FOUNDATIONS OF COLLEGE CHEMISTRY

Seventh Edition

MORRIS HEIN
MOUNT SAN ANTONIO COLLEGE

BROOKS/COLE PUBLISHING COMPANY
PACIFIC GROVE, CALIFORNIA

To stop learning is to wither on the vine.

Dedicated to
Mt. San Antonio College . . .
where it all began.

Sponsoring Editor: Harvey C. Pantzis
Production Services Manager: Joan Marsh
Editorial Assistant: Jennifer Kehr
Production: Julie Kranhold, Ex Libris
Design: Nancy Benedict Design
Illustrations: Cyndie Clarke-Huegel, John Foster, Lotus Art
Photo Researcher: Marion Hansen
Cover Photographer: Richard Fukuhara, West Light

Cover Photograph:
In the process of making fiber optics, a quartz tube is filled with layers of germanium-doped silicon dioxide, which is then solidified. As the solid in the tube is melted at 2100° Celsius, an optical fiber is formed. This photo shows the first drop formed.

Color insert photo credits appear on p. xvi.

Brooks/Cole Publishing Company
A Division of Wadsworth, Inc.
© 1990 by Wadsworth, Inc., Belmont, California 94002.

Printed in the United States of America
10 9 8 7 6 5 4 3 2 1

Library of Congress Cataloging in Publication Data
Hein, Morris.
  Foundations of college chemistry/Morris Hein.—7th ed.
    p. cm.
  ISBN 0-534-12966-8
  1. Chemistry.   I. Title.
  QD33.H45   1990
  540—dc20   89-70890
                                           CIP

PREFACE

To begin the preface of the seventh edition of *Foundations of College Chemistry*, I want to express my pleasure at the wide, favorable reception of the previous six editions and to thank my colleagues for the many helpful and constructive suggestions they have made.

This book is intended for use in a one-semester course in beginning or preparatory chemistry. Many students entering college may not have had a previous course in chemistry, or they may have an inadequate background preparation in mathematics or basic chemistry. The increased intensity of the general college chemistry course today is resulting in more students needing to take a preparatory course.

My purpose in writing *Foundations of College Chemistry* is to instruct students in the basic concepts of chemistry so that they will be qualified to enter the general college chemistry course. While presenting the technical phases of the subject matter, I attempt to create, hold, and motivate students' interest so that they may seek further scientific knowledge and, perhaps, a career in biology, medicine, engineering, or chemistry.

In selecting the subject matter for this book, I have drawn upon my own and the experience of my colleagues from teaching preparatory and college chemistry courses. I have included material that gives the most desirable background for the various levels of college chemistry courses. I have built the text around the principles of atomic and molecular structure and the properties of substances. In addition, I have stressed the quantitative relationships of these principles and the meanings of chemical equations.

Although a number of changes have been made in this edition, the level of the material remains the same. The entire text was carefully reviewed for further improvement in clarity of expression and to provide students with greater assistance in solving problems. Some of the major changes are as follows.

NEW FEATURES   Each chapter begins with an introduction that relates chemistry to some aspect of modern living. This is followed by a "Chapter Preview" which lists the sections covered in that chapter. Chapter objectives are now located at the end of the chapter in a section called "Concepts in Review." "Key Terms" defined in the chapter are also listed here. End-of-chapter "Exercises" have been expanded and are now numbered consecutively. As an aid to reinforce student learning, "Practice Problems" (with answers) immediately follow most "Examples" in the text.

We received so many compliments on the photo essay in the sixth edition that a new series of four photo essays are now included in this edition. These photo essays relate chemistry to nature, to the laboratory, to industry, and to our environment.

CHANGES IN ORGANIZATION

Chapter 1 has an added section (1.7) that describes the key pedagogical features of the text and explains their role in the study process. Chapters 3 and 4 have been reorganized so that Chapter 3 covers the classification of matter and Chapter 4 focuses on the properties of matter. The section "Heat: Quantitative Measurement" has been moved from Chapter 2 to Chapter 4 to separate heat measurement from temperature measurement and to associate heat more closely with energy.

I have observed over the years that students in this course have better success with the practical material than with theoretical or abstract material. Consequently, a major change in the chapter sequence has been made in this edition. Chapter 5 (atomic theory) has been split in two, "Early Atomic Theory and Structure" (Chapter 5) and "Modern Atomic Theory" (Chapter 10). Also moved to follow "Modern Atomic Theory" are the chapters on "The Periodic Arrangement of the Elements" (Chapter 11) and "Chemical Bonds: The Formation of Compounds from Atoms" (Chapter 12). Should any instructor prefer the original sequence of topics, moving Chapters 10, 11, and 12 to follow Chapter 5 will give them the same sequence of topics as in the previous editions.

Some of the benefits of the sequence changes are an earlier treatment of "Nomenclature of Inorganic Compounds" (Chapter 6), "Quantitative Composition of Compounds" (Chapter 7), "Chemical Equations" (Chapter 8), and "Calculations from Chemical Equations" (Chapter 9). Studying these chapters earlier enhances student success with naming compounds both in lecture and in the laboratory. It also increases student confidence in handling chemicals by name, introduces the mole sooner and provides an earlier understanding and use of chemical equations and calculations based on equations. I feel that this new sequence will better meet students' needs by allowing them to mature a bit in their chemical thinking before tackling the more abstract parts of the course.

LEARNING AIDS

Because they are inexperienced in the use of technical and abstract scientific content, students often feel somewhat apprehensive about undertaking the study of chemistry. As an experienced teacher of these students, I understand their feelings and therefore have used several helpful aids to ease students into confidently facing the technical subject matter.

- A list of "Concepts in Review" is given at the end of every chapter to guide students in studying the most important concepts in the chapter.
- Important "Key Terms" are set in boldface type where they are defined, and they are also printed in color in the margin. These new terms are collected in an alphabetical list at the end of each chapter.
- The many "Examples" in the text are solved in a step-by-step fashion using the conversion-factor dimensional analysis method.
- Most "Examples" are followed by a "Practice Problem" (with answer) for immediate reinforcement of student learning.
- Answers to mathematical problems are given in Appendix VI.
- Complete answers to all end-of-chapter "Exercises" are given in the *Solutions Manual.*
- All chapters contain at least one self-evaluation question. Answers to these questions are also given in Appendix VI.
- Seven sets of Review Exercises are provided for student self-testing and are interspersed at appropriate points throughout the text.
- A comprehensive glossary is given in Appendix V.

SUPPLEMENTS TO TEXT

Materials that may be helpful to students and to their instructors have been developed to accompany the text. A short description of them follows.

Study Guide by Peter C. Scott of Linn-Benton Community College includes a self-evaluation section for students to check their understanding of each chapter's objectives, a recap section, and answers to the self-evaluation section.

Solutions Manual includes answers and solutions to all end-of-chapter questions and problems.

Instructor's Supplement includes the answers to chapter review exercises, a set of objective test questions, and answers to the test questions.

Foundations of Chemistry in the Laboratory, 7th edition, by Morris Hein, Leo R. Best, and Robert L. Miner, includes twenty-eight experiments for a laboratory program that may accompany the lecture course. Also included are study aids and exercises.

Instructor's Manual to accompany the lab manual includes information on the management of the lab, evaluation of experiments, notes for individual experiments, a list of reagents needed, and answer keys to each experiment's report form.

EXP Test, the test generation system, Version 3.0 from Brooks/Cole for IBM PCs or compatibles.

ACKNOWLEDGMENTS

I am deeply grateful to the following reviewers who were kind enough to read and give their professional comments on the previous edition of the text: Barbara Andrews, Central Piedmont Community College; Harry Barnet, Palomar College; Hal Bender, Clackamas Community College; Orville Boge, Mesa College; Irvin M. Citron, Fairleigh Dickinson University; William Davis, St. Philip's College; Jerry A. Driscoll, University of Utah; Claudia Eley, University of Colorado; Don Glover, Bradley University; Howard Guyer, Fullerton College; Aniruddh Hathi, Texas State Technical Institute; Linda Hobart, Community College of the Fingerlakes; Kirk Hunter, Texas States Technical Institute; Christine S. Kerr, Montgomery College; Lawrence S. Lynn, St. Louis Community College at Meramec; Ivonna McCabe, Tacoma Community College; William A. Meena, Rock Valley College; Robin Monroe, Southeast Community College; Don Pon, Foothill Community College; Terri Prichard, Merced Community College; Barbara Rainard, Community College of Allegheny County; Patricia Redden, St. Peter's College; Keith Ries, Spokane Community College; Dale Rochnell, Rochester Institute of Technology; Michael Sady, Western Nevada Community College; Barbara Schowen, University of Kansas; Steven Thompson, Colorado State University; Frina Toby, Rutgers University; Katherine Weissmann, Mott Community College; and Peter K. Wong, Queensboro Community College.

I am particularly grateful to two of my colleagues at Mt. San Antonio College, Susan L. Arena and Eileen M. DiMauro, who worked closely with me and contributed many new ideas. Without their incredible effort this edition would not exist.

No textbook can be completed with the untiring effort of many publishing professionals. Special thanks to the talented staff at Brooks/Cole, especially Joan Marsh, who directed the book to its successful completion. Much credit must also be given to my editor, Harvey C. Pantzis, for his guidance, patience, flexibility, and strong desire to bring this edition to press.

Morris Hein

CONTENTS

CHAPTER THREE   51

Classification of Matter

CHAPTER FOUR   73

Properties of Matter

CHAPTER FIVE   95

Early Atomic Theory and Structure

CHAPTER SIX   114

Nomenclature of Inorganic Compounds

CHAPTER SEVEN   139

Quantitative Composition of Compounds

CHAPTER EIGHT   166

Chemical Equations

CHAPTER NINE   190

Calculations from Chemical Equations

CHAPTER TEN   214

Modern Atomic Theory

CHAPTER ELEVEN   236

The Periodic Arrangement of the Elements

CHAPTER TWELVE   256

Chemical Bonds: The Formation of Compounds from Atoms

Review Exercises for Chapters Ten through Twelve   285

CHAPTER THIRTEEN   290

The Gaseous State of Matter

CHAPTER FOURTEEN   336

Water and the Properties of Liquids

CHAPTER FIFTEEN   371

Solutions

CHAPTER SIXTEEN   416

Ionization: Acids, Bases, Salts

CHAPTER SEVENTEEN   448

Chemical Equilibrium

CHAPTER EIGHTEEN   484

Oxidation–Reduction

CHAPTER NINETEEN   516

Nuclear Chemistry

CHAPTER TWENTY   549

Introduction to Organic Chemistry

CHAPTER TWENTY-ONE   607

Introduction to Biochemistry

APPENDICES   A-1

INDEX   I-1

Start at top leftmost photo and read clockwise:

Chemistry in Nature

Page 1 Matt Bradley/Tom Stack & Associates; Jeff Foott/Bruce Coleman; Fred Ward; Fred McConnaughey/Photo Researchers. **Page 2** Sargent-Welch Scientific Company, a VWR Company. **Page 3** John Shaw/Bruce Coleman; Kenneth Lucas/Biological Photo Service; John Nakata/Terraphotographics/BPS; Jim Zuckerman/West Light. **Page 4** E. R. Degginger; Marvin Wax; Ken Davis/Tom Stack & Associates. **Page 5** E. R. Degginger, Photo Researchers; Gernological Institute of America; Fred Ward/Black Star. **Page 6** Roger Werth/Woodfin Camp & Associates; David Cavagnaro; Dick Rowan/Photo Researchers; Ken Lucas/Biological Photo Service. **Page 7** E. R. Degginger; Gordon Garradd/Photo Researchers; Spencer Swanger/Tom Stack & Associates. **Page 8** Ken Lucas/Biological Photo Service; Barbara J. Miller/Biological Photo Service; Fred Bavendam/Peter Arnold; D. J. Wrobel, Monterey Bay Aquarium/BPS; E. R. Degginger.

Chemistry in the Laboratory

Page 1 Richard Pasley/Stock Boston; Daemmrioh/Stock Boston; Richard Megna/Fundamental Photographs, NY; Rita Amaya. **Page 2** E. R. Degginger (3 photos); Yoav/Phototake; **Page 3** E. R. Degginger; Yoav/Phototake (2 photos); E. R. Degginger. **Page 4** Stacey Pick/Stock Boston; E. R. Degginger; Rita Amaya; E. R. Degginger. **Page 5** E. R. Degginger (4 photos). **Page 6** E. R. Degginger; American Petroleum Institute; E. R. Degginger (2 photos). **Page 7** Alexander Tsiaras/Photo Researchers; Nelson L. Max/Lawrence Livermore National Laboratory; Genentech; Lawrence Migdale/Stock Boston. **Page 8** Paulette Brunner/Tom Stack & Associates; Digital Instruments (3 photos).

Chemistry in Industry

Page 1 Philips and Du Pont Optical Company; S. L. Craig, Jr./Bruce Coleman; Corning, Inc; Gayle Cureton. **Page 2** American Petroleum Institute; Craig Aurness/West Light; Amoco. **Page 3** U.S. Department of Energy; Richard D. Hansen/UCLA-IDAEH Project NAKBE; Phil Brodatz; PG & E. **Page 4** Burroughs Wellcome; Genentech; Phil Hart/Burroughs Wellcome; Ellis Herwig/Stock Boston. **Page 5** Washington University School of Medicine in St. Louis (2 photos); Flex-Foot, Inc., Laguna Hills, CA (2 photos). **Page 6** Sara Hunsaker; Baltimore Spice Company; Richard Megna/Fundamental Photographs, NY; Sara Hunsaker. **Page 7** George Hall/Woodfin Camp & Associates; Tom Hollyman/Photo Researchers; Chesapeake Corporation; Christopher Springmann/After-Image. **Page 8** SCM Corporation; Allied Corporation; Armor All Products; Philip Jon Bailey/Taurus Photos.

Chemistry in the Environment

Page 1 Ellis Herwig/Stock Boston; Scott Kranhold; Philip Jon Bailey/Stock Boston. **Page 2** Victor Englebert/Photo Researchers; USDA Forest Service; Gary Milburn/Tom Stack & Associates; Jan Halaska/Photo Researchers; **Page 3** Du Pont; NASA; Russell Wood/Taurus Photos; National Science Foundation. **Page 4** T. Kitchin/Tom Stack & Associates; Du Pont; David M. Doody/Tom Stack & Associates; Fred Ward/Black Star. **Page 5** Philip Jon Bailey/Stock Boston; Mark Stephenson/West Light; Lawrence Kolozak/SCAQMD; Hans Silvester/Photo Researchers. **Page 6** Sara Hunsaker; Reynolds Aluminum Recycling Company (4 photos). **Page 7** Sara Hunsaker; Rubbermaid Commercial Products; Rita Amaya; Polystyrene Recycling; Du Pont. **Page 8** Peter J. Bryant/Biological Photo Service; Franklin Gress (2 photos); S. K. Webster, Monterey Bay Aquarium/BPS; Franklin Gress.

Calcium Carbonate

A container labeled "calcium carbonate" ($CaCO_3$) contains a white insoluble powder that is used principally for mortar. Nature has found much more spectacular uses for this compound. Calcium carbonate is a major component of nacre, lime, and limestone, which have served as building matrices for many very elaborate structures.

◀ This snail sports a beautifully colored shell, which gets its brilliance from nacre and crystalline $CaCO_3$.

The pearl is formed by many thin layers of nacre around an irritating substance that invades the oyster's shell.
▼

▲
Stalactites and stalagmites are haunting structures that are formed over hundreds of years by rainwater, saturated with dissolved limestone, dripping through the cracks in cave ceilings.

Brilliantly colored coral reefs form when ▶ a living layer of coral polyps synthesize their lime skeletons on top of those shed by millions of their ancestors.

Nature's Palette

Many chemical compounds exhibit brilliant colors that rival the pigments found in the biological world. Compounds of transition elements are typically very colorful and are useful as paint pigments. Compounds of elements in Groups IA and IIA reflect all wavelengths of light, which gives them a white color. Carbon compounds take on a variety of colors.

1. Cobalt (II) chloride, $CoCl_2$
2. Sodium chloride, NaCl
3. Lead sulfide, PbS
4. Sulfur, S
5. Zinc, Zn
6. Marble chips, $CaCO_3$
7. Logwood chips
8. Charcoal, C
9. Mercury (II) iodide, HgI_2
10. Pyrite, FeS_2
11. Chromium (III) oxide, Cr_2O_3
12. Iron (II) sulfate, $FeSO_4$
13. Sodium sulfite, Na_2SO_3
14. Rosen
15. Sodium thiosulfate, $Na_2S_2O_3$
16. Iron, Fe
17. Aluminum, Al
18. Potassium hexacyanoferrate, $K_3Fe(CN)_6$
19. Potassium chromium sulfate, $KCr(SO_4)_2$
20. Menthol, $C_{10}H_{19}OH$
21. Potassium permanganate, $KMnO_4$
22. Ammonium nickel sulfate, $(NH_4)_2Ni(SO_4)_2$
23. Copper (II) sulfate 5-hydrate, $CuSO_4 \cdot 5H_2O$
24. Sodium chromate, Na_2CrO_4
25. Trilead tetraoxide, Pb_3O_4
26. Hydroquinone, $C_6H_4(OH)_2$
27. Copper, Cu

Nature's Palette

Artists over the centuries have spent countless hours attempting to recreate the myriad of colors found in nature. These beautiful hues are the result of pigment molecules that produce color by reflecting specific wavelengths of light. Many pigments are found in the plant and animal kingdoms and serve a biological function as well as making the planet a beautiful place.

Tropical fish that dwell in coral reefs display dazzling colors and patterns, making them inconspicuous to predators.
▼

▲
Trees teach a brilliant lesson in photosynthesis. Chlorophyll traps energy from the sun to turn leaves green; as the plant prepares for inactivity, the chlorophyll degrades to reveal the normal red, yellow, and orange pigments that have been seasonally masked by chlorophyll.

▲
The breathtaking colors of flowers ensure their survival. Their pigments attract insects to pollinate them, which thereby continue the species.

▲
The brilliant coloring of birds serves as camouflage, as a recognition signal for members of the same species, and for attraction during mating season.

Natural Resources

We have renewable and nonrenewable resources. Renewable resources include all forms of plant and animal life as well as water and air. The earth has the machinery to regenerate renewable resources, and if this mechanism is not disturbed, they will be unlimited. Nonrenewable resources, such as minerals and fossil fuels, exist in finite quantities and have been formed over eons.

Salt flats are formed by the evaporation ▶ of water from large inland saline seas, leaving behind deposits of salt crystals.

Elemental sulfur is recovered from the earth by melting sulfur deposits with superheated steam. The molten sulfur is forced to the surface with compressed air where it is returned to the solid state. ▼

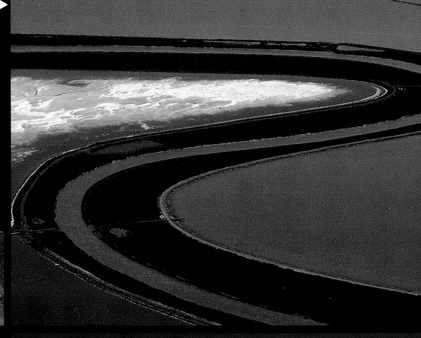

These strange, mineral formations, ▶ called "tufa," exist at Mono Lake in California—an extraordinary inland saline body of water. Tufa are formed by water bubbling through the sand, which is saturated with sodium chloride (NaCl), sodium carbonate (Na_2CO_3), and sodium sulfate (Na_2SO_4).

Gemstones

Gemstones are opaque, translucent minerals that possess great beauty. Opaque stones obtain their beauty from stunning color, whereas translucent gems are valued for their fire and brilliance. Particularly remarkable and rare stones, such as diamonds, emeralds, and rubies, are precious stones. Ameythyst, opal, lapis, and agate are examples of semiprecious stones.

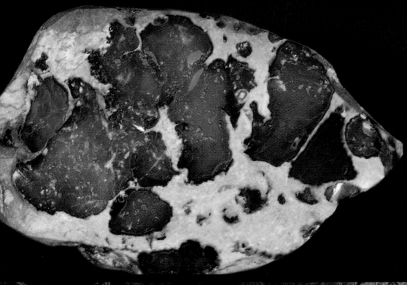

◀ Dazzling colors gleam inside this rock, which holds pockets of semiprecious opal (hydrated amorphous silica).

Its unusual violet color makes amethyst a prized form of quartz (SiO_2). Iron impurities (Fe^{3+}) create the beautiful color. ▼

▲ Its unequaled brilliance makes the diamond our most beloved and precious gemstone. One of the hardest substances on earth, it is made of pure carbon.

◀ The most precious gemstones are gouged out of the earth in mining operations, such as this diamond mine.

Chemical Phenomena

Nature is capable of producing an array of shows that are as brilliant as any fireworks display. Some of these shows are common occurrences familiar to everyone, while others are witnessed by only a select few. These events are diverse in nature having biological, geological, and atmospheric origins. Yet, each phenomenon can be explained at the chemical level.

A volcanic eruption occurs when intense pressure pushes molten lava upward. Lava is a solution of silicates, oxides, and sulfides combined with superheated steam and other gases.
▼

Elemental sulfur can be deposited on the rim of volcanos when gases, such as sulfur dioxide (SO_2) and hydrogen sulfide (H_2S) are oxidized or reduced.
▼

▲
The flashlight fish demonstrates the phenomenon of bioluminescence. The large organs of these bright fish contain bacteria that produce a luminous substance called "luciferin."

▲
Red tides are produced by "blooms" of certain microorganisms that give the water the appearance of tomato soup, and when disturbed they luminesce to produce a green-tinged flash.

◄ Nature's spectacular light show, the aurora borealis, is displayed when sunspots bombard nitrogen (N_2) and oxygen (O_2) molecules in the atmosphere with an increased number of charged particles.

▲ The unmistakable crack of lightning splits the air from an exchange of electrons between negatively charged water particles in a cloud and positively charged particles in another cloud or on the earth.

◄ Great drama is created by differences in the boiling points of water at sea level (100°C) and deep within the earth's crust (106°C). This is the driving force that releases the superheated water at geysers.

Ecosystems

An ecosystem is a community of organisms, nonliving matter, and radiant energy. Its characteristics are primarily detected by temperature, precipitation, and soil chemistry. The presence of trace elements helps to determine which types of vegetation and animals will flourish. A careful balance of all ecosystems must be maintained to guarantee our future.

Tropical rain forests exhibit the most abundant and diverse plant and animal life anywhere. The trees absorb CO_2 and produce vast amounts of life-sustaining O_2 by photosynthesis. ▶

A myriad of species, such as this Poison Arrow frog, exist in rain forests and have potential to produce useful foods and drugs. ▶

The ocean dissolves huge quantities of CO_2 from the atmosphere. It also supports a multitude of free-floating plankton, which produce much of the O_2 in the atmosphere by photosynthesis. ▼

◀ Kelp and coral are just two of the many useful organisms found in the ocean. Besides being sources of food and shelter, they have uses in medicine and industry, and are being researched even further.

FOUNDATIONS OF COLLEGE CHEMISTRY

Seventh Edition

CHAPTER ONE

Introduction

Have you ever strolled through a spring garden and been amazed at the diversity of colors in the flowers? Or perhaps you have curled up in front of a winter fire and become fascinated watching the flames. And think of those times when you have dropped a beverage container on a hard floor, and were relieved to find that it was plastic instead of glass. All of these phenomena are the result of chemistry—not in the laboratory but rather in our everyday lives. Chemical changes can bring us beautiful colors, warmth and light, or new and exciting products. Chemists seek to understand, explain, and utilize the diversity of materials we find around us.

Chapter Preview

1.1   The Nature of Chemistry

chemistry

What is chemistry? A popular dictionary gives this definition: **Chemistry** is the science of the composition, structure, properties, and reactions of matter, especially of atomic and molecular systems. Another, somewhat simpler dictionary definition is: Chemistry is the science dealing with the composition of *matter* and the changes in composition that matter undergoes. Neither of these definitions is entirely adequate. Chemistry, along with the closely related science of physics, is a fundamental branch of knowledge. Chemistry is also closely related to biology, not only because living organisms are made of material substances but also because life itself is essentially a complicated system of interrelated chemical processes.

The scope of chemistry is extremely broad: It includes the whole universe and everything, animate and inanimate, in it. Chemistry is concerned not only with the composition and changes in composition of matter, but also with the energy and energy changes associated with matter. Through chemistry we seek to learn and to understand the general principles that govern the behavior of all matter.

The chemist, like other scientists, observes nature and attempts to understand its secrets: What makes a rose red? Why is sugar sweet? What is occurring when iron rusts? Why is carbon monoxide poisonous? Why do people wither with age? Problems such as these—some of which have been solved, some of which are still to be solved—are part of what we call chemistry.

A chemist may interpret natural phenomena, devise experiments that will reveal the composition and structure of complex substances, study methods for improving natural processes, or, sometimes, synthesize substances unknown in nature. Ultimately, the efforts of successful chemists advance the frontiers of knowledge and at the same time contribute to the well-being of humanity. Chemistry can help us to understand nature; however, one need not be a professional chemist or scientist to enjoy natural phenomena. Nature and its beauty, its simplicity within complexity, are for all to appreciate.

The body of chemical knowledge is so vast that no one can hope to master it all, even in a lifetime of study. However, many of the basic concepts can be learned in a relatively short period of time. These basic concepts have become part of the education required for many professionals, including agriculturists, biologists, dental hygienists, dentists, medical technologists, microbiologists, nurses, nutritionists, pharmacists, physicians, and veterinarians, to name a few.

1.2   History of Chemistry

People have practiced empirical chemistry from the earliest times. Ancient civilizations were practicing the art of chemistry in such processes as wine-making, glass-making, pottery-making, dyeing, and elementary metallurgy. The early Egyptians, for example, had considerable knowledge of certain chemical processes. Excavations into ancient tombs dated about 3000 B.C. have uncovered workings of gold, silver, copper, and iron, pottery from clay, glass beads, and beautiful dyes and paints, as well as bodies of Egyptian kings in remarkably well-preserved states. Many other cultures made significant developments in chemistry. However, all these developments were empirical; that is, they were achieved by trial and error and did not rest on any valid theory of matter.

Philosophical ideas relating to the properties of matter (chemistry) did not develop as early as those relating to astronomy and mathematics. The Greek philosophers made great strides in philosophical speculation concerning materialistic ideas about chemistry. They led the way to placing chemistry on an intellectual, scientific basis. They introduced the concepts of elements, atoms, shapes of atoms, and chemical combination. They believed that all matter was derived from four elements: earth, air, fire, and water. The Greek philosophers had keen minds and perhaps came very close to establishing chemistry on a sound basis similar to the one that was to develop about 2000 years later. The main shortcoming of the Greek approach to scientific work was a failure to carry out systematic experimentation.

Greek civilization was succeeded ·by the Roman civilization. The Romans were outstanding in military, political, and economic affairs. They practiced empirical chemical arts such as metallurgy, enameling, glass-making, and pottery-making, but they did very little to advance new and theoretical knowledge. Eventually the Roman civilization was succeeded in Europe by the Dark Ages. During this period European civilization and learning were at a very low ebb.

In the Middle East and in North Africa, knowledge did not decline during the Dark Ages as it did in Western Europe. At this time Arabic cultures made contributions that were of great value to the development of modern chemistry. In particular, the Arabic number system, including the use of zero, gained acceptance; the branch of mathematics known as *algebra* was developed; and alchemy, a sort of pseudochemistry, was practiced extensively.

One of the more interesting periods in the history of chemistry was that of the alchemists (500–1600 A.D.). People have long had a lust for gold, and in those

FIGURE 1.1

**Medieval alchemy laboratory ("The Alchemist" by David
Teniers, Jr.** *Courtesy of the Fisher Collection of Pittsburgh.*)

days gold was considered the ultimate, most perfect metal formed in nature. The
principal goals of the alchemists were to find a method of prolonging human life
indefinitely and to change the base metals, such as iron, zinc, and copper, into
gold. They searched for a universal solvent to transmute base metals into gold
and for the "philosopher's stone" to rid the body of all diseases and to renew life.
In the course of their labors they learned a great deal of chemistry. Unfortunately, much of their work was done secretly because of the mysticism that
shrouded their activity, and very few records remain (See Figure 1.1).

Although the alchemists were not guided by sound theoretical reasoning and
were clearly not in the intellectual class of the Greek philosophers, they did
something that the philosophers had not considered worthwhile. They subjected
various materials to prescribed treatments under what might be loosely described

FIGURE 1.2

Robert Boyle, 1627–1691. (*Courtesy of AIP Niels Bohr Library.*)

as laboratory methods. These manipulations, carried out in alchemical laboratories, not only uncovered many facts of nature but paved the way for the systematic experimentation that is characteristic of modern science.

Alchemy began to decline in the 16th century when Paracelsus (1493–1541), a Swiss physician and outspoken revolutionary leader in chemistry, strongly advocated that the objectives of chemistry be directed toward the needs of medicine and the curing of human ailments. He openly condemned the mercenary efforts of alchemists to convert cheaper metals to gold.

But the real beginning of modern science can be traced to astronomy during the Renaissance. Nicolaus Copernicus (1473–1543), a Polish astronomer, began the downfall of the generally accepted belief in a geocentric universe. Although not all the Greek philosophers had believed that the sun and the stars revolved about the earth, the geocentric concept had come to be generally accepted. The heliocentric (sun-centered) universe concept of Copernicus was based on direct astronomical observation and represented a radical departure from the concepts handed down from Greek and Roman times. The ideas of Copernicus and the invention of the telescope stimulated additional work in astronomy. This work, especially that of Galileo Galilei (1564–1642) and Johannes Kepler (1571–1630), led directly to a rational explanation by Sir Isaac Newton (1642–1727) of the general laws of motion, which he formulated between about 1665 and 1685.

Modern chemistry was slower to develop than astronomy and physics; it began in the 17th and 18th centuries when Joseph Priestley (1733–1804), who

FIGURE 1.3

John Dalton, 1766–1844. (*Courtesy of the Smithsonian Institution.*)

discovered oxygen in 1774, and Robert Boyle (1627–1691) began to record and publish the results of their experiments and to discuss their theories openly. Boyle, who has been called the founder of modern chemistry, was one of the first to practice chemistry as a true science (Figure 1.2). He believed in the experimental method. In his most important book, *The Sceptical Chymist*, he clearly distinguished between an element and a compound or mixture. Boyle is best known today for the gas law that bears his name. A French chemist, Antoine Lavoisier (1743–1794), placed the science on a firm foundation with experiments in which he used a chemical balance to make quantitative measurements of the weights of substances involved in chemical reactions.

The use of the chemical balance by Lavoisier and others later in the 18th century was almost as revolutionary in chemistry as the use of the telescope had been in astronomy. Thereafter, chemistry became a quantitative experimental science. Lavoisier also contributed greatly to the organization of chemical data, to chemical nomenclature, and to the establishment of the Law of Conservation of Mass in chemical changes. During the period from 1803 to 1810, John Dalton (1766–1844), an English schoolteacher, advanced his atomic theory. See Figure 1.3. This theory (see Section 5.2) placed the atomistic concept of matter on a valid rational basis. It remains today as a tremendously important general concept of modern science.

Since the time of Dalton, knowledge of chemistry has advanced in great strides, with the most rapid advancement occurring at the end of the 19th century and during the 20th century. Especially outstanding achievements have been made in determining the structure of the atom, understanding the biochemical fundamentals of life, developing chemical technology, and the mass production of chemicals and related products.

1.3   The Branches of Chemistry

Chemistry may be broadly classified into two main branches: *organic* chemistry and *inorganic* chemistry. Organic chemistry is concerned with compounds containing the element carbon. The term *organic* was originally derived from the chemistry of living organisms: plants and animals. Inorganic chemistry deals with all the other elements as well as with some carbon compounds. Substances classified as inorganic are derived mainly from mineral sources rather than from animal or vegetable sources.

Other subdivisions of chemistry, such as analytical chemistry, physical chemistry, biochemistry, electrochemistry, geochemistry, and nuclear chemistry, may be considered specialized fields of, or auxiliary fields to, the two main branches.

Chemical engineering is the branch of engineering that deals with the development, design, and operation of chemical processes. A chemical engineer usually begins with a chemist's laboratory-scale process and develops it into an industrial-scale operation.

1.4   Relationship of Chemistry to Other Sciences and Industry

Besides being a science in its own right, chemistry is the servant of other sciences and industry. Chemical principles contribute to the study of physics, biology, agriculture, engineering, medicine, space research, oceanography, and many other sciences (Figure 1.4). Chemistry and physics are overlapping sciences, since both are based on the properties and behavior of matter. Biological processes are chemical in nature. The metabolism of food to provide energy to living organisms is a chemical process. Knowledge of molecular structure of proteins, hormones, enzymes, and the nucleic acids is assisting biologists in their investigations of the composition, development, and reproduction of living cells.

Chemistry is playing an important role in alleviating the growing shortage of food in the world. Agricultural production has been increased with the use of chemical fertilizers, pesticides, and improved varieties of seeds. Chemical refrigerants make possible the frozen food industry, which preserves large amounts of food that might otherwise spoil. Chemistry is also producing synthetic nutrients, but much remains to be done as the world population

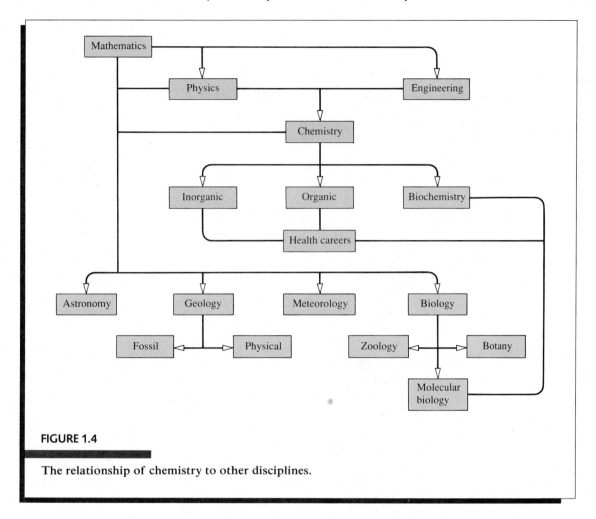

FIGURE 1.4

The relationship of chemistry to other disciplines.

increases relative to the land available for cultivation. Expanding energy needs have brought about difficult environmental problems in the form of air and water pollution. Chemists and other scientists are working diligently to alleviate these problems.

Advances in medicine and chemotherapy, through the development of new drugs, have contributed to prolonged life and the relief of human suffering. More than 90% of the drugs and pharmaceuticals being used in the United States today have been developed commerically within the past 50 years. The plastics and polymer industry, unknown 60 years ago, has revolutionized the packaging and textile industries and is producing durable and useful construction materials. Energy derived from chemical processes is used for heating, lighting, and transportation. Virtually every industry is dependent on chemicals—for example, the petroleum, steel, rubber, pharmaceutical, electronic, transportation, cosmetic, space, polymer, garment, aircraft, and television industries. (The list could go on.)

1.5   Scientific Method

Chemistry, as a science or field of knowledge, is concerned with ideas and concepts relating to the behavior of matter. Although these concepts are abstract, their application has had a concrete impact on human culture. This impact is due to modern technology, which may be said to have begun about 200 years ago and which has grown at an accelerating rate ever since.

An important difference between science and technology is that science represents an abstract body of knowledge, and technology represents the physical application of this knowledge to the world in which we live.

Why has the science of chemistry and its associated technology flourished in the last two centuries? Is it because we are growing more intelligent? No, we have absolutely no reason to believe that the general level of human intelligence is any higher today than it was in the Dark Ages. The use of the scientific method is usually credited with being the most important single factor in the amazing development of chemistry and technology. Although complete agreement is lacking on exactly what is meant by "using the scientific method," the general approach is as follows:

1. Collect facts or data that are relevant to the problem or question at hand, which is usually done by planned experimentation.

2. Analyze the data to find trends (regularities) that are pertinent to the problem. Formulate a hypothesis that will account for the data that have been accumulated and that can be tested by further experimentation.

3. Plan and do additional experiments to test the hypothesis. Such experiments extend beyond the range that is covered in Step 1.

4. Modify the hypothesis as necessary so that it is compatible with all the pertinent experimental data.

hypothesis

theory

scientific laws

Confusion sometimes arises regarding the exact meanings of the words *hypothesis, theory,* and *law.* A **hypothesis** is a tentative explanation of certain facts that provides a basis for further experimentation. A well-established hypothesis is often called a **theory.** Thus a theory is an explanation of the general principles of certain phenomena with considerable evidence or facts to support it. Hypotheses and theories explain natural phenomena, whereas **scientific laws** are simple statements of natural phenomena to which no exceptions are known under the given conditions.

Although the four steps listed in the preceding paragraph are a broad outline of the general procedure that is followed in much scientific work, they are not a

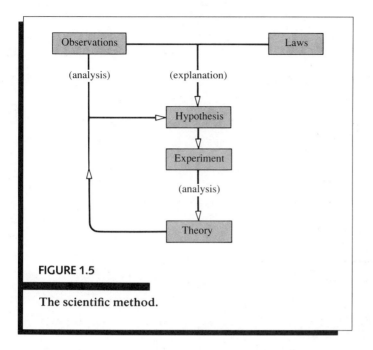

FIGURE 1.5

The scientific method.

recipe for doing chemistry or any other science (Figure 1.5). But chemistry is an experimental science, and much of its progress has been due to application of the scientific method through systematic research. Occasionally a great discovery is made by accident, but the majority of scientific achievements are accomplished by well-planned experiments.

Many theories and laws are studied in chemistry. They make the study of chemistry or any science easier, because they summarize particular aspects of the science. The student will note that some of the theories advanced by great scientists in the past have since been substantially altered and modified. Such changes do not mean that the discoveries of the past are less significant than those of today. Modification of existing theories in the light of new experimental evidence is essential to the growth and evolution of scientific knowledge.

1.6   How to Study Chemistry

How do you as a student approach a subject such as chemistry with its unfamiliar terminology, symbols, formulas, theories, and laws? All the generally accepted habits of good study are applicable to the study of chemistry. Budget your study time and spend it wisely. In particular, you can spend your time more profitably in regular, relatively short periods of study rather than in one prolonged cram session. Figure 1.6 shows how all the parts of the study process fit together.

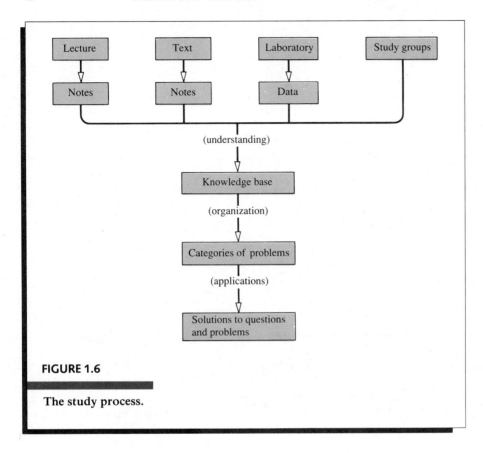

FIGURE 1.6

The study process.

Chemistry has its own language, and learning this language is of prime importance to the successful study of chemistry. Chemistry is a subject of many facts. At first you will simply have to memorize some of them. However, you will also learn these facts by referring to them frequently in your studies and by repetitive use. For example, you must learn the symbols of 30 or 40 common elements in order to be able to write chemical formulas and equations. As with the alphabet, repetitive use of these symbols will soon make them part of your vocabulary.

The need for careful reading of assigned material cannot be overemphasized. You should read each chapter at least twice. The first time, read the chapter rapidly, noting especially topic headings, diagrams, and other outstanding features. Then read more thoroughly and deliberately for better understanding. It may be profitable to underline and abstract material during the second reading. Isolated reading may be sufficient for some subjects, but it is not sufficient for learning chemistry. During the lectures, become an active mental participant and try to think along with your instructor, do not just occupy a seat. Lecture and laboratory sessions will be much more meaningful if you have already read the assigned material.

Your studies must include a good deal of *written* chemistry. Chemical symbolism, equations, problem solving, and so on, require much written practice for proficiency. One does not become an accomplished pianist by merely reading or listening to music—it takes practice. One does not become a good baseball player by reading the rules and watching baseball games—it takes practice. So it is with chemistry. One does not become proficient in chemistry by only reading about it—it takes practice.

You will encounter many mathematical problems as you progress through this text. To solve a numerical problem, you should read the problem carefully to determine what is being asked. Then develop a plan for solving the problem. It is a good idea to start by writing down the pertinent material—a formula, a diagram, an equation, the data given in the problem. This information will give you something to work with, to think about, to modify, and finally to expand into an answer. When you have arrived at an answer, consider it carefully to make sure that it is a reasonable one. The solutions to problems should be recorded in a neat, orderly, stepwise fashion. Fewer errors and saved time are the rewards of a neat and orderly approach to problem solving. If you need to read and study still further for complete understanding, do it!

1.7 Key Textbook Features

This new edition of *Foundations of College Chemistry* includes many tools to help you in the study process. You must be an active participant in the study of chemistry. This requires you to read the text and to ask yourself questions. You must address these questions as you read. To do this effectively, you must read with pencil and paper. There are many calculations and problems in chemistry, hence you will need a scientific calculator to help with the arithmetic, as real numbers can often be cumbersome. To study without the benefits of paper, pencil, and calculator is *inefficient* at best.

The text is designed to assist you in the study process. Each chapter begins with a "Chapter Preview" of the topics covered, and a short description of some practical aspects of the topic in your daily life. This description serves as a bridge to reveal how chemistry is an integral part of our lives.

Each chapter includes "Examples" which are worked out in sequence. These will provide you with an illustration of the problem-solving techniques necessary to understand a concept. The "skills" of chemistry are developed in the "Examples." Often, more than one method of solving a problem may be appropriate, thus alternative methods of solution are suggested. Select the method that makes you comfortable and use it to solve the "Practices" that follow the example. Be sure to work the practice problems while you are reading. The answers are provided so you may check them immediately.

At the end of most chapters, you will find a set of "Exercises." Some are less challenging, requiring basic information; while others are more demanding, requiring analysis or application of concepts already learned. The most challenging problems are marked with an asterisk (*). The tables and figures from the chapter, as well as the appendix and endpapers, will assist you with the problems. "Answers to Selected Problems" are found in the appendix. Complete solutions are found in the *Student Solutions Manual*.

If you have difficulty solving a problem or answering a question, refer back to the chapter for a similar situation. Use the examples as a guide. If you still can't solve the problem within a reasonable time, skip it for the moment. Often, leaving a frustrating problem and returning to it later will provide a fresh approach to a solution. Bring troublesome problems to your study group or your instructor for assistance.

At the end of each chapter, there are two important review sections. "Concepts in Review" is an overview of the important ideas and skills introduced in each chapter. If you accomplish each item in this section, you will understand the fundamental concepts from the chapter. "Key Terms in Review" provides a list of important terms introduced in the chapter. Chemistry is a new language to you, so the understanding of vocabulary is fundamental to a good foundation. Just as you do not memorize the definition of each new word you encounter, you should not attempt to memorize these terms. Rather, you should seek to understand and incorporate them in your speaking and writing. Also, "Review Exercises" appear throughout the text, which provide an opportunity to study larger blocks of material in an objective format.

The appendix of the text contains valuable information for use in study, including a review of mathematical concepts to assist you in problem solving and a "Glossary" that provides an accessible reference for terms that have been introduced elsewhere in the text. Other tables of useful information are found in the appendix as well. Consider it a handy reference library.

With these tools to assist you in your study of chemistry, you will be prepared to begin a great adventure. Surely, you will feel confusion, frustration, elation, and success along the way. Most certainly you will gain a new appreciation and understanding of the role of chemistry in the world around you.

Key Terms in Review

The terms listed here have been defined within the chapter. Review the definitions of each. Use the glossary and the margin notations within the chapter as study aids.

chemistry	theory
hypothesis	scientific laws

Exercises

1. Was the concept of a geocentric universe necessarily based on incorrect astronomical observations? Explain.
2. What were the principal goals of the alchemists?
3. What instrument, when first used by chemists, can be considered to be analogous to the use of the telescope by early astronomers? Explain the analogy.
4. Classify the following statements as observation, law, hypothesis, or theory:
   (a) When the pressure remains constant, the volume of a gas is directly proportional to the absolute temperature.
   (b) The water in a closed test tube boiled at 83°C.
   (c) Iron gets heavier when it rusts because it attracts particles of rust from the air.
   (d) All matter is composed of tiny particles called atoms.
   (e) As it approaches its melting point, glass turns a flame yellow.
   (f) Molecules in a gas are always moving.
   (g) When wood burns it decomposes to its elements, which all escape as gas.
5. Which of the following statements are correct? Rewrite the incorrect statements to make them correct.
   (a) Chemistry is the science that deals with the composition of substances and the transformations they undergo.
   (b) Robert Boyle, in the 17th century, clearly distinguished between an element and a compound or mixture.
   (c) Both the knowledge and intellectual capacity of Western European people decreased markedly during the Dark Ages.
   (d) From 1803 to 1810, John Dalton advanced his atomic theory.

   (e) Most of the drugs and pharmaceuticals used in the United States today have been available for at least a century.
   (f) A key feature of the scientific method is to plan and do additional experiments to test a hypothesis.
   (g) Scientific laws are simple statements of natural phenomena to which no exceptions are known.
   (h) Antoine Lavoisier was one of the first chemists to make quantitative measurements using a chemical balance.
   (i) The two main branches of chemistry are organic and biochemistry.
6. List ten examples of matter which you use daily.
7. Distinguish between theory and law.
8. If you perform an experiment and do not get the result you expect based upon your previous knowledge, what would you do next?
9. Design a weekly study plan to use in this course. Include:
   (a) all hours you are enrolled in all classes
   (b) study time for each course (plan for 2 hours per class hour)
   (c) work hours
   (d) relaxation time (life is not all academic)
10. Using the three numbers given and a combination of $+$, $-$, $\times$, and $\div$ symbols to yield the boxed number on the right. Remember that trial and error is an acceptable procedure for solution of a problem.

    $$7 \quad 5 \quad 4 = \boxed{16}$$
    $$6 \quad 2 \quad 8 = \boxed{12}$$
    $$3 \quad 8 \quad 6 = \boxed{4}$$

CHAPTER TWO

Standards for Measurement

Doing an experiment in chemistry is very much like cooking a meal in the kitchen. It is important to know the ingredients *and* the amounts of each in order to have a tasty product. Working on your car requires specific tools, in exact sizes. Buying new carpeting or draperies is an exercise in precise and accurate measurement for a good fit. A small difference in the concentration or amount of medication a pharmacist gives you may have significant effects on your well-being. In all of these cases, a strong foundation in the language, measurement, and use of numbers provides the basis for success. In chemistry, we begin by learning the metric system and the proper units for measuring mass, length, volume, and temperature.

Chapter Preview

2.1   Mass and Weight

Chemistry is an experimental science. The results of experiments are usually determined by making measurements. In elementary experiments the quantities that are commonly measured are mass, length, volume, pressure, temperature, and time. Measurements of electrical and optical quantities may also be needed in more sophisticated experimental work.

mass

Although mass and weight are often used interchangeably, the two words have quite different meanings. The **mass** of a body is defined as the amount of matter in that body. The mass of an object is a fixed and unvarying quantity that is independent of the object's location. The mass of an object may be measured on a balance, by comparison with other known masses.

weight

The **weight** of a body is the measure of the earth's gravitational attraction for that body. Weight is measured on a device called a scale, which measures force against a spring. Unlike mass, weight varies in relation to (1) the position of an object on or its distance from the earth and (2) whether the rate of motion of the object is changing with respect to the motion of the earth.

Consider an astronaut of mass 70.0 kilograms (154 pounds) who is being shot into orbit. At the instant before blast-off the weight of the astronaut is also 70.0 kilograms. As the distance from the earth increases and the rocket turns into an orbiting course, the gravitational pull on the astronaut's body decreases until a state of weightlessness (zero weight) is attained. However, the mass of the astronaut's body has remained constant at 70.0 kilograms during the entire event.

2.2   Measurement and Significant Figures

To understand certain aspects of chemistry it is necessary to set up and solve problems. Problem solving requires an understanding of the elementary mathematical operations used to manipulate numbers. Numerical values or data are obtained from measurements made in an experiment. A chemist may use these data to calculate the extent of the physical and chemical changes occurring in the substances that are being studied. By appropriate calculations the results of an experiment may be compared with those of other experiments and summarized in ways that are meaningful.

The result of a measurement is expressed by a numerical value together with a unit of that measurement. For example,

$$\overbrace{70.0 \text{ kilograms}} = 1\overbrace{54}^{} \text{ pounds}$$

numerical value

unit

Numbers obtained from a measurement are never exact values. They always have some degree of uncertainty due to the limitations of the measuring instrument and the skill of the individual making the measurement. The numerical value recorded for a measurement should give some indication of its reliability (precision). To express maximum precision this number should contain all the digits that are known plus one digit that is estimated. This last estimated digit introduces some uncertainty. Because of this uncertainty every number that expresses a measurement can have only a limited number of digits. These digits, **significant figures**   used to express a measured quantity, are known as **significant figures**, or **significant digits**.

Suppose we measure temperature on a thermometer calibrated in degrees and we observe that the mercury stops between 21 and 22 (see Figure 2.1a). We then know that the temperature is at least 21 degrees and is less than 22 degrees. To express the temperature with greater precision, we estimate that the mercury is about four-tenths the distance between 21 and 22. The temperature is, therefore, 21.4 degrees. The last digit (4) has some uncertainty, because it is an estimated value. The recorded temperature, 21.4 degrees, is said to have three significant figures. If the mercury stopped exactly on the 22 (Figure 2.1b), the temperature would be recorded as 22.0 degrees. The zero is used to indicate that the temperature was estimated to a precision of one-tenth degree. Finally, look at Figure 2.1c. On this thermometer, the temperature is recorded as $22.15°C$ (four significant figures). Since the thermometer is calibrated to tenths of a degree, the first estimated digit is the hundredths.

Some numbers are exact and have an infinite number of significant figures. Exact numbers occur in simple counting operations; when you count 25 dollars, you have exactly 25 dollars. Defined numbers, such as 12 inches in 1 foot, 60 minutes in 1 hour, and 100 centimeters in 1 meter, are also considered to be exact numbers. Exact numbers have no uncertainty.

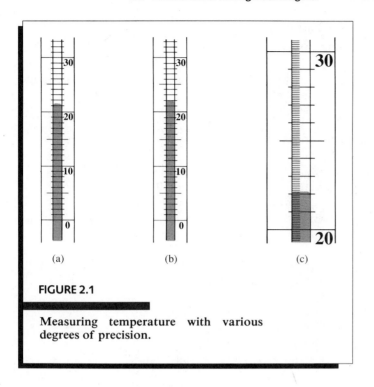

(a)   (b)   (c)

FIGURE 2.1

Measuring temperature with various degrees of precision.

Evaluating Zero

In any measurement all nonzero numbers are significant. However, zeros may or may not be significant, depending on their position in the number. Rules for determining when zero is significant in a measurement follow.

1. Zeros between nonzero digits are significant:

205 has three significant figures.
2.05 has three significant figures.
61.09 has four significant figures.

2. Zeros that precede the first nonzero digit are not significant. These zeros are used to locate a decimal point:

0.0025 has two significant figures (2, 5).
0.0108 has three significant figures (1, 0, 8).

3. Zeros at the end of a number that include a decimal point are significant:

0.500 has three significant figures (5, 0, 0).
25.160 has five significant figures.

3.00 has three significant figures.

20. has two significant figures.

4. Zeros at the end of a number without a decimal point are ambiguous and are not considered significant:

1000 (The zeros may or may not be significant.)

590 (The zero may or may not be significant.)

One way of indicating whether these zeros are significant is to write the number using a decimal point and a power of 10. Thus, if the value 1000 has been determined to four significant figures, it is written as 1.000×10^3. If 590 has only two significant figures, it is written as 5.9×10^2 (in this case the zero is not significant).

PRACTICE   How many significant figures are in each of these numbers?

(a) 4.5 inches   (e) 25.0 grams
(b) 3.025 feet   (f) 12.20 liters
(c) 125.0 meters   (g) 100,000 people
(d) 0.001 mile   (h) 205 birds

Answers: (a) 2   (b) 4   (c) 4   (d) 1   (e) 3   (f) 4   (g) 1   (h) 3

2.3   Rounding Off Numbers

In calculations we often obtain answers that have more digits than we are justified in using. It is necessary, therefore, to drop the nonsignificant digits in order to express the answer with the proper number of significant figures. When digits are dropped from a number, the value of the last digit retained is determined by a **rounding off** process known as **rounding off numbers**. Two rules will be used in this book for **numbers** rounding off numbers:

Rule 1   When the first digit after those you want to retain is 4 or less, that digit and all others to its right are dropped. The last digit retained is not changed.

Examples rounded off to four digits:

74.693 = 74.69
         This digit is dropped.

1.00629 = 1.006
          These two digits are dropped.

Rule 2   When the first digit after those you want to retain is 5 or greater, that digit and all others to the right of it are dropped and the last digit retained is increased by one.

Examples rounded off to four digits:

1.026868 = 1.027

— These three digits are dropped.
— This digit is changed to 7.

18.02500 = 18.03

— These three digits are dropped.
— This digit is changed to 3.

12.899 = 12.90

— This digit is dropped.
— These two digits are changed to 90.

PRACTICE   Round off these numbers to the number of significant digits indicated:
(a) 42.246 (four digits)   (d) 0.08965 (two digits)
(b) 88.015 (three digits)   (e) 225.3 (three digits)
(c) 0.08965 (three digits)   (f) 14.150 (three digits)
Answers: (a) 42.25 (Rule 2)   (b) 88.0 (Rule 1)   (c) 0.0897 (Rule 2)   (d) 0.090 (Rule 2)
(e) 225 (Rule 1)   (f) 14.2 (Rule 2)

2.4   Scientific Notation of Numbers

The age of the earth has been estimated as about 4,500,000,000 (4.5 billion) years. Because this is an estimated value, let us say to the nearest 0.1 billion years, we are justified in using only two significant figures to express it. To express this number with two significant figures we write it using a power of 10 as 4.5×10^9 years.

Very large and very small numbers are often used in chemistry. These numbers can be simplified and conveniently written using a power of 10. Writing a number as a power of 10 is called **scientific notation**.

scientific notation

To write a number in scientific notation, move the decimal point in the original number so that it is located after the first nonzero digit. This new number is multiplied by 10 raised to the proper power (exponent). The power of 10 is equal to the number of places that the decimal point has been moved. If the decimal was moved to the left, the power of 10 will be a positive number. If the decimal was moved to the right, the power of 10 will be a negative number.

The scientific notation of a number is the number written as a factor between 1 and 10 multiplied by 10 raised to a power. For example,

$$2468 = 2.468 \times 10^3$$

number   scientific notation
                of the number

Study the examples that follow.

EXAMPLE 2.1   Write 5283 in scientific notation.

5283.   Place the decimal between the 5 and the 2. Since the decimal was moved three
3 2 1   places to the left, the power of 10 will be 3, and the number 5.283 is multiplied
          by 10^3.

5.283×10^3 (Correct scientific notation)

EXAMPLE 2.2   Write 4,500,000,000 in scientific notation (two significant figures).

4,500,000,000.   Place the decimal between the 4 and the 5. Since the decimal was
9   1   moved nine places to the left, the power of 10 will be 9, and the
                    number 4.5 is multiplied by 10^9.

4.5×10^9 (Correct scientific notation)

EXAMPLE 2.3   Write 0.000123 in scientific notation.

0.000123   Place the decimal between the 1 and the 2. Since the decimal was moved
      4   four places to the right, the power of 10 will be -4, and the number 1.23 is
             multiplied by 10^{-4}.

1.23×10^{-4} (Correct scientific notation)

PRACTICE   Write the following numbers in scientific notation:
(a) 1200 (four digits)   (c) 0.0468
(b) 6,600,000 (two digits)   (d) 0.00003

Answers: (a) $1200 = 1.200 \times 10^3$ (b) $6\,600\,000 = 6.6 \times 10^6$
                   3   6

(c) $0.0468 = 4.68 \times 10^{-2}$ (d) $0.00003 = 3 \times 10^{-5}$
           2   5

2.5   Significant Figures in Calculations

The results of a calculation based on measurements cannot be more precise than
the least precise measurement.

Multiplication or Division

In calculations involving multiplication or division, the answer must contain the same number of significant figures as in the measurement that has the least number of significant figures. Consider the following examples:

EXAMPLE 2.4 $190.6 \times 2.3 = 438.38$

The value 438.38 was obtained with a hand calculator. The answer should have two significant figures, because 2.3, the number with the fewest significant figures, has only two significant figures. The answer must, therefore, be expressed in scientific notation.

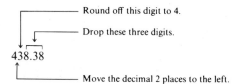

Round off this digit to 4.

Drop these three digits.

438.38

Move the decimal 2 places to the left.

The correct answer is 4.4×10^2.

EXAMPLE 2.5 $\dfrac{13.59 \times 6.3}{12} = 7.13475$

The value 7.13475 was obtained with a hand calculator. The answer should contain two significant figures because 6.3 and 12 have only two significant figures.

Drop these four digits.

7.13475

This digit remains the same.

The correct answer is 7.1.

PRACTICE   134 in. × 25 in. = ?
Answer: 3350 in.^{2} = 3.4×10^3 in.^{2}

PRACTICE $\dfrac{213 \text{ miles}}{4.20 \text{ hours}} = ?$
Answer: 50.7 miles/hour

PRACTICE $\dfrac{2.2 \times 273}{760} = ?$

Answer: 0.79

Addition or Subtraction

The results of an addition or a subtraction must be expressed to the same precision as the least precise measurement.

EXAMPLE 2.6   Add 125.17, 129.2, and 52.24.

125.17
129.2
 52.24
306.61   (306.6)

The number with the greatest uncertainty is 129.2. Therefore the answer is rounded off to the nearest tenth. 306.6

EXAMPLE 2.7   Subtract 14.1 from 132.56.

 132.56
−14.1
 118.46   (118.5)

14.1 is the number with the least precision. Therefore the answer is rounded off to the nearest tenth. 118.5

EXAMPLE 2.8   Subtract 120 from 1587.

 1587
−120
 1467   (1.47×10^3)

120 is the number with greatest uncertainty. The zero is not considered significant, therefore the answer must be rounded to the nearest ten. 1470 or 1.47×10^3.

EXAMPLE 2.9   Add 5672 and 0.00063.

5672
+ 0.00063
5672.00063   (5672)

The number with the greatest uncertainty is 5672. Therefore the answer is rounded off to the nearest unit. 5672

EXAMPLE 2.10 $$\frac{1.039 - 1.020}{1.039} = 0.018286814$$

The value 0.018286814 was obtained with a hand calculator. When the subtraction in the numerator is done,

$$1.039 - 1.020 = 0.019$$

the number of significant figures changes from four to two. Therefore the answer should contain two significant figures after the division is carried out:

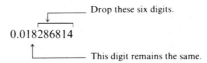

The correct answer is 0.018 or 1.8×10^{-2}.

PRACTICE How many significant figures should the answer contain in each of these calculations?
(a) 14.0×5.2 (b) 0.1682×8.2
(c) $\dfrac{160 \times 33}{4}$ (d) $8.2 + 0.125$
(e) $119.1 - 3.44$ (f) $\dfrac{94.5}{1.2}$
(g) $1200 + 6.34$ (h) $1.6 + 23 - 0.005$
Answers: (a) 2   (b) 2   (c) 1   (d) 2   (e) 4   (f) 2   (g) 2   (h) 2

Additional material on mathematical operations is given in the "Mathematical Review" in Appendix I. Review Appendix I and study carefully any portions that are not familiar to you. This study may be done at various times during the course as the need for additional knowledge of mathematical operations arises.

2.6 The Metric System

metric system
or SI

The **metric system**, or **International System (SI**, from *Système International*), is a decimal system of units for measurements of mass, length, time, and other physical quantities. It is built around a set of base units and uses factors of 10 to express larger or smaller numbers of these units. To express quantities that are larger or smaller than the base units, prefixes are added to the names of the units. These prefixes represent multiples of 10, making the metric system a decimal

...nd Numerical Values for SI Units*

Prefix	Symbol	Numerical value	Power of 10 equivalent
exa	E	1,000,000,000,000,000,000	10^{18}
peta	P	1,000,000,000,000,000	10^{15}
tera	T	1,000,000,000,000	10^{12}
giga	G	1,000,000,000	10^{9}
mega	M	1,000,000	10^{6}
kilo	k	1,000	10^{3}
hecto	h	100	10^{2}
deka	da	10	10^{1}
—	—	1	10^{0}
deci	d	0.1	10^{-1}
centi	c	0.01	10^{-2}
milli	m	0.001	10^{-3}
micro	μ	0.000001	10^{-6}
nano	n	0.000000001	10^{-9}
pico	p	0.000000000001	10^{-12}
femto	f	0.000000000000001	10^{-15}
atto	a	0.000000000000000001	10^{-18}

* The more commonly used prefixes are in color.

system of measurements. Table 2.1 shows the names, symbols, and numerical values of the prefixes. Some of the more commonly used prefixes are highlighted in color.

Examples are

1 kilometer	= 1000 meters
1 kilogram	= 1000 grams
1 millimeter	= 0.001 meter
1 microsecond	= 0.000001 second

The seven base units in the International System, their abbreviations, and the quantities they measure are given in Table 2.2. Other units are derived from these base units.

The metric system, or International System, is currently used by most of the countries in the world, not only for scientific and technical work, but also in commerce and industry. The United States is currently in the process of changing to the metric system of mass and measurements.

TABLE 2.2

International System Base Units of Measurement

Quantity	Name of unit	Abbreviation
Length	Meter	m
Mass	Kilogram	kg
Temperature	Kelvin	K
Time	Second	s
Amount of substance	Mole	mol
Electric current	Ampere	A
Luminous intensity	Candela	cd

2.7   Measurement of Length

Standards for the measurement of length have an interesting historical development. The Old Testament mentions such units as the *cubit* (the distance from a man's elbow to the tip of his outstretched hand). In ancient Scotland the inch was once defined as a distance equal to the width of a man's thumb.

meter

Reference standards of measurements have undergone continuous improvements in precision. The standard unit of length in the metric system is the **meter**. When the metric system was first introduced in the 1790s, the meter was defined as one ten-millionth of the distance from the equator to the North Pole measured along the meridian passing through Dunkirk, France. In 1889 the meter was redefined as the distance between two engraved lines on a platinum–iridium alloy bar maintained at 0° Celsius. This international meter bar is stored in a vault at Sèvres near Paris. Duplicate meter bars have been made and are used as standards by many nations.

By the 1950s length could be measured with such precision that a new standard was needed. Accordingly, the length of the meter was redefined in 1960 and again in 1983. The latest definition is: A meter is the distance that light travels in a vacuum during 1/299,792,458 of a second.

A meter is 39.37 inches, a little longer than 1 yard. One meter contains 10 decimeters, 100 centimeters, or 1000 millimeters (see Figure 2.2). A kilometer contains 1000 meters. Table 2.3 shows the relationships of these units.

The nanometer (10^{-9} m) is used extensively in expressing the wavelength of light and in atomic dimensions. See Appendix III for a complete table of common

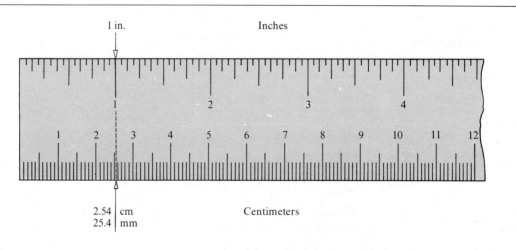

FIGURE 2.2

Comparison of the metric and American systems of length measurement: 2.54 cm = 1 in.

TABLE 2.3

Units of Length

Unit	Abbreviation	Meter equivalent	Exponential equivalent
Kilometer	km	1000 m	10^3 m
Meter	m	1 m	10^0 m
Decimeter	dm	0.1 m	10^{-1} m
Centimeter	cm	0.01 m	10^{-2} m
Millimeter	mm	0.001 m	10^{-3} m
Micrometer	μm	0.000001 m	10^{-6} m
Nanometer	nm	0.000000001 m	10^{-9} m
Angstrom	Å	0.0000000001 m	10^{-10} m

conversions. Other important relationships are:

$$1 \text{ m} = 100 \text{ cm} = 1000 \text{ mm} = 10^6 \ \mu\text{m} = 10^{10} \text{ Å}$$
$$1 \text{ cm} = 10 \text{ mm} = 0.01 \text{ m}$$
$$1 \text{ in.} = 2.54 \text{ cm}$$
$$1 \text{ mile} = 1.61 \text{ km}$$

2.8 Problem Solving

Many chemical principles are illustrated by mathematical concepts. Learning how to set up and solve numerical problems in a systematic fashion is *essential* in the study of chemistry. This skill, once acquired, is also very rewarding in other study areas. An electronic calculator will save you much time in computation.

Usually a problem can be solved by several methods. But in all methods it is best, especially for beginners, to use a systematic, orderly approach. The dimensional analysis, or factor-label, method is stressed in this book because

1. It provides a systematic, straightforward way to set up problems.
2. It gives a clear understanding of the principles involved.
3. It helps in learning to organize and evaluate data.
4. It helps to identify errors because unwanted units are not eliminated if the setup of the problem is incorrect.

The basic steps for solving problems are

1. Read the problem very carefully to determine what is to be solved for, and write it down.
2. Tabulate the data given in the problem. Even in tabulating data it is important to label all factors and measurements with the proper units.
3. Determine which principles are involved and which unit relationships are needed to solve the problem. Sometimes it is necessary to refer to tables for needed data.
4. Set up the problem in a neat, organized, and logical fashion, making sure that unwanted units cancel. Use sample problems in the text as guides for making setups.
5. Proceed with the necessary mathematical operations. Make certain that the answer contains the proper number of significant figures.
6. Check the answer to see if it is reasonable.

Just a few more words about problem solving. Don't allow any formal method of problem solving to limit your use of common sense and intuition. If a problem is clear to you and its solution seems simpler by another method, by all means use it. But in the long run you should be able to solve many otherwise difficult problems by using the *dimensional analysis method*.

The dimensional analysis method of problem solving converts one unit to another unit by the use of conversion factors.

$$\text{unit}_1 \times \text{conversion factor} = \text{unit}_2$$

If you want to know how many millimeters are in 2.5 meters, you need to convert meters (m) to millimeters (mm). Therefore, you start by writing

$$\text{m} \times \text{conversion factor} = \text{mm}$$

This conversion factor must accomplish two things. It must cancel, or eliminate, meters; and it must introduce millimeters, the unit wanted in the answer. Such a conversion factor will be in fractional form and have meters in the denominator and millimeters in the numerator:

$$\cancel{\text{m}} \times \frac{\text{mm}}{\cancel{\text{m}}} = \text{mm}$$

We know that 1 m = 1000 mm. From this relationship we can write two factors, 1 m per 1000 mm and 1000 mm per 1 m:

$$\frac{1 \text{ m}}{1000 \text{ mm}} \quad \text{and} \quad \frac{1000 \text{ mm}}{1 \text{ m}}$$

Using the factor 1000 mm/1 m, we can set up the calculation for the conversion of 2.5 m to millimeters,

$$2.5 \cancel{\text{m}} \times \frac{1000 \text{ mm}}{1 \cancel{\text{m}}} = 2500 \text{ mm} \qquad \text{or} \qquad 2.5 \times 10^3 \text{ mm}$$
$$\text{(two significant figures)}$$

Note that, in making this calculation, units are treated as numbers; meters in the numerator are canceled by meters in the denominator.

Now suppose you need to change 215 centimeters to meters. First you must determine that you need to convert centimeters to meters. We start with

$$\text{cm} \times \text{conversion factor} = \text{m}$$

The conversion factor must have centimeters in the denominator and meters in the numerator:

$$\cancel{\text{cm}} \times \frac{\text{m}}{\cancel{\text{cm}}} = \text{m}$$

From the relationship 100 cm = 1 m, we can write a factor that will accomplish this conversion:

$$\frac{1 \text{ m}}{100 \text{ cm}}$$

Now set up the calculation using all the data given.

$$215 \text{ cm} \times \frac{1 \text{ m}}{100 \text{ cm}} = \frac{215 \text{ m}}{100} = 2.15 \text{ m}$$

Some problems may require a series of conversions to reach the correct units in the answer. For example, suppose we want to know the number of seconds in 1 day. We need to go from the unit of days to seconds in this manner:

$$\text{day} \longrightarrow \text{hours} \longrightarrow \text{minutes} \longrightarrow \text{seconds}$$

This series requires three conversion factors, one for each step. We convert days to hours (hr), hours to minutes (min), and minutes to seconds (s). The conversions can be done individually or in a continuous sequence:

$$\text{day} \times \frac{\text{hr}}{\text{day}} \longrightarrow \text{hr} \times \frac{\text{min}}{\text{hr}} \longrightarrow \text{min} \times \frac{\text{s}}{\text{min}}$$

$$\text{day} \times \frac{\text{hr}}{\text{day}} \times \frac{\text{min}}{\text{hr}} \times \frac{\text{s}}{\text{min}} = \text{s}$$

Inserting the proper factors we calculate the number of seconds in 1 day to be

$$1 \text{ day} \times \frac{24 \text{ hr}}{1 \text{ day}} \times \frac{60 \text{ min}}{1 \text{ hr}} \times \frac{60 \text{ s}}{1 \text{ min}} = 86,400. \text{ s}$$

All five digits in 86,400 are significant, since all the factors in the calculation are exact numbers.

The dimensional analysis, or factor-label, method used in the preceding work shows how unit conversion factors are derived and used in calculations. After you become more proficient with the terms, you can save steps by writing the factors directly in the calculation. The problems that follow give examples of the conversion from American to metric units.

Label all factors with the proper units.

EXAMPLE 2.11

How many centimeters are in 2.00 ft?
The stepwise conversion of units from feet to centimeters may be done in this manner: Convert feet to inches; then convert inches to centimeters.

$$\text{ft} \longrightarrow \text{in.} \longrightarrow \text{cm}$$

The conversion factors needed are

$$\frac{12 \text{ in.}}{1 \text{ ft}} \quad \text{and} \quad \frac{2.54 \text{ cm}}{1 \text{ in.}}$$

$$2.00 \, \cancel{\text{ft}} \times \frac{12 \text{ in.}}{1 \, \cancel{\text{ft}}} = 24.0 \text{ in.}$$

$$24.0 \, \cancel{\text{in.}} \times \frac{2.54 \text{ cm}}{1 \, \cancel{\text{in.}}} = 61.0 \text{ cm} \quad \text{(Answer)}$$

Since 1 ft and 12 in. are exact numbers, the number of significant figures allowed in the answer is three, based on the number 2.00.

EXAMPLE 2.12   How many meters are in a 100. yd football field? The stepwise conversion of units from yards to meters may be done in this manner, using the proper conversion factors.

$$\text{yd} \longrightarrow \text{ft} \longrightarrow \text{in.} \longrightarrow \text{cm} \longrightarrow \text{m}$$

$$100. \, \cancel{\text{yd}} \times \frac{3 \text{ ft}}{1 \, \cancel{\text{yd}}} = 300 \text{ ft} \qquad (3 \text{ ft/yd})$$

$$300 \, \cancel{\text{ft}} \times \frac{12 \text{ in.}}{1 \, \cancel{\text{ft}}} = 3600 \text{ in.} \qquad (12 \text{ in./ft})$$

$$3600 \, \cancel{\text{in.}} \times \frac{2.54 \text{ cm}}{1 \, \cancel{\text{in.}}} = 9144 \text{ cm} \qquad (2.54 \text{ cm/in.})$$

$$9144 \, \cancel{\text{cm}} \times \frac{1 \text{ m}}{100 \, \cancel{\text{cm}}} = 91.4 \text{ m} \qquad (1 \text{ m/100 cm}) \qquad \text{(three significant figures)}$$

Examples 2.11 and 2.12 may be solved using a linear expression and writing down conversion factors in succession. This method often saves one or two calculation steps and allows numerical values to be reduced to simpler terms, leading to simpler calculations. The single linear expressions for Examples 2.11 and 2.12 are

$$2.00 \, \cancel{\text{ft}} \times \frac{12 \, \cancel{\text{in.}}}{1 \, \cancel{\text{ft}}} \times \frac{2.54 \text{ cm}}{1 \, \cancel{\text{in.}}} = 61.0 \text{ cm}$$

$$100. \, \cancel{\text{yd}} \times \frac{3 \, \cancel{\text{ft}}}{1 \, \cancel{\text{yd}}} \times \frac{12 \, \cancel{\text{in.}}}{1 \, \cancel{\text{ft}}} \times \frac{2.54 \, \cancel{\text{cm}}}{1 \, \cancel{\text{in.}}} \times \frac{1 \text{ m}}{100 \, \cancel{\text{cm}}} = 91.4 \text{ m}$$

Using the units alone (Example 2.12), we see that the stepwise cancellation proceeds in succession until the desired unit is reached.

$$\cancel{\text{yd}} \times \frac{\cancel{\text{ft}}}{\cancel{\text{yd}}} \times \frac{\cancel{\text{in.}}}{\cancel{\text{ft}}} \times \frac{\cancel{\text{cm}}}{\cancel{\text{in.}}} \times \frac{\text{m}}{\cancel{\text{cm}}} = \text{m}$$

PRACTICE   How many meters are in 1.00 mile?
Answer: 1.61×10^3 m

EXAMPLE 2.13   How many cubic centimeters (cm^3) are in a box that measures 2.20 in. by 4.00 in. by 6.00 in.? First we need to determine the volume of the box in cubic inches ($in.^3$) by multiplying together the length times the width times the height.

$$2.20 \text{ in.} \times 4.00 \text{ in.} \times 6.00 \text{ in.} = 52.8 \text{ in.}^3$$

Now we need to convert $in.^3$ to cm^3, which can be done by using the inches and centimeters relationship three times.

$$\cancel{in.^3} \times \frac{cm}{\cancel{in.}} \times \frac{cm}{\cancel{in.}} \times \frac{cm}{\cancel{in.}} = cm^3$$

$$52.8 \cancel{in.^3} \times \frac{2.54 \text{ cm}}{1 \cancel{in.}} \times \frac{2.54 \text{ cm}}{1 \cancel{in.}} \times \frac{2.54 \text{ cm}}{1 \cancel{in.}} = 865 \text{ cm}^3$$

PRACTICE   How many cubic meters are in a room measuring 8 ft $\times$ 10 ft $\times$ 12 ft?
Answer: $27.2 \text{ m}^3 = 3 \times 10^1 \text{ m}^3$

2.9   Measurement of Mass

kilogram

The gram is used as a unit of mass measurement, but it is a tiny amount of mass; for instance, a nickel has a mass of about 5 grams. Therefore the *standard unit* of mass in the SI system is the **kilogram** (equal to 1000 g). The amount of mass in a kilogram is defined by international agreement as exactly equal to the mass of a platinum–iridium cyclinder (international prototype kilogram) kept in a vault at Sèvres, France. Comparing this unit of mass to 1 lb (16 ozs), we find that 1 kg is equal to 2.2 lb. A pound is equal to 454 g (0.454 kg). The same prefixes used in length measurement are used to indicate larger and smaller gram units (see Table 2.4).

A balance is used to measure mass. Some balances will determine the mass of objects to the nearest microgram. The choice of balance depends on the accuracy required and the amount of material. Several balances are shown in Figure 2.3.

TABLE 2.4

Metric Units of Mass

Unit	Abbreviation	Gram equivalent	Exponential equivalent
Kilogram	kg	1000 g	10^3 g
Gram	g	1 g	10^0 g
Decigram	dg	0.1 g	10^{-1} g
Centigram	cg	0.01 g	10^{-2} g
Milligram	mg	0.001 g	10^{-3} g
Microgram	μg	0.000001 g	10^{-6} g

It is convenient to remember that

$1\,g = 1000\,mg$

$1\,kg = 1000\,g$

$1\,kg = 2.2\,lb$

$1\,lb = 454\,g$

To change grams to milligrams, multiply grams by the conversion factor 1000 mg/g. The setup for converting 25 g to milligrams is

$$25\,g \times \frac{1000\,mg}{1\,g} = 25{,}000\,mg \qquad (2.5 \times 10^4\,mg) \quad \text{(Answer)}$$

Note that multiplying a number by 1000 is the same as multiplying the number by 10^3 and can be done simply by moving the decimal point three places to the right

$$6.428 \times 1000 = 6428 \qquad (6.428)$$

To change milligrams to grams, multiply milligrams by the conversion factor 1 g/1000 mg. For example, to convert 155 mg to grams:

$$155\,mg \times \frac{1\,g}{1000\,mg} = 0.155\,g \quad \text{(Answer)}$$

Mass conversions from American to metric units are shown in Examples 2.14 and 2.15.

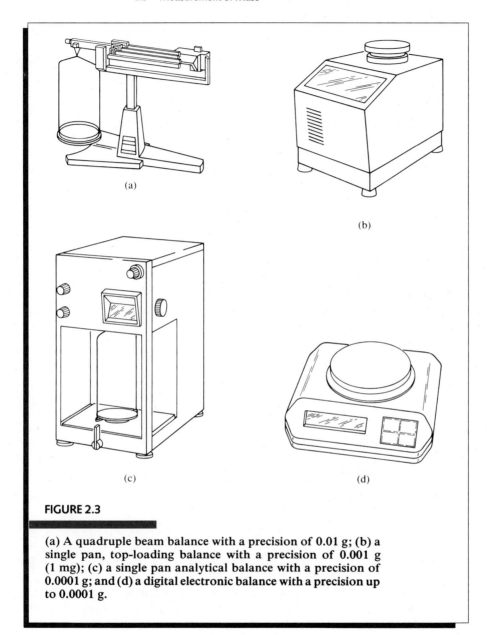

(a)

(b)

(c)   (d)

FIGURE 2.3

(a) A quadruple beam balance with a precision of 0.01 g; (b) a single pan, top-loading balance with a precision of 0.001 g (1 mg); (c) a single pan analytical balance with a precision of 0.0001 g; and (d) a digital electronic balance with a precision up to 0.0001 g.

EXAMPLE 2.14   A 1.50 lb package of sodium bicarbonate costs 80 cents. How many grams of this substance are in this package?

We are solving for the number of grams equivalent to 1.50 lb. Since 1 lb = 454 g, the factor to convert pounds to grams is 454 g/lb.

$$1.50 \cancel{lb} \times \frac{454 \text{ g}}{1 \cancel{lb}} = 681 \text{ g} \quad \text{(Answer)}$$

Note: The cost of the sodium bicarbonate has no bearing on the question asked in this problem.

EXAMPLE 2.15   Suppose four ostrich feathers weigh 1.00 lb. Assuming that each feather is equal in weight, how many milligrams does a single feather weigh? The unit conversion in this problem is from 1 lb/4 feathers to milligrams per feather. Since the unit *feathers* occurs in the denominator of both the starting unit and the desired unit, the unit conversions needed are

$$\text{lb} \longrightarrow \text{g} \longrightarrow \text{mg}$$

$$\frac{1.00 \cancel{lb}}{4 \text{ feathers}} \times \frac{454 \cancel{g}}{1 \cancel{lb}} \times \frac{1000 \text{ mg}}{1 \cancel{g}} = \frac{113,500 \text{ mg}}{1 \text{ feather}} \qquad (1.14 \times 10^5 \text{ mg/feather})$$

PRACTICE   You are traveling in Europe and wake up one morning to find your mass is 75.0 kg. Determine the American equivalent to see whether you need to go on a diet before you return home.

Answer: 165 lb

PRACTICE   A tennis ball has a mass of 65 g. Determine the American equivalent in pounds.

Answer: 0.14 lb

2.10   Measurement of Volume

volume

Volume, as used here, is the amount of space occupied by matter. The SI unit of volume is the *cubic meter* (m^3). However, the liter (pronounced *leeter* and abbreviated L) and the milliliter (abbreviated mL) are the standard units of volume used in most chemical laboratories.

The most common instruments or equipment for measuring liquids are the graduated cylinder, volumetric flask, buret, pipet, and syringe, which are illustrated in Figure 2.4. These pieces are usually made of glass and are available in various sizes.

It is convenient to remember that

$$1 \text{ L} = 1000 \text{ mL} = 1000 \text{ cm}^3$$
$$1 \text{ mL} = 1 \text{ cm}^3$$
$$1 \text{ L} = 1.057 \text{ qt}$$
$$946 \text{ mL} = 1 \text{ qt}$$

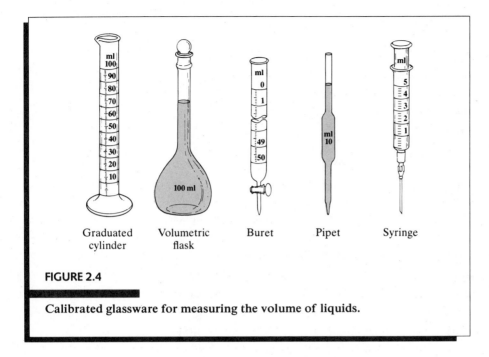

Graduated   Volumetric   Buret   Pipet   Syringe
cylinder   flask

FIGURE 2.4

Calibrated glassware for measuring the volume of liquids.

The volume of a cubic or rectangular container can be determined by multiplying its length times width times height. Thus a 10 cm square box has a volume of 10 cm × 10 cm × 10 cm = 1000 cm³.

The following examples illustrate volume conversions.

EXAMPLE 2.16   How many milliliters are contained in 3.5 liters?

The conversion factor to change liters to milliliters is 1000 mL/L.

$$3.5\,\cancel{L} \times \frac{1000\ \text{mL}}{\cancel{L}} = 3500\ \text{mL} \qquad (3.5 \times 10^3\ \text{mL})$$

Liters may be changed to milliliters by moving the decimal point three places to the right and changing the units to milliliters.

$$1.500\ \text{L} = 1500.\ \text{mL}$$

EXAMPLE 2.17   How many cubic centimeters are in a cube that is 11.1 inches on a side?

First change inches to centimeters. The conversion factor is 2.54 cm/in.

$$11.1\,\cancel{\text{in.}} \times \frac{2.54\ \text{cm}}{1\,\cancel{\text{in.}}} = 28.2\ \text{cm on a side}$$

Then change to cubic volume (length × width × height).

$$28.2 \text{ cm} \times 28.2 \text{ cm} \times 28.2 \text{ cm} = 22{,}426 \text{ cm}^3 \qquad (2.24 \times 10^4 \text{ cm}^3)$$

PRACTICE   An excellent bottle of chianti holds 750. mL. Determine the volume in quarts.

Answer: 0.793 qt

PRACTICE   Milk is often purchased by the half gallon. Determine the number of liters necessary to equal this amount.

Answer: 1.89 L (Number of significant figures is arbitrary.)

2.11   Measurement of Temperature

heat

Heat is a form of energy associated with the motion of small particles of matter. The term *heat* refers to the quantity of energy within a system or to a quantity of energy added to or taken away from a system. *System* as used here simply refers to the entity that is being heated or cooled. Depending on the amount of heat energy present, a given system is said to be hot or cold. **Temperature** is a measure of the intensity of heat, or how hot a system is, regardless of its size. Heat always flows from a region of higher temperature to one of lower temperature. The SI unit of temperature is the Kelvin. The common laboratory instrument for measuring temperature is a thermometer (see Figure 2.5).

temperature

The temperature of a system can be expressed by several different scales. Three commonly used temperature scales are the Celsius scale (pronounced *sell-see-us*), the Kelvin (absolute) scale, and the Fahrenheit scale. The unit of temperature on the Celsius and Fahrenheit scales is called a *degree*, but the size of the Celsius and the Fahrenheit degree is not the same. The symbol for the Celsius and Fahrenheit degrees is °, and it is placed as a superscript after the number and before the symbol for the scales. Thus, 100°C means 100 *degrees Celsius*. The degree sign is not used with Kelvin temperatures.

degrees Celsius = °C

Kelvin (absolute) = K

degrees Fahrenheit = °F

On the Celsius scale the interval between the freezing and boiling temperatures of water is divided into 100 equal parts, or degrees. The freezing point of water is assigned a temperature of 0°C and the boiling point of water a temperature of 100°C. The Kelvin temperature scale is also known as the absolute temperature scale, because 0 K is the lowest temperature theoretically attainable. The Kelvin zero is 273.16 degrees below the Celsius zero. A Kelvin is

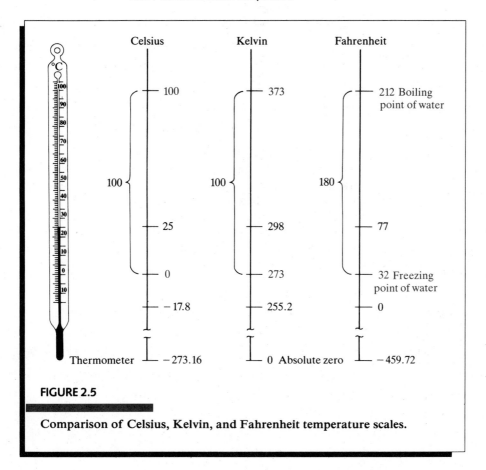

FIGURE 2.5

Comparison of Celsius, Kelvin, and Fahrenheit temperature scales.

equal in size to a Celsius degree. The freezing point of water on the Kelvin scale is 273.16 K (usually rounded to 273 K). The Fahrenheit scale has 180 degrees between the freezing and boiling temperatures of water. On this scale the freezing point of water is 32°F and the boiling point is 212°F.

$$0°C \cong 273 \text{ K} \cong 32°F$$

The three scales are compared in Figure 2.5. Although absolute zero (0 K) is the lower limit of temperature on these scales, temperature has no known upper limit. (Temperatures of several million degrees are known to exist in the sun and in other stars.)

By examining Figure 2.5 we can see that there are 100 Celsius degrees and 100 Kelvins between the freezing and boiling points of water. But there are 180 Fahrenheit degrees between these two temperatures. Hence, the size of a degree on the Celsius scale is the same as the size of one Kelvin, but one Celsius degree corresponds to 1.8 degrees on the Fahrenheit scale.

$$1°C = 1 \text{ K} = 1.8°F$$

From these data, mathematical formulas have been derived to convert a temperature on one scale to the corresponding temperature on another scale. These formulas are

$$K = {}^\circ C + 273 \qquad\qquad (1)$$

$${}^\circ F = (1.8 \times {}^\circ C) + 32 \qquad\qquad (2)$$

$${}^\circ C = \frac{({}^\circ F - 32)}{1.8} \qquad\qquad (3)$$

Interpretation: Formula (1) states that the addition of 273 to the degrees Celsius converts the temperature to Kelvins. Formula (2) states that to obtain the Fahrenheit temperature corresponding to a given Celsius temperature, we multiply the degrees Celsius by 1.8 and then add 32. Formula (3) states that to obtain the corresponding Celsius temperature, we subtract 32 from the degrees Fahrenheit and then divide this answer by 1.8. Examples of temperature conversions follow.

EXAMPLE 2.18

The temperature at which table salt (sodium chloride) melts is 800.°C. What is this temperature on the Kelvin and Fahrenheit scales?

We need to calculate K from °C, so we use formula (1) above. We also need to calculate °F from °C; for this calculation we use formula (2).

$$K = {}^\circ C + 273$$
$$K = 800.{}^\circ C + 273 = 1073 \text{ K}$$
$${}^\circ F = (1.8 \times {}^\circ C) + 32$$
$${}^\circ F = (1.8 \times 800.{}^\circ C) + 32$$
$${}^\circ F = 1440 + 32 = 1472{}^\circ F$$
$$800.{}^\circ C = 1073 \text{ K} = 1472{}^\circ F$$

EXAMPLE 2.19

The temperature for December 1 was 110.°C, a new record. Calculate this temperature in °C.

Formula (3) applies here.

$${}^\circ C = \frac{({}^\circ F - 32)}{1.8}$$

$${}^\circ C = \frac{(110. - 32)}{1.8} = \frac{78}{1.8} = 43{}^\circ C$$

EXAMPLE 2.20

What temperature on the Fahrenheit scale corresponds to $-8.0{}^\circ C$? (Be alert to the presence of the minus sign in this problem.)

$$°F = (1.8 \times °C) + 32$$
$$°F = [1.8 \times (-8.0)] + 32 = -14.4 + 32$$
$$°F = 17.6$$

Temperatures used in this book are in degrees Celsius (°C) unless specified otherwise. The temperature after conversion should be expressed to the same precision as the original measurement.

PRACTICE   Helium boils at 4 K. Convert this temperature to °C and then to °F.
Answer: $-269°C$; $-452°F$

PRACTICE   "Normal" human body temperature is 98.6°F. Convert this to °C and K.
Answer: 37.0°C; 310.0 K

2.12   Density

density

Density (d) is the ratio of the mass of a substance to the volume occupied by that mass; it is the mass per unit of volume and is given by the equation

$$d = \frac{\text{mass}}{\text{volume}}$$

Density is a physical characteristic of a substance and may be used as an aid to its identification. When the density of a solid or a liquid is given, the mass is usually expressed in grams and the volume in milliliters or cubic centimeters.

$$d = \frac{\text{mass}}{\text{volume}} = \frac{\text{g}}{\text{mL}} \qquad \text{or} \qquad d = \frac{g}{\text{cm}^3}$$

Since the volume of a substance (especially liquids and gases) varies with temperature, it is important to state the temperature along with the density. For example, the volume of 1.0000 g water at 4°C is 1.0000 mL, at 20°C it is 1.0018 mL, and at 80°C it is 1.0290 mL. Density, therefore, also varies with temperature.

The density of water at 4°C is 1.0000 g/mL, but at 80°C the density of water is 0.9718 g/mL.

$$d^{4°C} = \frac{1.0000 \text{ g}}{1.0000 \text{ mL}} = 1.0000 \text{ g/mL}$$

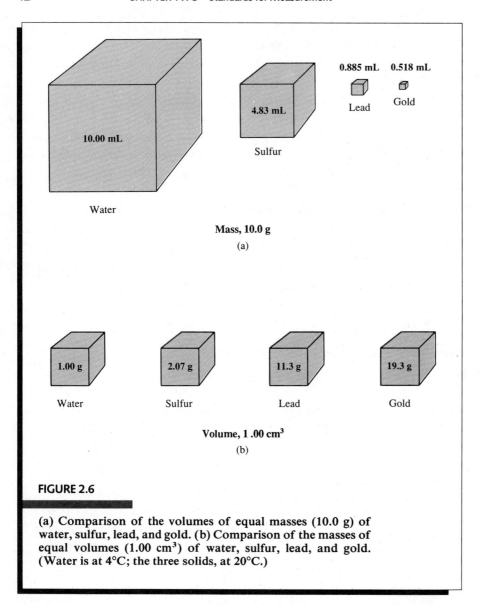

FIGURE 2.6

(a) Comparison of the volumes of equal masses (10.0 g) of water, sulfur, lead, and gold. (b) Comparison of the masses of equal volumes (1.00 cm³) of water, sulfur, lead, and gold. (Water is at 4°C; the three solids, at 20°C.)

$$d^{80°C} = \frac{1.0000 \text{ g}}{1.0290 \text{ mL}} = 0.9718 \text{ g/mL}$$

The density of iron at 20°C is 7.86 g/mL.

$$d^{20°C} = \frac{7.86 \text{ g}}{1.00 \text{ mL}} = 7.86 \text{ g/mL}$$

The densities of a variety of materials are compared in Figure 2.6

$$1.00 = \frac{1.09}{1.0ml}$$

TABLE 2.5

Densities of Some Selected Materials*

Liquids and solids		Gases	
Substance	**Density (g/mL at 20°C)**	**Substance**	**Density (g/L at 0°C)**
Wood (Douglas fir)	0.512	Hydrogen	0.090
Ethyl alcohol	0.789	Helium	0.178
Cottonseed oil	0.926	Methane	0.714
Water (4°C)	**1.0000**	Ammonia	0.771
Sugar	1.59	Neon	0.90
Carbon tetrachloride	1.595	Carbon monoxide	1.25
Magnesium	1.74	Nitrogen	1.251
Sulfuric acid	1.84	**Air**	**1.293**
Sulfur	2.07	Oxygen	1.429
Salt	2.16	Hydrogen chloride	1.63
Aluminum	2.70	Argon	1.78
Silver	10.5	Carbon dioxide	1.963
Lead	11.34	Chlorine	3.17
Mercury	13.55		
Gold	19.3		

* For comparing densities the density of water is the reference for solids and liquids; air is the reference for gases.

Densities for liquids and solids are usually represented in terms of grams per milliliter (g/mL) or grams per cubic centimeter (g/cm^3). The density of gases, however, is expressed in terms of grams per liter (g/L). Unless otherwise stated, gas densities are given for 0°C and 1 atmosphere pressure (discussed further in Chapter 13). Table 2.5 lists the densities of a number of common materials.

Suppose that water, carbon tetrachloride, and cottonseed oil are successively poured into a graduated cylinder. The result is a layered three-liquid system (Figure 2.7). Can we predict the order of the liquid layers? Yes, by looking up the densities in Table 2.5. Carbon tetrachloride has the greatest density (1.595 g/mL), and cottonseed oil has the lowest density (0.926 g/mL). Carbon tetrachloride will be the bottom layer, and cottonseed oil will be the top layer. Water, with a density between the other two liquids, will form the middle layer. This information can also be determined by experiment. Add a few milliliters of carbon tetrachloride to a beaker of water. The carbon tetrachloride, being more dense than the water, will sink. Cottonseed oil, being less dense than water, will float when added to the beaker. Direct comparisons of density in this manner can be made only with liquids that are *immiscible* (do not mix with or dissolve in one another).

The density of air at 0°C is approximately 1.293 g/L. Gases with densities

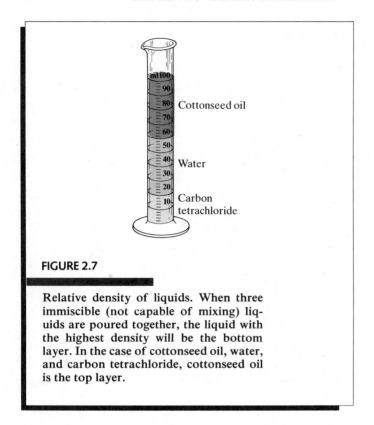

Cottonseed oil

Water

Carbon
tetrachloride

FIGURE 2.7

Relative density of liquids. When three immiscible (not capable of mixing) liquids are poured together, the liquid with the highest density will be the bottom layer. In the case of cottonseed oil, water, and carbon tetrachloride, cottonseed oil is the top layer.

less than this value are said to be "lighter than air." A helium-filled balloon will rise rapidly in air, because the density of helium is only 0.178 g/L.

When an insoluble solid object is dropped into water, it will sink or float, depending on its density. If the object is less dense than water, it will float, displacing a *mass* of water equal to the mass of the object. If the object is more dense than water, it will sink, displacing a *volume* of water equal to the volume of the object. This information can be utilized to determine the volume (and density) of irregularly shaped objects.

specific gravity   The **specific gravity** (sp gr) of a substance is a ratio of the density of that substance to the density of another substance. Water is usually used as the reference standard for solids and liquids. Air is usually used as the reference standard for gases. Specific gravity has no units because the density units cancel. The specific gravity tells us how many times as heavy a liquid, a solid, or a gas is as compared to the reference material.

$$\text{sp gr} = \frac{\text{density of a liquid or solid}}{\text{density of water}} \quad \text{or} \quad \frac{\text{density of a gas}}{\text{density of air}}$$

Sample calculations of density problems follow.

EXAMPLE 2.21 What is the density of a mineral if 427 g of the mineral occupy a volume of 35.0 mL? We need to solve for density, so we start by writing the formula for calculating density.

$$d = \frac{mass}{volume}$$

Then we substitute the data given in the problem into the equation and solve.

mass = 427 g   volume = 35.0 mL

$$d = \frac{mass}{volume} = \frac{427\ g}{35.0\ mL} = 12.2\ g/mL \quad (Answer)$$

EXAMPLE 2.22 The density of gold is 19.3 g/mL. What is the mass of 25.0 mL of gold?

Two ways to solve this problem are: (1) Solve the density equation for mass, then substitute the density and volume data into the new equation and calculate; (2) Solve by dimensional analysis.

Method 1 (a) Solve the density equation for mass:

$$d = \frac{mass}{volume} \qquad mass = d \times volume$$

(b) Substitute the data and calculate:

$$mass = \frac{19.3\ g}{mL} \times 25.0\ mL = 482\ g \quad (Answer)$$

Method 2 Dimensional analysis: Use density as a conversion factor, converting mL → g.

The conversion of units is

$$mL \times \frac{g}{mL} = g$$

$$25.0\ mL \times \frac{19.3\ g}{mL} = 482\ g \quad (Mass\ of\ gold)$$

EXAMPLE 2.23 Calculate the volume (in mL) of 100. g of ethyl alcohol.

From Table 2.5 we see that the density of ethyl alcohol is 0.789 g/mL. This density also means that 1 mL of the alcohol weighs 0.789 g (1 mL/0.789 g).

Method 1 This problem may be done by solving the density equation for volume and then substituting the data in the new equation.

$$d = \frac{mass}{volume} \qquad volume = \frac{mass}{d}$$

$$volume = \frac{100.\ g}{0.0789\ g/mL} = 127\ mL$$

Method 2 Dimensional analysis. For a conversion factor, we can use either

$$\frac{g}{mL} \quad \text{or} \quad \frac{mL}{g}$$

In this case the conversion is from g → mL, so we use mL/g. Substituting the data, we get

$$100\,\cancel{g} \times \frac{1\ mL}{0.789\,\cancel{g}} = 127\ mL \text{ of ethyl alchohol}$$

EXAMPLE 2.24   The water level in a graduated cylinder stands at 20.0 mL before and at 26.2 mL after a 16.74 g metal bolt is submerged in the water. (a) What is the volume of the bolt? (b) What is the density of the bolt?

(a) The bolt will displace a volume equal to its volume. Thus the increase in volume is the volume of the bolt.

$$26.2\ mL = \text{volume of water plus bolt}$$
$$\underline{-\,20.0\ mL = \text{volume of water}}$$
$$6.2\ mL = \text{volume of bolt} \quad \text{(Answer)}$$

(b) $d = \dfrac{\text{mass of bolt}}{\text{volume of bolt}} = \dfrac{16.74\ g}{6.2\ mL} = 2.7\ g/mL$ (Answer)

PRACTICE   Pure silver has a density of 10.5 g/mL. A ring sold as pure silver has a mass of 25.0 g. When placed in a graduated cylinder the water level rises 2.0 mL. Determine whether the ring is actually pure silver or if the customer should see the Better Business Bureau.
Answer: Density is 12.5 g/mL, therefore the ring is *not* pure silver.

PRACTICE   The water level in a metric measuring cup is 0.75 L before the addition of 150. g of shortening. The water level after the addition of the shortening is 0.92 L. Determine the density of the shortening.
Answer: 0.88 g/mL

Concepts in Review

1. Differentiate between mass and weight; include the instruments used to measure each.

2. Know the metric units of mass, length, and volume.

3. Know the numerical equivalent for the metric prefixes deci, centi, milli, micro, nano, kilo, and mega.

4. Express any number in scientific notation.

5. Express answers to calculations to the proper number of significant figures.

6. Set up and solve problems utilizing the method of dimensional analysis (factor-label method).

7. Convert measurements of mass, length, and volume from American units to metric units, and vice versa.

8. Make temperature conversions among Fahrenheit, Celsius, and Kelvin scales.

9. Differentiate between heat and temperature.

10. Calculate the density, mass, or volume of an object from the appropriate data.

Key Terms in Review

The terms listed here have all been defined within the chapter. Review the definitions of each. Use the glossary and the margin notations within the chapter as study aids.

density   scientific notation
heat   significant figures
kilogram   specific gravity
mass   temperature
meter   volume
metric system (SI)   weight
rounding off numbers

Exercises

An asterisk indicates a more challenging question or problem.

1. Use Table 2.3 to determine how many decimeters make up 1 km.

2. What is the temperature difference in Fahrenheit degrees between 25°C and 100°C? (see Figure 2.5.)

3. Use Figure 2.2 to determine the metric equivalent of 3 in.

4. Why do you suppose the neck of a 100 mL volumetric flask is narrower than the top of a 100 mL graduated cylinder? (See Figure 2.4.)

5. Refer to Table 2.5 and describe the arrangement that would be seen if these three immiscible substances were placed in a 100 mL graduated cylinder: 25 mL mercury, 25 mL carbon tetrachloride, and a cube of magnesium measuring 2.0 cm on an edge.

6. Arrange these materials in order of increasing density: salt, cottonseed oil, sulfur, aluminum, and ethyl alcohol.

7. Will an argon-filled balloon rise or sink in a methane atmosphere? Explain.

8. Ice floats in cottonseed oil and sinks in ethyl alcohol. The density of ice must therefore lie between what numerical values?

9. What are some of the important advantages of the metric system over the American system of weights and measurements?

10. What are the abbreviations for the following?
    (a) Gram
    (b) Kilogram
    (c) Milligram
    (d) Microgram
    (e) Centimeter
    (f) Millimeter
    (g) Micrometer
    (h) Angstrom
    (i) Milliliter
    (j) Microliter
    (k) Liter

11. For the following numbers, tell whether the zeros are significant or are not significant,
    (a) 503
    (b) 0.007
    (c) 4200
    (d) 3.0030
    (e) 100.00
    (f) 8.00×10^2

12. State the rules for rounding off numbers.

13. Distinguish between heat and temperature.

14. Distinguish between density and specific gravity.

15. Which of the following statements are correct? Rewrite the incorrect statements to make them correct.
    (a) The prefix *micro* indicates one-millionth of the unit expressed.
    (b) The length 10 cm is equal to 1000 mm.
    (c) The number 383.263 rounded to four significant figures becomes 383.3.
    (d) The number of significant figures in the number 29,004 is five.
    (e) The number 0.00723 contains three significant figures.
    (f) The sum of 32.276 + 2.134 should contain four significant figures.
    (g) The product of 18.42 cm × 3.40 cm should contain three significant figures.
    (h) One microsecond is 10^{-6} second.
    (i) One thousand meters is a longer distance than 1000 yards.
    (j) One liter is a larger volume than one quart.
    (k) One centimeter is longer than one inch.
    (l) One cubic centimeter (cm³) is equal to one milliliter.
    (m) The number 0.0002983 in exponential notation is 2.983×10^{-3}.
    (n) $3.0 \times 10^4 \times 6.0 \times 10^6 = 1.8 \times 10^{11}$
    (o) Temperature is a form of energy.
    (p) The density of water at 4°C is 1.00 g/mL.

(q) A pipet is a more accurate instrument for measuring 10.0 mL of water than is a graduated cylinder.

Significant Figures, Rounding, Exponential Notation

16. How many significant figures are in each of the following numbers?
    (a) 0.025
    (b) 40.0
    (c) 22.4
    (d) 0.0081
    (e) 0.0404
    (f) 129,042
    (g) 5.50×10^3
    (h) 4.090×10^{-3}

17. Round off the following numbers to three significant figures:
    (a) 93.246
    (b) 8.8726
    (c) 0.02854
    (d) 21.25
    (e) 4.644
    (f) 129.509
    (g) 34.250
    (h) 1.995×10^6

18. Express each of the following numbers in exponential notation:
    (a) 2,900,000
    (b) 0.0456
    (c) 0.58
    (d) 4082.2
    (e) 0.00840
    (f) 40.30
    (g) 12,000,000
    (h) 0.0000055

19. Solve the following mathematical problems, stating answers to the proper number of significant figures:
    (a) $12.62 + 1.5 + 0.25 =$
    (b) $4.68 \times 12.5 =$
    (c) $2.25 \times 10^3 \times 4.80 \times 10^4 =$
    (d) $\dfrac{182.6}{4.6} =$
    (e) $\dfrac{452 \times 6.2}{14.3} =$
    (f) $1986 + 23.48 + 0.012 =$
    (g) $0.0394 \times 12.8 =$
    (h) $2.92 \times 10^{-3} \times 6.14 \times 10^5 =$
    (i) $\dfrac{0.4278}{59.6} =$
    (j) $\dfrac{29.3}{284 \times 415} =$

20. Change these fractions to decimals. Express each answer to three significant figures.
    (a) $\dfrac{5}{6}$ (b) $\dfrac{3}{7}$ (c) $\dfrac{12}{16}$ (d) $\dfrac{9}{18}$

21. Solve each equation for X:
    (a) $3.42X = 6.5$
    (b) $\dfrac{X}{12.3} = 7.05$
    (c) $\dfrac{0.525}{X} = 0.25$
    (d) $0.298X = 15.3$

(e) $\dfrac{X}{0.819} = 10.9$ (f) $\dfrac{8.4}{X} = 282$

22. Solve each equation for the unknown:

(a) $°C = \dfrac{212 - 32}{1.8}$ (c) $K = 25 + 273$

(b) $°F = 1.8(22) + 32$ (d) $\dfrac{8.9\ g}{mL} = \dfrac{40.90\ g}{volume}$

Unit Conversions

23. Make the following conversions, showing mathematical setups:

(a) 28.0 cm to m
(b) 1000. m to km
(c) 9.28 cm to mm
(d) 150 mm to km
(e) 0.606 cm to km
(f) 4.5 cm to Å
(g) 6.5×10^{-7} m to Å
(h) 12.1 m to cm
(i) 8.0 km to m
(j) 315 mm to cm
(k) 25 km to mm
(l) 12 nm to cm
(m) 0.520 km to cm
(n) 3.884 Å to nm
(o) 42.2 in. to cm
(p) 0.64 mile to in.
(q) 504 miles to km
(r) 2.00 in.^{2} to cm^2
(s) 35.6 m to ft
(t) 16.5 km to miles
(u) 4.5 in.^{3} to mm^3
(v) 3.00 mile^3 to mm^3

24. Make the following conversions, showing mathematical setups:

(a) 10.68 g to mg
(b) 6.8×10^4 mg to kg
(c) 8.54 g to kg
(d) 42.8 kg to lb
(e) 164 mg to g
(f) 0.65 kg to mg
(g) 5.5 kg to g
(h) 95 lb to g

25. Make the following conversions, showing mathematical setups:

(a) 25.0 mL to L
(b) 22.4 L to mL
(c) 3.5 qt to mL
(d) 4.5×10^4 ft^3 to m^3
(e) 0.468 L to mL
(f) 35.6 L to gal
(g) 9.0 μL to mL
(h) 20.0 gal to L

26. An automobile traveling at 55 miles per hour is moving at what speed in (a) kilometers per hour and (b) feet per second?

27. Carl Lewis, a sprinter in the 1988 Olympic Games, ran the 100 m dash in 8.90 s. What was his speed in (a) feet per second and (b) miles per hour?

28. A lab experiment requires each student to use 6.55 g of sodium chloride. The instructor opens a new 1.00 lb jar of the salt. If 24 students each take exactly the assigned amount of salt, how much should be left in the bottle at the end of the lab period?

29. When the space probe *Galileo* reaches Jupiter in 1994, it will be traveling at an average speed of 27,000 miles per hour. What will be its speed in (a) miles per second and (b) kilometers per second?

30. How many kilograms does a 170 lb man weigh?

31. The usual aspirin tablet contains 5.0 grains of aspirin. How many grams of aspirin are in one tablet (1 grain $= \frac{1}{7000}$ lb)?

32. The sun is approximately 93 million miles from the earth. How many seconds will it take light to travel from the sun to the earth if the velocity of light is 3.00×10^{10} cm per second?

33. The average mass of the heart of a human baby is about 1 ounce. What is the mass in milligrams?

34. How much would 1.0 kg of potatoes cost if the price is $1.78 for 10 pounds?

35. The price of gold varies greatly and has been as high as $800 per ounce. What is the value of 227 g of gold at $345 per ounce? Gold is priced per troy ounce [1 lb (avoirdupois) = 14.58 oz (troy)].

36. An adult ruby-throated hummingbird has an average mass of 3.2 g, whereas an adult California condor may attain a weight of 21 lb. How many times heavier than the hummingbird is the condor?

37. At 35 cents per liter how much will it cost to fill a 15.8 gal tank with gasoline?

38. How many liters of gasoline will be used to drive 500 miles in a car that averages 34 miles per gallon?

39. Calculate the volume, in liters, of a box 75 cm long by 55 cm wide by 55 cm high.

*40. Assuming that there are 20 drops in 1.0 mL, how many drops are in one gallon?

41. How many liters of oil are in a 42 gallon barrel of oil?

42. How many milliliters will be delivered by a filled 5.0 μL syringe?

*43. Calculate the number of milliliters of water in 1.00 cubic foot of water.

*44. Oil spreads in a thin layer on water and is called an "oil slick." How much area in square meters (m^2) will 200 cm^3 of oil cover if it forms a layer 0.5 nm in thickness?

Temperature Conversions

45. Normal body temperature for humans is 98.6°F. What is this temperature on the Celsius scale?

46. Which is colder, $-100°C$ or $-138°F$?

47. Make the following conversions, showing mathematical setups:
    (a) 162°F to °C   (e) 32°C to °F
    (b) 0.0°F to °C   (f) −8.6°F to °C
    (c) 0.0°F to K   (g) 273°C to K
    (d) −18°C to °F   (h) 212 K to °C
*48. (a) At what temperature are the Fahrenheit and Celsius temperatures exactly equal?
     (b) At what temperature are they numerically equal but opposite in sign?

Density

49. Calculate the density of a liquid if 50.00 mL of the liquid has a mass of 78.26 g.
50. A 12.8 mL sample of bromine has a mass of 39.9 g. What is the density of bromine?
51. When a 32.7 g piece of chromium metal was placed into a graduated cylinder containing 25.0 mL of water, the water level rose to 29.6 mL. Calculate the density of the chromium.
52. Concentrated hydrochloric acid has a density of 1.19 g/mL. Calculate the mass of 500. mL of this acid.
53. An empty graduated cylinder has a mass of 42.817 g. When filled with 50.0 mL of an unknown liquid it has a mass of 106.773 g. What is the density of the liquid?
54. What mass of mercury (density 13.6 g/mL) will occupy a volume of 25.0 mL?
55. A 35.0 mL sample of ethyl alcohol (density 0.789 g/mL) is added to a graduated cylinder that has a mass of 49.28 g. What will be the mass of the cylinder plus the alcohol?
56. You are given three cubes, A, B, and C; one is magnesium, one is aluminum, and the third is silver. All three cubes have the same mass, but cube A has a volume of 25.9 mL, cube B has a volume of 16.7 mL, and cube C has a volume of 4.29 mL. Identify cubes A, B, and C.
*57. A cube of aluminum has a mass of 500 g. What will be the mass of a cube of gold of the same dimensions?
58. A 25.0 mL sample of water of 90°C has a mass of 24.12 g. Calculate the density of water at this temperature.
59. The mass of an empty container is 88.25 g. The mass of the container when filled with a liquid ($d = 1.25$ g/mL) is 150.50 g. What is the volume of the container?
60. Which liquid will occupy the greater volume, 50 g of water or 50 g of ethyl alcohol? Explain.
61. A gold bullion dealer advertised a bar of pure gold for sale. The gold bar had a mass of 3300 g and measured 2.00 cm by 15.0 cm by 6.00 cm. Was the gold bar pure gold? Show evidence for your answer.
62. The largest nugget of gold on record was found in 1872 in New South Wales, Australia, and had a mass of 93.3 kg. Assuming the nugget is pure gold, what is its volume in cubic centimeters? What is it worth by today's standards if gold is $345/oz? (See Exercise 35.)
*63. Forgetful Freddie placed 25.0 mL of a liquid in a graduated cylinder with a mass of 89.450 g when empty. When Freddie placed a metal slug with a mass of 15.434 g into the cylinder, the volume rose to 30.7 mL. Freddie was asked to calculate the density of the liquid and of the metal slug from his data, but he forgot to obtain the mass of the liquid. He was told that if he found the mass of the cylinder containing the liquid and the slug, he would have enough data for the calculations. He did so and found its mass to be 125.934 g. Calculate the density of the liquid and of the metal slug.

CHAPTER THREE

Classification of Matter

Throughout our lives we seek to bring order into the chaos that surrounds us. To do this, we classify things according to their similarities. In the library, we find books grouped according to the subject, and then by author. Our local department store organizes its merchandise by the size and style of clothing, as well as by the type of customer. The ball park or theater classifies its seat by price and location. The biologist divides the living world into plants and animals; this broad classification is further simplified into various phyla and on to specific genera. In chemistry, this classification process begins with pure substances (such as water or mercury) and mixtures (such as air or vinegar). This process continues and ultimately leads us to the fundamental building blocks of matter — the 109 elements.

Chapter Preview

3.1   Matter Defined

matter

The entire universe consists of matter and energy. Every day we come into contact with countless kinds of matter. Air, food, water, rocks, soil, glass, and this book are all different types of matter. Broadly defined, **matter** is *anything* that has mass and occupies space.

Matter may be quite invisible. If an apparently empty test tube is submerged mouth downward in a beaker of water, the water rises only slightly into the tube. The water cannot rise further because the tube is filled with invisible matter: air (see Figure 3.1).

To the eye matter appears to be continuous and unbroken. However, it is actually discontinuous and is composed of discrete, tiny particles called *atoms*. The particulate nature of matter will become evident when we study atomic structure and the properties of gases.

3.2   Physical States of Matter

solid

Matter exists in three physical states: solid, liquid, and gas. A **solid** has a definite shape and volume, with particles that cohere rigidly to one another. The shape of a solid can be independent of its container. For example, a crystal of sulfur has the same shape and volume whether it is placed in a beaker or simply laid on a glass plate.

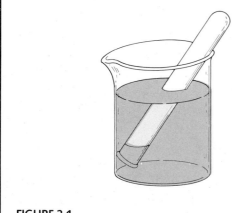

FIGURE 3.1

An apparently empty test tube is submerged, mouth downward, in water. Only a small volume of water rises into the tube, which is actually filled with air. This experiment proves that air, which is matter, occupies space.

Most commonly occurring solids, such as salt, sugar, quartz, and metals, are *crystalline*. Crystalline materials exist in regular, repeating, three-dimensional, geometric patterns. Solids such as plastics, glass, and gels, because they do not have any particular regular internal geometric pattern, are called **amorphous** solids. (*Amorphous* means without shape or form.)

amorphous

A **liquid** has a definite volume but not a definite shape, with particles that cohere firmly but not rigidly. Although the particles are held together by strong attractive forces and are in close contact with one another, they are able to move freely. Particle mobility gives a liquid fluidity and causes it to take the shape of the container in which it is stored.

liquid

A **gas** has indefinite volume and no fixed shape, with particles that are moving independently of one another. Particles in the gaseous state have gained enough energy to overcome the attractive forces that held them together as liquids or solids. A gas presses continuously in all directions on the walls of any container. Because of this quality a gas completely fills a container. The particles of a gas are relatively far apart compared with those of solids and liquids. The actual volume of the gas particles is usually very small in comparison with the volume of the space occupied by the gas. A gas therefore may be compressed into a very small volume or expanded almost indefinitely. Liquids cannot be compressed to any great extent, and solids are even less compressible than liquids.

gas

TABLE 3.1

Common Materials in the Solid, Liquids and Gaseous States of Matter

Solids	Liquids	Gases
Aluminum	Alcohol	Acetylene
Copper	Blood	Air
Gold	Gasoline	Butane
Polyethylene	Honey	Carbon dioxide
Salt	Mercury	Chlorine
Sand	Oil	Helium
Steel	Vinegar	Methane
Sulfur	Water	Oxygen

When a bottle of ammonia solution is opened in one corner of the laboratory, one can soon smell its familiar odor in all parts of the room. The ammonia gas escaping from the solution demonstrates that gaseous particles move freely and rapidly and tend to permeate the entire area into which they are released.

Although matter is discontinuous, attractive forces exist that hold the particles together and give matter its appearance of continuity. These attractive forces are strongest in solids, giving them rigidity; they are weaker in liquids but still strong enough to hold liquids to definite volumes. In gases the attractive forces are so weak that the particles of a gas are practically independent of one another. Table 3.1 lists a number of common materials that exist as solids, liquids, and gases. Table 3.2 summarizes comparative properties of solids, liquids, and gases.

3.3   Substances and Mixtures

substance

The term *matter* refers to all materials or material things that make up the universe. Many thousands of different and distinct kinds of matter or substances exist. A **substance** is a particular kind of matter with a definite, fixed composition. A substance, sometimes known as a *pure substance*, is either an element or a compound. Familiar examples of elements are copper, gold, and oxygen. Familiar compounds are salt, sugar, and water.

homogeneous

We can classify a sample of matter as either *homogeneous* or *heterogeneous* by examining it. **Homogeneous** matter is uniform in appearance and has the same properties throughout. Matter consisting of two or more physically distinct

---

TABLE 3.2

Physical Properties of Solids, Liquids and Gases

State	Shape	Volume	Particles	Compressibility
Solid	Definite	Definite	Rigidly cohering; tightly packed	Very slight
Liquid	Indefinite	Definite	Mobile; cohering	Slight
Gas	Indefinite	Indefinite	Independent of each other and relatively far apart	High

---

heterogeneous

phase

phases is **heterogeneous**. A **phase** is a homogeneous part of a system separated from other parts by physical boundaries. A system is simply the body of matter under consideration. Whenever we have a system in which visible boundaries exist between the parts or components, that system has more than one phase and is heterogeneous. It does not matter whether the components are in the solid, liquid, or gaseous states.

An important fact to keep in mind is that a pure substance, an element or compound, is always *homogeneous* in composition. However, a pure substance may exist as different phases in a heterogeneous system. Ice floating in water, for example, is a two-phase system made up of solid water and liquid water. The water in each phase is *homogeneous* in composition; but because two phases are present, the system is *heterogeneous*.

mixture

A **mixture** is a material containing two or more substances and can be either heterogeneous or homogeneous. Mixtures are variable in composition. If we add a tablespoonful of sugar to a glass of water, a heterogeneous mixture is formed immediately. The two phases are a solid (sugar) and a liquid (water). But upon stirring the sugar dissolves to form a homogeneous mixture or solution. Both substances are still present: All parts of the solution are sweet and wet. The proportions of sugar and water can be varied simply by adding more sugar and stirring to dissolve.

Many substances do not form homogeneous mixtures. If we mix sugar and fine white sand, a heterogeneous mixture is formed. Careful examination may be needed to decide that the mixture is heterogeneous because the two phases (sugar and sand) are both white solids. Ordinary matter exists mostly as mixtures. If we examine soil, granite, iron ore, or other naturally occurring mineral deposits, we find them to be heterogeneous mixtures. Seawater is a homogeneous mixture (solution) containing many substances. Air is a homogeneous mixture (solution) of several gases. Figure 3.2 illustrates the relationships of substances and mixtures.

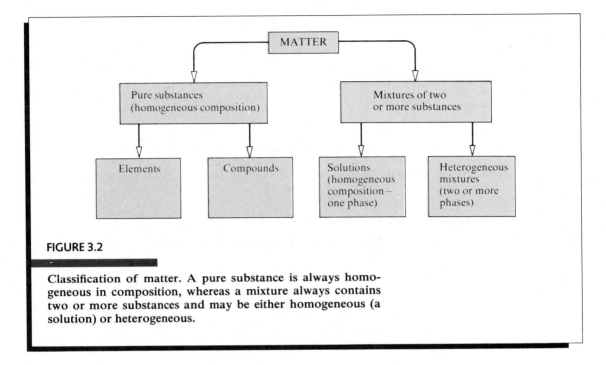

FIGURE 3.2

Classification of matter. A pure substance is always homo-geneous in composition, whereas a mixture always contains two or more substances and may be either homogeneous (a solution) or heterogeneous.

3.4   Elements

element

All the words in the English dictionary are formed from an alphabet consisting of only 26 letters. All known substances on earth—and most probably in the universe, too—are formed from a sort of "chemical alphabet" consisting of 109 presently known elements. An **element** is a fundamental or elementary substance that cannot be broken down by chemical means to simpler substances. Elements are the building blocks of all substances. The elements are numbered in order of increasing complexity beginning with hydrogen, number 1. Of the first 92 elements, 88 are known to occur in nature. The other four—technetium (43), promethium (61), astatine (85), and francium (87)—either do not occur in nature or have only transitory existences during radioactive decay. With the exception of number 94, plutonium, elements above number 92 are not known to occur naturally but have been synthesized, usually in very small quantities, in laboratories. The discovery of trace amounts of element 94 (plutonium) in nature has been reported recently. The syntheses of elements 107 and 109 were reported in 1981 and 1982. No elements other than those on the earth have been detected on other bodies in the universe.

Most substances can be decomposed into two or more simpler substances. Water can be decomposed into hydrogen and oxygen. Sugar can be decomposed into carbon, hydrogen, and oxygen. Table salt is easily decomposed into sodium and chlorine. An element, however, cannot be decomposed into simpler substances by ordinary chemical changes.

If we could take a small piece of an element, say copper, and divide it and subdivide it into smaller and smaller particles, we finally would come to a single unit of copper that we could no longer divide and still have copper. This ultimate

atom   particle, the smallest particle of an element that can exist, is called an **atom**. An atom is also the smallest unit of an element that can enter into a chemical reaction. Atoms are made up of still smaller subatomic particles. But these subatomic particles (described in Chapter 5) do not have the properties of elements.

3.5   Distribution of Elements

Elements are distributed very unequally in nature. At normal room temperature two of the elements, bromine and mercury, are liquids; eleven elements, hydrogen, nitrogen, oxygen, fluorine, chlorine, helium, neon, argon, krypton, xenon, and radon, are gases; all the other elements are solids.

Ten elements make up about 99% of the mass of the earth's crust, seawater, and atmosphere. Oxygen, the most abundant of these, constitutes about 50% of this mass. The distribution of the elements shown in Table 3.3 includes the earth's crust to a depth of about 10 miles, the oceans, fresh water, and the atmosphere but does not include the mantle and core of the earth, which are believed to consist of metallic iron and nickel. Because the atmosphere contains relatively little matter, its inclusion has almost no effect on the distribution shown in Table 3.3. But the inclusion of fresh and salt water does have an appreciable effect since water contains about 11.1% hydrogen. Nearly all of the 0.87% hydrogen shown is from water.

The average distribution of the elements in the human body is shown in Table 3.4. Note again the high percentage of oxygen.

3.6   Names of the Elements

The names of the elements came to us from various sources. Many are derived from early Greek, Latin, or German words that generally described some property of the element. For example, iodine is taken from the Greek word *iodes*,

TABLE 3.3

Distribution of the Elements in the Earth's Crust, Seawater, and Atmosphere

Element	Mass percent	Element	Mass percent
Oxygen	49.20	Chlorine	0.19
Silicon	25.67	Phosphorus	0.11
Aluminum	7.50	Manganese	0.09
Iron	4.71	Carbon	0.08
Calcium	3.39	Sulfur	0.06
Sodium	2.63	Barium	0.04
Potassium	2.40	Nitrogen	0.03
Magnesium	1.93	Fluorine	0.03
Hydrogen	0.87		
Titanium	0.58	All others	0.47

TABLE 3.4

Average Elemental Composition of the Human Body

Element	Mass percent
Oxygen	65.0
Carbon	18.0
Hydrogen	10.0
Nitrogen	3.0
Calcium	2.0
Phosphorus	1.0
Traces of several other elements	1.0

meaning violetlike. Iodine, indeed, is violet in the vapor state. The name of the metal bismuth had its origin from the German words *weisse masse*, which means white mass. Miners called it *wismat*; it was later changed to *bismat*, and finally to bismuth. Some elements are named for the location of their discovery—for example, germanium, discovered in 1886 by Winkler, a German chemist. Others are named in commemoration of famous scientists, such as einsteinium and curium, named for Albert Einstein and Marie Curie, respectively.

3.7 Symbols of the Elements

symbol

We all recognize Mr., N.Y., and Ave. as abbreviations for mister, New York, and avenue. In like manner chemists have assigned an abbreviation to each element; these are called **symbols** of the elements. Fourteen of the elements have a single letter as their symbol, six have three-letter symbols, and the rest have two letters. A symbol stands for the element itself, for one atom of the element, and (as we shall see later) for a particular quantity of the element.

Rules governing symbols of elements are as follows:

1. Symbols are composed of one, two, or three letters.
2. If one letter is used, it is capitalized.
3. If two or three letters are used, the first is capitalized and the others are lowercase letters.

Examples: Sulfur S   Barium Ba

The symbols and names of all the elements are given in the table on the inside back cover of this book. Table 3.5 lists the more commonly used symbols. If we examine this table carefully, we note that most of the symbols start with the same letter as the name of the element that is represented. A number of symbols, however, appear to have no connection with the names of the elements they

TABLE 3.5

Symbols of the Most Common Elements

Element	Symbol	Element	Symbol	Element	Symbol
Aluminum	Al	Fluorine	F	Phosphorus	P
Antimony	Sb	Gold	Au	Platinum	Pt
Argon	Ar	Helium	He	Potassium	K
Arsenic	As	Hydrogen	H	Radium	Ra
Barium	Ba	Iodine	I	Silicon	Si
Bismuth	Bi	Iron	Fe	Silver	Ag
Boron	B	Lead	Pb	Sodium	Na
Bromine	Br	Lithium	Li	Strontium	Sr
Cadmium	Cd	Magnesium	Mg	Sulfur	S
Calcium	Ca	Manganese	Mn	Tin	Sn
Carbon	C	Mercury	Hg	Titanium	Ti
Chlorine	Cl	Neon	Ne	Tungsten	W
Chromium	Cr	Nickel	Ni	Uranium	U
Cobalt	Co	Nitrogen	N	Zinc	Zn
Copper	Cu	Oxygen	O		

TABLE 3.6

Symbols of the Elements Derived from Early Names*

Present name	Symbol	Former name
Antimony	Sb	Stibium
Copper	Cu	Cuprum
Gold	Au	Aurum
Iron	Fe	Ferrum
Lead	Pb	Plumbum
Mercury	Hg	Hydrargyrum
Potassium	K	Kalium
Silver	Ag	Argentum
Sodium	Na	Natrium
Tin	Sn	Stannum
Tungsten	W	Wolfram

* These symbols are in use today even though they do not correspond to the current name of the element.

represent (see Table 3.6). These symbols have been carried over from earlier names (usually in Latin) of the elements and are so firmly implanted in the literature that their use is continued today.

Special care must be used in writing symbols. Begin each with a capital letter and use a lowercase second letter if needed. For example, consider Co, the symbol for the element cobalt. If through error CO (capital C and capital O) is written, the two elements carbon and oxygen (the *formula* for carbon monoxide) are represented instead of the single element cobalt. Another example of the need for care in writing symbols is the symbol Ca for calcium versus Co for cobalt. The letters must be distinct or else the symbol for the element may be misinterpreted.

Knowledge of symbols is essential for writing chemical formulas and equations. You should begin to learn the symbols immediately because they will be used extensively in the remainder of this book and in any future chemistry courses you may take. One way to learn the symbols is to practice a few minutes a day by making side-by-side lists of names and symbols and then covering each list alternately and writing the corresponding name or symbol. Initially it is a good plan to learn the symbols of the most common elements shown in Table 3.5.

The experiments of alchemists paved the way for the development of chemistry. Alchemists surrounded their work with mysticism, partly by devising a system of symbols known only to practitioners of alchemy (see Figure 3.3). The symbol ℞ (from the Latin *recipe*) is still used in medicine and was established during this time. In the early 1800s the Swedish chemist J. J. Berzelius (1779–1848) made a great contribution to chemistry by devising the present system of symbols using letters of the alphabet.

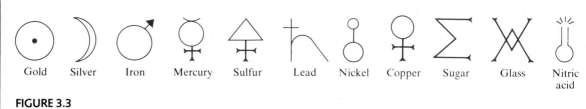

FIGURE 3.3

Some typical alchemists' symbols.

3.8   Metals, Nonmetals, and Metalloids

metal
nonmetal
metalloid

Three primary classifications of the elements are **metals, nonmetals,** and **metalloids.** Most of the elements are metals. We are familiar with metals because of their widespread use in tools, materials of construction, automobiles, and so on. But nonmetals are equally useful in our everyday life as major components of such items as clothing, food, fuel, glass, plastics, and wood.

The metallic elements are solids at room temperature (mercury is an exception). They have high luster, are good conductors of heat and electricity, are *malleable* (can be rolled or hammered into sheets), and are *ductile* (can be drawn into wires). Most metals have a high melting point and high density. Familiar metals are aluminum, chromium, copper, gold, iron, lead, magnesium, mercury, nickel, platinum, silver, tin, and zinc. Less familiar but still important metals are calcium, cobalt, potassium, sodium, uranium, and titanium.

Metals have little tendency to combine with each other to form compounds. But many metals readily combine with nonmetals such as chlorine, oxygen, and sulfur to form mainly ionic compounds such as metallic chlorides, oxides, and sulfides. In nature the more active metals are found combined with other elements as minerals. A few of the less active ones such as copper, gold, and silver are sometimes found in a native, or free, state.

Nonmetals, unlike metals, are not lustrous, have relatively low melting points and densities, and are generally poor conductors of heat and electricity. Carbon, phosphorus, sulfur, selenium, and iodine are solids; bromine is a liquid; the rest of the nonmetals are gases. Common nonmetals found uncombined in nature are carbon (graphite and diamond), nitrogen, oxygen, sulfur, and the noble gases (helium, neon, argon, krypton, xenon, and radon).

Nonmetals combine with one another to form molecular compounds such as carbon dioxide (CO_2), methane (CH_4), butane (C_4H_{10}), and sulfur dioxide (SO_2). Fluorine, the most reactive nonmetal, combines readily with almost all the other elements.

TABLE 3.7

Classification of the Elements into Metals, Metalloids, and Nonmetals

1 H																	2 He
3 Li	4 Be											5 B	6 C	7 N	8 O	9 F	10 Ne
11 Na	12 Mg											13 Al	14 Si	15 P	16 S	17 Cl	18 Ar
19 K	20 Ca	21 Sc	22 Ti	23 V	24 Cr	25 Mn	26 Fe	27 Co	28 Ni	29 Cu	30 Zn	31 Ga	32 Ge	33 As	34 Se	35 Br	36 Kr
37 Rb	38 Sr	39 Y	40 Zr	41 Nb	42 Mo	43 Tc	44 Ru	45 Rh	46 Pd	47 Ag	48 Cd	49 In	50 Sn	51 Sb	52 Te	53 I	54 Xe
55 Cs	56 Ba	57 La	72 Hf	73 Ta	74 W	75 Re	76 Os	77 Ir	78 Pt	79 Au	80 Hg	81 Tl	82 Pb	83 Bi	84 Po	85 At	86 Rn
87 Fr	88 Ra	89 Ac	104 Unq	105 Unp	106 Unh	107 Uns	108 Uno	109 Une									

Legend: ☐ Metals   ▨ Metalloids   ▨ Nonmetals

58 Ce	59 Pr	60 Nd	61 Pm	62 Sm	63 Eu	64 Gd	65 Tb	66 Dy	67 Ho	68 Er	69 Tm	70 Yb	71 Lu
90 Th	91 Pa	92 U	93 Np	94 Pu	95 Am	96 Cm	97 Bk	98 Cf	99 Es	100 Fm	101 Md	102 No	103 Lr

Several elements (boron, silicon, germanium, arsenic, antimony, tellurium, and polonium) are classified as *metalloids* and have properties that are intermediate between those of metals and those of nonmetals. The intermediate position of these elements is shown in Table 3.7, which lists and classifies all the elements as metals, nonmetals, or metalloids. Certain metalloids, such as boron, silicon, and germanium, are the raw materials for the semiconductor devices that make our modern electronics industry possible.

3.9   Compounds

compound

A **compound** is a distinct substance containing two or more elements chemically combined in definite proportions by mass. Compounds, unlike elements, can be decomposed chemically into simpler substances—that is, into simpler compounds and/or elements. Atoms of the elements in a compound are combined in whole-number ratios, never as fractional parts of atoms. Compounds fall into two general types, *molecular* and *ionic*.

molecule

A **molecule** is the smallest uncharged individual unit of a compound formed by the union of two or more atoms. Water is a typical molecular compound. If we divide a drop of water into smaller and smaller particles, we finally obtain a single molecule of water consisting of two hydrogen atoms bonded to one oxygen atom. This molecule is the ultimate particle of water; it cannot be further subdivided without destroying the water and forming hydrogen and oxygen.

ion

An **ion** is a positively or negatively charged atom or group of atoms. An ionic compound is held together by attractive forces that exist between positively and negatively charged ions. A positively charged ion is called a **cation** (pronounced *cat-eye-on*); a negatively charged ion is called an **anion** (pronounced *an-eye-on*).

cation
anion

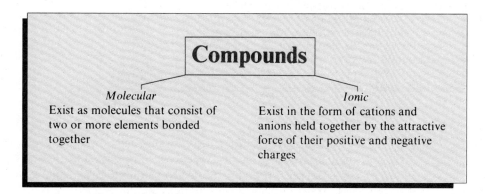

Sodium chloride is a typical ionic compound. The ultimate particles of sodium chloride are positively charged sodium ions and negatively charged chloride ions. Sodium chloride is held together in a crystalline structure by the attractive forces existing between these oppositely charged ions. Although ionic compounds consist of large aggregates of cations and anions, their formulas are normally represented by the simplest possible ratio of the atoms in the compound. For example in sodium chloride the ratio is one sodium ion to one chlorine ion, and the formula is NaCl. The two types of compounds, molecular and ionic, are illustrated in Figure 3.4.

There are more than 9 million known registered compounds, with no end in sight as to the number that will be prepared in the future. Each compound is unique and has characteristic properties. Let us consider two compounds, water and mercury(II) oxide, in some detail. Water is a colorless, odorless, tasteless liquid that can be changed to a solid (ice) at 0°C and to a gas (steam) at 100°C. Composed of two atoms of hydrogen and one atom of oxygen per molecule, water is 11.2% hydrogen and 88.8% oxygen by mass. Water reacts chemically with sodium to produce hydrogen gas and sodium hydroxide, with lime to produce calcium hydroxide, and with sulfur trioxide to produce sulfuric acid. No other compound has all these exact physical and chemical properties; they are characteristic of water alone.

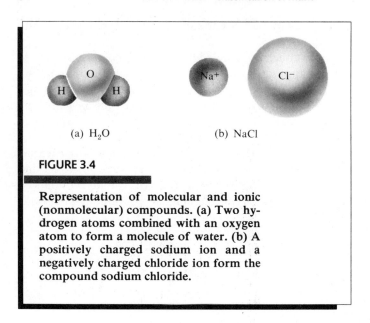

(a) H_2O (b) NaCl

FIGURE 3.4

Representation of molecular and ionic (nonmolecular) compounds. (a) Two hydrogen atoms combined with an oxygen atom to form a molecule of water. (b) A positively charged sodium ion and a negatively charged chloride ion form the compound sodium chloride.

Mercury(II) oxide is a dense, orange-red powder having a ratio of one atom of mercury to one atom of oxygen. Its composition by mass is 92.6% mercury and 7.4% oxygen. When it is heated to temperatures greater than 360°C, a colorless gas, oxygen, and a silvery, liquid metal, mercury, are produced. These specific properties belong to mercury(II) oxide and to no other substance. Thus, a compound may be identified and distinguished from all other compounds by its characteristic properties.

3.10   Elements that Exist as Diatomic Molecules

Seven of the elements (all nonmetals) occur as *diatomic molecules*. These elements and their symbols, formulas, and brief descriptions are listed in Table 3.8. Whether found free in nature or prepared in the laboratory, the molecules of these elements always contain two atoms. The formulas of the free elements are therefore always written to show this molecular composition: H_2, N_2, O_2, F_2, Cl_2, Br_2, and I_2.

It is important to see how symbols are used to designate either an atom or a molecule of an element. Consider hydrogen and oxygen. Hydrogen gas is present in volcanic gases and can be prepared by many chemical reactions. Regardless of their source, all samples of free hydrogen gas consist of diatomic molecules. Free hydrogen is designated and its composition is expressed by the formula H_2. Oxygen makes up about 21% by volume of the air that we breathe. This free

TABLE 3.8

Elements that Exist as Diatomic Molecules

Element	Symbol	Molecular formula	Normal state
Hydrogen	H	H_2	Colorless gas
Nitrogen	N	N_2	Colorless gas
Oxygen	O	O_2	Colorless gas
Fluorine	F	F_2	Pale yellow gas
Chlorine	Cl	Cl_2	Yellow-green gas
Bromine	Br	Br_2	Reddish-brown liquid
Iodine	I	I_2	Bluish-black solid

oxygen is constantly being replenished by photosynthesis; it can also be prepared in the laboratory by several reactions. All free oxygen is diatomic and is designated by the formula O_2. Now consider water, a compound designated by the formula H_2O (sometimes HOH). Water contains neither free hydrogen (H_2) nor free oxygen (O_2). The H_2 part of the formula H_2O simply indicates that two atoms of hydrogen are combined with one atom of oxygen to form water. Thus symbols are used to designate elements, show the composition of molecules of elements, and give the elemental composition of compounds.

3.11 Chemical Formulas

chemical formula

Chemical formulas are used as abbreviations for compounds. A **chemical formula** shows the symbols and the ratio of the atoms of the elements in a compound. Sodium chloride contains one atom of sodium per atom of chlorine; its formula is NaCl. The formula for water is H_2O; it shows that a molecule of water contains two atoms of hydrogen and one atom of oxygen.

The formula of a compound tells us which elements it is composed of and how many atoms of each element are present in a formula unit. For example, a molecule of sulfuric acid is composed of two atoms of hydrogen, one atom of sulfur, and four atoms of oxygen. We could express this compound as HHSOOOO, but the usual formula for writing sulfuric acid is H_2SO_4. The formula may be expressed verbally as "H-two-S-O-four." Numbers that appear partially below the line and to the right of a symbol of an element are called *subscripts.* Thus the 2 and the 4 in H_2SO_4 are subscripts. Characteristics of chemical formulas are

1. The formula of a compound contains the symbols of all the elements in the compound.
2. When the formula contains one atom of an element, the symbol of that element represents that one atom. The number one (1) is not used as a subscript to indicate one atom of an element.
3. When the formula contains more than one atom of an element, the number of atoms is indicated by a subscript written to the right of the symbol of that atom. For example, the two (2) in H_2O indicates two atoms of H in the formula.
4. When the formula contains more than one of a group of atoms that occurs as a unit, parentheses are placed around the group, and the number of units of the group are indicated by a subscript placed to the right of the parentheses. Consider the nitrate group, NO_3^-. The formula for sodium nitrate, $NaNO_3$, has only one nitrate group; therefore no parentheses are needed. Calcium nitrate, $Ca(NO_3)_2$, has two nitrate groups, as indicated by the use of parentheses and the subscript 2. $Ca(NO_3)_2$ has a total of nine atoms: one Ca, two N, and six O atoms. The formula $Ca(NO_3)_2$ is read as "C-A [pause] N-O-three-taken twice."
5. Formulas written as H_2O, H_2SO_4, $Ca(NO_3)_2$, and $C_{12}H_{22}O_{11}$ show only the number and kind of each atom contained in the compound; they do not show the arrangement of the atoms in the compound or how they are chemically bonded to one another.

Figure 3.5 illustrates how symbols and numbers are used in chemical formulas.

EXAMPLE 3.1

Write formulas for the following compounds, the atom composition of which is given. (a) Hydrogen chloride: 1 atom hydrogen + 1 atom chlorine; (b) Methane: 1 atom carbon + 4 atoms hydrogen; (c) Glucose: 6 atoms carbon + 12 atoms hydrogen + 6 atoms oxygen.

(a) First write the symbols of the atoms in the formula: H   Cl   Since the ratio of atoms is one to one, we merely bring the symbols together to give the formula for hydrogen chloride as HCl.
(b) Write the symbols of the atoms: C   H   Now bring the symbols together and place a subscript 4 after the hydrogen atom. The formula is CH_4.
(c) Write the symbols of the atoms: C   H   O   Now write the formula, bringing together the symbols followed by the correct subscripts according to the data given (six C, twelve H, six O). The formula is $C_6H_{12}O_6$.

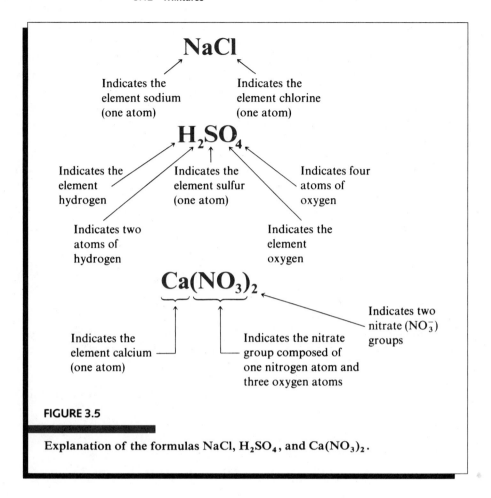

FIGURE 3.5

Explanation of the formulas $NaCl$, H_2SO_4, and $Ca(NO_3)_2$.

3.12 Mixtures

Single substances—elements or compounds—seldom occur naturally in the pure state. Air is a mixture of gases; seawater is a mixture containing a variety of dissolved minerals; ordinary soil is a complex mixture of minerals and various organic materials.

How is a mixture distinguished from a pure substance? A mixture always contains two or more substances that can be present in varying concentrations. Let us consider an example of a homogeneous mixture and an example of a heterogeneous mixture. Homogeneous mixtures (solutions) containing either 5% or 10% salt in water can be prepared simply by mixing the correct amounts of salt and water. These mixtures can be separated by boiling away the water, leaving the salt as a residue. The composition of a heterogeneous mixture of sulfur crystals and iron filings can be varied by merely blending in either more sulfur or more iron filings. This mixture can be separated physically by using a magnet to attract the iron or by adding carbon disulfide to dissolve the sulfur.

Iron(II) sulfide (FeS) contains 63.5% Fe and 36.5% S by mass. If we mix iron and sulfur in this proportion, do we have iron(II) sulfide? No, it is still a mixture; the iron is still attracted by a magnet. But if this mixture is heated strongly, a chemical change (reaction) occurs in which the reacting substances, iron and sulfur, form a new substance, iron(II) sulfide. Iron(II) sulfide, FeS, is a compound of iron and sulfur and has properties that are different from those of either iron or sulfur: It is neither attracted by a magnet nor dissolved by carbon disulfide. The differences between the iron and sulfur *mixture* and the iron(II) sulfide *compound* are as follows:

Mixture of iron and sulfur	Compound of iron and sulfur
Formula: mixture has no definite formula; consists of Fe and S.	Formula: FeS
Composition: mixture contains Fe and S in any proportion by mass	Composition: compound contains 63.5% Fe and 36.5% S by mass.
Separation: Fe and S can be separated by physical means.	Separation: Fe and S can be separated only by chemical change.

The general characteristics of mixtures and compounds are compared in Table 3.9.

TABLE 3.9

Comparison of Mixtures and Compounds

	Mixture	Compound
Composition	May be composed of elements, compounds, or both in variable composition.	Composed of two or more elements in a definite, fixed proportion by mass.
Separation of components	Separation may be made by physical or mechanical means.	Elements can be separated by chemical changes only.
Identification of components	Components do not lose their identity.	A compound does not resemble the elements from which it is formed.

Concepts in Review

1. Identify the three physical states of matter.
2. Distinguish between substances and mixtures.
3. Classify common materials as elements, compounds, or mixtures.
4. Write the symbols when given the names, or write the names when given the symbols, of the common elements listed in Table 3.5.
5. Understand how symbols, including subscripts and parentheses, are used to write chemical formulas.
6. Differentiate among atoms, molecules, and ions.
7. List the characteristics of metals, nonmetals, and metalloids.
8. List the elements that occur as diatomic molecules.

Key Terms in Review

The terms listed here have all been defined within the chapter. Review the definitions of each. Use the glossary and the margin notations within the chapter as study aids.

amorphous
anion
atom
cation
chemical formula
compound
element
gas
heterogeneous
homogeneous
ion

liquid
matter
metal
metalloid
mixture
molecule
nonmetal
phase
solid
substance
symbol

Exercises

1. List four different substances in each of the three states of matter.
2. In terms of the properties of the ultimate particles of a substance, explain
   (a) Why a solid has a definite shape but a liquid does not.
   (b) Why a liquid has a definite volume but a gas does not.
   (c) Why a gas can be compressed rather easily but a solid cannot be compressed appreciably.

3. What evidence can you find in Figure 3.1 that gases occupy space?

4. Which three liquids listed in Table 3.1 are not mixtures?

5. Which of the gases listed in Table 3.1 are not pure substances?

6. When the stopper is removed from a partly filled bottle containing solid and liquid acetic acid at 16.7°C, a strong vinegarlike odor is noticeable immediately. How many acetic acid phases must be present in the bottle? Explain.

7. Is the system enclosed in the bottle of Exercise 6 homogeneous or heterogeneous? Explain.

8. Is a system that contains only one substance necessarily homogeneous? Explain.

9. Is a system that contains two or more substances necessarily heterogeneous? Explain.

10. Are there more atoms of silicon or hydrogen in the earth's crust, seawater, and atmosphere? Use Table 3.3 and the fact that the mass of the silicon atom is about 28 times that of a hydrogen atom.

11. What does the symbol of an element stand for?

12. Write down what you believe to be the symbols for the elements phosphorus, aluminum, hydrogen, potassium, magnesium, sodium, nitrogen, nickel, silver, and plutonium. Now look up the correct symbols and rewrite them, comparing the two sets.

13. Interpret the difference in meanings for each of these pairs:
(a) Si and SI   (b) Pb and PB   (c) 4 P and P_4

14. List six elements and their symbols in which the first letter of the symbol is different from that of the name.

15. Write the names and symbols for the fourteen elements that have only one letter as their symbol. (See table on inside back cover.)

16. Distinguish between an element and a compound.

17. How many metals are there? Nonmetals? Metalloids? (See Table 3.7)

18. Of the ten most abundant elements in the earth's crust, seawater, and atmosphere, how many are metals? Nonmetals? Metalloids? (Table 3.3)

19. Of the six most abundant elements in the human body, how many are metals? Nonmetals? Metalloids?

20. Why is the symbol for gold Au rather than G or Go?

21. Give the names of (a) the solid diatomic nonmetal and (b) the liquid diatomic nonmetal.

22. Distinguish between a compound and a mixture.

23. What are the two general types of compounds? How do they differ from each other?

24. What is the basis for distinguishing one compound from another?

25. Given the following list of compounds and their formulas, what elements are present in each compound?
(a) Potassium iodide   KI
(b) Sodium carbonate Na_2CO_3
(c) Aluminum oxide Al_2O_3
(d) Calcium bromide $CaBr_2$
(e) Carbon tetrachloride CCl_4
(f) Magnesium bromide $MgBr_2$
(g) Nitric acid HNO_3
(h) Barium sulfate $BaSO_4$
(i) Aluminum phosphate $AlPO_4$
(j) Acetic acid $HC_2H_3O_2$

26. Write the formula for each of the following compounds, the composition of which is given after each name:
(a) Zinc oxide   1 atom Zn,
                        1 atom O
(b) Potassium chlorate 1 atom K,
                        1 atom Cl,
                        3 atoms O
(c) Sodium hydroxide   1 atom Na,
                        1 atom O,
                        1 atom H
(d) Aluminum bromide   1 atom Al,
                        3 atoms Br
(e) Calcium fluoride   1 atom Ca,
                        2 atoms F
(f) Lead(II) chromate   1 atom Pb,
                        1 atom Cr,
                        4 atoms O
(g) Ethyl alcohol   2 atoms C,
                        6 atoms H,
                        1 atom O
(h) Benzene   6 atoms C,
                        6 atoms H

27. Explain the meaning of each symbol and number in the following formulas:
(a) H_2O
(b) $AlBr_3$
(c) Na_2SO_4
(d) $Ni(NO_3)_2$
(e) $C_{12}H_{22}O_{11}$ (sucrose)

28. How many atoms are represented in each of these formulas?
    - (a) KF
    - (b) $CaCO_3$
    - (c) N_2
    - (d) $Ba(ClO_3)_2$
    - (e) $K_2Cr_2O_7$
    - (f) $NaC_2H_3O_2$
    - (g) CCl_2F_2 (Freon)
    - (h) $Al_2(SO_4)_3$
    - (i) $(NH_4)_2C_2O_4$

29. How many atoms are contained in (a) one molecule of hydrogen, (b) one molecule of water, and (c) one molecule of sulfuric acid?

30. What is the major difference between a cation and an anion?

31. Write the names and formulas of the elements that exist as diatomic molecules.

32. How many atoms of oxygen are represented in each formula?
    - (a) H_2O
    - (b) $CuSO_4$
    - (c) H_2O_2
    - (d) $Fe(OH)_3$
    - (e) $Al(ClO_3)_3$

33. How many atoms of hydrogen are represented in each formula?
    - (a) H_2
    - (b) $Ba(C_2H_3O_2)_2$
    - (c) $C_6H_{12}O_6$
    - (d) $HC_2H_3O_2$

34. Distinguish between homogeneous and heterogeneous mixtures.

35. Classify each of the following materials as an element, compound, or mixture:
    - (a) Air
    - (b) Oxygen
    - (c) Sodium chloride
    - (d) Platinum
    - (e) Wine
    - (f) Iodine
    - (g) Sulfuric acid
    - (h) Crude oil

36. Classify each of the following materials as an element, compound, or mixture:
    - (a) Paint
    - (b) Salt
    - (c) Copper
    - (d) Beer
    - (e) Sulfuric acid
    - (f) Silver
    - (g) Milk
    - (h) Sodium hydroxide

37. A white solid, on heating, formed a colorless gas and a yellow solid. Assuming that there was no reaction with the air, is the original solid an element or a compound? Explain.

38. Tabulate the properties that characterize metals and nonmetals.

39. Which of the following are diatomic molecules?
    - (a) H_2
    - (b) SO_2
    - (c) HCl
    - (d) H_2O
    - (e) NO
    - (f) NO_2
    - (g) $MgCl_2$

40. Which of the following statements are correct? Rewrite the incorrect statements to make them correct. (Try to answer this question without referring to the text.)

(a) Liquids are the least compact state of matter.

(b) Liquids have a definite volume and a definite shape.

(c) Matter in the solid state is discontinuous; that is, it is made up of discrete particles.

(d) Wood is homogeneous.

(e) Wood is a substance.

(f) Dirt is a mixture.

(g) Seawater, although homogeneous, is a mixture.

(h) Any system made up of only one substance is homogeneous.

(i) Any system containing two or more substances is heterogeneous.

(j) A solution, although it contains dissolved material, is homogeneous.

(k) The smallest unit of an element that can exist and enter into a chemical reaction is called a molecule.

(l) The basic building blocks of all substances, which cannot be decomposed into simpler substances by ordinary chemical change, are compounds.

(m) The most abundant element in the earth's crust, seawater, and atmosphere by mass is oxygen.

(n) The most abundant element in the human body, by mass, is carbon.

(o) Most of the elements are represented by symbols consisting of one or two letters.

(p) The symbol for copper is Co.

(q) The symbol for sodium is Na.

(r) The symbol for potassium is P.

(s) The symbol for lead is Le.

(t) Early names for some elements led to unlikely symbols, such as Fe for iron.

(u) A compound is a distinct substance that contains two or more elements combined in a definite proportion by mass.

(v) The smallest uncharged individual unit of a compound formed by the union of two or more atoms is called a substance.

(w) An ion is a positive or negative electrically charged atom or group of atoms.

(x) Bromine is an element that occurs as a diatomic molecule, Br_2.

(y) The formula Na_2CO_3 indicates a total of six atoms, including three oxygen atoms.

(z)   A general property of nonmetals is that they are good conductors of heat and electricity.

(aa) Metals have the properties of ductility and malleability.

(bb) *Malleable* means that when struck a hard blow the substance will shatter.

(cc) Elements that have properties intermediate between metals and nonmetals are called mixtures.

(dd) More of the elements are metals than nonmetals.

*41. What would be the density of a solution made by mixing 2.50 mL of carbon tetrachloride (CCl_4, $d = 1.595$ g/mL) and 3.50 mL of carbon tetrabromide (CBr_4, $d = 3.420$ g/mL)? Assume that the volume of the mixed liquids is the sum of the two volumes used.

*42. Only two elements, Br_2 ($d = 3.12$ g/mL) and Hg ($d = 13.6$ g/mL), are liquids at room temperature. How many milliliters of Br_2 will have the same mass as 12.5 mL Hg?

43. Pure gold is too soft a metal for many uses, so it is alloyed to give it more mechanical strength. One particular alloy is made by mixing 60 g of gold, 8.0 g of silver, and 12 g of copper. What carat gold is this alloy if pure gold is considered to be 24 carat?

*44. Methane, the chief component of natural gas, has the formula CH_4. Each atom of carbon has a mass 12 times greater than an atom of hydrogen. Calculate the mass percent of carbon in methane.

*45. White gold is a homogeneous solution of 90% gold and 10% palladium. How much gold is present in a bar of white gold with a mass of 8420 g?

*46. The metal used to make the U.S. nickel coin is an alloy of 75% copper and 25% nickel. What maximum mass of alloy could be produced if only 450 kg of nickel and 1180 kg of copper were on hand?

CHAPTER FOUR

Properties of Matter

The world we live in is a myriad of sights, sounds, smells, and tastes. Our senses help us to describe the objects in our lives. For example, the smell of freshly baked cinnamon rolls along with the sight of a bakery, combine to create a mouth-watering desire to gobble down a fresh-baked sample. And so it is with each substance—its own unique properties allow us to identify it and predict its interactions.

These interactions produce both physical and chemical changes. When you eat an apple, it results in the release of carbon dioxide and water. These same products are achieved by burning logs. Not only does a chemical change occur in these cases, but an energy change as well. Some reactions release energy (as does the apple or the log) whereas others require energy, such as the production of steel or the melting of ice. Over 90% of our current energy comes from chemical reactions.

Chapter Preview

4.1 Properties of Substances

properties

physical properties

chemical properties

How do we recognize substances? Each substance has a set of **properties** that is characteristic of that substance and gives it a unique identity. Properties are the personality traits of substances and are classified as either physical or chemical. **Physical properties** are the inherent characteristics of a substance that can be determined without altering its composition; they are associated with its physical existence. Common physical properties are color, taste, odor, state of matter (solid, liquid, or gas), density, melting point, and boiling point. **Chemical properties** describe the ability of a substance to form new substances, either by reaction with other substances or by decomposition.

We can select a few of the physical and chemical properties of chlorine as an example. Physically, chlorine is a gas about 2.4 times heavier than air. It is yellowish-green in color and has a disagreeable odor. Chemically, chlorine will not burn but will support the combustion of certain other substances. It can be used as a bleaching agent, as a disinfectant for water, and in many chlorinated substances such as refrigerants and insecticides. When chlorine combines with the metal sodium, it forms a salt called sodium chloride. These properties, among others, help to characterize and identify chlorine.

Substances, then, are recognized and differentiated by their properties. Table 4.1 lists four substances and tabulates several of their common physical properties. Information about common physical properties, such as that given in Table 4.1, is available in handbooks of chemistry and physics. Scientists do not pretend to know all the answers or to remember voluminous amounts of data, but it is important for them to know where to look for data in the literature. Handbooks are one of the most widely used resources for scientific data.*

---

* Two such handbooks are Robert C. Weast, ed., *Handbook of Chemistry and Physics*, 70th ed. (Cleveland: Chemical Rubber Company, 1989) and Norbert A. Lange, comp., *Handbook of Chemistry*, 13th ed. (New York: McGraw-Hill, 1985).

TABLE 4.1

Physical Properties of Chlorine, Water, Sugar, and Acetic Acid

Substance	Color	Odor	Taste	Physical state	Boiling point (°C)	Melting point (°C)
Chlorine	Yellowish-green	Sharp, suffocating	Sharp, sour	Gas	−34.6	−101.6
Water	Colorless	Odorless	Tasteless	Liquid	100.0	0.0
Sugar	White	Odorless	Sweet	Solid	Decomposes 170–186	—
Acetic acid	Colorless	Like vinegar	Sour	Liquid	118.0	16.7

> **No two substances will have identical physical and chemical properties.**

4.2 Physical Changes

physical change

Matter can undergo two types of changes, physical and chemical. **Physical changes** are changes in physical properties (such as size, shape, and density) or changes in state of matter without an accompanying change in composition. The changing of ice into water and water into steam are physical changes from one state of matter into another. No new substances are formed in these physical changes.

When a clean platinum wire is heated in a burner flame, the appearance of the platinum changes from silvery metallic to glowing red. This change is physical because the platinum can be restored to its original metallic appearance by cooling and, more importantly, because the composition of the platinum is not changed by heating and cooling.

4.3 Chemical Changes

chemical change

In a **chemical change**, new substances are formed that have different properties and composition from the original material. The new substances need not in any way resemble the initial material.

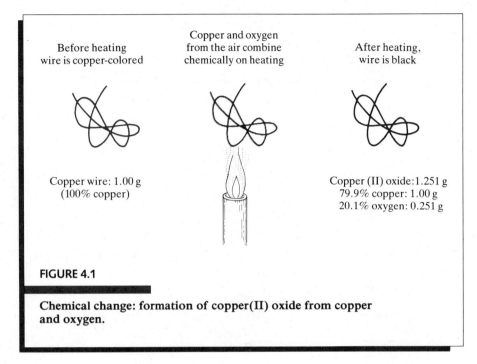

Before heating
wire is copper-colored

Copper and oxygen
from the air combine
chemically on heating

After heating,
wire is black

Copper wire: 1.00 g
(100% copper)

Copper (II) oxide:1.251 g
79.9% copper: 1.00 g
20.1% oxygen: 0.251 g

FIGURE 4.1

**Chemical change: formation of copper(II) oxide from copper
and oxygen.**

When a clean copper wire is heated in a burner flame, the appearance of the copper changes from coppery metallic to glowing red. Unlike the platinum previously mentioned, the copper is not restored to its original appearance by cooling but has become a black material. This black material is a new substance called copper(II) oxide. It was formed by chemical change when copper combined with oxygen in the air during the heating process. The unheated wire was essentially 100% copper, but the copper(II) oxide is 79.9% copper and 20.1% oxygen. One gram of copper will yield 1.251 g of copper(II) oxide (see Figure 4.1). The platinum was changed only physically when heated, but the copper was changed both physically and chemically when heated.

When 1.00 g of copper reacts with oxygen to yield 1.251 g of copper(II) oxide, the copper must have combined with 0.251 g of oxygen. The percentage of copper and oxygen can be calculated from this data, the copper and oxygen each being a percent of the total mass of copper(II) oxide.

$$1.00 \text{ g copper} + 0.251 \text{ g oxygen} \longrightarrow 1.251 \text{ g copper(II) oxide}$$

$$\frac{1.00 \text{ g copper}}{1.251 \text{ g copper(II) oxide}} \times 100 = 79.9\% \text{ copper}$$

$$\frac{0.251 \text{ g oxygen}}{1.251 \text{ g copper(II) oxide}} \times 100 = 20.1\% \text{ oxygen}$$

Mercury(II) oxide is an orange-red powder that, when subjected to high temperature (500–600°C), decomposes into a colorless gas (oxygen) and a silvery,

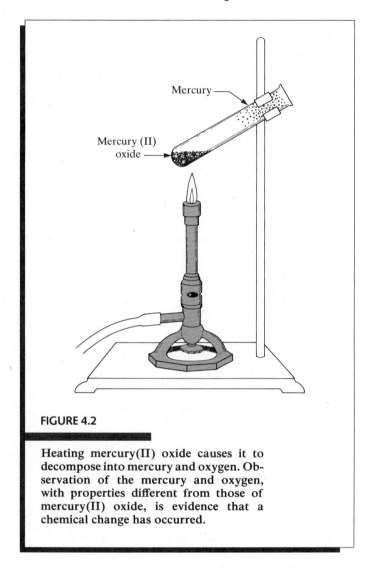

Mercury

Mercury (II) oxide

FIGURE 4.2

Heating mercury(II) oxide causes it to decompose into mercury and oxygen. Observation of the mercury and oxygen, with properties different from those of mercury(II) oxide, is evidence that a chemical change has occurred.

liquid metal (mercury). The composition and physical appearance of each product are noticeably different from those of the starting compound. When mercury(II) oxide is heated in a test tube (see Figure 4.2), small globules of mercury are observed collecting on the cooler part of the tube. Evidence that oxygen forms is observed when a glowing wood splint, lowered into the tube, bursts into flame. Oxygen supports and intensifies the combustion of the wood. From these observations we conclude that a chemical change has taken place.

Chemists have devised *chemical equations* as a shorthand method for expressing chemical changes. The two previous examples of chemical changes

TABLE 4.2

Examples of Processes Involving Physical or Chemical Changes

Process taking place	Type of change	Accompanying observations
Rusting of iron	Chemical	Shiny, bright metal changes to reddish-brown rust.
Boiling of water	Physical	Liquid changes to vapor.
Burning of sulfur in air	Chemical	Yellow solid sulfur changes to gaseous, choking sulfur dioxide.
Boiling an egg	Chemical	Liquid white and yolk change to solids.
Combustion of gasoline	Chemical	Liquid gasoline burns to gaseous carbon monoxide, carbon dioxide, and water.
Digesting food	Chemical	Food changes to liquid nutrients and partially solid wastes.
Sawing of wood	Physical	Smaller pieces of wood and sawdust are made from a larger piece of wood.
Burning of wood	Chemical	Wood burns to ashes, gaseous carbon dioxide, and water.
Heating of glass	Physical	Solid becomes pliable during heating, and the glass may change its shape.

can be represented by the following word equations:

$$\text{copper} + \text{oxygen} \xrightarrow{\Delta} \text{copper(II) oxide} \quad \text{(cupric oxide)} \tag{1}$$

$$\text{mercury(II) oxide} \xrightarrow{\Delta} \text{mercury} + \text{oxygen} \tag{2}$$

Equation (1) states: Copper plus oxygen when heated produce copper(II) oxide. Equation (2) states: Mercury(II) oxide when heated produces mercury plus oxygen. The arrow means "produces"; it points to the products. The Greek letter delta (Δ) represents heat. The starting substances (copper, oxygen, and mercury(II) oxide) are called the *reactants*, and the substances produced (copper(II) oxide, mercury, and oxygen) are called the *products*. In later chapters equations are presented in a still more abbreviated form, with symbols to represent substances.

Physical change usually accompanies a chemical change. Table 4.2 lists some common physical and chemical changes. In the examples given in the table, you will note that wherever a chemical change occurs, a physical change occurs also. However, wherever a physical change is listed, only a physical change occurs.

4.4 Conservation of Mass

Law of
Conservation
of Mass

The **Law of Conservation of Mass** states that no detectable change is observed in the total mass of the substances involved in a chemical change. This law, tested by extensive laboratory experimentation, is the basis for the quantitative mass relationships among reactants and products.

The decomposition of mercury(II) oxide into mercury and oxygen illustrates this law. One hundred grams of mercury(II) oxide decomposes into 92.6 g of mercury and 7.39 g of oxygen.

mercury(II) oxide $\longrightarrow$ mercury + oxygen

100. g   92.6 g   7.39 g

| 100. g Reactant | | 100. g Products |

Operation of the ordinary photographic flashcube also illustrates the law (see Figure 4.3). Sealed within the flashcube are fine wires of magnesium (a metal)

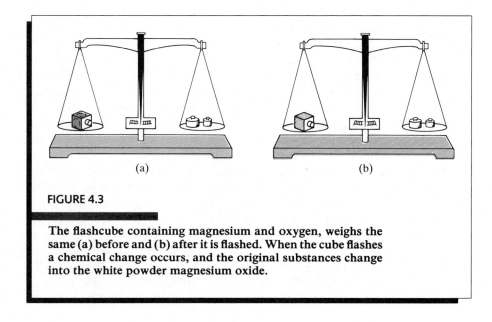

(a)   (b)

FIGURE 4.3

The flashcube containing magnesium and oxygen, weighs the same (a) before and (b) after it is flashed. When the cube flashes a chemical change occurs, and the original substances change into the white powder magnesium oxide.

and oxygen (a gas). When these reactants are energized, they combine chemically, producing magnesium oxide, a blinding white light, and considerable heat. The chemical change may be represented by this equation

$$\text{magnesium} + \text{oxygen} \longrightarrow \text{magnesium oxide} + \text{heat} + \text{light}$$

When the mass of the flashcube is determined before and after the chemical change, as illustrated in Figure 4.3, the cube shows no increase or decrease in mass.

mass of reactants = mass of products

4.5   Energy

From the prehistoric discovery that fire could be used to warm shelters and cook food, to the modern-day discovery that nuclear reactors can be used to produce vast amounts of controlled energy, our technical progress has been directed by the ability to produce, harness, and utilize energy. **Energy** is the capacity of matter to do work. Energy exists in several forms; some of the more familiar forms are mechanical, chemical, electrical, heat, nuclear, and radiant or light energy. Matter can have both potential and kinetic energy.

energy

potential energy

Potential energy is stored energy, or energy an object possesses due to its relative position. For example, a ball located 20 ft above the ground has more potential energy than when located 10 ft above the ground and will bounce higher when allowed to fall. Water backed up behind a dam represents potential energy that can be converted into useful work in the form of electrical or mechanical energy. Gasoline is a source of chemical potential energy. When gasoline burns (combines with oxygen), the heat released is associated with a decrease in potential energy. The new substances formed by burning have less chemical potential energy than the gasoline and oxygen did.

kinetic energy

Kinetic energy is the energy that matter possesses due to its motion. When the water behind the dam is released and allowed to flow, its potential energy is changed into kinetic energy, which can be used to drive generators and produce electricity. All moving bodies possess kinetic energy. The pressure exerted by a confined gas is due to the kinetic energy of rapidly moving gas particles. We all know the results when two moving vehicles collide: Their kinetic energy is expended in the crash that occurs.

Energy can be converted from one form to another form. Some kinds of energy can be converted to other forms easily and efficiently. For example, mechanical energy can be converted to electrical energy with an electric generator at better than 90% efficiency. On the other hand, solar energy has thus far been directly converted to electrical energy at an efficiency of only about 15%. In chemistry, energy is most frequently expressed as heat.

4.6 Heat: Quantitative Measurement

The SI-derived unit for energy is the joule (pronounced *jool* and abbreviated J). Another unit for heat energy, which has been used for many years, is the calorie (abbreviated cal). The relationship between joules and calories is

$$4.184 \text{ J} = 1 \text{ cal} \qquad \text{(exactly)}$$

joule
calorie

To give you some idea of the magnitude of these heat units, 4.184 **joule** or 1 **calorie** is the quantity of heat energy required to change the temperature of 1 g of water by 1°C, usually measured from 14.5°C to 15.5°C.

Since joule and calorie are rather small units, kilojoules (kJ) and kilocalories (kal) are used to express heat energy in many chemical processes. The kilocalorie is also known as the nutritional or large Calorie (spelled with a capital *C* and abbreviated Cal). In this book heat energy will be expressed in joules with parenthetical values in calories.

$$1 \text{ kJ} = 1000 \text{ J}$$
$$1 \text{ kcal} = 1000 \text{ cal}$$

The difference in the meanings of the terms *heat* and *temperature* can be seen by this example: Visualize two beakers, A and B. Beaker A contains 100 g of water at 20°C, and beaker B contains 200 g of water also at 20°C. The beakers are heated until the temperature of the water in each reaches 30°C. The temperature of the water in the beakers was raised by exactly the same amount, 10°C. But twice as much heat (8368 J or 2000 cal) was required to raise the temperature of the water in beaker B as was required in beaker A (4184 J or 1000 cal).

In the middle of the 18th century Joseph Black, a Scottish chemist, heated and cooled equal masses of iron and lead through the same temperature range. Black noted that much more heat was needed for the iron than for the lead. He had discovered a fundamental property of matter; namely, that every substance has a characteristic heat capacity. Heat capacities may be compared in terms of specific heats. The **specific heat** of a substance is the quantity of heat (lost or gained) required to change the temperature of 1 g of that substance by 1°C. It follows then that the specific heat of water is 4.184 J/g°C (1 cal/g°C). The specific heat of water is high compared with that of most substances. Aluminum and copper, for example, have specific heats of 0.900 and 0.385 J/g°C, respectively (see Table 4.3). The relation of mass, specific heat, temperature change (Δt), and

specific heat

TABLE 4.3

Specific Heat of Selected Substances

Substance	Specific heat J/g°C	Specific heat cal/g°C
Water	4.184	1.00
Ethyl alcohol	2.138	0.511
Ice	2.059	0.492
Aluminum	0.900	0.215
Iron	0.473	0.113
Copper	0.385	0.0921
Gold	0.131	0.0312
Lead	0.128	0.0305

quantity of heat lost or gained by a system is expressed by this general equation:

$$\begin{pmatrix} \text{mass of} \\ \text{substance} \end{pmatrix} \times \begin{pmatrix} \text{specific heat} \\ \text{of substance} \end{pmatrix} \times \Delta t = \text{energy (heat)} \tag{3}$$

Thus, the amount of heat needed to raise the temperature of 200 g of water by 10°C can be calculated as follows:

$$200 \ \cancel{g} \times \frac{4.184 \text{ J}}{\cancel{g}\cancel{°C}} \times 10 \cancel{°C} = 8368 \text{ J (2000 cal)}$$

Examples of specific-heat problems follow.

EXAMPLE 4.1

Calculate the specific heat of a solid in J/g°C and cal/g°C if 1638 J raise the temperature of 125 g of the solid from 25.0°C to 52.6°C.

First solve equation (3) to obtain an equation for specific heat.

$$\text{specific heat} = \frac{\text{J}}{\text{g} \times \Delta t}$$

Now substitute in the data:

$$\text{energy} = 1638 \text{ J} \qquad \text{mass} = 125 \text{ g} \qquad \Delta t = 52.6 - 25.0°C = 27.6°C$$

$$\text{specific heat} = \frac{1638 \text{ J}}{125 \text{ g} \times 27.6°C} = 0.475 \text{ J/g°C}$$

Now convert joules to calories using 1 cal/4.184 J:

$$\text{specific heat} = \frac{0.475\ \cancel{J}}{g°C} \times \frac{1\ \text{cal}}{4.184\ \cancel{J}} = 0.114\ \text{cal/g°C}$$

EXAMPLE 4.2   A sample of a metal with a mass of 212 g is heated to 125.0°C and then dropped into 375 g water at 24.0°C. If the final temperature of the water is 34.2°C, what is the specific heat of the metal? (Assume no heat losses to the surroundings.)

  When the metal enters the water it begins to cool, losing heat to the water. At the same time the temperature of the water rises. This process continues until the temperature of the metal and the temperature of the water are equal, at which point (34.2°C) no net flow of heat occurs.

  The heat lost or gained by a system is given by equation (3). We use this equation first to calculate the heat gained by the water and then to calculate the specific heat of the metal.

$$\text{temperature rise of the water } (\Delta t) = 34.2°C - 24.0°C = 10.2°C$$

$$\text{heat gained by the water} = 375\ g \times \frac{4.184\ J}{g°C} \times 10.2°C = 1.60 \times 10^4\ J$$

The metal dropped into the water must have a final temperature the same as the water (34.2°C).

$$\text{temperature drop by the metal } (\Delta t) = 125.0°C - 34.2°C = 90.8°C$$

$$\text{heat lost by the metal} = \text{heat gained by the water} = 1.60 \times 10^4\ J$$

Rearranging equation (3) we get

$$\text{specific heat} = \frac{J}{g \times \Delta t}$$

$$\text{specific heat of the metal} = \frac{1.60 \times 10^4\ J}{212\ g \times 90.8°C} = 0.831\ J/g°C$$

PRACTICE   Calculate the quantity of heat needed to heat 8.0 grams of water from 42°C to 45°C.
Answer: $1.0 \times 10^2\ J = 24\ \text{cal}$

PRACTICE   A 110.0 g sample of iron at 55.5°C raises the temperature of 150.0 mL of water from 23.0°C to 25.5°C. Determine the specific heat of the iron in calories/g°C.
Answer: 0.11 cal/g°C

4.7   Energy in Chemical Changes

In all chemical changes matter either absorbs or releases energy. Chemical changes can produce different forms of energy. Electrical energy to start automobiles is produced by chemical changes in the lead storage battery. Light energy for photographic purposes occurs as a flash during the chemical change in the magnesium flashbulb. Heat and light energies are released from the combustion of fuels. All the energy needed for our life processes—breathing, muscle contraction, blood circulation, and so on—is produced by chemical changes occurring within the cells of our bodies.

Conversely, energy is used to cause chemical changes. For example, a chemical change occurs in the electroplating of metals when electrical energy is passed through a salt solution in which the metal is submerged. A chemical change also occurs when radiant energy from the sun is used by green plants in the process of photosynthesis. And, as we saw, a chemical change occurs when heat causes mercury(II) oxide to decompose into mercury and oxygen. Chemical changes are often used primarily to produce energy rather than to produce new substances. The heat or thrust generated by the combustion of fuels is more important than the new substances formed.

4.8   Conservation of Energy

An energy transformation occurs whenever a chemical change occurs. If energy is absorbed during the change, the products will have more chemical potential energy than the reactants. Conversely, if energy is given off in a chemical change, the products will have less chemical potential energy than the reactants. Water, for example, can be decomposed in an electrolytic cell. Electrical energy is absorbed in the decomposition, and the products, hydrogen and oxygen, have a greater chemical potential energy level than that of water. This potential energy is released in the form of heat and light when the hydrogen and oxygen are burned to form water again (see Figure 4.4). Thus, energy can be changed from one form to another or from one substance to another and therefore is not lost.

The energy changes occurring in many systems have been thoroughly studied by many investigators. No system has been found to acquire energy except at the expense of energy possessed by another system. This principle is stated in other words as the **Law of Conservation of Energy**: Energy can be neither created nor destroyed, though it can be transformed from one form to another.

Law of
Conservation
of Energy

4.9   Interchangeability of Matter and Energy

Sections 4.4–4.8 dealt with matter and energy, which are clearly related; any attempt to deal with one inevitably involves the other. The nature of this relationship eluded the most able scientists until the beginning of the 20th

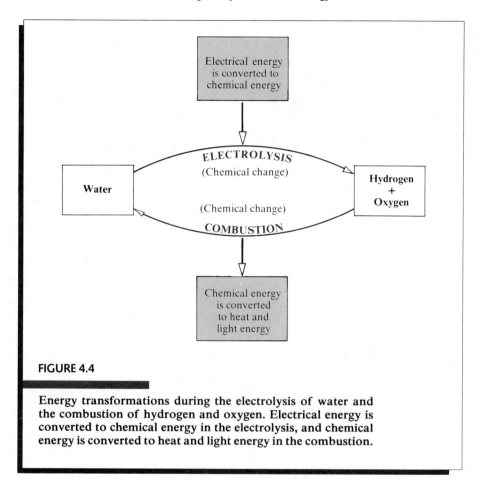

FIGURE 4.4

Energy transformations during the electrolysis of water and the combustion of hydrogen and oxygen. Electrical energy is converted to chemical energy in the electrolysis, and chemical energy is converted to heat and light energy in the combustion.

century. Then, in 1905, Albert Einstein (Figure 4.5) presented one of the most original scientific concepts ever devised.

Einstein stated that the quantity of energy E equivalent to the mass m could be calculated by the equation $E = mc^2$. In this equation the energy units of E are joules, m is in grams, and c is the velocity of light (3.0×10^8 m/s). According to Einstein's equation, whenever energy is absorbed or released by a substance, mass is lost or gained. Although the energy changes in chemical reactions are measurable and may appear to be large, the amounts are relatively small. The accompanying difference in mass between reactants and products in chemical changes is so small that it cannot be detected by available measuring instruments. According to Einstein's equation, 9.0×10^7 J of energy are equivalent to 0.0000010 g (1.0 μg) of mass.

FIGURE 4.5

Albert Einstein (1879–1955), world-renowned physicist and author of the theory of relativity and the interrelationship between matter and energy: $E = mc^2$. (*Courtesy of The Bettmann Archives.*)

$$1 \ \mu\text{g mass} = 9.0 \times 10^7 \text{ J energy}$$

In a more practical sense, when 2.8×10^3 g of carbon are burned to carbon dioxide, 9.0×10^7 J of energy are released. Of this very large amount of carbon, only about one-millionth of a gram, which is $3.6 \times 10^{-8}\%$ of the starting mass, is converted to energy. Therefore, in actual practice we may treat the reactants and products of chemical changes as having constant mass. However, because mass and energy are interchangeable, the two laws dealing with the conservation of matter may be combined into a single and generally more accurate statement:

The total amount of mass and energy remains constant during chemical change.

Concepts in Review

1. List the physical properties used to characterize a substance.
2. Distinguish between the physical and chemical properties of matter.
3. Classify changes undergone by matter as either physical or chemical.
4. Distinguish between kinetic and potential energy.
5. State the Law of Conservation of Mass.
6. State the Law of Conservation of Energy.
7. Differentiate clearly between heat and temperature.
8. Make calculations using the equation:

   energy = (mass) × (specific heat) × (Δt)

9. Explain why the laws dealing with the conservation of mass and energy may be combined into a single more accurate general statement.

Key Terms in Review

The terms listed here have all been defined within the chapter. Review the definitions of each, and use the glossary and the margin notations within the chapter as study aids.

calorie
chemical change
chemical properties
energy
joule
kinetic energy
Law of Conservation of Energy

Law of Conservation of Mass
physical change
physical properties
potential energy
properties
specific heat

Exercises

An asterisk indicates a more challenging question or problem.

1. In what physical state does acetic acid exist at 10°C? (See Table 4.1.)
2. In what physical state does chlorine exist at 102 K? (See Table 4.1.)

3. What evidence of chemical change is visible when mercury(II) oxide is heated as shown in Figure 4.2?
4. What chemical changes occur to the matter in the flashcube of Figure 4.3 when it is flashed?
5. What physical changes occur during the electrolysis of water? (See Figure 4.4.)

6. Distinguish between physical and chemical properties.

7. What is the fundamental difference between a chemical change and a physical change?

8. Classify the following as being primarily physical or primarily chemical changes:
   (a) Formation of a snowflake
   (b) Freezing ice cream
   (c) Boiling water
   (d) Boiling an egg
   (e) Churning cream to make butter
   (f) Souring milk

9. Classify the following as being primarily physical or chemical changes:
   (a) Lighting a candle
   (b) Stirring cake batter
   (c) Dissolving sugar in water
   (d) Decomposition of limestone by heat
   (e) A leaf turning yellow
   (f) Gas escaping from a freshly opened bottle of soda pop

10. Cite the evidence that indicated that only physical changes occurred when a platinum wire was heated in a burner flame.

11. Cite the evidence that indicated that both physical and chemical changes occurred when a copper wire was heated in a burner flame.

12. Cite the evidence that heating mercury(II) oxide brings about a chemical change.

13. Identify the reactants and products for each of the following:
    (a) Heating a copper wire in a burner flame
    (b) Heating mercury(II) oxide as shown in Figure 4.2
    (c) Flashing the flashcube shown in Figure 4.3

14. In a chemical change why can we consider that mass is neither lost nor gained (for practical purposes)?

15. Distinguish between potential and kinetic energy.

16. What happens to the kinetic energy of a speeding automobile when the automobile is braked to a stop?

17. What energy transformation is responsible for the fiery reentry of a returning space vehicle?

18. When the flashcube of Figure 4.3 is flashed, energy is given off to the surroundings. Explain why the mass of the flashcube appears to be the same after flashing as it was before, even though (according to Einstein) energy is equivalent to mass.

19. Which of the following statements are correct? Rewrite the incorrect ones to make them correct.
    (a) An automobile rolling down a hill possesses both kinetic and potential energy.
    (b) When heated in the air, a platinum wire gains mass.
    (c) When heated in the air, a copper wire loses mass.
    (d) The two kinds of pure substances are elements and compounds.
    (e) Boiling water represents a chemical change, because a change of state occurs.
    (f) All the following represent chemical changes: baking a cake, frying an egg, leaves changing color, iron changing to rust.
    (g) All of the following represent physical changes: breaking a stick, melting wax, folding a napkin, burning hydrogen to form water.
    (h) Chemical changes can produce electrical energy.
    (i) Electrical energy can produce chemical changes.
    (j) A stretched rubber band possesses kinetic energy.

20. Calculate the boiling point of acetic acid in (a) Kelvins and (b) degrees Fahrenheit. (See Table 4.1)

21. What is the percentage of iron in a mixture that contains 15.0 g of iron, 16.0 g of sulfur, and 18.5 g of sand?

22. How many grams of copper(II) oxide can be obtained from 1.80 g of copper? (See Figure 4.1.)

23. How many grams of copper will combine with 5.50 g of oxygen to form copper(II) oxide?

24. How many grams of mercury can be obtained from 65.0 g of mercury(II) oxide?

25. If 40.0 g of a meat sample contains 3.40 g of fat, what percentage fat is present?

26. When 10.5 g of magnesium was heated in air, 17.4 g of magnesium oxide was produced. Given the chemical reaction

magnesium + oxygen $\longrightarrow$ magnesium oxide

   (a) What mass of oxygen has combined with the magnesium?

(b) What percent of the magnesium oxide is magnesium?

27. When aluminum combines with chlorine gas, they produce the substance aluminum chloride. If 4.94 g of aluminum chloride is formed from 1.00 g of aluminum, how many grams of chlorine will combine with 5.50 g of aluminum?

28. How many joules of heat are required to raise the temperature of 80. g of water from 20.°C to 70°C?

29. How many joules are required to raise the temperature of 80. g of iron from 20.°C to 70.°C? How many calories?

30. A 250. g metal bar requires 5.866 kJ to change its temperature from 22°C to 100°C. What is the specific heat of the metal in $J/g°C$?

*31. A 20.0 g piece of a metal at 203°C is dropped into 100. g of water at 25.0°C. The water temperature rises to 29.0°C. Calculate the specific heat $(J/g°C)$ of the metal. Assume that all the heat lost by the metal is transferred to the water and no heat is lost to the surroundings.

*32. Assuming no heat losses by the system, what will be the final temperature when 50 g of water at 10°C are mixed with 10 g of water at 50°C?

*33. A 325 g piece of gold at 427°C is dropped into 200. mL of water at 22.0°C. The specific heat of gold is 0.131 $J/g°C$. Calculate the final temperature of the mixture. (Assume no heat losses to the surroundings.)

*34. The specific heat of zinc is 0.096 cal/g°C. Determine the energy required to raise the temperature of 250. g of zinc from room temperature (24.°C) to 150.°C

*35. If 40,000. J were absorbed by 500. g H_2O at 10.0°C, what would be the final temperature of the H_2O?

*36. Antimony (1.00 kg) absorbed 30.7 kJ, thus raising the temperature of the antimony from 20.°C to its melting point of 630.°C. Calculate the specific heat of antimony.

37. Three 500. g pans of iron, aluminum, and copper were used to fry an egg. Which pan would fry the egg (105°C) the quickest? Explain.

*38. What would be the change in temperature of 2.0 L of water at 24.0°C, if a 500. g iron bar at 212°C is placed in the water and the temperature stops changing?

*39. The heat of combustion of a sample of coal is 5500 cal/g. What quantity of this coal must be burned to heat 500. g of water from 20.°C to 90.°C?

40. The mass of a U.S. 25-cent coin is about 5.5 g.
    (a) How many joules would be released by the complete conversion of a 25-cent coin to energy? How many calories?
    (b) The energy calculated in (a) could heat how many gallons of water from room temperature to the boiling point if 1.27×10^6 J are needed to heat a gallon of water from room temperature (20°C) to boiling (100°C)?

Review Exercises for Chapters 1–4

CHAPTER ONE   Introduction

True–False. *Answer the following as either true or false.*

1. Chemistry is the science that deals with the composition of substances and the transformations they undergo.
2. Scientific laws are simple statements of natural phenomena to which no exceptions are known.
3. From 1803 to 1810, John Dalton advanced his atomic theory.
4. A key feature of the scientific method is to plan and do additional experiments to test a hypothesis.
5. An explanation of many observations that have been proven by many tests is called a hypothesis.
6. One of the principal goals of the alchemists was to change metals such as iron into gold.
7. Oxygen was discovered by Robert Boyle in 1774.
8. The use of the chemical balance revolutionized quantitative measurements in chemical reactions.
9. The two main branches of chemistry are organic and inorganic chemistry.
10. Only through well-planned experimentation can great scientific discoveries be derived.

Multiple Choice. *Choose the correct answer to each of the following.*

1. Early Greek philosophers believed that all matter was derived from each of the following *except*:
   (a) Earth   (d) Fire
   (b) Air   (e) Water
   (c) Metals
2. The origin of modern science is usually traced back to
   (a) Chemistry   (c) Biology
   (b) Physics   (d) Astronomy
3. If scientific experiment produced consistent results over a long period of time, it would verify a
   (a) Theory   (c) Hypothesis
   (b) Law   (d) Rule
4. Before any idea or hypothesis can be presented, a scientist must gather significant
   (a) Ideas   (c) Scientists
   (b) Data   (d) Equipment
5. The language of chemistry includes
   (a) Symbols   (c) Laws
   (b) Theories   (d) All of these

CHAPTER TWO   Standards for Measurement

True–False. *Answer the following as either true or false.*

1. A milligram is 0.001 g.
2. A centimeter is longer than a millimeter.
3. If 1 mL = 0.001 liter, then we can use the factor 10^3 mL/liter to convert liters to milliliters.
4. The density of water at 4°C is 1.00 g/mL.
5. As a metric prefix, *kilo* means 1000, or 10^3.
6. One milliliter equals 1 cm^3 exactly.
7. The joule is a unit of temperature.
8. The measurement 12.200 g contains three significant figures.
9. The answer to 25.2×0.1465 should contain three significant figures.
10. The number 14.0667 rounded off to four digits is 14.07.
11. The answer to $16.215 - 2.32$ should contain three digits.
12. A liter contains 100 mL.
13. 90°C is hotter than 210°F.
14. The units of specific gravity are g/mL.

15. The mass of an object is fixed and independent of its location.
16. The prefix *milli* means one-hundredth of.
17. The number 0.002040 written in scientific notation is 2.04×10^{-3}.
18. Two cubes of the same size but different masses will have different densities.
19. The density of liquid A is 2.20 g/mL, and that of liquid B is 1.44 g/mL. When equal volumes of these two immiscible liquids are mixed, liquid A will float on liquid B.
20. At $-20°C$ the Fahrenheit and Celsius temperatures are equal.

Multiple Choice. *Choose the correct answer to each of the following.*

1. 1.00 cm is equal to how many meters?
   (a) 2.54 (b) 100 (c) 10 (d) 0.01
2. 1.00 cm is equal to how many inches?
   (a) 0.394 (b) 0.10 (c) 12 (d) 2.54
3. 4.50 ft is how many centimeters?
   (a) 11.4 (b) 21.3 (c) 454 (d) 137
4. The number 0.0048 contains how many significant figures?
   (a) 1 (b) 2 (c) 3 (d) 4
5. Express 0.00382 in scientific notation.
   (a) 3.82×10^3 (c) 3.82×10^{-2}
   (b) 3.8×10^{-3} (d) 3.82×10^{-3}
6. 42.0°C is equivalent to:
   (a) 273 K (b) 5.55°F (c) 108°F (d) 53.3°F
7. 267°F is equivalent to:
   (a) 404 K (b) 116°C (c) 540 K (d) 389 K
8. An object has a mass of 62 g and a volume of 4.6 mL. Its density is:
   (a) 0.074 mL/g (c) 7.4 g/mL
   (b) 285 g/mL (d) 13 g/mL
9. The mass of a block is 9.43 g and its density is

2.35 g/mL. The volume of the block is:
   (a) 4.01 mL (c) 22.2 mL
   (b) 0.249 mL (d) 2.49 mL
10. The density of copper is 8.92 g/mL. The mass of a piece of copper that has a volume of 9.5 mL is:
   (a) 2.58 g (b) 85 g (c) 0.94 g (d) 1.07 g
11. An empty graduated cylinder has a mass of 54.772 g. When filled with 50.0 mL of an unknown liquid it has a mass of 101.074 g. The density of the liquid is:
   (a) 0.926 g/mL (c) 2.02 g/mL
   (b) 1.00 g/mL (d) 1.845 g/mL
12. The conversion factor to change grams to milligrams is:
   (a) $\dfrac{100 \text{ mg}}{1 \text{g}}$ (c) $\dfrac{1 \text{ g}}{1000 \text{ mg}}$
   (b) $\dfrac{1 \text{ g}}{100 \text{ mg}}$ (d) $\dfrac{1000 \text{ mg}}{1 \text{ g}}$
13. What Fahrenheit temperature is twice the Celsius temperature?
   (a) 64°F (b) 320°F (c) 200°F (d) 746°F
14. A gold alloy has a density of 12.41 g/mL and contains 75.0% gold by mass. The volume of this alloy that can be made from 255 g of pure gold is
   (a) 4.22×10^3 mL (c) 27.4 mL
   (b) 2.37×10^3 mL (d) 15.4 mL
15. A lead cylinder ($V = \pi r^2 h$) 12.0-cm in radius and 44.0-cm long has a density of 11.4 g/mL. The mass of the cylinder is:
   (a) 2.27×10^5 g (c) 1.78×10^3 g
   (b) 1.89×10^5 g (d) 3.50×10^5 g
16. All of the following units can be used for density *except*:
   (a) g/cm^3 (b) kg/m^3 (c) g/L (d) kg/m^2
17. 37.4 cm $\times$ 2.2 cm equals
   (a) 82.28 cm^2 (c) 82 cm^2
   (b) 82.3 cm^2 (d) 82.2 cm^2

CHAPTER THREE Classification of Matter

True–False. *Answer the following as either true or false.*

1. A substance is homogeneous but does not have a fixed composition.
2. A system having more than one phase is heterogeneous.

3. Matter that has identical properties throughout is homogeneous.
4. A gas is the least compact of the three states of matter.
5. Plastics, glass, and gels are examples of amorphous solids.

6. The most abundant element in the earth's crust, seawater, and atmosphere is nitrogen.
7. The symbol for cobalt can be written as Co or CO.
8. Metalloids are elements that have properties intermediate between those of metals and nonmetals.
9. Compounds exist as either molecules or ions.
10. The main characteristic of a mixture is that it has a definite composition.
11. A mixture is a combination of two or more substances in which the substances retain their identity.
12. A liquid has both a definite shape and a definite volume.
13. Pure substances occur in two forms, elements and ions.
14. Elements that have properties resembling both metals and nonmetals are called mixtures.
15. A molecule is a small, uncharged individual unit of a compound formed by the union of two or more atoms.
16. The symbols of the elements have two letters.
17. The present system of symbols for the elements was devised by J. J. Berzelius in the early 1800s.
18. The basic building blocks of all substances, which cannot be decomposed into simpler substances by ordinary chemical change, are compounds.
19. The smallest particle of an element that can exist and still retain the properties of the element is called an atom.
20. The symbol for silver is Ag.
21. The symbol for nitrogen is Ni.
22. The smallest uncharged unit of a compound is an atom.
23. All elements are expressed as diatomic molecules in the gaseous state.
24. A positively charged atom or group of atoms is called an *anion*.
25. An ion is an electrically charged atom or group of atoms.

Multiple Choice. *Choose the correct answer to each of the following.*

1. Which of the following is not one of the five most abundant elements by mass in the earth's crust, seawater, and atmosphere?

(a) Oxygen   (c) Silicon
(b) Hydrogen   (d) Aluminum

2. Which of the following is a compound?
(a) Lead   (c) Potassium
(d) Wood   (d) Water

3. Which of the following is a mixture?
(a) Water   (c) Wood
(b) Chromium   (d) Sulfur

4. How many atoms are represented in the formula Na_2CrO_4?
(a) 3   (b) (5)   (c) (7)   (d) 8

5. Which of the following is a characteristic of metals?
(a) Ductile   (c) Extremely strong
(b) Easily shattered   (d) Dull

6. Which of the following is a characteristic of nonmetals?
(a) Always a gas
(b) Poor conductor of electricity
(c) Shiny
(d) Combine only with metals

7. When a pure substance was analyzed, it was found to contain carbon and chlorine. This substance must be classified as:
(a) An element
(b) A mixture
(c) A compound
(d) Both a mixture and a compound

8. Chromium, fluorine, and magnesium have the symbols
(a) Ch, F, Ma   (c) Cr, F, Mg
(b) Cr, Fl, Mg   (d) Cr, F, Ma

9. Sodium, carbon, and sulfur have the symbols
(a) Na, C, S   (c) Na, Ca, Su
(b) So, C, Su   (d) So, Ca, Su

10. Coffee is an example of
(a) An element
(b) A compound
(c) A homogeneous mixture
(d) A heterogeneous mixture

11. The number of oxygen atoms in $Al(C_2H_3O_2)_3$ is
(a) 2   (b) 3   (c) 5   (d) 6

12. Which of the following is a mixture?
(a) Water   (c) Sugar solution
(b) Iron(II) oxide   (d) Iodine

13. Which is the most compact state of matter?
(a) Solid   (c) Gas
(b) Liquid   (d) Amorphous

14. Which is not characteristic of a solution?
    (a) A homogeneous mixture
    (b) A heterogeneous mixture
    (c) Contains two or more substances
    (d) Has a variable composition
15. Which of the following is an amorphous solid?
    (a) Quartz   (c) Glass
    (b) Silver   (d) Sodium

16. A chemical formula is a combination of
    (a) Symbols   (c) Elements
    (b) Atoms   (d) Compounds
17. The number of nonmetal atoms in $Al_2(SO_3)_3$ is
    (a) 5   (b) 7   (c) 12   (d) 14

CHAPTER FOUR   Properties of Matter

True–False. *Answer the following as either true or false.*

1. In a chemical change, substances are formed that are entirely different, having different properties and composition from the original material.
2. The Law of Conservation of Energy says that, because of the energy shortage, anyone wasting energy can be arrested.
3. The Law of Conservation of Mass states that no detectable change is observed in the total mass of the substances involved in a chemical change.
4. The starting substances in a chemical reaction are called the reactants.
5. When a clean copper wire is heated in a burner flame it gains mass.
6. The energy released when hydrogen and oxygen react to form water was stored in the hydrogen and oxygen as chemical or kinetic energy.
7. In a physical change the composition of matter does not change.
8. Physical properties describe the ability of a substance to form new substances.
9. All matter can undergo both physical and chemical changes.
10. Energy is the capacity of matter to do work.
11. 4.184 calories is the equivalent of 1.0 joule of energy.
12. The specific heat of a substance is the quantity of heat lost when the temperature of 1 g of that substance drops 1.0°C.
13. The specific heat of water is high compared to most other substances.
14. Iron (specific heat = 0.473 J/g°C) will absorb more heat energy per gram than gold (specific heat = 0.0131 J/g°C).

15. The Einstein equation relating mass and energy is $E = mc^2$.
16. Both mass and energy are conserved in a chemical change.
17. When mercury(II) oxide is heated it decomposes to give mercury and oxygen.
18. In a chemical change, the mass of the products is less than the mass of the reactants.

Multiple Choice. *Choose the correct answer to each of the following.*

1. Which of the following is not a physical property?
   (a) Boiling point   (c) Bleaching action
   (b) Physical state   (d) Color
2. Which of the following is a physical change?
   (a) A piece of sulfur is burned.
   (b) A firecracker explodes.
   (c) A rubber band is stretched.
   (d) A nail rusts.
3. Which of the following is a chemical change?
   (a) Water evaporates.
   (b) Ice melts.
   (c) Rocks are ground to sand.
   (d) A penny tarnishes.
4. When 9.44 g of calcium are heated in air, 13.22 g of calcium oxide are formed. The percent by mass of oxygen in the compound is:
   (a) 28.6%   (b) 40.0%   (c) 71.4%   (d) 13.2%
5. Barium iodide, BaI_2, contains 35.1% barium by mass. An 8.50 g sample of barium iodide contains what mass of iodine?
   (a) 5.52 g   (b) 2.98 g   (c) 3.51 g   (d) 6.49 g
6. Mercury(II) sulfide, HgS, contains 86.2% mercury by mass. The grams of HgS that can be

made from 30.0 of mercury are:

(a) 2586 g   (b) 2.87 g   (c) 25.9 g   (d) 34.8 g

7. The changing of liquid water to ice is known as a
   (a) Chemical change
   (b) Heterogeneous change
   (c) Homogeneous change
   (d) Physical change

8. Which of the following does not represent a chemical change?
   (a) Heating of copper in air
   (b) Combustion of gasoline
   (c) Cooling of red-hot iron
   (d) Digestion of food

9. To heat 30 g of water from 20°C to 50°C will require:
   (a) 30 cal   (c) 3.8×10^3 J
   (b) 50 cal   (d) 6.3×10^3 J

10. The specific heat of aluminum is 0.900 J/g°C. How many joules of energy are required to raise the temperature of 20.0 g of Al from 10.0°C to 15.0°C?
    (a) 79 J   (b) 90 J   (c) 100 J   (d) 112 J

11. A 100. g iron ball (specific heat = 0.473 J/g°C) is heated to 125°C and is placed in a calorimeter holding 200. g of water at 25.0°C. What will be the highest temperature reached by the water?
    (a) 43.7°C   (c) 65.3°C
    (b) 30.4°C   (d) 35.4°C

12. Which has the highest specific heat?
    (a) Ice   (b) Lead   (c) Water   (d) Aluminum

13. When 20.0 g of mercury is heated from 10.0°C to 20.0°C, 27.6 J of energy are absorbed. What is the specific heat of mercury?
    (a) 0.725 J/g°C   (c) 2.76 J/g°C
    (b) 0.138 J/g°C   (d) No correct answer given

14. Changing hydrogen and oxygen into water is a
    (a) Physical change
    (b) Chemical change
    (c) Conservation reaction
    (d) No correct answer given

CHAPTER FIVE

Early Atomic Theory and Structure

Since ancient times, we have sought to turn substances into gold. The Rumpelstilskin fairy tale and many other legends surround "alchemy." Yet, the alchemists discovered many elements, and founded the sciences of medicine, pharmacology, and metallurgy.

Pure substances are classified into elements and compounds. But just what makes a substance possess its unique properties? Salt tastes salty, but how small a piece of salt will retain this property? Carbon dioxide puts out fires, is used by plants to produce oxygen, and forms "dry" ice when solidified. But how small a mass of this material still behaves like carbon dioxide?

When substances finally reach the atomic, ionic, or molecular level they are in simplest identifiable form. Further division produces a loss of characteristic properties. What particles lie within an atom or ion? How are these tiny particles alike? How do they differ? How far can we continue to divide them? Alchemists began the quest, early chemists laid the foundation, and the modern chemist continues to build and expand on models of the atom.

Chapter Preview

5.1   Early Thoughts

The structure of matter has long intrigued and engaged the minds of people. The seed of modern atomic theory was sown during the time of the ancient Greek philosophers. About 440 B.C. Empedocles stated that all matter was composed of four "elements"—earth, air, water, and fire. Democritus (about 470–370 B.C.), one of the early atomistic philosophers, thought that all forms of matter were finitely divisible into invisible particles, which he called atoms. He held that atoms were in constant motion and that they combined with one another in various ways. This purely speculative hypothesis was not based on scientific observations. Shortly thereafter, Aristotle (384–322 B.C.) opposed the theory of Democritus and endorsed and advanced the Empedoclean theory. So strong was the influence of Aristotle that his theory dominated the thinking of scientists and philosophers until the beginning of the 17th century. The term *atom* is derived from the Greek word *atomos*, meaning indivisible.

5.2   Dalton's Atomic Theory

More than 2000 years after Democritus, the English schoolmaster John Dalton (1766–1844) revived the concept of atoms and proposed an atomic theory based on facts and experimental evidence. This theory, described in a series of papers published during the period 1803–1810, rested on the idea of a different kind of

Dalton's atomic
theory

atom for each element. The essence of **Dalton's atomic theory** may be summed up as follows:

1. Elements are composed of minute, indivisible particles called atoms.
2. Atoms of the same element are alike in mass and size.
3. Atoms of different elements have different masses and sizes.
4. Chemical compounds are formed by the union of two or more atoms of different elements.
5. Atoms combine to form compounds in simple numerical ratios such as one to one, two to one, two to three, and so on.
6. Atoms of two elements may combine in different ratios to form more than one compound.

Dalton's atomic theory stands as a landmark in the development of chemistry. The major premises of his theory are still valid today. However, some of the statements must be modified or qualified because investigations since Dalton's time have shown that (1) atoms are composed of subatomic particles; (2) not all the atoms of a specific element have the same mass; and (3) atoms, under special circumstances, can be decomposed.

5.3 Composition of Compounds

A large number of experiments extending over a long period of time have established the fact that a particular compound always contains the same elements in the same proportions by mass. For example, water will always contain 11.2% hydrogen and 88.8% oxygen by mass. The fact that water contains hydrogen and oxygen in this particular ratio does not mean that hydrogen and oxygen cannot combine in some other ratio. However, a compound with a different ratio would not be water. In fact, hydrogen peroxide is made up of two atoms of hydrogen and two atoms of oxygen per molecule and contains 5.9% hydrogen and 94.1% oxygen by mass; its properties are markedly different from those of water.

	Water	Hydrogen peroxide
Percent H	11.2	5.9
Percent O	88.8	94.1
Atomic composition	2 H + 1 O	2 H + 2 O

Law of Definite
Composition

The **Law of Definite Composition** states: A compound always contains two or more elements combined in a definite proportion by mass.

Let us consider two elements, oxygen and hydrogen, that form more than one compound. In water there are 8.0 g of oxygen for each gram of hydrogen. In hydrogen peroxide there are 16.0 g of oxygen for each gram of hydrogen. The masses of oxygen are in the ratio of small whole numbers, 16:8 or 2:1. Hydrogen peroxide has twice as much hydrogen (by mass) as does water. Using Dalton's atomic theory, we deduce that hydrogen peroxide has twice as many oxygens per hydrogen as water. In fact, we now write the formulas for water as H_2O and for hydrogen peroxide as H_2O_2.

Law of Multiple Proportions

The **Law of Multiple Proportions** states: Atoms of two or more elements may combine in different ratios to produce more than one compound. The reliability of this law and the law of Definite Composition is the cornerstone of the science of chemistry. In essence these laws state that (1) the composition of a particular substance will always be the same no matter what its origin or how it is formed, and (2) the composition of different compounds formed from the same elements will always be unique.

5.4   The Nature of Electric Charge

Many of us have received a shock after walking across a carpeted area on a dry day. We have also experienced the static associated with combing our hair, and have had our clothing cling to us. All of these phenomena result from an accumulation of *electric charge*. This charge may be transferred from one object to another. The properties of electric charge follow:

1. Charge may be of two types, positive and negative.
2. Unlike charges attract (positive attracts negative) and like charges repel (negative repels negative and positive repels positive).
3. Charge may be transferred from one object to another, by contact or induction.
4. The less distance between two charges the greater the force of attraction between unlike charges (or repulsion between identical charges).

5.5   Discovery of Ions

The great English scientist Michael Faraday (1791–1867) made the discovery that certain substances when dissolved in water could conduct an electric current. He also noticed that certain compounds could be decomposed into their elements by passing an electric current through the compound. Atoms of some elements were attracted to the positive electrode, while atoms of other elements were attracted to the negative electrode. Faraday concluded that these atoms were electrically charged. He called them *ions* after the Greek word, meaning "wanderer".

Any moving charge is an electric current. The electrical charge must travel through a substance known as a conducting medium. The most familiar conducting media are metals that are used in the form of wires.

The Swedish scientist Svante Arrhenius (1859–1927) extended Faraday's work. Arrhenius reasoned that an ion was an atom carrying a positive or negative charge. When a compound, such as sodium chloride (NaCl), was melted it conducted electricity. Water was unnecessary. Arrhenius' explanation of this conductivity was that upon melting the sodium chloride dissociated, or broke up, into charged ions, Na^+ and Cl^-. The Na^+ ions moved toward the negative electrode (cathode), whereas the Cl^- migrated toward the positive electrode (anode). Thus, positive ions are cations, and negative ions are anions.

From Faraday's and Arrhenius' work with ions, Irish physicist G. J. Stoney (1826–1911) realized there must be some fundamental unit of electricity associated with atoms. He named this unit the *electron* in 1891. Unfortunately he had no means of supporting his idea with experimental proof. Evidence remained elusive until 1897 when English physicist J. J. Thomson was able to show the existence of the electron experimentally.

5.6   Subatomic Parts of the Atom

The concept of the atom—a particle so small that it could not be seen even with the most powerful microscope—and the subsequent determination of its structure stand among the very greatest creative intellectual human achievements.

Any visible quantity of an element contains a vast number of identical atoms. But when we refer to an atom of an element, we isolate a single atom from the multitude in order to present the element in its simplest form. Figure 5.1 illustrates the hypothetical isolation of a single copper atom from its crystalline lattice.

Let us examine this tiny particle we call the atom. The diameter of a single atom ranges from 0.1 to 0.5 nanometers (1 nm = 1×10^{-9} m). Hydrogen, the smallest atom, has a diameter of about 0.1 nm. To arrive at some idea of how small an atom is, consider this dot (•), which has a diameter of about 1 mm, or 1×10^6 nm. It would take 10 million hydrogen atoms to form a line of atoms across this dot. As inconceivably small as atoms are, they contain smaller

subatomic particles

particles, the **subatomic particles**, such as electrons, protons, neutrons.

The development of atomic theory was helped in large part by the invention of new instruments. The Crookes tube, developed by Sir William Crookes in 1875, opened the door to the subatomic structure of the atom (Figure 5.2). The emissions generated in a Crookes tube are called *cathode rays*. Thomson (1856–1940) demonstrated in 1897 that cathode rays (1) travel in straight lines, (2) are negative in charge, (3) are deflected by electric and magnetic fields, (4) produce sharp shadows, and (5) are capable of moving a small paddle wheel. This was the experimental discovery of the fundamental unit of charge, the electron.

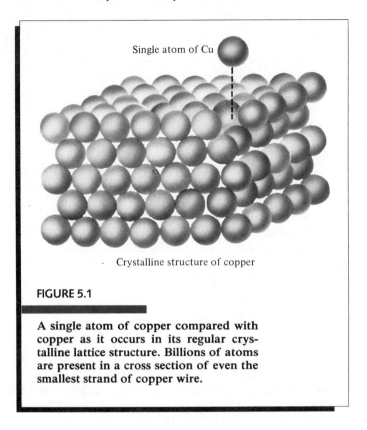

Single atom of Cu

Crystalline structure of copper

FIGURE 5.1

A single atom of copper compared with copper as it occurs in its regular crystalline lattice structure. Billions of atoms are present in a cross section of even the smallest strand of copper wire.

electron

The **electron** (e^-) is a particle with a negative electrical charge and a mass of 9.110×10^{-28} g. This mass is 1/1837 the mass of a hydrogen atom and corresponds to 0.0005486 atomic mass unit (amu) (defined in Section 5.11). One atomic mass unit has a mass of 1.6606×10^{-24} g. Although the actual charge of an electron is known, its value is too cumbersome for practical use and therefore has been assigned a relative electrical charge of -1. The size of an electron has not been determined exactly, but its diameter is believed to be less than 10^{-12} cm.

proton

Protons were first observed by German physicist E. Goldstein (1850–1930) in 1886. However, it was Thomson who discovered the nature of the proton. He showed that the proton was a particle, and he calculated its mass to be about 1837 times that of an electron. The **proton** (p) is a particle with a relative mass of 1 amu and an actual mass of 1.673×10^{-24} g. Its relative charge ($+1$) is equal in magnitude, but of opposite sign, to the charge on the electron. The mass of a proton is only very slightly less than that of a hydrogen atom.

Thomson had shown that atoms contained both negatively and positively charged particles. Clearly, the Dalton model of the atom was no longer acceptable. Atoms were not indivisible. They were composed of smaller parts. Thomson proposed a new model of the atom.

Thomson model of the atom

In the **Thomson model of the atom** the electrons were negatively charged particles imbedded in the atomic sphere. Since atoms were electrically neutral the

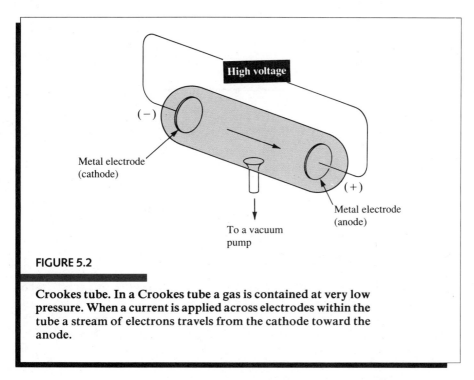

FIGURE 5.2

Crookes tube. In a Crookes tube a gas is contained at very low pressure. When a current is applied across electrodes within the tube a stream of electrons travels from the cathode toward the anode.

sphere also contained an equal number of protons, or positive charges. A neutral atom could become an ion by gaining or losing electrons.

Positive ions were explained by assuming that the neutral atom had lost electrons. An ion with a net charge of $+1$ (for example, H^+ or Li^+) would have lost one electron. An ion with a net charge of $+3$ (for example, Al^{3+}) would have lost three electrons.

Negative ions were explained by assuming that extra electrons had been added to the atoms. A net charge of -1 (for example, Cl^- or F^-) would be produced by the addition of one electron. A net charge of -2 (for example, O^{2-} or S^{2-}) would be explained as the addition of two electrons.

neutron

The third major subatomic particle was discovered in 1932 by James Chadwick (1891–1974). This particle, the **neutron** (n), bears neither a positive nor a negative charge and has a relative mass of about 1 amu. Its actual mass $(1.675 \times 10^{-24}$ g) is only very slightly greater than that of a proton. The properties of these three subatomic particles are summarized in Table 5.1.

Nearly all the ordinary chemical properties of matter can be explained in terms of atoms consisting of electrons, protons, and neutrons. The discussion of atomic structure that follows is based on the assumption that atoms contain only these principal subatomic particles. Many other subatomic particles such as mesons, positrons, neutrinos, and antiprotons have been discovered. At this time it is not clear whether all these particles are actually present in the atom or whether they are produced by reactions occurring within the nucleus. The fields of atomic and particle or high-energy physics are fascinating and have attracted

TABLE 5.1

Electrical Charge and Relative Mass of Electrons, Protons, and Neutrons

Particle	Symbol	Relative electrical charge	Relative mass (amu)	Actual mass (g)
Electron	e^-	-1	$\dfrac{1}{1837}$	9.110×10^{-28}
Proton	p	$+1$	1	1.673×10^{-24}
Neutron	n	0	1	1.675×10^{-24}

many young scientists in recent years. This interest has resulted in a great deal of research that is producing a long list of subatomic particles. Descriptions of the properties of many of these particles are to be found in recent physics textbooks and in various articles appearing in current issues of *Scientific American.*

EXAMPLE 5.1   The mass of a helium atom is 6.65×10^{-24} g. How many atoms are in a 4.0 g sample of helium?

$$4.0 \text{ g} \times \frac{1 \text{ atom}}{6.65 \times 10^{-24} \text{ g}} = 6.0 \times 10^{23} \text{ atoms}$$

PRACTICE   The mass of an atom of hydrogen is 1.673×10^{-24} g. How many atoms are in a 10.0 g sample of hydrogen?
Answer: 5.98×10^{24} atoms

5.7   The Nuclear Atom

The discovery that positively charged particles were present in atoms came soon after the discovery of radioactivity by Henri Becquerel in 1896. Radioactive elements spontaneously emit alpha, beta, and gamma rays from their nuclei (see Chapter 19).

Ernest Rutherford (Figure 5.3) had, by 1907, established that the positively charged alpha particles emitted by certain radioactive elements were ions of the element helium. Rutherford used these alpha particles to establish the nuclear

E. Rutherford

FIGURE 5.3

Ernest Rutherford (1871–1937), British physicist who identified two of the three principal rays emanating from radioactive
substances. His experiments with alpha particles led to the first
laboratory transmutation of an element and to his formulation
of the nuclear atom. Rutherford was awarded the Nobel prize
in 1908 for his work on transmutation. (*Courtesy Rutherford
Museum, McGill University.*)

nature of atoms. In some experiments performed in 1911, he directed a stream of
positively charged helium ions (alpha particles) at a very thin sheet of gold foil
(about 1000 atoms thick). See Figure 5.4(a). He observed that most of the alpha
particles passed through the foil with little or no deflection; but a few of the
particles were deflected at large angles, and occasionally one even bounced back
from the foil (Figure 5.4b). It was known that like charges repel each other and
that an electron with a mass of 1/1837 amu could not possibly have an appreciable effect on the path of a 4 amu alpha particle, which is about 7350 times
more massive than an electron. Rutherford therefore reasoned that each gold

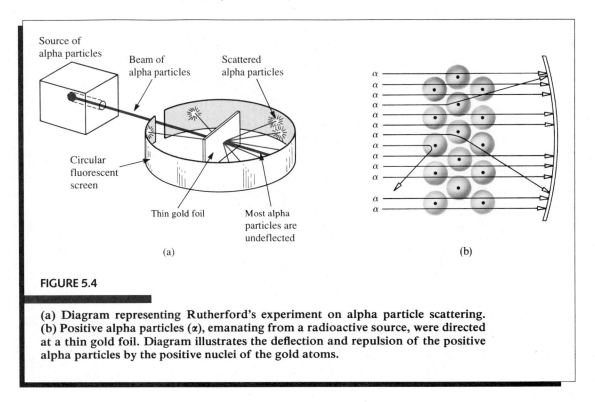

FIGURE 5.4

(a) Diagram representing Rutherford's experiment on alpha particle scattering. (b) Positive alpha particles (α), emanating from a radioactive source, were directed at a thin gold foil. Diagram illustrates the deflection and repulsion of the positive alpha particles by the positive nuclei of the gold atoms.

atom must contain a positively charged mass occupying a relatively tiny volume and that, when an alpha particle approached close enough to this positive mass, it was deflected. Rutherford spoke of this positively charged mass as the *nucleus* of the atom. Because alpha particles have relatively high masses, the extent of the deflections (some actually bounced back) indicated to Rutherford that the nucleus is relatively very heavy and dense. (The density of the nucleus of a hydrogen atom is about 10^{12} g/cm^3, about one trillion times the density of water.) Because most of the alpha particles passed through the thousand or so gold atoms without any apparent deflection, he further concluded that most of an atom consists of empty space.

When we speak of the mass of an atom, we are, for practical purposes, referring primarily to the mass of the nucleus. The nucleus contains all the protons and neutrons, which represent more than 99.9% of the total mass of any atom (see Table 5.1). By way of illustration, the largest number of electrons known to exist in an atom is 109. The mass of even 109 electrons is only about 1/17 of the mass of a single proton or neutron. The mass of an atom, therefore, is primarily determined by the combined masses of its protons and neutrons.

5.8 General Arrangement of Subatomic Particles

nucleus

The alpha particle scattering experiments of Rutherford established that the atom contains a dense, positively charged nucleus. The later work of Chadwick demonstrated that the atom contains neutrons, which are particles with mass but no charge. He also noted that light, negatively charged electrons are present and offset the positive charges in the nucleus. Based on this experimental evidence, a general description of the atom and the location of its subatomic particles was devised. Each atom consists of a **nucleus** surrounded by electrons. The nucleus contains protons and neutrons but not electrons. In a neutral atom the positive charge of the nucleus (due to protons) is exactly offset by the negative electrons. Because the charge of an electron is equal but of opposite sign to the charge of a proton, a neutral atom must contain exactly the same number of electrons as protons. However, this generalized picture of atomic structure provides no information on the arrangement of electrons within the atom. Electron configuration is discussed in Chapter 10.

> **A neutral atom contains the same number of protons and electrons.**

5.9 Atomic Numbers of the Elements

atomic number

The **atomic number** of an element is the number of protons in the nucleus of an atom of that element. The atomic number determines the identity of an atom. For example, every atom with an atomic number of 1 is a hydrogen atom; it contains one proton in its nucleus. Every atom with an atomic number of 8 is an oxygen atom; it contains 8 protons in its nucleus. Every atom with an atomic number of 92 is a uranium atom; it contains 92 protons in its nucleus. The atomic number tells us not only the number of positive charges in the nucleus but also the number of electrons in the neutral atom.

> **atomic number = number of protons in the nucleus**

There is no need to memorize the atomic numbers of the elements as the periodic table is commonly provided in texts, laboratories, and on examinations. The atomic numbers of all elements are shown in the Periodic Table on the inside front cover of this book and are also listed in the Table of Atomic Masses on the inside back cover.

5.10   Isotopes of the Elements

Shortly after Rutherford's conception of the nuclear atom, experiments were performed to determine the masses of individual atoms. These experiments showed that the masses of nearly all atoms were greater than could be accounted for by simply adding up the masses of all the protons and electrons that were known to be present in an atom. This fact led to the concept of the neutron, a particle with no charge but with a mass about the same as that of a proton. Because this particle has no charge, it was very difficult to detect, and the existence of the neutron was not proven experimentally until 1932. All atomic nuclei except that of the simplest hydrogen atom are now believed to contain neutrons.

All atoms of a given element have the same number of protons, but experimental evidence has shown that, in most cases, all atoms of a given element do not have identical masses because atoms of the same element may have different numbers of neutrons in their nuclei.

isotopes

Atoms of an element having the same atomic number but different atomic masses are called **isotopes** of that element. Atoms of the various isotopes of an element, therefore, have the same number of protons and electrons but different numbers of neutrons.

Three isotopes of hydrogen (atomic number 1) are known. Each has one proton in the nucleus and one electron. The first isotope (protium), without a neutron, has a mass of 1; the second isotope (deuterium), with one neutron in the nucleus, has a mass of 2; the third isotope (tritium), with two neutrons, has a mass of 3 (see Figure 5.5).

The three isotopes of hydrogen may be represented by the symbols $_1^1H$, $_1^2H$, $_1^3H$, indicating an atomic number of 1 and mass numbers of 1, 2, and 3, respectively. This method of representing atoms is called *isotopic notation*. The

mass number

subscript (Z) is the atomic number; the superscript (A) is the **mass number**, which is the sum of the number of protons and the number of neutrons in the nucleus.

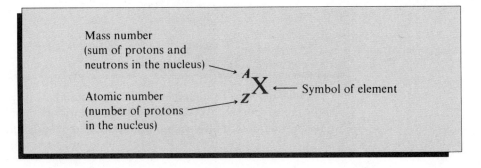

Mass number
(sum of protons and
neutrons in the nucleus) → $_Z^A X$ ← Symbol of element

Atomic number
(number of protons
in the nucleus) →

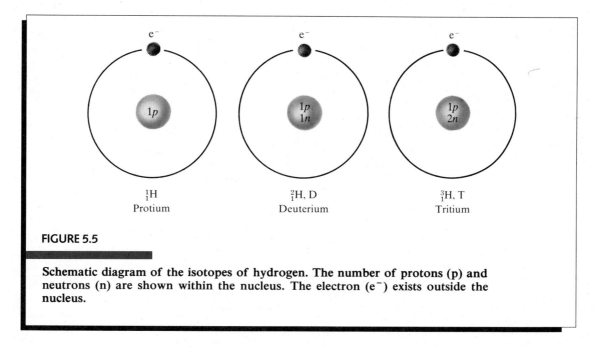

FIGURE 5.5

Schematic diagram of the isotopes of hydrogen. The number of protons (p) and neutrons (n) are shown within the nucleus. The electron (e⁻) exists outside the nucleus.

The hydrogen isotopes may also be referred to as hydrogen-1, hydrogen-2, and hydrogen-3.

Most of the elements occur in nature as mixtures of isotopes. However, not all isotopes are stable; some are radioactive and are continuously decomposing to form other elements. For example, of the seven known isotopes of carbon, only two, carbon-12 and carbon-13, are stable. Of the seven known isotopes of oxygen, only three, $^{16}_{8}O$, $^{17}_{8}O$, and $^{18}_{8}O$, are stable. Of the fifteen known isotopes of arsenic, $^{75}_{33}As$ is the only one that is stable.

5.11   Atomic Mass (Atomic Weight)

The mass of a single atom is far too small to measure individually on a balance. But fairly precise determinations of the masses of individual atoms can be made with an instrument called a *mass spectrometer* (see Figure 5.6). The mass of a single hydrogen atom is 1.6736×10^{-24} g. However, it is neither convenient nor practical to compare the actual masses of atoms expressed in grams; therefore a table of relative atomic masses using *atomic mass units* was devised. (The term *atomic weight* is often used instead of atomic mass.) The carbon isotope, having six protons and six neutrons and designated carbon-12, or $^{12}_{6}C$, was chosen as the standard for atomic masses. This reference isotope was assigned a value of exactly 12 atomic mass units (amu). Thus one **atomic mass unit** is defined as equal

atomic mass unit

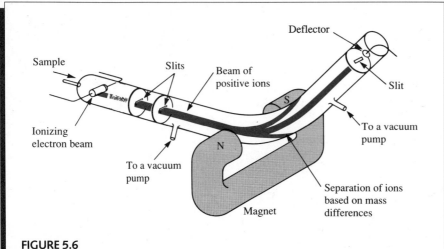

FIGURE 5.6

Schematic diagram of a modern mass spectrometer. A beam of positive ions is produced from the sample as it enters the chamber. The positive ions are then accelerated as they pass through slits in an electric field. When the ions enter the magnetic field, they are deflected differently, depending on mass and charge. The ions are then detected at the end of the tube. From intensity and position of the lines on the mass spectrogram, the different isotopes of the elements and their relative amounts can be determined.

to exactly 1/12 of the mass of a carbon-12 atom. The actual mass of a carbon-12 atom is 1.9927×10^{-23} g, and that of one atomic mass unit is 1.6606×10^{-24} g. In the table of atomic masses all elements then have values that are relative to the mass assigned to the reference isotope carbon-12.

A table of atomic masses is given on the inside back cover of this book. Hydrogen atoms, with a mass of about 1/12 that of a carbon atom, have an average atomic mass of 1.00797 amu on this relative scale. Magnesium atoms, which are about twice as heavy as carbon, have an average mass of 24.305 amu. The average atomic mass of oxygen is 15.9994 amu (usually rounded off to 16.0).

Since most elements occur as mixtures of isotopes with different masses, the atomic mass determined for an element represents the average relative mass of all the naturally occurring isotopes of that element. The atomic masses of the individual isotopes are approximately whole numbers, because the relative masses of the protons and neutrons are approximately 1.0 amu each. Yet we find that the atomic masses given for many of the elements deviate considerably from whole numbers. For example, the atomic mass of rubidium is 85.4678 amu, that of copper is 63.546 amu, and that of magnesium is 24.305 amu. The deviation of an atomic mass from a whole number is due mainly to the unequal occurrence of the various isotopes of an element. It is also due partly to the difference between the mass of a free proton or neutron and the mass of these same particles in the nucleus. For example, the two principal isotopes of copper are $^{63}_{29}Cu$ and $^{65}_{29}Cu$. It is apparent that copper-63 atoms are the more abundant isotope, since the atomic mass of copper, 63.546 amu, is closer to 63 than to 65 amu. The actual values of the copper isotopes observed by mass spectra determination are shown here:

Isotope	Isotopic mass (amu)	Abundance (%)	Average atomic mass (amu)
$^{63}_{29}Cu$	62.9298	69.09	63.55
$^{65}_{29}Cu$	64.9278	30.91	

The average atomic mass can be calculated by multiplying the atomic mass of each isotope by the fraction of each isotope present and adding the results. The calculation for copper is

$$62.9298 \text{ amu} \times 0.6909 = 43.48 \text{ amu}$$
$$64.9278 \text{ amu} \times 0.3091 = \underline{20.07} \text{ amu}$$
$$63.55 \text{ amu}$$

atomic mass

The **atomic mass** of an element is the average relative mass of the isotopes of that element referred to the atomic mass of carbon-12 (exactly 12.0000 amu).

The relationship between mass number and atomic number is such that, if we subtract the atomic number from the mass number of a given isotope, we obtain the number of neutrons in the nucleus of an atom of that isotope. Table 5.2 shows the application of this method of determining the number of neutrons. For example, the fluorine atom ($^{19}_{9}F$), atomic number 9, having a mass of 19 amu, contains 10 neutrons:

$$\text{Mass number} - \text{Atomic number} = \text{Number of neutrons}$$
$$19 \quad - \quad 9 \quad = \quad 10$$

TABLE 5.2

Determination of the Number of Neutrons in an Atom by Subtracting Atomic Number from Mass Number

	Hydrogen ($^{1}_{1}H$)	Oxygen ($^{16}_{8}O$)	Sulfur ($^{32}_{16}S$)	Fluorine ($^{19}_{9}F$)	Iron ($^{56}_{26}Fe$)
Mass number	1	16	32	19	56
Atomic number	−1	− 8	−16	− 9	−26
Number of neutrons	0	8	16	10	30

The atomic masses given in the table on the inside back cover of this book are values accepted by international agreement. You need not memorize atomic masses. In most of the calculations needed in this book, use of atomic masses to the first decimal place will give results of sufficient accuracy.

EXAMPLE 5.2

How many protons, neutrons, and electrons are found in an atom of $^{14}_{6}C$?

The element is carbon, atomic number 6. The number of protons or electrons equals the atomic number and is 6. The number of neutrons is determined by subtracting the atomic number from the mass number: $14 - 6 = 8$.

PRACTICE　How many protons, neutrons, and electrons are in each of the following?

(a) $^{16}_{8}O$　(b) $^{80}_{35}Br$　(c) $^{235}_{92}U$　(d) $^{64}_{29}Cu$

		protons	neutrons	electrons
Answers:	(a)	8	8	8
	(b)	35	45	35
	(c)	92	143	92
	(d)	29	35	29

EXAMPLE 5.3

Chlorine is found in nature as two isotopes, $^{37}_{17}Cl$ (24.47%) and $^{35}_{17}Cl$ (75.53%). The atomic masses are 36.96590 amu and 34.96885 amu, respectively. Determine the average atomic mass of chlorine.

Multiply each mass by its percentage and add the results to find the average.

$$(0.2447) \times (36.96590 \text{ amu}) + (0.7553) \times (34.96885 \text{ amu})$$

$$= 35.4575 \text{ amu}$$
$$= 35.46 \text{ amu (4 significant figures)}$$

PRACTICE　Silver is found in two isotopes with atomic masses of 106.9041 and 108.9047 amu, respectively. The first isotope represents 51.82% and the second 48.18%. Determine the average atomic mass of silver.

Answer:　107.9 amu

Concepts in Review

1. State the major provisions of Dalton's atomic theory.
2. State the Law of Definite Composition and indicate its significance.

3. State the Law of Multiple Proportions and indicate its significance.

4. Give the names, symbols, and relative masses of the three principal subatomic particles.

5. Describe the Thomson model of the atom.

6. Describe the atom as conceived by Ernest Rutherford after his alpha-scattering experiment.

7. Determine the atomic number, mass number, or number of neutrons of an isotope when given the values of any two of these three items.

8. Name and distinguish among the three isotopes of hydrogen.

9. Calculate the average atomic mass of an element, given the isotopic masses and the abundance of its isotopes.

Key Terms in Review

The terms listed here have all been defined within the chapter. Review the definitions of each, and use the glossary and margin notations within the chapter as study aids.

atomic mass   Law of Multiple Proportions
atomic mass unit   mass number
atomic number   neutron
Dalton's atomic theory   nucleus
electron   proton
isotopes   subatomic particles
Law of Definite Composition   Thomson model of the atom

Exercises

An asterisk indicates a more challenging question or problem.

1. What are the atomic numbers of (a) copper, (b) nitrogen, (c) phosphorus, (d) radium, and (e) zinc?

2. A neutron is approximately how many times heavier than an electron?

3. Explain why, in Rutherford's experiments, some alpha particles were scattered at large angles by the gold foil or even bounced back from it?

Atomic Structure

4. From the point of view of a chemist what are the essential differences among a proton, a neutron, and an electron?

5. Describe the general arrangement of subatomic particles in the atom.

6. What part of the atom contains practically all its mass?

7. What experimental evidence led Rutherford to conclude each of the following?
   (a) The nucleus of the atom contains most of the atomic mass.

(b) The nucleus of the atom is positively charged.

(c) The atom consists of mostly empty space.

8. What contribution did each of the following scientists make to the atomic theory?
   (a) Dalton   (c) Rutherford
   (b) Thomson

9. Which of the following statements are correct? Rewrite the incorrect statements to make them correct.
   (a) John Dalton developed an important atomic theory in the early 1800s.
   (b) Dalton said that elements are composed of minute indivisible particles called *atoms*.
   (c) Dalton said that when atoms combine to form compounds, they do so in simple numerical ratios.
   (d) Dalton said that atoms are composed of protons, neutrons, and electrons.
   (e) All of Dalton's theory is still considered valid today.
   (f) Hydrogen is the smallest atom.
   (g) A proton is about 1837 times as heavy as an electron.
   (h) The nucleus of an atom contains protons, neutrons, and electrons.

10. What are the numbers of protons, neutrons, and electrons in each of the following?
    (a) $^{79}_{35}Br$ (b) $^{131}_{56}Ba$ (c) $^{238}_{92}U$ (d) $^{56}_{26}Fe$

11. What letters are used to designate atomic number and mass number in isotopic notation of atoms?

12. Write isotopic notation symbols for the following:
    (a) $Z = 26$, $A = 55$ (d) $Z = 14$, $A = 29$
    (b) $Z = 12$, $A = 26$ (e) $Z = 79$, $A = 188$
    (c) $Z = 3$, $A = 6$

13. Give the isotopic notation ($^{73}_{32}Ge$, for example) for:
    (a) An atom containing 27 protons, 32 neutrons, and 27 electrons.
    (b) An atom containing 110 neutrons, 74 electrons, and 74 protons.

14. Which of the following statements are correct? Rewrite the incorrect statements to make them correct.
    (a) An element with an atomic number of 29 has 29 protons, 29 neutrons, and 29 electrons.
    (b) An atom of the isotope $^{60}_{26}Fe$ has 34 neutrons in its nucleus.

(c) $^{2}_{1}H$ is a symbol for the isotope deuterium.

(d) An atom of $^{31}_{15}P$ contains 15 protons, 16 neutrons, and 31 electrons.

(e) In the isotope $^{6}_{3}Li$, $Z = 3$ and $A = 3$.

(f) Isotopes of a given element have the same number of protons but differ in the number of neutrons.

(g) The three isotopes of hydrogen are called protium, deuterium, and tritium.

(h) $^{23}_{11}Na$ and $^{24}_{11}Na$ are isotopes.

(i) $^{24}_{11}Na$ has one more electron than $^{23}_{11}Na$.

(j) $^{24}_{11}Na$ has one more proton than $^{23}_{11}Na$.

(k) $^{24}_{11}Na$ has one more neutron than $^{23}_{11}Na$.

(l) Only a few of the elements exist in nature as mixtures of isotopes.

Atomic mass

15. Explain why the atomic masses of elements are not whole numbers.

16. Is the isotopic mass of a given isotope ever an exact whole number? Is it always? In answering, consider the masses of $^{12}_{6}C$ and $^{63}_{29}Cu$.

17. Which of the isotopes of calcium in Exercise 23 is the most abundant isotope? Can you be sure? Explain your choice.

18. In what ways are isotopes alike? In what ways are they different?

19. What special names are given to the isotopes of hydrogen?

20. List the similarities and differences in the three isotopes of hydrogen.

21. What is the symbol and name of the element that has an atomic number of 24 and a mass number of 52.

22. An atom of an element has a mass number of 201 and has 121 neutrons in its nucleus.
    (a) What is the electrical charge of the nucleus?
    (b) What is the symbol and name of the element?

23. What is the nuclear composition of the six naturally occurring isotopes of calcium having mass numbers of 40, 42, 43, 44, 46, and 48?

24. Lanthanum exists in four naturally occurring isotopes. Explain how these isotopes affect the atomic mass of lanthanum.

25. Distinguish between an atom and an ion.

26. Define a diatomic molecule and list seven examples.

27. Define an isotope and state two examples.

28. Which of the following statements are correct? Rewrite the incorrect statements to make them correct.
    (a) One atomic mass unit has a mass 12 times as much as one carbon-12 atom.
    (b) The atomic masses of protium and deuterium differ by about 100%.
    (c) $^{23}_{11}Na$ and $^{24}_{11}Na$ have the same atomic masses.
    (d) The atomic mass of an element represents the average relative atomic mass of all the naturally occurring isotopes of that element.
29. Change the following to powers of 10 (scientific notation):
    (a) 510,000   (c) $(0.001)^2$
    (b) 0.000274   (d) $(8.0)^3$
30. Complete the following table with the appropriate data for each isotope given:

Atomic number	Mass number	Symbol of element	Number of protons	Number of neutrons
(a) 8	16			
(b)		Ni		30
(c)	199		80	

*31. The actual mass of one atom of an unknown isotope is 2.18×10^{-22} amu. Calculate the atomic mass of this isotope.
32. Naturally occurring silver exists as two stable isotopes, ^{107}Ag with a mass of 106.9041 amu (51.82%) and ^{109}Ag with a mass of 108.9047 amu (48.18%). Calculate the average atomic mass of silver.
33. Naturally occurring magnesium consists of three stable isotopes: ^{24}Mg, 23.985 amu (78.99%); ^{25}Mg, 24.986 amu (10.00%); and ^{26}Mg, 25.983 amu (11.01%). Calculate the average atomic mass of magnesium
34. The mass of an atom of argon is 6.63×10^{-24} g. How many atoms are in a 40.0-g sample of argon?
35. 68.9257 amu is the mass of 60.4% of the atoms of an element with only two naturally occurring isotopes. The atomic mass of the other isotope is 70.9249 amu. Determine the average atomic mass of the element. Identify the element.
36. Complete the following table:

Element	Symbol	Atomic number	Number of protons	Number of neutrons	Number of electrons
(a) Platinum					
(b)	^{30}P	53			
(c)			36		
(d)				45	34
(e)	^{40}Ca				
(f)					

CHAPTER SIX

Nomenclature of Inorganic Compounds

As children, we begin to communicate with other people in our lives by learning the names of objects around us. As we continue to develop, we learn to speak and use language to complete a wide variety of tasks. As we enter school we begin to learn of other languages—the languages of mathematics, of other cultures, of computers. In each case, we begin by learning the names of the building blocks, and then proceed to more abstract concepts. In chemistry, a new language beckons also—a whole new way of describing the objects so familiar to us in our daily lives—the nomenclature of compounds. Only after learning the language are we able to understand the complexities of the modern model of the atom and its applications to the various fields we have chosen as our careers.

Chapter Preview

6.1   Common and Systematic Names

Chemical nomenclature is the system of names that chemists use to identify compounds. When a new substance is formulated, it must be named in order to distinguish it from all other substances. In this chapter, we will restrict our discussion to the nomenclature of inorganic compounds, compounds which do not generally contain carbon. The naming of organic compounds, carbon-containing compounds, will be covered separately in Chapter 20.

Common names are arbitrary names that are not based on the chemical composition of compounds. Before chemistry was systematized, a substance was given a name that generally associated it with one of its outstanding physical or chemical properties. For example, *quicksilver* was a common name for mercury, and nitrous oxide (N_2O), used as an anesthetic in dentistry, has been called *laughing* gas because it induces laughter when inhaled. Water and ammonia also are common names because neither provides any information about the chemical composition of the compounds. If every substance were assigned a common name, the amount of memorization required to learn over nine million names would be astronomical.

Common names have distinct limitations, but they remain in frequent use. Often common names continue to be used in industry because the systematic name is too long or too technical for everyday use. For example, calcium oxide (CaO) is called lime by plasterers; photographers refer to *hypo* rather than sodium thiosulfate ($Na_2S_2O_3$); and nutritionists use *vitamin D$_3$*, instead of *9,10 secocholesta-5,7,10(19)-trien-3- β-o1* ($C_{27}H_{44}O$). Table 6.1 lists the common names, formulas, and systematic names of some familiar substances.

Chemists prefer systematic names that precisely identify the chemical composition of chemical compounds. The system for inorganic nomenclature was devised by the International Union of Pure and Applied Chemistry (IUPAC), which first began in 1921. They continue to meet regularly and constantly review and update the system.

TABLE 6.1

Common Names, Formulas, and Chemical Names of Familiar Substances

Common name	Formula	Chemical name
Acetylene	C_2H_2	Ethyne
Lime	CaO	Calcium oxide
Slaked lime	$Ca(OH)_2$	Calcium hydroxide
Water	H_2O	Water
Galena	PbS	Lead(II) sulfide
Alumina	Al_2O_3	Aluminum oxide
Baking soda	$NaHCO_3$	Sodium hydrogen carbonate
Cane or beet sugar	$C_{12}H_{22}O_{11}$	Sucrose
Blue stone, blue vitriol	$CuSO_4 \cdot 5\,H_2O$	Copper(II) sulfate pentahydrate
Borax	$Na_2B_4O_7 \cdot 10\,H_2O$	Sodium tetraborate decahydrate
Brimstone	S	Sulfur
Calcite, marble, limestone	$CaCO_3$	Calcium carbonate
Cream of tartar	$KHC_4H_4O_6$	Potassium hydrogen tartrate
Epsom salts	$MgSO_4 \cdot 7\,H_2O$	Magnesium sulfate heptahydrate
Gypsum	$CaSO_4 \cdot 2\,H_2O$	Calcium sulfate dihydrate
Grain alcohol	C_2H_5OH	Ethanol, ethyl alochol
Hypo	$Na_2S_2O_3$	Sodium thiosulfate
Laughing gas	N_2O	Dinitrogen monoxide
Litharge	PbO	Lead(II) oxide
Lye, caustic soda	NaOH	Sodium hydroxide
Milk of magnesia	$Mg(OH)_2$	Magnesium hydroxide
Muriatic acid	HCl	Hydrochloric acid
Oil of vitriol	H_2SO_4	Sulfuric acid
Plaster of paris	$CaSO_4 \cdot \frac{1}{2}\,H_2O$	Calcium sulfate hemihydrate
Potash	K_2CO_3	Potassium carbonate
Pyrite (fool's gold)	FeS_2	Iron disulfide
Quicksilver	Hg	Mercury
Sal ammoniac	NH_4Cl	Ammonium chloride
Saltpeter (chile)	$NaNO_3$	Sodium nitrate
Table salt	NaCl	Sodium chloride
Washing soda	$Na_2CO_3 \cdot 10\,H_2O$	Sodium carbonate decahydrate
Wood alcohol	CH_3OH	Methanol, methyl alcohol

In the IUPAC system, the compound is considered to be composed of two parts, one positive and the other negative. The positive portion is named and written first. The negative portion, generally nonmetallic, follows. The names of the elements are modified with prefixes and suffixes to identify the classes of compounds.

Before we can discuss the specifics of inorganic nomenclature, it is necessary to learn a method for keeping track of the positive, negative, or zero value that an element has within a compound or ion.

6.2   Oxidation Numbers

oxidation number

oxidation state

The **oxidation number** or **oxidation state** of an element is an integer value assigned to each element in a compound or ion. This value allows us to keep track of electrons associated with each atom. Oxidation numbers have a variety of uses in chemistry—from writing formulas, to predicting properties of compounds, and assisting in the balancing of oxidation-reduction reactions in which electrons are transferred (Chapter 19).

As a starting point, the oxidation number of an uncombined element, regardless of whether it is monatomic or diatomic, is zero. Other oxidation numbers are assigned by the following somewhat arbitrary set of rules:

1. Any element in its free state has an oxidation number of zero (Examples: Na, Mg, H_2, O_2, Cl_2).
2. Metals generally have positive oxidation numbers.
3. The oxidation number of hydrogen in a compound or an ion is generally $+1$. An exception is a metal hydride when hydrogen is second in the formula and has an oxidation number of -1 (Examples: NaH, H is -1; HCl, H is $+1$).
4. The oxidation number of oxygen in a compound or an ion is generally -2, with the exception of peroxides where it is -1 (Examples: H_2O, O is -2; H_2O_2, O is -1).
5. The oxidation number of a monatomic ion is the same as the charge on the ion (Examples: Cl^-, Mg^{2+}).
6. The algebraic sum of the oxidation numbers for all the atoms in a compound must equal zero.
7. The algebraic sum of the oxidation numbers for all the atoms in a polyatomic ion (ions containing more than one atom) must equal the charge on the ion.

The oxidation numbers of many elements are predictable from their position in the periodic table. In Figure 6.1, groups of elements are labeled with their oxidation number at the top of selected columns. This figure also shows the oxidation numbers for selected common ions.

The names, formulas, and ionic charges of some common polyatomic ions are given in Table 6.2. A list of both monatomic and polyatomic ions is given on the inside back cover of this book. Writing formulas of compounds and chemical equations is facilitated by a knowledge of oxidation numbers and ionic charges.

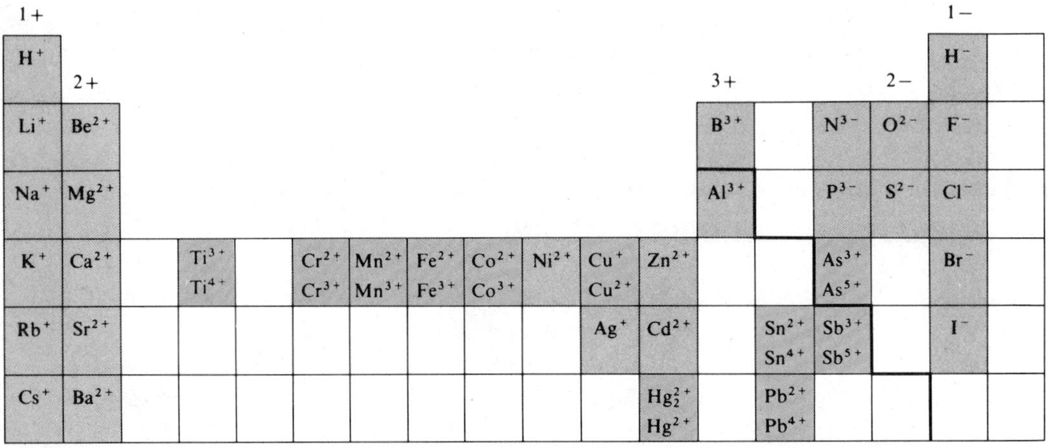

1+																1−
H^+	2+											3+			2−	H^-
Li^+	Be^{2+}											B^{3+}		N^{3-}	O^{2-}	F^-
Na^+	Mg^{2+}											Al^{3+}		P^{3-}	S^{2-}	Cl^-
K^+	Ca^{2+}	Ti^{3+} Ti^{4+}		Cr^{2+} Cr^{3+}	Mn^{2+} Mn^{3+}	Fe^{2+} Fe^{3+}	Co^{2+} Co^{3+}	Ni^{2+}	Cu^+ Cu^{2+}	Zn^{2+}			As^{3+} As^{5+}			Br^-
Rb^+	Sr^{2+}								Ag^+	Cd^{2+}		Sn^{2+} Sn^{4+}	Sb^{3+} Sb^{5+}			I^-
Cs^+	Ba^{2+}								Hg_2^{2+} Hg^{2+}			Pb^{2+} Pb^{4+}				

FIGURE 6.1

Oxidation numbers of selected elements in the periodic table. The elements with multiple oxidation numbers are highlighted in color. Hg_2^{2+} is a diatomic ion, indicating a +1 charge from each atom in the ion.

TABLE 6.2

Names, Formulas, and Charges of Some Common Polyatomic Ions

Name	Formula	Charge	Name	Formula	Charge
Acetate	$C_2H_3O_2$	−1	Cyanide	CN^-	−1
Ammonium	NH_4^+	+1	Dichromate	$Cr_2O_7^{2-}$	−2
Arsenate	AsO_4^{3-}	−3	Hydroxide	OH^-	−1
Bicarbonate	HCO_3^-	−1	Nitrate	NO_3^-	−1
Bisulfate	HSO_4^-	−1	Nitrite	NO_2^-	−1
Bromate	BrO_3^-	−1	Permanganate	MnO_4^-	−1
Carbonate	CO_3^{2-}	−2	Phosphate	PO_4^{3-}	−3
Chlorate	ClO_3^-	−1	Sulfate	SO_4^{2-}	−2
Chromate	CrO_4^{2-}	−2	Sulfite	SO_3^{2-}	−2

Use the following steps to find the oxidation number for an element within a compound.

Step 1 Write the oxidation number of each known atom below the atom in the formula.

Step 2 Multiply each oxidation number by the number of atoms of that element in the compound.

Step 3 Write an equation indicating the sum of all the oxidation numbers in the compound. Remember that the sum of all the oxidation numbers in a compound must equal zero.

EXAMPLE 6.1   Determine the oxidation number of carbon in carbon dioxide, CO_2.

$$CO_2$$

Step 1 -2
Step 2 $(-2)2$
Step 3 $C + (-4) = 0$
            $C = +4$ (oxidation number for carbon)

EXAMPLE 6.2   Determine the oxidation number for sulfur in sulfuric acid, H_2SO_4:

$$H_2S\ O_4$$

Step 1 $+1$ -2
Step 2 $2(+1) = +2$ $4(-2) = -8$
Step 3 $+2 + S + (-8) = 0$
                      $S = +6$

PRACTICE   Determine the oxidation number of (a) S in Na_2SO_4, (b) As in K_3AsO_4, (c) C in $CaCO_3$.

Answers: (a) $S = +6$ (b) $As = +5$ (c) $C = +4$

Oxidation numbers in a polyatomic ion are determined in a similar fashion, remembering that in a polyatomic ion the sum of the oxidation numbers must equal the charge on the ion instead of zero.

EXAMPLE 6.3   Determine the oxidation number of manganese in the permanganate ion MnO_4^-.

$$MnO_4^-$$

Step 1 -2
Step 2 $(-2)4$
Step 3 $Mn + (-8) = -1$
           $Mn = +7$ (oxidation number for manganese)

EXAMPLE 6.4   Determine the oxidation number of carbon in the oxalate ion $C_2O_4^{2-}$

$$C_2O_4^{2-}$$

          -2
Step 1 $(-2)4$
Step 2 $2C + (-8) = -2$
Step 3 $2C = +6$
           $C = +3$ (oxidation number for C)

PRACTICE   Determine the oxidation numbers of (a) N in NH_4^+, (b) Cr in $Cr_2O_7^{2-}$, (c) P in PO_4^{3-}.

Answers: (a) $N = -3$ (b) $Cr = +6$ (c) $P = +5$

(Note: H is $+1$ in (a) even though it comes second in the formula. N is not a metal.)

6.3   Using Ions to Write Formulas of Compounds

The sum of the oxidation numbers of all the atoms in a compound is zero. This statement applies to all substances. For ionic compounds the sum of the charges of all the ions in the compound must be zero. Hence the formulas of ionic substances can easily be determined and written by simply combining the ions in the simplest proportion that makes the sum of the ionic charges add up to zero.

To illustrate: Sodium chloride consists of Na^+ and Cl^- ions. Since $(+1) + (-1) = 0$, these ions combine in a one-to-one ratio, and the formula is written

NaCl. Calcium fluoride is made up of Ca^{2+} and F^- ions; one Ca^{2+} ion $(+2)$ and two F^- ions (-2) are needed to make zero, so the formula is CaF_2. Aluminum oxide is a bit more complicated, because it consists of Al^{3+} and O^{2-} ions. Since 6 is the lowest common multiple of 3 and 2, we have $2(+3) + 3(-2) = 0$; that is, two Al^{3+} ions $(+6)$ and three O^{2-} ions (-6) are needed; therefore, the formula is Al_2O_3.

The foregoing compounds all are made up of monatomic ions. The same procedure is used for polyatomic ions. Consider calcium hydroxide, which is made up of Ca^{2+} and OH^- ions. Since $(+2) + 2(-1) = 0$, one Ca^{2+} and two OH^- ions are needed, so the formula is $Ca(OH)_2$. The parentheses are used to enclose the OH^- so that two hydroxide ions can be shown. It is not correct to write CaO_2H_2 in place of $Ca(OH)_2$ because the identity of the compound would be lost by so doing. Note that the positive ion is written first in formulas. The following table provides examples of formula writing for ionic compounds.

Name of compound	Ions	Lowest common multiple	Sum of charges on ions	Formula
Sodium bromide	Na^+, Br^-	1	$(+1) + (-1) = 0$	NaBr
Potassium sulfide	K^+, S^{2-}	2	$2(+1) + (-2) = 0$	K_2S
Zinc sulfate	Zn^{2+}, SO_4^{2-}	2	$(+2) + (-2) = 0$	$ZnSO_4$
Ammonium phosphate	NH_4^+, PO_4^{3-}	3	$3(+1) + (-3) = 0$	$(NH_4)_3PO_4$
Aluminum chromate	Al^{3+}, CrO_4^{2-}	6	$2(+3) + 3(-2) = 0$	$Al_2(CrO_4)_3$

> **The sum of the charges on the ions of an ionic compound must equal zero.**

EXAMPLE 6.5   Write formulas for (a) calcium chloride; (b) iron (III) sulfide; (c) aluminum sulfate.

(a) Use the following steps for calcium chloride.

Step 1 From the name we know that calcium chloride is composed of calcium and chloride ions. First write down the formulas of these ions.

Ca^{2+} and Cl^-

Step 2 To write the formula of the compound, combine the smallest numbers of Ca^{2+} and Cl^- ions to give a charge sum equal to zero. In this case the lowest common multiple of the charges is 2:

$$(Ca^{2+}) + 2(Cl^-) = 0$$
$$(+2) + 2(-1) = 0$$

Therefore, the formula is $CaCl_2$.

(b) Use the same procedure for iron(III) sulfide.

Step 1 Write down the formulas for the iron(III) and sulfide ions.

$$Fe^{3+} \quad \text{and} \quad S^{2-}$$

Step 2 Use the smallest numbers of these ions required to give a charge sum equal to zero. The lowest common multiple of the charges is 6:

$$2(Fe^{3+}) + 3(S^{2-}) = 0$$
$$2(+3) + 3(-2) = 0$$

Therefore, the formula is Fe_2S_3.

(c) Use the same procedure for aluminum sulfate.

Step 1 Write down the formulas for the aluminum and sulfate ions.

$$Al^{3+} \quad \text{and} \quad SO_4^{2-}$$

Step 2 Use the smallest numbers of these ions required to give a charge sum equal to zero. The lowest common multiple of the charges is 6:

$$2(Al^{3+}) + 3(SO_4^{2-}) = 0$$
$$2(+3) + 3(-2) = 0$$

Therefore, the formula is $Al_2(SO_4)_3$. Note the use of parentheses around the SO_4^{2-} ion.

6.4   Binary Compounds

Binary compounds contain only two different elements. Their names have two parts: the name of the more positive element followed by the name of the more negative element modified to end in *ide*. [Some nonbinary compounds have names ending in *ide*, but they are exceptions to the rule and are discussed in part (d) of this section.]

(a) Binary compounds in which the positive element has a fixed oxidation state
Most of these compounds contain a metal and a nonmetal. The chemical name is composed of the name of the metal followed by the name of the nonmetal, which has been modified to an identifying stem plus the suffix *ide*.

For example, sodium chloride, NaCl, is composed of one atom each of sodium and chlorine. The name of the metal, sodium, is written first and is not modified. The second part of the name is derived from the nonmetal, chlorine, by using the stem *chlor* and adding the ending *ide*; it is named *chloride*. The compound name is sodium chloride.

NaCl

Elements:	Sodium (metal)
	Chlorine(nonmetal)
	name modified to the stem *chlor* + *ide*
Name of compound:	Sodium chloride

Stems of the more common negative-ion forming elements are shown in Table 6.3. Table 6.4 shows some compounds with names ending in *ide*.

Compounds may contain more than one atom of the same element, but as long as they contain only two different elements and if only one compound of these two elements exists, the name follows the rule for binary compounds:

Examples: $CaBr_2$ Mg_3N_2 Ag_2O

Calcium bromide   Magnesium nitride   Silver oxide

TABLE 6.3

Examples of Elements Forming Negative Ions

Symbol	Element	Stem	Binary name ending
B	Boron	Bor	Boride
Br	Bromine	Brom	Bromide
Cl	Chlorine	Chlor	Chloride
F	Fluorine	Fluor	Fluoride
H	Hydrogen	Hydr	Hydride
I	Iodine	Iod	Iodide
N	Nitrogen	Nitr	Nitride
O	Oxygen	Ox	Oxide
P	Phosphorus	Phosph	Phosphide
S	Sulfur	Sulf	Sulfide

TABLE 6.4

Examples of Compounds with Names Ending in *ide*

Formula	Name	Formula	Name
$AlCl_3$	Aluminum chloride	ZnS	Zinc sulfide
Al_2O_3	Aluminum oxide	LiI	Lithium iodide
CaC_2	Calcium carbide	$MgBr_2$	Magnesium bromide
HCl	Hydrogen chloride	NaH	Sodium hydride
HI	Hydrogen iodide	Na_2O	Sodium oxide

(b) Binary compounds containing metals of variable oxidation numbers   Two systems are commonly used for compounds in this category. The official system, designated by the International Union of Pure and Applied Chemistry (IUPAC), is known as the *Stock System*. In the Stock System, when a compound contains a metal that can have more than one oxidation number, the oxidation number of the metal is designated by a Roman numeral placed in parentheses immediately following the name of the metal. The negative element is treated in the usual manner for binary compounds.

Oxidation number	+1	+2	+3	+4	+5	+6
Roman numeral	(I)	(II)	(III)	(IV)	(V)	(VI)

Examples:
$FeCl_2$ Iron(II) chloride Fe^{2+}
$FeCl_3$ Iron(III) chloride Fe^{3+}
$CuCl$ Copper(I) chloride Cu^+
$CuCl_2$ Copper(II) chloride Cu^{2+}

The fact that $FeCl_2$ has two chloride ions, each with a -1 charge, establishes that the oxidation number of Fe is $+2$. To distinguish between the two iron chlorides, $FeCl_2$ is named iron(II) chloride and $FeCl_3$ is named iron(III) chloride.

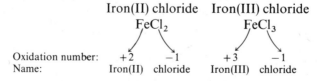

When a metal has only one possible oxidation state, we need not distinguish one oxidation state from another, so Roman numerals are not needed. Thus we do not say calcium(II) chloride for $CaCl_2$, but rather calcium chloride, since the oxidation number of calcium is understood to be $+2$.

In classical nomenclature, when the metallic ion has only two oxidation numbers, the name of the metal (usually the Latin name) is modified with the suffixes *ous* and *ic* to distinguish between the two. The lower oxidation state is given the *ous* ending, and the higher one, the *ic* ending.

Examples: $FeCl_2$ Ferrous chloride Fe^{2+} (lower oxidation state)

$FeCl_3$ Ferric chloride Fe^{3+} (higher oxidation state)

$CuCl$ Cuprous chloride Cu^+ (lower oxidation state)

$CuCl_2$ Cupric chloride Cu^{2+} (higher oxidation state)

Table 6.5 lists some common metals that have more than one oxidation number.

Notice that the *ous–ic* naming system does not give the oxidation state of an element but merely indicates that at least two oxidation states exist. The Stock System avoids any possible uncertainty by clearly stating the oxidation number.

(c) Binary compounds containing two nonmetals   In a compound between two nonmetals, the element that occurs first in the series is written and named first.

Si, B, P, H, C, S, I, Br, N, Cl, O, F

TABLE 6.5

Names and Oxidation Numbers of Some Common Metal Ions that Have More than One Oxidation Number

Formula	Stock system name	Classical name
Cu^{1+}	Copper(I)	Cuprous
Cu^{2+}	Copper(II)	Cupric
$Hg^{1+}\ (Hg_2)^{2+}$	Mercury(I)	Mercurous
Hg^{2+}	Mercury(II)	Mercuric
Fe^{2+}	Iron(II)	Ferrous
Fe^{3+}	Iron(III)	Ferric
Sn^{2+}	Tin(II)	Stannous
Sn^{4+}	Tin(IV)	Stannic
Pb^{2+}	Lead(II)	Plumbous
Pb^{4+}	Lead(IV)	Plumbic
As^{3+}	Arsenic(III)	Arsenous
As^{5+}	Arsenic(V)	Arsenic
Ti^{3+}	Titanium(III)	Titanous
Ti^{4+}	Titanium(IV)	Titanic

The name of the second element retains the modified binary ending. A Latin or Greek prefix (*mono, di, tri,* and so on) is attached to the name of each element to indicate the number of atoms of that element in the molecule. The prefix *mono* is generally omitted except when needed to distinguish between two or more compounds, such as carbon monoxide, CO, and carbon dioxide, CO_2. Some common prefixes and their numerical equivalences follow.

Mono = 1	*Hexa* =	6
Di = 2	*Hepta* =	7
Tri = 3	*Octa* =	8
Tetra = 4	*Nona* =	9
Penta = 5	*Deca* =	10

Here are some examples of compounds that illustrate this system:

CO	Carbon monoxide		N_2O	Dinitrogen monoxide
CO_2	Carbon dioxide		N_2O_4	Dinitrogen tetroxide
PCl_3	Phosphorus trichloride		NO	Nitrogen oxide
PCl_5	Phosphorus pentachloride		N_2O_3	Dinitrogen trioxide
P_2O_5	Diphosphorus pentoxide		S_2Cl_2	Disulfur dichloride
CCl_4	Carbon tetrachloride		S_2F_{10}	Disulfur decafluoride

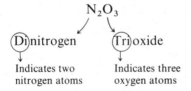

N_2O_3

(Di)nitrogen — Indicates two nitrogen atoms

(Tri)oxide — Indicates three oxygen atoms

Some common names, such as ammonia for NH_3, are exceptions to the system.

(d) Exceptions that use *ide* endings   Three notable exceptions that use the *ide* ending are hydroxides (OH^-), cyanides (CN^-), and ammonium (NH_4^+) compounds. These polyatomic ions, when combined with another element, take the ending *ide*, even though more than two elements are present in the compound.

NH_4I	Ammonium iodide
$Ca(OH)_2$	Calcium hydroxide
KCN	Potassium cyanide

(e) Acids derived from binary compounds   Certain binary hydrogen compounds, when dissolved in water, form solutions that have *acid* properties. Because of this property, these compounds are given acid names in addition to their regular *ide* names. For example, HCl is a gas and is called *hydrogen chloride*, but its water

TABLE 6.6

Names and Formulas of Selected Binary Acids

Formula	Acid name
HF	Hydrofluoric acid
HCl	Hydrochloric acid
HBr	Hydrobromic acid
HI	Hydriodic acid
H_2S	Hydrosulfuric acid
H_2Se	Hydroselenic acid

solution is known as *hydrochloric acid*. Binary acids are composed of hydrogen and one other nonmetallic element. However, not all binary hydrogen compounds are acids. To express the formula of a binary acid it is customary to write the symbol of hydrogen first, followed by the symbol of the second element (for example, HCl, HBr, H_2S). When we see formulas such as CH_4 or NH_3, we understand that these compounds are not normally considered to be acids.

To name a binary acid, place the prefix *hydro* in front of, and the suffix *ic* after, the stem of the nonmetal name. Then add the word *acid*.

HCl H_2S

Examples: *Hydro* chlor/*ic acid*   *Hydro* sulfur/*ic acid*
(hydrochloric acid)   (hydrosulfuric acid)

Acids are hydrogen-containing substances that liberate hydrogen ions when dissolved in water. The same formula is often used to express binary hydrogen compounds such as HCl, regardless of whether they are dissolved in water. Table 6.6 shows several examples of binary acids.

EXAMPLE 6.6

Name the compound CaS.

Step 1 From the formula it is a two-element compound and follows the rules for binary compounds.

Step 2 The compound is composed of Ca, a metal, and S, a nonmetal. Elements in the periodic table under the +2 column have only only one oxidation state. Thus, we name the positive part of the compound *calcium*.

Step 3 Modify the name of the second element to the identifying stem *sulf* and add the binary ending *ide* to form the name of the negative part, *sulfide*.

Step 4 The name of the compound, therefore, is *calcium sulfide*.

EXAMPLE 6.7   Name the compound FeS.

Step 1 This compound follows the rules for a binary compound and, like CaS, must be a sulfide.

Step 2 It is a compound of Fe, a metal, and S, a nonmetal. In the oxidation number tables we see that Fe has two oxidation numbers. In sulfides, the oxidation number of S is -2. Therefore, the oxidation number of Fe must be $+2$, and the name of the positive part of the compound is *iron(II)*, or *ferrous*.

Step 3 We have already determined that the name of the negative part of the compound will be *sulfide*.

Step 4 The name of FeS is *iron(II) sulfide*, or *ferrous sulfide*.

EXAMPLE 6.8   Name the compound PCl_5

Step 1 Phosphorus and chlorine are nonmetals, so the rules for naming binary compounds containing two nonmetals apply. Phosphorus is named first (see Section 6.4c).Therefore the compound is a chloride.

Step 2 No prefix is needed for phosphorus because each molecule has only one atom of phosphorus. The prefix *penta* is used with chloride to indicate the five chlorine atoms. (PCl_3 is also a known compound.)

Step 3 The name for PCl_5 is *phosphorus pentachloride*.

PRACTICE   Name each of the following binary compounds: (a) KBr, (b) Ca_3N_2, (c) SO_3, (d) SnF_2, (e) $CuCl_2$, (f) N_2O_4.

Answers: (a) potassium bromide   (d) tin(II) fluoride
                                                       stannous fluoride
              (b) calcium nitride   (e) copper(II) chloride
                                                       cupric chloride
              (c) sulfur trioxide   (f) dinitrogen tetraoxide

6.5 Ternary Compounds

Ternary compounds contain three elements and usually consist of a cation (either hydrogen or a metal), combined with a negative polyatomic ion. In general, when naming ternary compounds, the positive group is given first, followed by the name of the negative ion.

In order to name ternary compounds the names of the polyatomic ions must be known. Many polyatomic ions contain oxygen and generally have the suffix *ate* or *ite*. Unfortunately, the suffix does not indicate the number of oxygen atoms present. The *ate* form contains more oxygen atoms than the *ite* form. Examples include sulfate (SO_4^{2-}), sulfite (SO_3^{2-}), nitrate (NO_3^-), and nitrite (NO_2^-).

Techniques

A chemistry laboratory is a truly exciting and intriguing place where concepts that have only been seen on a chalkboard or in pages of a textbook can be observed first hand. A chance to work in a chemistry laboratory provides an opportunity to gain exposure to techniques and develop skills that can serve as a foundation for a career in science or carry over into other disciplines.

◀ Using a balance to determine the mass of objects is a fundamental skill learned early in a chemist's career.

A graduated cylinder, the measuring cup of the chemist, is used for volume determinations.
▼

▲
A pipet is used when more accurate measurements of volume are critical.

▲
Proper adjustment of a Bunsen burner flame results in a safe and rapid reaction.

Reactions

Chemical reactions form new substances with different compositions. Atoms, molecules, or ions interact and rearrange themselves to form new products. A reaction is generally accompanied by a physical change, which is often very discreet, requiring keen skills of observation. Others are intense, to the point of explosive. Proper safety prodecures are important in a laboratory because of violent or unexpected effects.

Combustion is a chemical reaction in which heat and light result. This purple glow signals that sulfur dioxide (SO_2) has been created through the combustion of sulfur in a container of oxygen (O_2). ▶

▲
(*Left*) Oxidation-reduction is a process in which the oxidation number of an element is changed by the transfer of electrons. Here, zinc metal is placed in a beaker of aqueous copper (II) sulfate ($CuSO_4$).
(*Right*) Electrons from the zinc metal are transferred to the copper ions (Cu^{2+}) to form solid copper, which can be seen as a deposit on the zinc metal.

A burst of UV light is emitted as a ribbon of metallic magnesium combusts in air, producing magnesium oxide (MgO). ▶

◀ Pure sodium is so reactive that contact with water creates enough heat to actually melt metal. Hydrogen gas (H_2) evolves, and can be ignited by sparking produced during the reaction.

Metallic zinc transfers electrons to hydrogen ions (H^+) to form bubbles of insoluble hydrogen gas (H_2).
▼

▲
Hydrated salts contain H_2O molecules in their crystalline structure. Heating cobalt chloride hexahydrate ($CoCl_2 \cdot 6H_2O$), blue, removes H_2O molecules to form anhydrous cobalt chloride ($CoCl_2$), red.

A double replacement reaction results ▶ from mixing aqueous solutions of potassium hydroxide (KOH) and iron (II) nitrate [$Fe(NO_3)_3$], which produces insoluble iron (II) hydroxide [$Fe(OH)_2$],

Separations

The primary goal of many chemical reactions that take place in the laboratory is to obtain an isolated, pure compound. Techniques used to separate compounds or mixtures depend on individual chemical or physical characteristics. Among the most used characteristics are boiling point, solubility, polarity, electrical charge, and size.

Separation and quantitation of compounds ▶ in a liquid chromatography instrument is based on the differential interaction of the compounds in a sample between a liquid mobile phase and a solid stationary phase.

Molecular iodine (I_2) and ionic iodine (I^1) can be separated on the basis of their differing solubilities in two immiscible solvents, such as water (H_2O) and carbon tetrachloride (CCl_4). Nonpolar I_2 dissolves in nonpolar CCl_4 as indicated by the pink color, whereas I^{1-} remains in H_2O, which is a polar solvent.
▼

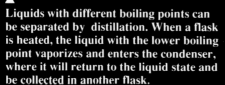

▲
Liquids with different boiling points can be separated by distillation. When a flask is heated, the liquid with the lower boiling point vaporizes and enters the condenser, where it will return to the liquid state and be collected in another flask.

▲
The process of filtration depends on the insolubility of a solid in a given solvent. Compounds that are not dissolved are trapped in the filter while dissolved compounds pass through.

The Colors of Chemistry

Color in a chemistry laboratory can provide a wealth of information. The color of a solution can reveal the oxidation state of the ions dissolved. Colored indicators can determine the concentration of an acid or a base. In addition, solutions of chemical substances can differentially absorb colors of light, and this property can be used to assess concentration and purity.

◀ The pH of a solution can be determined by adding a small quantity of the universal indicator to the solution and comparing the color to a standard chart.

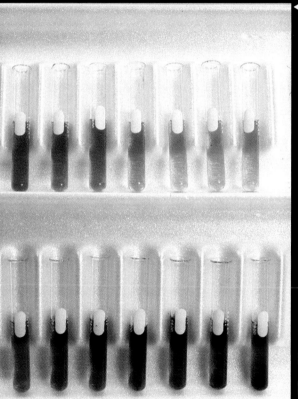

(Left) An indicator can be selected for a titration so that it will signal the endpoint by changing to a different color at the instant of chemical equivalence. For example, phenolphthalein is colorless in acid solution. *(Right)* Phenolphthalein turns pink at the endpoint of an acid-base titration when the equivalents of a base just exceed those of an acid.

◀ Principal oxidation states of chromium are +2, +3, and +6. Oxidation numbers are indicated by the color of a chromium solution: +2 is gray-blue, +3 is violet or green, and +6 is yellow or orange.

Polymers

Polymers are macromolecules formed by bonding together thousands of repeated smaller molecules called "monomers." Starch, glycogen, cellulose, proteins, and DNA are naturally occurring polymers essential to life processes. Dacron, orlon, nylon, polyethylene, polyester, and polystyrene are some of the synthetic polymers that we use on a daily basis.

Polyurethanes are highly cross-linked thermosetting polymers. Here, the unpolymerized reagents are poured into molds and polymerization is initiated. ▶

(Left) Nylon is formed at the interface of solutions of hexamethylene diamine dissolved in water (bottom) and adipoyl chloride dissolved in hexane (top). *(Right)* The length of the strand will continue to grow until the supply of one of the reagents is exhausted.
▼

▲
New applications are being found for polymers. This chemist is studying synthetic resins to find better ways of using them in the manufacture of adhesives.

Genetic Engineering

Genetic engineering involves reprogramming the genetic material of a cell so that it will produce a new or different protein. The ability to isolate specific genes and synthesize nucleic-acid sequences allows researchers to modify the genomes of bacteria so that they manufacture proteins that are beneficial to human beings.

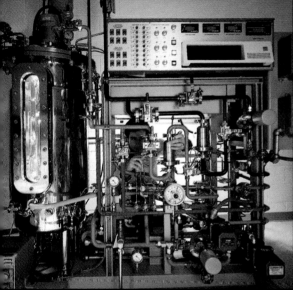

◀ The temperature, oxygen flow, and pH balance of an organism can be monitored in a biological fermenter. This sterilized, controlled environment allows genetically engineered organisms to flourish.

DNA is made up of four nucleotides and arranged in a double helix.
▼

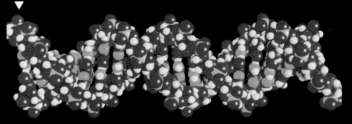

▲
A researcher uses a mass spectrometer to analyze a protein.

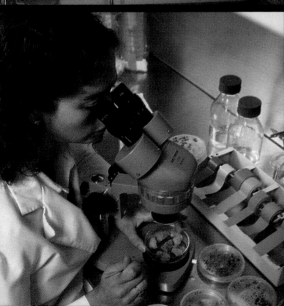

◀ This technician is examining and recording growth patterns of genetically altered bacteria in petri dishes.

Microscopes

Everyday, chemists observe physical and chemical properties attributed to the atom. The scanning electron microscope (SEM) allowed visualization of the shapes of large molecules. And now, a recent breakthrough in resolution, using a scanning tunneling microscope (STM), has revealed structural details of molecules as small as a single atom. At last, the atom has been seen. This instrument has the potential for thousands of uses.

The SEM gives a wealth of information ▶
about subcellular shapes and structures.

STM image of a DNA. One exciting use
of the STM is to actually see structures
of individual molecules and how they
interact to form complex compounds.
▼

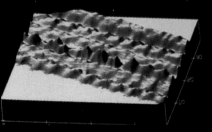

STM image of iodine adsorbed on a
platinum crystal surface.
▼

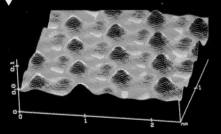

The STM uses a sharp-tipped needle to ▶
scan across a sample. A small current
that "tunnels" across the gap from
needle to sample is translated into three-
dimensional images.

Some elements form more than two different polyatomic ions containing oxygen. To name these ions, prefixes are used in addition to the suffix. To indicate more oxygen than in the *ate* form we add the prefix *per*, which is a short form of *hyper*, meaning more. The prefix *hypo*, meaning less (oxygen in this case), is used for the ion containing less oxygen than the *ite* form. An example of this system is shown for the chlorine, oxygen polyatomic ions in Table 6.7. The prefixes are also used with other similar ions, such as iodate (IO_3^-), bromate (BrO_3^-), and phosphate (PO_4^{3-}).

Only two of the common negatively charged polyatomic ions do not use the *ate/ite* system. These exceptions are hydroxide (OH^-) and cyanide (CN^-). Care must be taken with these, as their endings can easily be confused with the *ide* ending for binary compounds (Section 6.4d).

There are two common positively charged polyatomic ions as well—the ammonium and the hydronium ions. The ammonium ion (NH_4^+) is frequently found in ternary compounds (Section 6.4), whereas the hydronium ion (H_3O^+) is usually seen in aqueous solutions of acids (Chapter 16).

(a) Ternary oxy-acids Inorganic ternary compounds containing hydrogen, oxygen, and one other element are called *oxy-acids*. The element other than hydrogen or oxygen in these acids is often a nonmetal, but can be a metal.

The first step in naming these acids is to determine that the compound in question is really an oxy-acid. The keys to identification are that (1) hydrogen is the first element in the compound's formula; and (2) the second portion of the formula consists of a polyatomic ion containing oxygen.

Hydrogen in a ternary oxy-acid is not specifically designated in the acid name. The presence of hydrogen in the compound is indicated by the use of the word *acid* in the name of the substance. To determine the particular type of acid the polyatomic ion following hydrogen must be examined. The name of the polyatomic ion is modified in the following manner: (1) *ate* changes to an *ic* ending; (2) *ite* changes to an *ous* ending. Note also that this rule is similar to the earlier rule used in the Stock system for naming binary compounds. The *ic* ending represents a higher oxidation state, whereas the *ous* ending represents a lower oxidation state. Consider the following examples:

H_2SO_4 sulf*ate* ⟶ sulfur*ic* acid   (S is +6)
H_2SO_3 sulf*ite* ⟶ sulfur*ous* acid   (S is +4)
HNO_3 nitr*ate* ⟶ nitr*ic* acid   (N is +5)
HNO_2 nitr*ite* ⟶ nitr*ous* acid   (N is +3)

The complete system for naming ternary oxy-acids is shown for the various acids containing chlorine in Table 6.7.

Examples of other ternary oxy-acids and their names are shown in Table 6.8.

(b) Salts of ternary acids A *salt* is formed when the hydrogen of an acid is replaced by a metal or ammonium ion. The name of the cation is given first,

---

TABLE 6.7

Oxy-Acids and Oxy-Anions of Chlorine

Acid formula	Acid name	Chlorine oxidation number	Anion formula	Anion name
HClO	*Hypo*chlor*ous* acid	+1	ClO^-	*Hypo*chlor*ite*
$HClO_2$	Chlor*ous* acid	+3	ClO_2^-	Chlor*ite*
$HClO_3$	Chlor*ic* acid	+5	ClO_3^-	Chlor*ate*
$HClO_4$	*Per*chlor*ic* acid	+7	ClO_4^-	*Per*chlor*ate*

TABLE 6.8

Formulas and Names of Selected Ternary Oxy-Acids

Formula	Acid name	Formula	Acid name
H_2SO_3	Sulfurous acid	$HC_2H_3O_2$	Acetic acid
H_2SO_4	Sulfuric acid	$H_2C_2O_4$	Oxalic acid
HNO_2	Nitrous acid	H_2CO_3	Carbonic acid
HNO_3	Nitric acid	$HBrO_3$	Bromic acid
H_3PO_2	Hypophosphorous acid	HIO_3	Iodic acid
H_3PO_3	Phosphorous acid	H_3BO_3	Boric acid
H_3PO_4	Phosphoric acid		

followed by the name of the polyatomic anion. The names of the anions are derived from the names of the corresponding ternary acids by changing the *ous* and *ic* endings to *ite* and *ate*, respectively. The stem portion of the acid name is not changed.

Ternary oxy-acid		*Ternary oxy-salt*
ous ending of acid	becomes	*ite* ending in salt
ic ending of acid	becomes	*ate* ending in salt

Thus the sulf*ite* ion (SO_3^{2-}) is derived from sulfur*ous* acid (H_2SO_3) and the sulf*ate* ion (SO_4^{2-}), from sulfur*ic* acid (H_2SO_4). The names and formulas of the sodium, calcium, and aluminum salts of sulfurous and sulfuric acids are as follows:

$$Na_2SO_3 \qquad CaSO_3 \qquad Al_2(SO_3)_3$$

Sodium sulfite   Calcium sulfite   Aluminum sulfite

$$Na_2SO_4 \qquad CaSO_4 \qquad Al_2(SO_4)_3$$

Sodium sulfate   Calcium sulfate   Aluminum sulfate

The complete system for naming ternary acids and their anions, using the oxy-acids of chlorine as a model, is shown in Table 6.7. A comparison of the acid, anion, and salt names for a number of ternary oxy-compounds is presented in Table 6.9.

The endings *ous*, *ic*, *ite*, and *ate* are part of classical nomenclature; they are not used in the Stock System to indicate different oxidation states of the elements. These endings are still used, however, in naming many common compounds. The Stock name for H_2SO_4 is tetraoxosulfuric(VI) acid, and that for H_2SO_3 is trioxosulfuric(IV) acid. These Stock names are not commonly used.

TABLE 6.9

Comparison of Acid, Anion and Salt Names for Selected Ternary Oxy-Compounds

Acid	Anion	Formula and names of representative salts	
H_2SO_4 Sulfuric acid	SO_4^{2-} Sulfate ion	$CaSO_4$ $Fe_2(SO_4)_3$	Calcium sulfate Iron(III) sulfate or ferric sulfate
H_2SO_3 Sulfurous acid	SO_3^{2-} Sulfite ion	Na_2SO_3 Ag_2SO_3	Sodium sulfite Silver sulfite
HNO_3 Nitric acid	NO_3^- Nitrate ion	KNO_3 $Hg(NO_3)_2$	Potassium nitrate Mercury(II) nitrate or mercuric nitrate
HNO_2 Nitrous acid	NO_2^- Nitrite ion	KNO_2 $Co(NO_2)_2$	Potassium nitrite Cobalt(II) nitrite or cobaltous nitrie
H_2CO_3 Carbonic acid	CO_3^{2-} Carbonate ion	Li_2CO_3 $BaCO_3$	Lithium carbonate Barium carbonate
H_3PO_4 Phosphoric acid	PO_4^{3-} Phosphate ion	$AlPO_4$ $Zn_3(PO_4)_2$	Aluminum phosphate Zinc phosphate
H_3PO_3 Phosphorous acid	PO_3^{3-} Phosphite ion	Na_3PO_3 $Zn_3(PO_3)_2$	Sodium phosphite Zinc phosphite
HIO_3 Iodic acid	IO_3^- Iodate ion	$AgIO_3$ $Cu(IO_3)_2$	Silver iodate Copper(II) iodate or cupric iodate
$HC_2H_3O_2$ Acetic acid	$C_2H_3O_2^-$ Acetate ion	$Pb(C_2H_3O_2)_2$ $NH_4C_2H_3O_2$	Lead(II) acetate Ammonium acetate
$H_2C_2O_4$ Oxalic acid	$C_2O_4^{2-}$ Oxalate ion	CaC_2O_4 $(NH_4)_2C_2O_4$	Calcium oxalate Ammonium oxalate

EXAMPLE 6.9   (a) Name the salt KNO_3 and (b) name the acid HNO_3 from which this salt can be derived.

> **Step 1** The compound contains three elements and follows the rules for ternary compounds.
> **Step 2** The salt is composed of a K^+ ion and a NO_3^- ion. The name of the positive part of the compound is *potassium*.
> **Step 3** Since it is a ternary salt, the name will end in *ite* or *ate*. In the oxidation number tables, we see that the name of the NO_3^- ion is *nitrate*.
> **Step 4** The name of the compound is *potassium nitrate*.

(b) The name of the acid follows the rules for ternary oxy-acids. Because the name of the salt KNO_3 ends in *ate*, the name of the corresponding acid will end in *ic acid*. Change the *ate* ending of nitrate to *ic*. Thus *nitrate* becomes *nitric*, and the name of the acid is *nitric acid*.

PRACTICE   Name the following salts and the acids from which they are derived.
(a) $CaSO_4$, H_2SO_4; (b) Mg_3PO_4, H_3PO_4; (c) Na_2CO_3, H_2CO_3.
Answers: (a) calcium sulfate, sulfuric acid
           (b) magnesium phosphate, phosphoric acid
           (c) sodium carbonate, carbonic acid

6.6   Salts with More than One Kind of Positive Ion

Salts can be formed from acids that contain two or more acid hydrogen atoms by replacing only one of the hydrogen atoms with a metal or by replacing both hydrogen atoms with different metals. Each positive group is named first, and then the appropriate salt ending is added.

Acid	Salt	Name of salt
H_2CO_3	$NaHCO_3$	Sodium hydrogen carbonate or sodium bicarbonate
H_2S	$NaHS$	Sodium hydrogen sulfide or sodium bisulfide
H_3PO_4	$MgNH_4PO_4$	Magnesium ammonium phosphate
H_2SO_4	$NaKSO_4$	Sodium potassium sulfate

Note the name *sodium bicarbonate* given in the table. The prefix *bi* is commonly used to indicate a compound in which one of two acid hydrogen atoms has been replaced by a metal. The HCO_3^- ion is known as the hydrogen carbonate ion or the bicarbonate ion. Another example is sodium bisulfate, which

TABLE 6.10

Names of Selected salts that contain More than One Kind of Positive Ion

Acid	Salt	Name of salt
H_2SO_4	$KHSO_4$	Potassium hydrogen sulfate or potassium bisulfate
H_2SO_3	$Ca(HSO_3)_2$	Calcium hydrogen sulfite or calcium bisulfite
H_2S	NH_4HS	Ammonium hydrogen sulfide or ammonium bisulfide
H_3PO_4	$MgNH_4PO_4$	Magnesium ammonium phosphate
H_3PO_4	NaH_2PO_4	Sodium dihydrogen phosphate
H_3PO_4	Na_2HPO_4	Disodium hydrogen phosphate
$H_2C_2O_4$	KHC_2O_4	Potassium hydrogen oxalate or potassium binoxalate
H_2SO_4	$KAl(SO_4)_2$	Potassium aluminum sulfate
H_2CO_3	$Al(HCO_3)_3$	Aluminum hydrogen carbonate or aluminum bicarbonate

has the formula $NaHSO_4$. Table 6.10 shows examples of other salts that contain more than one kind of positive ion.

Note that prefixes are also used in chemical nomenclature to give special clarity or emphasis to certain compounds as well as to distinguish between two or more compounds.

Examples: Na_3PO_4 Trisodium phosphate
Na_2HPO_4 Disodium hydrogen phosphate
NaH_2PO_4 Sodium dihydrogen phosphate

6.7   Bases

Inorganic bases contain the hydroxide ion, OH^-, in chemical combination with a metal ion. These compounds are called *hydroxides*. The OH^- group is named as a single ion and is given the ending *ide*. Several common bases are listed below:

NaOH   Sodium hydroxide
KOH   Potassium hydroxide
NH_4OH Ammonium hydroxide

$Ca(OH)_2$ Calcium hydroxide

$Ba(OH)_2$ Barium hydroxide

We have now looked at ways of naming a variety of inorganic compounds—binary compounds consisting of a metal and a nonmetal and of two nonmetals, and binary and ternary acids, salts, and bases. These compounds are just a small part of the classified chemical compounds. Most of the remaining classes are in the broad field of organic chemistry under such categories as hydrocarbons, alcohols, ethers, aldehydes, ketones, phenols, and carboxylic acids.

Concepts in Review

1. State the rules for assigning oxidation numbers.
2. Write the formulas of compounds formed by combining the ions from Figure 6.1 and Table 6.2 (or from the inside back cover of this book) in the correct ratios.
3. Assign the oxidation number to each element in a compound or ion.
4. Write the names or formulas for inorganic binary compounds in which the metal has only one common oxidation state.
5. Write the names or formulas for inorganic binary compounds that contain metals of variable oxidation state, using either the Stock System or classical nomenclature.
6. Write the names or formulas for inorganic binary compounds that contain two nonmetals.
7. Write the names or formulas for binary acids.
8. Write the names or formulas for ternary oxy-acids.
9. Write the names or formulas for ternary salts.
10. Given the formula of a salt, write the name and formula of the acid from which the salt may be derived.
11. Write the names or formulas for salts that contain more than one kind of positive ion.
12. Write the names or formulas for inorganic bases.

Key Terms in Review

The terms listed here have all been defined within the chapter. Review the definitions of each, and use the glossary and margin notations within the chapter as study aids.

oxidation number   oxidation state

Exercises

1. Using the oxidation numbers shown in Figure 6.1, write formulas for:
   (a) The oxygen compounds of Ca, Ni, Ag, Al, Na
   (b) The iodine compounds of Li, Al, B, Sr, Mg
2. Use the oxidation number table on the inside back cover of your book to determine the formulas for compounds composed of the following ions:
   (a) Sodium and chlorate
   (b) Hydrogen and sulfate
   (c) Tin(II) and acetate
   (d) Copper(I) and oxide
   (e) Zinc and bicarbonate
   (f) Iron(III) and carbonate
3. Determine the oxidation number of the element underlined in each formula:
   (a) $M\underline{n}CO_3$ (d) $KM\underline{n}O_4$ (g) $\underline{W}Cl_5$
   (b) $\underline{Sn}F_4$ (e) $Ba\underline{C}O_3$ (h) $K_2\underline{Cr}_2O_7$
   (c) $K\underline{N}O_3$ (f) $\underline{P}Cl_3$
4. Determine the oxidation number of the element underlined in each formula:
   (a) $\underline{In}I_3$ (d) $\underline{C}_2H_5OH$ (g) $\underline{Fe}_2(CO_3)_3$
   (b) $K\underline{Cl}O_3$ (e) $Mg(\underline{N}O_3)_2$ (h) $Na\underline{Cl}O_4$
   (c) $Na_2\underline{S}O_4$ (f) $\underline{Sn}O_2$

5. Write the formula of the compound that would be formed between the given elements:
   (a) Na and I   (c) Al and O   (e) Cs and Cl
   (b) Ba and F   (d) K and S   (f) Sr and Br
6. Write the formula of the compound that would be formed between the given elements:
   (a) Ba and O   (c) Al and Cl   (e) Li and Si
   (b) H and S   (d) Be and Br   (f) Mg and P
7. Does the fact that two elements combine in a one-to-one atomic ratio mean that their oxidation numbers are both 1? Explain.
8. Write formulas for the following cations (do not forget to include the charges): sodium, magnesium, aluminum, copper(II), iron(II), ferric, lead(II), silver, cobalt(II), barium, hydrogen, mercury(II), tin(II), chromium(III), stannic, manganese(II) bismuth(III).
9. Write formulas for the following anions (do not forget to include the charges): chloride, bromide, fluoride, iodide, cyanide, oxide, hydroxide, sulfide, sulfate, bisulfate, bisulfite, chromate, carbonate, bicarbonate, acetate, chlorate, permanganate, oxalate.
10. Complete the table, filling in each box with the proper formula.

		Anions			
Cations	Br^-	O^{2-}	NO_3^-	PO_4^{3-}	CO_3^{2-}
K^+	KBr				
Mg^{2+}					
Al^{3+}					
Zn^{2+}				$Zn_3(PO_4)_2$	
H^+					

11. Complete the table, filling in each box with the proper formula.

Anions

Cations	SO_4^{2-}	Cl^-	AsO_4^{3-}	$C_2H_3O_2^-$	CrO_4^{2-}
NH_4^+			$(NH_4)_3AsO_4$		
Ca^{2+}					
Fe^{3+}	$Fe_2(SO_4)_3$				
Ag^+					
Cu^{2+}					

12. State how each of the following is used in naming inorganic compounds: *ide, ous, ic, hypo, per, ite, ate,* Roman numerals.

13. Write formulas for the following binary compounds, all of which are composed of non-metals:
    (a) Carbon monoxide
    (b) Sulfur trioxide
    (c) Carbon tetrabromide
    (d) Phosphorus trichloride
    (e) Nitrogen dioxide
    (f) Dinitrogen pentoxide

14. Name the following binary compounds, all of which are composed of nonmetals:
    (a) CO_2 (e) SO_2 (i) NF_3
    (b) N_2O (f) N_2O_4 (j) CS_2
    (c) PCl_5 (g) P_2O_5
    (d) CCl_4 (h) OF_2

15. Name the following compounds:
    (a) K_2O (d) $BaCO_3$ (g) $Zn(NO_3)_2$
    (b) NH_4Br (e) Na_3PO_4 (h) Ag_2SO_4
    (c) CaI_2 (f) Al_2O_3

16. Write formulas for the following compounds:
    (a) Sodium nitrate
    (b) Magnesium fluoride
    (c) Barium hydroxide
    (d) Ammonium sulfate
    (e) Silver carbonate
    (f) Calcium phosphate

17. Name each of the following compounds by both the Stock (IUPAC) System, and the *ous–ic* system:
    (a) $CuCl_2$ (d) $FeCl_3$ (g) $As(C_2H_3O_2)_3$
    (b) $CuBr$ (e) SnF_2 (h) TiI_3
    (c) $Fe(NO_3)_2$ (f) $HgCO_3$

18. Write formulas for the following compounds:
    (a) Tin(IV) bromide   (d) Mercuric nitrite
    (b) Copper(I) sulfate   (e) Titanic sulfide
    (c) Ferric carbonate   (f) Iron(II) acetate

19. Write formulas for the following acids:
    (a) Hydrochloric acid   (d) Carbonic acid
    (b) Chloric acid   (e) Sulfurous acid
    (c) Nitric acid   (f) Phosphoric acid

20. Name the following acids:
    (a) HNO_2 (d) HBr (g) HF
    (b) H_2SO_4 (e) H_3PO_3 (h) $HBrO_3$
    (c) $H_2C_2O_4$ (f) $HC_2H_3O_2$

21. Write formulas for the following acids:
    (a) Acetic acid   (d) Boric acid
    (b) Hydrofluoric acid   (e) Nitrous acid
    (c) Hypochlorous acid   (f) Hydrosulfuric acid

22. Name the following acids:
    (a) H_3PO_4 (c) HIO_3 (e) $HClO$ (g) HI
    (b) H_2CO_3 (d) HCl (f) HNO_3 (h) $HClO_4$

23. Name the following compounds:
    (a) $Ba(NO_3)_2$ (d) $MgSO_4$ (g) NiS
    (b) $NaC_2H_3O_2$ (e) $CdCrO_4$ (h) $Sn(NO_3)_4$
    (c) PbI_2 (f) $BiCl_3$ (i) $Ca(OH)_2$

24. Write formulas for the following compounds:
    (a) Silver sulfite
    (b) Cobalt(II) bromide
    (c) Tin(II) hydroxide
    (d) Aluminum sulfate
    (e) Manganese(II) fluoride
    (f) Ammonium carbonate
    (g) Chromium(III) oxide
    (h) Cupric chloride
    (i) Potassium permanganate
    (j) Barium nitrite
    (k) Sodium peroxide
    (l) Ferrous sulfate
    (m) Potassium dichromate
    (n) Bismuth(III) chromate

25. Write formulas for the following compounds:
    (a) Sodium chromate
    (b) Magnesium hydride
    (c) Nickel(II) acetate
    (d) Calcium chlorate
    (e) Lead(II) nitrate
    (f) Potassium dihydrogen phosphate
    (g) Manganese(II) hydroxide
    (h) Cobalt(II) bicarbonate
    (i) Sodium hypochlorite
    (j) Arsenic(V) carbonate
    (k) Chromium(III) sulfite
    (l) Antimony(III) sulfate
    (m) Sodium oxalate
    (n) Potassium thiocyanate

26. Write the name of each salt and the formula and name of the acid from which the salt may be derived. [*Example*: NiC_2O_4, nickel(II) oxalate; $H_2C_2O_4$, oxalic acid.]

    (a) $ZnSO_4$ (k) $Ca(HSO_4)_2$
    (b) $HgCl_2$ (l) $As_2(SO_3)_3$
    (c) $CuCO_3$ (m) $Sn(NO_2)_2$
    (d) $Cd(NO_3)_2$ (n) $FeBr_3$
    (e) $Al(C_2H_3O_2)_3$ (o) $KHCO_3$
    (f) CoF_2 (p) $BiAsO_4$
    (g) $Cr(ClO_3)_3$ (q) $Fe(BrO_3)_2$
    (h) Ag_3PO_4 (r) $(NH_4)_2HPO_4$
    (i) NiS (s) $NaClO$
    (j) $BaCrO_4$ (t) $KMnO_4$

27. Write the chemical formula for each of the following substances:
    (a) Baking soda   (c) Epsom salts
    (b) Lime   (d) Muriatic acid

    (e) Vinegar   (k) Limestone
    (f) Potash   (l) Cane sugar
    (g) Lye   (m) Milk of magnesia
    (h) Quicksilver   (n) Washing soda
    (i) Fool's gold   (o) Grain alcohol
    (j) Saltpeter

28. Give the name and formula of three salts with *ide* endings that are not binary compounds. Do not use the exact salts used as examples in the chapter.

29. Which of these statements are correct? Rewrite each incorrect statement to make it correct.
    (a) An oxidation number can be positive, negative, or zero.
    (b) The oxidation number of Li, Na, and K in compounds is -1.
    (c) The sum of the oxidation numbers for all the atoms in a polyatomic ion is zero.
    (d) The formula for calcium hydride is CaH_2.
    (e) All the following compounds are acids: H_2SO_4, HCl, HNO_3, $NaC_2H_3O_2$.
    (f) The ions of all the following metals have an oxidation number of $+2$: Ca, Ba, Sr, Cd, Zn.
    (g) The formulas for nitrous and sulfurous acids are HNO_2 and H_2SO_3.
    (h) The formula for the compound between Fe^{3+} and O^{2-} is Fe_3O_2.
    (i) The oxidation number of Cr in K_2CrO_4 is $+6$.
    (j) The oxidation number of Sn in $SnCl_4$ is $+4$.
    (k) The oxidation number of Co in $CoCl_2^{-1}$ is $+4$.
    (l) The name for $NaNO_2$ is sodium nitrite.
    (m) The name for $Ca(ClO_3)_2$ is calcium chlorate.
    (n) The name for CuO is copper(I) oxide.
    (o) The name for SO_4^{2-} is sulfate ion.
    (p) The name for N_2O_4 is dinitrogen tetroxide.
    (q) The name for Na_2O is disodium oxide.
    (r) If the name of an anion ends with *ide*, the name of the corresponding acid will start with *hydro*.
    (s) If the name of an anion ends with *ite*, the corresponding acid name will end with *ic*.
    (t) If the name of an acid ends with *ous*, the corresponding salt name will end with *ate*.

(u) In FeI_2, the iron is iron(II) because it is combined with two I^- ions.

(v) In Cu_2SO_4, the copper is copper(II) because there are two copper ions.

(w) In Ru_2O_3, we can deduce the oxidation state of Ru as $+3$ because two ions are combined with three oxide ions.

(x) When two nonmetals combine, prefixes of *di*, *tri*, *tetra*, and so on are used to specify how many atoms of each element are in a molecule.

(y) N_2O_3 is called dinitrogen trioxide.

(z) $Sn(CrO_4)_2$ is called tin dichromate.

(aa) In the Stock System of nomenclature, when a compound contains a metal that can have more than one oxidation number, the oxidation number of the metal is designated by a Roman numeral written immediately after the name of the metal.

CHAPTER SEVEN

Quantitative Composition of Compounds

Knowing the substances contained in a product is not sufficient information to produce the product. An artist can create an incredible array of colors from a limited number of pigments. A pharmacist can combine the same drugs in various amounts to produce different effects in patients. Cosmetics, cereals, cleaning products, and pain remedies all show a list of ingredients on the labels of the packages.

In each of these products the key to the production lies in the amount of each ingredient. The pharmaceutical industry maintains strict regulations on the amounts of the ingredients in the medicines we purchase. The formulas for soft drinks and most cosmetics are trade secrets. Small deviations in the composition of these products can result in large losses or lawsuits for these organizations.

The composition of compounds is an important concept in chemistry. The numerical relationships among elements within compounds and the measurement of exact quantities of particles, too small to be seen, are fundamental tasks for the chemist.

Chapter Preview

7.1   The Mole

In the laboratory we normally determine the mass of substances on a balance in units of grams. But, when we run a chemical reaction, the reaction occurs between atoms and molecules. For example, in the reaction between magnesium and sulfur, one atom of sulfur reacts with one atom of magnesium.

$$Mg + S \longrightarrow MgS$$

However, when we measure the masses of these two elements that react, we find that 24.305 g of Mg are required to react with 32.06 g of S. Because magnesium and sulfur react in a 1:1 atom ratio, we can conclude from this experiment that 24.305 g of Mg contain the same number of atoms as 32.06 g of S. How many atoms are in 24.305 g of Mg or 32.06 g of S? These two amounts each contain one mole of atoms.

The mole (abbreviated mol) is one of the seven base units in the International System and is the unit for an amount of substance. The mole is a counting unit as in other things that we count, such as a dozen (12) eggs or a gross (144) of pencils. But a mole is a much larger number of things, namely 6.022×10^{23}. Thus one mole contains 6.022×10^{23} entities of anything. In reference to our reaction between magnesium and sulfur, 1 mol Mg (24.305 g) contains 6.022×10^{23} atoms of magnesium and 1 mol S (32.06 g) contains 6.022×10^{23} atoms of sulfur.

Avogadro's number

The number 6.022×10^{23} is known as **Avogadro's number** in honor of Amedeo Avogadro (1776–1856), an Italian physicist. Avogadro's number is an important constant in chemistry and physics and has been experimentally determined by several independent methods.

> **Avogadro's number** $= 6.022 \times 10^{23}$

It is difficult to imagine how large Avogadro's number really is, but perhaps the following analogy will help express it: If 10,000 people started to count Avogadro's number and each counted at the rate of 100 numbers per minute each minute of the day, it would take them over 1 trillion (10^{12}) years to count the total number. So, even the most minute amount of matter contains extremely large numbers of atoms. For example, 1 mg (0.001 g) of sulfur contains 2×10^{19} atoms of sulfur.

Avogadro's number is the basis for the amount of substance that is used to express a particular number of chemical species, such as atoms, molecules, formula units, ions, or electrons. This amount of substance is the mole. We define a **mole** as an amount of a substance containing the same number of formula units as there are atoms in exactly 12 g of carbon-12. (Recall that carbon-12 is the reference isotope for atomic masses.) Other definitions are used, but they all relate to a mole being Avogadro's number of formula units of a substance. A **formula unit** is the atom or molecule indicated by the formula of the substance under consideration—for example, Mg, MgS, H_2O, O_2, $^{75}_{33}As$.

From the above definition we can say that the atomic mass in grams of any element contains one mole of atoms.

The term *mole* is so commonplace in chemical jargon that chemists use it as freely as the words *atom* and *molecule*. The mole is used in conjunction with many different particles, such as atoms, molecules, ions, and electrons, to represent Avogadro's number of these particles. If we can speak of a mole of atoms, we can also speak of a mole of molecules, a mole of electrons, a mole of ions, understanding that in each case we mean 6.022×10^{23} formula units of these particles.

mole

formula unit

> **1 mole of atoms = 6.022×10^{23} atoms**
>
> **1 mole of molecules = 6.022×10^{23} molecules**
>
> **1 mole of ions = 6.022×10^{23} ions**

molar mass

The atomic mass in grams of an element contains Avogadro's number of atoms and is defined as the **molar mass** of that element. To determine the molar mass of an element, change the units of the atomic mass found in the periodic table from amu to grams. Sulfur has an atomic mass of 32.06 amu. One molar mass of sulfur has a mass of 32.06 grams and contains 6.022×10^{23} atoms of sulfur. See the following table.

Element	Atomic mass	Molar mass	Number of atoms
H	1.0079 amu	1.0079 g	6.022×10^{23}
Mg	24.305 amu	24.305 g	6.022×10^{23}
Na	22.9898 amu	22.9898 g	6.022×10^{23}

> **1 molar mass (g) = 1 mole of atoms**
> **= Avogadro's number (6.022 × 10²³) of atoms**

We frequently encounter problems that require conversions involving quantities of mass, numbers, and moles of atoms of an element. Conversion factors that can be used for this purpose are

(a) Grams to atoms: $\dfrac{6.022 \times 10^{23} \text{ atoms of the element}}{1 \text{ molar mass of the element}}$

(b) Atoms to grams: $\dfrac{1 \text{ molar mass of the element}}{6.022 \times 10^{23} \text{ atoms of the element}}$

(c) Grams to moles: $\dfrac{1 \text{ mole of the element}}{1 \text{ molar mass of the element}}$

(d) Moles to grams: $\dfrac{1 \text{ molar mass of the element}}{1 \text{ mole of the element}}$

EXAMPLE 7.1   How many moles of iron does 25.0 g of Fe represent?

The problem requires that we change grams of Fe to moles of Fe. We look up the atomic mass of Fe in the table of atomic masses on the periodic table or table of atomic masses and find it to be 55.8. Then we use the proper conversion factor to obtain moles. The conversion factor is (c).

grams Fe $\longrightarrow$ moles Fe   grams Fe $\times \dfrac{1 \text{ mole Fe}}{1 \text{ molar mass Fe}}$

$25.0 \text{ g Fe} \times \dfrac{1 \text{ mol Fe}}{55.8 \text{ g Fe}} = 0.448 \text{ mol Fe}$ (Answer)

EXAMPLE 7.2   How many magnesium atoms are contained in 5.00 g of Mg?

The problem requires that we change grams of magnesium to atoms of magnesium.

grams Mg $\longrightarrow$ atoms Mg

We find the atomic mass of magnesium to be 24.3 and set up the calculation using conversion factor (a).

grams Mg $\times \dfrac{6.022 \times 10^{23} \text{ atoms Mg}}{1 \text{ molar mass Mg}}$

$5.00 \text{ g Mg} \times \dfrac{6.022 \times 10^{23} \text{ atoms Mg}}{24.3 \text{ g Mg}} = 1.24 \times 10^{23} \text{ atoms Mg}$ (Answer)

An alternative solution is first to convert grams of magnesium to moles of magnesium, which are then changed to atoms of magnesium.

grams Mg $\longrightarrow$ moles Mg $\longrightarrow$ atoms Mg

Use conversion factor (c) followed by (a). The calculation setup is

$$5.00 \text{ g Mg} \times \frac{1 \text{ mol Mg}}{24.3 \text{ g Mg}} \times \frac{6.022 \times 10^{23} \text{ atoms Mg}}{1 \text{ mol Mg}} = 1.24 \times 10^{23} \text{ atoms Mg}$$

Thus 1.24×10^{23} atoms of Mg are contained in 5.00 g of Mg.

EXAMPLE 7.3   What is the mass, in grams, of one atom of carbon?

The molar mass of carbon is 12.0 g. The factor needed to convert atoms to grams is conversion factor (b).

$$\text{atoms C} \longrightarrow \text{grams C} \qquad \text{atoms C} \times \frac{1 \text{ molar mass}}{6.022 \times 10^{23} \text{ atoms C}}$$

$$1 \text{ atom C} \times \frac{12.0 \text{ g C}}{6.022 \times 10^{23} \text{ atoms C}} = 1.99 \times 10^{-23} \text{ g C} \qquad \text{(Answer)}$$

EXAMPLE 7.4   What is the mass of 3.01×10^{23} atoms of sodium?

The information needed to solve this problem is the molar mass of Na (23.0 g) and conversion factor (b).

$$\text{atoms Na} \longrightarrow \text{grams Na} \qquad \text{atoms Na} \times \frac{1 \text{ molar mass Na}}{6.022 \times 10^{23} \text{ atoms Na}}$$

$$3.01 \times 10^{23} \text{ atoms Na} \times \frac{23.0 \text{ g Na}}{6.022 \times 10^{23} \text{ atoms Na}} = 11.5 \text{ g Na} \qquad \text{(Answer)}$$

EXAMPLE 7.5   What is the mass of 0.252 mol of Cu?

The information needed to solve this problem is the molar mass of Cu (63.5 g) and conversion factor (d).

$$\text{moles Cu} \longrightarrow \text{grams Cu} \qquad \text{moles Cu} \times \frac{1 \text{ molar mass Cu}}{1 \text{ mole Cu}}$$

$$0.252 \text{ mol Cu} \times \frac{63.5 \text{ g Cu}}{1 \text{ mol Cu}} = 16.0 \text{ g Cu} \qquad \text{(Answer)}$$

EXAMPLE 7.6   How many oxygen atoms are present in 1.00 mol of oxygen molecules?

Oxygen is a diatomic molecule with the formula O_2. Therefore a molecule of oxygen contains two atoms of oxygen.

$$\frac{2 \text{ atoms O}}{1 \text{ molecule O}_2}$$

The sequence of conversions is

$$\text{moles } O_2 \longrightarrow \text{molecules } O_2 \longrightarrow \text{atoms } O$$

Two conversion factors are needed; they are

$$\frac{6.022 \times 10^{23} \text{ molecules } O_2}{1 \text{ mol } O_2} \quad \text{and} \quad \frac{2 \text{ atoms } O}{1 \text{ molecule } O_2}$$

The calculation is

$$1.00 \text{ mol } O_2 \times \frac{6.022 \times 10^{23} \text{ molecules } O_2}{1 \text{ mol } O_2} \times \frac{2 \text{ atoms } O}{1 \text{ molecule } O_2} = 1.20 \times 10^{24} \text{ atoms } O$$

(Answer)

PRACTICE   What is the mass of 2.50 mol of helium?
Answer: 10.0 g of helium

PRACTICE   How many atoms are present in 0.025 mol of iron?
Answer: 1.51×10^{22} atoms

7.2   Molar Mass of Compounds

One mole of a compound contains Avogadro's number of formula units of that compound. The terms *molecular weight* and *formula weight* have been used in the past to refer to the mass of one mole of a compound. However, the term *molar mass* is more inclusive, since it can be used for all types of compounds.

If the formula of a compound is known, its molar mass may be determined by adding together the molar masses of all the atoms in the formula. If more than one atom of any element is present, its mass must be added as many times as it is used.

EXAMPLE 7.7   The formula for water is H_2O. What is its molar mass?
Proceed by looking up the molar masses of H (1.008) and O (15.999) and adding together the masses of all the atoms in the formula unit. Water contains two atoms of H and one atom of O. Thus,

$$2 H = 2 \times 1.008 \text{ g} = 2.016 \text{ g}$$
$$1 O = 1 \times 15.999 \text{ g} = \underline{15.999 \text{ g}}$$
$$18.015 \text{ g} = \text{molar mass}$$

EXAMPLE 7.8 Calculate the molar mass of calcium hydroxide, $Ca(OH)_2$.

The formula of this substance contains one atom of Ca and two atoms each of O and H. Proceed as in Example 7.7. Thus

$$1 \text{ Ca} = 1 \times 40.08 \text{ g} = 40.08 \text{ g}$$
$$2 \text{ O} = 2 \times 15.999 \text{ g} = 31.998 \text{ g}$$
$$2 \text{ H} = 2 \times 1.008 \text{ g} = \underline{2.016 \text{ g}}$$
$$74.094 \text{ g} = \text{molar mass}$$

The molar masses of elements are often rounded off to one decimal place to simplify calculations. (However, this simplification cannot be made in more exacting chemical work.) If we calculate the molar mass of $Ca(OH)_2$ on the basis of one decimal place, we find the value to be 74.1 g instead of 74.094 g. The molar mass of $Ca(OH)_2$ would then be calculated as follows:

$$1 \text{ Ca} = 1 \times 40.1 \text{ g} = 40.1 \text{ g}$$
$$2 \text{ O} = 2 \times 16.0 \text{ g} = 32.0 \text{ g}$$
$$2 \text{ H} = 2 \times 1.0 \text{ g} = \underline{2.0 \text{ g}}$$
$$74.1 \text{ g} = \text{molar mass}$$

PRACTICE   Calculate the molar mass of KNO_3.
Answer: 101.1 g KNO_3

A one molar mass of a compound is one mole of that compound and therefore contains Avogadro's number of formula units or molecules. Consider the compound hydrogen chloride, HCl. One atom of H combines with one atom of Cl to form one molecule of HCl. When 1 molar mass of H (1.0 g of H representing 1 mol or 6.022×10^{23} H atoms) combines with 1 molar mass of Cl (35.5 g of Cl representing 1 mol or 6.022×10^{23} Cl atoms), 1 molar mass of HCl (36.5 g of HCl representing 1 mol or 6.022×10^{23} HCl molecules) is produced. These relationships are summarized in the following table.

H	Cl	HCl
6.022×10^{23} H *atoms*	6.022×10^{23} Cl *atoms*	6.022×10^{23} HCl *molecules*
1 mol H *atoms*	1 mol Cl *atoms*	1 mol HCl *molecules*
1.0 g H	35.5 g Cl	36.5 g HCl
1 molar mass H	1 molar mass Cl	1 molar mass HCl

In dealing with diatomic elements (H_2, O_2, N_2, F_2, Cl_2, Br_2, and I_2), special care must be taken to distinguish between a mole of atoms and a mole of molecules. For example consider *one* mole of oxygen molecules, which has a mass of 32.0 g. This quantity is equal to *two* moles of oxygen atoms. The key concept is that one mole represents Avogadro's number of the particular chemical entity—atoms, molecules, formula units and so forth—that is under consideration.

$$1 \text{ mol } H_2O = 18.0 \text{ g } H_2O = 6.022 \times 10^{23} \text{ molecules}$$
$$1 \text{ mol NaCl} = 58.5 \text{ g NaCl} = 6.022 \times 10^{23} \text{ formula units}$$
$$1 \text{ mol } H_2 = 2.0 \text{ g } H_2 = 6.022 \times 10^{23} \text{ molecules}$$
$$1 \text{ mol } HNO_3 = 63.0 \text{ g } HNO_3 = 6.022 \times 10^{23} \text{ molecules}$$
$$1 \text{ mol } K_2SO_4 = 174.3 \text{ g } K_2SO_4 = 6.022 \times 10^{23} \text{ formula units}$$

---

1 mol = 6.022 × 10²³ formula units or molecules

= 1 molar mass of a compound

We often need to convert moles of a compound to grams, and grams of a compound to moles. The factors for these conversions are

grams to moles: $\dfrac{1 \text{ mol of a substance}}{1 \text{ molar mass of the substance}}$

moles to grams: $\dfrac{1 \text{ molar mass of a substance}}{1 \text{ mol of the substance}}$

EXAMPLE 7.9

What is the mass of 1 mol of sulfuric acid, H_2SO_4?

One mole of H_2SO_4 is one molar mass of H_2SO_4. The problem, therefore, is solved in a similar manner to Examples 7.7 and 7.8. Look up the molar masses of H, S, and O, and solve.

$$2 \text{ H} = 2 \times 1.0 \text{ g} = 2.0 \text{ g}$$
$$1 \text{ S} = 1 \times 32.1 \text{ g} = 32.1 \text{ g}$$
$$4 \text{ O} = 4 \times 16.0 \text{ g} = \underline{64.0 \text{ g}}$$
$$98.1 \text{ g} = \text{mass of 1 mol of } H_2SO_4$$

EXAMPLE 7.10

How many moles of NaOH are there in 1.00 kg of sodium hydroxide?

First we know that

1 molar mass = (23.0 g + 16.0 g + 1.0 g) or 40.0 g NaOH

To convert grams to moles we use the conversion factor

$$\dfrac{1 \text{ mol}}{1 \text{ molar mass}} \quad \text{or} \quad \dfrac{1 \text{ mol NaOH}}{40.0 \text{ g NaOH}}$$

Use this conversion sequence:

$$kg\ NaOH \longrightarrow g\ NaOH \longrightarrow mol\ NaOH$$

The calculation is

$$1.00\ \cancel{kg\ NaOH} \times \frac{1000\ \cancel{g\ NaOH}}{\cancel{kg\ NaOH}} \times \frac{1\ mol\ NaOH}{40.0\ \cancel{g\ NaOH}} = 25.0\ mol\ NaOH$$

$$1.00\ kg\ NaOH = 25.0\ mol\ NaOH$$

EXAMPLE 7.11   What is the mass of 5.00 mol of water?
First we know that

$$1\ mol\ H_2O = 18.0\ g \qquad (Example\ 7.7)$$

The conversion is

$$mol\ H_2O \longrightarrow g\ H_2O$$

To convert moles to grams use the conversion factor

$$\frac{1\ molar\ H_2O}{1\ mol\ H_2O} \qquad or \qquad \frac{18.0\ g\ H_2O}{1\ mol\ H_2O}$$

The calculation is

$$5.00\ \cancel{mol\ H_2O} \times \frac{18.0\ g\ H_2O}{1\ \cancel{mol\ H_2O}} = 90.0\ g\ H_2O \qquad (Answer)$$

EXAMPLE 7.12   How many molecules of HCl are there in 25.0 g of hydrogen chloride?
From the formula we find that the molar mass of HCl is 36.5 g (1.0 g + 35.5 g). The sequence of conversions is

$$g\ HCl \longrightarrow mol\ HCl \longrightarrow molecules\ HCl$$

using the conversion factors

$$\frac{1\ mol\ HCl}{36.5\ g\ HCl} \qquad and \qquad \frac{6.022 \times 10^{23}\ molecules\ HCl}{1\ mol\ HCl}$$

$$25.0\ \cancel{g\ HCl} \times \frac{1\ \cancel{mol\ HCl}}{36.5\ \cancel{g\ HCl}} \times \frac{6.022 \times 10^{23}\ molecules\ HCl}{1\ \cancel{mol\ HCl}} = 4.12 \times 10^{23}\ molecules\ HCl$$

$$(Answer)$$

PRACTICE   What is the mass of 0.150 mol of Na_2SO_4?
Answer: 21.3 g Na_2SO_4

PRACTICE   How many moles are there in 500.0 g of $AlCl_3$?

Answer: 3.745 mol of $AlCl_3$

7.3   Percent Composition of Compounds

percent
composition
of a compound

Percent means parts per one hundred parts. Just as each piece of pie is a percent of the whole pie, each element in a compound is a percent of the whole compound. The **percent composition of a compound** is the *mass percent* of each element in the compound. The molar mass represents the total mass, or 100%, of the compound. Thus the percent composition of water, H_2O, is 11.1% H and 88.9% O by mass. According to the Law of Definite Composition, the percent composition must be the same no matter what size sample is taken.

The percent composition of a compound can be determined if its formula is known or if the masses of two or more elements that have combined with each other are known or are experimentally determined.

If the formula is known, it is essentially a two-step process to determine the percent composition.

Step 1   Calculate the molar mass as was done in Section 7.2.

Step 2   Divide the total mass of each element in the formula by the molar mass and multiply by 100. This gives the percent composition.

$$\frac{\text{total mass of the element}}{\text{molar mass}} \times 100 = \text{percent of the element}$$

EXAMPLE 7.13

Calculate the percent composition of sodium chloride, NaCl

Step 1 Calculate the molar mass of NaCl:

$$1 \text{ Na} = 1 \times 23.0 \text{ g} = 23.0 \text{ g}$$
$$1 \text{ Cl} = 1 \times 35.5 \text{ g} = \underline{35.5 \text{ g}}$$
$$58.5 \text{ g (molar mass)}$$

Step 2 Now calculate the percent composition. We know there are 23.0 g Na and 35.5 g Cl in 58.5 g NaCl.

$$\text{Na:} \quad \frac{23.0 \text{ g Na}}{58.5 \text{ g NaCl}} \times 100 = \quad 39.3\% \text{ Na}$$

$$\text{Cl:} \quad \frac{35.5 \text{ g Cl}}{58.5 \text{ g NaCl}} \times 100 = \frac{60.7\% \text{ Cl}}{100.0\% \text{ total}}$$

In any two-component system, if one percent is known, the other is automatically defined by the difference; that is, if Na = 39.3%, then Cl = 100% − 39.3% = 60.7%. However, the calculation of the percent of each component should be carried out, since this provides a check against possible error. The percent composition data should add up to $100 \pm 0.2\%$.

EXAMPLE 7.14   Calculate the percent composition of potassium chloride, KCl.

Step 1 Calculate the molar mass of KCl:

$$1\ K = 1 \times 39.1\ g = 39.1\ g$$

$$1\ Cl = 1 \times 35.5\ g = \dfrac{35.5\ g}{74.6\ g} \quad \text{(molar mass)}$$

Step 2 Now calculate the percent composition. We know there are 39.1 g K and 35.5 g Cl in 74.6 g KCl.

K: $\dfrac{39.1\ \cancel{g}\ K}{74.6\ \cancel{g}\ KCl} \times 100 = \quad 52.4\%\ K$

Cl: $\dfrac{35.5\ \cancel{g}\ Cl}{74.6\ \cancel{g}\ KCl} \times 100 = \dfrac{47.6\%\ Cl}{100.0\%\ total}$

Comparing the data calculated for NaCl and for KCl, we see that NaCl contains a higher percentage of Cl by mass, although each compound has a one-to-one atom ratio of Cl to Na and Cl to K. The reason for this mass percent difference is that Na and K do not have the same atomic masses.

It is important to realize that, when we compare 1 mol of NaCl with 1 mol of KCl, each quantity contains the same number of Cl atoms—namely, 1 mol of Cl atoms. However, if we compare equal masses of NaCl and KCl, there will be more Cl atoms in the mass of NaCl since NaCl has a higher mass percent of Cl.

1 mole NaCl contains	100 g NaCl contains	1 mole KCl contains	100 g KCl contains
1 mol Na	39.3 g Na	1 mol K	52.4 g K
1 mol Cl	60.7 g Cl	1 mol Cl	47.6 g Cl
60.7% Cl		47.6% Cl	

EXAMPLE 7.15   Calculate the percent composition of potassium sulfate, K_2SO_4.

Step 1 Calculate the molar mass of K_2SO_4:

$$2\ K = 2 \times 39.1\ g = \quad 78.2\ g$$

$$1\ S = 1 \times 32.1\ g = \quad 32.1\ g$$

$$4\ O = 4 \times 16.0\ g = \dfrac{64.0\ g}{174.3\ g} \quad \text{(molar mass)}$$

Step 2 Now calculate the percent composition. We know there are 78.2 g of K, 32.1 g of S, and 64.0 g of O in 174.3 g of K_2SO_4.

$$K: \quad \frac{78.2 \text{ g K}}{174.3 \text{ g K}_2SO_4} \times 100 = \quad 44.9\% \text{ K}$$

$$S: \quad \frac{32.1 \text{ g S}}{174.3 \text{ g K}_2SO_4} \times 100 = \quad 18.4\% \text{ S}$$

$$O: \quad \frac{64.0 \text{ g O}}{174.3 \text{ g K}_2SO_4} \times 100 = \quad \frac{36.7\% \text{ O}}{100.0\% \text{ total}}$$

PRACTICE Calculate the percent composition of $Ca(NO_3)_2$
Answer: Ca = 24.4%, N = 17.1%, O = 58.5%

PRACTICE Calculate the percent composition of K_2CrO_4.
Answer: K = 40.3%, Cr = 26.8%, O = 33.0%

The percent composition can be determined from experimental data without knowing the formula of a compound. This determination is done by calculating the mass of each element in a compound as a percentage of the total mass of the compound formed.

EXAMPLE 7.16 When heated in the air, 1.63 g of zinc, Zn, combine with 0.40 g of oxygen, O_2, to form zinc oxide. Calculate the percent composition of the compound formed.
First, calculate the total mass of the compound formed.

$$\frac{\begin{array}{l} 1.63 \text{ g Zn} \\ 0.40 \text{ g O}_2 \end{array}}{2.03 \text{ g}} = \text{total mass of product}$$

Then divide the mass of each element by the total mass (2.03 g) and multiply by 100.

$$\frac{1.63 \text{ g}}{2.03 \text{ g}} \times 100 = \quad 80.3\% \text{ Zn}$$

$$\frac{0.40 \text{ g}}{2.03 \text{ g}} \times 100 = \frac{19.7\% \text{ O}}{100.0\% \text{ total}}$$

The compound formed contains 80.3% Zn and 19.7% O.

PRACTICE Aluminum chloride is formed by reacting 13.43 g aluminum with 53.18 g chlorine. What is the percent composition of the compound?
Answer: Al = 20.16%, Cl = 79.84%

7.4 Empirical Formula versus Molecular Formula

empirical formula

The **empirical formula**, or *simplest formula*, gives the smallest whole-number ratio of the atoms that are present in a compound. This formula gives the relative number of atoms of each element in the compound.

molecular formula

The **molecular formula** is the true formula, representing the total number of atoms of each element present in one molecule of a compound. It is entirely possible that two or more substances will have the same percent composition, yet be distinctly different compounds. For example, acetylene, C_2H_2, is a common gas used in welding; benzene, C_6H_6, is an important solvent obtained from coal tar and is used in the synthesis of styrene and nylon. Both acetylene and benzene contain 92.3% C and 7.7% H. The smallest ratio of C and H corresponding to these percents is CH (1:1). Therefore the empirical formula for both acetylene and benzene is CH, even though it is known that the molecular formulas are C_2H_2 and C_6H_6, respectively. It is not uncommon for the molecular formula to be the same as the empirical formula. If the molecular formula is not the same, it will be an integral (whole number) multiple of the empirical formula.

$$CH = \text{empirical formula}$$
$$(CH)_2 = C_2H_2 = \text{acetylene} \quad \text{(molecular formula)}$$
$$(CH)_6 = C_6H_6 = \text{benzene} \quad \text{(molecular formula)}$$

Table 7.1 summarizes the data concerning these CH formulas. Table 7.2 shows empirical and molecular formula relationships of other compounds.

TABLE 7.1

Molecular Formulas of Two Compounds Having an Empirical Formula with a 1:1 Ratio of Carbon and Hydrogen Atoms

| | Composition | | |
Formula	% C	% H	Molar mass
CH (empirical)	92.3	7.7	13.0 (empirical)
C_2H_2 (acetylene)	92.3	7.7	26.0 (2 × 13.0)
C_6H_6 (benzene)	92.3	7.7	78.0 (6 × 13.0)

TABLE 7.2

Some Empirical and Molecular Formulas

Compound	Empirical formula	Moleuclar formula	Compound	Empirical formula	Molecular formula
Acetylene	CH	C_2H_2	Diborane	BH_3	B_2H_6
Benzene	CH	C_6H_6	Hydrazine	NH_2	N_2H_4
Ethylene	CH_2	C_2H_4	Hydrogen	H	H_2
Formaldehyde	CH_2O	CH_2O	Chlorine	Cl	Cl_2
Acetic acid	CH_2O	$C_2H_4O_2$	Bromine	Br	Br_2
Glucose	CH_2O	$C_6H_{12}O_6$	Oxygen	O	O_2
Hydrogen chlorine	HCl	HCl	Nitrogen	N	N_2
Carbon dioxide	CO_2	CO_2			

7.5   Calculation of Empirical Formula

It is possible to establish an empirical formula because (1) the individual atoms in a compound are combined in whole-number ratios and (2) each element has a specific atomic mass.

In order to calculate the empirical formula we need to know (1) the elements that are combined, (2) their atomic masses, and (3) the ratio by mass or percentage in which they are combined. If elements A and B form a compound, we may represent the empirical formula as A_xB_y, where x and y are small whole numbers that represent the number of atoms of A and B. To write the empirical formula we must determine x and y.

The solution to this problem requires three or four steps.

Step 1   Assume a definite starting quantity (usually 100 g) of the compound, if not given, and express the mass of each element in grams.

Step 2   Multiply the mass (in grams) of each element by the factor 1 mol/1 molar mass to convert grams to moles. This conversion gives the number of moles of atoms of each element in the quantity assumed. At this point these numbers will usually not be whole numbers.

Step 3   Divide each of the values obtained in Step 2 by the smallest of these values. If the numbers obtained by this procedure are whole numbers, use them as subscripts in writing the empirical formula. If the numbers obtained are not whole numbers, go on to Step 4.

Step 4   Multiply the values obtained in Step 3 by the smallest number that will convert them to whole numbers. Use these whole numbers as the subscripts in the empirical formula. For example, if the ratio of A to B is 1.0:1.5, multiply both numbers by 2 to obtain a ratio of 2:3. The empirical formula then is A_2B_3.

EXAMPLE 7.17   Calculate the empirical formula of a compound containing 11.19% hydrogen, H, and 88.89% oxygen, O.

Step 1   Express each element in grams. If we assume that there are 100 g of material, then the percent of each element is equal to the grams of each element in 100 g, and the percent sign can be omitted.

$$H = 11.19 \text{ g}$$

$$O = 88.89 \text{ g}$$

Step 2   Multiply the grams of each element by the proper mol/molar mass factor to obtain the relative number of moles of atoms:

$$\text{H:}\quad 11.19 \text{ g H} \times \frac{1 \text{ mol H atoms}}{1.01 \text{ g H}} = 11.1 \text{ mol H atoms}$$

$$\text{O:}\quad 88.89 \text{ g O} \times \frac{1 \text{ mol O atoms}}{16.0 \text{ g O}} = 5.55 \text{ mol O atoms}$$

The formula could be expressed as $H_{11.1}O_{5.55}$. However, it is customary to use the smallest whole-number ratio of atoms. This ratio is calculated in Step 3.

Step 3   Change these numbers to whole numbers by dividing each of them by the smaller number.

$$H = \frac{11.1 \text{ mol}}{5.55 \text{ mol}} = 2 \qquad O = \frac{5.55 \text{ mol}}{5.55 \text{ mol}} = 1$$

In this step the ratio of atoms has not changed, because we divided the number of moles of each element by the same number.

The simplest ratio of H to O is 2:1.

Empirical formula $= H_2O$

EXAMPLE 7.18   The analysis of a salt showed that it contained 56.58% potassium, K, 8.68% carbon, C, and 34.73% oxygen, O. Calculate the empirical formula for this substance.

Steps 1 and 2   After changing the percent of each element to grams, find the relative number of moles of each element by multiplying by the proper mol/molar mass factor.

$$\text{K:}\quad 56.58\ \text{g}\cancel{K} \times \frac{1\ \text{mol K atoms}}{39.1\ \text{g}\cancel{K}} = 1.45\ \text{mol K atoms}$$

$$\text{C:}\quad 8.68\ \text{g}\cancel{C} \times \frac{1\ \text{mol C atoms}}{12.0\ \text{g}\cancel{C}} = 0.720\ \text{mol C atoms}$$

$$\text{O:}\quad 34.73\ \text{g}\cancel{O} \times \frac{1\ \text{mol O atoms}}{16.0\ \text{g}\cancel{O}} = 2.17\ \text{mol O atoms}$$

Step 3   Divide each number of moles by the smallest value.

$$\text{K} = \frac{1.45\ \text{mol}}{0.720\ \text{mol}} = 2.01$$

$$\text{C} = \frac{0.720\ \text{mol}}{0.720\ \text{mol}} = 1.00$$

$$\text{O} = \frac{2.17\ \text{mol}}{0.720\ \text{mol}} = 3.01$$

The simplest ratio of K:C:O is 2:1:3.

Empirical formula = K_2CO_3

EXAMPLE 7.19   A sulfide of iron was formed by combining 2.233 g of iron, Fe, with 1.926 g of sulfur, S. What is the empirical formula of the compound?

Steps 1 and 2   The grams of each element are given, so we use them directly in our calculations. Calculate the relative number of moles of each element by multiplying grams of each element by the proper mol/molar mass factor.

$$\text{Fe:}\quad 2.233\ \text{g}\cancel{Fe} \times \frac{1\ \text{mol Fe atoms}}{55.8\ \text{g}\cancel{Fe}} = 0.0400\ \text{mol Fe atoms}$$

$$\text{S:}\quad 1.926\ \text{g}\cancel{S} \times \frac{1\ \text{mol S atoms}}{32.1\ \text{g}\cancel{S}} = 0.0600\ \text{mol S atoms}$$

Step 3   Divide each number of moles by the smaller of the two numbers.

$$\text{Fe} = \frac{0.0400\ \text{mol}}{0.0400\ \text{mol}} = 1.00$$

$$\text{S} = \frac{0.0600\ \text{mol}}{0.0400\ \text{mol}} = 1.50$$

Step 4   We still have not reached a ratio that will give a formula containing whole numbers of atoms, so we must double each value to obtain a ratio of 2.00 atoms of Fe to 3.00 atoms of S. Doubling both values does not change the ratio of Fe and S atoms.

$$Fe: 1.00 \times 2 = 2.00$$

$$S: 1.50 \times 2 = 3.00$$

Empirical formula $= Fe_2S_3$

In many of these calculations results may vary somewhat from an exact whole number, which can be due to experimental errors in obtaining the data or from rounding off numbers. Calculations that vary by no more than ± 0.1 from a whole number can usually be rounded off to the nearest whole number. Deviations greater than about 0.1 unit usually mean that the calculated ratios need to be multiplied by a factor to make them all whole numbers. For example, an atom ratio of 1:1.33 should be multiplied by 3 to make the ratio 3:4.

PRACTICE Calculate the empirical formula of a compound containing 53.33% C, 11.11% H, and 35.53% O.
Answer: C_2H_5O

PRACTICE Calculate the empirical formula of a compound that contains 43.7% phosphorus and 56.3% O by mass.
Answer: P_2O_5

7.6   Calculation of the Molecular Formula from the Empirical Formula

The molecular formula can be calculated from the empirical formula if the molar mass, in addition to data for calculating the empirical formula, is known. The molecular formula, as stated in Section 7.4, will be equal to or some multiple of the empirical formula. For example, if the empirical formula of a compound of hydrogen and fluorine is HF, the molecular formula can be expressed as $(HF)_n$, where $n = 1, 2, 3, 4, \ldots$ This n means that the molecular formula could be $HF, H_2F_2, H_3F_3, H_4F_4$, and so on. To determine the molecular formula, we must evaluate n.

$$n = \frac{\text{molar mass}}{\text{mass of empirical formula}} = \text{number of empirical formula units}$$

What we actually calculate is the number of units of the empirical formula that is contained in the molecular formula.

EXAMPLE 7.20   A compound of nitrogen and oxygen with a molar mass of 92.0 g was found to have an empirical formula of NO_2. What is its molecular formula?

Step 1 Let n be the number of (NO_2) units in a molecule; then the molecular formula is (NO_2)$_n$.

Step 2 Each (NO_2) unit has a mass of $[14 \text{ g} + (2 \times 16 \text{ g})]$ or 46.0 g. The molar mass of (NO_2)$_n$ is 92.0 g and the number of 46.0 units in 92.0 is 2.

$$n = \frac{92.0 \text{ g}}{46.0 \text{ g}} = 2 \quad \text{(empirical formula units)}$$

Step 3 The molecular formula is (NO_2)$_2$, or N_2O_4.

EXAMPLE 7.21   The hydrocarbon propylene has a molar mass of 42.0 g and contains 14.3% H and 85.7% C. What is its molecular formula?

Step 1 First find the empirical formula:

$$\text{C:} \quad 85.7 \text{ g C} \times \frac{1 \text{ mol C atoms}}{12.0 \text{ g C}} = 7.14 \text{ mol C atoms}$$

$$\text{H:} \quad 14.3 \text{ g H} \times \frac{1 \text{ mol H atoms}}{1.0 \text{ g H}} = 14.3 \text{ mol H atoms}$$

Divide each value by the smaller number of moles.

$$\text{C} = \frac{7.14 \text{ mol}}{7.14 \text{ mol}} = 1.0$$

$$\text{H} = \frac{14.3 \text{ mol}}{7.14 \text{ mol}} = 2.0$$

Empirical formula = CH_2

Step 2 Determine the molecular formula from the empirical formula and the molar mass.

Molecular formula = (CH_2)$_n$

Molar mass = 42.0 g

Each CH_2 unit has a mass of $(12.0 \text{ g} + 2.0 \text{ g})$ or 14.0 g. The number of CH_2 units in 42.0 g is 3.

$$n = \frac{42.0 \text{ g}}{14.0 \text{ g}} = 3 \quad \text{(empirical formula units)}$$

The molecular formula is (CH_2)$_3$, or C_3H_6.

PRACTICE   Calculate the empirical and molecular formulas of a compound that contains 80.0% C, 20.0% H, and has a molar mass of 30.0 g.

Answers: The empirical formula is CH_3
            The molecular formula is C_2H_6

Concepts in Review

1. Explain the meaning of the mole.
2. Discuss the relationship between a mole and Avogadro's number.
3. Convert grams, atoms, molecules, and molar masses to moles, and vice versa.
4. Determine the molar mass of a compound from the formula.
5. Calculate the percent composition of a compound from its formula.
6. Calculate the percent composition of a compound from experimental data on combining masses.
7. Explain the relationship between an empirical formula and a molecular formula.
8. Determine the empirical formula for a compound from its percent composition.
9. Calculate the molecular formula of a compound from its percent composition and molar mass.

Key Terms in Review

The terms listed here have all been defined within the chapter. Review the definitions of each, and use the glossary and margin notations within the chapter as study aids.

Avogadro's number   mole
empirical formula   molecular formula
formula unit   percent composition of a compound
molar mass

Exercises

An asterisk indicates a more challenging question or problem.

1. What is a mole?
2. Which would have a higher mass: a mole of potassium atoms, or a mole of gold atoms?
3. Which would contain more atoms: a mole of potassium atoms, or a mole of gold atoms?
4. Which would contain more electrons: a mole of potassium atoms, or a mole of gold atoms?

*5. If the atomic mass scale had been defined differently, with an atom of $^{12}_{6}C$ being defined as weighing 50 amu, would this have any effect on the value of Avogadro's number? Explain.
6. What is the numerical value of Avogadro's number?
7. What is the relationship between Avogadro's number and the mole?
8. Complete the following statements, supplying the proper quantity.

(a) A mole of oxygen atoms (O) contains _____ atoms.

(b) A mole of oxygen molecules (O_2) contains _____ molecules.

(c) A mole of oxygen molecules (O_2) contains _____ atoms.

(d) A mole of oxygen atoms (O) has a mass _____ grams.

(e) A mole of oxygen molecules (O_2) has a mass _____ grams.

9. Which of the following statements are correct? Rewrite the incorrect statements to make them correct.

(a) One atomic mass of any element contains 6.022×10^{23} atoms.

(b) The mass of one atom of chlorine is
$$\frac{35.5 \text{ g}}{6.022 \times 10^{23} \text{ atoms}}.$$

(c) A mole of magnesium atoms (24.3 g) contains the same number of atoms as a mole of sodium atoms (23.0 g).

(d) A mole of bromine atoms contains 6.022×10^{23} atoms of bromine.

(e) A mole of chlorine molecules (Cl_2) contains 6.022×10^{23} atoms of chlorine.

(f) A mole of aluminum atoms has the same mass as a mole of tin atoms.

(g) A mole of H_2O contains 6.022×10^{23} atoms.

(h) A mole of hydrogen molecules (H_2) contains 1.204×10^{24} electrons.

10. How many molecules are present in 1 molar mass of sulfuric acid, H_2SO_4? How many atoms are present?

11. In calculating the empirical formula of a compound from its percent composition, why do we choose to start with 100 g of the compound?

12. Which of the following statements are correct? Rewrite the incorrect statements to make them correct.

(a) A mole of sodium and a mole of sodium chloride contain the same number of sodium atoms.

(b) One mole of nitrogen gas (N_2) has a mass of 14.0 g.

(c) The percent of oxygen is higher in K_2CrO_4 than it is in Na_2CrO_4.

(d) The number of Cr atoms is the same in a mole of K_2CrO_4 as it is in a mole of Na_2CrO_4.

(e) Both K_2CrO_4 and Na_2CrO_4 contain the same percent by mass of Cr.

(f) A molar mass of sucrose, $C_{12}H_{22}O_{11}$, contains 1 mol of sucrose molecules.

(g) Two moles of nitric acid, HNO_3, contain 6 moles of oxygen atoms.

(h) The empirical formula of sucrose, $C_{12}H_{22}O_{11}$, is CH_2O.

(i) A hydrocarbon that has a molar mass of 280 and an empirical formula of CH_2 has a molecular formula of $C_{22}H_{44}$.

(j) The empirical formula is often called the simplest formula.

(k) The empirical formula of a compound gives the smallest whole-number ratio of the atoms that are present in a compound.

(l) If the molecular formula and the empirical formula of a compound are not the same, the empirical formula will be an integral multiple of the molecular formula.

(m) The empirical formula of benzene, C_6H_6, is CH.

(n) A compound having an empirical formula of CH_2O, and a molar mass of 60, has a molecular formula of $C_3H_6O_3$.

13. Determine the molar masses of the following compounds:

(a) KBr   (f) Fe_3O_4

(b) Na_2SO_4 (g) $C_{12}H_{22}O_{11}$

(c) $Pb(NO_3)_2$ (h) $Al_2(SO_4)_3$

(d) C_2H_5OH (i) $(NH_4)_2HPO_4$

(e) $HC_2H_3O_2$

14. Determine the molar mass of the following compounds:

(a) NaOH   (f) C_6H_5COOH

(b) Ag_2CO_3 (g) $C_6H_{12}O_6$

(c) Cr_2O_3 (h) $K_4Fe(CN)_6$

(d) $(NH_4)_2CO_3$ (i) $BaCl_2 \cdot 2\,H_2O$

(e) $Mg(HCO_3)_2$

Moles and Avogadro's number

15. How many moles of atoms are contained in the following?

(a) 22.5 g Zn

(b) 0.688 g Mg

(c) 4.5×10^{22} atoms Cu

(d) 382 g Co

(e) 0.055 g Sn

(f) 8.5×10^{24} molecules N_2

16. How many moles are contained in the following?
    (a) 25.0 g NaOH   (d) 14.8 g CH_3OH
    (b) 44.0 g Br_2 (e) 2.88 g Na_2SO_4
    (c) 0.684 g $MgCl_2$ (f) 4.20 lb ZnI_2

17. Calculate the number of grams in each of the following:
    (a) 0.550 mol Au
    (b) 15.8 mol H_2O
    (c) 12.5 mol Cl_2
    (d) 3.15 mol NH_4NO_3
    (e) 4.25×10^{-4} mol H_2SO_4
    (f) 4.5×10^{22} molecules CCl_4
    (g) 0.00255 mol Ti
    (h) 1.5×10^{16} atoms S

18. How many molecules are contained in each of the following:
    (a) 1.26 mol O_2
    (b) 0.56 mol C_6H_6
    (c) 16.0 g CH_4
    (d) 1000. g HCl

19. Calculate the mass in grams of each of the following:
    (a) 1 atom Pb
    (b) 1 atom Ag
    (c) 1 molecule H_2O
    (d) 1 molecule $C_3H_5(NO_3)_3$

20. Make the following conversions:
    (a) 8.66 mol Cu to grams Cu
    (b) 125 mol Au to kilograms Au
    (c) 10 atoms C to moles C
    (d) 5000 molecules CO_2 to moles CO_2
    (e) 28.4 g S to moles S
    (f) 2.50 kg NaCl to moles NaCl
    (g) 42.4 g Mg to atoms Mg
    (h) 485 mL Br_2 (d = 3.12 g/mL) to moles Br_2

21. One mole of carbon disulfide (CS_2) contains
    (a) How many carbon disulfide molecules?
    (b) How many carbon atoms?
    (c) How many sulfur atoms?
    (d) How many total atoms of all kinds?

22. White phosphorus is one of several forms of phosphorus and exists as a waxy solid consisting of P_4 molecules. How many atoms are present in 0.350 mol of P_4?

23. How many grams of sodium contain the same number of atoms as 10.0 g of potassium?

24. One atom of an unknown element is found to have a mass of 1.79×10^{-23} g. What is the molar mass of this element?

*25. If a stack of 500 sheets of paper is 4.60 cm high, what will be the height, in meters, of a stack of Avogadro's number of sheets of paper?

26. There are about 5.0 billion (5.0×10^9) people on earth. If 1 mole of dollars were distributed equally among these people, how many dollars would each person receive?

*27. If 20 drops of water equal 1.0 mL (1.0 cm^3),
    (a) How many drops of water are there in a cubic mile of water?
    (b) What would be the volume in cubic miles of a mole of drops of water?

*28. Silver has a density of 10.5 g/cm^3. If 1.00 mol of silver were shaped into a cube,
    (a) What would be the volume of the cube?
    (b) What would be the length of one side of the cube?

29. How many atoms of oxygen are contained in each of the following?
    (a) 16.0 g O_2
    (b) 0.622 mol MgO
    (c) 6.00×10^{22} molecules $C_6H_{12}O_6$
    (d) 5.0 mol MnO_2
    (e) 250 g $MgCO_3$
    (f) 5.0×10^{18} molecules H_2O

30. Calculate the number of:
    (a) Grams of silver in 25.0 g AgBr
    (b) Grams of chlorine in 5.00 g $PbCl_2$
    (c) Grams of nitrogen in 6.34 mol $(NH_4)_3PO_4$
    (d) Grams of oxygen in 8.45×10^{22} molecules SO_3
    (e) Grams of hydrogen in 45.0 g C_3H_8O

*31. A sulfuric acid solution contains 65.0% H_2SO_4 by mass and has a density of 1.55 g/mL. How many moles of the acid are present in 1.00 L of the solution?

*32. A nitric acid solution containing 72.0% HNO_3 by mass has a density of 1.42 g/mL. How many moles of HNO_3 are present in 100 mL of the solution?

33. Given 1.00 g samples of each of the following compounds, CO_2, O_2, H_2O, and CH_3OH,
    (a) Which sample will contain the largest number of molecules?
    (b) Which sample will contain the largest number of atoms?
    Show proof for your answers.

34. How many grams of Fe_2S_3 will contain a total number of atoms equal to Avogadro's number?

Percent composition

35. Calculate the percent composition by mass of the following compounds:
    (a) NaBr   (c) $FeCl_3$ (e) $Al_2(SO_4)_3$
    (b) $KHCO_3$ (d) $SiCl_4$ (f) $AgNO_3$

36. Calculate the percent composition of the following compounds:
    (a) $ZnCl_2$
    (b) $NH_4C_2H_3O_2$
    (c) MgP_2O_7
    (d) $(NH_4)_2SO_4$
    (e) $Fe(NO_3)_3$
    (f) ICl_3

37. Calculate the percent of iron, Fe, in the following compounds:
    (a) FeO
    (b) Fe_2O_3
    (c) Fe_3O_4
    (d) $K_4Fe(CN)_6$

38. Which of the following chlorides has the highest and which has the lowest percentage of chlorine, Cl, by mass, in its formula?
    (a) KCl (b) $BaCl_2$ (c) $SiCl_4$ (d) LiCl

39. A 6.20 g sample of phosphorus was reacted with oxygen to form an oxide with a mass of 14.20 g. Calculate the percent composition of the compound.

40. A sample of ethylene chloride was analyzed to contain 6.00 g of C, 1.00 g of H, and 17.75 g of Cl. Calculate the percent composition of ethylene chloride.

41. How many grams of lithium will combine with 20.0 grams of sulfur to form the compound Li_2S?

42. Calculate the percentage of
    (a) Mercury in $HgCO_3$
    (b) Oxygen in $Ca(ClO_3)_2$
    (c) Nitrogen in $C_{10}H_{14}N_2$ (nicotine)
    (d) Mg in $C_{55}H_{72}MgN_4O_5$ (chlorophyll)

43. Answer the following by examination of the formulas. Check your answers by calculations if you wish. Which compound has the:
    (a) Higher percent by mass of hydrogen, H_2O or H_2O_2?
    (b) Lower percent by mass of nitrogen, NO or N_2O_3?
    (c) Higher percent by mass of oxygen, NO_2 or N_2O_4?
    (d) Lower percent by mass of chlorine, $NaClO_3$ or $KClO_3$?

(e) Higher percent by mass of sulfur, $KHSO_4$ or K_2SO_4?
(f) Lower percent by mass of chromium, Na_2CrO_4 or $Na_2Cr_2O_7$?

Empirical and molecular formulas

44. Calculate the empirical formula of each compound from the percent compositions given.
    (a) 63.6% N, 36.4% O
    (b) 46.7% N, 53.3% O
    (c) 25.9% N, 74.1% O
    (d) 43.4% Na, 11.3% C, 45.3% O
    (e) 18.8% Na, 29.0% Cl, 52.3% O
    (f) 72.02% Mn, 27.98% O

45. Calculate the empirical formula of each compound from the percent compositions given.
    (a) 64.1% Cu, 35.9% Cl
    (b) 47.2% Cu, 52.8% Cl
    (c) 51.9% Cr, 48.1% S
    (d) 55.3% K, 14.6% P, 30.1% O
    (e) 38.9% Ba, 29.4% Cr, 31.7% O
    (f) 3.99% P, 82.3% Br, 13.7% Cl

46. A sample of tin (Sn) having a mass of 3.996 g was oxidized and found to have combined with 1.077 g of oxygen. Calculate the empirical formula of this oxide of tin.

47. A 3.054 g sample of vanadium (V) combined with oxygen to form 5.454 g of product. Calculate the empirical formula for this compound.

*48. Zinc and sulfur react to form zinc sulfide, ZnS. If we mix 19.5 g of zinc and 9.40 g of sulfur, have we added sufficient sulfur to fully react all the zinc? Show evidence for your answer.

49. Hydroquinone is an organic compound commonly used as a photographic developer. It has a molar mass of 110 g/mol and a composition of 65.45% C, 5.45% H, and 29.09% O. Calculate the molecular formula of hydroquinone.

50. Fructose is a very sweet natural sugar that is present in honey, fruits, and fruit juices. It has a molar mass of 180 g/mol and a composition of 40.0% C, 6.7% H, and 53.3% O. Calculate the molecular formula of fructose.

51. Aspirin is well known as a pain reliever (analgesic) and as a fever reducer (anti-pyretic). It has a molar mass of 180 g/mol and a composition of 60.0% C, 4.48% H, and 35.5% O. Calculate the molecular formula of aspirin.

52. How many grams of oxygen are contained in 8.50 g $Al_2(SO_4)_3$?

53. Gallium arsenide is one of the newer materials used to make semiconductor chips for use in supercomputers. Its composition is 48.2% Ga and 51.8% As. What is the empirical formula?

54. Listed below are the compositions of four different compounds of carbon and chlorine. Determine both the empirical formula and the molecular formula for each.

Percentage C	Percentage Cl	Molar mass
(a) 7.79	92.21	154
(b) 10.13	89.87	237
(c) 25.26	74.74	285
(d) 11.25	88.75	320

Review Exercises for Chapters 5–7

CHAPTER 5   Early Atomic Theory and Structure

True–False.   *Answer the following as either true or false.*

1. Dalton's atomic theory states that all atoms are composed of protons, neutrons, and electrons.
2. The electron was discovered by J. J. Thomson.
3. The proton was discovered by James Chadwick in 1932.
4. The Law of Definite Composition states that a compound contains two or more elements combined in a definite proportion by mass.
5. An atom of $^{108}_{47}Ag$ contains 47 protons, 47 electrons, and 108 neutrons.
6. One atomic mass unit is defined as one-twelfth the mass of a carbon-12 atom.
7. The proton and the neutron have approximately equal charge.
8. The listed atomic mass of an element represents the average relative mass of all the naturally occurring isotopes of that element.
9. The atoms of two different elements must have different mass numbers.
10. $^{35}_{17}Cl$ and $^{37}_{17}Cl$ are isotopes of chlorine.
11. In the isotope $^{112}_{47}Ag$, $Z = 112$ and $A = 47$.
12. All the isotopes of an element have the same number of electrons.
13. The reason the atomic mass of Mg is 24.305 rather than almostly exactly 24 is that protons and neutrons do not have exactly the same mass.
14. The lightest element is helium.
15. The compounds HgO and Hg_2O illustrate the Law of Multiple Proportions.
16. Electric charge can be either positive or negative.
17. Positive ions are called *cations*; negative ions are called *anions*.
18. The nucleus of an atom contains protons, neutrons, and electrons.
19. All isotopes of the same element have the same number of neutrons.
20. The element represented by $^{51}_{21}X$ is antimony, Sb.

Multiple Choice.   *Choose the correct answer to each of the following.*

1. The concept of the positive charge and most of the mass concentrated in a small nucleus surrounded by the electrons was the contribution of:
   (a) Dalton   (c) Thomson
   (b) Rutherford   (d) Chadwick
2. The neutron was discovered in 1932 by
   (a) Dalton   (c) Thomson
   (b) Rutherford   (d) Chadwick
3. An atom of atomic number 53 and mass number 127 contains how many neutrons?
   (a) 53   (c) 127
   (b) 74   (d) 180
4. How many electrons are in an atom of $^{40}_{18}Ar$?
   (a) 20   (c) 40
   (b) 22   (d) No correct answer given
5. The number of neutrons is an atom of $^{139}_{56}Ba$ is:
   (a) 56   (c) 139
   (b) 83   (d) No correct answer given
6. The name of the isotope containing one proton and two neutrons is:
   (a) Protium   (c) Deuterium
   (b) Tritium   (d) Helium
7. Each atom of a specific element has the same
   (a) Number of protons
   (b) Atomic mass
   (c) Number of neutrons
   (d) No correct answer given
8. Which pair of symbols represents isotopes?
   (a) $^{23}_{11}Na$ and $^{23}_{12}Na$ (c) $^{63}_{29}Cu$ and $^{29}_{64}Cu$
   (b) $^{7}_{3}Li$ and $^{6}_{3}Li$ (d) $^{12}_{24}Mg$ and $^{12}_{26}Mg$
9. Two naturally occurring isotopes of an element have masses and abundance as follows: 54.00 amu (20.00%) and 56.00 amu (80.00%). What is the relative atomic mass of the element?
   (a) 54.20   (c) 54.80
   (b) 54.40   (d) 55.60

10. Substance X has 13 protons, 14 neutrons, and 10 electrons. Determine its identity.
    (a) ^{27}Mg (c) ^{27}Al^{+3} (b) ^{27}Ne (d) ^{27}Al

11. The mass of a chlorine atom is 5.90×10^{-23} g. How many atoms are in a 42.0 g sample of chlorine?
    (a) 2.48×10^{-21} (c) 1.40×10^{-24}
    (b) 7.12×10^{23} (d) no correct answer given

12. The number of neutrons in an atom of $^{108}_{47}$Ag is
    (a) 47 (c) 155
    (b) 108 (d) no correct answer given

13. The number of electrons in an atom of $^{27}_{13}$Al is
    (a) 13 (c) 27
    (b) 14 (d) 40

14. The number of protons in an atom of $^{65}_{30}$Zn is
    (a) 65 (c) 30
    (b) 35 (d) 95

15. The number of electrons in the nucleus of an atom of $^{24}_{12}$Mg is
    (a) 12 (c) 36
    (b) 24 (d) no correct answer given

CHAPTER 6   Nomenclature of Inorganic Compounds

True–False. *Answer the following as either true or false.*

1. The oxidation number of an element can have a positive, negative, or zero value.
2. The sum of oxidation numbers of all the elements in a compound equals zero.
3. The oxidation number of an ion in an ionic compound is equal to the charge of the ion.
4. The oxidation numbers of the elements boron and aluminum are 0 or +3.
5. The oxidation number of oxygen in a compound is usually −1.
6. In KMnO$_4$ the oxidation number of Mn is +6.
7. Binary compounds have names ending in *ide*.
8. The oxidation number of oxygen in peroxides is −2.
9. The compound formed from Ga^{3+} and O^{2-} is Ga$_3$O$_2$.
10. The compound formed from NH$_4^+$ and SO$_4^{2-}$ is (NH$_4$)$_2$SO$_4$.
11. The nitrite ion has three oxygen atoms and the nitrate ion has four oxygen atoms.
12. The prefixes *tetra* and *penta* mean four and five, respectively.
13. If the name of an acid ends in *ous*, the corresponding salt name will end in *ate*.
14. The lower and higher oxidation states of iron (Fe) are called *ferrous* and *ferric*, respectively.
15. The common name for sulfur is brimstone.
16. The formula for muriatic acid is HNO$_3$.
17. The formula for cane or beet sugar is C$_6$H$_{12}$O$_6$.
18. The name for Cl$_2$O$_3$ is dichloroheptoxide.
19. The formula for copper(II) oxide is Cu$_2$O.
20. The formula for barium hydroxide is BaOH.
21. An ion is a positive or negative electrically charged atom or group of atoms.
22. A chemical formula is a shorthand expression for a chemical reaction.
23. The name of ZnBr$_2$ is zinc bromide.

Names and Formulas. *In which of the following is the formula correct for the name given?*

1. Copper(II) sulfate, CuSO$_4$
2. Ammonium hydroxide, NH$_4$OH
3. Mercury(I) carbonate, HgCO$_3$
4. Phosphorus triiodide, PI$_3$
5. Calcium acetate, Ca(C$_2$H$_3$O$_2$)$_2$
6. Hypochlorous acid, HClO
7. Dichlorine heptoxide, Cl$_2$O$_7$
8. Magnesium iodide, MgI
9. Sulfurous acid, H$_2$SO$_3$
10. Potassium manganate, KMnO$_4$
11. Lead(II) chromate, PbCrO$_4$
12. Ammonium bicarbonate, NH$_4$HCO$_3$
13. Iron(II) phosphate, FePO$_4$
14. Calcium hydrogen sulfate, CaHSO$_4$
15. Mercury(II) sulfate, HgSO$_4$
16. Dinitrogen pentoxide, N$_2$O$_5$
17. Sodium hypochlorite, NaClO
18. Sodium dichromate, Na$_2$Cr$_2$O$_7$
19. Cadmium cyanide, Cd(CN)$_2$
20. Bismuth(III) oxide, Bi$_3$O$_2$
21. Carbonic acid, H$_2$CO$_3$
22. Silver oxide, Ag$_2$O
23. Ferric iodide, FeI$_2$
24. Tin(II) fluoride, TiF$_2$

25. Carbon monoxide, CO
26. Phosphoric acid, H_3PO_3
27. Sodium bromate, Na_2BrO_3
28. Hydrosulfuric acid, H_2S
29. Potassium hydroxide, POH
30. Sodium carbonate, Na_2CO_3
31. Zinc sulfate, $ZnSO_3$
32. Sulfur trioxide, SO_3

33. Tin(IV) nitrate, $Sn(NO_3)_4$
34. Ferrous sulfate, $FeSO_4$
35. Chloric acid, HCl
36. Aluminum sulfide, Al_2S_3
37. Cobalt(II) chloride, $CoCl_2$
38. Acetic acid, $HC_2H_3O_2$
39. Zinc oxide, ZnO_2
40. Stannous fluoride, SnF_2

CHAPTER 7   Quantitative Composition of Compounds

True–False.   *Answer the following as either true or false.*

1. A mole contains Avogadro's number of atoms, molecules, or formula units.
2. A mole of Ag (107.9 g) contains the same number of atoms as a mole of Na (23.0 g).
3. One mole of chlorine molecules contains 2 mol of chlorine atoms.
4. One mole of glucose, $C_6H_{12}O_6$, contains 24 mol of atoms.
5. The mass of 2 mol of hydrogen molecules is 4.0 g.
6. One gram of sulfur contains 6.022×10^{23} atoms.
7. The mass of a mole of NaCl is less than the mass of a mole of KCl.
8. $CaCl_2$ has a higher percentage of chlorine than $MgCl_2$.
9. A compound has an empirical formula of C_2H_2O and a molar mass of 168.0. The molecular formula is $C_6H_6O_3$.
10. The percent composition of a compound is the mass percent of each element in the compound.
11. One mole of $HC_2H_3O_2$ has a mass of 60.0 g.
12. If the molecular formula and empirical formula of a compound are not the same, the empirical formula will be an integral multiple of the molecular formula.
13. The empirical formula of a compound gives the smallest ratio of the atoms that are present in the compound.
14. A mole of magnesium and a mole of magnesium oxide, MgO, contain the same number of magnesium atoms.
15. The number of sulfur atoms is the same in 1 mol of Na_2SO_4 as in 1 mol of K_2SO_4.

16. The number of sulfur atoms is the same in 1 g of Na_2SO_4 as in 1 g of K_2SO_4.
17. There are 14 mol of chlorine atoms in 3.5 mol of CCl_4.
18. A compound that has a carbon to hydrogen ratio of 1:2 can have a molar mass of 48.0.
19. The molar mass of a compound is the sum of the molar masses of all the atoms in the formula of the compound.

Multiple Choice.   *Choose the correct answer to each of the following.*

1. 4.0 g of oxygen contains:
   (a) 1.5×10^{23} atoms of oxygen
   (b) 4.0 molar masses of oxygen
   (c) 0.50 mol of oxygen
   (d) 6.022×10^{23} atoms of oxygen
2. One mole of hydrogen atoms contains:
   (a) 2.0 g of hydrogen
   (b) 6.022×10^{23} atoms of hydrogen
   (c) 1 atom of hydrogen
   (d) 12 g of carbon-12
3. The mass of one atom of magnesium is
   (a) 24.3 g   (c) 12.0 g
   (b) 54.9 g   (d) 4.035×10^{-23} g
4. Avogadro's number of magnesium atoms:
   (a) Has a mass of 1.0 g
   (b) Has the same mass as Avogadro's number of sulfur atoms
   (c) Has a mass of 12.0 g
   (d) Is 1 mol of magnesium atoms
5. Which of the following contains the largest number of moles?
   (a) 1.0 g Li   (c) 1.0 g Al
   (b) 1.0 g Na   (d) 1.0 g Ag

6. The number of moles in 112 g of acetylsalicylic acid (aspirin), $C_9H_8O_4$, is:
   (a) 1.61   (b) 0.622   (c) 112   (d) 0.161

7. How many moles of aluminum hydroxide are in one antacid tablet containing 400 mg of $Al(OH)_3$?
   (a) 5.13×10^{-3} (c) 5.13
   (b) 0.400   (d) 9.09×10^{-3}

8. How many grams of Au_2S can be obtained from 1.17 mol of Au?
   (a) 182 g   (b) 249 g   (c) 364 g   (d) 499 g

9. The molar mass of $Ba(NO_3)_2$ is:
   (a) 199.3   (b) 261.3   (c) 247.3   (d) 167.3

10. A 16 g sample of O_2:
    (a) Is 1 mol of O_2
    (b) Contains 6.022×10^{23} molecules of O_2
    (c) Is 0.50 molecule of O_2
    (d) Is 0.50 molar mass of O_2

11. What is the percent composition for a compound formed from 8.15 g of zinc and 2.00 g of oxygen?
    (a) 80.3% Zn 19.7% O   (c) 70.3% Zn 29.7% O
    (b) 80.3% O 19.7% Zn   (d) 65.3% Zn 34.7% O

12. Which of the following compounds contains the largest percentage of oxygen?
    (a) SO_2 (c) N_2O_3
    (b) SO_3 (d) N_2O_5

13. 2.00 mol of CO_2:
    (a) Has a mass of 56.0 g
    (b) Contain 1.20×10^{24} molecules
    (c) Has a mass of 56.0 g
    (d) Contain 6.00 molar masses of CO_2

14. In Ag_2CO_3, the percent by mass of:
    (a) Carbon is 43.5%   (c) Oxygen is 17.4%
    (b) Silver is 64.2%   (d) Oxygen is 21.9%

15. The empirical formula of the compound containing 31.0% Ti and 69.0% Cl is:
    (a) TiCl   (c) $TiCl_3$
    (b) $TiCl_2$ (d) $TiCl_4$

16. A compound contains 54.3% C, 5.6% H, and 40.1% Cl. The empirical formula is:
    (a) CH_3Cl (c) $C_2H_4Cl_2$
    (b) C_2H_5Cl (d) C_4H_5Cl

17. A compound contains 40.0% C, 6.7% H, and 53.3% O. The molar mass is 60.0 g/mol. The molecular formula is:
    (a) $C_2H_3O_2$ (c) C_2HCl
    (b) C_3H_8O (d) $C_2H_4O_2$

18. How many chlorine atoms are in 4.0 mol of PCl_3?
    (a) 3   (c) 12.0
    (b) 7.2×10^{24} (d) 2.4×10^{24}

19. What is the mass of 4.53 mol of Na_2SO_4?
    (a) 142.1 g   (c) 31.4 g
    (b) 644 g   (d) 3.19×10^{-2} g

20. The percent composition of Mg_3N_2 is:
    (a) 72.2% Mg, 27.8% N
    (b) 63.4% Mg, 36.6% N
    (c) 83.9% Mg, 16.1% N
    (d) No correct answer given

21. How many grams of oxygen are contained in 0.500 mol of Na_2SO_4?
    (a) 16.0 g   (c) 64.0 g
    (b) 32.0 g   (d) No correct answer given

22. The empirical formula of a compound is CH. If the molar mass of this compound is 78.0, then the molecular formula is:
    (a) C_2H_2 (c) C_6H_6
    (b) C_5H_{18} (d) No correct answer given

CHAPTER EIGHT

Chemical Equations

In today's world much of our energy is directed toward expressing information in a concise, useful manner. From our earliest days in childhood, we are taught to translate ideas and desires into sentences. In mathematics, we learn to translate numerical relationships and situations into mathematical expressions and equations. An historian translates a thousand years of history into a 500-page textbook. A secretary translates an entire letter or document into a few lines of shorthand. A film maker translates an entire event, such as the Olympics, into several hours of entertainment.

And so it is in chemistry. A chemist uses a chemical equation to translate the reactions that are observed over widely varying timeframes in the laboratory or in nature. Chemical equations provide the necessary means (1) to summarize the reaction, (2) to determine the substances that are reacting, (3) to predict the products, and (4) to indicate the amounts of all component substances in the reaction.

Chapter Preview

8.1   The Chemical Equation

In a chemical reaction the substances entering the reaction are called reactants and the substances formed are called the products. During a chemical reaction atoms, molecules, or ions interact and rearrange themselves to form the products. During this process chemical bonds are broken and new bonds are formed. The reactants and products may be in the solid, liquid, or gaseous state, or in solution.

chemical equation

word equation

A **chemical equation** is a shorthand expression for a chemical change or reaction. It shows, among other things, the rearrangement of the atoms that are involved in the reaction. A **word equation** states in words, in equation form, the substances involved in a chemical reaction. For example, when mercury(II) oxide is heated, it decomposes to form mercury and oxygen. The word equation for this decomposition is

$$\text{mercury(II) oxide} + \text{heat} \longrightarrow \text{mercury} + \text{oxygen}$$

From the chemist's point of view this method of describing a chemical reaction is inadequate. It is bulky and cumbersome to use and does not give quantitative information. The chemical equation, using symbols and formulas, is a far better way to describe the decomposition of mercury(II) oxide:

$$2\,HgO \xrightarrow{\Delta} 2\,Hg + O_2\uparrow$$

This equation gives all the information from the word equation plus formulas, composition, reactive amounts of all the substances involved in the reaction, and much additional information (see Section 8.4). Even though a chemical equation provides much quantitative information, it is still not a complete description; it does not tell us how much energy is needed to cause decomposition, what we observe during the reaction, or anything about the rate of reaction. This information must be obtained from other sources or from experimentation.

167

8.2 Format for Writing Chemical Equations

A chemical equation uses the chemical symbols and formulas of the reactants and products and other symbolic terms to represent a chemical reaction. Equations are written according to this general format:

1. The reactants are separated from the products by an arrow ($\rightarrow$) that indicates the direction of the reaction. A double arrow ($\rightleftharpoons$) indicates that the reaction goes in both directions and establishes an equilibrium between the reactants and the products.
2. The reactants are placed to the left and the products to the right of the arrow. A plus sign ($+$) is placed between reactants and between products when needed.
3. Conditions required to carry out the reaction may, if desired, be placed above or below the arrow or equality sign. For example, a delta sign placed over the arrow ($\xrightarrow{\Delta}$) indicates that heat is supplied to the reaction.
4. Coefficients (integral numbers) are placed in front of substances (for example, $2 H_2O$) to balance the equation and to indicate the number of formula units (atoms, molecules, moles, ions) of each substance reacting or being produced. When no number is shown, it is understood that one formula unit of the substance is indicated.
5. The physical state of a substance is indicated by the following symbols: (s) for solid state; (l) for liquid state; (g) for gaseous state; and (aq) for substances in aqueous solution.

Symbols commonly used in equations are given in Table 8.1.

TABLE 8.1

Symbols Commonly Used in Chemical Equations

Symbol	Meaning
$\rightarrow$	Yields; produces (points to products)
$\rightleftharpoons$	Reversible reaction; equilibrium between reactants and products
$\uparrow$	Gas evolved (written after a substance)
$\downarrow$	Solid or precipitate formed (written after a substance)
(s)	Solid state (written after a substance)
(l)	Liquid state (written after a substance)
(g)	Gaseous state (written after a substance)
(aq)	Aqueous solution (substance dissolved in water)
Δ	Heat
$+$	Plus or added to (placed between substances)

8.3   Writing and Balancing Equations

balanced equation

To represent the quantitative relationships of a reaction, the chemical equation must be balanced. A **balanced equation** contains the same number of each kind of atom on each side of the equation. The balanced equation, therefore, obeys the Law of Conservation of Mass.

The ability to balance equations must be acquired by every chemistry student. Simple equations are easy to balance, but some care and attention to detail are required. The way to balance an equation is to adjust the number of atoms of each element so that it is the same on each side of the equation. But we must not change a correct formula in order to achieve a balanced equation. Each equation must be treated on its own merits; we have no simple "plug in" formula for balancing equations. The following outline gives a general procedure for balancing equations. Study this outline and refer to it as needed when working examples. There is no substitute for practice in learning to write and balance chemical equations.

1. Identify the reaction for which the equation is to be written   Formulate a description or word equation for the reaction if needed (e.g., mercury(II) oxide decomposes yielding mercury and oxygen).

2. Write the unbalanced, or skeleton, equation   Make sure that the formula for each substance is correct and that the reactants are written to the left and the products to the right of the arrow (e.g., $HgO \rightarrow Hg + O_2$). The correct formulas must be known or ascertained from the periodic table, oxidation numbers, lists of ions, or experimental data.

3. Balance the equation   Use the following steps as necessary:
   (a) Count and compare the number of atoms of each element on each side of the equation and determine those that must be balanced.
   (b) Balance each element, one at a time, by placing whole numbers (coefficients) in front of the formulas containing the unbalanced element. It is usually best to balance metals first, then nonmetals, then hydrogen and oxygen. Select the smallest coefficients that will give the same number of atoms of the element on each side. A coefficient placed before a formula multiplies every atom in the formula by that number (for example, $2\,H_2SO_4$ means two molecules of sulfuric acid and also means four H atoms, two S atoms, and eight O atoms.)
   (c) Check all other elements after each individual element is balanced to see whether, in balancing one, other elements have become unbalanced. Make adjustments as needed.
   (d) Balance polyatomic ions such as SO_4^{2-}, which remain unchanged from

one side of the equation to the other, in the same way as individual atoms.

(e) Do a final check, making sure that each element and/or polyatomic ion is balanced and that the smallest possible set of whole-number coefficients has been used.

$$4\,HgO \longrightarrow 4\,Hg + 2\,O_2 \quad \text{(incorrect form)}$$

$$2\,HgO \longrightarrow 2\,Hg + O_2 \quad \text{(correct form)}$$

Not all chemical equations can be balanced by the simple method of inspection just described. The balancing of more complex equations such as oxidation-reduction equations is described in Chapter 18.

The following examples show stepwise sequences leading to balanced equations. Study each example carefully.

EXAMPLE 8.1

Write the balanced equation for the reaction that takes place when magnesium metal is burned in air to produce magnesium oxide.

1. *Word equation*

   magnesium + oxygen $\longrightarrow$ magnesium oxide

2. *Skeleton equation*

   $$Mg + O_2 \longrightarrow MgO \quad \text{(unbalanced)}$$

3. *Balance*
   (a) Oxygen is not balanced. Two O atoms appear on the left side and one on the right side.
   (b) Place the coefficient 2 before MgO.

   $$Mg + O_2 \longrightarrow 2\,MgO \quad \text{(unbalanced)}.$$

   (c) Now Mg is not balanced. One Mg atom appears on the left side and two on the right side. Place a 2 before Mg.

   $$2\,Mg + O_2 \longrightarrow 2\,MgO \quad \text{(balanced)}$$

   (d) *Check*: Each side has two Mg and two O atoms.

EXAMPLE 8.2

When methane, CH_4, undergoes complete combustion, it reacts with oxygen to produce carbon dioxide and water. Write the balanced equation for this reaction.

1. *Word equation*

   methane + oxygen $\longrightarrow$ carbon dioxide + water

2. *Skeleton equation*

   $$CH_4 + O_2 \longrightarrow CO_2 + H_2O \quad \text{(unbalanced)}$$

3. *Balance*
(a) Carbon is balanced. Hydrogen and oxygen are not balanced.
(b) Balance H atoms by placing a 2 before H_2O.

$$CH_4 + O_2 \longrightarrow CO_2 + 2\,H_2O \quad \text{(unbalanced)}$$

Each side of the equation has four H atoms; oxygen is still not balanced. Place a 2 before O_2 to balance the oxygen atoms.

$$CH_4 + 2\,O_2 \longrightarrow CO_2 + 2\,H_2O \quad \text{(balanced)}$$

(c) *Check*: The equation is correctly balanced; it has one C, four O, and four H atoms on each side.

EXAMPLE 8.3

Oxygen and potassium chloride are formed by heating potassium chlorate. Write a balanced equation for this reaction.

1. *Word equation*

potassium chlorate $\xrightarrow{\Delta}$ potassium chloride + oxygen

2. *Skeleton equation*

$$KClO_3 \xrightarrow{\Delta} KCl + O_2 \quad \text{(unbalanced)}$$

3. *Balance*
(a) Oxygen is unbalanced (three O atoms on the left and two on the right side).
(b) How many oxygen atoms are needed? The subscripts of oxygen (3 and 2) in $KClO_3$ and O_2 have a least common multiple of 6. Therefore coefficients for $KClO_3$ and O_2 are needed to give six oxygen atoms on each side. Place a 2 before $KClO_3$ and a 3 before O_2 to give six O atoms on each side.

$$2\,KClO_3 \xrightarrow{\Delta} KCl + 3\,O_2 \quad \text{(unbalanced)}$$

Now K and Cl are not balanced. Place a 2 before KCl, which balances both K and Cl at the same time.

$$2\,KClO_3 \xrightarrow{\Delta} 2\,KCl + 3\,O_2 \quad \text{(balanced)}$$

(c) *Check*: Each side now contains two K, two Cl, and six O atoms.

EXAMPLE 8.4

Balance by starting with the word equation given.

1. *Word equation*

silver nitrate + hydrogen sulfide $\longrightarrow$ silver sulfide + nitric acid

2. *Skeleton equation*

$$AgNO_3 + H_2S \longrightarrow Ag_2S + HNO_3 \quad \text{(unbalanced)}$$

3. *Balance*
(a) Ag and H are unbalanced.
(b) Place a 2 in front of $AgNO_3$ to balance Ag.

$$2\ AgNO_3 + H_2S \longrightarrow Ag_2S + HNO_3 \quad \text{(unbalanced)}$$

(c) H and NO_3^- are still unbalanced. Balance by placing a 2 in front of HNO_3.

$$2\ AgNO_3 + H_2S \longrightarrow Ag_2S + 2\ HNO_3 \quad \text{(balanced)}$$

(d) In this example N and O atoms are balanced by balancing the NO_3^- ion as a unit.
(e) *Check*: Each side has two Ag, two H, and one S atom. Also, each side has two NO_3^- ions.

EXAMPLE 8.5   Balance by starting with the word equation given.

1. *Word equation*

aluminum hydroxide + sulfuric acid $\longrightarrow$ aluminum sulfate + water

2. *Skeleton equation*

$$Al(OH)_3 + H_2SO_4 \longrightarrow Al_2(SO_4)_3 + H_2O \quad \text{(unbalanced)}$$

3. *Balance*
(a) All elements are unbalanced.
(b) Balance Al by placing a 2 in front of $Al(OH)_3$. Treat the unbalanced SO_4^{2-} ion as a unit and balance by placing a 3 before H_2SO_4.

$$2\ Al(OH)_3 + 3\ H_2SO_4 \longrightarrow Al_2(SO_4)_3 + H_2O \quad \text{(unbalanced)}$$

Balance the unbalanced H and O by placing a 6 in front of H_2O.

$$2\ Al(OH)_3 + 3\ H_2SO_4 \longrightarrow Al_2(SO_4)_3 + 6\ H_2O \quad \text{(balanced)}$$

(c) *Check*: Each side has two Al, twelve H, three S, and eighteen O atoms.

EXAMPLE 8.6   When the fuel in a butane gas stove undergoes complete combustion, it reacts with oxygen to form carbon dioxide and water. Write the balanced equation for this reaction.

1. *Word equation*

butane + oxygen $\longrightarrow$ carbon dioxide + water

2. *Skeleton equation*

$$C_4H_{10} + O_2 \longrightarrow CO_2 + H_2O \quad \text{(unbalanced)}$$

3. *Balance*
(a) All elements are unbalanced.
(b) Balance C by placing a 4 in front of CO_2.

$$C_4H_{10} + O_2 \longrightarrow 4\,CO_2 + H_2O \quad \text{(unbalanced)}$$

Balance H by placing a 5 in front of H_2O.

$$C_4H_{10} + O_2 \longrightarrow 4\,CO_2 + 5\,H_2O \quad \text{(unbalanced)}$$

Oxygen remains unbalanced. The oxygen atoms on the right side are fixed, because $4\,CO_2$ and $5\,H_2O$ are derived from the single C_4H_{10} molecule on the left. When we try to balance the O atoms, we find that there is no integer (whole number) that can be placed in front of O_2 to bring about a balance. The equation can be balanced if we use $6\frac{1}{2}\,O_2$ and then double the coefficients of each substance, including the $6\frac{1}{2}\,O_2$, to obtain the balanced equation.

$$C_4H_{10} + 6\tfrac{1}{2}\,O_2 \longrightarrow 4\,CO_2 + 5\,H_2O \quad \text{(balanced—incorrect form)}$$
$$2\,C_4H_{10} + 13\,O_2 \longrightarrow 8\,CO_2 + 10\,H_2O \quad \text{(balanced)}$$

(c) *Check*: Each side now has eight C, twenty H, and twenty-six O atoms.

PRACTICE   Balance the following word equation.

Aluminum + oxygen $\longrightarrow$ Aluminum oxide

Answer: the coefficients are 4, 2 $\longrightarrow$ 2

PRACTICE   Balance the following word equation.

Magnesium hydroxide + phosphoric acid $\longrightarrow$ magnesium phosphate + water

Answer: the coefficients are 3, 2 $\longrightarrow$ 1, 6

8.4   What Information Does an Equation Tell Us?

Depending on the particular context in which it is used, a formula can have different meanings. The meanings refer either to an individual chemical entity (atom, ion, molecule, or formula unit) or to a mole of that chemical entity. For example, the formula H_2O can be used to indicate any of the following:

1. 2 H atoms and 1 O atom
2. 1 molecule of water
3. 1 mol of water
4. 6.022×10^{23} molecules of water
5. 1 molar mass of water
6. 18.0 g of water

Formulas used in equations can be expressed in units of individual chemical entities or as moles, the latter being more commonly used. For example, in the reaction of hydrogen and oxygen to form water,

$$2\,H_2 + O_2 \longrightarrow 2\,H_2O$$

the $2\,H_2$ can represent 2 molecules or 2 moles of hydrogen; the O_2, 1 molecule or 1 mole of oxygen; and the $2\,H_2O$, 2 molecules or 2 moles of water. In terms of moles, this equation is stated: 2 moles of H_2 react with 1 mole of O_2 to give 2 moles of H_2O.

As indicated earlier, a chemical equation is a shorthand description of a chemical reaction. Interpretation of the balanced equation gives us the following information:

1. What the reactants are and what the products are
2. The formulas of the reactants and products
3. The number of molecules or formula units of reactants and products in the reaction
4. The number of atoms of each element involved in the reaction
5. The number of molar masses of each substance used or produced
6. The number of moles of each substance
7. The number of grams of each substance used or produced

Consider the equation

$$H_2(g) + Cl_2(g) \longrightarrow 2\,HCl(g)$$

This equation states that hydrogen gas reacts with chlorine gas to produce hydrogen chloride, also a gas. Let us summarize all the information relating to the equation. The information that can be stated about the relative amount of each substance, with respect to all other substances in the balanced equation, is written below its formula in the following equation:

$$H_2(g) \qquad + \qquad Cl_2(g) \qquad \longrightarrow \qquad 2\,HCl(g)$$

Hydrogen	Chlorine	Hydrogen Chloride
1 molecule	1 molecule	2 molecules
2 atoms	2 atoms	2 atoms H + 2 atoms Cl
1 molar mass	1 molar mass	2 molar masses
1 mole	1 mole	2 moles
2.0 g	71.0 g	2×36.5 g or 73.0 g

These data are very useful in calculating quantitative relationships that exist among substances in a chemical reaction. For example, if we react 2 moles of hydrogen (twice as much as is indicated by the equation) with 2 moles of chlorine, we can expect to obtain 4 moles, or 146 g, of hydrogen chloride as a product. We will study this phase of using equations in more detail in the next chapter.

Let us try another equation. When propane gas is burned in air, the products are carbon dioxide, CO_2, and water, H_2O. The balanced equation and its interpretation are as follows:

$$C_3H_8(g) \quad + \quad 5\,O_2(g) \quad \longrightarrow \quad 3\,CO_2(g) \quad + \quad 4\,H_2O(g)$$

Propane	Oxygen	Carbon dioxide	Water
1 molecule	5 molecules	3 molecules	4 molecules
3 atoms C	10 atoms O	3 atoms C	8 atoms H
8 atoms H		6 atoms O	4 atoms O
1 molar mass	5 molar masses	3 molar masses	4 molar masses
1 mole	5 moles	3 moles	4 moles
44.0 g	5×32.0 g (160.0 g)	3×44.0 g (132.0 g)	4×18.0 g (72.0 g)

8.5 Types of Chemical Equations

Chemical equations represent chemical changes or reactions. To be of any significance an equation must represent an actual or possible reaction. Part of the problem of writing equations is determining the products formed. We have no sure method of predicting products, nor do we have time to carry out experimentally all the reactions we may wish to consider. Therefore we must use data reported in the writings of other workers, certain rules to aid in our predictions, and the atomic structure and combining capacities of the elements to help us predict the formulas of the products of a chemical reaction. The final proof of the existence of any reaction, of course, is in the actual observation of the reaction in the laboratory (or elsewhere).

Reactions are classified into types to assist in writing equations and to aid in predicting other reactions. Many chemical reactions fit one or another of the four principal reaction types that are discussed in the following paragraphs. Reactions are also classified as oxidation–reduction. Special methods are used to balance complex oxidation–reduction equations (see Chapter 18).

combination or
synthesis reaction

1. Combination or Synthesis Reaction In a **combination reaction**, two reactants combine to give one product. The general form of the equation is

$$A + B \longrightarrow AB$$

in which A and B are either elements or compounds and AB is a compound. The formula of the compound in many cases can be determined from a knowledge of the oxidation numbers of the reactants in their combined states. Some reactions that fall into this category are the following:

(a) metal + oxygen ⟶ metal oxide

$$2\,Mg(s) + O_2(g) \xrightarrow{\Delta} 2\,MgO(s)$$

$$4\,Al(s) + 3\,O_2(g) \xrightarrow{\Delta} 2\,Al_2O_3(s)$$

(b) nonmetal + oxygen ⟶ nonmetal oxide

$$S(s) + O_2(g) \xrightarrow{\Delta} SO_2(g)$$

$$N_2(g) + O_2(g) \xrightarrow{\Delta} 2\,NO(g)$$

(c) metal + nonmetal ⟶ salt

$$2\,Na(s) + Cl_2(g) \longrightarrow 2\,NaCl(s)$$

$$2\,Al(s) + 3\,Br_2(l) \longrightarrow 2\,AlBr_3(s)$$

(d) metal oxide + water ⟶ base (metal hydroxide)

$$Na_2O(s) + H_2O(l) \longrightarrow 2\,NaOH(aq)$$

$$CaO(s) + H_2O(l) \longrightarrow Ca(OH)_2(aq)$$

(e) nonmetal oxide + water ⟶ oxy-acid

$$SO_3(g) + H_2O(l) \longrightarrow H_2SO_4(aq)$$

$$N_2O_5(s) + H_2O(l) \longrightarrow 2\,HNO_3(aq)$$

---

decomposition
reaction

2. Decomposition Reaction   In a **decomposition reaction** a single substance is decomposed or broken down to give two or more different substances. The reaction may be considered the reverse of combination. The starting material must be a compound, and the products may be elements or compounds. The general form of the equation is

$$AB \longrightarrow A + B$$

Predicting the products of a decomposition reaction can be difficult and requires an understanding of each individual reaction. Heating oxygen-containing compounds often results in decomposition. Some reactions that fall into this category are the following:

(a) Metal oxides. Some metal oxides decompose to yield the free metal plus oxygen, others give a lower oxide, and some are very stable, resisting decomposition by heating.

$$2\,HgO(s) \xrightarrow{\Delta} 2\,Hg(l) + O_2(g)$$

$$2\,PbO_2(s) \xrightarrow{\Delta} 2\,PbO(s) + O_2(g)$$

(b) Carbonates and bicarbonates decompose to yield CO_2 when heated.

$$CaCO_3(s) \xrightarrow{\Delta} CaO(s) + CO_2(g)$$

$$2\,NaHCO_3(s) \xrightarrow{\Delta} Na_2CO_3(s) + H_2O(g) + CO_2(g)$$

(c) Miscellaneous.

$$2\,KClO_3(s) \xrightarrow{\Delta} 2\,KCl(s) + 3\,O_2(g)$$

$$2\,NaNO_3(s) \xrightarrow{\Delta} 2\,NaNO_2(s) + O_2(g)$$

$$2\,H_2O_2(l) \xrightarrow{\Delta} 2\,H_2O(l) + O_2(g)$$

single-displacement reaction

3. Single-Displacement Reaction In a **single-displacement reaction** one element reacts with a compound to take the place of one of the elements of that compound. A different element and a different compound are formed. The general form of the equation is

$$A + BC \longrightarrow B + AC \qquad \text{or} \qquad A + BC \longrightarrow C + BA$$

metal   halogen

If A is a metal, A will replace B to form AC, providing A is a more reactive metal than B. If A is a halogen, it will replace C to form BA, providing A is a more reactive halogen than C.

A brief activity series of selected metals (and hydrogen) and halogens are shown in Table 8.1. The series are listed in descending order of chemical reactivity, with the most active metals and halogens at the top. From such series it is possible to predict many chemcial reactions. Any metal on the list will replace the ions of those metals that appear anywhere underneath it on the list. For example, zinc metal will replace hydrogen from a hydrochloric acid solution. But copper metal, which is underneath hydrogen on the list and thus less reactive than hydrogen, will not replace hydrogen from a hydrochloric acid solution. Some reactions that fall into this category follow.

(a) metal + acid $\longrightarrow$ hydrogen + salt

$$Zn(s) + 2\,HCl(aq) \longrightarrow H_2(g) + ZnCl_2(aq)$$

$$2\,Al(s) + 3\,H_2SO_4(aq) \longrightarrow 3\,H_2(g) + Al_2(SO_4)_3(aq)$$

(b) metal + water $\longrightarrow$ hydrogen + metal hydroxide or metal oxide

$$2\,Na(s) + 2\,H_2O \longrightarrow H_2(g) + 2\,NaOH(aq)$$

$$Ca(s) + 2\,H_2O \longrightarrow H_2(g) + Ca(OH)_2(aq)$$

$$3\,Fe(s) + 4\,H_2O(g) \longrightarrow 4\,H_2(g) + Fe_3O_4(s)$$

Steam

TABLE 8.2

Activity Series

Metals	Halogens
K	F_2
Ca	Cl_2
Na	Br_2
Mg	I_2
Al	
Zn	
Fe	
Ni	
Sn	
Pb	
H	
Cu	
Ag	
Hg	
Au	

increasing activity →

(c)   metal + salt $\longrightarrow$ metal + salt

$$Fe(s) + CuSO_4(aq) \longrightarrow Cu(s) + FeSO_4(aq)$$
$$Cu(s) + 2\,AgNO_3(aq) \longrightarrow 2\,Ag(s) + Cu(NO_3)_2(aq)$$

(d)   halogen + halogen salt $\longrightarrow$ halogen + halogen salt

$$Cl_2(g) + 2\,NaBr(aq) \longrightarrow Br_2(l) + 2\,NaCl(aq)$$
$$Cl_2(g) + 2\,KI(aq) \longrightarrow I_2(s) + 2\,KCl(aq)$$

A common chemical reaction is the displacement of hydrogen from water or acids. This reaction is a good illustration of the relative reactivity of metals and the use of the activity series.

K, Ca, and Na displace hydrogen from cold water, steam, and acids.

Mg, Al, Zn, and Fe displace hydrogen from steam and acids.

Ni, Sn, and Pb displace hydrogen only from acids.

Cu, Ag, Hg, and Au do not displace hydrogen.

EXAMPLE 8.7   Will a reaction occur between (a) nickel metal and hydrochloric acid and (b) tin metal and a solution of aluminum chloride? Write balanced equations for the reactions.

(a) Nickel is more reactive than hydrogen so it will displace hydrogen from hydrochloric acid. The products are hydrogen gas and a salt of Ni^{2+} and Cl^- ions.

$$Ni(s) + 2\,HCl(aq) \longrightarrow H_2(g) + NiCl_2(aq)$$

(b) According to the activity series, tin is less reactive than aluminum, so no reaction will occur.

$$Sn(s) + AlCl_3(aq) \longrightarrow no\ reaction$$

PRACTICE   Write balanced equations for the reactions:
(a) iron metal and a solution of magnesium chloride
(b) zinc metal and a solution of lead nitrate
Answers: (a) $Fe + MgCl_2 \longrightarrow$ no reaction
                (b) $Zn + Pb(NO_3)_2 \longrightarrow Pb\downarrow + Zn(NO_3)_2$

double-displacement or metathesis reaction

4. Double-Displacement or Metathesis Reaction   In a **double-displacement reaction**, two compounds exchange partners with each other to produce two different compounds. The general form of the equation is

$$AB + CD \longrightarrow AD + CB$$

This reaction may be thought of as an exchange of positive and negative groups, in which A combines with D, and C combines with B. In writing the formulas of the products, we must take into account the oxidation numbers or charges of the combining groups.

It is possible to write an equation in the form of a double displacement reaction when a reaction has not occurred. For example, when solutions of sodium chloride and potassium nitrate are mixed, the following equation can be written:

$$NaCl + KNO_3 \longrightarrow NaNO_3 + KCl$$

When the procedure is carried out, no physical changes are observed, indicating that no chemcial reaction has taken place.

A double displacement reaction is accompanied by evidence of such reactions as the evolution of heat, the formation of an insoluble precipitate, or the production of bubbles of a gas. Some of the reactions that fall into these categories follow.

(a) Neutralization of an acid and a base   The production of a molecule of water from an H^+ and an OH^- ion will be accompanied by a release of heat, which can be detected by touching the reaction container.

$$\text{acid} + \text{base} \longrightarrow \text{salt} + \text{water}$$

$$HCl(aq) + NaOH(aq) \longrightarrow NaCl(aq) + H_2O$$

$$H_2SO_4(aq) + Ba(OH)_2(aq) \longrightarrow BaSO_4\downarrow + 2\ H_2O$$

(b) Formation of an insoluble precipitate   The solubilities of the products can be determined by consulting the Solubility Table in Appendix IV. One or both of the products may be insoluble.

$$BaCl_2(aq) + 2\ AgNO_3(aq) \longrightarrow 2\ AgCl\downarrow + Ba(NO_3)_2(aq)$$

$$FeCl_3(aq) + 3\ NH_4OH(aq) \longrightarrow Fe(OH)_3\downarrow + 3\ NH_4Cl(aq)$$

(c) Metal oxide + acid   Heat is released by the production of a molecule of water.

$$\text{metal oxide} + \text{acid} \longrightarrow \text{salt} + \text{water}$$

$$CuO(s) + 2\ HNO_3(aq) \longrightarrow Cu(NO_3)_2(aq) + H_2O$$

$$CaO(s) + 2\ HCl(aq) \longrightarrow CaCl_2(aq) + H_2O$$

(d) Formation of a gas   A gas such as HCl or H_2S may be produced directly as in these two examples:

$$H_2SO_4(l) + NaCl(s) \longrightarrow NaHSO_4(s) + HCl\uparrow$$

$$2\ HCl(aq) + ZnS(s) \longrightarrow ZnCl_2(aq) + H_2S\uparrow$$

A gas can be produced indirectly. Some unstable compounds formed in a double displacement reaction, such as H_2CO_3, H_2SO_3, and NH_4OH, will decompose to form water and a gas:

$$2\ HCl(aq) + Na_2CO_3(aq) \longrightarrow 2\ NaCl(aq) + H_2CO_3 \longrightarrow 2\ NaCl(aq) + H_2O + CO_2\uparrow$$

$$2\ HNO_3(aq) + K_2SO_3(aq) \longrightarrow 2\ KNO_3(aq) + H_2SO_3(aq) \longrightarrow 2\ KNO_3(aq) + H_2O + SO_2\uparrow$$

$$NH_4Cl(aq) + NaOH(aq) \longrightarrow NaCl(aq) + NH_4OH \longrightarrow NaCl(aq) + H_2O + NH_3\uparrow$$

EXAMPLE 8.8   Write the equation for the reaction between aqueous solutions of hydrobromic acid and potassium hydroxide.

First write the formulas for the reactants. They are HBr and KOH. Then classify the type of reaction that would occur between them. Because the reactants are compounds and one is an acid and the other is a base, the reaction will be of the neutralization type:

$$\text{acid} + \text{base} \longrightarrow \text{salt} + \text{water}$$

Now rewrite the equation by putting down the formulas for the known substances:

$$HBr(aq) + KOH(aq) \longrightarrow salt + H_2O$$

In this reaction, which is a double-displacement type, the H^+ from the acid combines with the OH^- from the base to form water. The salt must be composed of the other two ions, K^+ and Br^-. We determine the formula of the salt to be KBr from the fact that K is a $+1$ cation and Br is a -1 anion. The final balanced equation is

$$HBr(aq) + KOH(aq) \longrightarrow KBr(aq) + H_2O(l)$$

EXAMPLE 8.9 Complete and balance the equation for the reaction between aqueous solutions of barium chloride and sodium sulfate.

First determine the formulas for the reactants. They are $BaCl_2$ and Na_2SO_4. Then classify these substances as acids, bases, or salts. Both substances are salts. Since both substances are compounds, the reaction looks as though it will be of the double-displacement type. Start writing the equation with the reactants:

$$BaCl_2(aq) + Na_2SO_4(aq) \longrightarrow$$

If the reaction is double-displacement, Ba^{2+} will be written combined with SO_4^{2-}, and Na^+ with Cl^- as the products. The balanced equation is

$$BaCl_2(aq) + Na_2SO_4(aq) \longrightarrow BaSO_4 + 2\,NaCl$$

The final step is to determine the nature of the products, which controls whether or not the reaction will take place. If both products are soluble, we may merely have a mixture of all the possible products in solution. But if an insoluble precipitate is formed, the reaction will definitely occur. We know from experience that NaCl is fairly soluble in water, but what about $BaSO_4$? The Solubility Table in Appendix IV can give us this information. From this table we see that $BaSO_4$ is insoluble in water, so it will be a precipitate in the reaction. Thus the reaction will occur, forming a white precipitate. The equation is

$$BaCl_2(aq) + Na_2SO_4(aq) \longrightarrow BaSO_4\downarrow + 2\,NaCl(aq)$$

PRACTICE Complete and balance the equations for the reactions:
(a) potassium phosphate + barium chloride
(b) hydrochloric acid + nickel carbonate
(c) ammonium chloride + sodium nitrate
Answers: (a) $2\,K_3PO_4(aq) + 3\,BaCl_2(aq) \longrightarrow Ba_3(PO_4)_2\downarrow + 2\,KCl(aq)$
(b) $2HCl(aq) + NiCO_3 \longrightarrow NiCl_2(aq) + H_2O + CO_2\uparrow$
(c) $NH_4Cl(aq) + NaNO_3(aq) \longrightarrow$ no reaction

Some reactions we attempt may fail because the substances are not reactive, or the proper conditions for reaction may not be present. For example, mercury(II) oxide does not decompose until it is heated; magnesium does not burn in air or oxygen until the temperature is raised to the point at which it begins to react. When silver is placed in a solution of copper(II) sulfate, no reaction takes place; however, when a strip of copper is placed in a solution of silver nitrate, the single-displacement reaction (3c on p. 178) takes place because copper is a more reactive metal than silver. The successful prediction of the products of a reaction is not always easy. The ability to predict products correctly comes with knowledge and experience. Although you may not be able to predict many reactions at this point, as you continue you will find that reactions can be categorized, and that prediction of the products thereby becomes easier, if not always certain.

We have a great deal yet to learn about which substances react with each other, how they react, and what conditions are necessary to bring about their reaction. It is possible to make accurate predictions concerning the occurrence of proposed reactions. Such predictions require, in addition to appropriate data, a good knowledge of thermodynamics, a subject usually reserved for advanced courses in chemistry and physics. But even without the formal use of thermodynamics your knowledge of such generalities as the four reaction types just cited, the periodic table, atomic structure, oxidation numbers, and so on, can be put to good use in predicting reactions and in writing equations. Indeed, such applications serve to make chemistry an interesting and fascinating study.

8.6   Heat in Chemical Reactions

Energy changes always accompany chemical reactions. One reason why reactions occur is that the products attain a lower, more stable energy state than the reactants. For the products to attain this more stable state, energy must be liberated and given off to the surroundings as heat (or as heat and work). When a solution of a base is neutralized by the addition of an acid, the liberation of heat energy is signaled by an immediate rise in the temperature of the solution. When an automobile engine burns gasoline, heat is certainly liberated; at the same time, part of the liberated energy does the work of moving the automobile.

exothermic reaction

endothermic reaction

Reactions are either exothermic or endothermic. **Exothermic reactions** liberate heat; **endothermic reactions** absorb heat. In an exothermic reaction heat is a product and may be written on the right side of the equation for the reaction. In an endothermic reaction heat can be regarded as a reactant and is written on the left side of the equation. Examples indicating heat in an exothermic and an endothermic reaction follow.

$$H_2(g) + Cl_2(g) \longrightarrow 2\ HCl(g) + 185\ kJ\ (44.2\ kcal)\quad \text{(exothermic)}$$
$$N_2(g) + O_2(g) + 181\ kJ\ (43.2\ kcal) \longrightarrow 2\ NO(g)\quad \text{(endothermic)}$$

heat of reaction

The quantity of heat produced by a reaction is known as the **heat of reaction**. The units used can be kilojoules or kilocalories. Consider the reaction represented by this equation

$$C(s) + O_2(g) \longrightarrow CO_2(g) + 393 \text{ kJ (94.0 kcal)}$$

When the heat liberated is expressed as part of the equation, the substances are expressed in units of moles. Thus, when 1 mol (12.0 g) of C combines with 1 mol (32.0 g) of O_2, 1 mol (44.0 g) of CO_2 is formed and 393 kJ (94.0 kcal) of heat are liberated. In this reaction, as in many others, the heat energy is more useful than the chemical products.

Aside from relatively small amounts of energy from nuclear processes, the sun is the major provider of energy for life on earth. The sun maintains the temperature necessary for life and also supplies light energy for the endothermic photosynthetic reactions carried on by green plants. In photosynthesis carbon dioxide and water are converted to free oxygen and glucose.

$$6\ CO_2 + 6\ H_2O + 2519 \text{ kJ (673 kcal)} \longrightarrow C_6H_{12}O_6 + 6\ O_2$$
$$\text{Glucose}$$

Nearly all of the chemical energy used by living organisms is obtained from glucose or compounds derived from glucose. Modern technology depends for energy on fossil fuels—coal, petroleum, and natural gas. The energy is obtained from the combustion (burning) of these fuels, which are converted to carbon

combustion   dioxide and water. **Combustion** is the term for a chemical reaction in which heat and light are given off.

Fossil fuels constitute a huge energy reservoir. Some coal is about 90% carbon. Since 393 kJ are obtained from the combustion of 1 mol (12.0 g) of carbon, the combustion of a single ton of coal yields about 2.68×10^{10} J (6.40×10^9 cal) of energy. At 4.184 J/g°C, this energy is enough to heat about 21,000 gallons of water from room temperature to the boiling point (20° to 100°C).

Natural gas is primarily methane, CH_4. Petroleum is a mixture of hydrocarbons (compounds of carbon and hydrogen). Liquefied petroleum gas (LPG) is a mixture of propane (C_3H_8) and butane (C_4H_{10}).

The combustion of these fuels releases a tremendous amount of energy, but reactions won't occur to a significant extent at ordinary temperatures. A spark or a flame must be present before methane will ignite. The amount of energy that

activation   must be supplied to start a chemical reaction is called the **activation energy**. Once
energy   the activation energy has been provided, enough energy is generated to keep the reaction going.

$$CH_4(g) + 2\ O_2(g) \longrightarrow CO_2(g) + 2\ H_2O(g) + 890 \text{ kJ (213 kcal)}$$
$$C_3H_8(g) + 5\ O_2(g) \longrightarrow 3\ CO_2(g) + 4\ H_2O(g) + 2200 \text{ kJ (526 kcal)}$$
$$2\ C_8H_{18}(l) + 25\ O_2(g) \longrightarrow 16\ CO_2(g) + 18\ H_2O(g) + 10{,}900 \text{ kJ (2606 kcal)}$$

Be careful not to confuse an exothermic reaction that merely requires heat (activation energy) to get it started with a truly endothermic process. The

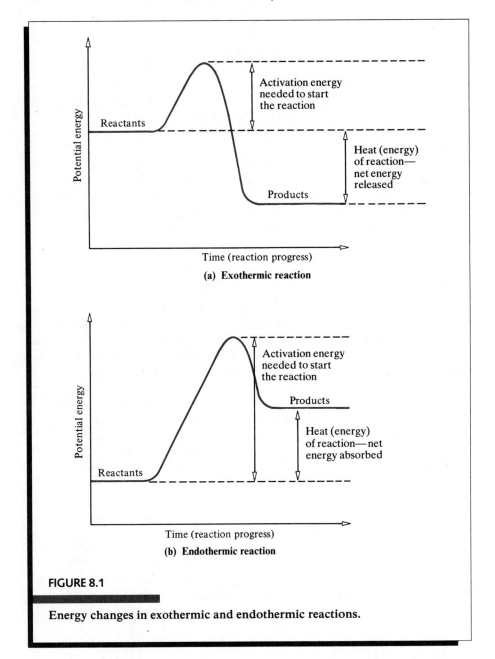

(a) **Exothermic reaction**

(b) **Endothermic reaction**

FIGURE 8.1

Energy changes in exothermic and endothermic reactions.

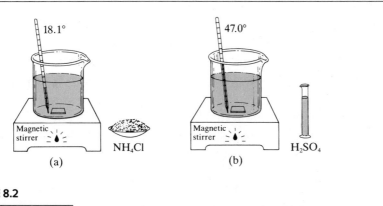

FIGURE 8.2

In the endothermic process (a) dissolving 10 g of NH_4Cl in 100 mL of H_2O caused the temperature of the solution to decrease from 24.5°C to 18.1°C. In the exothermic process (b) dissolving 10 mL of concentrated H_2SO_4 in 100 mL of H_2O caused the temperature of the solution to increase from 24.8°C to 47.0°C.

combustion of magnesium is highly exothermic, yet magnesium must be heated to a fairly high temperature in air before combustion begins. Once started, however, the combustion reaction goes very vigorously until either the magnesium or the available supply of oxygen is exhausted. The electrolytic decomposition of water to hydrogen and oxygen is highly endothermic. If the electric current is shut off when this process is going on, the reaction stops instantly. The relative energy levels of reactants and products in exothermic and in endothermic processes are presented graphically in Figure 8.1.

In reaction (a) of Figure 8.1, the products are at a lower energy level than the reactants. Energy (heat) is given off, and the reaction is exothermic. In reaction (b) the products are at a higher energy level than the reactants. Energy has therefore been absorbed, and the reaction is endothermic.

Examples of endothermic and exothermic processes that can be demonstrated easily in the laboratory are shown in Figure 8.2. When dissolving ammonium chloride, NH_4Cl, in water, we observe an endothermic process. The temperature changes from 24.5°C to 18.1°C when 10 g of NH_4Cl are added to 100 mL of water. This energy, in the form of heat, is taken from the immediate surroundings, the water, causing the salt solution to become colder. In the second example, we observe a temperature change from 24.8°C to 47.0°C when 10 mL of concentrated sulfuric acid, H_2SO_4, are dissolved in 100 mL of water. This reaction is an exothermic process. In both examples the temperature changes are large enough to be detected by touching the containers.

Concepts in Review

1. Know the format used in setting up chemical equations.

2. Recognize the various symbols commonly used in writing chemical equations.

3. Balance simple chemical equations.

4. Interpret a balanced equation in terms of the relative numbers or amounts of molecules, atoms, grams, or moles of each substance represented.

5. Classify equations as combination, decomposition, single-displacement, or double-displacement reactions.

6. Use the activity series to predict whether a single-displacement reaction will occur.

7. Complete and balance equations for simple combination, decomposition, single-displacement, and double-displacement reactions when given the reactants.

8. Distinguish between exothermic and endothermic reactions, and relate the quantity of heat to the amounts of substances involved in the reaction.

9. Identify the major sources of chemical energy and their uses.

Key Terms in Review

The terms listed here have all been defined within the chapter. Review the definitions of each, and use the glossary and margin notations within the chapter as study aids.

activation energy
balanced equation
chemical equation
combination or synthesis reaction
combustion
decomposition reaction
double-displacement, or
    metathesis, reaction

endothermic reaction
exothermic reaction
heat of reaction
single-displacement reaction
word equation

Exercises

1. Balance the following equations:
   (a) $H_2 + O_2 \longrightarrow H_2O$
   (b) $H_2 + Br_2 \longrightarrow HBr$
   (c) $C + Fe_2O_3 \longrightarrow Fe + CO$
   (d) $H_2O_2 \longrightarrow H_2O + O_2$
   (e) $Ba(ClO_3)_2 \xrightarrow{\Delta} BaCl_2 + O_2$
   (f) $H_2SO_4 + NaOH \longrightarrow H_2O + Na_2SO_4$
   (g) $NH_4I + Cl_2 \longrightarrow NH_4Cl + I_2$
   (h) $CrCl_3 + AgNO_3 \longrightarrow Cr(NO_3)_3 + AgCl$
   (i) $Al_2(CO_3)_3 \xrightarrow{\Delta} Al_2O_3 + CO_2$
   (j) $Al + C \xrightarrow{\Delta} Al_4C_3$

2. Classify the reactions in Exercise 1. as combination, decomposition, single displacement, or double displacement.

3. Balance the following equations:
   (a) $SO_2 + O_2 \longrightarrow SO_3$
   (b) $Al + MnO_2 \xrightarrow{\Delta} Mn + Al_2O_3$
   (c) $Na + H_2O \longrightarrow NaOH + H_2$
   (d) $AgNO_3 + Ni \longrightarrow Ni(NO_3)_2 + Ag$
   (e) $Bi_2S_3 + HCl \longrightarrow BiCl_3 + H_2S$
   (f) $PbO_2 \xrightarrow{\Delta} PbO + O_2$
   (g) $LiAlH_4 \xrightarrow{\Delta} LiH + Al + H_2$
   (h) $KI + Br_2 \longrightarrow KBr + I_2$
   (i) $K_3PO_4 + BaCl_2 \longrightarrow KCl + Ba_3(PO_4)_2$

4. Balance the following equations.
   (a) $MnO_2 + CO \longrightarrow Mn_2O_3 + CO_2$
   (b) $Mg_3N_2 + H_2O \longrightarrow Mg(OH)_2 + NH_3$
   (c) $C_3H_5(NO_3)_3 \longrightarrow CO_2 + H_2O + N_2 + O_2$
   (d) $FeS + O_2 \longrightarrow Fe_2O_3 + SO_2$
   (e) $Cu(NO_3)_2 \longrightarrow CuO + NO_2 + O_2$
   (f) $NO_2 + H_2O \longrightarrow HNO_3 + NO$
   (g) $Al + H_2SO_4 \longrightarrow Al_2(SO_4)_3 + H_2$
   (h) $HCN + O_2 \longrightarrow N_2 + CO_2 + H_2O$
   (i) $B_5H_9 + O_2 \longrightarrow B_2O_3 + H_2O$
   (j) $NH_3 + O_2 \longrightarrow NO + H_2O$

5. Change the following word equations into formula equations and balance them:
   (a) Copper + Sulfur $\xrightarrow{\Delta}$ Copper(I) sulfide
   (b) Phosphoric acid + Calcium hydroxide $\longrightarrow$ Calcium phosphate + Water
   (c) Silver oxide $\xrightarrow{\Delta}$ Silver + Oxygen
   (d) Iron(III) chloride + Sodium hydroxide $\longrightarrow$ Iron(III) hydroxide + Sodium chloride
   (e) Nickel(II) phosphate + Sulfuric acid $\longrightarrow$ Nickel(II) sulfate + Phosphoric acid
   (f) Zinc carbonate + Hydrochloric acid $\longrightarrow$ Zinc chloride + Water + Carbon dioxide
   (g) Silver nitrate + Aluminum chloride $\longrightarrow$ Silver chloride + Aluminum nitrate

6. Change the following word equations into formula equations and balance them:
   (a) Water $\longrightarrow$ Hydrogen + Oxygen
   (b) Acetic acid + Potassium hydroxide $\longrightarrow$ Potassium acetate + Water
   (c) Phosphorus + Iodine $\longrightarrow$ Phosphorus triiodide
   (d) Aluminum + Copper(II) sulfate $\longrightarrow$ Copper + Aluminum sulfate
   (e) Ammonium sulfate + Barium chloride $\longrightarrow$ Ammonium chloride + Barium sulfate
   (f) Sulfur tetrafluoride + Water $\longrightarrow$ Sulfur dioxide + Hydrogen fluoride
   (g) Chromium(III) carbonate $\xrightarrow{\Delta}$ Chromium(III) oxide + Carbon dioxide

7. Complete and balance the equations for these combination reactions:
   (a) $K + O_2 \longrightarrow$ (c) $CO_2 + H_2O \longrightarrow$
   (b) $Al + Cl_2 \longrightarrow$ (d) $CaO + H_2O \longrightarrow$

8. Complete and balance the equations for these decomposition reactions:
   (a) $HgO \xrightarrow{\Delta}$ (c) $MgCO_3 \xrightarrow{\Delta}$
   (b) $NaClO_3 \xrightarrow{\Delta}$ (d) $PbO_2 \xrightarrow{\Delta} PbO +$

9. Complete and balance the equations for these single-displacement reactions:
   (a) $Zn + H_2SO_4 \longrightarrow$
   (b) $AlI_3 + Cl_2 \longrightarrow$
   (c) $Mg + AgNO_3 \longrightarrow$
   (d) $Al + CoSO_4 \longrightarrow$

10. Use the activity series to predict which of the following reactions will occur. Complete and balance the equations. Where no reaction will occur, write "no reaction" as the product.
    (a) $Ag + H_2SO_4(aq) \longrightarrow$
    (b) $Cl_2 + NaBr(aq) \longrightarrow$
    (c) $Mg + ZnCl_2(aq) \longrightarrow$
    (d) $Pb + AgNO_3(aq) \longrightarrow$
    (e) $Cu + FeCl_3(aq) \longrightarrow$

(f) $H_2 + Al_2O_3(s) \xrightarrow{\Delta}$

(g) $Al + HBr(aq) \longrightarrow$

(h) $I_2 + HCl(aq) \longrightarrow$

11. Complete and balance the equations for these double-displacement reactions:

(a) $ZnCl_2 + KOH \longrightarrow$

(b) $CuSO_4 + H_2S \longrightarrow$

(c) $Ca(OH)_2 + H_3PO_4 \longrightarrow$

(d) $(NH_4)_3PO_4 + Ni(NO_3)_2 \longrightarrow$

(e) $Ba(OH)_2 + HNO_3 \longrightarrow$

(f) $(NH_4)_2S + HCl \longrightarrow$

12. Predict which of the following double-displacement reactions will occur. Complete and balance the equations. Where no reaction will occur, write "no reaction" as the product.

(a) $AgNO_3(aq) + KCl(aq) \longrightarrow$

(b) $Ba(NO_3)_2(aq) + MgSO_4(aq) \longrightarrow$

(c) $H_2SO_4(aq) + Mg(OH)_2(aq) \longrightarrow$

(d) $MgO(s) + H_2SO_4(aq) \longrightarrow$

(e) $Na_2CO_3(aq) + NH_4Cl(aq) \longrightarrow$

13. Complete and balance the equations for these reactions. All reactions yield products:

(a) $H_2 + I_2 \longrightarrow$

(b) $CaCO_3 \xrightarrow{\Delta}$

(c) $Mg + H_2SO_4 \longrightarrow$

(d) $FeCl_2 + NaOH \longrightarrow$

(e) $SO_2 + H_2O \longrightarrow$

(f) $SO_3 + H_2O \longrightarrow$

(g) $Ca + H_2O \longrightarrow$

(h) $Bi(NO_3)_3 + H_2S \longrightarrow$

14. Complete and balance the equations for the following reactions. All reactions yield products:

(a) $Ba + O_2 \longrightarrow$

(b) $NaHCO_3 \xrightarrow{\Delta} Na_2CO_3 +$

(c) $Ni + CuSO_4 \longrightarrow$

(d) $MgO + HCl \longrightarrow$

(e) $H_3PO_4 + KOH \longrightarrow$

(f) $C + O_2 \longrightarrow$

(g) $Al(ClO_3)_3 \xrightarrow{\Delta} O_2 +$

(h) $CuBr_2 + Cl_2 \longrightarrow$

(i) $SbCl_3 + (NH_4)_2S \longrightarrow$

(j) $NaNO_3 \xrightarrow{\Delta} NaNO_2 +$

15. What is the purpose of balancing equations?

16. What is represented by the numbers (coefficients) that are placed in front of the formulas in a balanced equation?

17. Interpret the following chemical reactions in terms of the number of moles of each reactant and product:

(a) $MgBr_2 + 2\,AgNO_3 \longrightarrow$
$$Mg(NO_3)_2 + 2\,AgBr$$

(b) $N_2 + 3\,H_2 \longrightarrow 2\,NH_3$

(c) $2\,C_3H_7OH + 9\,O_2 \longrightarrow 6\,CO_2 + 8\,H_2O$

18. Interpret each of the following equations in terms of the relative number of moles of each substance involved and indicate whether the reaction is exothermic or endothermic:

(a) $2\,Na + Cl_2 \longrightarrow$
$$2\,NaCl + 822\ kJ\ (196.4\ kcal)$$

(b) $PCl_5 + 92.9\ kJ\ (22.2\ kcal) \longrightarrow PCl_3 + Cl_2$

19. Write balanced equations for each of these reactions, including the heat term:

(a) Lime, CaO, is converted to slaked lime, $Ca(OH)_2$, by reaction with water. The reaction liberates 65.3 kJ (15.6 kcal) of heat for each mole of lime reacted.

(b) The industrial production of aluminum metal from aluminum oxide is an endothermic electrolytic process requiring 1630 kJ per mole of Al_2O_3. Oxygen is also a product.

20. Write balanced equations for the combustion of the following hydrocarbons.

(a) ethane, C_2H_6 (c) heptane, C_7H_{16}

(b) benzene, C_6H_6

21. Which of the following statements are correct? Rewrite the incorrect ones to make them correct.

(a) The coefficients in front of the formulas in a balanced chemical equation give the relative number of moles of the reactants and products in the reaction.

(b) A balanced chemical equation is one that has the same number of moles on each side of the equation.

(c) In a chemical equation, the symbol $\xrightarrow{\Delta}$ indicates that the reaction is exothermic.

(d) A chemical change that absorbs heat energy is said to be endothermic.

(e) In the reaction $H_2 + Cl_2 \longrightarrow 2\,HCl$, 100 molecules of HCl are produced for every 50 molecules of H_2 reacted.

(f) The symbol (aq) after a substance in an equation means that the substance is in a water solution.

(g) The equation $H_2O \longrightarrow H_2 + O_2$ can be balanced by placing a 2 in front of H_2O.

(h) In the equation $3 H_2 + N_2 \longrightarrow 2 NH_3$ there are fewer moles of product than there are moles of reactants.

(i) The total number of moles of reactants and products represented by this equation is 5 moles:
$$Mg + 2 HCl \longrightarrow MgCl_2 + H_2$$

(j) One mole of glucose, $C_6H_{12}O_6$, contains 6 moles of carbon atoms.

(k) The reactants are the substances produced by the chemical reaction.

(l) In a balanced equation each side of the equation contains the same number of atoms of each element.

(m) When a precipitate is formed in a chemical reaction, it can be indicated in the equation with the symbol ↓ or (s) immediately before the formula of the substance precipitated.

(n) When a gas is evolved in a chemical reaction, it can be indicated in the equation with the symbol ↑ or (g) immediately following the formula of the gas.

(o) According to the equation $3 H_2 + N_2 \longrightarrow 2 NH_3$, 4 mol of NH_3 will be formed when 6 mol of H_2 and 2 mol of N_2 react.

(p) The products of an exothermic reaction are at a lower energy state than the reactants.

(q) The combustion of hydrocarbons produces carbon dioxide and water as products.

CHAPTER NINE

Calculations from Chemical Equations

Often, the old adage "waste not, want not" is appropriate in our daily life. For example, a hostess determines the quantity of food and beverage necessary to serve all her guests. These amounts are defined by specific recipes and a knowledge of the particular likes and dislikes of the guests. A seamstress determines the amount of material, lining, and trim necessary to produce a gown for her client by relying on a pattern or her own experience to guide the selection. A carpet layer determines the correct amount of carpet and padding necessary to recarpet a customer's house by calculating the floor area. The IRS determines the correct deduction for federal income taxes from your paycheck based on your expected annual income.

The chemist also finds it necessary to calculate amounts of products or reactants using a balanced chemical equation. With these calculations, the chemist can control the amount of product by scaling the reaction up or down to fit the needs of the laboratory, and thereby minimize waste or excess materials formed during the reaction.

Chapter Preview

9.1   A Short Review

Molar Mass   The molar mass is the sum of the atomic masses of all the atoms in a molecule. The molar mass also applies to the mass of any formula unit—atoms, molecules, or ions; it is the atomic mass of an atom, or the sum of the atomic masses in a molecule or an ion.

Relationship Between Molecule and Mole   A molecule is the smallest unit of a molecular substance (for example, Cl_2), and a mole is Avogadro's number, 6.022×10^{23}, of molecules of that substance. A mole of chlorine (Cl_2) has the same number of molecules as a mole of carbon dioxide, a mole of water, or a mole of any other molecular substance. When we relate molecules to molar mass 1 molar mass = 1 mol, or 6.022×10^{23} molecules.

In addition to referring to molecular substances, the term *mole* may refer to any chemical species. It represents a quantity in grams equal to the molar mass and may be applied to atoms, ions, electrons, and formula units of nonmolecular substances.

$$1 \text{ mole} = \begin{cases} 6.022 \times 10^{23} \text{ molecules} \\ 6.022 \times 10^{23} \text{ formula units} \\ 6.022 \times 10^{23} \text{ atoms} \\ 6.022 \times 10^{23} \text{ ions} \end{cases}$$

Other useful mole relationships are

$$\text{number of moles} = \frac{\text{grams of a substance}}{\text{molar mass of the substance}}$$

$$\text{number of moles} = \frac{\text{grams of a monatomic element}}{\text{molar mass of the element}}$$

$$\text{number of moles} = \frac{\text{number of molecules}}{6.022 \times 10^{23} \text{ molecules/mole}}$$

Two other useful equalities can be derived algebraically from each of these mole relationships. What are they?

Balanced Equations   When using chemical equations for calculations of mole–mass–volume relationships between reactants and products, the equations must be balanced. Remember that the number in front of a formula in a balanced chemical equation can represent the number of moles of that substance in the chemical reaction.

9.2   Introduction to Stoichiometry: The Mole-Ratio Method

It is often necessary to calculate the amount of a substance that is produced from, or needed to react with, a given quantity of another substance. The area of chemistry that deals with the quantitative relationships among reactants and products is known as **stoichiometry** (*stoy-key-ah-meh-tree*). Although several methods are known, we firmly believe that the *mole* or *mole-ratio* method is generally best for solving problems in stoichiometry. This method is straightforward and, in our opinion, makes it easy to see and understand the relationships of the reacting species.

stoichiometry

mole ratio

A **mole ratio** is a ratio between the number of moles of any two species involved in a chemical reaction. For example, in the reaction

$$2\,H_2 \;+\; O_2 \;\longrightarrow\; 2\,H_2O$$
$$\text{2 mol} \qquad \text{1 mol} \qquad \text{2 mol}$$

six mole ratios apply only to this reaction:

$$\frac{2\text{ mol }H_2}{1\text{ mol }O_2} \qquad \frac{2\text{ mol }H_2}{2\text{ mol }H_2O} \qquad \frac{1\text{ mol }O_2}{2\text{ mol }H_2}$$

$$\frac{1\text{ mol }O_2}{2\text{ mol }H_2O} \qquad \frac{2\text{ mol }H_2O}{2\text{ mol }H_2} \qquad \frac{2\text{ mol }H_2O}{1\text{ mol }O_2}$$

The mole ratio is a conversion factor used to convert the number of moles of one substance to the corresponding number of moles of another substance in a chemical reaction. For example, if we want to calculate the number of moles of H_2O that can be obtained from 4.0 mol of O_2, we use the mole ratio $2\text{ mol }H_2O/1\text{ mol }O_2$.

$$4.0 \;\cancel{\text{mol }O_2} \times \frac{2\text{ mol }H_2O}{1\;\cancel{\text{mol }O_2}} = 8.0\text{ mol }H_2O$$

Since stoichiometric problems are encountered throughout the entire field of chemistry, it is profitable to master this general method for their solution. The mole-ratio method makes use of three simple basic operations.

1. Convert the quantity of starting substance to moles (if it is not given in moles).
2. Convert the moles of starting substance to moles of desired substance.
3. Convert the moles of desired substance to the units specified in the problem.

Like learning to balance chemical equations, learning to make stoichiometric calculations require practice. A detailed step-by-step description of the general method, together with a variety of worked examples, is given in the following paragraphs. Study this material and apply the method to the problems at the end of this chapter.

Step 1 Use a balanced equation. Write a balanced equation for the chemical reaction in question or check to see that the equation given is balanced.

Step 2 Determine the number of moles of starting substance. Identify the starting substance from the data given in the statement of the problem. When the starting substance is given in moles, use it in that form. If it is not in moles, convert the quantity of the starting substance to moles. For example, if grams of a starting substance are given,

$$\left(\begin{array}{c}\text{moles of starting} \\ \text{substance}\end{array}\right) = \left(\begin{array}{c}\text{grams of starting} \\ \text{substance}\end{array}\right) \times \dfrac{\left(\begin{array}{c}1 \text{ mol of} \\ \text{starting substance}\end{array}\right)}{\left(\begin{array}{c}\text{molar mass of} \\ \text{starting substance}\end{array}\right)}$$

Step 3 Determine the mole ratio of the desired substance to the starting substance. The number of moles of each substance in the balanced equation is indicated by the coefficient in front of each substance. Use these coefficients to set up the mole ratio:

$$\text{mole ratio} = \dfrac{\text{moles of desired substance in the equation}}{\text{moles of starting substance in the equation}}$$

Step 4 Calculate the number of moles of the desired substance. Multiply the number of moles of starting substance (from Step 2) by the mole ratio (from Step 3) to obtain the number of moles of desired substance:

$$\left(\begin{array}{c}\text{moles of desired} \\ \text{substance}\end{array}\right) = \underset{\text{From Step 2}}{\left(\begin{array}{c}\text{moles of starting} \\ \text{substance}\end{array}\right)} \times \underset{\text{Mole ratio from Step 3}}{\dfrac{\left(\begin{array}{c}\text{moles of desired substance} \\ \text{in the equation}\end{array}\right)}{\left(\begin{array}{c}\text{moles of starting substance} \\ \text{in the equation}\end{array}\right)}}$$

Note that the units of moles of starting substance cancel out in the numerator and the denominator.

Step 5   Calculate the desired substance in the units specified in the problem. If the answer is to be in moles, the problem is finished in Step 4. If units other than moles are wanted, multiply the moles of the desired substance (from Step 4) by the appropriate factor to convert moles to the units required.

For example, if grams of the desired substance are wanted,

$$\begin{pmatrix} \text{grams of desired} \\ \text{substance} \end{pmatrix} = \underset{\text{From Step 4}}{\begin{pmatrix} \text{moles of desired} \\ \text{substance} \end{pmatrix}} \times \frac{\begin{pmatrix} \text{molar mass of} \\ \text{desired substance} \end{pmatrix}}{\begin{pmatrix} \text{one mole of} \\ \text{desired substance} \end{pmatrix}}$$

Use the conversion factors 6.022×10^{23} atoms/mol or 6.022×10^{23} molecules/mol when the problem asks for the answer in atoms or molecules.

The steps for converting the mass of starting substance A to either mass, atoms, or molecules of desired substance B are summarized below:

$$\text{grams of A} \xrightarrow{\text{Step 2}} \text{moles of A} \xrightarrow{\text{Steps 3 and 4}} \text{moles B} \begin{array}{c} \nearrow \text{grams of B} \\ \underset{\text{Step 5}}{} \\ \searrow \text{atoms or molecules of B} \end{array}$$

9.3   Mole-Mole Calculations

The first application of the mole-ratio method of solving stoichiometric problems is that of mole-mole calculations.

The quantity of starting substance is given in moles and the quantity of desired substance is requested in moles. Steps 2 and 5 of the mole-ratio method (Section 9.2) are not required. Understanding the use of mole ratios will be very helpful in solving later problems. Some illustrative problems follow.

EXAMPLE 9.1   How many moles of carbon dioxide will be produced by the complete oxidation of 2.0 mol of glucose ($C_6H_{12}O_6$), according to the following reaction?

$$C_6H_{12}O_6 + 6\,O_2 \longrightarrow 6\,CO_2 + 6\,H_2O$$
$$1\text{ mol}6\text{ mol}6\text{ mol}6\text{ mol}$$

The balanced equation states that 6 mol of CO_2 will be produced from 1 mol of $C_6H_{12}O_6$. Even though we can readily see that 12 mol of CO_2 will be formed from 2.0 mol of $C_6H_{12}O_6$, the mole-ratio method of solving the problem is shown here.

Step 1 The equation given is balanced.
Step 2 The number of moles of starting substance is 2.0 mol $C_6H_{12}O_6$.
Step 3 The conversion needed is

$$\text{moles } C_6H_{12}O_6 \longrightarrow \text{moles } CO_2$$

Step 4 From the balanced equation, set up the mole ratio between the two substances in question, placing the moles of the substance being sought in the numerator and the moles of the starting substance in the denominator. The number of moles, in each case, is the same as the coefficient in front of the substance in the balanced equation.

$$\text{mole ratio} = \frac{6 \text{ mol } CO_2}{1 \text{ mol } C_6H_{12}O_6} \quad \text{(from equation)}$$

Step 5 Multiply 2.0 mol of glucose (given in the problem) by this mole ratio.

$$2.0 \text{ mol } C_6H_{12}O_6 \times \frac{6 \text{ mol } CO_2}{1 \text{ mol } C_6H_{12}O_6} = 12 \text{ mol } CO_2 \quad \text{(Answer)}$$

Again note the use of units. The moles of $C_6H_{12}O_6$ cancel, leaving the answer in units of moles of CO_2.

EXAMPLE 9.2

How many moles of ammonia can be produced from 8.00 mol of hydrogen reacting with nitrogen?

Step 1 First we need the balanced equation

$$3 \text{ H}_2 + \text{N}_2 \longrightarrow 2 \text{ NH}_3$$

Step 2 The starting substance is 8.00 mol of hydrogen.
Step 3 The conversion needed is

$$\text{moles } H_2 \longrightarrow \text{moles } NH_3$$

The balanced equation states that we get 2 mol of NH_3 for every 3 mol of H_2 that react. Set up the mole ratio of desired substance (NH_3) to starting substance (H_2):

$$\text{mole ratio} = \frac{2 \text{ mol } NH_3}{3 \text{ mol } H_2} \quad \text{(from equation)}$$

Step 4 Multiplying the 8.00 mol of starting H_2 by this mole ratio, we get

$$8.00 \text{ mol } H_2 \times \frac{2 \text{ mol } NH_3}{3 \text{ mol } H_2} = 5.33 \text{ mol } NH_3 \quad \text{(Answer)}$$

EXAMPLE 9.3   Given the balanced equation

$$K_2Cr_2O_7 + 6\,KI + 7\,H_2SO_4 \longrightarrow Cr_2(SO_4)_3 + 4\,K_2SO_4 + 3\,I_2 + 7\,H_2O$$

1 mol   6 mol   3 mol

calculate (a) the number of moles of potassium dichromate ($K_2Cr_2O_7$) that will react with 2.0 mol of potassium iodide (KI); (b) the number of moles of iodine (I_2) that will be produced from 2.0 mol of potassium iodide.

After the equation is balanced, we are concerned only with $K_2Cr_2O_7$, KI, and I_2, and we can ignore all the other substances. The equation states that 1 mol of $K_2Cr_2O_7$ will react with 6 mol of KI to produce 3 mol of I_2.

(a) Calculate the number of moles of $K_2Cr_2O_7$.

    Step 1 The equation given is balanced.
    Step 2 The starting substance is 2.0 mol of KI.
    Step 3 The conversion needed is

$$\text{moles KI} \longrightarrow \text{moles } K_2Cr_2O_7$$

    Set up the mole ratio of desired substance to starting substance:

$$\text{mole ratio} = \frac{1 \text{ mol } K_2Cr_2O_7}{6 \text{ mol KI}} \quad \text{(from equation)}$$

    Step 4 Multiply the moles of starting material by this ratio to obtain the answer.

$$2.0 \text{ mol KI} \times \frac{1 \text{ mol } K_2Cr_2O_7}{6 \text{ mol KI}} = 0.33 \text{ mol } K_2Cr_2O_7 \quad \text{(Answer)}$$

(b) Calculate the number of moles of I_2.

    Steps 1 and 2 The equation given is balanced and the moles of starting substance are 2.0 mol KI as in part (a).
    Step 3 The conversion needed is

$$\text{moles KI} \longrightarrow \text{moles } I_2$$

    Set up the mole ratio of desired substance to starting substance:

$$\text{mole ratio} = \frac{3 \text{ mol } I_2}{6 \text{ mol KI}} \quad \text{(from equation)}$$

    Step 4 Multiply the moles of starting material by this ratio to obtain the answer.

$$2.0 \text{ mol KI} \times \frac{3 \text{ mol } I_2}{6 \text{ mol KI}} = 1.0 \text{ mol } I_2 \quad \text{(Answer)}$$

EXAMPLE 9.4   How many molecules of water can be produced by reacting 0.010 mol of oxygen with hydrogen?

The sequence of conversions needed in the calculation is

$$\text{moles } O_2 \longrightarrow \text{moles } H_2O \longrightarrow \text{molecules } H_2O$$

Step 1 First we write the balanced equation:

$$2 \, H_2 + O_2 \longrightarrow 2 \, H_2O$$

$$\underset{1 \text{ mol}}{} \qquad \underset{2 \text{ mol}}{}$$

Step 2 The starting substance is 0.010 mol O_2.

Step 3 The conversion needed is moles $O_2 \longrightarrow$ moles H_2O. Set up the mole ratio of desired substance to starting substance:

$$\text{mole ratio} = \frac{2 \text{ mol } H_2O}{1 \text{ mol } O_2} \quad \text{(from equation)}$$

Step 4 Multiplying the 0.010 mol of oxygen by this ratio, we obtain

$$0.010 \text{ mol } O_2 \times \frac{2 \text{ mol } H_2O}{1 \text{ mol } O_2} = 0.020 \text{ mol } H_2O$$

Step 5 Since the problem asks for molecules instead of moles of H_2O, we must convert moles to molecules. Use the conversion factor $(6.022 \times 10^{23} \text{ molecules})$/mole.

$$0.020 \text{ mol } H_2O \times \frac{6.022 \times 10^{23} \text{ molecules}}{1 \text{ mol}} = 1.2 \times 10^{22} \text{ molecules } H_2O \quad \text{(Answer)}$$

Note that 0.020 mol is still quite a large number of water molecules.

PRACTICE How many moles of aluminum oxide will be produced from 0.50 mol of oxygen?

$$4 \, Al + 3 \, O_2 \longrightarrow 2 \, Al_2O_3$$

Answer: 0.33 mol Al_2O_3

PRACTICE How many moles of aluminum hydroxide are required to produce 22.0 mol of water?

$$2 \, Al(OH)_3 + 3 \, H_2SO_4 \longrightarrow Al_2(SO_4)_3 + 6 \, H_2O$$

Answer: 7.33 mol $Al(OH)_3$

9.4 Mole-Mass Calculations

The object of this type of problem is to calculate the mass of one substance that reacts with or is produced from a given number of moles of another substance in a chemical reaction. If the mass of the starting substance is given, it is necessary to convert it to moles. The mole ratio is used to convert from moles of starting substance to moles of desired substance. Moles of desired substance can then be changed to mass if required.

As you are now familiar with the mole-ratio method, Steps 3 and 4 (Section 9.2) are combined in the examples. Each example is solved in a continuous calculation sequence, after the step-by-step method is demonstrated.

EXAMPLE 9.5

What mass of hydrogen can be produced by reacting 6.0 mol of aluminum with hydrochloric acid?

First calculate the moles of hydrogen produced, using the mole-ratio method, and then calculate the mass of hydrogen by multiplying the moles of hydrogen by its grams per mole. The sequence of conversions in the calculation is

$$\text{moles Al} \longrightarrow \text{moles } H_2 \longrightarrow \text{grams } H_2$$

Step 1 The balanced equation is

$$2 \text{ Al}(s) + 6 \text{ HCl}(aq) \longrightarrow 2 \text{ AlCl}_3(aq) + 3 \text{ H}_2(g)$$

    2 mol   3 mol

Step 2 The starting substance is 6.0 mol of aluminum.
Steps 3 and 4 Calculate moles of H_2 by the mole-ratio method.

$$6.0 \text{ mol Al} \times \frac{3 \text{ mol } H_2}{2 \text{ mol Al}} = 9.0 \text{ mol } H_2$$

Step 5 Convert moles of H_2 to grams [g = mol × (g/mol)]:

$$9.0 \text{ mol } H_2 \times \frac{2.0 \text{ g } H_2}{1 \text{ mol } H_2} = 18 \text{ g } H_2 \quad \text{(Answer)}$$

We see that 18 g of H_2 can be produced by reacting 6.0 mol of Al with HCl. The following setup combines all the above steps into one continuous calculation:

$$6.0 \text{ mol Al} \times \frac{3 \text{ mol } H_2}{2 \text{ mol Al}} \times \frac{2.0 \text{ g } H_2}{1 \text{ mol } H_2} = 18 \text{ g } H_2$$

EXAMPLE 9.6

How many moles of water can be produced by burning 325 g of octane (C_8H_{18})? The sequence of conversions in the calculation is

$$\text{grams } C_8H_{18} \longrightarrow \text{moles } C_8H_{18} \longrightarrow \text{moles } H_2O$$

Step 1 The balanced equation is

$$2 \text{ C}_8H_{18}(g) + 25 \text{ O}_2(g) \longrightarrow 16 \text{ CO}_2(g) + 18 \text{ H}_2O(g)$$

    2 mol   18 mol

Step 2 The starting substance is 325 g C_8H_{18}. Convert 325 g of C_8H_{18} to moles:

$$325 \text{ g } C_8H_{18} \times \frac{1 \text{ mol } C_8H_{18}}{114.0 \text{ g } C_8H_{18}} \longrightarrow 2.85 \text{ mol } C_8H_{18}$$

Steps 3 and 4 Calculate the moles of water by the mole-ratio method:

$$2.85 \text{ mol } C_8H_{18} \times \frac{18 \text{ mol } H_2O}{2 \text{ mol } C_8H_{18}} = 25.7 \text{ mol } H_2O \quad \text{(Answer)}$$

The calculation in a continuous sequence is

$$325 \text{ g } C_8H_{18} \times \frac{1 \text{ mol } C_8H_{18}}{114.0 \text{ g } C_8H_{18}} \times \frac{18 \text{ mol } H_2O}{2 \text{ mol } C_8H_{18}} = 25.7 \text{ mol } H_2O$$

PRACTICE How many moles of potassium chloride can be produced from 100.0 g of potassium chlorate?

$$2 \text{ KClO}_3 \longrightarrow 2 \text{ KCl} + 3 \text{ O}_2$$

Answer: 0.8157 mol KCl

PRACTICE How many grams of silver nitrate are required to produce 0.25 mol of silver sulfide?

$$2 \text{ AgNO}_3 + H_2S \longrightarrow Ag_2S + 2 \text{ HNO}_3$$

Answer: 85 g $AgNO_3$

9.5 Mass-Mass Calculations

Solving mass-mass stoichiometry problems requires all steps of the mole-ratio method. The mass of starting substance is converted to moles. The mole ratio is then used to determine moles of desired substance which in turn is converted to mass.

Steps 3 and 4 (Section 9.2) are combined in the examples. At the end of each example the problem is solved in a continuous calculation sequence.

EXAMPLE 9.7 What mass of carbon dioxide is produced by the complete combustion of 100. g of the hydrocarbon pentane, C_5H_{12}?

The sequence of conversions in the calculation is

$$\text{grams } C_5H_{12} \longrightarrow \text{moles } C_5H_{12} \longrightarrow \text{moles } CO_2 \longrightarrow \text{grams } CO_2$$

Step 1 The balanced equation is

$$C_5H_{12} + 8\,O_2 \longrightarrow 5\,CO_2 + 6\,H_2O$$

$$\quad\text{1 mol} \qquad\qquad\qquad \text{5 mol}$$

Step 2 The starting substance is 100. g of C_5H_{12}. Convert 100. g of C_5H_{12} to moles:

$$100.\ \cancel{\text{g } C_5H_{12}} \times \frac{1 \text{ mol } C_5H_{12}}{72.0 \ \cancel{\text{g } C_5H_{12}}} = 1.39 \text{ mol } C_5H_{12}$$

Steps 3 and 4 Calculate the moles of CO_2 by the mole-ratio method:

$$1.39\ \cancel{\text{mol } C_5H_{12}} \times \frac{5 \text{ moles } CO_2}{1 \ \cancel{\text{mol } C_5H_{12}}} = 6.95 \text{ mol } CO_2$$

Step 5 Convert moles of CO_2 to grams:

$$\text{mol } CO_2 \times \frac{\text{molar mass } CO_2}{1 \text{ mol } CO_2} = \text{grams } CO_2$$

$$6.95 \ \cancel{\text{mol } CO_2} \times \frac{44.0 \text{ g } CO_2}{1 \ \cancel{\text{mol } CO_2}} = 306 \text{ g } CO_2 \quad \text{(Answer)}$$

The calculation in a continuous sequence is

$$100.\ \cancel{\text{g } C_5H_{12}} \times \frac{1 \ \cancel{\text{mol } C_5H_{12}}}{72.0 \ \cancel{\text{g } C_5H_{12}}} \times \frac{5 \ \cancel{\text{mol } CO_2}}{1 \ \cancel{\text{mol } C_5H_{12}}} \times \frac{44.0 \text{ g } CO_2}{1 \ \cancel{\text{mol } CO_2}} = 306 \text{ g } CO_2$$

EXAMPLE 9.8   How many grams of nitric acid, HNO_3, are required to produce 8.75 g of dinitrogen monoxide, N_2O, according to the following equation?

$$4\,Zn(s) + 10\,HNO_3(aq) \longrightarrow 4\,Zn(NO_3)_2(aq) + N_2O(g) + 5\,H_2O(l)$$

$$\qquad\qquad \text{10 mol} \qquad\qquad\qquad\qquad\qquad \text{1 mol}$$

The sequence of conversions in the calculation is

$$\text{grams } N_2O \longrightarrow \text{moles } N_2O \longrightarrow \text{moles } HNO_2 \longrightarrow \text{grams } HNO_3$$

Step 1 The balanced equation is given.
Step 2 The starting substance is 8.75 g of N_2O. Convert 8.75 g of N_2O to moles:

$$8.75 \text{ g } N_2O \times \frac{1 \text{ mole } N_2O}{44.0 \text{ g } N_2O} = 0.199 \text{ mol } N_2O$$

Steps 3 and 4 Calculate the moles of CO_2 by the mole-ratio method:

$$0.199 \text{ mole } N_2O \times \frac{10 \text{ mol } HNO_3}{1 \text{ mol } N_2O} = 1.99 \text{ mol } HNO_3$$

Step 5 Convert moles of HNO_3 to grams:

$$1.99 \text{ mol } HNO_3 \times \frac{63.0 \text{ g } HNO_3}{1 \text{ mol } HNO_3} = 125 \text{ g } HNO_3 \quad \text{(Answer)}$$

The calculation in a continuous sequence is

$$8.75 \text{ g } N_2O \times \frac{1 \text{ mol } N_2O}{44.0 \text{ g } N_2O} \times \frac{10 \text{ mol } HNO_3}{1 \text{ mol } N_2O} \times \frac{63.0 \text{ g } HNO_3}{1 \text{ mol } HNO_3} = 125 \text{ g } HNO_3$$

PRACTICE   How many grams of chromium (III) chloride are required to produce 75.0 g of silver chloride?

$$CrCl_3 + 3 \, AgNO_3 \longrightarrow Cr(NO_3)_3 + 3 \, AgCl$$

Answer: 27.6 g $CrCl_3$

PRACTICE   What mass of water is produced by the complete combustion of 225.0 g of butane (C_4H_{10})?

$$2 \, C_4H_{10} + 13 \, O_2 \longrightarrow 8 \, CO_2 + 10 \, H_2O$$

Answer: 349.1 g H_2O

9.6   Limiting Reactant and Yield Calculations

In many chemical processes the quantities of the reactants used are such that the amount of one reactant is in excess of the amount of a second reactant in the reaction. The amount of the product(s) formed in such a case will depend on the reactant that is not in excess. Thus the reactant that is not in excess is

limiting reactant

known as the **limiting reactant** (sometimes called the limiting reagent), because it limits the amount of product that can be formed.

As an example, consider the case in which solutions containing 1.0 mol of sodium hydroxide and 1.5 mol of hydrochloric acid are mixed:

$$NaOH + HCl \longrightarrow NaCl + H_2O$$

1 mol   1 mol   1 mol   1 mol

According to the equation it is possible to obtain 1.0 mol of NaCl from 1.0 mol of NaOH, and 1.5 mol of NaCl from 1.5 mol of HCl. However, we cannot have two different yields of NaCl from the reaction. When 1.0 mol of NaOH and 1.5 mol of HCl are mixed, there is insufficient NaOH to react with all of the HCl. Therefore, HCl is the reactant in excess and NaOH is the limiting reactant. Since the amount of NaCl formed is dependent on the limiting reactant, only 1.0 mol of NaCl will be formed. Because 1.0 mol of NaOH reacts with 1.0 mol of HCl, 0.5 mol of HCl remains unreacted.

$$\left.\begin{array}{c} 1.0 \text{ mol NaOH} \\ 1.5 \text{ mol HCl} \end{array}\right\} \longrightarrow \begin{array}{c} 1.0 \text{ mol NaCl} \\ 1.0 \text{ mol H}_2\text{O} \end{array} + 0.5 \text{ mol HCl unreacted}$$

Problems giving the amounts of two reactants are generally of the limiting-reactant type, and there are several methods used to identity them in a chemical reaction. In the most direct method two steps are needed to determine the limiting reactant and the amount of product formed.

1. Calculate the amount of product (moles or grams, as needed) that can be formed from each reactant.
2. Determine which reactant is limiting. (The reactant that gives the least amount of product is the limiting reactant and the other reactant is in excess. The limiting reactant will determine the amount of product formed in the reaction.)

Sometimes it is necessary to calculate the amount of excess reactant. This can be done first by calculating the amount of excess reactant required to react with the limiting reactant. Then subtract the amount that reacts from the starting quantity of the reactant in excess. This result is the amount of that substance that remains unreacted.

EXAMPLE 9.9

How many moles of Fe_3O_4 can be obtained by reacting 16.8 g Fe with 10.0 g H_2O? Which substance is the limiting reactant? Which substance is in excess?

$$3 \text{ Fe}(s) + 4 \text{ H}_2\text{O}(g) \xrightarrow{\Delta} \text{Fe}_3\text{O}_4(s) + 4 \text{ H}_2(g)$$

Calculate the moles of Fe_3O_4 that can be formed from each reactant.

$$\text{g reactant} \longrightarrow \text{mol reactant} \longrightarrow \text{mol Fe}_3\text{O}_4$$

Step 1 Calculate the moles of Fe_3O_4 that can be formed from each reactant.

$$16.8 \text{ g Fe} \times \frac{1 \text{ mol Fe}}{55.8 \text{ g Fe}} \times \frac{1 \text{ mol Fe}_3\text{O}_4}{3 \text{ mol Fe}} = 0.100 \text{ mol Fe}_3\text{O}_4$$

$$10.0 \text{ g H}_2\text{O} \times \frac{1 \text{ mol H}_2\text{O}}{18.0 \text{ g H}_2\text{O}} \times \frac{1 \text{ mol Fe}_3\text{O}_4}{4 \text{ mol H}_2\text{O}} = 0.139 \text{ mol Fe}_3\text{O}_4$$

Step 2 Determine the limiting reactant. The limiting reactant is Fe because it produces less Fe_3O_4; the H_2O is in excess. The yield of product is 0.100 mol of Fe_3O_4.

EXAMPLE 9.10 How many grams of silver bromide, AgBr, can be formed when solutions containing 50.0 g of $MgBr_2$ and 100. g of $AgNO_3$ are mixed together? How many grams of the excess reactant remain unreacted?

$$MgBr_2(aq) + 2\,AgNO_3(aq) \longrightarrow 2\,AgBr\downarrow + Mg(NO_3)_2(aq)$$

Step 1 Calculate the grams of AgBr that can be formed from each reactant.

$$\text{g reactant} \longrightarrow \text{mol reactant} \longrightarrow \text{mol AgBr} \longrightarrow \text{g AgBr}$$

$$50.0\,\text{g MgBr}_2 \times \frac{1\,\text{mol MgBr}_2}{184.1\,\text{g MgBr}_2} \times \frac{2\,\text{mol AgBr}}{1\,\text{mol MgBr}_2} \times \frac{187.8\,\text{g AgBr}}{1\,\text{mol AgBr}} = 102\,\text{g AgBr}$$

$$100.\,\text{g AgNO}_3 \times \frac{1\,\text{mol AgNO}_3}{169.9\,\text{g AgNO}_3} \times \frac{2\,\text{mol AgBr}}{2\,\text{mol AgNO}_3} \times \frac{187.8\,\text{g AgBr}}{1\,\text{mol AgBr}} = 111\,\text{g AgBr}$$

Step 2 Determine the limiting reactant. The limiting reactant is $MgBr_2$ because it gives less AgBr. $AgNO_3$ is in excess. The yield is 102 g AgBr.

Step 3 Calculate the grams of unreacted $AgNO_3$.
   Calculate the grams of $AgNO_3$ that will react with 50.0 g of $MgBr_2$.

$$\text{g MgBr}_2 \longrightarrow \text{mol MgBr}_2 \longrightarrow \text{mol AgNO}_3 \longrightarrow \text{g AgNO}_3$$

$$50.0\,\text{g MgBr}_2 \times \frac{1\,\text{mol MgBr}_2}{184.1\,\text{g MgBr}_2} \times \frac{2\,\text{mol AgNO}_3}{1\,\text{mol MgBr}_2} \times \frac{169.9\,\text{g AgNO}_3}{1\,\text{mol AgNO}_3} = 92.3\,\text{g AgNO}_3$$

Thus 92.3 g of $AgNO_3$ react with 50.0 g of $MgBr_2$. The amount of $AgNO_3$ that remains unreacted is

$$100.\,\text{g AgNO}_3 - 92.3\,\text{g AgNO}_3 = 7.7\,\text{g AgNO}_3 \text{ unreacted}$$

The final mixture will contain 102 g AgBr(s) and 7.7 g $AgNO_3$ in solution.

PRACTICE How many grams of hydrogen chloride can be produced from 0.490 g of hydrogen and 50.0 g of chlorine?

$$H_2(g) + Cl_2(g) \longrightarrow 2\,HCl(g)$$

Answer: 17.9 g HCl

PRACTICE How many grams of barium sulfate will be formed from 200.0 g of barium nitrate and 100.0 grams of sodium sulfate?

$$Ba(NO_3)_2 + Na_2SO_4 \longrightarrow BaSO_4 + 2\,NaNO_3$$

Answer: 144.3 g $BaSO_4$

The quantities of the products that we have been calculating from equations represent the maximum yield (100%) of product according to the reaction represented by the equation. Many reactions, especially those involving organic substances, fail to give a 100% yield of product. The main reasons for this failure are the side reactions that give products other than the main product and the fact that many reactions are reversible. In addition, some product may be lost in handling and transferring from one vessel to another. The **theoretical yield** of a reaction is the calculated amount of product that can be obtained from a given amount of reactant, according to the chemical equation. The **actual yield** is the amount of product that we finally obtain.

theoretical yield

actual yield

percent yield

The **percent yield** is the ratio of the actual yield to the theoretical yield multiplied by 100. Both yields must have the same units.

$$\frac{\text{actual yield}}{\text{theoretical yield}} \times 100 = \text{percent yield}$$

For example, if the theoretical yield calculated for a reaction is 14.8 g, and the amount of product obtained is 9.25 g, the percent yield is

$$\text{percent yield} = \frac{9.25 \text{ g}}{14.8 \text{ g}} \times 100 = 62.5\%$$

EXAMPLE 9.11   Carbon tetrachloride was prepared by reacting 100. g of carbon disulfide and 100. g of chlorine. Calculate the percent yield if 65.0 g of CCl_4 was obtained from the reaction.

$$CS_2 + 3 Cl_2 \longrightarrow CCl_4 + S_2Cl_2$$

In this problem we need to determine the limiting reactant in order to calculate the quantity of CCl_4 (theoretical yield) that can be formed. Then we can compare this amount with the 65.0 g of CCl_4 actual yield to calculate the percent yield.

Step 1 Determine the theoretical yield. Calculate the grams of CCl_4 that can be formed from each reactant.

$$\text{g reactant} \longrightarrow \text{mol reactant} \longrightarrow \text{mol } CCl_4 \longrightarrow \text{g } CCl_4$$

$$100. \text{ g } CS_2 \times \frac{1 \text{ mol } CS_2}{76.2 \text{ g } CS_2} \times \frac{1 \text{ mol } CCl_4}{1 \text{ mol } CS_2} \times \frac{154 \text{ g } CCl_4}{1 \text{ mol } CCl_4} = 202 \text{ g } CCl_4$$

$$100. \text{ g } Cl_2 \times \frac{1 \text{ mol } Cl_2}{71.0 \text{ g } Cl_2} \times \frac{1 \text{ mol } CCl_4}{3 \text{ mol } Cl_2} \times \frac{154 \text{ g } CCl_4}{1 \text{ mol } CCl_4} = 72.3 \text{ g } CCl_4$$

Determine the limiting reactant. The limiting reactant is Cl_2 because it gives less CCl_4. The CS_2 is in excess. The theoretical yield is 72.3 g CCl_4.

Step 2 Calculate the percent yield.

According to the equation, 72.3 g of CCl_4 is the maximum amount or theoretical yield of CCl_4 possible from 100. g of Cl_2. Actual yield is 65.0 g of CCl_4.

$$\text{percent yield} = \frac{65.0 \text{ g}}{72.3 \text{ g}} \times 100 = 89.9\%$$

EXAMPLE 9.12 Silver bromide was prepared by reacting 200.0 g of magnesium bromide and an adequate amount of silver nitrate. Calculate the percent yield if 375.0 g of silver bromide was obtained from the reaction.

$$MgBr_2 + 2 AgNO_3 \longrightarrow Mg(NO_3)_3 + 2 AgBr$$

Step 1 Determine the theoretical yield. Calculate the grams of AgBr that can be formed.

$$200.0 \text{ g } MgBr_2 \times \frac{1 \text{ mol } MgBr_2}{184.1 \text{ g } MgBr_2} \times \frac{2 \text{ mol } AgBr}{1 \text{ mol } MgBr_2} \times \frac{187.8 \text{ g } AgBr}{1 \text{ mol } AgBr} = 408.0 \text{ g } AgBr$$

The theoretical yield is 408.0 g AgBr

Step 2 Calculate the percent yield. According to the equation, 408.0 g AgBr is the maximum amount of AgBr possible from 200.0 g $MgBr_2$. Actual yield is 375.0 g AgBr.

$$\text{Percent yield} = \frac{375.0 \text{ g } AgBr}{408.0 \text{ g } AgBr} \times 100 = 91.91\%$$

PRACTICE Aluminum oxide was prepared by heating 225 g of chromium II oxide with 125 g of aluminum. Calculate the percent yield if 100.0 g of aluminum oxide was obtained.

$$2 Al + 3 CrO \longrightarrow Al_2O_3 + 3 Cr$$

Answer: 88.5% yield

When solving problems, you will achieve better results if at first you do not try to take shortcuts. Write the data and numbers in a logical, orderly manner. Make certain that the equations are balanced and that the computations are accurate and expressed to the correct number of significant figures. Remember that units are very important; a number without units has little meaning. Finally, a scientific calculator can save you many hours of tedious computations.

Concepts in Review

1. Write mole ratios for any two substances involved in chemical reaction.
2. Outline the mole or mole-ratio method for making stoichiometric calculations.
3. Calculate the number of moles of a desired substance obtainable from a given number of moles of a starting substance in a chemical reaction (mole-to-mole calculations).
4. Calculate the mass of a desired substance obtainable from a given number of

moles of a starting substance in a chemical reaction, and vice versa (mole-to-mass and mass-to-mole calculations).

5. Calculate the mass of a desired substance involved in a chemical reaction from a given mass of a starting substance (mass-to-mass calculation).

6. Deduce the limiting reactant or reagent when given the amounts of starting substances, and then calculate the moles or mass of desired substance obtainable from a given chemical reaction (limiting reactant calculation).

7. Apply theoretical yield or actual yield to any of the foregoing types of problems, or calculate theoretical and actual yields of a chemical reaction.

Key Terms in Review

The terms listed here have all been defined within the chapter. Review the definitions of each, and use the glossary and margin notations within the chapter as study aids.

actual yield   percent yield
limiting reactant   stoichiometry
mole ratio   theoretical yield

Exercises

An asterisk indicates a more challenging question or problem.

Mole Review Problems

1. Calculate the number of moles in each of the following quantities:
   (a) 25.0 g KNO_3
   (b) 10.8 g $Ca(NO_3)_2$
   (c) 5.4×10^2 g $(NH_4)_2C_2O_4$
   (d) 2.10 kg $NaHCO_3$
   (e) 525 mg $ZnCl_2$
   (f) 56 millimol NaOH
   (g) 9.8×10^{24} molecules CO_2
   (h) 250 mL ethyl alcohol, C_2H_2OH
       ($d = 0.789$ g/mL)
   *(i) 16.8 mL H_2SO_4 solution ($d = 1.727$ g/mL, 80.0% H_2SO_4 by mass)

2. Calculate the number of grams in each of the following quantities:
   (a) 2.55 mol $Fe(OH)_3$
   (b) 0.00844 mol $NiSO_4$
   (c) 125 kg $CaCO_3$
   (d) 0.0600 mol $HC_2H_3O_2$
   (e) 10.5 mol NH_3
   (f) 0.725 mol Bi_2S_3
   (g) 72 millimol HCl
   (h) 4.50×10^{21} molecules glucose, $C_6H_{12}O_6$
   (i) 500 mL of liquid Br_2 ($d = 3.119$ g/mL)
   *(j) 75 mL K_2CrO_4 solution ($d = 1.175$ g/mL, 20.0% K_2CrO_4 by mass)

3. Which contains the larger number of molecules?
   (a) 10.0 g H_2O or 10.0 g H_2O_2
   (b) 25.0 g HCl or 85.0 g $C_6H_{12}O_6$

Mole-Ratio Problems

4. Given the equation for the combustion of iso-propyl alcohol:

$$2 C_3H_7OH + 9 O_2 \longrightarrow 6 CO_2 + 8 H_2O$$

what is the mole ratio of:
(a) CO_2 to C_3H_7OH (d) H_2O to C_3H_7OH
(b) C_3H_7OH to O_2 (e) CO_2 to H_2O
(c) O_2 to CO_2 (f) H_2O to O_2

5. For the reaction

$$3 CaCl_2 + 2 H_3PO_4 \longrightarrow Ca_3(PO_4)_2 + 6 HCl$$

set up the mole ratio of
(a) $CaCl_2$ to $Ca_3(PO_4)_2$
(b) HCl to H_3PO_4
(c) $CaCl_2$ to H_3PO_4
(d) $Ca_3(PO_4)_2$ to H_3PO_4
(e) HCl to $Ca_3(PO_4)_2$
(f) H_3PO_4 to HCl

6. How many moles of Cl_2 can be produced from 5.60 mol HCl?

$$4 HCl + O_2 \longrightarrow 2 Cl_2 + 2 H_2O$$

7. How many grams of sodium hydroxide can be produced from 500 g of calcium hydroxide according to this equation?

$$Ca(OH)_2 + Na_2CO_3 \longrightarrow 2 NaOH + CaCO_3$$

8. Given the equation

$$Al_4C_3 + 12 H_2O \longrightarrow 4 Al(OH)_3 + 3 CH_4$$

(a) How many moles of water are needed to react with 100. g of Al_4C_3?
(b) How many moles of $Al(OH)_3$ will be produced when 0.600 mol of CH_4 is formed?

9. How many grams of zinc phosphate, $Zn_3(PO_4)_2$, are formed when 10.0 g of Zn are reacted with phosphoric acid?

10. Given the equation

$$4 FeS_2 + 11 O_2 \longrightarrow 2 Fe_2O_3 + 8 SO_2$$

(a) How many moles of Fe_2O_3 can be made from 1.00 mol of FeS_2?
(b) How many moles of O_2 are required to react with 4.50 mol of FeS_2?

(c) If the reaction produces 1.55 mol of Fe_2O_3, how many moles of SO_2 are produced?
(d) How many grams of SO_2 can be formed from 0.512 mol of FeS_2?
(e) If the reaction produces 40.6 g of SO_2, how many moles of O_2 were reacted?
(f) How many grams of FeS_2 are needed to produce 221 g of Fe_2O_3?

11. An early method of producing chlorine was by the reaction of pyrolusite, MnO_2, and hydrochloric acid. How many moles of HCl will react with 1.05 mol of MnO_2? (Balance the equation first.)

$$MnO_2(s) + HCl(aq) \longrightarrow$$
$$Cl_2(g) + MnCl_2(aq) + H_2O$$

12. 180 g of zinc were dropped into a beaker of hydrochloric acid. After the reaction ceased, 35 g of unreacted zinc remained in the beaker:

$$Zn + HCl \longrightarrow ZnCl_2 + H_2$$

(a) How many moles of hydrogen gas were produced?
(b) How many grams of HCl were reacted?

13. In a blast furnace, iron(III) oxide reacts with coke (carbon) to produce molten iron and carbon monoxide:

$$Fe_2O_3 + 3 C \longrightarrow 2 Fe + 3 CO$$

How many kilograms of iron would be formed from 125 kg of Fe_2O_3?

14. How many grams of steam and iron must react to produce 375 g of magnetic iron oxide, Fe_3O_4?

$$3 Fe(s) + 4 H_2O(g) \longrightarrow Fe_3O_4(s) + 4 H_2(g)$$

15. Ethane gas, C_2H_6, burns in air (i.e., reacts with the oxygen in air) to form carbon dioxide and water:

$$2 C_2H_6 + 7 O_2 \longrightarrow 4 CO_2 + 6 H_2O$$

(a) How many moles of O_2 are needed for the complete combustion of 15.0 mol of ethane?
(b) How many grams of CO_2 are produced for each 8.00 g of H_2O produced?
(c) How many grams of CO_2 will be produced by the combustion of 75.0 g of C_2H_6?

Limiting-Reactant and Percent-Yield Problems

16. In the following equations, determine which reactant is the limiting reactant and which reactant is in excess. The amounts mixed together are shown below each reactant. Show evidence for your answers.

    (a) $KOH + HNO_3 \longrightarrow KNO_3 + H_2O$
       16.0 g   12.0 g

    (b) $2 NaOH + H_2SO_4 \longrightarrow Na_2SO_4 + 2 H_2O$
       10.0 g   10.0 g

    (c) $2 Bi(NO_3)_3 + 3 H_2S \longrightarrow Bi_2S_3 + 6 HNO_3$
       50.0 g   6.00 g

    (d) $3 Fe + 4 H_2O \longrightarrow Fe_3O_4 + 4 H_2$
       40.0 g   16.0 g

17. Given the equation

    $$Fe(s) + CuSO_4(aq) \longrightarrow Cu(s) + FeSO_4(aq)$$

    (a) When 2.0 mol of Fe and 3.0 mol of $CuSO_4$ are reacted, what substances will be present when the reaction is over? How many moles of each substance are present?

    (b) When 20.0 g of Fe and 40.0 g of $CuSO_4$ are reacted, what substances will be present when the reaction is over? How many grams of each substance are present?

· 18. The reaction for the combustion of propane, C_3H_8, is:

    $$C_3H_8 + 5 O_2 \longrightarrow 3 CO_2 + 4 H_2O$$

    (a) If 5.0 mol of C_3H_8 and 5.0 mol of O_2 are reacted, how many moles of CO_2 can be produced?

    (b) If 3.0 mol of C_3H_8 and 20.0 mol of O_2 are reacted, how many moles of CO_2 can be produced?

    (c) If 20.0 mol of C_3H_8 and 3.0 mol of O_2 are reacted, how many moles of CO_2 can be produced?

    (d) If 2.0 mol of C_3H_8 and 14.0 mol of O_2 are placed in a closed container, and they react to completion (until one reactant is completely used up), what compounds are present in the container after the reaction, and how many moles of each compound are present?

    (e) If 20.0 g of C_3H_8 and 20.0 g of O_2 are reacted, how many grams of CO_2 can be produced?

    (f) If 20.0 g of C_3H_8 and 80.0 g of O_2 are reacted, how many grams of CO_2 can be produced?

    (g) If 20.0 g C_3H_8 and 200 g of O_2 are reacted, how many grams of CO_2 can be produced?

19. Aluminum reacts with bromine to form aluminum bromide:

    $$2 Al + 3 Br_2 \longrightarrow 2 AlBr_3$$

    If 25.0 g of Al and 100. g of Br_2 are reacted, and 64.2 g of $AlBr_3$ product are recovered, what is the percent yield for the reaction?

20. Methyl alcohol, CH_3OH, is made by reacting carbon monoxide and hydrogen in the presence of certain metal oxide catalysts. How much alcohol can be obtained by reacting 40.0 g of CO and 10.0 g of H_2? How many grams of excess reactant remain unreacted?

    $$CO(g) + 2 H_2(g) \longrightarrow CH_3OH(l)$$

· 21. Iron was reacted with a solution containing 400. g of copper(II) sulfate. The reaction was stopped after 1 hour, and 151 g of copper were obtained. Calculate the percent yield of copper obtained.

    $$Fe(s) + CuSO_4(aq) \longrightarrow Cu(s) + FeSO_4(aq)$$

*22. Ethyl alcohol, C_2H_5OH, also called grain alcohol, can be made by the fermentation of sugar, which often comes from starch in grain:

    $$C_6H_{12}O_6 \longrightarrow 2 C_2H_5OH + 2 CO_2$$
    Glucose   Ethyl alcohol

    If an 84.6% yield of ethyl alcohol is obtained,
    (a) What mass of ethyl alcohol will be produced from 750 g of glucose?
    (b) What mass of glucose should be used to produce 475 g of C_2H_5OH?

*23. Carbon disulfide, CS_2, can be made from coke, C, and sulfur dioxide, SO_2:

    $$3 C + 2 SO_2 \longrightarrow CS_2 + 2 CO_2$$

If the actual yield of CS_2 is 86.0% of the theoretical yield, what mass of coke is needed to produce 950 g of CS_2?

*24. Acetylene, C_2H_2, can be manufactured by the reaction of water and calcium carbide, CaC_2:

$$CaC_2 + 2\,H_2O \longrightarrow C_2H_2(g) + Ca(OH)_2$$

When 44.5 g of commercial grade (impure) calcium carbide were reacted, 0.540 mol of C_2H_2 was produced. Assuming that all of the CaC_2 was reacted to C_2H_2, what is the percent of CaC_2 in the commercial grade material?

Additional Problems

*25. Both $CaCl_2$ and $MgCl_2$ react with $AgNO_3$ to precipitate $AgCl$. When solutions containing equal masses of $CaCl_2$ and $MgCl_2$ are reacted, which salt will produce the larger amount of $AgCl$? Show proof.

*26. An astronaut excretes about 2500 g of water a day. If lithium oxide, Li_2O, is used in the spaceship to absorb this water, how many kilograms of Li_2O must be carried for a 30-day space trip for three astronauts?

$$Li_2O + H_2O \longrightarrow 2\,LiOH$$

*27. Much commerical hydrochloric acid is prepared by the reaction of concentrated sulfuric acid with sodium chloride:

$$H_2SO_4 + 2\,NaCl \longrightarrow Na_2SO_4 + 2\,HCl$$

How many kilograms of concentrated H_2SO_4, 96% H_2SO_4 by mass, are required to produce 20.0 L of concentrated hydrochloric acid ($d = 1.20$ g/mL, 42.0% HCl by mass)?

*28. Gastric juice contains about 3.0 g of HCl per liter. If a person produces about 2.5 L of gastric juice per day, how many antacid tablets, each containing 400. mg of $Al(OH)_3$, are needed to neutralize all the HCl produced in one day?

$$Al(OH)_3(s) + 3\,HCl(aq) \longrightarrow$$
$$AlCl_3(aq) + 6\,H_2O(l)$$

*29. 12.82 g of a mixture of $KClO_3$ and NaCl are heated strongly. The $KClO_3$ reacts according to

the following equation:

$$2\,KClO_3(s) \longrightarrow 2\,KCl(s) + 3\,O_2(g)$$

The NaCl does not undergo any reaction. After the heating, the mass of the residue (KCl and NaCl) is 9.45g. Assuming that all the loss of mass represents loss of oxygen gas, calculate the percent of $KClO_3$ in the original mixture.

30. Phosphine, PH_3, can be prepared by the hydrolysis of calcium phosphide, Ca_3P_2:

$$Ca_3P_2 + 6\,H_2O \longrightarrow 3\,Ca(OH)_2 + 2\,PH_3$$

Based on this equation, which of the following statements are correct? Show evidence to support your answer.
(a) One mole of Ca_3P_2 produces 2 mol of PH_3.
(b) One gram of Ca_3P_2 produces 2 g of PH_3.
(c) Three moles of $Ca(OH)_2$ are produced for each 2 mol of PH_3 produced.
(d) The mole ratio between phosphine and calcium phosphide is

$$\frac{2 \text{ mol } PH_3}{1 \text{ mol } Ca_3P_2}$$

(e) When 2.0 mol of Ca_3P_2 and 3.0 mol of H_2O react, 4.0 mol of PH_3 can be formed.
(f) When 2.0 mol of Ca_3P_2 and 15.0 mol of H_2O react, 6.0 mol of $Ca(OH)_2$ can be formed.
(g) When 200. g of Ca_3P_2 and 100. g of H_2O react, Ca_3P_2 is the limiting reactant.
(h) When 200. g of Ca_3P_2 and 100. g of H_2O react, the theoretical yield of PH_3 is 57.4 g.

31. The equation representing the reaction used for the commercial preparation of hydrogen cyanide is

$$2\,CH_4 + 3\,O_2 + 2\,NH_3 \longrightarrow$$
$$2\,HCN + 6\,H_2O$$

Based on this equation, which of the statements below are correct? Rewrite incorrect statements to make them correct.
(a) Three moles of O_2 are required for 2 mol of NH_3.
(b) Twelve moles of HCN are produced for every 16 mol of O_2 that react.

(c) The mole ratio between H_2O and CH_4 is

$$\frac{6 \text{ mol } H_2O}{2 \text{ mol } CH_4}$$

(d) When 12 mol of HCN are produced, 4 mol of H_2O will be formed.

(e) When 10 mol of CH_4, 10 mol of O_2, and 10 mol of NH_3 are mixed and reacted, O_2 is the limiting reactant.

(f) When 3 mol each of CH_4, O_2, and NH_3 are mixed and reacted, 3 mol of HCN will be produced.

Review Exercises for Chapters 8–9

CHAPTER 8   Chemical Equations

True–False.   *Answer the following as either true or false.*

1. In balancing an equation, we change the formulas of compounds to make the number of atoms on each side of the equation balance.
2. The equation $N_2 + 3 H_2 \rightleftharpoons 2 NH_3$ can be interpreted as saying that 1 mol of N_2 reacts with 3 mol of H_2 to form 2 mol of NH_3.
3. The equation $N_2 + 3 H_2 \rightleftharpoons 2 NH_3$ can be interpreted as saying that 1 g of N_2 reacts with 3 g of H_2 to form 2 g of NH_3.
4. The substances on the right side of a chemical equation are called the products.
5. When a gas is evolved in a chemical reaction, it can be indicated in the equation with the symbol ↑ or (*g*) immediately following the formula of the gas.
6. A balanced chemical equation is one that has the same number of moles on each side of the equation.
7. The coefficients in front of the formulas in a balanced chemical equation give the relative number of moles of the reactants and products in the reaction.
8. Water is formed in a neutralization reaction.
9. When carbonates or bicarbonates react with acids, carbon monoxide is formed.
10. In the reaction

$$Cu(s) + 2 AgNO_3(aq) \longrightarrow$$
$$Cu(NO_3)_2(aq) + 2 Ag(s)$$

Cu replaces Ag^+ because copper is a more reactive element than silver.
11. The combustion of hydrocarbons is an exothermic reaction.
12. The addition of sulfuric acid to water is an endothermic process.
13. In a balanced chemical equation, the mass of the products is equal to the mass of the reactants.

14. The equation shown is balanced.

$$2 Cu + 3 HNO_3 \longrightarrow$$
$$Cu(NO_3)_2 + NO_2 + 2 H_2O$$

15. The products for the complete combustion of butane (C_4H_{10}) are CO_2 and H_2O.

Multiple Choice.   *Choose the correct answer to each of the following.*

1. The reaction $BaCl_2 + (NH_4)_2CO_3 \longrightarrow BaCO_3 + 2 NH_4Cl$ is an example of:
   (a) combination
   (b) decomposition
   (c) single displacement
   (d) double displacement
2. When the equation
   $Al + O_2 \longrightarrow Al_2O_3$ is properly balanced, which of the following terms appears?
   (a) 2 Al   (b) 2 Al_2O_3 (c) 3 Al   (d) 2 O_2
3. Which equation is *incorrectly* balanced?
   (a) $2 KNO_3 \xrightarrow{\Delta} 2 KNO_2 + O_2$
   (b) $H_2O_2 \longrightarrow H_2O + O_2$
   (c) $2 Na_2O_2 + 2 H_2O \longrightarrow 4 NaOH + O_2$
   (d) $2 H_2O \xrightarrow[H_2SO_4]{\text{Electrical energy}} 2 H_2 + O_2$
4. The reaction $2 Al + 3 Br_2 \longrightarrow 2 AlBr_3$ is an example of:
   (a) combination
   (b) single displacement
   (c) decomposition
   (d) double displacement.
5. When the equation
   $PbO_2 \xrightarrow{\Delta} PbO + O_2$ is balanced, one of the terms in the balanced equation is:
   (a) PbO_2 (b) $3 O_2$ (c) 3 PbO   (d) O_2
6. When the equation
   $Cr_2S_3 + HCl \longrightarrow CrCl_3 + H_2S$ is balanced, one of the terms in the balanced equation is:
   (a) 3 HCl   (b) $CrCl_3$ (c) $3 H_2S$ (d) $2 Cr_2S_3$

7. When the equation
$F_2 + H_2O \longrightarrow HF + O_2$ is balanced, a term in the balanced equation is:
(a) $2\,HF$ (b) $3\,O_2$ (c) $4\,HF$ (d) $4\,H_2O$

8. When the equation
$NH_4OH + H_2SO_4 \longrightarrow$
is completed and balanced, one of the terms in the balanced equation is:
(a) NH_4SO_4 (c) H_2OH
(b) $2\,H_2O$ (d) $2\,(NH_4)_2SO_4$

9. When the equation
$H_2 + V_2O_5 \longrightarrow V +$ is completed and balanced, a term in the balanced equation is:
(a) $2\,V_2O_5$ (b) $3\,H_2O$ (c) $2\,V$ (d) $8\,H_2$

10. When the equation
$Al(OH)_3 + H_2SO_4 \longrightarrow Al_2(SO_4)_3 + H_2O$ is balanced the sum of the coefficients will be:
(a) 9   (c) 13
(b) 11   (d) 15

11. When the equation
$H_3PO_4 + Ca(OH)_2 \longrightarrow H_2O + Ca_3(PO_4)$ is balanced the proper sequence of coefficients is:
(a) 3, 2, 1, 6   (c) 2, 3, 1, 6
(b) 2, 3, 6, 1   (d) 2, 3, 3, 1

12. When the equation
$Fe_2(SO_4)_3 + Ba(OH)_2 \longrightarrow$
is completed and balanced, a term in the balanced equation is:
(a) $Ba_2(SO_4)_3$ (c) $2\,Fe_2(SO_4)_3$
(b) $2\,Fe(OH)_2$ (d) $2\,Fe(OH)_3$

13. For the reaction
$2\,H_2 + O_2 \longrightarrow 2\,H_2O + 572.4\,kJ$, which of the following is not true?
(a) The reaction is exothermic.
(b) 572.4 kJ of heat are liberated for each mole of water formed.
(c) Two moles of hydrogen react with 1 mol of oxygen.
(d) 572.4 kJ of heat are liberated for each 2 mol of hydrogen reacted.

14. When a nonmetal oxide reacts with water,
(a) A base is formed
(b) An acid is formed
(c) A salt is formed
(d) A nonmetal oxide is formed

CHAPTER 9   Calculations from Chemical Equations

True–False. *Answer the following as either true or false.*

1. Stoichiometry is the section of chemistry involving calculations based on mass and mole relationships of substances in chemical reactions.

2. In a limiting-reactant problem, you determine which reactant has the fewest moles available.

3. The maximum amount of product that can be produced according to the equation, from the amounts of reactants supplied, is the theoretical yield.

4. A mole ratio is a ratio of the moles of products to the moles of reactants.

5. A mole ratio is used to convert the moles of starting substance to the moles of desired substance.

6. The limiting reactant of a chemical reaction limits the amount of product that can be formed.

7. The equation for a reaction should be balanced before doing stoichiometric calculations.

8. The mole ratio for converting moles of CO_2 to moles of O_2 in the reaction
$2\,C_2H_6 + 7\,O_2 \longrightarrow 4\,CO_2 + 6\,H_2O$
is 4 mol CO_2/7 mol O_2.

9. The mole ratio for converting moles of CH_4 to moles of H_2O in the reaction

$2\,CH_4 + 3\,O_2 + 2\,NH_3 \longrightarrow 2\,HCN + 6\,H_2O$

is 6 mol H_2O/2 mol CH_4.

10. The percent yield of a product is
$\dfrac{\text{actual yield}}{\text{theoretical yield}} \times 100.$

Multiple Choice. *Choose the correct answer to each of the following.*

1. 20.0 g of Na_2CO_3 is how many moles?
   (a) 1.89 mol   (c) 212 mol
   (b) 2.12×10^3 mol   (d) 0.189 mol
2. What is the mass in grams of 0.30 mol of $BaSO_4$?
   (a) 7.0×10^3 g   (c) 70 g
   (b) 0.13 g   (d) 700.20 g
3. How many molecules are in 5.8 g of acetone, C_3H_6O?
   (a) 0.10 molecules
   (b) 6.0×10^{22} molecules
   (c) 3.5×10^{24} molecules
   (d) 6.022×10^{23} molecules

Problems 4–10 refer to the reaction

$$2\,C_2H_4 + 6\,O_2 \longrightarrow 4\,CO_2 + 4\,H_2O$$

4. If 6.0 mol of CO_2 are produced, how many moles of O_2 were reacted?
   (a) 4.0 mol   (c) 9.0 mol
   (b) 7.5 mol   (d) 15.0 mol
5. How many moles of O_2 are required for the complete reaction of 45 g of C_2H_4?
   (a) 1.3×10^2 mol   (c) 112.5 mol
   (b) 0.64 mol   (d) 4.8 mol
6. If 18.0 g of CO_2 are produced, how many grams of H_2O are produced?
   (a) 7.36 g   (b) 3.68 g   (c) 9.0 g   (d) 14.7 g
7. How many moles of CO_2 can be produced by the reaction of 5.0 mol of C_2H_4 and 12.0 mol of O_2?
   (a) 4.0 mol   (c) 8.0 mol
   (b) 5.0 mol   (d) 10.0 mol
8. How many moles of CO_2 can be produced by the reaction of 0.480 mol of C_2H_4 and 1.08 mol of O_2?
   (a) 0.240 mol   (c) 0.720 mol
   (b) 0.960 mol   (d) 0.864 mol

9. How many grams of CO_2 could be produced from 2.0 g of C_2H_4 and 5.0 g of O_2?
   (a) 5.5 g   (b) 4.6 g   (c) 7.6 g   (d) 6.3 g
10. If 14.0 g of C_2H_4 is reacted, and the actual yield of H_2O is 7.84 g, the percent yield in the reaction is:
   (a) 0.56%   (b) 43.6%   (c) 87.1%   (d) 56.0%

Problems 11–13 refer to the equation

$$H_3PO_4 + MgCO_3 \longrightarrow$$
$$Mg_3(PO_4)_2 + CO_2 + H_2O$$

11. The sequence of coefficients for the balanced equation is
   (a) 2, 3, 1, 3, 3   (c) 2, 2, 1, 2, 3
   (b) 3, 1, 3, 2, 3   (d) 2, 3, 1, 3, 2
12. If 20.0 g of carbon dioxide is produced, the number of moles of magnesium carbonate used is
   (a) 0.228 mol   (c) 0.910 mol
   (b) 1.37 mol   (d) 0.455 mol
13. If 50.0 g of magnesium carbonate react completely with H_3PO_4, the number of grams of carbon dioxide produced is
   (a) 52.2 g   (c) 13.1 g
   (b) 26.1 g   (d) 50.0 g
14. $MgCl_2 + Na_2CO_3 \longrightarrow MgCO_3 + 2\,NaCl$. When 10.0 g of $MgCl_2$ and 10.0 g of Na_2CO_3 are reacted, the limiting reactant is:
   (a) $MgCl_2$ (c) $MgCO_3$
   (b) Na_2CO_3 (d) NaCl.
15. $Cu + 2\,AgNO_3 \longrightarrow Cu(NO_3)_2 + 2\,Ag$. When 50.0 g of copper was reacted with silver nitrate solution, 148 g of silver was obtained. What is the percent yield of silver obtained?
   (a) 87.1%   (c) 55.2%
   (b) 84.9%   (d) No correct answer given

CHAPTER TEN

Modern Atomic Theory

How do we go about studying an object too small to see? Think back to that birthday when you had a present you could look at but not yet open. Simply judging from the wrapping and size of the box was not very useful. Shaking, turning, and lifting the package all gave clues, indirectly, to the contents. After all the experiments were done on the package, a fairly good hypothesis of the contents could be made. But was it correct? The only means of absolute verification would be opening the package.

The same is true for chemists in their study of the atom. Atoms are so very small that it is not possible to use the normal senses and rules to describe them. Chemists are working within a package essentially in the dark. Yet, as such instruments as x-ray machines and scanning-tunneling microscopes and such measuring devices as spectrophotometers and magnetic resonance imaging (mri) have progressed along with man's skill in mathematics and probability, the secrets of the atom are beginning to be unraveled.

Chapter Preview

10.1   Introduction

In the last 200 years, vast amounts of data have been accumulated to support atomic theory. When atoms were originally suggested by the early Greeks, no evidence existed to physically support their ideas. Lavoisier, Faraday, Arrhenius, and others did a variety of experiments, which culminated in the Dalton theory of the atom. Because of the limitations of Dalton's model, modifications were proposed by Thomson and then by Rutherford to evolve eventually into the nuclear atom. These early models of the atom work reasonably well—in fact, we continue to use them today to visualize a variety of chemical concepts. There remain, however, questions that these models cannot answer, including an explanation of how atomic structure relates to the periodic table. In this chapter, we will look at our modern model of the atom to see how it varies from and improves upon the earlier atomic theories.

10.2   The Bohr Atom

At high temperatures, or when subjected to high voltages, elements in the gaseous state give off colored light. Brightly colored neon signs illustrate this property of matter very well. When passed through the prism or the grating of a spectroscope, the light emitted by a gas appears as a set of brightly colored lines and is called a line spectrum. These colored lines indicate that the light is being emitted only at certain wavelengths, or frequencies, that correspond to specific colors. Each element possesses a unique set of these spectral lines that is different from the sets of all the other elements.

In 1912–1913, while studying the line spectra of hydrogen, Niels Bohr (1885–1962), a Danish physicist, made a significant contribution to the rapidly growing knowledge of atomic structure. His research led him to believe that electrons in an atom exist in specific regions at various distances from the nucleus. He also visualized the electrons as rotating in orbits around the nucleus, like planets rotating around the sun.

Bohr's first paper in this field dealt with the hydrogen atom, which he described as a single electron rotating in an orbit about a relatively heavy nucleus. He applied the concept of energy quanta, proposed in 1900 by the German physicist Max Planck (1858–1947), to the observed line spectra of hydrogen. Planck stated that energy is never emitted in a continuous stream but only in small discrete packets called quanta (Latin, *quantus*, how much). Bohr theorized that electrons have several possible orbits at different distances from the nucleus and that an electron had to be in one specific orbit (energy level) or another; it could not exist between orbits. In other words, the energy of the electron is said to be quantized. Bohr also stated that, when a hydrogen atom absorbed one or more quanta of energy, its electron would "jump" to an orbit at a greater distance from the nucleus.

Bohr was able to account for spectral lines this way. A number of orbits are available, each corresponding to a different energy level. The orbit closest to the nucleus is the lowest, or ground state, energy level; orbits at increasing distances are the second, third, fourth, etc., energy levels. When an electron falls from a high-energy orbit to a lower one (say, from the fourth to the second), a quantum of energy is emitted as light at a specific frequency, or wavelength. This light corresponds to one of the lines visible in the hydrogen spectrum (see Figure 10.1). Several lines are visible in this spectrum. Each line corresponds to a specific electron energy-level shift within the hydrogen atom.

The chemical properties of an element and its position in the periodic table (Chapter 11) depend on electron behavior within the atoms. In turn, much of our knowledge of the behavior of electrons within atoms is based on spectroscopy. Niels Bohr contributed a great deal to our knowledge of atomic structure by (1) suggesting quantized energy levels for electrons and (2) showing that spectral lines result from the radiation of small increments of energy (Planck's quanta) when electrons shift from one energy level to another. Bohr's calculations succeeded very well in correlating the experimentally observed spectral lines with electron energy levels for the hydrogen atom. However, Bohr's methods of calculation did not succeed for heavier atoms. More theoretical work on atomic structure was needed.

In 1924 the French physicist, Louis de Broglie, suggested that moving electrons had properties of waves as well as mass. In 1926 Erwin Schrödinger, an Austrian physicist, introduced a new method of calculation—quantum mechanics, or wave mechanics. By Schrödinger's method electrons are described in mathematical terms as having dual characteristics; that is, some electron properties are best described in terms of waves (like light) and others in terms of particles having mass.

Since the late 1920's quantum mechanical concepts have generally replaced

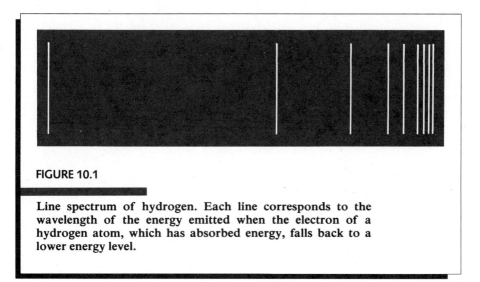

FIGURE 10.1

Line spectrum of hydrogen. Each line corresponds to the wavelength of the energy emitted when the electron of a hydrogen atom, which has absorbed energy, falls back to a lower energy level.

Bohr's ideas in theoretical considerations of atomic structure. One important difference between the Bohr and the quantum mechanics concepts is in the treatment of electrons. Quantum mechanics retains the concept of electrons being in specific energy levels. But the electrons are treated not as revolving about the nucleus in *orbits* but as being located in *orbitals*. An **orbital** is simply a region in space about the nucleus where there is a high probability of finding a given electron.

orbital

10.3   Energy Levels of Electrons

Not all the electrons in an atom are located the same distance from the nucleus. As pointed out by both the Bohr theory and quantum mechanics, the probability of finding electrons is greatest at certain specified distances from the nucleus, called **energy levels**. Energy levels are also referred to as **electron shells** and may contain only a limited number of electrons. The main or principal energy levels (**n**) are numbered, starting with **n** = 1 as the energy level nearest to the nucleus and going to **n** = 7, for the known elements. (Theoretically, the number of energy levels is infinite.) Each succeeding energy level is located farther from the nucleus. The electrons in energy levels at increasing distances from the nucleus have increasingly higher energies. The order of energy for the principal energy levels **n** is

energy levels

electron shells

Principal energy levels: $1 < 2 < 3 < 4 < 5 < 6 < 7$

TABLE 10.1

Maximum Number of Electrons that Can Occupy Each Principal Energy Level

Principal energy level, n	Maximum number of electrons in each energy level, $2n^2$
1	$2 \times 1^2 = 2$
2	$2 \times 2^2 = 8$
3	$2 \times 3^2 = 18$
4	$2 \times 4^2 = 32$
5	$2 \times 5^2 = 50^a$

[a] The theoretical value of 50 electrons in energy level 5 has never been attained in any element known to date.

The number of electrons that can exist in each energy level is limited. The maximum number of electrons for a specific energy level can be calculated from the formula $2n^2$, where **n** is the number of the principal energy level. For example, energy level 1 (**n** = 1) can have a maximum of two electrons ($2 \times 1^2 = 2$); energy level 2 (**n** = 2) can have a maximum of eight electrons ($2 \times 2^2 = 8$), and so on. Table 10.1 shows the maximum number of electrons that can exist in each of the first five energy levels.

10.4   Energy Sublevels

The principal energy levels contain sublevels designated by the letters *s*, *p*, *d*, and *f*. These orbitals are the ones in which electrons are located. The *s* sublevel consists of one orbital; the *p* sublevel consists of three orbitals; the *d* sublevel consists of five orbitals; and the *f* sublevel consists of seven orbitals. An electron spins on its own axis in one of only two directions, clockwise or counterclockwise. As a result, only two electrons can occupy the same orbital, one spinning clockwise and the other spinning counterclockwise. When an orbital contains two electrons, the electrons are said to be paired. Because no more than two electrons can exist in an orbital, the maximum numbers of electrons that can exist in the sublevels are 2 in the *s* orbital, 6 in the three *p* orbitals, 10 in the five *d* orbitals, and 14 in the seven *f* orbitals.

Type of sublevel	Number of orbitals possible	Number of electrons possible
s	1	2
p	3	6
d	5	10
f	7	14

The order of energy of the sublevels within a principal energy level is the following: s electrons are lower in energy than p electrons, which are lower than d electrons, which are lower than f electrons. This order may be expressed in the following manner:

Sublevel energy: $s < p < d < f$

Not all principal energy levels contain each and every type of sublevel. To determine what types of sublevels occur in any given energy level, we need to know the maximum number of electrons possible in that energy level (see Table 10.1), and we need to use these three rules:

1. No more than two electrons can occupy one orbital.
2. Electrons occupy the lowest possible energy sublevels; they enter a higher sublevel only when the lower sublevels are filled.
3. Orbitals in a given sublevel of equal energy are each occupied by a single electron before a second electron enters them. For example, all three p orbitals must contain one electron before a second electron enters a p orbital.

The maximum number of electrons in the first energy level is two; both are s orbital electrons, designated as $1s^2$. (The s orbital in the second energy level ($\mathbf{n} = 2$) is written as $2s$, in the third energy level as $3s$, and so on.) The second energy level, with a maximum of eight electrons, contains only s and p electrons— namely, a maximum of two s and six p electrons, designated as $2s^2 2p^6$. The following diagram shows how to read these electron designations:

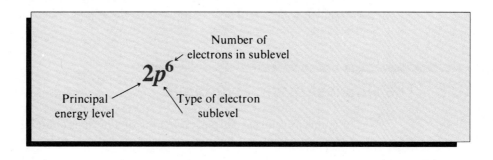

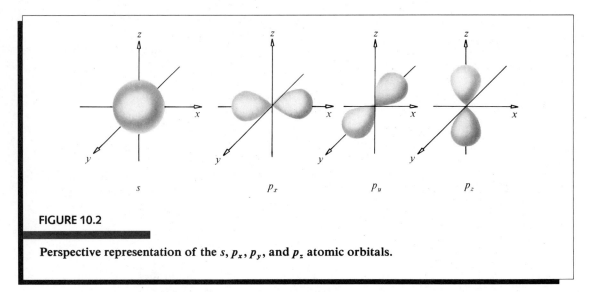

FIGURE 10.2

Perspective representation of the s, p_x, p_y, and p_z atomic orbitals.

If each orbital contains two electrons, the second energy level can have four orbitals (8 electrons): one s orbital and three individual p orbitals. These three p orbitals are energetically equivalent to each other and are labeled $2p_x$, $2p_y$, and $2p_z$ to indicate their orientation in space (see Figure 10.2). The symbols $3s^2$, $3p^6$, and $3d^{10}$ show the sublevel breakdown of electrons in the third energy level. From this line of reasoning, we can see that, if an atom has sufficient electrons, f electrons first appear in the fourth energy level. Table 10.2 shows the type of sublevel electron orbitals and the maximum number of orbitals and electrons in each energy level. No elements in the fifth, sixth and seventh energy levels contain the calculated maximum number of electrons.

Since the $spdf$ atomic orbitals have definite orientations in space, they are represented by particular spatial shapes. At this time we will consider only the s and p orbitals. The s orbitals are spherically symmetrical about the nucleus, as illustrated in Figure 10.2. A $2s$ orbital is a larger sphere than a $1s$ orbital. The p orbitals (p_x, p_y, p_z) are dumbbell-shaped and are oriented at right angles to each other along the x, y, and z axes in space. An electron has an equal probability of being located in either lobe of the p orbital. In illustrations such as Figure 10.2, the boundaries of the orbitals enclose the region of the greatest probability (about a 90% chance) of finding an electron. In the ground state, or lowest energy level, of a hydrogen atom, this region is a sphere having a radius of 0.053 nm.

10.5   The Simplest Atom: Hydrogen

The common hydrogen atom, consisting of a nucleus containing one proton and an electron orbital containing one electron, is the simplest known atom. (Some

TABLE 10.2

Sublevel Electron Orbitals in Each Principal Energy Level and the Maximum Number of Orbitals and Electrons in Each Energy Level

Principal energy level	Sublevel electron	Maximum number of orbitals	Maximum number of electrons
1	s	1	2
2	s, p	4	8
3	s, p, d	9	18
4	s, p, d, f	16	32
5	s, p, d, f	Incomplete[a]	(50)[a]
6	s, p, d	Incomplete[a]	(72)[a]
7	s	Incomplete[a]	(96)[a]

[a] Insufficient electrons to complete the shell.

hydrogen atoms are known to contain one or two neutrons in their nucleus. (See Section 5.10 on isotopes.) The electron in hydrogen occupies an s orbital in the first energy level. This electron does not move in any definite path but rather in a rapid random motion throughout its entire orbital, forming an electron "cloud" about the nucleus. The diameter of the nucleus is believed to be about 10^{-13} cm, and the diameter of the electron orbital to be about 10^{-8} cm. Hence, the diameter of the electron orbital of a hydrogen atom is about 100,000 times greater than the diameter of the nucleus.

What we have, then, is a positive nucleus surrounded by an electron cloud formed by an electron in an s orbital. The net electrical charge on the hydrogen atom is zero; it is called a *neutral atom*. Figure 10.3 shows two methods of representing a hydrogen atom.

10.6   Atomic Structures of the First Twenty Elements

Starting with hydrogen and progressing in order of increasing atomic number to helium, lithium, beryllium, and so on, the atoms of each successive element contain one more proton and one more electron than do the atoms of the preceding element. This sequence continues, without exception, through the entire list of known elements. It is one of the most impressive examples of order in nature.

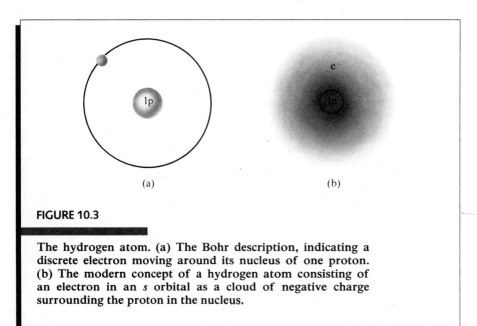

(a)   (b)

FIGURE 10.3

The hydrogen atom. (a) The Bohr description, indicating a discrete electron moving around its nucleus of one proton. (b) The modern concept of a hydrogen atom consisting of an electron in an *s* orbital as a cloud of negative charge surrounding the proton in the nucleus.

The number of neutrons also increases as we progress through the list of elements. But this number, unlike the number of protons and electrons, does not increase in a perfectly uniform manner. Furthermore, atoms of the same element may contain different numbers of neutrons; such atoms are called *isotopes*. For example, three different hydrogen isotopes are described in Section 5.10. The most abundant helium isotope (element number 2) contains two neutrons, but two other helium isotopes exist, containing one and four neutrons, respectively.

The ground-state electron structures of the first 20 elements fall into a regular pattern. The one hydrogen electron is in the first energy level as are both helium electrons. The electron structures for hydrogen and helium are $1s^1$ and $1s^2$. The maximum number of electrons in the first energy level is two ($2\mathbf{n}^2 = 2 \times 1^2 = 2$; see Section 10.3), so the two electrons fill the first energy level of helium.

An atom with three electrons will have its third electron in the second energy level because the first level can contain only two electrons. Thus in lithium (atomic number 3) the third electron is in the $2s$ sublevel of the second energy level. Lithium has the electron structure $1s^2 2s^1$.

In succession, the atoms of beryllium (4), boron (5), carbon (6), nitrogen (7), oxygen (8), fluorine (9), and neon (10) have one more proton and one more electron than the preceding element until, in neon, both the first and second energy levels are filled to capacity, with 2 and 8 electrons, respectively.

H	$1s^1$	Li	$1s^2 2s^1$	B	$1s^2 2s^2 2p^1$	N	$1s^2 2s^2 2p^3$	F	$1s^2 2s^2 2p^5$
He	$1s^2$	Be	$1s^2 2s^2$	C	$1s^2 2s^2 2p^2$	O	$1s^2 2s^2 2p^4$	Ne	$1s^2 2s^2 2p^6$

Element 11, sodium (Na), has two electrons in the first energy level and eight electrons in the second energy level, with the remaining electron occupying the $3s$ orbital in the third energy level. The electron structure of sodium is $1s^2 2s^2 2p^6 3s^1$. Magnesium (12), aluminum (13), silicon (14), phosphorus (15), sulfur (16), chlorine (17), and argon (18) follow in order. Each of these elements will add one electron to the third energy level up to argon, which has eight electrons in the third energy level.

Up through the $3p$ level the sequence of filling the sublevels is exactly as expected, based on the increasing principal and sublevel energy levels. However, after the $3p$ level is filled, variations occur. The third energy level might logically be expected to fill to its capacity of 18 electrons with $3d$ electrons before electrons enter the $4s$ sublevel. However, this order of filling the third energy level does not occur because the $4s$ sublevel is at a lower energy than the $3d$ sublevel (see Figure 10.4). Consequently, because the sublevels fill in order of increasing energy, the last electron in potassium (19) and the last two electrons in calcium (20) are in the $4s$ sublevel. The electron structures for potassium and calcium are

$$K \qquad 1s^2 2s^2 2p^6 3s^2 3p^6 4s^1$$
$$Ca \qquad 1s^2 2s^2 2p^6 3s^2 3p^6 4s^2$$

This break in sequence does not invalidate the formula $2\mathbf{n}^2$, which prescribes the maximum number of electrons that each shell can contain but not the order in which the shells are filled. Table 10.3 shows the electron structure of the first 20 elements.

The relative energies of the electron orbitals are shown in Figure 10.4. The order given can be used to determine the electron distribution in the atoms of the elements, although some exceptions to the pattern are known. Suppose we wish to determine the electron structure of a chlorine atom (atomic number 17), which has 17 electrons. Following the order in Figure 10.4, we begin by placing two electrons in the $1s$ orbital, then two electrons in the $2s$ orbital, and then six electrons in the $2p$ orbitals. We now have used ten electrons.

$$1s^2 2s^2 2p^6$$

Finally we place the next two electrons in the $3s$ orbital and the remaining five electrons in the $3p$ orbitals, which uses all 17 electrons, giving the electron structure for a chlorine atom as $1s^2 2s^2 2p^6 3s^2 3p^5$. The sum of the superscripts equals 17, the number of electrons in the atom. This procedure is summarized below.

Order of orbitals to be filled: $\qquad 1s2s2p3s3p$

Distribution of the 17 electrons
in a chlorine atom: $\qquad\qquad 1s^2 2s^2 2p^6 3s^2 3p^5$

TABLE 10.3

Electron Structure of the First Twenty Elements

Element	Number of protons (atomic number)	Number of electrons	Electron structure
H	1	1	$1s^1$
He	2	2	$1s^2$
Li	3	3	$1s^2 2s^1$
Be	4	4	$1s^2 2s^2$
B	5	5	$1s^2 2s^2 2p^1$
C	6	6	$1s^2 2s^2 2p^2$
N	7	7	$1s^2 2s^2 2p^3$
O	8	8	$1s^2 2s^2 2p^4$
F	9	9	$1s^2 2s^2 2p^5$
Ne	10	10	$1s^2 2s^2 2p^6$
Na	11	11	$1s^2 2s^2 2p^6 3s^1$
Mg	12	12	$1s^2 2s^2 2p^6 3s^2$
Al	13	13	$1s^2 2s^2 2p^6 3s^2 3p^1$
Si	14	14	$1s^2 2s^2 2p^6 3s^2 3p^2$
P	15	15	$1s^2 2s^2 2p^6 3s^2 3p^3$
S	16	16	$1s^2 2s^2 2p^6 3s^2 3p^4$
Cl	17	17	$1s^2 2s^2 2p^6 3s^2 3p^5$
Ar	18	18	$1s^2 2s^2 2p^6 3s^2 3p^6$
K	19	19	$1s^2 2s^2 2p^6 3s^2 3p^6 4s^1$
Ca	20	20	$1s^2 2s^2 2p^6 3s^2 3p^6 4s^2$

EXAMPLE 10.1   What is the electron distribution in a phosphorus atom?

First determine the number of electrons contained in a phosphorus atom. The atomic number of phosphorus is 15; therefore, each atom contains 15 protons and 15 electrons. Now tabulate the number of electrons in each principal and subenergy level until all 15 electrons are assigned.

Sublevel	Number of e^-	Total e^-
$1s$ orbital	$2e^-$	2
$2s$ orbital	$2e^-$	4
$2p$ orbital	$6e^-$	10
$3s$ orbital	$2e^-$	12
$3p$ orbital	$3e^-$	15

Therefore the electron distribution in phosphorus is $1s^2 2s^2 2p^6 3s^2 3p^3$.

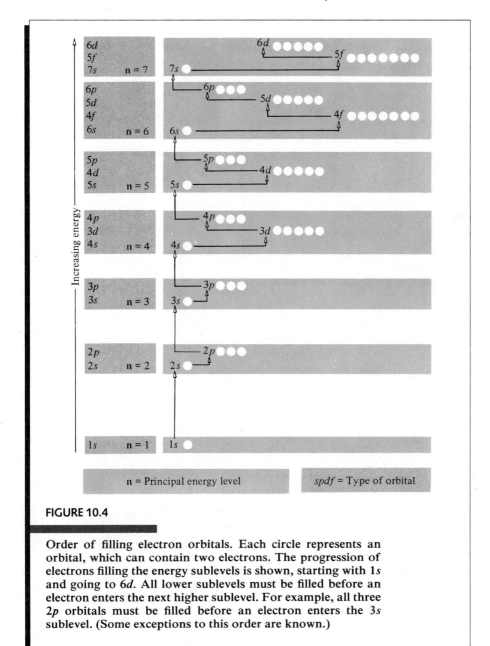

FIGURE 10.4

Order of filling electron orbitals. Each circle represents an orbital, which can contain two electrons. The progression of electrons filling the energy sublevels is shown, starting with 1*s* and going to 6*d*. All lower sublevels must be filled before an electron enters the next higher sublevel. For example, all three 2*p* orbitals must be filled before an electron enters the 3*s* sublevel. (Some exceptions to this order are known.)

PRACTICE Determine the electron configuration of each of the following elements:
(a) Al (b) Si (c) Be
Answers: (a) $1s^2 2s^2 2p^6 3s^2 3p^1$ (b) $1s^2 2s^2 2p^6 3s^2 3p^2$ (c) $1s^2 2s^2$

10.7   Electron Structures of the Elements Beyond Calcium

The elements following calcium have a less regular pattern of adding electrons. The lowest energy level available for the 21st electron is the $3d$ sublevel. Scandium (21) has one more electron than calcium (20). Its electron structure will be the same as calcium plus one electron in the $3d$ sublevel. The electron structure

TABLE 10.4

Electron Structures of the Elements

Element	Atomic number	Electron structure	Element	Atomic number	Electron structure
H	1	$1s^1$	C	6	$1s^2 2s^2 2p^2$
He	2	$1s^2$	N	7	$1s^2 2s^2 2p^3$
Li	3	$1s^2 2s^1$	O	8	$1s^2 2s^2 2p^4$
Be	4	$1s^2 2s^2$	F	9	$1s^2 2s^2 2p^5$
B	5	$1s^2 2s^2 2p^1$	Ne	10	$1s^2 2s^2 2p^6$
Na	11	$[Ne]\,3s^1$	Pr	59	$[Xe]\,6s^2 4f^3$
Mg	12	$[Ne]\,3s^2$	Nd	60	$[Xe]\,6s^2 4f^4$
Al	13	$[Ne]\,3s^2 3p^1$	Pm	61	$[Xe]\,6s^2 4f^5$
Si	14	$[Ne]\,3s^2 3p^2$	Sm	62	$[Xe]\,6s^2 4f^6$
P	15	$[Ne]\,3s^2 3p^3$	Eu	63	$[Xe]\,6s^2 4f^7$
S	16	$[Ne]\,3s^2 3p^4$	Gd	64	$[Xe]\,6s^2 4f^7 5d^1$
Cl	17	$[Ne]\,3s^2 3p^5$	Tb	65	$[Xe]\,6s^2 4f^9$
Ar	18	$[Ne]\,3s^2 3p^6$	Dy	66	$[Xe]\,6s^2 4f^{10}$
K	19	$[Ar]\,4s^1$	Ho	67	$[Xe]\,6s^2 4f^{11}$
Ca	20	$[Ar]\,4s^2$	Er	68	$[Xe]\,6s^2 4f^{12}$
Sc	21	$[Ar]\,4s^2 3d^1$	Tm	69	$[Xe]\,6s^2 4f^{13}$
Ti	22	$[Ar]\,4s^2 3d^2$	Yb	70	$[Xe]\,6s^2 4f^{14}$
V	23	$[Ar]\,4s^2 3d^3$	Lu	71	$[Xe]\,6s^2 4f^{14} 5d^1$
Cr	24	$[Ar]\,4s^1 3d^5$	Hf	72	$[Xe]\,6s^2 4f^{14} 5d^2$
Mn	25	$[Ar]\,4s^2 3d^5$	Ta	73	$[Xe]\,6s^2 4f^{14} 5d^3$
Fe	26	$[Ar]\,4s^2 3d^6$	W	74	$[Xe]\,6s^2 4f^{14} 5d^4$
Co	27	$[Ar]\,4s^2 3d^7$	Re	75	$[Xe]\,6s^2 4f^{14} 5d^5$
Ni	28	$[Ar]\,4s^2 3d^8$	Os	76	$[Xe]\,6s^2 4f^{14} 5d^6$
Cu	29	$[Ar]\,4s^1 3d^{10}$	Ir	77	$[Xe]\,6s^2 4f^{14} 5d^7$
Zn	30	$[Ar]\,4s^2 3d^{10}$	Pt	78	$[Xe]\,6s^1 4f^{14} 5d^9$
Ga	31	$[Ar]\,4s^2 3d^{10} 4p^1$	Au	79	$[Xe]\,6s^1 4f^{14} 5d^{10}$
Ge	32	$[Ar]\,4s^2 3d^{10} 4p^2$	Hg	80	$[Xe]\,6s^2 4f^{14} 5d^{10}$

Table continues ▶

for scandium is $1s^2 2s^2 2p^6 3s^2 3p^6 4s^2 3d^1$. The elements following scandium—titanium (22) through copper (29)—continue to add d electrons until the third energy level has its maximum of 18. Two exceptions in the orderly electron addition are chromium (24) and copper (29), the structures of which are given in Table 10.4. The third energy level of electrons is first completed in the element copper. Table 10.4 shows the order of filling of the electron orbitals and the electron configuration for the elements.

TABLE 10.4 (*continued*)

Element	Atomic number	Electron structure	Element	Atomic number	Electron structure
As	33	[Ar] $4s^2 3d^{10} 4p^3$	Tl	81	[Xe] $6s^2 4f^{14} 5d^{10} 6p^1$
Se	34	[Ar] $4s^2 3d^{10} 4p^4$	Pb	82	[Xe] $6s^2 4f^{14} 5d^{10} 6p^2$
Br	35	[Ar] $4s^2 3d^{10} 4p^5$	Bi	83	[Xe] $6s^2 4f^{14} 5d^{10} 6p^3$
Kr	36	[Ar] $4s^2 3d^{10} 4p^6$	Po	84	[Xe] $6s^2 4f^{14} 5d^{10} 6p^4$
Rb	37	[Kr] $5s^1$	At	85	[Xe] $6s^2 4f^{14} 5d^{10} 6p^5$
Sr	38	[Kr] $5s^2$	Rn	86	[Xe] $6s^2 4f^{14} 5d^{10} 6p^6$
Y	39	[Kr] $5s^2 4d^1$	Fr	87	[Rn] $7s^1$
Zr	40	[Kr] $5s^2 4d^2$	Ra	88	[Rn] $7s^2$
Nb	41	[Kr] $5s^1 4d^4$	Ac	89	[Rn] $7s^2 6d^1$
Mo	42	[Kr] $5s^1 4d^5$	Th	90	[Rn] $7s^2 6d^2$
Tc	43	[Kr] $5s^2 4d^5$	Pa	91	[Rn] $7s^2 5f^2 6d^1$
Ru	44	[Kr] $5s^1 4d^7$	U	92	[Rn] $7s^2 5f^3 6d^1$
Rh	45	[Kr] $5s^1 4d^8$	Np	93	[Rn] $7s^2 5f^4 6d^1$
Pd	46	[Kr] $4d^{10}$	Pu	94	[Rn] $7s^2 5f^6$
Ag	47	[Kr] $5s^1 4d^{10}$	Am	95	[Rn] $7s^2 5f^7$
Cd	48	[Kr] $5s^2 4d^{10}$	Cm	96	[Rn] $7s^2 5f^7 6d^1$
In	49	[Kr] $5s^2 4d^{10} 5p^1$	Bk	97	[Rn] $7s^2 5f^9$
Sn	50	[Kr] $5s^2 4d^{10} 5p^2$	Cf	98	[Rn] $7s^2 5f^{10}$
Sb	51	[Kr] $5s^2 4d^{10} 5p^3$	Es	99	[Rn] $7s^2 5f^{11}$
Te	52	[Kr] $5s^2 4d^{10} 5p^4$	Fm	100	[Rn] $7s^2 5f^{12}$
I	53	[Kr] $5s^2 4d^{10} 5p^5$	Md	101	[Rn] $7s^2 5f^{13}$
Xe	54	[Kr] $5s^2 4d^{10} 5p^6$	No	102	[Rn] $7s^2 5f^{14}$
Cs	55	[Xe] $6s^1$	Lr	103	[Rn] $7s^2 5f^{14} 6d^1$
Ba	56	[Xe] $6s^2$	Unq	104	[Rn] $7s^2 5f^{14} 6d^2$
La	57	[Xe] $6s^2 5d^1$	Unp	105	[Rn] $7s^2 5f^{14} 6d^3$
Ce	58	[Xe] $6s^2 4f^1 5d^1$	Unh	106	[Rn] $7s^2 5f^{14} 6d^4$

Note: For simplicity of expression, symbols of the chemically stable noble gases are used as a portion of the electron structure for the elements beyond neon. For example, the electron structure of a sodium atom, Na, consists of ten electrons, as in neon [Ne], plus a $3s^1$ electron. Detailed electron structures for the noble gases are given in Table 10.5.

EXAMPLE 10.2   What is the electron structure for a sulfur atom (atomic number 16) and for an iron atom (atomic number 26)?

Look in Table 10.4 for the element with atomic number 16 and write down its structure, [Ne] $3s^2 3p^4$. [Ne] is an abbreviated structure for neon, which is $1s^2 2s^2 2p^6$. Therefore, the electron structure for a sulfur atom is $1s^2 2s^2 2p^6 3s^2 3p^4$.

For an iron atom, Table 10.4 shows a structure of [Ar] $4s^2 3d^6$. This notation means that the electron structure of iron consists of the electron structure for argon plus $4s^2 3d^6$. The table shows that the structure for [Ar] is [Ne] $3s^2 3p^6$, which is equal to $1s^2 2s^2 2p^6 3s^2 3p^6$. Therefore, the electron structure for iron is $1s^2 2s^2 2p^6 3s^2 3p^6 4s^2 3d^6$.

If Table 10.4 is not available, the structure can be determined by tabulating electrons as in Example 10.1 and using Figure 10.4. Structures for all the noble gases, He, Ne, Ar, and so on, are also given in Table 10.5.

PRACTICE   Write the electron configuration for each of the following elements, using Table 10.4.
(a) Co   (b) Cd   (c) Pt
Answers: (a) $[Ar]4s^2 3d^7$ (b) $[Kr]5s^2 4d^{10}$ (c) $[Xe]6s^2 4f^{14} 5d^7$

10.8   Diagramming Atomic Structures

We can use several methods to diagram the atomic structures of atoms, depending on what we are trying to illustrate. When we want to show both the nuclear makeup and the electron structure of each energy level (without orbital detail), we can use a diagram such as Figure 10.5.

A method of diagramming subenergy levels is shown in Figure 10.6. Each orbital is represented by a square □. When the orbital contains one electron, an arrow (↑) is placed in the square. A second arrow, pointing downward (↓), indicates the second electron in that orbital.

The diagram for hydrogen is ↑. Helium, with two electrons, is drawn as ↑↓; both electrons are 1s electrons. The diagram for lithium shows three electrons in two energy levels, $1s^2 2s^1$. All four electrons of beryllium are s electrons, $1s^2 2s^2$. Boron has the first p electron, which is located in the $2p_x$ orbital. Because it is energetically more difficult for the next p electron to pair up with the electron in the p_x orbital than to occupy a second p orbital, the second p electron in carbon is located in the $2p_y$ orbital. The third p electron in nitrogen is still unpaired and is found in the $2p_z$ orbital. The next three electrons pair with each of the 2p electrons through the element neon. Also shown in Figure 10.6 are the equivalent linear expressions for these orbital electron structures.

The electrons in successive elements are found in sublevels of increasing energy. The general sequence of increasing energy of sublevels and the order of

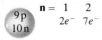 **n** = 1   2
2e^- 7e^-

Fluorine atom

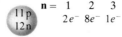 **n** = 1   2   3
2e^- 8e^- 1e^-

Sodium atom

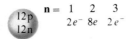 **n** = 1   2   3
2e^- 8e 2e^-

Magnesium atom

FIGURE 10.5

Atomic structure diagrams of fluorine, sodium, and magnesium atoms. The numbers of protons and neutrons are shown in the nucleus. The number of electrons is shown in each principal energy level outside the nucleus.

Element	Orbital electron structure						Linear expression of electron structure
	$1s$	$2s$	$2p_x$	$2p_y$	$2p_z$	$3s$	
H	↑						$1s^1$
He	↑↓						$1s^2$
Li	↑↓	↑					$1s^2 2s^1$
Be	↑↓	↑↓					$1s^2 2s^2$
B	↑↓	↑↓	↑				$1s^2 2s^2 2p_x^1$
C	↑↓	↑↓	↑	↑			$1s^2 2s^2 2p_x^1 2p_y^1$
N	↑↓	↑↓	↑	↑	↑		$1s^2 2s^2 2p_x^1 2p_y^1 2p_z^1$
O	↑↓	↑↓	↑↓	↑	↑		$1s^2 2s^2 2p_x^2 2p_y^1 2p_z^1$
F	↑↓	↑↓	↑↓	↑↓	↑		$1s^2 2s^2 2p_x^2 2p_y^2 2p_z^1$
Ne	↑↓	↑↓	↑↓	↑↓	↑↓		$1s^2 2s^2 2p_x^2 2p_y^2 2p_z^2$
Na	↑↓	↑↓	↑↓	↑↓	↑↓	↑	$1s^2 2s^2 2p_x^2 2p_y^2 2p_z^2 3s^1$

FIGURE 10.6

Subenergy-level electron structure of hydrogen through sodium atoms. Each electron is indicated by an arrow placed in the square, which represents the orbital.

filling sublevels with electrons is

$$1s\ 2s\ 2p\ 3s\ 3p\ 4s\ 3d\ 4p\ 5s\ 4d\ 5p\ 6s\ 4f\ 5d\ 6p\ 7s\ 5f\ 6d$$

Minor variations from the electron structure predicted by the foregoing general sequence are found in a number of atoms. Table 10.4 shows the accepted ground-state electron structure for the elements.

EXAMPLE 10.3   Diagram the electron structure of a zinc atom and a rubidium atom. Use the $1s^2 2s^2 2p^6$, etc., method.

 The atomic number of zinc is 30; therefore it has 30 protons and 30 electrons in a neutral atom. Using Figure 10.4 tabulate the 30 electrons as follows:

Orbital	Number of e^-	Total e^-
$1s$	$2e^-$	2
$2s$	$2e^-$	4
$2p$	$6e^-$	10
$3s$	$2e^-$	12
$3p$	$6e^-$	18
$4s$	$2e^-$	20
$3d$	$10e^-$	30

 The electron structure of a zinc atom is $1s^2 2s^2 2p^6 3s^2 3p^6 4s^2 3d^{10}$. Check by adding the superscripts, which should equal 30.

 The atomic number of rubidium is 37; therefore it has 37 protons and 37 electrons in a neutral atom. With a little practice, and using Figure 10.4, the electron structure may be written directly in the linear form. The electron structure of a rubidium atom is $1s^2 2s^2 2p^6 3s^2 3p^6 4s^2 3d^{10} 4p^6 5s^1$. Check by adding the superscripts, which should equal 37.

10.9   Lewis-Dot Representation of Atoms

The Lewis-dot (or electron-dot) method of representing atoms, proposed by the American chemist G. N. Lewis (1875–1946), uses the symbol for the element and dots for electrons. The number of dots placed around the symbol equals the number of s and p electrons in the outermost energy level of the atom. Paired dots represent paired electrons; unpaired dots represent unpaired electrons. For example, $\mathbf{H}\cdot$ is the Lewis symbol for a hydrogen atom, $1s^1$; $:\mathbf{\dot{B}}$ is the Lewis symbol for a boron atom, $1s^2 2s^2 2p^1$; $:\mathbf{\ddot{I}}\cdot$ is an iodine atom, which has seven

$$\text{H}\cdot \quad \text{He:} \quad \text{Li}\cdot \quad \text{Be:} \quad :\overset{\cdot}{\text{B}} \quad :\overset{\cdot}{\text{C}}\cdot \quad :\overset{\cdot}{\underset{\cdot}{\text{N}}}\cdot \quad \cdot\overset{\cdot\cdot}{\underset{\cdot}{\text{O}}}: \quad :\overset{\cdot\cdot}{\underset{\cdot}{\text{F}}}: \quad :\overset{\cdot\cdot}{\underset{\cdot\cdot}{\text{Ne}}}:$$

$$\text{Na}\cdot \quad \text{Mg:} \quad :\overset{\cdot}{\text{Al}} \quad :\overset{\cdot}{\text{Si}}\cdot \quad :\overset{\cdot}{\underset{\cdot}{\text{P}}}\cdot \quad \cdot\overset{\cdot\cdot}{\underset{\cdot}{\text{S}}}: \quad :\overset{\cdot\cdot}{\underset{\cdot}{\text{Cl}}}: \quad :\overset{\cdot\cdot}{\underset{\cdot\cdot}{\text{Ar}}}:$$

$$\text{K}\cdot \quad \text{Ca:}$$

FIGURE 10.7

Lewis-dot diagrams of the first twenty elements. Dots represent electrons in the outermost energy level only.

electrons in its outermost energy level. In the case of boron, the symbol B represents the boron nucleus and the $1s^2$ electrons; the dots represent only the $2s^2 2p^1$ electrons.

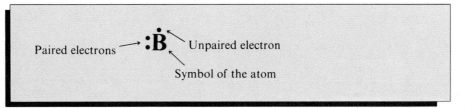

Paired electrons → $:\mathbf{B}$ ← Unpaired electron

Symbol of the atom

The Lewis-dot method is often used, not only because of its simplicity of expression, but also because much of the chemistry of the atom is directly associated with the electrons in the outermost energy level. This association is especially true for the first 20 elements and the remaining Group A elements of the periodic table (see Chapter 11). Figure 10.7 shows Lewis-dot diagrams for the elements hydrogen through calcium.

EXAMPLE 10.4   Write the Lewis-dot structure for a phosphorus atom.

First establish the electron structure for a phosphorus atom. It is $1s^2 2s^2 2p^6 3s^2 3p^3$. Note that there are five electrons in the outermost principal energy level; they are $3s^2 3p^3$.

Write the symbol for phosphorus and place the five electrons as dots around it:

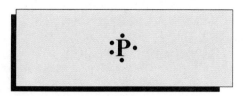

The $3s^2$ electrons are paired and are represented by the paired dots. The $3p^3$ electrons, which are unpaired, are represented by the single dots.

PRACTICE   Write the Lewis-dot structure for the following elements:
(a) N   (b) Al   (c) Sr   (d) Br
Answers: (a) :N·   (b) :Al·   (c) Sr: (d) :Br:

10.10   The Octet Rule

noble gases

The family of elements consisting of helium, neon, argon, krypton, xenon, and radon is known as the **noble gases**. These elements, formerly called *inert gases*, have almost no chemical reactivity; in fact, no compounds of any of them were known before 1962.

TABLE 10.5

Arrangement of Electrons in the Noble Gases[a]

Noble gas	Symbol	Electron structure					
		$n = 1$	2	3	4	5	6
Helium	He	$1s^2$					
Neon	Ne	$1s^2$	$2s^2 2p^6$				
Argon	Ar	$1s^2$	$2s^2 2p^6$	$3s^2 3p^6$			
Krypton	Kr	$1s^2$	$2s^2 2p^6$	$3s^2 3p^6 3d^{10}$	$4s^2 4p^6$		
Xenon	Xe	$1s^2$	$2s^2 2p^6$	$3s^2 3p^6 3d^{10}$	$4s^2 4p^6 4d^{10}$	$5s^2 5p^6$	
Radon	Rn	$1s^2$	$2s^2 2p^6$	$3s^2 3p^6 3d^{10}$	$4s^2 4p^6 4d^{10} 4f^{14}$	$5s^2 5p^6 5d^{10}$	$6s^2 6p^6$

[a] Each gas except helium has eight electrons in its outermost energy level.

octet rule

Each of these elements, except helium, has an outer shell of eight electrons, two *s* and six *p* (see Table 10.5). The only shell of helium is an *s* orbital filled with two electrons. The electron structure of the noble gases is such that the outer shell *s* and *p* orbitals are filled with paired electrons. This arrangement is very stable and makes the atoms of the noble gases chemically unreactive. Recognition of the extraordinary stability of this structure led to the **octet rule**: Through chemical changes many of the elements tend to attain an electron structure of eight electrons in their outermost energy level, identical to that of the chemically stable noble gases. Although the octet rule is useful and applies to the behavior of many elements and compounds, it is not universally applicable; some elements do not obey this rule. Applications of the octet rule are given in Chapter 12.

Concepts in Review

1. Describe the atom as conceived by Niels Bohr.

2. Discuss the contributions to atomic theory made by Dalton, Thomson, Rutherford, Bohr, Chadwick, and Schrödinger.

3. Describe what is meant by an electron orbital.

4. Determine the maximum number of electrons that can exist in the principal energy levels and sublevels.

5. Write the electron configuration for any of the first 56 elements.

6. Explain what is represented by the Lewis-dot (electron-dot) structure of an element.

7. Write the Lewis-dot (electron-dot) symbols for the first twenty elements.

8. Understand the basis for the octet rule.

Key Terms in Review

The terms listed here have all been defined within the chapter. Review the definitions of each, and use the glossary and margin notations within the chapter as study aids.

electron shells	octet rule
energy levels	orbital
noble gases	

Exercises

1. What is an electron orbital?
2. Under which conditions can a second electron enter an orbital already containing one electron?
3. What is meant when we say that the electron structure of an atom is in its ground state?
4. How do 1s and 2s orbitals differ? How are they alike?
5. What letters are used to designate the energy sublevels?
6. List the following electron sublevels in order of increasing energy: 2s, 2p, 4s, 1s, 3d, 3p, 4p, 3s.
7. How many s electrons, p electrons, and d electrons are possible in any electron shell?
8. How many protons are in the nucleus of an atom of each of these elements: H, B, F, Sc, Ag, U, Br, Sb, and Pb?
9. Give the electron structure ($1s^2 2s^2 2p^6 \ldots$) for B, Ti, Zn, Br, and Sr.
10. Why is the eleventh electron of the sodium atom located in the third energy level rather than in the second energy level?
11. What is the major difference between an orbital and a Bohr orbit?
12. Explain how the spectral lines of hydrogen occur.
13. Explain how Bohr used the data of the hydrogen bright-line spectrum to support his solar system model of atomic structure.
14. Explain why and how Bohr's solar system model of the atom was modified to the cloud model of the atom.
15. Using the formula $2n^2$, calculate the number of electrons that can exist in principal energy levels: **n** = 1, 2, 3, 4, 5, and 6.
16. How many electrons can be present in the fourth energy level?
17. How many orbitals can exist in the third energy level? What are they?
18. Sketch the s, p_x, p_y, and p_z orbitals.
19. Diagram the atomic structures of the following atoms:
    (a) $^{14}_{7}N$ (d) $^{91}_{40}Zr$
    (b) $^{35}_{17}Cl$ (e) $^{127}_{53}I$
    (c) $^{65}_{30}Zn$

20. Write out the electron configuration (long form) for each of the following elements:
    (a) Chlorine (d) Iron
    (b) Silver (e) Iodine
    (c) Lithium
21. Show the Lewis-dot structures for C, Mg, Al, Cl, and K.
22. In the designation $3d^7$, give the significance of the 3, the d, and the 7.
23. Using the method shown in Figure 10.6, show the orbital electron structure for an atom of:
    (a) Si (b) S (c) Ar (d) V
24. What electron structure do the noble gases have in common?
25. Why is the last electron in potassium located in the fourth energy level rather than in the third energy level?
26. Which atoms have the following electron structures?
    (a) $1s^2 2s^2 2p^6 3s^2$
    (b) $1s^2 2s^2 2p^5$
    (c) $1s^2 2s^2 2p^6 3s^2 3p^6 4s^2 3d^8$
    (d) $1s^2 2s^2 2p^6 3s^2 3p^6 4s^2 3d^5$
    (e) $1s^2 2s^2 2p^6 3s^2 3p^6 4s^2 3d^{10} 4p^6 5s^1 4d^5$
27. Show the electron structures ($1s^2 2s^2 2p^6 \ldots$) for elements of atomic numbers:
    (a) 8 (b) 11 (c) 17 (d) 23 (e) 28 (f) 34
28. Using only Table 10.4 show the electron structures ($1s^2 2s^2 2p^6 \ldots$) for the elements having the following numbers of electrons:
    (a) 9 (b) 26 (c) 31 (d) 39 (e) 52
29. Which elements have the following electron structures?
    (a) $[Ar]4s^2 3d^1$ (c) $[Kr]5s^2 4d^{10} 5p^2$
    (b) $[Ar]4s^2 3d^{10} 4p^6$ (d) $[Xe]6s^1$
30. Identify these atoms from their atomic structure diagrams:

    (a) $\left(\begin{array}{c}16p\\16n\end{array}\right)$ $2e^-$ $8e^-$ $6e^-$

    (b) $\left(\begin{array}{c}28p\\32n\end{array}\right)$ $2e^-$ $8e^-$ $16e^-$ $2e^-$

31. Diagram the atomic structures (as in Exercise 30) for these atoms:
    (a) $^{27}_{13}Al$ (b) $^{51}_{23}V$
32. State the octet rule and its relationship to the noble gases.
33. Write Lewis-dot symbols for these atoms: He, B, O, Na, Si, Ar, Ga, Ca, Br, and Kr.
34. Which of the following statements are correct? Rewrite each incorrect statement to make it correct.
    (a) In the ground state, electrons tend to occupy orbitals having the lowest possible energy.
    (b) The maximum number of p electrons in the first energy level is six.
    (c) A $2s$ electron is in a lower energy state than a $2p$ electron.
    (d) The electron structure for a carbon atom is $1s^2 2s^2 2p^2$.
    (e) The $2p_x$, $2p_y$, and $2p_z$ electron orbitals are all in the same energy state.
    (f) The energy level of a $3d$ electron is higher than that of a $4s$ electron.
    (g) The electron structure for a calcium atom is $1s^2 2s^2 2p^6 3s^2 3p^6 3d^2$.
    (h) There are seven principal energy levels for the known elements.
    (i) The third energy level can have a maximum of 18 electrons.
    (j) The number of possible d electrons in the third energy level is ten.
    (k) The first f electron occurs in the fourth principal energy level.
    (l) The Lewis-dot symbol for nitrogen is $:\overset{\cdot}{N}\cdot$
    (m) The Lewis-dot symbol for potassium is $P\cdot$
    (n) Atoms of all the noble gases (except helium) have eight electrons in their outermost energy level.
    (o) A p orbital is spherically symmetrical around the nucleus.
    (p) An atom of nitrogen has two electrons in a $1s$ orbital, two electrons in a $2s$ orbital, and one electron in each of three different $2p$ orbitals.
    (q) The maximum number of electrons that can occupy a specific energy level $\mathbf{n}$ is given by $2\mathbf{n}^2$.
    (r) $^{12}_{6}C$ is an isotope of carbon that is used as the reference standard for the atomic mass system.
    (s) The Lewis-dot symbol for the noble gas helium is $:\overset{\cdot\cdot}{\underset{\cdot\cdot}{He}}:$
    (t) When an orbital contains two electrons, the electrons have parallel spins.
    (u) The Bohr theory proposed that electrons move around the nucleus in circular orbits.
    (v) Bohr concluded from his experiment that the positive charge and almost all the mass were concentrated in a very small nucleus.

CHAPTER ELEVEN

The Periodic Arrangement of the Elements

Organizing and classifying large volumes of information for easy use and convenience is a necessary task in modern society. In the hardware store, paint colors are sorted into color chips—first by major color group, then by shades or hues within the specific color. In the local winery, wines are classified as red, white, or rosé—then further grouped by type of grape, origin, and year. The use of personal computers has soared in recent years, and allows us to classify many things, such as addresses, birthdays, or household finances. Even the calendar itself is broken into rows and columns to permit us to organize our time and activities by day, week, and month.

So, too, chemical information is classified in a concise and useful way. The periodic table provides the structure for this classification scheme. It permits the prediction of chemical properties and interactions in much the same manner as the calendar or computer spreadsheet.

11.1   Early Attempts to Classify the Elements

Chemists of the early 19th century had sufficient knowledge of the properties of elements to recognize similarities among groups of elements. As early as 1817 J. W. Döbereiner (1780–1849), professor at the University of Jena in Germany, observed the existence of *triads* of similarly behaving elements, in which the middle element had an atomic mass approximating the average of the other two elements. He also noted that for many other properties the value for the central element was approximately the average of the values for the other two elements. Table 11.1 presents comparative data on atomic mass and density for two sets of Döbereiner's triads.

In 1864, J. A. R. Newlands (1837–1898), an English chemist, reported his *Law of Octaves*. In his studies Newlands observed that, when the elements were arranged according to increasing atomic masses, every eighth element had similar properties. (The noble gases were not yet discovered at that time.) Newlands' theory was ridiculed by his contemporaries in the Royal Chemical Society, and they refused to publish his work. Many years later, however, Newlands was awarded the highest honor of the society for this important contribution to the development of the periodic law.

In 1869 Dmitri Ivanovitch Mendeleev (1834–1907) of Russia and Lothar Meyer (1830–1895) of Germany independently published their periodic arrangements of the elements that were based on increasing atomic masses. Because his arrangement was published slightly earlier and was in a somewhat more useful form than that of Meyer, Mendeleev's name is usually associated with the modern periodic table.

TABLE 11.1

Döbereiner's Triads

Triads	Atomic mass	Density (g/mL at 4°C)
Chlorine	35.5	1.56[a]
Bromine	79.9	3.12
Iodine	126.9	4.95
Average of chlorine and iodine	81.2	3.26
Calcium	40.1	1.55
Strontium	87.6	2.6
Barium	137.4	3.5
Average of calcium and barium	88.8	2.52

[a] Density at −34°C (liquid)

11.2   The Periodic Law

Only about 63 elements were known when Mendeleev constructed his table. He arranged these elements so that those with similar chemical properties fitted into columns to form family groups. The arrangement left many gaps between elements, and Mendeleev predicted that these spaces would be filled as new elements were discovered. For example, spaces for undiscovered elements were left after calcium, under aluminum, and under silicon. He called these unknown elements eka-boron, eka-aluminum, and eka-silicon. The term *eka* comes from Sanskrit meaning "one" and was used to indicate that the missing element was one place away in the table from the element indicated. Mendeleev even went so far as to predict with high accuracy the physical and chemical properties of these undiscovered elements. The three elements, scandium (atomic number 21), gallium (31), and germanium (32), were in fact discovered during Mendeleev's lifetime and were found to have properties agreeing very closely with the predictions that he had made for eka-boron, eka-aluminum, and eka-silicon. The amazing way in which Mendeleev's predictions were fulfilled is illustrated in Table 11.2, which compares the predicted properties of eka-silicon with those of germanium, discovered by the German chemist C. Winkler in 1886.

Two major additions have been made to the periodic table since Mendeleev's time: (1) A new family of elements, the noble gases, was discovered and added; and (2) elements having atomic numbers greater than 92 have been discovered and fitted into the table.

TABLE 11.2

Comparison of Properties of Eka-Silicon (predicted by Mendeleev) with Germanium

Property	Eka-silicon, Es (predicted)	Germanium, Ge (observed)
Atomic mass	72	72.6
Color of metal	Dirty gray	Grayish-white
Density	5.5 g/mL	5.35 g/mL
Oxide formula	EsO_2	GeO_2
Oxide density	4.7 g/mL	4.70 g/mL
Chloride formula	$EsCl_4$	$GeCl_4$
Chloride density	1.9 g/mL	1.87 g/mL
Boiling temperature of chloride	Under 100°C	86°C

The term *periodic* means recurring at regular intervals. The original periodic tables were based on the premise that the properties of the elements are periodic functions of their atomic masses. However, this basic premise had some disturbing discrepancies. For example, the atomic mass for argon is greater than that of potassium, yet argon had to be placed before potassium, because argon is certainly a noble gas, and potassium behaves like the other alkali metals. These discrepancies were resolved by the work of the British physicist, H. G. J. Moseley (1887–1915) and by the discovery of the existence of isotopes. Moseley noted that the x-ray emission frequencies of the elements increased in a regular, stepwise fashion each time the nuclear charge (atomic number) increased by one unit. This observation meant that the periodic table must be based on a revised premise—namely, that the properties of the elements are a periodic function of their *atomic numbers*. Under this revised premise argon (18) properly comes before potassium (19), even though the atomic mass of argon is greater than the atomic mass of potassium. The current statement of the **periodic law** is

periodic law

> **The properties of the chemical elements recur periodically when the elements are arranged in increasing order of their atomic numbers.**

As the format of the periodic table is studied, it becomes evident that the periodicity of the properties of the elements is due to the recurring similarities of their electron structures.

TABLE 11.3

Periodic Table of the Elements

Group IA — 1 / Period

Legend:
- Atomic number — 11
- Name — Sodium
- Symbol — Na
- Electron structure — 2 8 1
- Atomic mass — 22.98977

[a] Mass number of most stable or best-known isotope
[b] Mass of the isotope of longest half-life

Transition elements

Period	IA 1	IIA 2	IIIB 3	IVB 4	VB 5	VIB 6	VIIB 7	VIII 8	VIII 9
1	1 Hydrogen **H** 1.0079 (1)								
2	3 Lithium **Li** 6.941 (2,1)	4 Beryllium **Be** 9.01218 (2,2)							
3	11 Sodium **Na** 22.98977 (2,8,1)	12 Magnesium **Mg** 24.305 (2,8,2)							
4	19 Potassium **K** 39.098 (2,8,8,1)	20 Calcium **Ca** 40.08 (2,8,8,2)	21 Scandium **Sc** 44.9559 (2,8,9,2)	22 Titanium **Ti** 47.90 (2,8,10,2)	23 Vanadium **V** 50.9414 (2,8,11,2)	24 Chromium **Cr** 51.996 (2,8,13,1)	25 Manganese **Mn** 54.9380 (2,8,13,2)	26 Iron **Fe** 55.847 (2,8,14,2)	27 Cobalt **Co** 58.9332 (2,8,15,2)
5	37 Rubidium **Rb** 85.4678 (2,8,18,8,1)	38 Strontium **Sr** 87.62 (2,8,18,8,2)	39 Yttrium **Y** 88.9059 (2,8,18,9,2)	40 Zirconium **Zr** 91.22 (2,8,18,10,2)	41 Niobium **Nb** 92.9064 (2,8,18,12,1)	42 Molybdenum **Mo** 95.94 (2,8,18,13,1)	43 Technetium **Tc** 98.9062[b] (2,8,18,14,1)	44 Ruthenium **Ru** 101.07 (2,8,18,15,1)	45 Rhodium **Rh** 102.9055 (2,8,18,16,1)
6	55 Cesium **Cs** 132.9054 (2,8,18,18,8,1)	56 Barium **Ba** 137.34 (2,8,18,18,8,2)	57 Lanthanum **La*** 138.9055 (2,8,18,18,9,2)	72 Hafnium **Hf** 178.49 (2,8,18,32,10,2)	73 Tantalum **Ta** 180.9479 (2,8,18,32,11,2)	74 Wolfram (Tungsten) **W** 183.85 (2,8,18,32,12,2)	75 Rhenium **Re** 186.2 (2,8,18,32,13,2)	76 Osmium **Os** 190.2 (2,8,18,32,14,2)	77 Iridium **Ir** 192.22 (2,8,18,32,17,0)
7	87 Francium **Fr** (223)[a] (2,8,18,32,18,8,1)	88 Radium **Ra** 226.0254[b] (2,8,18,32,18,8,2)	89 Actinium **Ac**** (227)[a] (2,8,18,32,18,9,2)	104 Unnilquadium **Unq** (261)[a] (2,8,18,32,32,10,2)	105 Unnilpentium **Unp** (262)[a] (2,8,18,32,32,11,2)	106 Unnilhexium **Unh** (263)[a] (2,8,18,32,32,12,2)	107 Unnilseptium **Uns** (262)[a]	108 Unniloctium **Uno** (265)[a]	109 Unnilennium **Une** (266)[a]

Lanthanide series 6 *

58 Cerium **Ce** 140.12 (2,8,18,20,8,2)	59 Praseodymium **Pr** 140.9077 (2,8,18,21,8,2)	60 Neodymium **Nd** 144.24 (2,8,18,22,8,2)	61 Promethium **Pm** (145)[a] (2,8,18,23,8,2)	62 Samarium **Sm** 150.4 (2,8,18,24,8,2)

Actinide series 7 **

90 Thorium **Th** 232.0381[b] (2,8,18,32,18,10,2)	91 Protactinium **Pa** 231.0359[b] (2,8,18,32,20,9,2)	92 Uranium **U** 238.029 (2,8,18,32,21,9,2)	93 Neptunium **Np** 237.0482 (2,8,18,32,22,9,2)	94 Plutonium **Pu** (242)[a] (2,8,18,32,23,9,2)

Table continues ▶

Noble gases 18

IIIA 13	IVA 14	VA 15	VIA 16	VIIA 17	
					2 — Helium — **He** — 4.00260 — (2)
5 — Boron — **B** — 10.81 — (2,3)	6 — Carbon — **C** — 12.011 — (2,4)	7 — Nitrogen — **N** — 14.0067 — (2,5)	8 — Oxygen — **O** — 15.9994 — (2,6)	9 — Fluorine — **F** — 18.99840 — (2,7)	10 — Neon — **Ne** — 20.179 — (2,8)
13 — Aluminum — **Al** — 26.98154 — (2,8,3)	14 — Silicon — **Si** — 28.086 — (2,8,4)	15 — Phosphorus — **P** — 30.97376 — (2,8,5)	16 — Sulfur — **S** — 32.06 — (2,8,6)	17 — Chlorine — **Cl** — 35.453 — (2,8,7)	18 — Argon — **Ar** — 39.948 — (2,8,8)

	IB 11	IIB 12	IIIA 13	IVA 14	VA 15	VIA 16	VIIA 17	18
10								

28 — Nickel — **Ni** — 58.71 — (2,8,16,2)	29 — Copper — **Cu** — 63.546 — (2,8,18,1)	30 — Zinc — **Zn** — 65.38 — (2,8,18,2)	31 — Gallium — **Ga** — 69.72 — (2,8,18,3)	32 — Germanium — **Ge** — 72.59 — (2,8,18,4)	33 — Arsenic — **As** — 74.9216 — (2,8,18,5)	34 — Selenium — **Se** — 78.96 — (2,8,18,6)	35 — Bromine — **Br** — 79.904 — (2,8,18,7)	36 — Krypton — **Kr** — 83.80 — (2,8,18,8)
46 — Palladium — **Pd** — 106.4 — (2,8,18,18,0)	47 — Silver — **Ag** — 107.868 — (2,8,18,18,1)	48 — Cadmium — **Cd** — 112.40 — (2,8,18,18,2)	49 — Indium — **In** — 114.82 — (2,8,18,18,3)	50 — Tin — **Sn** — 118.69 — (2,8,18,18,4)	51 — Antimony — **Sb** — 121.75 — (2,8,18,18,5)	52 — Tellurium — **Te** — 127.60 — (2,8,18,18,6)	53 — Iodine — **I** — 126.9045 — (2,8,18,18,7)	54 — Xenon — **Xe** — 131.30 — (2,8,18,18,8)
78 — Platinum — **Pt** — 195.09 — (2,8,18,32,17,1)	79 — Gold — **Au** — 196.9665 — (2,8,18,32,18,1)	80 — Mercury — **Hg** — 200.59 — (2,8,18,32,18,2)	81 — Thallium — **Tl** — 204.37 — (2,8,18,32,18,3)	82 — Lead — **Pb** — 207.2 — (2,8,18,32,18,4)	83 — Bismuth — **Bi** — 208.9804 — (2,8,18,32,18,5)	84 — Polonium — **Po** — (210)^a — (2,8,18,32,18,6)	85 — Astatine — **At** — (210)^a — (2,8,18,32,18,7)	86 — Radon — **Rn** — (222)^a — (2,8,18,32,18,8)

Inner transition elements

63 — Europium — **Eu** — 151.96 — (2,8,18,25,8,2)	64 — Gadolinium — **Gd** — 157.25 — (2,8,18,25,9,2)	65 — Terbium — **Tb** — 158.9254 — (2,8,18,27,8,2)	66 — Dysprosium — **Dy** — 162.50 — (2,8,18,28,8,2)	67 — Holmium — **Ho** — 164.9304 — (2,8,18,29,8,2)	68 — Erbium — **Er** — 167.26 — (2,8,18,30,8,2)	69 — Thulium — **Tm** — 168.9342 — (2,8,18,31,8,2)	70 — Ytterbium — **Yb** — 173.04 — (2,8,18,32,8,2)	71 — Lutetium — **Lu** — 174.97 — (2,8,18,32,9,2)
95 — Americium — **Am** — (243)^a — (2,8,18,32,25,8,2)	96 — Curium — **Cm** — (247)^a — (2,8,18,32,25,9,2)	97 — Berkelium — **Bk** — (249)^a — (2,8,18,32,26,9,2)	98 — Californium — **Cf** — (251)^a — (2,8,18,32,27,9,2)	99 — Einsteinium — **Es** — (254)^a — (2,8,18,32,28,9,2)	100 — Fermium — **Fm** — (253)^a — (2,8,18,32,29,9,2)	101 — Mendelevium — **Md** — (256)^a — (2,8,18,32,30,9,2)	102 — Nobelium — **No** — (254)^a — (2,8,18,32,31,9,2)	103 — Lawrencium — **Lr** — (257)^a — (2,8,18,32,32,9,2)

11.3   Arrangement of the Periodic Table

periodic table

periods of
elements

groups or families
of elements

The modern **periodic table** is shown in Table 11.3 and on the inside front cover of this book. In this table the elements are arranged horizontally in numerical sequence according to their atomic numbers; the result is seven horizontal rows called **periods**. Each period, with the exception of the first, starts with an alkali metal and ends with a noble gas. This arrangement forms vertical columns of elements that have identical or similar outer-shell electron structures and thus similar chemical properties. These columns are known as **groups** or **families of elements**.

The heavy zigzag line starting at boron and running diagonally down the table separates the elements into metals and nonmetals. The elements to the right of the line are nonmetallic, and those to the left are metallic. The elements bordering the zigzag line are the metalloids, which show both metallic and nonmetallic properties. With some exceptions the characteristic electron arrangement of metals is that their atoms have one, two, or three electrons in the outer energy level, whereas nonmetals have five, six, or seven electrons in the outer energy level.

In this periodic arrangement the elements also fall into blocks according to the sublevel of electrons that is being filled. The grouping of the elements into *spdf* blocks is shown in Table 11.4. The *s* and *p* blocks are made up of Group A elements and noble gases. The *s* block consists of the Group IA and IIA elements and helium; each element in this block has one or two *s* electrons in its outer energy level. The *p* block contains the Group IIIA through VIIA elements and the noble gases (except helium). In the *p* block, electrons are filling the *p* sublevel orbitals. Each *p* block element has two *s* electrons and from one to six *p* electrons in its outermost energy level.

The *d* block includes the transition elements of Groups IB through VIIB and Group VIII (see Section 11.7). In this block electrons are filling the *d* sublevel orbitals. Each period beginning with period 4 can have ten *d* block elements. Periods 4, 5, and 6 each have ten; period 7 is incomplete (see Table 11.4).

The *f* block consists of the inner transition elements and includes two series of 14 elements in periods 6 and 7. The first series, found in period 6, is the lanthanide series (Ce to Lu); the second, found in period 7, is the actinide series (Th to Lr). Electrons are filling the $4f$ sublevel orbitals in the lanthanide series and the $5f$ orbitals in the actinide series.

11.4   Periods of Elements

Of the seven periods of elements (see Tables 11.3 and 11.4), the first three are known as *short periods*, and the remaining four as *long periods*. The first period contains 2 elements, hydrogen and helium. Period 2 contains 8 elements, starting

TABLE 11.4

Arrangement of Elements According to the Sublevel of Electrons Being Filled in Their Atomic Structure

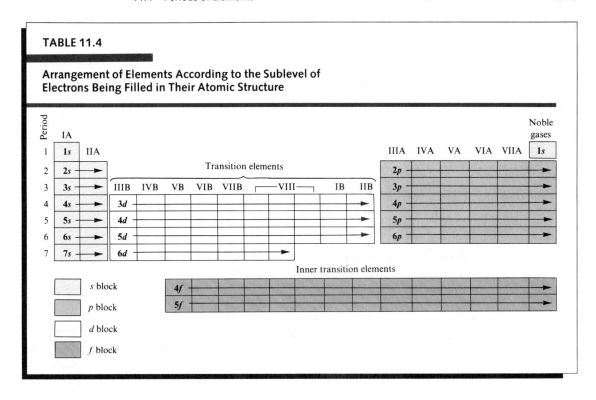

with lithium and ending with neon. Period 3 also contains 8 elements, sodium to argon. Periods 4 and 5 each contain 18 elements; period 6 has 32 elements; and period 7, which is incomplete, contains the remaining elements.

The number of each period corresponds to the number of the outermost energy level that contains electrons of the elements in that period. For example, the period 1 elements contain electrons in the first energy level only; period 2 elements contain electrons in the first and second energy levels; period 3 elements contain electrons in the first, second, and third levels; and so on.

From the standpoint of properties, we find that the two elements in the first period are gases; hydrogen (1) is reactive, and helium (2) is a noble gas. Thereafter, as we move from left to right across the successive periods from 2 to 6, we find that the elements vary from strongly metallic at the beginning to nonmetallic at the end of each period. Period 7 is incomplete but begins with a strongly metallic element, francium (87).

Starting with the third element of the long periods, 4, 5, 6, and 7 (scandium, Sc; yttrium, Y; lanthanum, La; actinium, Ac), the inner shells of d and f orbital electrons begin to fill in, forming the transition and inner transition elements (d and f blocks). Both the transition and inner transition elements are metallic.

TABLE 11.5

The Number of Elements in Each Period

Period number	Number of elements	Electron orbitals being filled in each period
1	2	$1s$
2	8	$2s2p$
3	8	$3s3p$
4	18	$4s3d4p$
5	18	$5s4d5p$
6	32	$6s4f5d6p$
7	22	$7s5f6d$

In general the atomic radii of the elements within a period decrease with increasing nuclear charge. This decrease occurs because, as the positive charge on the nucleus increases, it exerts a greater attractive force on the electrons, causing the atom to become smaller. Therefore the size of the atoms becomes progressively smaller from left to right within each period (see Figure 11.1). Because the noble gases do not readily combine with other elements to form compounds, the radii of their atoms are not determined in the same comparative manner as are those of the other elements. However, calculations have shown that the radii of the noble gas atoms are about the same as, or slightly smaller than, the element immediately preceding them. Slight deviations in atomic radii occur in the middle of the long periods of the elements.

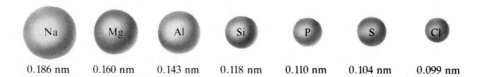

Na	Mg	Al	Si	P	S	Cl
0.186 nm	0.160 nm	0.143 nm	0.118 nm	0.110 nm	0.104 nm	0.099 nm

FIGURE 11.1

Radii of period 3 elements. In general the size of the atoms in a period decreases with increasing nuclear charge.

11.5 Groups or Families of Elements

representative
elements

transition
elements

The groups, or families, of elements are numbered IA through VIIA, IB through VIIB, VIII, and noble gases. Group A elements are often referred to as the **representative**, or *main*, groups of **elements**. They are filling in electrons in the *s* and *p* orbitals. Group B and Group VIII elements are known as the **transition elements**. They are filling electrons in the *d* and *f* orbitals.

The elements comprising each family have similar outer energy-level electron structures. In the atoms of Group A elements the number of electrons in the outer energy level is identical to the group number. Group IA is known as the *alkali metals*. Each atom of this family of elements has one *s* electron in its outer energy level. Group IIA atoms, known as the *alkaline earth metals*, each have two *s* electrons in their outer energy level. Group VIIA is known as the halogens; their atoms each have seven electrons (two *s* and five *p*) in their outer energy level. The noble gases (except helium) have eight electrons in their outer energy level.

Each alkali metal starts a new period of elements in which the *s* electron occupies a principal energy level one greater than in the previous period. As a result the size of the atoms of this and other families of elements increases from the top to the bottom of the family. (This generality has some exceptions.) The relative size of the alkali metal atoms is illustrated in Figure 11.2.

One major distinction between groups is the energy level to which the last electron is added. In the elements of Groups IA through VIIA, IB, IIB, and the noble gases, the last electron is added either to an *s* or to a *p* orbital located in the outermost energy level. (This rule has some exceptions; see Table 10.4.) In the elements of Groups IIIB through VIIB and VIII, the last electron goes to a *d* or to an *f* orbital located in an inner energy level. For examples see Table 11.6.

The general characteristics of group properties are as follows:

1. The number of electrons in the outer energy level of Groups IA through VIIA, IB, and IIB elements is the same as the group number. The other B groups and Group VIII do not show this characteristic. Each noble gas except helium has eight electrons in its outer energy level.
2. The groups on the left and in the middle sections of the table tend to be metallic in nature. The groups on the right tend to be nonmetallic.
3. The radii of the elements increase from top to bottom within a particular group (for example, from lithium to francium).
4. Elements at the bottom of a group tend to be more metallic in their properties than those at the top. This tendency is especially noticeable in Groups IVA through VIIA.
5. Elements within an A group have the same number of electrons in their outer shell and show closely related chemical properties.
6. Elements within a B group have some similarity in electron structure and also show some similarities in chemical properties.

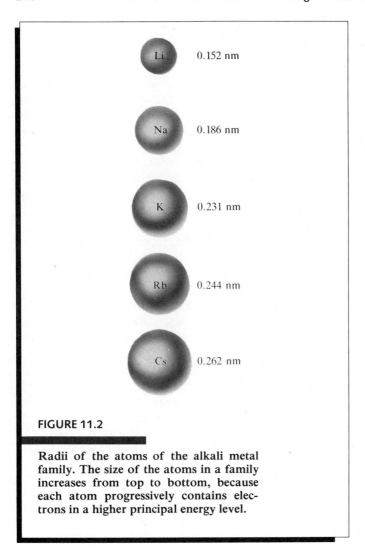

Li   0.152 nm

Na   0.186 nm

K   0.231 nm

Rb   0.244 nm

Cs   0.262 nm

FIGURE 11.2

Radii of the atoms of the alkali metal family. The size of the atoms in a family increases from top to bottom, because each atom progressively contains electrons in a higher principal energy level.

A new format for numbering the groups in the periodic table has been recommended by the American Chemical Society Committee on Nomenclature and by the International Union of Pure and Applied Chemistry (IUPAC) Commission on the Nomenclature of Inorganic Chemistry. The new format numbers the groups from 1 to 18 going from left to right. A periodic table using

---

TABLE 11.6

Comparison of Placement of the Last Electron in Group A and Group B Elements*

Group	Noble gas	IA	IIA	IIIB	IVB
Element	Ar	K	Ca	Sc	Ti
Electron structure	2,8,8	2,8,8,1	2,8,8,2	2,8,9,2	2,8,10,2
Energy level	1,2,3	1,2,3,4	1,2,3,4	1,2,3,4	1,2,3,4
Last electron added	↑	↑	↑	↑	↑

* The energy level to which the last electron is added is indicated by the arrow, an outer level for Group A elements and an inner level for Group B elements.

---

this numbering method is shown in Table 11.3. Because the new method of numbering is still controversial, we will continue to use the A and B system in this book.

EXAMPLE 11.1   Write Lewis-dot symbols for the representative elements Na, K, Sr, Se, and Br.

First we need to locate these elements in the periodic table. They are all Group A elements. Then, knowing that Group A numbers are the same as the number of outer-shell electrons and that Lewis-dot symbols show only the outer-shell electrons, we can proceed to write the Lewis-dot symbols.

Na, K	Group IA	$(1\,e^-)$	Na· K·
Sr	Group IIA	$(2\,e^-)$	Sr:
Se	Group VIA	$(6\,e^-)$	:S̈e·
Br	Group VIIA	$(7\,e^-)$	:B̈r·

11.6   Predicting Formulas by Use of the Periodic Table

The periodic table can be used to predict the formulas of simple compounds. As we shall see in Chapter 12, the chemical properties of the elements are dependent on their electrons. The representative (Group A) elements ordinarily form compounds using only the electrons in their outer energy level. If we examine

TABLE 11.7

Formulas of Compounds Formed by Alkali Metals

Lewis-dot structure	Monoxides	Chlorides	Bromides	Sulfates
Li·	Li_2O	LiCl	LiBr	Li_2SO_4
Na·	Na_2O	NaCl	NaBr	Na_2SO_4
K·	K_2O	KCl	KBr	K_2SO_4
Rb·	Rb_2O	RbCl	RbBr	Rb_2SO_4
Cs·	Cs_2O	CsCl	CsBr	Cs_2SO_4

Group IA, the alkali metals, we see that all of them have one electron in their outer energy level; all follow a noble gas in the table; and all, except lithium, have eight electrons in their next inner shell. These likenesses suggest that these metals should have a great deal of similarity in their chemistry, since their chemical properties are vested primarily in their outer-shell electron. And they do have similarity: All readily lose their outer electron and attain a noble gas electron structure. In doing so they form compounds with similar atomic compositions. For example, all the monoxides of Group IA contain two atoms of the alkali metal to one atom of oxygen. Their formulas are Li_2O, Na_2O, K_2O, Rb_2O, Cs_2O, and Fr_2O.

How can we use the table to predict formulas of other compounds? Because of similar electron structures, the elements in a family generally form compounds with the same atomic ratios, as was shown in the preceding paragraph for the oxides of the Group IA metals. In general, if we know the atomic ratio of a particular compound, say sodium chloride (NaCl), we can predict the atomic ratios and formulas of the other alkali metal chlorides. These formulas are LiCl, KCl, RbCl, CsCl, and FrCl (see Table 11.7).

In a similar way, if we know that the formula of the oxide of hydrogen is H_2O, we can predict that the formula of the sulfide will be H_2S, because sulfur has the same outer-shell electron structure as oxygen. It must be recognized, however, that these are only predictions; it does not necessarily follow that every element in a group will behave like the others or even that a predicted compound will actually exist. Knowing the formulas for potassium chlorate, bromate, and iodate to be $KClO_3$, $KBrO_3$, and KIO_3, we can correctly predict the corresponding sodium compounds to have the formulas $NaClO_3$, $NaBrO_3$, and $NaIO_3$. Fluorine belongs to the same family of elements (Group VIIA) as chlorine, bromine, and iodine. So we can predict that potassium and sodium fluorates will have the formulas KFO_3 and $NaFO_3$. But this prediction would not be correct, because potassium and sodium fluorates are not known to exist. However, if they did exist, the formulas could very well be correct, for these predictions are based on comparisons with known formulas and similar electron structures.

Predicting formulas using this type of analogy is not as reliable for the transition (Group B) elements.

EXAMPLE 11.2   The formula for calcium sulfate is $CaSO_4$ and that for lithium carbonate is Li_2CO_3. Predict formulas for (a) magnesium sulfate, (b) potassium carbonate, and (c) magnesium selenate.

(a) Look in the periodic table for calcium and magnesium. They are both in Group IIA. Since the formula for calcium sulfate is $CaSO_4$, it is reasonable to predict that the formula for magnesium sulfate is $MgSO_4$.

(b) Find lithium and potassium in the periodic table. They are in Group IA. Since the formula for lithium carbonate is Li_2CO_3, it is reasonable to predict that K_2CO_3 is the formula for potassium carbonate.

(c) Find selenium in the periodic table. It is in Group VIA just below sulfur. Therefore it is reasonable to assume that selenium forms selenate in the same way that sulfur forms sulfate. Since $MgSO_4$ was the predicted formula for magnesium sulfate in part (a), it is reasonable to assume that the formula for magnesium selenate is $MgSeO_4$.

11.7 Transition Elements

Elements in Groups IB, IIIB through VIIB, and VIII are known as the transition elements. Transition elements occur in periods 4, 5, 6, and 7. The transition elements are characterized by an increasing number of d or f electrons in an inner shell; they all have either one or two electrons in their outer shell. In period 4, electrons enter the $3d$ sublevel. In period 5, electrons enter the $4d$ sublevel. The transition elements in period 6 include the lanthanide series (or *rare earth elements*, Ce to Lu), in which electrons are entering the $4f$ sublevel. The $5d$ sublevel also fills up in the sixth period. The seventh period of elements is an incomplete period. It includes the actinide series (Th to Lr) in which electrons are entering the $5f$ and $6d$ sublevels.

All of the transition elements are metals. Their oxidation states may be variable; that is, in the formation of compounds electrons may come from more than one energy level. For this reason many of these metals form multiple series of compounds. These compounds often occur in the form of some of the most beautifully colored crystals to be found in nature.

11.8   Noble Gases

The noble gases are the last group in the periodic table and are sometimes called the zero group. These gases are characterized by extremely low chemical reactivity and were, in fact, formerly known as the inert gases because they were believed to be chemically inert. Their low reactivity is associated with a stable electron outer shell consisting of filled s orbitals and (except helium) filled p orbitals (see Section 10.11).

All the noble gases except radon are normally present in the atmosphere. Argon is the most abundant, about 1% by volume. The others are present only in trace amounts. Argon was discovered in 1894 by Lord Raleigh (1842–1919) and Sir William Ramsay (1852–1916). Helium was first observed in the spectrum of the sun during an eclipse in 1868. It was not until 1894 that Ramsay recognized that helium exists in the earth's atmosphere. In 1898 he and his co-worker, Morris W. Travers (1872–1961), announced the discovery of neon, krypton, and xenon, having isolated them from liquid air. Friedrich E. Dorn (1848–1916) first identified radon, the heaviest member of the noble gases, as a radioactive gas emanating from the element radium.

Because of its low density and nonflammability, helium has been used for filling balloons and dirigibles. Only hydrogen surpasses helium in lifting power. Helium mixed with oxygen is used by deep-sea divers for breathing. This mixture reduces the danger of acquiring the "bends" (caisson disease), pains and paralysis suffered by divers on returning from the ocean depths to normal atmospheric pressure. Helium is also used in heliarc welding, to supply an inert atmosphere for the welding of active metals such as magnesium. Helium is found in some natural gas wells in the southwestern United States. As a liquid it is used to study the properties of substances at very low temperatures. The boiling point of liquid helium, 4.2 K, is not far above absolute zero.

We are all familiar with the neon sign, in which a characteristic red color is produced when an electric discharge is passed through a tube filled with neon. This color may be modified by mixing the neon with other gases or by changing the color of the glass. Argon is used primarily in gas-filled electric light bulbs and other types of electronic tubes to provide an inert atmosphere for prolonging tube life. Argon is also used in some welding applications where an inert atmosphere is needed. Krypton and xenon have not been used extensively because of their limited availability. Radon is radioactive; it has been used medicinally in the treatment of cancer.

For many years it was believed that the noble gases could not be made to combine chemically with any other element. Then, in 1962, Neil Bartlett at the University of British Columbia, Vancouver, synthesized the first noble gas compound, xenon hexafluoroplatinate, $XePtF_6$. This outstanding discovery opened a new field in the techniques of preparing noble gas compounds and investigating their chemical bonding and properties. Other compounds of xenon and compounds of krypton and radon have been prepared. Some of these are XeF_2, XeF_4, XeF_6, $XeOF_4$, $Xe(OH)_6$, and KrF_2.

11.9   New Elements

Mendeleev allowed gaps in his orderly periodic table for elements whose discovery he predicted. These were actually discovered, as were all the elements up to atomic number 92 that occur naturally on the earth. Seventeen elements beyond uranium (atomic numbers 93–109) have been discovered or synthesized since 1939. All these elements have unstable nuclei and are radioactive. Beyond element 101 the isotopes synthesized thus far have such short lives that chemical identification has not been accomplished.

Intensive research is continuing on the synthesis of still heavier elements to extend the periodic table beyond the presently known elements. It is predicted that elements 110 to 118 will be very stable but still radioactive. Element 114 will lie below lead (82) in the periodic table and should be exceptionally stable. Element 118 should be a member of the noble gas family. Elements 119 and 120 should be in Groups IA and IIA, respectively, and have electrons in the $8s$ sublevel.

11.10   Value of the Periodic Table

The arrangement of the elements into the modern periodic table has been and continues to be of great value to chemists and chemistry students. This value increases with your knowledge of chemistry. For a given element the following data can usually be obtained directly from the table: name, symbol, atomic number, atomic mass, electron configuration, group number, period number, and whether it is a metal, nonmetal, or metalloid. From the location of the element in the periodic table, one can estimate and compare many of its properties—such as ionization energy, density, atomic radius, atomic volume, oxidation states, electrical conductance, and electronegativity—with those of other elements.

The periodic table is still used as a guide in predicting the synthesis of possible new elements. It presents a very large amount of chemical information in compact form and correlates the properties and relationships of all the elements. The table is so useful that a copy hangs in nearly every chemistry lecture hall and laboratory in the world. Refer to it often.

Concepts in Review

1. Describe briefly the contributions of Döbereiner, Newlands, Mendeleev, Meyer, and Moseley to the development of the periodic law.

2. State the periodic law in its modern form.

3. Indicate the locations of the metals, nonmetals, metalloids, and noble gases in the periodic table.

4. Indicate in the periodic table the areas where the s, p, d, and f orbitals are being filled.

5. Describe how atomic radii vary (a) from left to right in a period, and (b) from top to bottom in a group.

6. Distinguish between representative elements and transition elements.

7. Identify groups of elements by their special names.

8. Describe the changes in outer-shell electron structure (a) when moving from left to right in a period, and (b) when going from top to bottom in a group.

9. Explain the relationship between group number and the number of outer-shell electrons for the representative elements.

10. Indicate how the changes in the electron structure between adjacent transition elements differs from that between adjacent representative elements.

11. Write Lewis-dot symbols for the representative elements from their position in the periodic table.

12. Predict formulas of simple compounds formed between the representative (Group A) elements using the periodic table.

Key Terms in Review

The terms listed here have all been defined within the chapter. You should review the definitions of each. Remember to use the glossary and the margin notations within the chapter to help you.

groups or families of elements
periodic law
periodic table

periods of elements
representative elements
transition elements

Exercises

1. From the standpoint of electron structure, what do the elements in the s block have in common?

2. Write the symbols for the elements having atomic numbers 8, 16, 34, 52, and 84. What do these elements have in common?

3. Mendeleev described a then-undiscovered element as eka-silicon. When this element was discovered, (a) what was it named and (b) how did the actual density agree with that predicted by Mendeleev?

4. How does the size of the atoms of the elements in the third period vary from left to right?

5. Write the symbols of the family of elements that have seven electrons in their outer energy level.

6. Write the symbols of the alkali metal family in order of increasing atomic size.

7. What is the greatest number of elements to be found in any period? Which periods have this number?

8. From the standpoint of energy level, how does the placement of the last electron differ in Group A elements from that of Group B elements?

9. What were Döbereiner's triads? In what way did they lead to later developments in periodicity?

10. What do you think is the basis for Newlands' Law of Octaves?

11. Given that the noble gases were not discovered at the time of Newlands' Law of Octaves, could his law be extended as far as the element bromine? Explain.

12. What is meant by the term *periodicity* as applied to the elements?

13. Why are some missing elements in Mendeleev's periodic table considered a victory for his table rather than a defeat?

14. How does our modern periodic table differ from Mendeleev's?

15. What additional understanding of periodic properties was added by the work of H. G. J. Moseley?

16. State the periodic law in its current form. How is this law different from the law as stated by Mendeleev?

17. Find the places in the modern periodic table where elements are not in proper sequence according to atomic mass.

18. How are elements in a period related to one another?

19. How are elements in a group related to one another?

20. What is common about the electron structures of the alkali metals?

21. Why would you expect the elements zinc, cadmium, and mercury to be in the same chemical family?

22. Draw the Lewis-dot symbols for Cs, Ba, Tl, Pb, Po, At, and Rn. How do these structures correlate with the group in which each element occurs?

23. Pick the electron structures below that represent elements in the same chemical family:
    (a) $1s^2 2s^1$
    (b) $1s^2 2s^2 2p^4$
    (c) $1s^2 2s^2 2p^2$
    (d) $1s^2 2s^2 2p^6 3s^2 3p^4$
    (e) $1s^2 2s^2 2p^6 3s^2 3p^6$
    (f) $1s^2 2s^2 2p^6 3s^2 3p^6 4s^2$
    (g) $1s^2 2s^2 2p^6 3s^2 3p^6 4s^1$
    (h) $1s^2 2s^2 2p^6 3s^2 3p^6 4s^2 3d^1$

24. Pick the electron structures below that represent elements in the same chemical family:
    (a) $[He]2s^2 2p^6$
    (b) $[Ne]3s^1$
    (c) $[Ne]3s^2$
    (d) $[Ne]3s^2 3p^3$
    (e) $[Ar]4s^1 3d^{10}$
    (f) $[Ar]4s^2 3d^{10} 4p^6$
    (g) $[Ar]4s^2 3d^5$
    (h) $[Kr]5s^1 4d^{10}$

25. In the periodic table, calcium, element 20, is surrounded by elements 12, 19, 21, and 38. Which of these have physical and chemical properties most resembling calcium?

26. Oxygen is a gas. Sulfur is a solid. What is it about their electron structures that causes them to be grouped in the same chemical family?

27. Classify each of the following elements as metals, nonmetals, or metalloids:
    (a) Potassium
    (b) Plutonium
    (c) Sulfur
    (d) Antimony
    (e) Iodine
    (f) Tungsten
    (g) Molybdenum
    (h) Germanium

28. In which period and group does an electron first appear in a *d* orbital?

29. How many electrons occur in the outer shell of Group IIIA and IIIB elements? Why are they different?

30. In which groups are transition elements located?

31. How do the electron structures of the transition elements differ from representative elements?

32. Which element in each of the following pairs has the larger atomic radius?
    (a) Na or K (c) O or F (e) Ti or Zr
    (b) Na or Mg (d) Br or I

33. Which element in each of Groups IA–VIIA has the smallest atomic radius?

34. Why does the atomic size increase in going down any family of the periodic table?

35. All the atoms within each Group A family of elements can be represented by the same Lewis-dot symbol. Complete the table, expressing the Lewis-dot symbol for each group. Use E to represent the elements.

Group	IA	IIA	IIIA	IVA	VA	VIA	VIIA
	E·						

36. Let E be any representative element. Following the pattern in the table, write the formulas for the hydrogen and oxygen compounds of
(a) Na (c) Al (e) Sb (g) Cl
(b) Ca (d) Sn (f) Se

Group IA	IIA	IIIA	IVA	VA	VIA	VIIA
EH	EH_2	EH_3	EH_4	EH_3	H_2E	HE
E_2O	EO	E_2O_3	EO_2	E_2O_5	EO_3	E_2O_7

37. Group IB elements have one electron in their outer shell, as do Group IA elements. Would you expect them to form compounds such as CuCl, AgCl, and AuCl? Explain.

38. The formula for lead(II) bromide is $PbBr_2$; predict formulas for tin(II) and germanium(II) bromides. (If needed, see Section 6.4, part (b), for the use of Roman numerals in naming compounds.)

39. The formula for sodium sulfate is Na_2SO_4. Write the names and formulas for the other alkali metal sulfates.

40. The formula for calcium bromide is $CaBr_2$. Write formulas for magnesium bromide, strontium bromide, and barium bromide.

41. Why should the discovery of the existence of isotopes have any bearing on the fact that the periodicity of the elements is a function of their atomic numbers and not their atomic masses?

Try to answer Exercises 42–44 without referring to the periodic table.

42. The atomic numbers of the noble gases are 2, 10, 18, 36, 54, and 86. What are the atomic numbers for the elements with six electrons in their outer electron shells?

43. Element number 87 is in Group IA, period 7. Describe its outermost energy level. How many energy levels of electrons does it have?

44. If element 36 is a noble gas, in which groups would you expect elements 35 and 37 to occur?

45. Write a paragraph describing the general features of the periodic table.

46. Rank the following five elements according to the radii of their atoms, from smallest to largest: Na, Mg, Cl, K, and Rb.

47. Which of the following statements are correct? Rewrite the incorrect statements to make them correct.

(a) Properties of the elements are periodic functions of their atomic numbers.

(b) There are more nonmetallic elements than metallic elements.

(c) Metallic properties of the elements increase from left to right across a period.

(d) Metallic properties of the elements increase from top to bottom in a family of elements.

(e) Calcium is a member of the alkaline earth metal family.

(f) Iron belongs to the alkali metal family.

(g) Bromine belongs to the halogen family.

(h) Neon is a noble gas.

(i) Group A elements do not contain partially filled d or f sublevels.

(j) An atom of oxygen is larger than an atom of lithium.

(k) An atom of sulfur is larger than an atom of oxygen.

(l) An atom of aluminum (Group IIIA) has five electrons in its outer shell.

(m) If the formula for calcium iodide is CaI_2, then the formula for cesium iodide is CsI_2.

(n) If the formula for aluminum oxide is Al_2O_3, then the formula for gallium oxide is Ga_2O_3.

(o) Uranium is an inner transition element.

(p) The element $[Ar]4s^2 3d^{10}4p^5$ is a halogen.

(q) The element $[Kr]5s^2$ is a nonmetal.

(r) The element with $Z = 12$ forms compounds similar to the element with $Z = 37$.

(s) Nitrogen, fluorine, neon, gallium, and bromine are all nonmetals.

(t) The Lewis-dot symbol for tin is $:\overset{\cdot}{S}n\cdot$

(u) The atom having an outer-shell electron structure of $5s^2 5p^2$ would be in period 6, Group IVA.

(v) The yet-to-be-discovered element with an atomic number of 118 would be a noble gas.

(w) The most metallic element of Group VIIA is iodine.

CHAPTER TWELVE

Chemical Bonds: The Formation of Compounds from Atoms

For centuries, we have been aware that certain metals cling to a magnet. Balloons may stick to a wall. Why? Recently, the activity of a superconductor floating in air has been the subject of television commercials. High-speed levitation trains are heralded to be the wave of the future. How do they function? In each case, forces of attraction and repulsion are at work.

Interestingly, human interactions suggest that "opposites attract" and "likes repel." Attractions draw us into friendships and significant relationships, whereas repulsive forces may produce debate and antagonism. We form and break apart interpersonal bonds throughout our lives.

In chemistry, we also see this phenomenom. Nonliving substances form chemical bonds as a result of electrical attractions and repulsions. Chemical bonds provide the tremendous diversity of compounds found in nature.

Chapter Preview

12.1   Chemical Bonds

chemical bonds

Except in very rare instances matter does not fly apart spontaneously. It is prevented from doing so by forces acting at the ionic and molecular levels. Through chemical reactions atoms tend to attain more stable states at lower chemical potential energy levels. Atoms react chemically by losing, gaining, or sharing electrons. Forces arise from electron transferring and electron sharing interactions. Those forces that hold oppositely charged ions together or that bind atoms together in molecules are called **chemical bonds**. The two principal types of bonds are the ionic bond and the covalent bond.

12.2   Ionization Energy and Electron Affinity

ionization energy

According to Niels Bohr's concept, electrons exist at various discrete energy levels depending on the amount of energy that the atom has absorbed. If sufficient energy is applied to an atom, it is possible to remove completely (or "knock out") one or more electrons from its structure, thereby forming a positive ion:

atom + energy $\longrightarrow$ positive ion + electron (e^-)

The amount of energy required to remove one electron from an atom or an ion is called the **ionization energy**. The *first* ionization energy is the amount of energy required to remove the first electron from an atom, the *second* ionization energy is the amount needed to remove the second electron from that atom, and so on.

TABLE 12.1

Ionization Energies for Selected Elements*

Element	Required amounts of energy (kJ/mol)				
	1st e⁻	2nd e⁻	3rd e⁻	4th e⁻	5th e⁻
H	1,314				
He	2,372	5,247			
Li	520	7,297	11,810		
Be	900	1,757	14,845	21,000	
B	800	2,430	3,659	25,020	32,810
C	1,088	2,352	4,619	6,222	37,800
Ne	2,080	3,962	6,276	9,376	12,190
Na	496	4,565	6,912	9,540	13,355

* Values are expressed in kilojoules per mole, showing energies required to remove 1 to 5 electrons per atom. Color indicates the energy needed to remove an electron from a noble-gas electron structure.

Ionization energies may be expressed in terms of various energy units such as electron volts, kilojoules, or kilocalories per mole. In the following discussion, ionization energy is expressed as kilojoules per mole (kJ/mol), indicating the number of kilojoules required to remove one electron from each atom in a mole of atoms.

$$\underset{\substack{\text{1 mol} \\ \text{sodium atoms}}}{Na} \quad + 494\ kJ \longrightarrow \quad \underset{\substack{\text{1 mol} \\ \text{sodium ions}}}{Na^+} \quad + \quad \underset{\substack{\text{1 mol} \\ \text{electrons}}}{e^-}$$

Table 12.1 gives the ionization energies for the removal of one to five electrons from several elements. Thus, 1314 kJ are required to remove one electron from a mole of hydrogen atoms, but only 494 kJ are needed to remove the first electron from a mole of sodium atoms. The table shows that increasingly higher amounts of energy are required to remove the second, third, fourth, and fifth electrons. This sequence is logical, because the removal of electrons does not decrease the number of protons, or the amount of charge, in the nucleus. Therefore the remaining electrons are held more tightly. The data of Table 12.1 also show that an extra-large ionization energy is needed whenever an electron is pulled away from a noble-gas electron structure, clearly demonstrating the high stability of this structure.

First ionization energies have been experimentally determined for most of the elements. Figure 12.1 is a graphic plot of the first ionization energies of the first 56 elements, H through Ba.

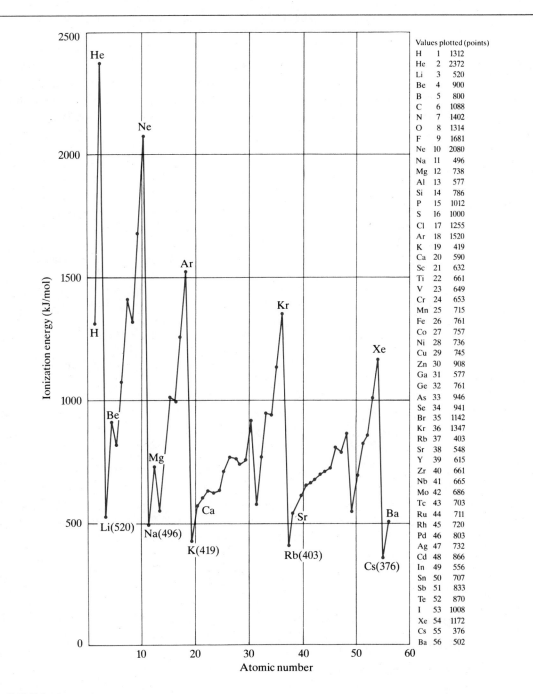

FIGURE 12.1

Periodic relationship of the first ionization energy to the atomic number of the elements.

Certain periodic relationships are evident in Figure 12.1. In the Group A, or representative elements, the first ionization energy generally decreases from top to bottom in groups or families. For example, in alkali metals (Group IA) the first ionization energy decreases from 520 kJ/mol for lithium to 376 kJ/mol for cesium. The two main reasons for this family trend are that (1) as we go from top to bottom in the family, the electron being removed is farther away from its nucleus, and (2) the electron is shielded from its nucleus by more inner shells of electrons. Furthermore all the alkali metals have relatively low first ionization energies, indicating that each has one electron that is easily removed.

From left to right within a period, the ionization energy gradually increases despite some irregularities. The noble gases have relatively high values, confirming the nonreactivity of this family and the stability of an eight-electron structure in the outer energy level.

electron affinity

Atoms also have *electron affinity*—that is, the ability to attract electrons and form negative ions. **Electron affinity** may be defined as the amount of energy released or absorbed when an electron is added to an atom to form a negative ion. For most of the elements, heat is released when an atom adds an electron.

$$\text{atom} + \text{electron} \longrightarrow \text{negative ion} + \text{energy}$$

Electron affinity is a measure of the attraction of an atom for an electron—in other words, of the tendency to form a negative ion. Chlorine, for example, is a nonmetallic element and has a strong tendency to form negative ions. Consequently the electron affinity of chlorine is high.

$$\text{Cl} \quad + \quad e^- \quad \longrightarrow \quad \text{Cl}^- \quad + 348 \text{ kJ (83 kcal)}$$

1 mol	1 mol	1 mol
chlorine atoms	electrons	chloride ions

Electron affinity tends to be high for nonmetals (particularly for the halogens, oxygen, and sulfur) and low for metals. The general trend of electron affinity is to increase from left to right in any period and to decrease from top to bottom in a family of elements.

12.3   Electrons in the Outer Shell: Valence Electrons

One outstanding property of the elements is their tendency to form a stable outer-shell electron structure. For many elements this stable outer shell contains eight electrons (two *s* and six *p*) identical to the outer-shell electron structure of the noble gases. Atoms undergo rearrangements of electron structure to attain a state of greater stability. These rearrangements are accomplished by losing, gaining, or sharing electrons with other atoms. For example, a hydrogen atom has a tendency to accept another electron and thus attain an electron structure like that of the stable noble gas helium; a fluorine atom can acquire one more electron to attain a stable electron structure like neon; a sodium atom tends to lose one electron to attain a stable electron structure like neon.

$$\text{Na} + \text{energy} \longrightarrow \text{Na}^+ + 1\,e^-$$

valence electrons

The electrons in the outermost shell of an atom are responsible for most of this electron activity and are called the **valence electrons**. In Lewis-dot symbols of atoms, the dots represent the outer-shell electrons and thus also represent the valence electrons. For example, hydrogen has one valence electron; sodium, one; aluminum, three; and oxygen, six. When a rearrangement of these electrons takes place between atoms, a chemical change occurs.

H·	Äl·	·Ö:
One valence electron	Three valence electrons	Six valence electrons

12.4   The Ionic Bond: Transfer of Electrons from One Atom to Another

The chemistry of the elements, especially the representative ones, is to attain an outer-shell electron structure like that of the chemically stable noble gases. With the exception of helium, this stable structure consists of eight electrons in the outer shell (see Table 10.5).

Let us look at the electron structures of sodium and chlorine to see how each element can attain a structure of 8 electrons in its outer shell. A sodium atom has 11 electrons: 2 in the first energy level, 8 in the second energy level, and 1 in the third energy level. A chlorine atom has 17 electrons: 2 in the first energy level, 8 in the second energy level, and 7 in the third energy level. If a sodium atom transfers or loses its $3s$ electron, its third energy level becomes vacant, and it becomes a sodium ion with an electron configuration identical to that of the noble gas neon. This process absorbs energy.

$$11+ \quad 2e^-\ 8e^-\ 1e^- \quad \longrightarrow \quad 11+^{1+} \quad 2e^-\ 8e^- + 1e^-$$

Na atom $(1s^2 2s^2 2p^6 3s^1)$ Na$^+$ ion $(1s^2 2s^2 2p^6)$

An atom that has lost or gained electrons will have a plus or minus electric charge, depending on which charged particles, protons or electrons, are in excess. Recall that a charged atom or group of atoms is called an *ion*.

By losing a negatively charged electron, the sodium atom becomes a positively charged particle known as a sodium ion. The charge, $+1$, results because the nucleus still contains 11 positively charged protons, and the electron orbitals contain only 10 negatively charged electrons. The charge is indicated by a plus sign (+) and is written as a superscript after the symbol of the element (Na^+).

A chlorine atom with 7 electrons in the third energy level needs 1 electron to pair up with its one unpaired $3p$ electron to attain the stable outer-shell electron structure of argon. By gaining 1 electron the chlorine atom becomes a chloride ion (Cl^-), a negatively charged particle containing 17 protons and 18 electrons. This process releases energy.

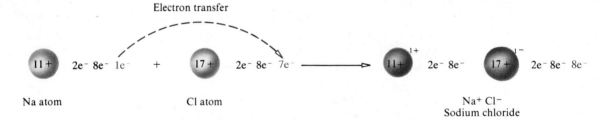

$$17+ \quad 2e^- \; 8e^- \; 7e^- \quad + \quad 1e^- \longrightarrow \quad 17+ \quad 2e^- \; 8e^- \; 8e^-$$

Cl atom ($1s^2 2s^2 2p^6 3s^2 3p^5$)   Cl$^-$ ion ($1s^2 2s^2 2p^6 3s^2 3p^6$)

Consider the case in which sodium and chlorine atoms react with each other. The $3s$ electron from the sodium atom transfers to the half-filled $3p$ orbital in the chlorine atom to form a positive sodium ion and a negative chloride ion. The compound sodium chloride results because the Na^+ and Cl^- ions are strongly attracted to each other by their opposite electrostatic charges. The force holding the oppositely charged ions together is an ionic bond.

Electron transfer

$$11+ \quad 2e^- \; 8e^- \; 1e^- \quad + \quad 17+ \quad 2e^- \; 8e^- \; 7e^- \longrightarrow \quad 11+ \quad 2e^- \; 8e^- \qquad 17+ \quad 2e^- \; 8e^- \; 8e^-$$

Na atom   Cl atom Na^+ Cl^-
                                                                        Sodium chloride

The Lewis-dot representation of sodium chloride formation is shown below.

$$Na\cdot + \cdot\ddot{\underset{\cdot\cdot}{Cl}}: \longrightarrow Na^+ : \ddot{\underset{\cdot\cdot}{Cl}}:^-$$

The chemical reaction between sodium and chlorine is a very vigorous one, producing considerable heat in addition to the salt formed. When energy is released in a chemical reaction, the products are more stable than the reactants. Note that in NaCl both atoms attained a noble-gas electron structure.

Sodium chloride is made up of cubic crystals in which each sodium ion is surrounded by six chloride ions and each chloride ion by six sodium ions, except at the crystal surface. A visible crystal is a regularly arranged aggregate of millions of these ions, but the ratio of sodium to chloride ions is one-to-one, hence the formula NaCl. The cubic crystalline lattice arrangement of sodium chloride is shown in Figure 12.2.

Figure 12.3 contrasts the relative sizes of sodium and chlorine atoms with those of their ions. The sodium ion is smaller than the atom due primarily to two

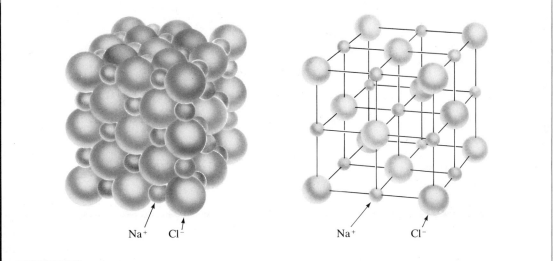

FIGURE 12.2

Sodium chloride crystal. Diagram represents a small fragment of sodium chloride, which forms cubic crystals. Each sodium ion is surrounded by six chloride ions, and each chloride ion is surrounded by six sodium ions.

factors: (1) The sodium atom has lost its outer shell of 1 electron, thereby reducing its size; and (2) the 10 remaining electrons are now attracted by 11 protons and are thus drawn closer to the nucleus. Conversely the chloride ion is larger than the atom because it has 18 electrons but only 17 protons; the nuclear attraction on each electron is thereby decreased, allowing the chlorine atom to expand as it forms an ion.

We have seen that when sodium reacts with chlorine, each atom becomes an electrically charged ion. Sodium chloride, like all ionic substances, is held together by the attraction existing between positive and negative charges. An **ionic bond** is the attraction between oppositely charged ions.

ionic bond

Ionic bonds are formed whenever one or more electrons are transferred from one atom to another. The metals, which have relatively little attraction for their valence electrons, tend to form ionic bonds when they combine with nonmetals.

It is important to recognize that substances with ionic bonds do not exist as molecules. In sodium chloride, for example, the bond does not exist solely between a single sodium ion and a single chloride ion. Each sodium ion in the crystal attracts six near-neighbor negative chloride ions; in turn, each negative chloride ion attracts six near-neighbor positive sodium ions (see Figure 12.2).

A metal will usually have one, two, or three electrons in its outer energy level. In reacting, metal atoms characteristically lose these electrons, attain the electron

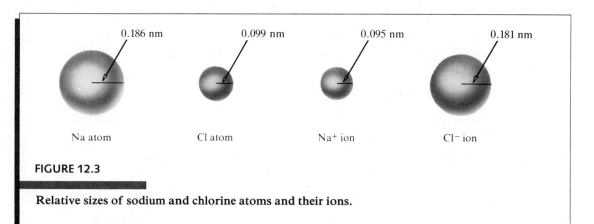

0.186 nm   0.099 nm   0.095 nm   0.181 nm

Na atom   Cl atom   Na⁺ ion   Cl⁻ ion

FIGURE 12.3

Relative sizes of sodium and chlorine atoms and their ions.

TABLE 12.2

Change in Atomic Radii of Selected Metals and Nonmetals*

Atomic radius (nm)	Ionic radius (nm)		Atomic radius (nm)	Ionic radius (nm)	
Li 0.152	Li⁺	0.060	F 0.071	F⁻	0.136
Na 0.186	Na⁺	0.095	Cl 0.099	Cl⁻	0.181
K 0.227	K⁺	0.133	Br 0.114	Br⁻	0.195
Mg 0.160	Mg²⁺	0.065	O 0.074	O²⁻	0.140
Al 0.143	Al³⁺	0.050	S 0.103	S²⁻	0.184

* The metals shown lose electrons to become positive ions. The nonmetals gain electrons to become negative ions.

structure of a noble gas, and become positive ions. A nonmetal, on the other hand, is only a few electrons short of having a complete octet in its outer energy level and thus has a tendency to gain electrons (electron affinity). In reacting with metals, nonmetal atoms characteristically gain one, two, or three electrons, attain the electron structure of a noble gas, and become negative ions. The ions formed by loss of electrons are much smaller than the corresponding metal atoms; the ions formed by gaining electrons are larger than the corresponding nonmetal atoms. The actual dimensions of the atomic and ionic radii of several metals and nonmetals are given in Table 12.2.

Study the following examples. Note the loss and gain of electrons between atoms; also note that the ions in each compound have a noble-gas electron structure.

EXAMPLE 12.1   Explain how magnesium and chlorine combine to form magnesium chloride, $MgCl_2$.

A magnesium atom of electron structure $1s^2 2s^2 2p^6 3s^2$ must lose two electrons or gain six electrons to reach a stable electron structure. If magnesium reacts with chlorine and each chlorine atom can accept only one electron, two chlorine atoms will be needed for the two electrons from one magnesium atom. The compound formed will contain one magnesium ion and two chloride ions. The magnesium atom, having lost two electrons, becomes a magnesium ion with a $+2$ charge. Each chloride ion will have a -1 charge. The transfer of electrons from a magnesium atom to two chlorine atoms is shown in the following illustration.

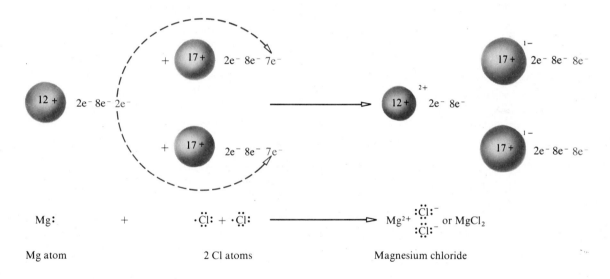

EXAMPLE 12.2   Explain the formation of sodium fluoride, NaF, from its elements.

The fluorine atom, with seven electrons in its outer shell, behaves similarly to the chlorine atom.

EXAMPLE 12.3   Explain the formation of aluminum chloride, $AlCl_3$ from its elements.

$$\cdot \ddot{\underset{\cdot\cdot}{Cl}} : \qquad\qquad : \ddot{\underset{\cdot\cdot}{Cl}} :^-$$

$$\ddot{Al}\cdot \;+\; \cdot \ddot{\underset{\cdot\cdot}{Cl}} : \longrightarrow \; Al^{3+} : \ddot{\underset{\cdot\cdot}{Cl}} :^- \qquad or \qquad AlCl_3$$

$$\cdot \ddot{\underset{\cdot\cdot}{Cl}} : \qquad\qquad : \ddot{\underset{\cdot\cdot}{Cl}} :^-$$

Aluminum atom	Chlorine atoms	Aluminum chloride

Each chlorine atom can accept only one electron. Therefore three chlorine atoms are needed to combine with the three outer-shell electrons of one aluminum atom. The aluminum atom has lost three electrons to become an aluminum ion, Al^{3+}, with a $+3$ charge.

EXAMPLE 12.4   Explain the formation of magnesium oxide, MgO from its elements.

$$\boxed{12+}\; 2e^-\; 8e^-\; 2e^- \quad + \quad \boxed{8+}\; 2e^-\; 6e^- \longrightarrow \overset{2+}{\boxed{12+}}\; 2e^-\; 8e^- \quad \overset{2-}{\boxed{8+}}\; 2e^-\; 8e^-$$

$$Mg : \qquad\qquad + \quad \cdot \ddot{\underset{\cdot}{O}} : \qquad\qquad\qquad\qquad Mg^{2+} \; : \ddot{\underset{\cdot\cdot}{O}} :^{2-}$$

Magnesium atom	Oxygen atom	Magnesium oxide

The magnesium atom, with two electrons in the outer energy level, exactly fills the need of one oxygen atom for two electrons. The resulting compound has a ratio of one atom of magnesium to one atom of oxygen. The oxygen (oxide) ion has a -2 charge, having gained two electrons. In combining with oxygen, magnesium behaves the same way as when combining with chlorine; it loses two electrons.

EXAMPLE 12.5   Explain the formation of sodium sulfide, Na_2S, from its elements.

$$\begin{array}{c} Na\cdot \\ \\ Na\cdot \end{array} \;+\; \cdot \ddot{\underset{\cdot}{S}} : \longrightarrow \begin{array}{c} Na^+ \\ :\ddot{\underset{\cdot\cdot}{S}} :^{2-} \\ Na^+ \end{array} \qquad or \qquad Na_2S$$

Sodium atoms	Sulfur atom	Sodium sulfide

Two sodium atoms supply the electrons that one sulfur atom needs to make eight in its outer shell.

EXAMPLE 12.6   Explain the formation of aluminum oxide, Al_2O_3, from its elements.

$$\begin{array}{c} \ddot{Al}\cdot \\ \\ \\ \ddot{Al}\cdot \end{array} \begin{array}{c} \cdot \ddot{\underset{\cdot}{O}} : \\ \\ +\; \cdot \ddot{\underset{\cdot}{O}} : \\ \\ \cdot \ddot{\underset{\cdot}{O}} : \end{array} \longrightarrow \begin{array}{c} : \ddot{\underset{\cdot\cdot}{O}} :^{2-} \\ Al^{3+} \\ : \ddot{\underset{\cdot\cdot}{O}} :^{2-} \\ Al^{3+} \\ : \ddot{\underset{\cdot\cdot}{O}} :^{2-} \end{array} \qquad or \qquad Al_2O_3$$

Aluminum atoms	Oxygen atoms	Aluminum oxide

One oxygen atom, needing two electrons, cannot accommodate the three electrons from one aluminum atom. One aluminum atom falls one electron short of the four electrons needed by two oxygen atoms. A ratio of two atoms of aluminum to three atoms of oxygen, involving the transfer of six electrons (two to each oxygen atom), gives each atom a stable electron configuration.

Note that in each of the examples above, outer shells containing eight electrons were formed in all the negative ions. This formation resulted from the pairing of all the *s* and *p* electrons in these outer shells.

Chemistry would be considerably simpler if all compounds were made by the direct formation of ions as outlined in the examples just given. However, this method is only one of the two general methods of compound formation. The second general method will be outlined in the sections that follow.

12.5   The Covalent Bond: Sharing Electrons

covalent bond

Some atoms do not transfer electrons from one atom to another to form ions. Instead they form a chemical bond by sharing pairs of electrons between them.

A **covalent bond** consists of a pair of electrons shared between two atoms. This bonding concept was introduced in 1916 by G. N. Lewis of the University of California at Berkeley. In the millions of compounds that are known, the covalent bond is the predominant chemical bond.

True molecules exist in substances in which the atoms are covalently bonded. Hence, it is proper to refer to molecules of such substances as hydrogen, chlorine, hydrogen chloride, carbon dioxide, water, or sugar. These substances contain only covalent bonds and exist as aggregates of molecules. We do not use the term *molecule* when talking about ionically bonded compounds such as sodium chloride, because such substances exist as large aggregates of positive and negative ions, not as molecules.

A study of the hydrogen molecule will give an insight into the nature of the covalent bond and its formation. The formation of a hydrogen molecule, H_2, involves the overlapping and pairing of $1s$ electron orbitals from two hydrogen atoms. This overlapping and pairing is shown in Figure 12.4. Each atom contributes one electron of the pair that is shared jointly by two hydrogen nuclei. The orbital of the electrons now includes both hydrogen nuclei, but probability factors show that the most likely place to find the electrons (the point of highest electron density) is between the two nuclei. The two nuclei are shielded from each other by the pair of electrons, allowing the two nuclei to be drawn very close to each other. (The average bond length between the hydrogen nuclei is 7.4×10^{-9} cm.)

The tendency for hydrogen atoms to form a molecule is very strong. In the molecule, each electron is attracted by two positive nuclei. This attraction gives

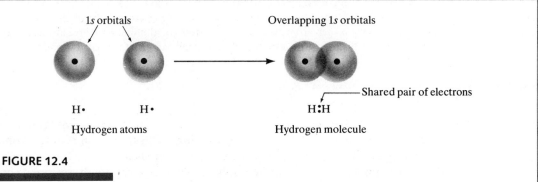

FIGURE 12.4

The formation of a hydrogen molecule from two hydrogen atoms. The two 1s orbitals overlap, forming the H₂ molecule. In this molecule the two electrons are shared between the atoms, forming a covalent bond.

the hydrogen molecule a more stable structure than the individual hydrogen atoms had. Energy is released when a bond forms between two atoms. Consequently, the same amount of energy is needed to break that bond. Experimental evidence of stability is shown by the fact that 436 kJ are needed to break the bonds between the hydrogen atoms in 1 mole of hydrogen molecules. The strength of a bond may be determined by the energy required to break it. The energy required to break a covalent bond is known as the *bond dissociation energy*. The following bond dissociation energies illustrate relative bond strengths. (All substances are considered to be in the gaseous state and to form neutral atoms.)

Reaction	Bond dissociation energy (kJ/mol)
$H_2 \longrightarrow 2\,H$	436
$N_2 \longrightarrow 2\,N$	946
$O_2 \longrightarrow 2\,O$	595
$F_2 \longrightarrow 2\,F$	153
$Cl_2 \longrightarrow 2\,Cl$	243
$Br_2 \longrightarrow 2\,Br$	193
$I_2 \longrightarrow 2\,I$	151

The formula for chlorine gas is Cl_2. When the two atoms of chlorine combine to form this molecule, the electrons must interact in a manner that is similar to that shown in the preceding example. Each chlorine atom would be more stable with eight electrons in its outer shell. But chlorine atoms are identical, and neither

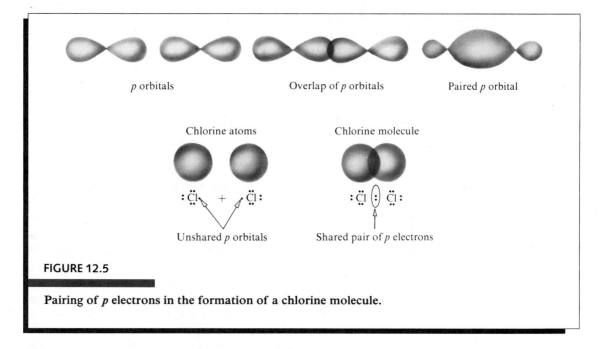

p orbitals   Overlap of *p* orbitals   Paired *p* orbital

Chlorine atoms   Chlorine molecule

Unshared *p* orbitals   Shared pair of *p* electrons

FIGURE 12.5

Pairing of *p* electrons in the formation of a chlorine molecule.

is able to pull an electron away from the other. What happens is this: The unpaired $3p$ electron orbital of one chlorine atom overlaps the unpaired $3p$ electron orbital of the other atom, resulting in a pair of electrons that are mutually shared between the two atoms. Each atom furnishes one of the pair of shared electrons. Thus, each atom attains a stable structure of eight electrons by sharing an electron pair with the other atom. The pairing of the *p* electrons and formation of a chlorine molecule are illustrated in Figure 12.5. Neither chlorine atom has a positive or negative charge, since both contain the same number of protons and have equal attraction for the pair of electrons being shared. Other examples of molecules in which electrons are equally shared between two atoms are hydrogen, H_2; oxygen, O_2; nitrogen, N_2; fluorine, F_2; bromine, Br_2; and iodine, I_2. Note that more than one pair of electrons may be shared between atoms.

$$H:H \qquad :\ddot{F}:\ddot{F}: \qquad :\ddot{Br}:\ddot{Br}: \qquad :\ddot{I}:\ddot{I}: \qquad :\ddot{O}::\ddot{O}: \qquad :N:::N:$$

Hydrogen   Fluorine   Bromine   Iodine   Oxygen   Nitrogen

The Lewis structure given for oxygen does not adequately account for all the properties of the oxygen molecule. Other theories explaining the bonding in oxygen molecules have been advanced; however, they are complex and beyond the scope of this book.

A common practice in writing structures is to replace the pair of dots used to represent a shared pair of electrons by a dash (—). But you must remember that a dash represents a shared pair of electrons. One dash represents a single bond; two dashes, a double bond; and three dashes, a triple bond. The six structures just shown may be written thus:

$$\text{H—H} \qquad :\!\ddot{\text{F}}\!\!-\!\!\ddot{\text{F}}\!: \qquad :\!\ddot{\text{B}}\text{r}\!-\!\ddot{\text{B}}\text{r}\!: \qquad :\!\ddot{\text{I}}\!-\!\ddot{\text{I}}\!: \qquad :\!\ddot{\text{O}}\!\!=\!\!\ddot{\text{O}}\!: \qquad :\!\text{N}\!\!\equiv\!\!\text{N}\!:$$

12.6   Electronegativity

**electro-
negativity**

When two different kinds of atoms share a pair of electrons, one atom assumes a partial positive charge and the other a partial negative charge with respect to each other. This difference in charge occurs because the two atoms exert unequal attraction for the pair of shared electrons. The attractive force that an atom of an element has for shared electrons in a molecule is known as its **electronegativity**. Elements differ in their electronegativities. For example, both hydrogen and chlorine need one electron to form stable electron configurations. They share a pair of electrons in hydrogen chloride, HCl. Chlorine is more electronegative and therefore has a greater attraction for the shared electrons than does hydrogen. As a result the pair of electrons is displaced toward the chlorine atom, giving it a partial negative charge and leaving the hydrogen atom with a partial positive charge. It should be understood that the electron is not transferred entirely to the chlorine atom, as in the case of sodium chloride, and no ions are formed. The entire molecule, HCl, is electrically neutral. A partial charge is usually indicated by the Greek letter delta, δ. Thus, a partial positive charge is represented by $\delta+$ and a partial negative charge by $\delta-$.

Hydrogen chloride

The pair of shared electrons in HCl is closer to the more electronegative chlorine atom than to the hydrogen atom, giving chlorine a partial negative charge with respect to the hydrogen atom.

The electronegativity, or ability of an atom to attract an electron, depends on several factors: (1) the charge on the nucleus, (2) the distance of the outer electrons from the nucleus, and (3) the amount the nucleus is shielded from the outer-shell electrons by intervening shells of electrons. A scale of relative electronegativities, in which the most electronegative element, fluorine, is assigned a value of 4.0, was developed by the Nobel laureate (1954 and 1962) Linus Pauling (b. 1901). Table 12.3 shows that the relative electronegativity of the nonmetals is high and that of the metals is low. These electronegativities indicate that atoms of metals

TABLE 12.3

Relative Electronegativity of the Elements *

1 **H** 2.1																	2 **He** —
3 **Li** 1.0	4 **Be** 1.5											5 **B** 2.0	6 **C** 2.5	7 **N** 3.0	8 **O** 3.5	9 **F** 4.0	10 **Ne** —
11 **Na** 0.9	12 **Mg** 1.2											13 **Al** 1.5	14 **Si** 1.8	14 **P** 2.1	16 **S** 2.5	17 **Cl** 3.0	18 **Ar** —
19 **K** 0.8	20 **Ca** 1.0	21 **Sc** 1.3	22 **Ti** 1.4	23 **V** 1.6	24 **Cr** 1.6	25 **Mn** 1.5	26 **Fe** 1.8	27 **Co** 1.8	28 **Ni** 1.8	29 **Cu** 1.9	30 **Zn** 1.6	31 **Ga** 1.6	32 **Ge** 1.8	33 **As** 2.0	34 **Se** 2.4	35 **Br** 2.8	36 **Kr** —
37 **Rb** 0.8	38 **Sr** 1.0	39 **Y** 1.2	40 **Zr** 1.4	41 **Nb** 1.6	42 **Mo** 1.8	43 **Tc** 1.9	44 **Ru** 2.2	45 **Rh** 2.2	46 **Pd** 2.2	47 **Ag** 1.9	48 **Cd** 1.7	49 **In** 1.7	50 **Sn** 1.8	51 **Sb** 1.9	52 **Te** 2.1	53 **I** 2.5	54 **Xe** —
55 **Cs** 0.7	56 **Ba** 0.9	57–71 **La–Lu** 1.1–1.2	72 **Hf** 1.3	73 **Ta** 1.5	74 **W** 1.7	75 **Re** 1.9	76 **Os** 2.2	77 **Ir** 2.2	78 **Pt** 2.2	79 **Au** 2.4	80 **Hg** 1.9	81 **Tl** 1.8	82 **Pb** 1.8	83 **Bi** 1.9	84 **Po** 2.0	85 **At** 2.2	86 **Rn** —
87 **Fr** 0.7	88 **Ra** 0.9	89– **Ac–** 1.1–1.7	104 **Unq**	105 **Unp**	106 **Unh**	107 **Uns**	108 **Uno**	109 **Une**									

Key: 9 — Atomic number; F — Symbol; 4.0 — Electronegativity

* The electronegativity value is given below the symbol of each element.

have a greater tendency to lose electrons than do atoms of nonmetals and that nonmetals have a greater tendency to gain electrons than do metals. The higher the electronegativity value, the greater the attraction for electrons.

12.7 Writing Lewis Structures

As we have seen, Lewis structures are a convenient way of showing the covalent bonds in many molecules or ions of the representative elements. In writing Lewis structures the object is to connect the atoms in a molecule with covalent bonds by rearranging the valence electrons of the atoms so that each atom has eight outer-shell electrons around it. Exceptions to this rule are hydrogen, which requires only two electrons, and several other elements such as lithium, beryllium, and boron.

The most difficult part of writing Lewis structures is determining the arrangement of the atoms in a molecule or an ion. In simple molecules with more than two atoms, one atom will be the central atom surrounded by the other atoms. Thus Cl_2O has two possible arrangements, Cl—Cl—O or Cl—O—Cl. Usually, but not always, the single atom in the formula (except H) will be the central atom.

Although Lewis structures for many molecules and ions can be written by inspection, the following procedure will be helpful for learning to write these structures:

1. Obtain the total number of valence electrons to be used in the structure by adding the number of valence electrons in all of the atoms in the molecule or ion. If you are writing the structure of an ion, add one electron for each negative charge or subtract one electron for each positive charge on the ion. Remember, the number of valence electrons of Group A elements is the same as their group number in the periodic table.

2. Write down the skeletal arrangement of the atoms and connect them with a single covalent bond (two dots or one dash). Hydrogen, which contains only one bonding electron, can form only one covalent bond. Oxygen atoms are not normally bonded to each other, except in compounds known to be peroxides. Oxygen atoms normally have a maximum of two covalent bonds, two single bonds or one double bond.

3. Subtract two electrons for each single bond you used in Step 2 from the total number of electrons calculated in Step 1. This calculation gives you the net number of electrons available for completing the structure.

4. Distribute pairs of electrons (pairs of dots) around each atom (except hydrogen) to give each atom a total of eight electrons around it.

5. If there are not enough electrons to give these atoms eight electrons, change single bonds between atoms to double or triple bonds by shifting unbonded pairs of electrons as needed. Check to see that each atom (except H) has eight electrons around it. A double bond counts as four electrons for each atom to which it is bonded.

EXAMPLE 12.7   How many valence electrons are in each of these atoms: Cl, H, C, O, N, S, P, I?

You can look in Table 10.4 for the electron structure, or, if the element is in an A Group of the periodic table, the number of valence electrons is equal to the group number.

Atom	Periodic group	Valence electrons
Cl	VIIA	7
H	IA	1
C	IVA	4
O	VIA	6
N	VA	5
S	VIA	6
P	VA	5
I	VIIA	7

EXAMPLE 12.8   Write the Lewis structure for water, H_2O.

Step 1   The total number of valence electrons is 8, 2 from the two hydrogen atoms and 6 from the oxygen atom.

Step 2   The two hydrogen atoms are connected to the oxygen atom. Write the skeletal structure:

H   O

    H

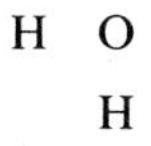

Place two dots between the hydrogen and oxygen atoms to form the covalent bonds:

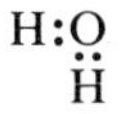

H:O
  ̈H

Step 3   Subtract the 4 electrons used in Step 2 from 8 to obtain 4 electrons yet to be used.

Step 4   Distribute the four electrons around the oxygen atom. Hydrogen atoms cannot accommodate any more electrons.

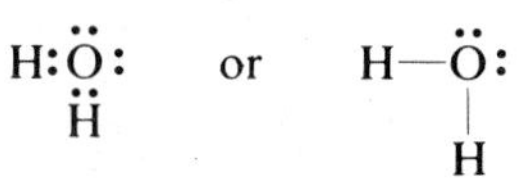

H:Ö: or   H—Ö:
 ̈H |
               H

This arrangement is the Lewis structure. Each atom has a noble-gas electron structure.

EXAMPLE 12.9   Write Lewis structures for a molecule of (a) methane, CH_4, and (b) carbon tetrachloride, CCl_4.

Part (a)

Step 1   The total number of valence electrons is 8, 1 from each hydrogen atom and 4 from the carbon atom.

Step 2   The skeletal structure contains four H atoms around a central C atom. Place 2 electrons between the C and each H.

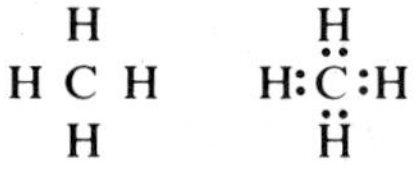

   H   H
              ̈
H C H   H:C:H
   H   ̈H

Step 3   Subtract the 8 electrons used in Step 2 from 8 to obtain zero electrons yet to be placed. Therefore, the Lewis structure must be as written in Step 2.

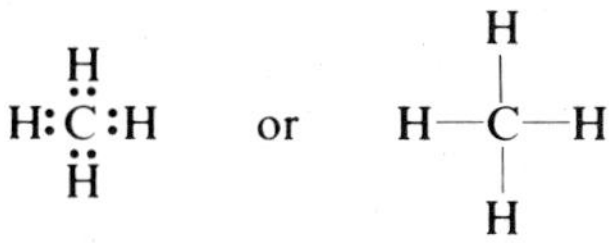

                H
  ̈ |
H:C:H   or   H—C—H
  ̈H |
                H

Part (b)

Step 1   The total number of valence electrons to be used is 32, 4 from the carbon atom and 7 from each of the four chlorine atoms.

Step 2 The skeletal structure contains the four CI atoms around a central C atom. Place 2 electrons between the C and each Cl.

$$\begin{array}{ccc} & \text{Cl} & \\ \text{Cl} & \text{C} & \text{Cl} \\ & \text{Cl} & \end{array} \qquad \begin{array}{c} \text{Cl} \\ \text{Cl:C:Cl} \\ \text{Cl} \end{array}$$

Step 3 Subtract the 8 electrons used in Step 2 from 32, to obtain 24 electrons yet to be placed.

Step 4 Distribute the 24 electrons (12 pairs) around the Cl atoms so that each Cl atom has 8 electrons around it.

$$\begin{array}{c} :\ddot{C}l: \\ :\ddot{C}l:\ddot{C}:\ddot{C}l: \\ :\ddot{C}l: \end{array} \quad \text{or} \quad \begin{array}{c} :\ddot{C}l: \\ | \\ :\ddot{C}l—C—\ddot{C}l: \\ | \\ :\ddot{C}l: \end{array}$$

This arrangement is the Lewis structure; CCl_4 contains four covalent bonds.

EXAMPLE 12.10   Write Lewis structures for (a) carbon dioxide, CO_2, and (b) nitric acid, HNO_3.

Part (a)

Step 1 The total number of valence electrons is 16, 4 from the carbon atom and 6 from each oxygen atom.

Step 2 The two O atoms are bonded to a central C atom. Write the skeletal structure and place 2 electrons between the C and each O atom.

O:C:O

Step 3 Subtract the 4 electrons used in Step 2 from 16 to obtain 12 electrons yet to be placed.

Step 4 Distribute the 12 electrons around the C and O atoms. Several possibilities exist:

$$:\ddot{O}:C:\ddot{O}: \qquad :\ddot{O}:C:\ddot{O}: \qquad :\ddot{O}:\ddot{C}:\ddot{O}:$$
$$\quad\;\; \text{I} \qquad\qquad\quad \text{II} \qquad\qquad\quad \text{III}$$

Step 5 All the atoms do not have 8 electrons around them. Move one pair of unbonded electrons from each O atom in structure I and place one additional pair of electrons between each C and O atom, forming two double bonds:

$$:\ddot{O}::C::\ddot{O}: \qquad \text{or} \qquad :\ddot{O}=C=\ddot{O}:$$

Each atom now has 8 electrons around it. Carbon is sharing four pairs of electrons, and each oxygen is sharing two pairs.

Part (b)

Step 1 The total number of valence electrons is 24, 1 from the hydrogen atom, 5 from the nitrogen atom, and 6 from each oxygen atom.

Step 2 The three O atoms are bonded to a central N atom. The H atom is bonded to one of
the O atoms. Write the skeletal structure and place 2 electrons between each pair of
atoms:

$$\begin{array}{c} O \\ H\!:\!O\!:\!\overset{..}{N}\!:\!O \end{array}$$

Step 3 Subtract the 8 electrons used in Step 2 from 24 to obtain 16 electrons yet to
be placed.

Step 4 Distribute the 16 electrons around the N and O atoms:

$$\begin{array}{c} :\overset{..}{\underset{..}{O}}: \longleftarrow \text{electron deficient} \\ H\!:\!\overset{..}{\underset{..}{O}}\!:\!\overset{..}{\underset{..}{N}}\!:\!\overset{..}{\underset{..}{O}}: \end{array}$$

Step 5 One pair of electrons is still needed to give all the N and O atoms 8 electrons. Move
the unbonded pair of electrons from the N atom and place it between the N and the
electron-deficient O atom, making a double bond:

$$\begin{array}{c} :\overset{..}{O} \\ :: \\ H\!:\!\overset{..}{\underset{..}{O}}\!:\!\overset{..}{\underset{..}{N}}\!:\!\overset{..}{\underset{..}{O}}: \end{array} \quad \text{or} \quad \begin{array}{c} :\overset{..}{O} \\ \| \\ H\!-\!\overset{..}{\underset{..}{O}}\!-\!N\!-\!\overset{..}{\underset{..}{O}}: \end{array}$$

This arrangement is the Lewis structure.

EXAMPLE 12.11   Write the Lewis structure for a sulfate ion, SO_4^{2-}.

Step 1 These five atoms have 30 valence electrons (6 in each atom) plus 2 electrons from
the -2 charge, which makes 32 electrons to be placed.

Step 2 In the sulfate ion the sulfur is the central atom surrounded by the four oxygen
atoms. Write the skeletal structure and place two electrons between each pair of
atoms:

$$\begin{array}{c} O \\ \overset{..}{} \\ O\!:\!\overset{..}{S}\!:\!O \\ \overset{..}{O} \end{array}$$

Step 3 Subtract the 8 electrons used in Step 2 from 32 to give 24 electrons yet to be placed.

Step 4 Distribute the 24 electrons around the four O atoms and indicate that the sulfate
ion has a -2 charge:

$$\begin{array}{c} :\overset{..}{\underset{..}{O}}: \\ :\overset{..}{\underset{..}{O}}\!:\!\overset{..}{\underset{..}{S}}\!:\!\overset{..}{\underset{..}{O}}\!:^{2-} \\ :\overset{..}{\underset{..}{O}}: \end{array} \quad \text{or} \quad \begin{array}{c} :\overset{..}{O}: \\ | \\ :\overset{..}{O}\!-\!S\!-\!\overset{..}{O}:^{2-} \\ | \\ :\overset{..}{O}: \end{array}$$

This arrangement is the Lewis structure. Each atom has 8 electrons around it.

PRACTICE   Write the Lewis structures for each of the following: (a) NH_3 (b) SiO_2 (c) SO_3 (d) ClO_3^- (e) H_3O^+

Answers: (a) H:N̈:H   (b) O::Si::O   (c) :Ö::S̈
                   Ḧ   :Ö:

(d) $\left[:\ddot{O}:\ddot{C}l:\ddot{O}:\right]^-$ (e) $\left[H:\ddot{O}:H\right]^+$
          :Ö: Ḧ

Although many compounds follow the octet rule for covalent bonding, the rule has numerous exceptions. Sometimes it is impossible to write a structure in which each atom has eight electrons around it. For example, in BF_3 the boron atom has only 6 electrons around it, and in SF_6 the sulfur atom has 12 electrons around it.

12.8   Nonpolar and Polar Covalent Bonds

nonpolar covalent bond

We have considered bonds to be either ionic or covalent depending on whether electrons are transferred from one atom to another or are shared between two atoms. In a **nonpolar covalent bond** the shared pair of electrons is attracted equally by the two atoms. Nonpolar covalent bonds form between the same kind of atoms. For example, the covalent bond in a hydrogen molecule or in a chlorine molecule is nonpolar because the electronegativity difference between identical atoms is zero. (See Figure 12.6.)

polar covalent bond

A covalent bond between two different kinds of atoms has a partial ionic character and is known as a **polar covalent bond**. This bond is polar because the two atoms have different electronegativities. The polarity is due to unequal sharing of the electrons because the more electronegative atom has a greater attraction for the shared pair of electrons. The more electronegative atom thus acquires a partial negative charge, and the other atom acquires a partial positive charge. When a polar covalent bond is present in a diatomic molecule, such as HCl, the molecule is polar. By *polar* we mean an unequal distribution of electrical charge. Overall, however, the HCl molecule is neutral because the partial negative charge on the chlorine atom is exactly balanced by the partial positive charge on the hydrogen atom.

dipole

A **dipole** is a molecule that is electrically unsymmetrical, causing it to be oppositely charged at two points. A dipole is often written as ⊕⊖. A hydrogen chloride molecule is polar and behaves as a small dipole. The HCl dipole may be written as H ⟶ Cl. The arrow points toward the negative end of the dipole. Molecules of H_2O, HBr, and ICl are polar.

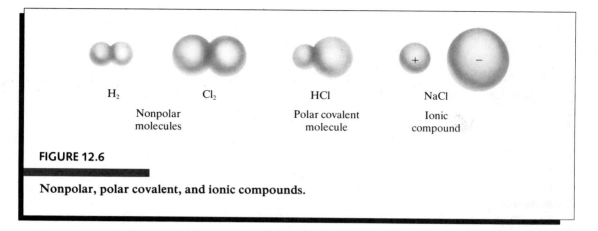

FIGURE 12.6

Nonpolar, polar covalent, and ionic compounds.

$$\overset{\delta+}{H} \longrightarrow \overset{\delta-}{Cl} \qquad \overset{\delta+}{H} \longrightarrow \overset{\delta-}{Br} \qquad \overset{\delta+}{I} \longrightarrow \overset{\delta-}{Cl} \qquad \overset{\delta-}{\underset{\underset{H \qquad H}{\delta+ \qquad \delta+}}{O}}$$

How do we know whether a bond between two atoms is ionic or covalent? The difference in electronegativity between two atoms determines the character of the bond formed between them. As the difference in electronegativity increases, the polarity of the bond (or percent ionic character) increases. As a rule, if the electronegativity difference between two bonded atoms is greater than 1.7–1.9, the bond will be more ionic than covalent. If the electronegativity difference is greater than 2.0, the bond is strongly ionic. If the electronegativity difference is less than 1.5, the bond is strongly covalent.

Care must be taken to distinguish between polar bonds and polar molecules. A covalent bond between different kinds of atoms is always polar. But a molecule containing different kinds of atoms may or may not be polar, depending on its shape or geometry. Molecules of HF, HCl, HBr, HI, and ICl are all polar because each contains a single polar bond. However, CO_2, CH_4, and CCl_4 are nonpolar molecules despite the fact that all three contain polar bonds. The carbon dioxide molecule, O=C=O, is nonpolar because the carbon-oxygen dipoles cancel each other by acting in opposite directions.

$$\overset{\longleftarrow \quad + \quad \longrightarrow}{O=C=O}$$

Dipoles in opposite directions

Methane (CH_4) and carbon tetrachloride (CCl_4) are nonpolar because the four C—H and C—Cl polar bonds are identical, and, because these bonds emanate

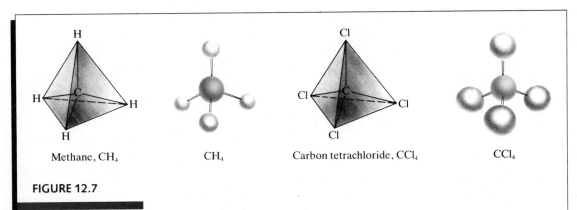

FIGURE 12.7

Ball-and-stick models of methane and carbon tetrachloride. Methane and carbon tetrachloride are nonpolar molecules because their polar bonds cancel each other in the tetrahedral arrangement of their atoms. The carbon atoms are located in the centers of the tetrahedrons.

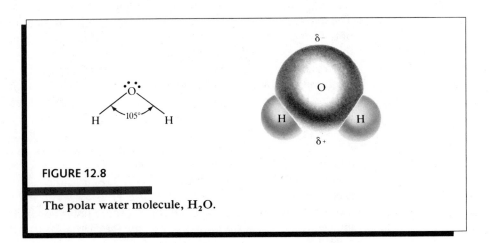

FIGURE 12.8

The polar water molecule, H_2O.

from the center to the corners of a tetrahedron in the molecule, the effect of their polarities cancel one another (Figure 12.7).

We have said that water is a polar molecule. If the atoms in water were linear like those in carbon dioxide, the two O—H dipoles would cancel each other, and the molecule would be nonpolar. However, water is definitely polar and has a nonlinear (bent) structure with an angle of $105°$ between the two O—H bonds (see Figure 12.8).

12.9   Coordinate Covalent Bonds

coordinate covalent bond

In Section 12.5 we saw that a covalent bond was formed by the overlapping of electron orbitals between two atoms. The two atoms each furnish an electron to make a pair that is shared between them.

Covalent bonds can also be formed by one atom furnishing both electrons that are shared between the two atoms. The bond so formed is called a **coordinate covalent bond**. This bond is often designated by an arrow pointing away from the electron-pair donor (for example, A $\longrightarrow$ B). Once formed, a coordinate covalent bond cannot be distinguished from any other covalent bond; it is simply a pair of electrons shared between two atoms.

The Lewis structures of sulfurous and sulfuric acids show coordinate covalent bonds between the sulfur and the oxygen atoms that are not bonded to hydrogen atoms. The colored dots indicate the electrons of the sulfur atom.

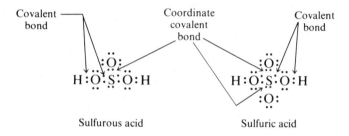

Sulfurous acid   Sulfuric acid

The lone (unbonded) pair of electrons on the sulfur atom in sulfurous acid allows room for another oxygen atom with six valence electrons to fit perfectly into its structure to form sulfuric acid. Other atoms with six valence electrons in their outer shell, such as sulfur, could also fit into this pattern. The coordinate covalent bond explains the formation of many complex molecules.

12.10   Polyatomic Ions

A polyatomic ion is a stable group of atoms that has either a positive or a negative charge and behaves as a single unit in many chemical reactions. Sodium sulfate, Na_2SO_4, contains two sodium ions and a sulfate ion. The sulfate ion, SO_4^{2-}, is a polyatomic ion composed of one sulfur atom and four oxygen atoms and has a charge of -2. One sulfur and four oxygen atoms have a total of 30 electrons in their outer shells. The sulfate ion contains 32 outer-shell electrons and therefore has a charge of -2. In this case the two additional electrons come from the two sodium atoms, which are now sodium ions.

$$Na^+ \begin{bmatrix} :\ddot{O}: \\ :\ddot{O}:S:\ddot{O}: \\ :\ddot{O}: \end{bmatrix}^{2-} Na^+ \qquad \begin{bmatrix} :\ddot{O}: \\ :\ddot{O}:S:\ddot{O}: \\ :\ddot{O}: \end{bmatrix}^{2-}$$

            Sodium sulfate   Sulfate ion

Sodium sulfate has both ionic and covalent bonds. Ionic bonds exist between each of the sodium ions and the sulfate ion. Covalent bonds are present between the sulfur and oxygen atoms within the sulfate ion. One important difference between the ionic and covalent bonds in this compound can be demonstrated by dissolving sodium sulfate in water. It dissolves in water forming three charged particles, two sodium ions and one sulfate ion, per formula unit of sodium sulfate:

$$Na_2SO_4 \xrightarrow{\text{Water}} 2\,Na^+ + SO_4^{2-}$$

    Sodium sulfate   Sodium ions   Sulfate ion

The SO_4^{2-} ion remains as a unit, held together by covalent bonds; whereas, where the bonds were ionic, dissociation of the ions took place. Do not think, however, that polyatomic ions are so stable that they cannot be altered. Chemical reactions by which polyatomic ions can be changed to other substances do exist.

The Lewis structures for several common polyatomic ions are shown below:

$$\begin{bmatrix} H \\ H:\ddot{N}:H \\ H \end{bmatrix}^{+} \qquad \begin{bmatrix} :\ddot{O}: \\ N::\ddot{O}: \\ :\ddot{O}: \end{bmatrix}^{-} \qquad \begin{bmatrix} :\ddot{O}: \\ :\ddot{O}:P:\ddot{O}: \\ :\ddot{O}: \end{bmatrix}^{3-} \qquad \begin{bmatrix} :\ddot{O}:H \end{bmatrix}^{-}$$

Ammonium ion, NH_4^+ Nitrate ion, NO_3^- Phosphate ion, PO_4^{3-} Hydroxide ion, OH^-

Concepts in Review

1. Describe how the ionization energies of the elements vary with respect to (a) position in the periodic table and (b) the removal of successive electrons.

2. Determine the number of valence electrons in an atom for any Group A element.

3. Describe (a) the formation of ions by electron transfer and (b) the nature of the chemical bond formed by electron transfer.

4. Show by means of Lewis-dot structures the formation of an ionic compound from atoms.

5. Describe a crystal of sodium chloride.

6. Predict the formulas of the monatomic ions of Group A metals and nonmetals.

7. Predict the relative sizes of an atom and a monatomic ion for a given element.

8. Describe the covalent bond and predict whether a given covalent bond would be polar or nonpolar.

9. Draw Lewis-dot structures for the diatomic elements.

10. Identify single, double, and triple covalent bonds.

11. Describe the changes in electronegativity in (1) moving across a period and (2) moving down a group in the periodic table.

12. Describe the effect of electronegativity on the type of chemical bonds in a compound.

13. Draw Lewis-dot structures for (a) the molecules of covalent compounds and (b) polyatomic ions.

14. Describe the difference between polar and nonpolar bonds.

15. Distinguish clearly between ionic and molecular substances.

16. Predict whether the bonding in a compound will be primarily ionic or covalent.

17. Distinguish coordinate-covalent from covalent bonds in a Lewis-dot structure.

Key Terms in Review

The terms listed here have all been defined within the chapter. Review the definitions of each. Use the glossary, and the margin notations within the chapter as study aids.

chemical bonds	ionic bond
coordinate covalent bond	ionization energy
covalent bond	nonpolar covalent bond
dipole	polar covalent bond
electron affinity	valence electrons
electronegativity	

Exercises

1. Explain why much more ionization energy is required to remove the first electron from neon than from sodium.

2. Explain the large increase in ionization energy needed to remove the third electron from beryllium compared with that needed for the second electron.

3. Does the first ionization energy increase or decrease from top to bottom in the periodic table for the alkali metal family? Explain.

4. Does the first ionization energy increase or decrease from top to bottom in the periodic table for the noble gas family? Explain.

5. In which general areas of the periodic table are the elements with (a) the highest and (b) the lowest electronegativities found?

6. Which is larger, a magnesium atom or a magnesium ion? Explain.

7. Which is smaller, a bromine atom or a bromide ion? Explain.

8. Using the table of electronegativity values (Table 12.3), indicate which element is positive and which is negative in the following compounds:
(a) H_2O (e) NO   (i) CCl_4
(b) NaF   (f) CH_4 (j) IBr
(c) NH_3 (g) HCl   (k) MgH_2
(d) PbS   (h) LiH   (l) OF_2

9. Classify the bond between the following pairs of elements as principally ionic or principally covalent (use Table 12.3):
(a) Sodium and chlorine
(b) Carbon and hydrogen
(c) Chlorine and carbon
(d) Calcium and oxygen
(e) Hydrogen and sulfur
(f) Barium and oxygen
(g) Fluorine and fluorine
(h) Potassium and fluorine

10. Explain why, in general, the ionization energy decreases from top to bottom in a family of elements.

11. Why does barium (Ba) have a lower ionization energy than beryllium (Be)?

12. Why is there such a large increase in the ionization energy required to remove the second electron from a sodium atom as opposed to the first?

13. Explain how electron affinity differs from ionization energy.

14. Explain what happens to the electron structures of Mg and Cl atoms when they react to form $MgCl_2$.

15. Write an equation representing (a) the change of a fluorine atom to a fluoride ion (F^-) and (b) the change of a calcium atom to a calcium ion (Ca^{2+}).

16. Use Lewis-dot symbols to show the electron transfer for the formation of the following ionic compounds from the atoms.
(a) MgF_2 (b) K_2O (c) CaO   (d) NaBr

17. What are valence electrons?

18. How many valence electrons are in each of these atoms? H, K, Mg, He, Al, Si, N, P, O, Cl

19. How many electrons must be gained or lost for each of the following to achieve a noble-gas electron structure?
(a) A calcium atom   (d) A chloride ion
(b) A sulfur atom   (e) A nitrogen atom
(c) A helium atom   (f) A potassium atom

20. Explain why potassium forms a K^+ ion but not a K^{2+} ion.

21. Why does an aluminum ion have a $+3$ charge?

22. Which is larger? Explain.
(a) A potassium atom or a potassium ion
(b) A bromine atom or a bromide ion
(c) A magnesium ion or an aluminum ion
(d) Fe^{2+} or Fe^{3+}

23. Write Lewis-dot structures for
(a) Na (b) Br^- (c) O^{2-} (d) Ga (e) Ga^{3+}

24. Why is it not proper to speak of sodium chloride molecules?

25. What is a covalent bond? How does it differ from an ionic bond?

26. What is different about the formation of a coordinate covalent bond from that of an ordinary covalent bond?

27. Classify the bonding in each compound as ionic or covalent:
(a) H_2O (c) MgO   (e) HCl   (g) NH_3
(b) NaCl   (d) Br_2 (f) $BaCl_2$ (h) SO_2

28. Predict the type of bond that would be formed between each of the following pairs of atoms:
(a) Na and N   (d) H and Si
(b) N and S   (e) O and F
(c) Br and I

29. Draw Lewis structures for:
(a) H_2 (b) N_2 (c) Cl_2 (d) O_2 (e) Br_2

30. Draw Lewis structures for:
(a) NCl_3 (c) H_3PO_4 (e) H_2S
(b) H_2CO_3 (d) C_2H_6 (f) CS_2

31. Briefly comment on the structure Na$:\overset{\cdot\cdot}{\underset{\cdot\cdot}{O}}:$Na for the compound Na_2O.

32. What are the four most electronegative elements?

33. Draw Lewis structures for:
(a) Ba^{2+} (e) SO_4^{2-} (h) CO_3^{2-}
(b) Al^{3+} (f) SO_3^{2-} (i) ClO_3^-
(c) I^- (g) CN^- (j) NO_3^-
(d) S^{2-}

34. The Lewis structure for chloric acid is

$$H:\overset{..}{\underset{..}{O}}:\overset{..}{\underset{..}{Cl}}:\overset{..}{\underset{..}{O}}:$$
$$:\overset{..}{\underset{..}{O}}:$$

Point out the covalent and coordinate covalent bonds in this structure.

35. Draw the Lewis structure for ammonia (NH_3). What type of bond is present? Can this molecule form coordinate covalent bonds? If so, how many?

36. Rank these elements from highest electronegativity to lowest: Mg, S, F, H, O, Cs.

37. Classify the following molecules as polar or nonpolar:
(a) H_2O (c) CF_4 (e) CO_2
(b) HBr (d) F_2 (f) NH_3

38. Is it possible for a molecule to be nonpolar even though it contains polar covalent bonds? Explain.

39. Why is CO_2 a nonpolar molecule, whereas CO is a polar molecule?

40. Which of these statements are correct? (Try to answer this exercise using only the periodic table.) Rewrite each incorrect statement to make it correct.

(a) The amount of energy required to remove one electron from an atom is known as the ionization energy.

(b) Metallic elements tend to have relatively low electronegativities.

(c) Elements with a high ionization energy tend to have very metallic properties.

(d) Sodium and chlorine react to form molecules of NaCl.

(e) A chlorine atom has fewer electrons than a chloride ion.

(f) The noble gases have a tendency to lose one electron to become positively charged ions.

(g) The chemical bonds in a water molecule are ionic.

(h) The chemical bonds in a water molecule are polar.

(i) Valence electrons are those electrons in the outermost shell of an atom.

(j) An atom with eight electrons in its outer shell has all its s and p orbitals filled.

(k) Fluorine has the lowest electronegativity of all the elements.

(l) Oxygen has a greater electronegativity than carbon.

(m) A cation is larger than its corresponding neutral atom.

(n) Cl_2 is more ionic in character than HCl.

(o) A neutral atom with eight electrons in its valence shell will likely be an atom of a noble gas.

(p) A nitrogen atom has four valence electrons.

(q) An aluminum atom must lose three electrons to become an aluminum ion, Al^{3+}

(r) A stable group of atoms that has either a positive or a negative charge and behaves as a single unit in many chemical reactions is called a polyatomic ion.

(s) Sodium sulfate, Na_2SO_4, has covalent bonds between sulfur and the oxygen atoms, and ionic bonds between the sodium ions and the sulfate ion.

(t) The water molecule is a dipole.

(u) In an ethylene molecule, C_2H_4,

$$\underset{H}{\overset{H}{\diagdown}}C=C\underset{\diagdown H}{\overset{\diagup H}{}}$$

two pairs of electrons are shared between the carbon atoms.

(v) The octet rule is mainly useful for atoms where only s and p electrons enter into bonding.

(w) When electrons are transferred from one atom to another, the resulting compound contains ionic bonds.

(x) A phosphorus atom, $\cdot\overset{..}{P}\cdot$, needs three additional electrons to attain a stable octet of electrons.

(y) The simplest compound between oxygen, $\cdot\overset{..}{O}\cdot$, and fluorine, $:\overset{..}{F}\cdot$, atoms is FO_2.

(z) The $H:\overset{..}{\underset{..}{Cl}}:$ molecule has three unshared pairs of electrons.

(aa) The smaller the difference in electronegativity between two atoms, the more ionic the bond between them will be.

(bb) Lewis structures are mainly useful for the representative elements.

(cc) The correct Lewis structure for NH_3 is

$$H:\overset{..}{N}:H$$
$$\overset{..}{H}$$

(dd) The correct Lewis structure for CO_2 is

$$:\ddot{O}:C:\ddot{O}:$$

(ee) The correct Lewis structure for SO_4^{2-} is

$$:\ddot{O}:\ddot{O}:\ddot{S}:\ddot{O}:\ddot{O}:^{2-}$$

(ff) In period 4 of the periodic table, the element having the lowest ionization energy is Xe.

(gg) An atom having an electron structure of $1s^2 2s^2 2p^6 3s^2 3p^2$ has four valence electrons.

(hh) When an atom of bromine becomes a bromide ion, its size increases.

(ii) The structures that show that H_2O is a dipole and that CO_2 is not a dipole are

$$H—\ddot{O}—H \text{ and } :\ddot{O}=C=\ddot{O}:$$

(jj) The Cl^- and S^{2-} ions have the same electron structure.

Review Exercises for Chapters 10–12

CHAPTER TEN   Modern Atomic Theory

True–False.   *Answer the following as either true or false.*

1. A *p* orbital is spherically symmetrical around the nucleus.
2. An atom of nitrogen has two electrons in a 1*s* orbital, two electrons in a 2*s* orbital, and one electron in each of three different 2*p* orbitals.
3. The Lewis-dot symbol for calcium is Ca:
4. In the ground state, electrons tend to occupy orbitals of the lowest possible energy.
5. The Lewis-dot symbol for potassium is ·P·
6. The fourth principal energy level can contain a total of 32 electrons.
7. The second principal energy level contains only *s* and *p* sublevels.
8. The first *f* sublevel electrons are in the fourth principal energy level.
9. The 4*s* energy sublevel fills before the 3*d* sublevel because the 4*s* is at a lower energy level.
10. A *p* type energy sublevel can contain six electrons.
11. A *d* type energy sublevel can contain five orbitals.
12. In a Lewis-dot representation of an element, the symbol of the element is shown surrounded by the number of electrons in the outermost energy level of the atom.
13. The element represented by $1s^2 2s^2 2p^6 3s^2 3p^4$ has an atomic number of 16.
14. The maximum number of electrons in an *f* orbital is 14.
15. An atom with eight electrons in its outer shell is a noble gas.
16. An atom of $_{16}S$ has nine sublevel orbitals that contain one or more electrons.
17. The electron configuration of $_{22}Ti$ is $1s^2 2s^2 2p^6 3s^2 3p^6 3d^4$.
18. An atom of an element of atomic number 21 has electrons in 11 orbitals.
19. An atom of an element of atomic number 22 has electrons in 11 orbitals.
20. The maximum number of electrons that can occupy a specific energy level **n** is given by **n²**.

Multiple Choice.   *Choose the correct answer to each of the following.*

1. The concept of electrons existing in specific orbits around the nucleus was the contribution of:
   (a) Thomson   (c) Bohr
   (b) Rutherford   (d) Schrödinger
2. The correct electron structure for a fluorine atom, F, is:
   (a) $1s^2 2s^2 2p^5$ (c) $1s^2 2s^2 2p^4 3s^1$
   (b) $1s^2 2s^2 2p^2 3s^2 3p^1$ (d) $1s^2 2s^2 2p^3$
3. The correct electron structure for $_{48}Cd$ is:
   (a) $1s^2 2s^2 2p^6 3s^2 3p^6 4s^2 3d^{10}$
   (b) $1s^2 2s^2 2p^6 3s^2 3p^6 4s^2 3d^{10} 4p^6 5s^2 4d^{10}$
   (c) $1s^2 2s^2 2p^6 3s^2 3p^6 4s^2 3d^{10} 4p^6 4d^4$
   (d) $1s^2 2s^2 2p^6 3s^2 3p^6 4s^2 4p^6 4d^{10} 5s^2 5d^{10}$
4. The correct electron structure of $_{23}V$ is:
   (a) $[Ar] 4s^2 3d^3$ (c) $[Ar] 4s^2 4d^3$
   (b) $[Ar] 4s^2 4p^3$ (d) $[Kr] 4s^2 3d^3$
5. Which of the following is the correct atomic structure for $_{22}^{48}Ti$?

   (a) $\overset{22p}{\underset{26n}{\bigcirc}}$ 2 8 10 2   (c) $\overset{26p}{\underset{22n}{\bigcirc}}$ 2 8 8 4

   (b) $\overset{22p}{\underset{48n}{\bigcirc}}$ 2 8 8 4   (d) $\overset{22p}{\underset{26n}{\bigcirc}}$ 2 8 8 4

6. The number of orbitals in a *d* sublevel is
   (a) 3   (c) 7
   (b) 5   (d) No correct answer given
7. The number of electrons in the third principal energy level in an atom having the electron structure $1s^2 2s^2 2p^6 3s^2 3p^2$ is:
   (a) 2   (b) 4   (c) 6   (d) 8
8. The total number of orbitals that contain at least one electron in an atom having the structure

$1s^2 2s^2 2p^6 3s^2 3p^2$ is:
(a) 5   (c) 14
(b) 8   (d) No correct answer given

9. Which of these elements has two s and six p electrons in its outer energy level?
(a) He   (c) Ar
(b) O   (d) No correct answer given

10. Which is not a correct Lewis-dot symbol for phosphorus, $_{15}P$?

(a) :P̈·   (b) :P̈·   (c) ·P̈·   (d) ·P̈:

11. Which element is not a noble gas?
(a) Ra   (b) Xe   (c) He   (d) Ar

12. Which of the following elements has the largest number of unpaired electrons?
(a) F   (b) S   (c) Cu   (d) N

13. Which of the following is the correct Lewis-dot symbol for bromine:
(a) Br·   (b) B̈r·   (c) :B̈r·   (d) :B̈r:

14. How many unpaired electrons are in the electron structure of $_{24}Cr$, $[Ar]\,4s^1 3d^5$?
(a) 2   (b) 4   (c) 5   (d) 6

CHAPTER ELEVEN   The Periodic Arrangement of the Elements

True–False. *Answer the following as either true or false.*

1. The periodic law states that the atomic masses of the elements are a periodic function of their atomic number.

2. The atomic radius of phosphorus, P, will be larger than that of sodium, Na.

3. The atomic radius of barium, Ba, will be larger than that of magnesium, Mg.

4. Elements at the bottom of a group in the periodic table tend to be more metallic in their properties than those at the top.

5. Elements within a group will have similar electron structures and will show some similarities in their chemical properties.

6. The transition elements are characterized by an increasing number of d or f electrons in an inner shell.

7. Based on the periodic table, the formula for aluminum oxide would be Al_2O_3.

8. J. A. R. Newlands observed that, when the elements were arranged according to increasing atomic masses, every eighth element had similar properties.

9. The horizontal rows of elements in the periodic table are called periods.

10. The p block of elements contains transition elements.

11. Group IIIA elements all have three electrons in their outer energy level.

12. Group B elements are referred to as the representative elements.

13. The element in period 3 having the largest radius is argon.

14. The element in period 3 that is classified as a metalloid is silicon.

15. The halogen family is located in Group VIIA.

16. There are seven periods of elements, two short and five long periods.

17. The representative elements do not contain d or f electrons.

18. Group VIA elements are called the alkaline earth metals.

19. The Lewis-dot symbol for $_{84}Po$ is ·P̈o:

20. Oxygen has the smallest radius of the elements in Group VIA.

Multiple Choice. *Choose the correct answer to each of the following.*

1. The scientist who noticed the existence of triads of elements with similar properties was:
(a) Döbereiner   (c) Mendeleev
(b) Newlands   (d) Meyer

2. The chemist who developed a periodic table with the known elements arranged by atomic mass, left spaces where elements seemed to be missing, and predicted the properties of the missing elements was:
(a) Newlands   (c) Meyer
(b) Mendeleev   (d) Moseley

3. The physicist who, through his work in X-ray emission frequencies of elements, established that atomic number was the correct basis for

arranging the elements was:

(a) Döbereiner   (c) Meyer

(b) Mendeleev   (d) Moseley

4. The German scientist who in 1869 independently published a periodic arrangement of elements based on increasing atomic masses similar to Mendeleev's arrangement was:

(a) Döbereiner   (c) Newlands

(b) Meyer   (d) Moseley

5. Periods IIIA through VIIA plus the noble gases form the area of the periodic table where the electron sublevels being filled are:

(a) p sublevels   (c) d sublevels

(b) s and p sublevels   (d) f sublevels

6. In moving down an A group on the periodic table, the number of electrons in the outermost energy level:

(a) Increases regularly

(b) Remains constant

(c) Decreases regularly

(d) Changes in an unpredictable manner

7. Which of the following would be an incorrect formula?

(a) NaCl   (b) K_2O (c) AlO   (d) BaO

8. Elements of the noble gas family:

(a) Form no compounds at all

(b) Have full outer electron shells

(c) Have an outer-shell electron structure of ns^2np^6 (helium excepted), where n is the period number

(d) No correct answer given

9. The lanthanide and actinide series of elements are:

(a) Representative elements

(b) Transition elements

(c) Filling in d level electrons

(d) No correct answer given

10. The element having the structure $1s^2 2s^2 2p^6 3s^2 3p^2$ is in Group:

(a) IIA   (b) IIB   (c) IVA   (d) IVB

11. In Group VA, the element having the smallest atomic radius is:

(a) Bi   (b) P   (c) As   (d) N

12. In Group IVA, the most metallic element is:

(a) C   (b) Si   (c) Ge   (d) Sn

13. Which group in the periodic table contains the least reactive elements?

(a) IA   (b) IIA   (c) IIIA   (d) Noble gases

14. Which group in the periodic table contains the alkali metals?

(a) IA   (b) IIA   (c) IIIA   (d) IVA

15. An atom of fluorine is smaller than an atom of oxygen. One possible explanation is that, compared to oxygen, fluorine has:

(a) A larger mass number

(b) A smaller atomic number

(c) A greater nuclear charge

(d) More unpaired electrons

16. If the size of the fluorine atom is compared to the size of the fluoride ion,

(a) They would both be the same size.

(b) The atom is larger than the ion.

(c) The ion is larger than the atom.

(d) The size difference depends on the reaction

17. Sodium is a very active metal because

(a) It has a low ionization energy

(b) It has only one outermost electron

(c) It has a relatively small atomic mass

(d) All of the above

CHAPTER TWELVE   Chemical Bonds: The Formation of Compounds from Atoms

True–False. *Answer the following as either true or false.*

1. The amount of energy required to remove one electron from an atom is known as the ionization energy.

2. It requires less energy to remove a second electron from an atom than to remove the first electron.

3. The first ionization energy of potassium will be greater than that for sodium.

4. The first ionization energy of sulfur will be greater than that for sodium.

5. The electrons in the outermost shell of an element are called the valence electrons.

6. When an atom loses an electron it becomes a negative ion.

7. A bromide ion is smaller than a bromine atom.
8. Crystals of sodium chloride consist of arrays of NaCl molecules.
9. The sharing of a pair of electrons between a positive ion and a negative ion is called an ionic bond.
10. When elements combine by ionic bonding, they normally form molecules.
11. When a single atom furnishes both electrons that are shared between two atoms, the bond is called a coordinate covalent bond.
12. The NH_4^+ ion contains at least one coordinate covalent bond.
13. Oxygen has a greater electronegativity than carbon.
14. Bromine has a greater electronegativity than chlorine.
15. A dipole is a molecule that is electrically unsymmetrical, causing it to be oppositely charged at two points.
16. The water molecule is a dipole.
17. The formula of the aluminum ion is Al^+.
18. A stable group of atoms that has either a positive or a negative charge and behaves as a single unit is called a polyatomic ion.
19. A phosphate ion, PO_4^{3-}, is a polyatomic ion.
20. A sodium ion is smaller than a sodium atom.
21. In crystals of sodium chloride, each sodium ion is surrounded by six chloride ions, and each chloride ion is surrounded by six sodium ions.
22. Boron has a greater electronegativity than barium.
23. Metals tend to lose their valence electrons, forming positively charged ions.
24. When metals with low electronegativity combine with nonmetals of high electronegativity, they tend to form ionic compounds.
25. A pair of electrons shared between two atoms constitutes a covalent bond.
26. The ionic bond is the electrostatic attraction existing between oppositely charged ions.
27. Two atoms with different electronegativities that form a polar covalent bond will have partial ionic charges.
28. The covalent bond between a B atom and an N atom will be nonpolar.
29. The covalent bond between two carbon atoms will be nonpolar.
30. A covalent bond formed by sharing one pair of electrons is called a single bond.

Multiple Choice. *Choose the correct answer to each of the following.*

1. Which of the following formulas is not correct?
   (a) Na^+ (b) S^- (c) Al^{3+} (d) F^-
2. Which of the following does not have a polar covalent bond?
   (a) CH_4 (b) H_2O (c) CH_3OH (d) Cl_2
3. Which of the following molecules is a dipole?
   (a) HBr (b) CH_4 (c) H_2 (d) CO_2
4. Which of the following has bonding that is ionic?
   (a) H_2 (b) MgF_2 (c) H_2O (d) CH_4
5. Which of the following is a correct Lewis structure?

   (a) $:\overset{..}{O}:C:\overset{..}{O}:$ (b) $:\overset{..}{Cl}:\overset{\overset{..}{:}\overset{..}{Cl}:}{\underset{:\overset{..}{Cl}:}{C}}:\overset{..}{Cl}:$

   (c) $\overset{..}{Cl}::\overset{..}{Cl}$ (d) $:\overset{..}{N}:\overset{..}{N}:$

6. Which of the following is an incorrect Lewis structure?

   (a) $H:\overset{..}{N}:H$ (b) $:\overset{..}{O}:H$
         $\quad\; H$ $\quad\; H$

   (c) $H:\overset{H}{\underset{H}{\overset{..}{C}}}:H$ (d) $:N:::N:$

7. The correct Lewis structure for SO_2 is:

   (a) $:\overset{..}{O}:\overset{..}{S}:\overset{..}{O}:$ (b) $:\overset{..}{O}:\overset{..}{S}::\overset{..}{O}:$

   (c) $:\overset{..}{O}::\overset{..}{S}::\overset{..}{O}:$ (d) $:\overset{..}{O}:\overset{..}{S}:\overset{..}{O}:$

8. Which element has seven valence electrons?
   (a) S (b) Ne (c) Br (d) Ag
9. Carbon dioxide, CO_2, is a nonpolar molecule because:
   (a) Oxygen is more electronegative than carbon.
   (b) The two oxygen atoms are bonded to the carbon atom.
   (c) The molecule has a linear structure with the carbon atom in the middle.
   (d) The carbon–oxygen bonds are polar covalent.

10. When a magnesium atom participates in a chemical reaction, it is most likely to:
    (a) Lose 1 electron   (c) Lose 2 electrons
    (b) Gain 1 electron   (d) Gain 2 electrons

11. If X represents an element of Group IIIA, what is the general formula for its oxide?
    (a) X_3O_4 (b) X_3O_2 (c) XO (d) X_2O_3

12. Which of the following has the same electron structure as an argon atom?
    (a) Ca^{2+} (b) Cl^0 (c) Na^+ (d) K^0

13. As the difference in electronegativity between two elements decreases, the tendency for the elements to form a covalent bond:
    (a) Increases
    (b) Decreases
    (c) Remains the same
    (d) Sometimes increases and sometimes decreases

14. Which compound forms a tetrahedral molecule?
    (a) NaCl   (b) CO_2 (c) CH_4 (d) $MgCl_2$

15. Which compound has a bent (V-shaped) molecular structure?
    (a) NaCl   (b) CO_2 (c) CH_4 (d) H_2O

16. Which compound has double bonds within its molecular structure?
    (a) NaCl   (b) CO_2 (c) CH_4 (d) H_2O

17. The total number of valence electrons in a nitrate ion, NO_3^-, is:
    (a) 12   (b) 18   (c) 23   (d) 24

18. The number of electrons in a triple bond is:
    (a) 3   (b) 4   (c) 6   (d) 8

19. The number of lone (unbonded) pairs of electrons in H_2O is:
    (a) 0   (b) 1   (c) 2   (d) 4

20. Which of the following does not have a noble-gas electron structure?
    (a) Na   (b) Sc^{3+} (c) Ar   (d) O^{2-}

CHAPTER THIRTEEN

The Gaseous State of Matter

Our atmosphere is composed of a mixture of gases, including nitrogen, oxygen, carbon dioxide, ozone, and trace amounts of other gases. Carbon dioxide is taken in by plants and converted to carbohydrates. Yet it also is associated with the potentially hazardous greenhouse effect. Ozone surrounds the earth at high altitudes and protects us from harmful ultraviolet rays, but it also destroys rubber and plastics. We require air to live, yet scuba divers must be concerned about oxygen poisoning, nitrogen narcosis, and the bends.

In chemistry, the study of the behavior of gases, allows us to lay a foundation for understanding our atmosphere and the effects that gases have on our lives.

Chapter Preview

13.1   General Properties of Gases

In Chapter 3, solids, liquids, and gases are described in a brief outline. In this chapter we shall consider the behavior of gases in greater detail.

Gases are the least compact and most mobile of the three states of matter. A solid has a rigid structure, and its particles remain in essentially fixed positions. When a solid absorbs sufficient heat, it melts and changes into a liquid. Melting occurs because the molecules (or ions) have absorbed enough energy to break out of the rigid crystal lattice structure of the solid. The molecules or ions in the liquid are more energetic than they were in the solid, as shown by their increased mobility. Molecules in the liquid state are *coherent*; that is, they cling to one another. When the liquid absorbs additional heat, the more energetic molecules break away from the liquid surface and go into the gaseous state. Gases represent the most mobile state of matter. Gas molecules move with very high velocities and have high kinetic energy (KE). The average velocity of hydrogen molecules at 0°C is over 1600 meters (1 mile) per second. Because of the high velocities of their molecules, mixtures of gases are uniformly distributed within the container in which they are confined.

A quantity of a substance occupies a much greater volume as a gas than it does as a liquid or a solid. For example, 1 mole of water (18 g) has a volume of

18 mL at 4°C. This same amount of water would occupy about 22,400 mL in the gaseous state—more than a 1200-fold increase in volume. We may assume from this difference in volume that (1) gas molecules are relatively far apart, (2) gases can be greatly compressed, and (3) the volume occupied by a gas is mostly empty space.

13.2   The Kinetic-Molecular Theory

Careful scientific studies of the behavior and properties of gases were begun in the 17th century by Robert Boyle (1627–1691). This work was carried forward by many investigators after Boyle. The accumulated data were used in the second half of the 19th century to formulate a general theory to explain the behavior and properties of gases. This theory is called the **Kinetic-Molecular Theory (KMT)**. The KMT has since been extended to cover, in part, the behavior of liquids and solids. It ranks today with the atomic theory as one of the greatest generalizations of modern science.

Kinetic-Molecular
Theory (KMT)

The KMT is based on the motion of particles, particularly gas molecules. A gas that behaves exactly as outlined by the theory is known as an **ideal**, or **perfect**, **gas**. Actually no ideal gases exist, but under certain conditions of temperature and pressure gases approach ideal behavior, or at least show only small deviations from it. Under extreme conditions, such as very high pressure and low temperature, real gases may deviate greatly from ideal behavior. For example, at low temperature and high pressure many gases become liquids.

ideal, or perfect,
gas

The principal assumptions of the Kinetic-Molecular Theory are

1. Gases consist of tiny (submicroscopic) molecules.
2. The distance between molecules is large compared with the size of the molecules themselves. The volume occupied by a gas consists mostly of empty space.
3. Gas molecules have no attraction for one another.
4. Gas molecules move in straight lines in all directions, colliding frequently with one another and with the walls of the container.
5. No energy is lost by the collision of a gas molecule with another gas molecule or with the walls of the container. All collisions are perfectly elastic.
6. The average kinetic energy for molecules is the same for all gases at the same temperature, and its value is directly proportional to the Kelvin temperature.

The kinetic energy (KE) of a molecule is one-half its mass times its velocity squared. It is expressed by the equation

$$KE = \frac{1}{2} mv^2$$

where m is the mass and v is the velocity of the molecule.

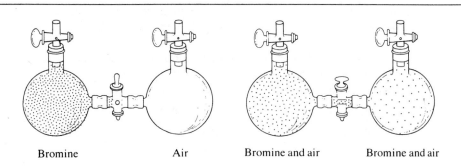

Bromine   Air   Bromine and air   Bromine and air

FIGURE 13.1

Diffusion of gases. When the stopcock between the two flasks is opened, colored bromine molecules can be seen diffusing into the flask containing air.

All gases have the same kinetic energy at the same temperature. Therefore, from the kinetic energy equation we can see that, if we compare the velocities of the molecules of two gases, the lighter molecules will have a greater velocity than the heavier ones. For example, calculations show that the velocity of a hydrogen molecule is four times the velocity of an oxygen moelcule.

diffusion

Due to their molecular motion gases have the property of **diffusion**, the ability of two or more gases to mix spontaneously until they form a uniform mixture. The diffusion of gases may be illustrated by use of the apparatus shown in Figure 13.1. Two large flasks, one containing reddish-brown bromine vapors and the other dry air, are connected by a side tube. When the stopcock between the flasks is opened, the bromine and air will diffuse into each other. After standing awhile, both flasks will contain bromine and air.

effusion

If we put a pinhole in a balloon, the gas inside will effuse or flow out of the balloon. **Effusion** is a process by which gas molecules pass through a very small orifice (opening) from a container at higher pressure to one at lower pressure.

Graham's Law of Effusion

Thomas Graham (1805–1869), a Scottish chemist, observed that the rate of effusion was dependent on the density of a gas. This observation led to **Graham's Law of Effusion**: The rates of effusion of two gases at the same temperature and pressure are inversely proportional to the square roots of their densities or molar masses,

$$\frac{\text{rate of effusion of gas A}}{\text{rate of effusion of gas B}} = \sqrt{\frac{d\text{B}}{d\text{A}}} = \sqrt{\frac{\text{molar mass B}}{\text{molar mass A}}}$$

A major application of Graham's law occurred during World War II with the separation of the isotopes of uranium-235 (U-235) and uranium-238 (U-238). Naturally occurring uranium consists of 0.7% U-235, 99.3% U-238, and a trace of U-234. However, only U-235 is useful as fuel for nuclear reactors and atomic bombs, so the concentration of U-235 in the mixture of isotopes had to be increased.

Uranium was first changed to uranium hexafluoride, UF_6, a white solid that readily goes into the gaseous state. The gaseous mixture of $^{235}UF_6$ and $^{238}UF_6$ was then allowed to effuse through porous walls. Although the effusion rate of the lighter gas is only slightly faster than that of the heavier one,

$$\frac{\text{effusion rate } ^{235}UF_6}{\text{effusion rate } ^{238}UF_6} = \sqrt{\frac{\text{molar mass } ^{238}UF_6}{\text{molar mass } ^{235}UF_6}} = \sqrt{\frac{352}{349}} = 1.0043$$

the separation and enrichment of U-235 was accomplished by subjecting the gaseous mixture to several thousand stages of effusion.

The properties of an ideal gas are independent of the molecular constitution of the gas. Mixtures of gases also obey the Kinetic-Molecular Theory if the gases in the mixture do not enter into a chemical reaction with one another.

13.3   Measurement of Pressure of Gases

pressure

Pressure is defined as force per unit area. Do gases exert pressure? Yes. When a rubber balloon is inflated with air, it stretches and maintains its larger size because the pressure on the inside is greater than that on the outside. Pressure results from the collisions of gas molecules with the walls of the balloon (see Figure 13.2). When the gas is released, the force or pressure of the air escaping from the small neck propels the balloon in a rapid, irregular flight. If the balloon is inflated until it bursts, the gas escaping all at once causes an explosive noise. This pressure that gases display can be measured; it can also be transformed into useful work. Steam under pressure, used in the steam locomotive, played an important role in the early development of the United States. Compressed steam is used today to generate at least part of the electricity for many cities. Compressed air is used to operate many different kinds of mechanical equipment.

The mass of air surrounding the earth is called the *atmosphere*. It is composed of about 78% nitrogen, 21% oxygen, and 1% argon, and other minor constituents by volume (see Table 13.1). The outer boundary of the atmosphere is not known precisely, but more than 99% of the atmosphere is below an altitude of 20 miles (32 km). Thus, the concentration of gas molecules in the atmosphere decreases with altitude, and at about 4 miles the amount of oxygen is insufficient to sustain human life. The gases in the atmosphere exert a pressure known as

atmospheric pressure

atmospheric pressure. The pressure exerted by a gas depends on the number of molecules of gas present, the temperature, and the volume in which the gas is confined. Gravitational forces hold the atmosphere relatively close to the earth and prevent air molecules from flying off into outer space. Thus, the atmospheric pressure at any point is due to the mass of the atmosphere pressing downward at that point.

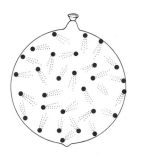

FIGURE 13.2

Here we see the pressure resulting from the collisions of gas molecules with the walls of the balloon. This pressure keeps the balloon inflated.

TABLE 13.1

Average Composition of Normal Dry Air

Gas	Percent by volume	Gas	Percent by volume
N_2	78.08	He	0.0005
O_2	20.95	CH_4	0.0002
Ar	0.93	Kr	0.0001
CO_2	0.033	Xe, H_2, and N_2O	Trace
Ne	0.0018		

barometer

The pressure of a gas can be measured with a pressure gauge, a manometer, or a **barometer**. A mercury barometer is commonly used in the laboratory to measure atmospheric pressure. A simple barometer of this type may be prepared by completely filling a long tube with pure, dry mercury and inverting the open end into an open dish of mercury. If the tube is longer than 760 mm, the mercury level will drop to a point at which the column of mercury in the tube is just supported by the pressure of the atmosphere. If the tube is properly prepared, a vacuum will exist above the mercury column. The mass of mercury, per unit area,

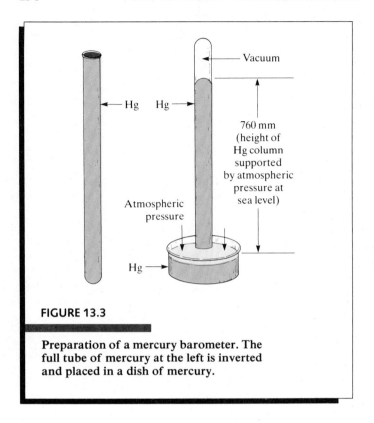

Vacuum

Hg   Hg

760 mm
(height of
Hg column
supported
by atmospheric
pressure at
sea level)

Atmospheric
pressure

Hg

FIGURE 13.3

**Preparation of a mercury barometer. The
full tube of mercury at the left is inverted
and placed in a dish of mercury.**

is equal to the pressure of the atmosphere. The column of mercury is supported
by the pressure of the atmosphere, and the height of the column is a measure of
this pressure (see Figure 13.3). The mercury barometer was invented in 1643 by
the Italian physicist E. Torricelli (1608–1647), for whom the unit of pressure *torr*
was named.

Air pressure is measured and expressed in many units. The standard
atmospheric pressure, or simply **1 atmosphere**, is the pressure exerted by a column of mercury 760 mm high at a temperature of 0°C. The abbreviation for
atmosphere is atm. The normal pressure of the atmosphere at sea level is 1 atm
or 760 torr or 760 mm Hg. The SI unit for pressure is the pascal (Pa), where
1 atm = 101,325 Pa or 101.3 kPa. Other units for expressing pressure are inches
of mercury, centimeters of mercury, the millibar (mbar), and pounds per square
inch (lb/in.^{2} or psi). The meteorologist uses inches of mercury in reporting atmospheric pressure. The values of these units equivalent to 1 atm are summarized in Table 13.2 (1 atm ≡ 760 torr ≡ 760 mm Hg ≡ 76 cm Hg ≡ 101,325 Pa ≡
1013 mbar ≡ 29.9 in. Hg ≡ 14.7 lb/in.^{2}). (The symbol ≡ means *identical with*.)

Atmospheric pressure varies with altitude. The average pressure at Denver,

1 atmosphere

TABLE 13.2

Pressure Units Equivalent to 1 Atmosphere

1 atm
760 torr
760 mm Hg
76 cm Hg
101,325 Pa
1013 mbar
29.9 in. Hg
14.7 lb/in.^{2}

Colorado, 1.61 km (1 mile) above sea level, is 630 torr (0.83 atm). Atmospheric pressure is 0.5 atm at about 5.5 km (3.4 miles) altitude.

Other liquids besides mercury may be employed for barometers, but they are not as useful as mercury because of the difficulty of maintaining a vacuum above the liquid and because of impractical heights of the liquid column. For example, a pressure of 1 atm will support a column of water about 10,336 mm (33.9 ft) high.

Pressure is often measured by reading the heights of mercury columns in millimeters on barometers and manometers. Thus pressure may be recorded as mm Hg. But in many applications the torr is superseding mm Hg as a unit of pressure. In problems dealing with gases it is necessary to make interconversions among the various pressure units. Since atm, torr, and mm Hg are common pressure units, we use illustrative problems involving all three of these units.

$$1 \text{ atm} = 760 \text{ torr} = 760 \text{ mm Hg}$$

EXAMPLE 13.1

The average atmospheric pressure at Walnut, California, is 740. mm Hg. Calculate this pressure in (a) torr and (b) atmospheres.

This problem can be solved using conversion factors relating one unit of pressure to another.

(a) To convert mm Hg to torr, use the conversion factor 760. torr/760. mm Hg (1 torr/1 mm Hg):

$$740. \text{ mm Hg} \times \frac{1 \text{ torr}}{1 \text{ mm Hg}} = 740. \text{ torr}$$

(b) To convert mm Hg to atm, use the conversion factor 1 atm/760. mm Hg:

$$740. \; \cancel{\text{mm Hg}} \times \frac{1 \text{ atm}}{760. \; \cancel{\text{mm Hg}}} = 0.934 \text{ atm}$$

PRACTICE　A barometer reads 1.12 atm. Calculate the corresponding pressure in (a) torr and (b) mm Hg.

Answer:　(a) 851 torr　(b) 851 mm Hg

13.4　Dependence of Pressure on Number of Molecules and Temperature

Pressure is produced by gas molecules colliding with the walls of a container. At a specific temperature and volume, the number of collisions depends on the number of gas molecules present. The number of collisions can be increased by increasing the number of gas molecules present. If we double the number of molecules, the frequency of collisions and the pressure should double. We find, for an ideal gas, that this doubling is actually what happens: When the temperature and volume are kept constant, the pressure is directly proportional to the number of moles or molecules of gas present. Figure 13.4 illustrates this concept.

A good example of this molecule-pressure relationship may be observed in an ordinary cylinder of compressed gas that is equipped with a pressure gauge. When the valve is opened, gas escapes from the cylinder. The volume of the cylinder is constant, and the decrease in quantity (moles) of gas is registered by a drop in pressure indicated on the gauge.

The pressure of a gas in a fixed volume also varies with temperature. When the temperature is increased, the kinetic energy of the molecules increases, causing more frequent collisions of the molecules with the walls of the container. This increase in collision frequency results in a pressure increase (see Figure 13.5, p. 300).

13.5　Boyle's Law

Boyle's law

Robert Boyle demonstrated experimentally that, at constant temperature (T), the volume (V) of a fixed mass of a gas is inversely proportional to the pressure (P). This relationship of P and V is known as **Boyle's law**. Mathematically, Boyle's

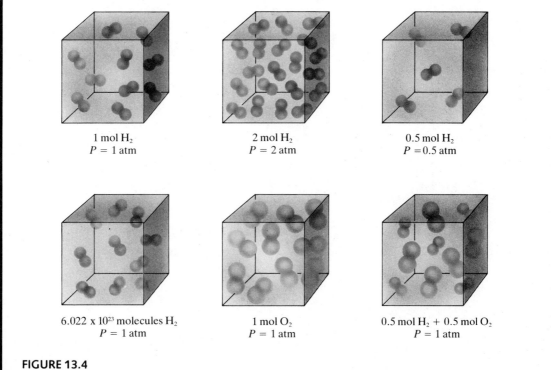

1 mol H₂
P = 1 atm

2 mol H₂
P = 2 atm

0.5 mol H₂
P = 0.5 atm

6.022 x 10²³ molecules H₂
P = 1 atm

1 mol O₂
P = 1 atm

0.5 mol H₂ + 0.5 mol O₂
P = 1 atm

FIGURE 13.4

The pressure exerted by a gas is directly proportional to the number of molecules present. In each case shown, the volume is 22.4 L and the temperature is 0°C.

law may be expressed

$$V \propto \frac{1}{P} \quad \text{(mass and temperature are constant)}$$

This equation says that the volume varies ($\propto$) inversely with the pressure, at constant mass and temperature. When the pressure on a gas is increased, its volume will decrease, and vice versa. The inverse relationship of pressure and volume is shown graphically in Figure 13.6 (p. 301).

Boyle demonstrated that, when he doubled the pressure on a specific quantity of a gas, keeping the temperature constant, the volume was reduced to one-half the original volume; when he tripled the pressure on the system, the new volume was one-third the original volume; and so on. His demonstration showed that the product of volume and pressure is constant if the temperature is not

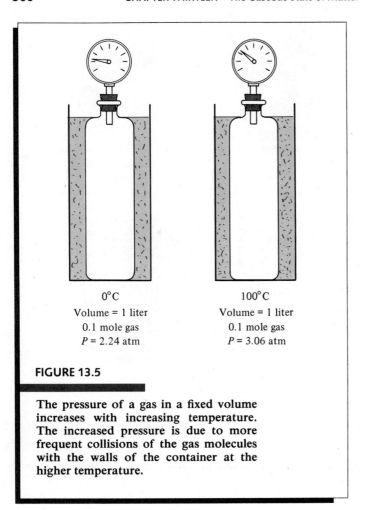

0°C
Volume = 1 liter
0.1 mole gas
P = 2.24 atm

100°C
Volume = 1 liter
0.1 mole gas
P = 3.06 atm

FIGURE 13.5

The pressure of a gas in a fixed volume increases with increasing temperature. The increased pressure is due to more frequent collisions of the gas molecules with the walls of the container at the higher temperature.

changed:

$$PV = \text{constant} \quad \text{or} \quad PV = k \quad \text{(mass and temperature are constant)}$$

Let us demonstrate this law by using a cylinder with a movable piston so that the volume of gas inside the cylinder may be varied by changing the external pressure (see Figure 13.7). We assume that the temperature and the number of gas molecules do not change. Let us start with a volume of 1000 mL and a pressure of 1 atm. When we change the pressure to 2 atm, the gas molecules are crowded closer together, and the volume is reduced to 500 mL. When we increase the pressure to 4 atm, the volume becomes 250 mL.

Note that the product of the pressure times the volume is the same number in each case, substantiating Boyle's law. We may then say that

$$P_1V_1 = P_2V_2$$

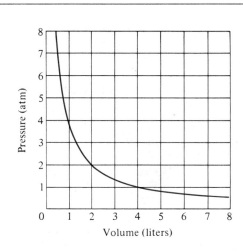

FIGURE 13.6

Graph of pressure versus volume showing the inverse PV relationship of an ideal gas

where P_1V_1 is the pressure–volume product at one set of conditions, and P_2V_2 is the product at another set of conditions. In each case the new volume may be calculated by multiplying the starting volume by a ratio of the two pressures involved. Of course, the ratio of pressures used must reflect the direction in which the volume should change. When the pressure is changed from 1 atm to 2 atm, the ratio to be used is 1 atm/2 atm. Now we can verify the results given in Figure 13.7:

(a) Starting volume, 1000 mL; pressure change, 1 atm $\longrightarrow$ 2 atm

$$1000 \text{ mL} \times \frac{1 \text{ atm}}{2 \text{ atm}} = 500 \text{ mL}$$

(b) Starting volume, 1000 mL; pressure change, 1 atm $\longrightarrow$ 4 atm

$$1000 \text{ mL} \times \frac{1 \text{ atm}}{4 \text{ atm}} = 250 \text{ mL}$$

(c) Starting volume, 500 mL; pressure change, 2 atm $\longrightarrow$ 4 atm

$$500 \text{ mL} \times \frac{2 \text{ atm}}{4 \text{ atm}} = 250 \text{ mL}$$

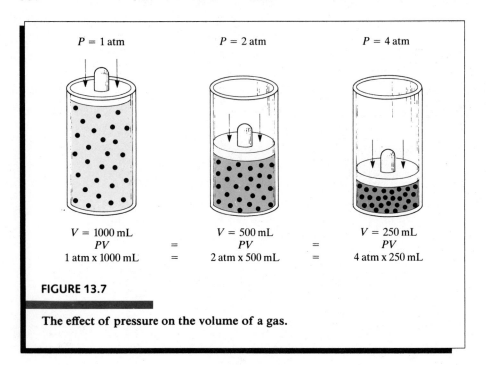

$P = 1$ atm $P = 2$ atm $P = 4$ atm

$V = 1000$ mL		$V = 500$ mL		$V = 250$ mL
PV	$=$	PV	$=$	PV
1 atm x 1000 mL	$=$	2 atm x 500 mL	$=$	4 atm x 250 mL

FIGURE 13.7

The effect of pressure on the volume of a gas.

In summary, a change in the volume of a gas due to a change in pressure can be calculated by multiplying the original volume by a ratio of the two pressures. If the pressure is increased, the ratio should have the smaller pressure in the numerator and the larger pressure in the denominator. If the pressure is decreased, the larger pressure should be in the numerator and the smaller pressure in the denominator.

new volume = original volume × ratio of pressures

Examples of problems based on Boyle's law follow. If no mention is made of temperature, assume that it remains constant.

EXAMPLE 13.2   What volume will 2.50 L of a gas occupy if the pressure is changed from 760. mm Hg to 630. mm Hg?

Method (a):   Conversion Factors

Step 1 Determine whether pressure is being increased or decreased.

pressure decreases ⟶ volume increases

Step 2 Multiply the original volume by a ratio of pressures that will result in an increase in volume.

$$V = 2.50 \text{ L} \times \frac{760 \text{ mm Hg}}{630 \text{ mm Hg}} = 3.02 \text{ L (new volume)}$$

Method (b):   Algebraic Equation

Step 1 Organize given information.

$$P_1 = 760. \text{ mm Hg} \qquad V_1 = 2.50 \text{ L}$$
$$P_2 = 630. \text{ mm Hg} \qquad V_2 = ?$$

Step 2 Write and solve equation for the unknown.

$$P_1V_1 = P_2V_2 \qquad V_2 = V_1 \times \frac{P_1}{P_2}$$

Step 3 Put given information into equation and calculate.

$$V_2 = 2.50 \text{ L} \times \frac{760 \text{ mm Hg}}{630 \text{ mm Hg}} = 3.02 \text{ L}$$

EXAMPLE 13.3   A given mass of hydrogen occupies 40.0 L at 700. torr. What volume will it occupy at 5.0 atm pressure?

Method (a):   Conversion Factors

Step 1 Determine whether the pressure is being increased or decreased. Notice that in order to compare the values the units must be the same.

$$700 \text{ torr} \times \frac{1 \text{ atm}}{760 \text{ torr}} = 0.921 \text{ atm}$$

pressure increases $\longrightarrow$ volume decreases

Step 2 Multiply the original volume by a ratio of pressures that will result in a decrease in volume.

$$V = 40.0 \text{ L} \times \frac{0.921 \text{ atm}}{5.00 \text{ atm}} = 7.37 \text{ L} \quad \text{(Answer)}$$

Method (b):   Algebraic Equation

Step 1 Organize the given information. Remember to make units the same.

$$P_1 = 700. \text{ torr} = 0.921 \text{ atm} \qquad V_1 = 40.0 \text{ L}$$
$$P_2 = 5.00 \text{ atm} \qquad\qquad\qquad V_2 = ?$$

Step 2 Write and solve equation for the unknown.

$$P_1 V_1 = P_2 V_2 \qquad V_2 = V_1 \times \frac{P_1}{P_2}$$

Step 3 Put given information into equation and calculate.

$$V_2 = 40.0 \text{ L} \times \frac{0.921 \text{ atm}}{5.00 \text{ atm}} = 7.37 \text{ L} \quad \text{(Answer)}$$

EXAMPLE 13.4   A gas occupies a volume of 200. mL at 400. torr pressure. To what pressure must the gas be subjected in order to change the volume to 75.0 mL?

Method (a):   Conversion Factors

Step 1 Determine whether volume is being increased or decreased.

volume decreases $\longrightarrow$ pressure increases

Step 2 Multiply the original pressure by a ratio of volumes that will result in an increase in pressure.

new pressure = original pressure × ratio of volumes

$$P = 400 \text{ torr} \times \frac{200 \text{ mL}}{75.0 \text{ mL}} = 1067 \text{ torr} \qquad \text{or} \qquad 1.07 \times 10^3 \text{ torr} \quad \text{(new pressure)}$$

Method (b):   Algebraic Equation

Step 1 Organize given information. Remember to make units the same.

$P_1 = 400. \text{ torr} \qquad V_1 = 200. \text{ mL}$
$P_2 = ? \qquad\qquad V_2 = 75.0 \text{ mL}$

Step 2 Write and solve equation for the unknown.

$$P_1 V_1 = P_2 V_2 \qquad P_2 = P_1 \times \frac{V_1}{V_2}$$

Step 3 Put given information into equation and calculate.

$$P_2 = P_1 \times \frac{V_1}{V_2} = 400 \text{ torr} \times \frac{200 \text{ mL}}{75.0 \text{ mL}} = 1.07 \times 10^3 \text{ torr}$$

PRACTICE   A gas occupies a volume of 3.86 L at 0.750 atm. What must the pressure be changed to in order for the volume to change to 4.86 L?

Answer: 0.600 atm

13.6 Charles' Law

The effect of temperature on the volume of a gas was observed in about 1787 by the French physicist J. A. C. Charles (1746–1823). Charles found that various gases expanded by the same fractional amount when heated through the same temperature interval. Later it was found that if a given volume of any gas initially at 0°C was cooled by 1°C, the volume decreased by $\frac{1}{273}$; if cooled by 2°C, by $\frac{2}{273}$; if cooled by 20°C, by $\frac{20}{273}$; and so on. Since each degree of cooling reduced the volume by $\frac{1}{273}$, it was apparent that any quantity of any gas would have zero volume, if it could be cooled to −273°C. Of course no real gas can be cooled to −273°C for the simple reason that it liquefies before that temperature is reached. However, −273°C (more precisely −273.16°C) is referred to as *absolute zero*; this temperature is the zero point on the Kelvin (absolute) temperature scale. It is the temperature at which the volume of an ideal, or perfect, gas would become zero.

The volume–temperature relationship for gases is shown graphically in Figure 13.8. Experimental data show the graph to be a straight line that, when extrapolated, crosses the temperature axis at −273.16°C, or absolute zero.

Charles' law

In modern form, **Charles' law** states that at *constant pressure* the volume of a fixed mass of any gas is directly proportional to the absolute temperature. Mathematically, Charles' law may be expressed as

$$V \propto T \quad (P \text{ is constant})$$

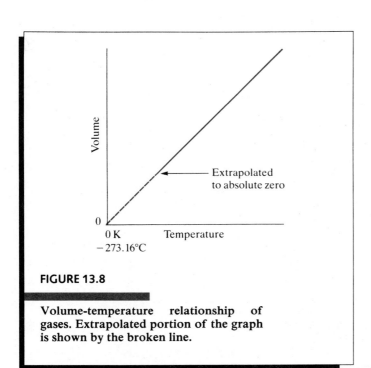

FIGURE 13.8

Volume-temperature relationship of gases. Extrapolated portion of the graph is shown by the broken line.

which means that the volume of a gas varies directly with the absolute temperature when the pressure remains constant. In equation form Charles' law may be written as

$$V = kT \qquad \text{or} \qquad \frac{V}{T} = k \quad \text{(at constant pressure)}$$

where k is a constant for a fixed mass of the gas. If the absolute temperature of a gas is doubled, the volume will double. (A capital T is usually used for absolute temperature, K, and a small t for °C.)

To illustrate, let us return to the gas cylinder with the movable or free-floating piston, (see Figure 13.9). Assume that the cylinder labeled (a) contains a quantity of gas and the pressure on it is 1 atm. When the gas is heated, the molecules move faster, and their kinetic energy increases. This action should increase the number of collisions per unit of time and therefore increase the pressure. However, the increased internal pressure will cause the piston to rise to a level at which the internal and external pressures again equal 1 atm, as we see in cylinder (b). The net result is an increase in volume due to an increase in temperature.

Another equation relating the volume of a gas at two different temperatures is

$$\frac{V_1}{T_1} = \frac{V_2}{T_2} \quad \text{(constant } P\text{)}$$

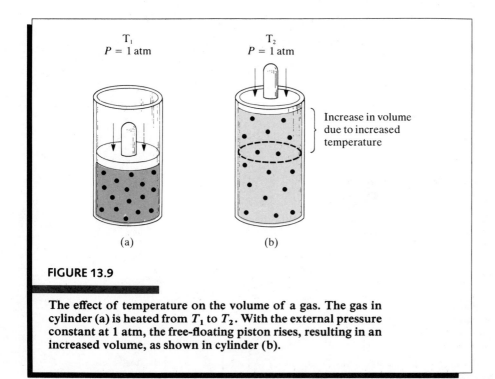

(a)   (b)

FIGURE 13.9

The effect of temperature on the volume of a gas. The gas in cylinder (a) is heated from T_1 to T_2. With the external pressure constant at 1 atm, the free-floating piston rises, resulting in an increased volume, as shown in cylinder (b).

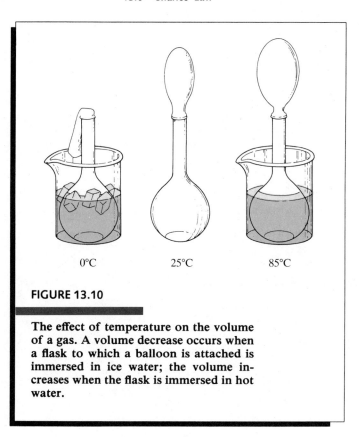

0°C   25°C   85°C

FIGURE 13.10

The effect of temperature on the volume of a gas. A volume decrease occurs when a flask to which a balloon is attached is immersed in ice water; the volume increases when the flask is immersed in hot water.

where V_1 and T_1 are one set of conditions and V_2 and T_2 are another set of conditions.

A simple experiment showing the variation of the volume of a gas with temperature is illustrated in Figure 13.10. A flask to which a balloon is attached is immersed in either ice water or hot water. In ice water the volume is reduced, as shown by the collapse of the balloon; in hot water the gas expands and the balloon increases in size.

EXAMPLE 13.5   Three liters of hydrogen at $-20.°C$ are allowed to warm to a room temperature of $27°C$. What is the volume at room temperature if the pressure remains constant?

Method (a):   Conversion Factors

Step 1 Determine whether temperature is being increased or decreased. Remember temperature must be changed to Kelvin in gas problems.

$$-20.°C + 273 = 253 \text{ K}$$
$$27°C + 273 = 300. \text{ K}$$

temperature increases $\longrightarrow$ volume increases

Step 2 Multiply the original volume by a ratio of temperatures that will result in an increase in volume.

$$V = 3.00 \text{ L} \times \frac{300.\,\cancel{K}}{253\,\cancel{K}} = 3.56 \text{ L} \text{ (new volume)}$$

Method (b): Algebraic Equation

Step 1 Organize given information. Remember to make units the same.

$$V_1 = 3.00 \text{ L}   T_1 = -20.°C = 253 \text{ K}$$
$$V_2 = ?   T_2 = 27°C = 300. \text{ K}$$

Step 2 Write and solve equation for the unknown.

$$\frac{V_1}{T_1} = \frac{V_2}{T_2}   V_2 = V_1 \times \frac{T_2}{T_1}$$

Step 3 Put given information into equation and calculate.

$$V_2 = V_1 \times \frac{T_2}{T_1} = 3.00 \text{ L} \times \frac{300\,\cancel{K}}{253\,\cancel{K}} = 3.56 \text{ L}$$

EXAMPLE 13.6   If 20.0 L of oxygen are cooled from 100°C to 0°C, what is the new volume?
Since no mention is made of pressure, assume that pressure does not change.

Method (a): Conversion Factors

Step 1 Change °C to K:

$$100°C + 273 = 373 \text{ K}$$
$$0°C + 273 = 273 \text{ K}$$

Step 2 The ratio of temperature to be used is 273 K/373 K, because the final volume should be smaller than the original volume. The calculation is

$$V = 20.0 \text{ L} \times \frac{273\,\cancel{K}}{373\,\cancel{K}} = 14.6 \text{ L} \text{ (new volume)}$$

Method (b): Algebraic Equation

Step 1 Organize the given information. Remember to make units coincide.

$$V_1 = 20.0 \text{ L}   T_1 = 100°C = 373 \text{ K}$$
$$V_2 = ?   T_2 = 0°C = 273 \text{ K}$$

Step 2 Write and solve equation for the unknown.

$$\frac{V_1}{T_1} = \frac{V_2}{T_2} \qquad V_2 = V_1 \times \frac{T_2}{T_1}$$

Step 3 Put given information into equation and calculate.

$$V_2 = V_1 \times \frac{T_2}{T_1} = 20.0 \text{ L} \times \frac{273 \text{ K}}{373 \text{ K}} = 14.6 \text{ L}$$

PRACTICE A 4.50 L container of nitrogen at 28.0°C is heated to 56.0°C. Assuming the the volume of the container can vary, what is the new volume of the gas?
Answer: 4.92 L

13.7 Gay-Lussac's Law

Three variables—pressure, P; volume, V; and temperature, T—are needed to describe a fixed amount of a gas. Boyle's law, $PV = k$, relates pressure and volume at constant temperature; Charles' law, $V = kT$, relates volume and temperature at constant pressure. A third relationship involving pressure and temperature at constant volume is stated: The pressure of a fixed mass of a gas, at constant volume, is directly proportional to the Kelvin temperature. In equation form the relationship is

$$P = kT \quad \text{(at constant volume)} \qquad \frac{P_1}{T_1} = \frac{P_2}{T_2}$$

Gay-Lussac's law

This relationship is a modification of Charles' law and is sometimes called **Gay-Lussac's law.**

EXAMPLE 13.7

The pressure of a container of helium is 650. torr at 25°C. If the sealed container is cooled to 0°C, what will the pressure be?

Method (a): Conversion Factors

Step 1 Determine whether temperature is being increased or decreased.

temperature decreases $\longrightarrow$ pressure decreases

Step 2 Multiply the original pressure by a ratio of Kelvin temperatures which will result in a decrease in pressure.

$$650. \text{ torr} \times \frac{273 \text{ K}}{298 \text{ K}} = 595 \text{ torr}$$

Method (b):   Algebraic Equation

Step 1 Organize given information. Remember to make units the same.

$$P_1 = 650. \text{ torr} \qquad T_1 = 25°C = 298 \text{ K}$$
$$P_2 = ? \qquad\qquad T_2 = 0°C = 273 \text{ K}$$

Step 2 Write and solve equation for the unknown.

$$\frac{P_1}{T_1} = \frac{P_2}{T_2}$$

$$P_2 = P_1 \times \frac{T_2}{T_1}$$

Step 3 Put given information into equation and calculate.

$$P_2 = 650. \text{ torr} \times \frac{273 \text{ K}}{298 \text{ K}} = 595 \text{ torr}$$

PRACTICE   A gas cylinder contains 40.0 L of gas at 45.0°C and has a pressure of 650. torr. What will the pressure be if the temperature is changed to 100.°C?

Answer: 762 torr

We may summarize the effects of changes in pressure, temperature, and quantity of a gas as follows:

1. In the case of a fixed or constant volume,
    (a) when the temperature is increased, the pressure increases.
    (b) when the quantity of a gas is increased, the pressure increases (T remaining constant).
2. In the case of a variable volume,
    (a) when the external pressure is increased, the volume decreases (T remaining constant).
    (b) when the temperature of a gas is increased, the volume increases (P remaining constant).
    (c) when the quantity of a gas is increased, the volume increases (P and T remaining constant).

13.8   Standard Temperature and Pressure

standard
conditions

In order to compare volumes of gases, common reference points of temperature and pressure were selected and called **standard conditions** or **standard temperature and pressure** (abbreviated **STP**). Standard temperature is 273 K (0°C), and

standard
temperature and
pressure (STP)

standard pressure is 1 atm or 760 torr or 760 mm Hg or 101.325 kPa. For purposes of comparison, volumes of gases are usually changed to STP conditions.

> **standard temperature = 273 K or 0°C**
> **standard pressure = 1 atm or 760 torr or 760 mm Hg or**
> **101.325 kPa**

13.9 Combined Gas Laws: Simultaneous Changes in Pressure, Volume, and Temperature

When temperature and pressure change at the same time, the new volume may be calculated by multiplying the initial volume by the correct ratios of both pressure and temperature, as follows:

$$\text{final volume} = \text{initial volume} \times \left(\begin{array}{c} \text{ratio of} \\ \text{pressures} \end{array} \right) \times \left(\begin{array}{c} \text{ratio of} \\ \text{temperatures} \end{array} \right)$$

This equation combines Boyle's and Charles' laws, and the same considerations for the pressure and the temperature ratios should be used in the calculation. The four possible variations are:

1. Both T and P cause an increase in volume.
2. Both T and P cause a decrease in volume.
3. T causes an increase and P causes a decrease in volume.
4. T causes a decrease and P causes an increase in volume.

The P, V, and T relationships for a given mass of any gas, in fact, may be expressed as a single equation, $PV/T = k$. For problem solving this equation is usually written

$$\frac{P_1 V_1}{T_1} = \frac{P_2 V_2}{T_2}$$

where P_1, V_1, and T_1 are the initial conditions and P_2, V_2, and T_2 are the final conditions.

This equation can be solved for any one of the six variables and is useful in dealing with the pressure–volume–temperature relationships of gases. Note that when T is constant ($T_1 = T_2$), Boyle's law is represented; when P is constant ($P_1 = P_2$), Charles' law is represented; and when V is constant ($V_1 = V_2$), the modified Charles' or Gay-Lussac's law is represented.

EXAMPLE 13.8   Given 20.0 L of ammonia gas at 5°C and 730. torr, calculate the volume at 50.°C and 800. torr.

Step 1 Organize the given information, putting temperatures in Kelvin.

$$P_1 = 730. \text{ torr} \qquad P_2 = 800. \text{ torr}$$

$$V_1 = 20.0 \text{ L} \qquad V_2 = ?$$

$$T_1 = 5°C = 278 \text{ K} \qquad T_2 = 50.°C = 323 \text{ K}$$

Method (a):   Conversion Factors

Step 2 Set up ratios of T and P.

$$T \text{ ratio} = \frac{323 \text{ K}}{278 \text{ K}} \quad (\text{increase in } T \text{ should increase } V)$$

$$P \text{ ratio} = \frac{730. \text{ torr}}{800. \text{ torr}} \quad (\text{increase in } P \text{ should decrease } V)$$

Step 3 Multiply the original pressure by the ratios.

$$V_2 = 20.0 \text{ L} \times \frac{730. \text{ torr}}{800. \text{ torr}} \times \frac{323 \text{ K}}{278 \text{ K}} = 21.2 \text{ L}$$

Method (b):   Algebraic Equation

Step 2 Write and solve equation for the unknown.

Solve $\dfrac{P_1 V_1}{T_1} = \dfrac{P_2 V_2}{T_2}$ for V_2 by multiplying both sides of the equation by T_2/P_2 and rearranging to obtain

$$V_2 = \frac{V_1 \times P_1 \times T_2}{P_2 \times T_1}$$

Step 3 Put given information into equation and calculate.

$$V_2 = \frac{20.0 \text{ L} \times 730. \text{ torr} \times 323 \text{ K}}{800. \text{ torr} \times 278 \text{ K}} = 21.2 \text{ L}$$

EXAMPLE 13.9   To what temperature (°C) must 10.0 L of nitrogen at 25°C and 700. torr be heated in order to have a volume of 15.0 L and a pressure of 760. torr?

Step 1 Organize the given information, putting temperatures in Kelvin.

$$P_1 = 700. \text{ torr} \qquad P_2 = 760. \text{ torr}$$

$$V_1 = 10.0 \text{ L} \qquad V_2 = 15.0 \text{ L}$$

$$T_1 = 25°C = 298 \text{ K} \qquad T_2 = ?$$

Method 1:   Conversion Factors

Step 2 Set up ratios of V and P.

$$P \text{ ratio} = \frac{760. \text{ torr}}{700. \text{ torr}} \quad \text{(increase in } P \text{ should increase } T\text{)}$$

$$V \text{ ratio} = \frac{15.0 \text{ L}}{10.0 \text{ L}} \quad \text{(increase in } V \text{ should increase } T\text{)}$$

Step 3 Multiply the original pressure by the ratios.

$$T_2 = \frac{298 \text{ K} \times 760.\cancel{\text{ torr}} \times 15.0 \cancel{L}}{700.\cancel{\text{ torr}} \times 10.0 \cancel{L}} = 485 \text{ K}$$

Method 2:   Algebraic Equation

Step 2 Write and solve equation for unknown.

$$\frac{P_1 V_1}{T_1} = \frac{P_2 V_2}{T_2} \qquad T_2 = \frac{T_1 \times P_2 \times V_2}{P_1 \times V_1}$$

Step 3 Put given information into equation and calculate.

$$T_2 = \frac{298 \text{ K} \times 760.\cancel{\text{ torr}} \times 15.0 \cancel{L}}{700.\cancel{\text{ torr}} \times 10.0 \cancel{L}} = 485 \text{ K}$$

In either method since the problem asks for °C, we must subtract 273 from the Kelvin answer:

$$485 \text{ K} - 273 = 212°\text{C} \quad \text{(Answer)}$$

EXAMPLE 13.10   The volume of a gas-filled balloon is 50.0 L at 20.°C and 742 torr. What volume will it occupy at standard temperature and pressure (STP)?

Step 1 Organize the given information, putting temperatures in Kelvin.

$P_1 = 742. \text{ torr}$ $P_2 = 760. \text{ torr (standard pressure)}$

$V_1 = 50.0 \text{ L}$ $V_2 = ?$

$T_1 = 20.°\text{C} = 293 \text{ K}$ $T_2 = 273 \text{ K (standard pressure)}$

Method 1:   Conversion Factors

Step 2 Set up ratios of T and P.

$$T \text{ ratio} = \frac{273 \text{ K}}{293 \text{ K}} \quad \text{[decrease in } T \text{ (293 K to 273 K) should decrease } V\text{]}$$

$$P \text{ ratio} = \frac{742 \text{ torr}}{760. \text{ torr}} \quad \text{[increase in } P \text{ (742 torr to 760. torr) should decrease } V\text{]}$$

Step 3 Multiply the original volume by the ratios.

$$V_2 = 50.0 \text{ L} \times \frac{273\,K}{293\,K} \times \frac{742 \text{ torr}}{760. \text{ torr}} = 45.5 \text{ L}$$

Method (b):   Algebraic Equation

Step 2 Write and solve equation for the unknown.

$$\frac{P_1 V_1}{T_1} = \frac{P_2 V_2}{T_2} \qquad V_2 = \frac{P_1 V_1 T_2}{P_2 T_1}$$

Step 3 Put given information into equation and calculate.

$$V_2 = \frac{742 \text{ torr} \times 50.0 \text{ L} \times 273\,K}{760. \text{ torr} \times 293\,K} = 45.5 \text{ L}$$

PRACTICE   15.00 L of gas at 45.0°C and 800. torr is heated to 400.°C, and the pressure changed to 300. torr. What is the new volume?

Answer: 84.7 L

PRACTICE   To what temperature must 5.0 L of oxygen at 50.°C and 600. torr be heated in order to have a volume of 10.0 L and a pressure of 800. torr?

Answer: 861 K (588°C)

13.10   Dalton's Law of Partial Pressures

Dalton's Law of Partial Pressures

partial pressures

If gases behave according to the Kinetic-Molecular Theory, there should be no difference in the pressure–volume–temperature relationships whether the gas molecules are all the same or different. This similarity in the behavior of gases is the basis for an understanding of **Dalton's Law of Partial Pressures**, which states that the total pressure of a mixture of gases is the sum of the partial pressures exerted by each of the gases in the mixture. Each gas in the mixture exerts a pressure that is independent of the other gases present. These pressures are called **partial pressures**. Thus, if we have a mixture of three gases, A, B, and C, exerting partial pressures of 50 torr, 150 torr, and 400 torr, respectively, the total pressure will be 600 torr.

$$P_{\text{Total}} = p_A + p_B + p_C$$
$$P_{\text{Total}} = 50 \text{ torr} + 150 \text{ torr} + 400 \text{ torr} = 600 \text{ torr}$$

We can see an application of Dalton's law in the collection of insoluble gases over water. When prepared in the laboratory, oxygen is commonly collected by the downward displacement of water. Thus the oxygen is not pure but is mixed with water vapor (see Figure 13.11). When the water levels are adjusted to the

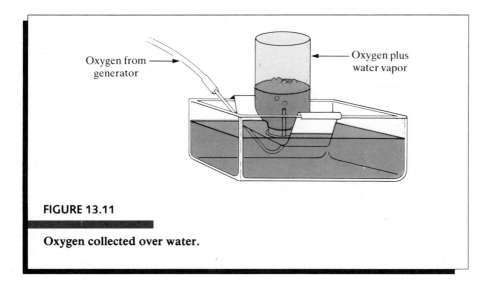

Oxygen from generator

Oxygen plus water vapor

FIGURE 13.11

Oxygen collected over water.

same height inside and outside the bottle, the pressure of the oxygen plus water vapor inside the bottle is equal to the atmospheric pressure:

$$P_{atm} = p_{O_2} + p_{H_2O}$$
$$p_{O_2} = P_{atm} - p_{H_2O}$$

To determine the amount of O_2 or any other gas collected over water, we must subtract the pressure of the water vapor from the total pressure of the gas. The vapor pressure of water at various temperatures is tabulated in Appendix II.

EXAMPLE 13.11   A 500. mL sample of oxygen was collected over water at 23°C and 760. torr. What volume will the dry O_2 occupy at 23°C and 760 torr? The vapor pressure of water at 23°C is 21.2 torr.

To solve this problem, we must first determine the pressure of the oxygen alone, by subtracting the pressure of the water vapor present.

Step 1 Determine the pressure of dry O_2.

$$P_{Total} = 760 \text{ torr} = p_{O_2} + p_{H_2O}$$
$$p_{O_2} = 760 \text{ torr} - 21.2 \text{ torr} = 739 \text{ torr} \quad (\text{dry } O_2)$$

Step 2 Organize the given information.

$$P_1 = 739 \text{ torr} \qquad P_2 = 760 \text{ torr}$$
$$V_1 = 500. \text{ mL} \qquad V_2 = ?$$

T is constant

Step 3 Solve as a Boyle's law problem:

$$V = 500 \text{ mL} \times \frac{739 \text{ torr}}{760. \text{ torr}} = 486 \text{ mL dry O}_2$$

PRACTICE   Hydrogen gas was collected by downward displacement of water. A volume of 600.0 mL of gas was collected at 25.0°C and 740.0 torr. What volume will the dry hydrogen occupy at STP?

Answer: 518 mL

13.11   Avogadro's Law

Gay-Lussac's Law
of Combining
Volumes of Gases

Early in the 19th century J. L. Gay-Lussac (1778–1850) of France studied the volume relationships of reacting gases. His results, published in 1809, were summarized in a statement known as **Gay-Lussac's Law of Combining Volumes of Gases**: *When measured at the same temperature and pressure, the ratios of the volumes of reacting gases are small whole numbers.* Thus, H_2 and O_2 combine to form water vapor in a volume ratio of 2 to 1 (Figure 13.12); H_2 and Cl_2 react to form HCl in a volume ratio of 1 to 1; and H_2 and N_2 react to form NH_3 in a volume ratio of 3 to 1.

Avogadro's law

Two years later, in 1811, Amedeo Avogadro used the Law of Combining Volumes of Gases to make a simple but significant and far-reaching generalization concerning gases. **Avogadro's law** states:

> **Equal volumes of different gases at the same temperature and pressure contain the same number of molecules.**

This law was a real breakthrough in understanding the nature of gases. (1) It offered a rational explanation of Gay-Lussac's Law of Combining Volumes of Gases and indicated the diatomic nature of such elemental gases as hydrogen, chlorine, and oxygen; (2) it provided a method for determining the molar masses of gases and for comparing the densities of gases of known molar mass (see Sections 13.12 and 13.13); and (3) it afforded a firm foundation for the development of the Kinetic-Molecular Theory.

On a volume basis hydrogen and chlorine react thus:

hydrogen + chlorine $\longrightarrow$ hydrogen chloride

1 volume   1 volume   2 volumes

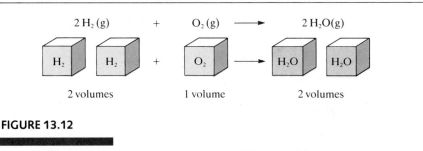

2 volumes   1 volume   2 volumes

FIGURE 13.12

Gay-Lussac's Law of Combining Volumes of Gases applied to the reaction of hydrogen and oxygen. When measured at the same temperature and pressure, hydrogen and oxygen react in a volume ratio of 2 to 1.

By Avogadro's law, equal volumes of hydrogen and chlorine contain the same number of molecules. Therefore, hydrogen molecules react with chlorine molecules in a 1:1 ratio. Since two volumes of hydrogen chloride are produced, one molecule of hydrogen and one molecule of chlorine must produce two molecules of hydrogen chloride. Therefore, each hydrogen molecule and each chlorine molecule must be made up of two atoms. The coefficients of the balanced equation for the reaction give the correct ratios for volumes, molecules, and moles of reactants and products:

$$H_2 \; + \; Cl_2 \; \longrightarrow \; 2\,HCl$$

1 volume	1 volume	2 volumes
1 molecule	1 molecule	2 molecules
1 mol	1 mol	2 mol

By like reasoning oxygen molecules also must contain at least two atoms because one volume of oxygen reacts with two volumes of hydrogen to produce two volumes of water vapor.

The volume of a gas depends on the temperature, the pressure, and the number of gas molecules. Different gases at the same temperature have the same average kinetic energy. Hence, if two different gases are at the same temperature, occupy equal volumes, and exhibit equal pressures, each gas must contain the same number of molecules. This statement is true because systems with identical PVT properties can be produced only by equal numbers of molecules having the same average kinetic energy.

13.12   Mole-Mass-Volume Relationships of Gases

Because a mole contains 6.022×10^{23} molecules (Avogadro's number), a mole of any gas will have the same volume as a mole of any other gas at the same temperature and pressure. It has been experimentally determined that the volume

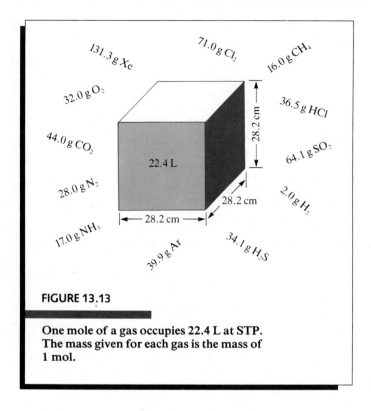

FIGURE 13.13

One mole of a gas occupies 22.4 L at STP.
The mass given for each gas is the mass of
1 mol.

molar volume

occupied by a mole of any gas is 22.4 L at standard temperature and pressure. This volume, 22.4 L, is known as the **molar volume** of a gas. The molar volume is a cube about 28.2 cm (11.1 in.) on a side (see Figure 13.13). The molar masses of several gases, each occupying 22.4 L at STP, are also shown in Figure 3.13.

One mole of a gas occupies 22.4 liters at STP.

The molar volume is useful for determining the molar mass of a gas or of substances that can be easily vaporized. If the mass and the volume of a gas at STP are known, we can calculate its molar mass. For example, 1 liter of pure oxygen at STP has a mass of 1.429 g. The molar mass of oxygen may be calculated by multiplying the mass of 1 liter by 22.4 L/mol.

$$\frac{1.429 \text{ g}}{1\,\cancel{L}} \times \frac{22.4\,\cancel{L}}{1 \text{ mol}} = 32.0 \text{ g/mol} \quad \text{(molar mass)}$$

If the mass and volume are at other than standard conditions, we change the volume to STP and then calculate the molar mass. Note that we do not correct the mass to standard conditions—only the volume.

The molar volume, 22.4 L/mol, is used as a conversion factor to convert grams per liter to grams per mole (molar mass) and also to convert liters to moles. The two conversion factors are

$$\frac{22.4 \text{ L}}{1 \text{ mol}} \quad \text{and} \quad \frac{1 \text{ mol}}{22.4 \text{ L}}$$

These conversions must be done at STP except under certain special circumstances. Examples follow.

EXAMPLE 13.12   If 2.00 L of a gas measured at STP have a mass of 3.23 g, what is the molar mass of the gas?
The unit of molar mass is g/mol; the conversion is from

$$\frac{\text{g}}{\text{L}} \longrightarrow \frac{\text{g}}{\text{mol}}$$

The starting amount is $\dfrac{3.23 \text{ g}}{2.00 \text{ L}}$.   The conversion factor is $\dfrac{22.4 \text{ L}}{1 \text{ mol}}$.

The calculation is $\dfrac{3.23 \text{ g}}{2.00 \, \cancel{\text{L}}} \times \dfrac{22.4 \, \cancel{\text{L}}}{1 \text{ mol}} = 36.2 \text{ g/mol}$ (molar mass)

EXAMPLE 13.13   Measured at 40 °C and 630. torr, the mass of 691 mL of ethyl ether is 1.65 g. Calculate the molar mass of ethyl ether.

Step 1 Organize the given information, converting temperatures to Kelvin. Note that we must change to STP in order to determine molar mass.

$P_1 = 630.$ torr $P_2 = 760.$ torr

$V_1 = 691$ mL $V_2 = ?$

$T_1 = 313$ K(40.°C) $T_2 = 273$ K

Step 2 Use either the conversion factor method or the algebraic method and the combined gas law to determine V_2:

$$V_2 = 691 \text{ mL} \times \frac{273 \, \cancel{K}}{313 \, \cancel{K}} \times \frac{630. \, \cancel{\text{torr}}}{760. \, \cancel{\text{torr}}} = 500 \text{ mL} = 0.500 \text{ L} \quad \text{(at STP)}$$

Step 3 In the example V_2 is the volume for 1.65 g of the gas, so we can now find the molar mass by converting g/L to g/mol.

$$\frac{1.65 \text{ g}}{0.500 \, \cancel{\text{L}}} \times \frac{22.4 \, \cancel{\text{L}}}{\text{mol}} = 73.9 \text{ g/mol}$$

PRACTICE   A gas with a mass of 86 g occupies 5.00 L at 25°C and 3.00 atm pressure. What is the molar mass of the gas?

Answer: 1.4×10^2 g/mol

13.13   Density of Gases

The density, d, of a gas is its mass per unit volume, which is generally expressed in grams per liter (g/L) as follows:

$$d = \frac{\text{mass}}{\text{volume}} = \frac{g}{L}$$

Because the volume of a gas depends on temperature and pressure, both should be given when stating the density of a gas. The volume of a solid or liquid is hardly affected by changes in pressure and is changed only slightly when the temperature is varied. Increasing the temperature from 0°C to 50°C will reduce the density of a gas by about 18% if the gas is allowed to expand, whereas a 50°C rise in the temperature of water (0°C $\longrightarrow$ 50°C) will change its density by less than 0.2%.

The density of a gas at any temperature and pressure can be determined by calculating the mass of gas present in 1 L. At STP, in particular, the density can be calculated by multiplying the molar mass of the gas by 1 mol/22.4 L.

$$d \text{ (at STP)} = \text{molar mass} \times \frac{1 \text{ mol}}{22.4 \text{ L}} \qquad \text{molar mass} = d \text{ (at STP)} \times \frac{22.4 \text{ L}}{1 \text{ mol}}$$

Table 13.3 lists the densities of some common gases.

EXAMPLE 13.14   Calculate the density of Cl_2 at STP.

First calculate the molar mass of Cl_2. It is 71.0 g/mol. Since $d = g/L$, the conversion is

$$\frac{g}{\text{mol}} \longrightarrow \frac{g}{L}$$

The conversion factor is $\dfrac{1 \text{ mol}}{22.4 \text{ L}}$.

$$d = \frac{71.0 \text{ g}}{1 \text{ mol}} \times \frac{1 \text{ mol}}{22.4 \text{ L}} = 3.17 \text{ g/L}$$

PRACTICE   The molar mass of a gas is 20. g. Calculate the density of the gas at STP.

Answer: 0.89 g/L

Pollution

Our technologically advanced society provides many products and services that save time as well as add convenience to our lives. Unfortunately, objects made from synthetic plastics are not readily broken down and can remain intact for more than 50 years. Chemistry will play an important role in solving these problems as they seek new types of plastic molecules that are biodegradable and new energy sources that release fewer pollutants.

◀ In the Los Angeles basin, a smoky mixture of sunlight and industrial and automotive emissions forms to create photochemical smog. Ozone (O_3) is the major photochemical oxidant.

▲
Researchers sample lakewater to determine the level of organic contamination.

◀ Increased dependence on nonbiodegradable products are exhausting our landfills at an alarming rate.

Acid Rain

The phenomena of acid rain became widely publicized in the 1980s. Major contributors to acid rain are oxides of sulfur and nitrogen which dissolve in drops of rainwater to form sulfuric acid (H_2SO_4) and nitric acid (HNO_3). Small quantities of sulfur oxides are released from natural sources, but the main source comes from burning sulfur-containing coal and oil.

High smoke stacks protect plant life in the immediate vicinity, but oxides of sulfur and nitrogen released higher up are eventually deposited in other locations.
▼

Stark landscapes are becoming increasingly visible as the decreased pH of rainwater dissolves nutrients essential to many forms of vegetation.
▼

▲
Over the past several years, industrial pollution has caused the pH of rainwater to decrease, resulting in increased acidity.

Marble masterpieces, sculpted to last forever, are slowly disappearing as acid rain dissolves the calcium carbonate ($CaCO_3$). ▶

Ozone

Ozone (O_3) is formed in the stratosphere by ultraviolet (UV) radiation from the sun on atmospheric oxygen (O_2). An equilibrium exists between O_3, O_2, and oxygen atoms, with UV light being continually absorbed. Pollutants remain unchanged in the lower atmosphere, but when carried to the stratosphere they can destroy O_3 molecules, which contributes to a depletion of the O_3 layer. Without the filtering capability, increased UV radiation will strike the earth resulting in climatic changes and an increase in skin cancer.

◀ Refrigerants, a staple used in air conditioners and refrigerators, add to the released atmospheric CFCs.

The O_3 hole over Antarctica gives us cause for concern. The opening contained 50% less O_3 than normal for springtime.
▼

▲
Balloons released into the sky from an Antarctica research station enable scientists to measure the amount of O_3.

▲
Chlorofluorocarbons (CFCs), commonly used as propellants in aerosol cans, contribute to the stratospheric imbalance by attacking and destroying O_3 molecules.

Industrial Waste

Modern industries create thousands of pounds of waste products that can be hazardous if they are not disposed of carefully and permanently. Radioactive isotopes, biohazardous wastes, and solvents, such as methylethylketone (MEK), are all materials that contribute to the industrial waste problem. Improper disposal of these items can lead to ecological contamination.

Chemical wastes sealed in leakproof containers are buried in special lined landfills to prevent chemical seepage from contaminating the groundwater. ▼

Medical wastes, such as containers for radioactive isotopes and used syringes, must be carefully handled to avoid contamination of medical personnel. ▶

▲
Illegal disposal of over 20,000 tons of hazardous chemicals occurred at Love Canal necessitating the evacuation of the entire population.

▲
Protected from exposure to dangerous chemicals by proper safety gear, a hazardous waste team works at a spill site.

Greenhouse Effect

Carbon compounds are essential to life on earth (for example, as in photosynthesis). But, more and more CO_2 is being released through fossil fuel burning, car exhaust, factory emissions, and deforestation. Carbon dioxide absorbs energy reflected from the earth's surface and converts it to heat. An increase in the amounts of CO_2 in the atmosphere could lead to a dramatic global warming and the melting of polar ice caps.

Large quantities of CO_2 are dissolved in ▶ the oceans, which act as sinks for the excesses produced over the years.

Research is very active in the area of alternate fuels that will burn cleaner and more efficiently. Gasohol, a blend of gasoline and alcohol, is one such fuel. ▼

▲
Electric powered cars emit no CO_2 and may be a long-term solution to our problem.

◀ The large-scale burning of rain forests is destroying plant life, which is essential in removing CO₂ from the atmosphere.

Aluminum Recycling

Recycling consumer products can relieve the heavy burden on landfills as well as conserve natural resources. The combination of paper, metal, and glass results in 50% of solid waste in landfills. These items can easily be reclaimed and manufactured into new consumer goods. In the future, recycling will be the primary method of solid waste disposal.

Everyone can help by returning reusable items to convenient neighborhood collection centers.
▼

Although the use of aluminum cans is widespread, it is one of the most commonly recycled items.
▼

▲
Reprocessing of aluminum cans begins by heating the cans to their melting point. The molten metal is then purified and cooled.

▲
After the cooled metal is formed into a large ingot, new items are ready to be manufactured.

▲
A recycled aluminum product.

Plastics Recycling

Plastics are manufactured, nonbiodegradable polymers, such as polyethylene, polystyrene, and polypropylene. And, they are responsible for about 20% of solid wastes in landfills. In an era of dwindling space for solid waste disposal, there is an urgent need for reusable and degradable plastics. Researchers are working diligently on this problem, and promising advances in technology are encouraging for the future of recycling plastics.

Unsightly litter left on beaches can be collected and sorted into recyclable items.
▼

Manufacturers purchase recycled polystyrene to produce durable consumer goods.
▼

Disposable diapers have a substantial impact on landfills. Now, through chemical engineering, biodegradable diapers are appearing in the marketplace.

▲

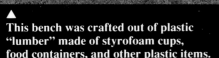

▲
This bench was crafted out of plastic "lumber" made of styrofoam cups, food containers, and other plastic items.

▲
This polystyrene plant recycles plastics into polystyrene flakes that can be manufactured into useful products.

Solutions Do Exist

Dichlorodiphenyltrichloroethane (DDT) is an insecticide that was used world-wide to protect crops and to prevent outbreaks of malaria. The insecticide accumulates in fish and becomes concentrated in fish-eating birds, such as the California brown pelican. The United States banned DDT in the 1970s, but almost one-third of the DDT used still exists in our oceans and lakes. Yet, much of the world uses DDT today.

DDT interferes with the normal reproduction of the California brown pelican.
▼

DDT is a nonpolar molecule that becomes concentrated in the fatty tissues of exposed pelicans.
▼

DDT causes shells of pelican eggs to become so fragile that they crush during nesting.
▼

▲
Since the banning of DDT in the 1970s, the concentration of the pesticide in pelicans is decreasing . . .

. . . and the pelican population is ▶ increasing.

TABLE 13.3

Density of Common Gases at STP

Gas	Molecular mass	Density (g/L at STP)
H_2	2.0	0.090
CH_4	16.0	0.714
NH_3	17.0	0.760
C_2H_2	26.0	1.16
HCN	27.0	1.21
CO	28.0	1.25
N_2	28.0	1.25
air	**(28.9)**	**1.29**
O_2	32.0	1.43
H_2S	34.1	1.52
HCl	36.5	1.63
F_2	38.0	1.70
CO_2	44.0	1.96
C_3H_8	44.0	1.96
O_3	48.0	2.14
SO_2	64.1	2.86
Cl_2	71.0	3.17

13.14   Ideal Gas Equation

We have used four variables in calculations involving gases: the volume, V; the pressure, P; the absolute temperature, T; and the number of molecules or moles, which is abbreviated n. Combining these variables into a single expression, we obtain

$$V \propto \frac{nT}{P} \quad \text{or} \quad V = \frac{nRT}{P}$$

where R is a proportionality constant known as the *ideal gas constant*. The equation is commonly written as

$$PV = nRT$$

ideal gas equation   and is known as the **ideal gas equation**. This equation states in a single expression what we have considered in our earlier discussions: The volume of a gas varies directly with the number of gas molecules and the absolute temperature, and varies inversely with the pressure. The value and units of R depend on the units of P, V, and T. We can calculate one value of R by taking 1 mol of a gas at STP conditions. Solve the equation for R:

$$R = \frac{PV}{nT} = \frac{1 \text{ atm} \times 22.4 \text{ L}}{1 \text{ mol} \times 273 \text{ K}} = 0.0821 \frac{\text{L-atm}}{\text{mol-K}}$$

The units of R in this case are liter-atmospheres (L-atm) per mol-K. When the value of $R = 0.0821$ L-atm/mol-K, P is in atmospheres, n is in moles, V is in liters, and T is in Kelvin.

The ideal gas equation can be used to calculate any one of the four variables if the other three are known.

EXAMPLE 13.15   What pressure will be exerted by 0.400 mol of a gas in a 5.00 L container at 17.0°C?

Step 1 Organize the given information, putting temperatures in Kelvin

$P = ?$

$V = 5.00$ L

$T = 290.$ K

$n = 0.400$ mol

Step 2 Write and solve equation for the unknown.

$$PV = nRT \qquad \text{or} \qquad P = \frac{nRT}{V}$$

Step 3 Put given information into equation and calculate.

$$P = \frac{0.400 \text{ mol} \times 0.0821 \text{ L-atm} \times 290. \text{ K}}{5.00 \text{ L} \times \text{mol-K}} = 1.90 \text{ atm} \quad \text{(Answer)}$$

EXAMPLE 13.16   How many moles of oxygen gas are in a 50.0-L tank at 22.0°C if the pressure gauge reads 2000. lb/in.^{2}?

Step 1 Organize the given information, putting the temperature in Kelvin and changing pressure to atmospheres.

$$P = \frac{2000 \text{ lb}}{\text{in.}^2} \times \frac{1 \text{ atm}}{14.7 \text{ lb/in.}^2} = 136 \text{ atm}$$

$V = 50.0$ L

$T = 295$ K

$n = ?$

Step 2 Write and solve equation for the unknown.

$$PV = nRT \quad \text{or} \quad n = \frac{PV}{RT}$$

Step 3 Put given information into equation and calculate.

$$n = \frac{136 \text{ atm} \times 50.0 \text{ L}}{(0.0821 \text{ L atm/mol-K}) \times 295 \text{ K}} = 281 \text{ mol O}_2 \quad \text{(Answer)}$$

PRACTICE   A 23.8 L cylinder contains oxygen gas at 20.0°C and 732 torr. How many moles of oxygen are in the cylinder?

Answer: 0.953 mol

The molar mass of a gaseous substance can be determined using the ideal gas equation. Since molar mass = g/mol, then mol = g/molar mass. Using M for molar mass we can substitute g/M for n (moles) in the ideal gas equation to get

$$PV = \frac{g}{M} RT \quad \text{or} \quad M = \frac{gRT}{PV} \quad \text{(modified ideal gas equation)}$$

which will allow us to calculate the molar mass, M, for any substance in the gaseous state.

EXAMPLE 13.17   Calculate the molar mass of butane gas, if 3.69 g occupy 1.53 L at 20°C and 1 atm pressure. Change 20°C to 293 K and substitute the data into the modified ideal gas equation:

$$M = \frac{gRT}{PV} = \frac{3.69 \text{ g} \times 0.0821 \text{ L atm} \times 293 \text{ K}}{1 \text{ atm} \times 1.53 \text{ L} \times \text{mol-K}} = 58.0 \text{ g/mol} \quad \text{(Answer)}$$

PRACTICE   A sample of 0.286 g of a certain gas occupies 50.0 mL at standard temperature and 76.0 cm Hg pressure. Determine the molar mass of the gas.

Answer: 128 g/mol

13.15   Calculations from Chemical Equations Involving Gases

Mole–Volume (Gas) and Mass–Volume (Gas) Calculations   Stoichiometric problems involving gas volumes can be solved by the general mole-ratio method outlined in Chapter 9. The factors 1 mol/22.4 L and 22.4 L/1 mol are used for converting volume to moles and moles to volume, respectively. These conversion factors are used under the assumption that the gases are at STP and behave as ideal gases. In practice, gases are measured at other than STP conditions, and the volumes are converted to STP for stoichiometric calculations.

In a balanced equation, the number preceding the formula of a gaseous substance represents the number of moles or molar volumes (22.4 L at STP) of that substance.

The following are examples of typical problems involving gases and chemical equations.

EXAMPLE 13.18   What volume of oxygen (at STP) can be formed from 0.500 mol of potassium chlorate?

Step 1 Write the balanced equation:

$$2 \text{ KClO}_3 \longrightarrow 2 \text{ KCl} + 3 \text{ O}_2\uparrow$$

$$\phantom{2 \text{ KClO}_3}\text{2 mol}\phantom{\longrightarrow 2 \text{ KCl} +}\text{3 mol}$$

Step 2 The starting amount is 0.500 mol $KClO_3$. The conversion is from

$$\text{moles KClO}_3 \longrightarrow \text{moles O}_2 \longrightarrow \text{liters O}_2$$

Step 3 Calculate the moles of O_2, using the mole-ratio method:

$$0.500 \text{ mol KClO}_3 \times \frac{3 \text{ mol O}_2}{2 \text{ mol KClO}_3} = 0.750 \text{ mol O}_2$$

Step 4 Convert moles of O_2 to liters of O_2. The moles of a gas at STP are converted to liters by multiplying by the molar volume, 22.4 L/mole:

$$0.750 \text{ mol O}_2 \times \frac{22.4 \text{ L}}{1 \text{ mol}} = 16.8 \text{ L O}_2 \quad \text{(Answer)}$$

Setting up a continuous calculation, we obtain

$$0.500 \text{ mol KClO}_3 \times \frac{3 \text{ mol O}_2}{2 \text{ mol KClO}_3} \times \frac{22.4 \text{ L}}{1 \text{ mol}} = 16.8 \text{ L O}_2$$

EXAMPLE 13.19   How many grams of aluminum must react with sulfuric acid to produce 1.25 L of hydrogen gas at STP?

Step 1 The balanced equation is

$$2 \text{ Al}(s) + 3 \text{ H}_2\text{SO}_4(aq) \longrightarrow \text{Al}_2(\text{SO}_4)_3(aq) + 3 \text{ H}_2(g)$$

$$\text{2 mol}\phantom{\text{Al}(s) + 3 \text{ H}_2\text{SO}_4(aq) \longrightarrow \text{Al}_2(\text{SO}_4)_3(aq) + }\text{3 mol}$$

Step 2 We first convert liters of H_2 to moles of H_2. Then the familiar stoichiometric calculation from the equation is used. The conversion is

$$L\ H_2 \longrightarrow mol\ H_2 \longrightarrow mol\ Al \longrightarrow g\ Al$$

$$1.25\ \cancel{L}\ \cancel{H_2} \times \frac{1\ \cancel{mol}}{22.4\ \cancel{L}} \times \frac{2\ \cancel{mol\ Al}}{3\ \cancel{mol\ H_2}} \times \frac{27.0\ g\ Al}{1\ \cancel{mol\ Al}} = 1.00\ g\ Al \quad \text{(Answer)}$$

EXAMPLE 13.20 What volume of hydrogen, collected at 30°C and 700. torr pressure, will be formed by reacting 50.0 g of aluminum with hydrochloric acid?

$$2\ Al(s) + 6\ HCl(aq) \longrightarrow 2\ AlCl_3(aq) + 3\ H_2(g)$$

2 mol   3 mol

In this problem the conditions are not at STP, so we cannot use the method shown in Example 13.18. Either we need to calculate the volume at STP from the equation and then convert this volume to the conditions given in the problem, or we can use the ideal gas equation. Let's use the ideal gas equation.

First calculate the moles of H_2 obtained from 50.0 g of Al. Then, using the ideal gas equation, calculate the volume of H_2 at the conditions given in the problem.

Step 1 Moles of H_2: The conversion is

$$grams\ Al \longrightarrow moles\ Al \longrightarrow moles\ H_2$$

$$50.0\ \cancel{g\ Al} \times \frac{1\ \cancel{mol\ Al}}{27.0\ \cancel{g\ Al}} \times \frac{3\ \cancel{mol}\ H_2}{2\ \cancel{mol\ Al}} = 2.78\ mol\ H_2$$

Step 2 Liters of H_2: Solve $PV = nRT$ for V and substitute the data into the equation.
Convert °C to K: 30°C + 273 = 303 K.
Convert torr to atm: 700. $\cancel{torr} \times 1\ atm/760\ \cancel{torr} = 0.921$ atm.

$$V = \frac{nRT}{P} = \frac{2.78\ \cancel{mol}\ H_2 \times 0.0821\ \text{L-}\cancel{atm} \times 303\ \cancel{K}}{0.921\ \cancel{atm} \times \cancel{mol\text{-}K}} = 75.1\ L\ H_2 \quad \text{(Answer)}$$

Note: The volume at STP is 62.3 L H_2.

PRACTICE If 10.0 g of sodium peroxide (Na_2O_2) react with water to produce sodium hydroxide and oxygen, how many liters of oxygen will be produced at 20°C and 750. torr?

$$2\ Na_2O_2 + 2\ H_2O \longrightarrow 4\ NaOH + O_2$$

Answer: 1.56 L O_2

Volume–Volume Calculations When all substances in a reaction are in the gaseous state, simplifications in the calculation can be made that are based on Avogadro's law that gases under identical conditions of temperature and pressure contain the same number of molecules and occupy the same volume. Using this same law, we can also state that, under the same conditions of temperature and pressure, the volumes of gases reacting are proportional to the numbers of moles of the gases in the balanced equation. Consider the reaction

$$H_2(g) \; + \; Cl_2(g) \; \longrightarrow \; 2\,HCl(g)$$

1 mol	1 mol	2 mol
22.4 L	22.4 L	2×22.4 L
1 volume	1 volume	2 volumes
Y volume	Y volume	$2\,Y$ volumes

In this reaction 22.4 L of hydrogen will react with 22.4 L of chlorine to give $2 \times 22.4 = 44.8$ L of hydrogen chloride gas. This statement is true because these volumes are equivalent to the number of reacting moles in the equation. Therefore, Y volume of H_2 will combine with Y volume of Cl_2 to give $2\,Y$ volumes of HCl. For example, 100 L of H_2 react with 100 L of Cl_2 to give 200 L of HCl; if the 100 L of H_2 and of Cl_2 are at 50°C, they will give 200 L of HCl at 50°C. When the temperature and pressure before and after a reaction are the same, volumes can be calculated without changing the volumes to STP.

For reacting gases: Volume–volume relationships are the same as mole–mole relationships.

EXAMPLE 13.21   What volume of oxygen will react with 150 L of hydrogen to form water vapor? What volume of water vapor will be formed?

Assume that both reactants and products are measured at the same conditions. Calculation by reacting volumes:

$$2\,H_2(g) \; + \; O_2(g) \longrightarrow 2\,H_2O(g)$$

2 mol	1 mol	2 mol
2×22.4 L	22.4 L	2×22.4 L
2 volumes	1 volume	2 volumes
150 L	75 L	150 L

For every two volumes of H_2 that react, one volume of O_2 reacts and two volumes of $H_2O(g)$ are produced.

$$150\,\text{L}\;H_2 \times \frac{1\ \text{volume}\ O_2}{2\ \text{volumes}\ H_2} = 75\,\text{L}\;O_2$$

$$150\,\text{L}\;H_2 \times \frac{1\ \text{volume}\ O_2}{1\ \text{volume}\ H_2} = 150\,\text{L}\;H_2$$

EXAMPLE 13.22   The equation for the preparation of ammonia is

$$3H_2 + N_2 \xrightarrow{\ 400°C\ } 2\,NH_3$$

Assuming that the reaction goes to completion,

(a) What volume of H_2 will react with 50 L of N_2?
(b) What volume of NH_3 will be formed from 50 L of N_2?
(c) What volume of N_2 will react with 100 mL of H_2?
(d) What volume of NH_3 will be produced from 100 mL of H_2?
(e) If 600 mL of H_2 and 400 mL of N_2 are sealed in a flask and allowed to react, what amounts of H_2, N_2, and NH_3 are in the flask at the end of the reaction?

The answers to parts (a)–(d) are shown in the boxes and can be determined from the equation by inspection, using the principle of reacting volumes.

$$3\ H_2\ +\ N_2\ \longrightarrow\ 2\ NH_3$$

3 volumes   1 volume   2 volumes

(a) $\boxed{150\ \text{L}}$ 50 L
(b)   50 L $\boxed{100\ \text{L}}$
(c) 100 mL $\boxed{33.3\ \text{mL}}$
(d) 100 mL $\boxed{66.7\ \text{mL}}$

(e) Volume ratio from the equation $= \dfrac{3 \text{ volumes } H_2}{1 \text{ volume } N_2}$

Volume ratio used $= \dfrac{600 \text{ mL } H_2}{400 \text{ mL } N_2} = \dfrac{3 \text{ volumes } H_2}{2 \text{ volumes } N_2}$

Comparing these two ratios, we see that an excess of N_2 is present in the gas mixture. Therefore, the reactant limiting the amount of NH_3 that can be formed is H_2:

$$3\ H_2\ +\ N_2\ \longrightarrow 2\ NH_3$$

600 mL   200 mL   400 mL

In order to have a 3:1 ratio of volumes reacting, 600 mL of H_2 will react with 200 mL of N_2 to produce 400 mL of NH_3, leaving 200 mL of N_2 unreacted. At the end of the reaction the flask will contain 400 mL of NH_3 and 200 mL of N_2.

PRACTICE   What volume of oxygen will react with 15.0 L of propane (C_3H_8) to form carbon dioxide and water? How much carbon dioxide will be formed? How much water?

$$C_3H_8(g) + 5\ O_2(g) \longrightarrow 3\ CO_2(g) + 4\ H_2O(g)$$

Answers: 75.0 L O_2 45.0 L CO_2 60.0 L H_2O

13.16   Real Gases

All the gas laws are based on the behavior of an ideal gas—that is, a gas with a behavior that is described exactly by the gas laws for all possible values of P, V, and T. Most real gases actually do behave very nearly as predicted by the gas laws over a fairly wide range of temperatures and pressures. However, when conditions are such that the gas molecules are crowded closely together (high pressure and/or low temperature), they show marked deviations from ideal behavior. Deviations occur because molecules have finite volumes and also have intermolecular attractions, which result in less compressibility at high pressures and greater compressibility at low temperatures than predicted by the gas laws. Many gases become liquids at high pressure and low temperature.

13.17   Air Pollution

Chemical reactions occur among the gases that are emitted into our atmosphere. In recent years, concern has been growing regarding the effect these reactions have upon our environment and our lives.

The outer portion of the atmosphere plays a significant role in determining the conditions for life at the surface of the earth. This stratosphere protects the surface from the intense radiation and particles bombarding our planet. Some of the high-energy radiation from the sun acts upon oxygen molecules in the stratosphere, converting them into ozone, O_3. Different molecular forms of an element are called **allotropes** of that element. Thus oxygen and ozone are allotropic forms of oxygen.

allotrope

$$O_2 \xrightarrow{\text{Sunlight}} O + O$$
$$\text{Oxygen atoms}$$
$$O_2 + O \longrightarrow O_3$$

Ultraviolet radiation from the sun is highly damaging to living tissues of plants and animals. The ozone layer, however, shields the earth by absorbing ultraviolet radiation and thus prevents most of this lethal radiation from reaching the earth's surface. The reaction that occurs is the reverse of the preceding one:

$$O_3 \xrightarrow[\text{radiation}]{\text{Ultraviolet}} O_2 + O + \text{heat}$$

Scientists have become concerned about a growing hazard to the ozone layer. Chlorofluorocarbon propellants, such as the Freons, CCl_3F and CCl_2F_2, which were used in aerosol spray cans and are used in refrigeration and air conditioning units, are stable compounds and remain unchanged in the lower atmosphere. When these chlorofluorocarbons are carried by convection currents to the stratosphere, they absorb ultraviolet radiation and produce chlorine atoms (chlorine free radicals) that in turn react with ozone. The following reaction sequence involving free radicals (see Section 17.15) has been proposed to explain the partial destruction of the ozone layer by chlorofluorocarbons.

$$CCl_3F \xrightarrow[\text{radiation}]{\text{Ultraviolet}} \cdot CCl_2F + Cl\cdot$$

| Fluorocarbon molecule | Fluorocarbon free radical | Chlorine free radical (atom) |

$$Cl\cdot + O_3 \longrightarrow ClO\cdot + O_2$$
$$ClO\cdot + O \longrightarrow O_2 + Cl\cdot$$

Because a chlorine atom is generated for each ozone molecule that is destroyed, a single chlorofluorocarbon molecule can be responsible for the destruction of many ozone molecules. During the past decade, scientists have discovered an annual thinning in the ozone layer over Antarctica. This is what we call the "hole" in the ozone layer. If this hole were to occur over populated regions of the world, severe effects would result, including a rise in the cancer rate, increased climatic temperatures, and vision problems.

Ozone can be prepared by passing air or oxygen through an electrical discharge:

$$3 O_2(g) + 286 \text{ kJ (68.4 kcal)} \xrightarrow[\text{discharge}]{\text{Electrical}} 2 O_3(g)$$

The characteristic pungent odor of ozone is noticeable in the vicinity of electrical machines and power transmission lines. Ozone is formed in the atmosphere during electrical storms and by the photochemical action of ultraviolet radiation on a mixture of nitrogen dioxide and oxygen. Areas with high air pollution are subject to high atmospheric ozone concentrations.

Ozone is not a desirable low-altitude constituent of the atmosphere, because it is known to cause extensive plant damage, cracking of rubber, and the formation of eye-irritating substances. Concentrations of ozone greater than 0.1 part per million (ppm) of air cause coughing, choking, headache, fatigue, and reduced resistance to respiratory infection. Concentrations between 10 and 20 ppm are fatal to humans.

In addition to ozone, the air in urban areas contains nitrogen oxides, substances which are components of smog. The term *smog* refers to air pollution in urban environments. Often the chemical reactions occur as part of a *photochemical process*. Nitric oxide (NO) is oxidized in the air or in automobile engines to produce nitrogen dioxide (NO_2). In the presence of light,

$$NO_2 \longrightarrow NO + O$$

In addition to nitrogen oxides, combustion of fossil fuels releases CO_2, CO, and sulfur oxides. Incomplete combustion releases unburned and partially burned hydrocarbons.

Society is continually attempting to discover, understand, and control emissions that contribute to this sort of atmospheric chemistry. It is a problem that each one of us faces as we look to the future if we want to continue to support life as we know it on our planet.

Concepts in Review

1. State the principal assumptions of the Kinetic Molecular Theory.
2. Estimate the relative rates of effusion of two gases of known molar mass.
3. Sketch and explain the operation of a mercury barometer.
4. List two factors that determine gas pressure in a vessel of fixed volume.
5. State Boyle's, Charles' and Gay-Lussac's laws. Use all of them in problems.
6. State the combined gas law. Indicate when it is used.
7. Use Dalton's Law of Partial Pressures and the combined gas law to determine the dry STP volume of a gas collected over water.
8. State Avogadro's law.
9. Understand the mole-mass-volume relationship of gases.
10. Determine the density of any gas at STP.
11. Determine the molar mass of a gas from its density at a known temperature and pressure.
12. Solve problems involving the ideal gas equation.
13. Make mole-volume, mass-volume, and volume-volume stoichiometric calculations from balanced chemical equations.
14. State two reasons why real gases may deviate from the behavior predicted for an ideal gas.

Key Terms in Review

The terms listed here have all been defined within the chapter. Review the definitions of each. Use the glossary and the margin notations within the chapter as study aids.

allotrope
1 atmosphere
atmospheric pressure
Avogadro's law
Barometer
Boyle's law
Charles' law
Dalton's Law of Partial Pressures
diffusion
effusion
Gay-Lussac's law
   (modified Charles' law)

Gay-Lussac's Law of Combining
   Volumes of Gases
Graham's Law of Effusion
ideal gas
ideal gas equation
Kinetic-Molecular Theory (KMT)
molar volume
partial pressures
pressure
standard conditions
standard temperature and
   pressure (STP)

Exercises

An asterisk indicates a more challenging question or problem

1. What evidence is used to show diffusion in Figure 13.1? If hydrogen, H_2, and oxygen, O_2, were in the two flasks, how could we prove that diffusion had taken place?

2. How does the air pressure inside the balloon shown in Figure 13.2 compare with the air pressure outside the balloon? Explain.

3. According to Table 13.1, what two gases are the major constituents of dry air?

4. How does the pressure represented by 1 torr compare in magnitude to the pressure represented by 1 mm Hg? See Table 13.2.

5. In which container illustrated in Figure 13.5 are the molecules of gas moving faster? Assume both gases to be hydrogen.

6. In Figure 13.6, what gas pressure corresponds to a volume of 4 L?

7. How do the data illustrated in Figure 13.6 substantiate Boyle's law?

8. What effect would you observe in Figure 13.9 if T_2 were lower than T_1?

9. In the diagram shown in Figure 13.11, is the pressure of the oxygen plus water vapor inside the bottle equal to, greater than, or less than the atmospheric pressure outside the bottle? Explain.

10. List five gases in Table 13.3 that are more dense than air. Explain the basis for your selections.

11. What are the basic assumptions of the Kinetic-Molecular Theory?

12. Arrange the following gases, all at standard temperature, in order of increasing relative molecular velocities: H_2, CH_4, Rn, N_2, F_2, He. What is your basis for determining the order?

13. List, in descending order, the average kinetic energies of the molecules in Exercise 12.

14. What are the four parameters used to describe the behavior of a gas?

15. What are the characteristics of an ideal gas?

16. Under what condition of temperature, high or low, is a gas least likely to exhibit ideal behavior? Explain.

17. Under what conditions of pressure, high or low, is a gas least likely to exhibit ideal behavior? Explain.

18. Compare, at the same temperature and pressure, equal volumes of H_2 and O_2 as to:
    (a) Number of molecules
    (b) Mass
    (c) Number of moles
    (d) Average kinetic energy of the molecules
    (e) Rate of effusion
    (f) Density

19. How does the Kinetic-Molecular Theory account for the behavior of gases as described by:
    (a) Boyle's law
    (b) Charles' law
    (c) Dalton's Law of Partial Pressures

20. Explain how the reaction $N_2(g) + O_2(g) \xrightarrow{\Delta} 2\ NO(g)$ proves that nitrogen and oxygen are diatomic molecules.

21. What is the reason for referring gases to STP?

22. Is the conversion of oxygen to ozone an exothermic or endothermic reaction? How do you know?

23. Write formulas for an oxygen atom, an oxygen molecule, and an ozone molecule. How many electrons are in an oxygen molecule?

24. When constant pressure is maintained, what effect does heating a mole of N_2 gas have on the following?
    (a) Its density
    (b) Its mass
    (c) The average kinetic energy of its molecules
    (d) The average velocity of its molecules
    (e) The number of N_2 molecules in the sample

25. Assuming ideal gas behavior, which of the following statements are correct? (Try to answer without referring to your text.) Rewrite each incorrect statement to make it correct.
    (a) The pressure exerted by a gas at constant volume is independent of the temperature of the gas.
    (b) At constant temperature, increasing the pressure exerted on a gas sample will cause a decrease in the volume of the gas sample.
    (c) At constant pressure, the volume of a gas is inversely proportional to the absolute temperature.

(d) At constant temperature, doubling the pressure on a gas sample will cause the volume of the gas sample to decrease to one-half its original volume.

(e) Compressing a gas at constant temperature will cause its density and mass to increase.

(f) Equal volumes of CO_2 and CH_4 gases at the same temperature and pressure contain
   (1) The same number of molecules
   (2) The same mass
   (3) The same densities
   (4) The same number of moles
   (5) The same number of atoms

(g) At constant temperature, the average kinetic energy of O_2 molecules at 200 atm pressure is greater than the average kinetic energy of O_2 molecules at 100 atm pressure.

(h) According to Charles' law, the volume of a gas becomes zero at $-273°C$.

(i) One liter of O_2 gas at STP has the same mass as 1 L of O_2 gas at 273°C and 2 atm pressure.

(j) The volume occupied by a gas depends only on its temperature and pressure.

(k) In a mixture containing O_2 molecules and N_2 molecules, the O_2 molecules, on the average, are moving faster than the N_2 molecules.

(l) $PV = k$ is a statement of Charles' law.

(m) If the temperature of a sample of gas is increased from 25°C to 50°C, the volume of the gas will increase by 100%.

(n) One mole of chlorine, Cl_2, at 20°C and 600 torr pressure contains 6.022×10^{23} molecules.

(o) One mole of H_2 plus 1 mole of O_2 in an 11.2 L container exert a pressure of 4 atm at 0°C.

(p) When the pressure on a sample of gas is halved, with the temperature remaining constant, the density of the gas is also halved.

(q) When the temperature of a sample of gas is increased at constant pressure, the density of the gas will decrease.

(r) According to the equation

$$2\ KClO_3(s) \xrightarrow{\Delta} 2\ KCl(s) + 3\ O_2(g),$$

1 mol of $KClO_3$ will produce 67.2 L of O_2 at STP.

(s) $PV = nRT$ is a statement of Avogadro's law.

(t) STP conditions are 1 atm and 0°C.

Pressure Units

26. The barometer reads 715 mm Hg. Calculate the corresponding pressure in
   (a) atmospheres (d) torrs
   (b) inches of Hg (e) millibars
   (c) lb/in.²   (f) kilopascals

27. Express the following pressures in atmospheres:
   (a) 28 mm Hg   (c) 795 torr
   (b) 6000. cm Hg   (d) 5.00 kPa

Boyle's, Charles', and Gay-Lussac's Laws

28. A gas occupies a volume of 400. mL at 500. mm Hg pressure. What will be its volume, at constant temperature, if the pressure is changed to (a) 760 mm Hg; (b) 250 torr; (c) 2.00 atm?

29. A 500. mL sample of a gas is at a pressure of 640. mm Hg. What must be the pressure, at constant temperature, if the volume is changed to (a) 855 mL and (b) 450 mL?

30. At constant temperature, what pressure would be required to compress 2500 L of hydrogen gas at 1.0 atm pressure into a 25 L tank?

31. Given 6.00 L of N_2 gas at $-25°C$, what volume will the nitrogen occupy at (a) 0.0°C; (b) 0.0°F; (c) 100.K; (d) 345 K? (Assume constant pressure.)

32. Given a sample of a gas at 27°C, at what temperature would the volume of the gas sample be doubled, the pressure remaining constant?

*33. A gas sample at 22°C and 740 torr pressure is heated until its volume is doubled. What pressure would restore the sample to its original volume?

34. A gas occupies 250 mL at 700. torr and 22°C. When the pressure is changed to 500. torr, what temperature (°C) is needed to maintain the same volume?

35. Hydrogen stored in a metal cylinder has a pressure of 252 atm at 25°C. What will be the pressure in the cylinder when the cylinder is lowered into liquid nitrogen at $-196°C$?

36. The tires on an automobile were filled with air to 30. psi at 71.0°F. When driving at high speeds, the tires become hot. If the tires have a bursting

pressure of 44 psi, at what temperature (°F) will the tires "blow out"?

Combined Gas Laws

37. A gas occupies a volume of 410 mL at 27°C and 740 mm Hg pressure. Calculate the volume the gas would occupy at (a) STP; (b) 250.°C and 680 mm Hg pressure.

38. What volume would 5.30 L of H_2 gas at STP occupy at 70°C and 830 torr pressure?

39. What pressure will 800. mL of a gas at STP exert when its volume is 250. mL at 30°C?

40. An expandable balloon contains 1400. L of He at 0.950 atm pressure and 18°C. At an altitude of 22 miles (temperature 2.0°C and pressure 4.0 torr), what will be the volume of the balloon?

41. A gas occupies 22.4 L at 2.50 atm and 27°C. What will be its volume at 1.50 atm and −5.00°C?

42. How many gas molecules are present in 600. mL of N_2O at 40°C and 400. torr pressure? How many atoms are present? What would be the volume of the sample at STP?

Dalton's Law of Partial Pressures

43. What would be the partial pressure of O_2 gas collected over water at 20°C and 720. torr pressure? (Check Appendix II for the vapor pressure of water.)

44. An equilibrium mixture contains H_2 at 600. torr pressure, N_2 at 200 torr pressure, and O_2 at 300. torr pressure. What is the total pressure of the gases in the system?

45. A sample of methane gas, CH_4, was collected over water at 25.0°C and 720. torr. The volume of the wet gas is 2.50 L. What will be the volume of the dry methane at standard pressure?

46. 5.00 L of CO_2 at 500. torr and 3.00 L of CH_4 at 400. torr are put into a 10.0 L container. What is the pressure exerted by the gases in the container?

Mole-Mass-Volume Relationships

47. What volume will 2.5 mol of Cl_2 occupy at STP?

48. A steel cylinder contains 60. mol of H_2 at a pressure of 1500 lb/in.². (a) How many moles of H_2 are in the cylinder when the pressure reads 850 lb/in.²? (b) How many grams of H_2 were initially in the cylinder?

49. How many grams of CO_2 are present in 2500 mL of CO_2 at STP?

50. At STP, 560 mL of a gas have a mass of 1.08 g. What is the molar mass of the gas?

51. What volume will each of the following occupy at STP?
    (a) 1.0 mol of NO_2
    (b) 17.05 g of NO_2
    (c) 1.20×10^{24} molecules of NO_2

52. How many molecules of NH_3 gas are present in a 1.00 L flask at STP?

*53. How many moles of Cl_2 are in one cubic meter (1.00 m³) at STP?

Density of Gases

54. A gas has a density at STP of 1.78 g/L. What is its molar mass?

55. Calculate the density of the following gases at STP:
    (a) Kr   (b) He   (c) SO_3 (d) C_4H_8

56. Calculate the density of:
    (a) F_2 gas at STP
    (b) F_2 gas at 27°C and 1 atm pressure

*57. At what temperature (°C) will the density of methane, CH_4, be 1.0 g/L at 1.0 atm pressure?

Ideal Gas Equation and Stoichiometry

58. Using the ideal gas equation, $PV = nRT$, calculate:
    (a) The volume of 0.510 mole of H_2 at 47°C and 1.6 atm pressure
    (b) The number of grams in 16.0 L of CH_4 at 27°C and 600 torr pressure
    *(c) The density of CO_2 at 4.00 atm pressure and −20°C
    *(d) The molar mass of a gas having a density of 2.58 g/L at 27°C and 1.00 atm pressure.
    Hints for (c) and (d): n = moles = m/molar mass, and $d = m/V$.

59. What is the molar mass of a gas if 1.15 g occupy 0.215 L at 0.813 atm and 30°C?

60. At 27°C and 750 torr pressure, what will be the volume of 2.3 mol of Ne?

61. What volume will a mixture of 5.00 mol of H_2 and 0.500 mol of CO_2 occupy at STP?

62. 4.50 mol of a gas occupy 0.250 L at 4.15 atm. What is the Kelvin temperature of the system?

63. How many moles of N_2 gas occupy 5.20 L at 250 K and 0.500 atm?

64. What volume of hydrogen at STP can be produced by reacting 8.30 mol of Al with sulfuric acid? The equation is

$$2\ Al(s) + 3\ H_2SO_4(aq) \longrightarrow$$
$$Al_2(SO_4)_3(aq) + 3\ H_2(g)$$

65. Given the equation

$$4\ NH_3(g) + 5\ O_2(g) \longrightarrow 4\ NO(g) + 6\ H_2O(g)$$

(a) How many moles of NH_3 are required to produce 5.5 mol of NO?

(b) How many moles of NH_3 will react with 7.0 mol of O_2?

(c) How many liters of NO can be made from 12 L of O_2 and 10 L of NH_3 at STP?

(d) At constant temperature and pressure how many liters of NO can be made by the reaction of 800. mL of O_2?

(e) At constant temperature and pressure, what is the maximum volume, in liters, of NO that can be made from 3.0 L of NH_3 and 3.0 L of O_2?

(f) How many grams of O_2 must react to produce 60. L of NO measured at STP?

*(g) How many grams of NH_3 must react to produce a total of 32 L of products, NO plus H_2O, measured at STP?

66. Given the equation

$$4\ FeS(s) + 7\ O_2(g) \xrightarrow{\Delta}$$
$$2\ Fe_2O_3(s) + 4\ SO_2(g)$$

(a) How many liters of O_2, measured at STP, will react with 0.600 kg of FeS_2?

(b) How many liters of SO_2, measured at STP, will be produced from 0.600 kg of FeS_2?

*67. Acetylene, C_2H_2, and hydrogen fluoride react to give difluoroethane.

$$C_2H_2(g) + 2\ HF(g) \longrightarrow C_2H_4F_2(g)$$

When 1.0 mol of C_2H_2 and 5.0 mol of HF are reacted in a 10.0 L flask, what will be the pressure in the flask at 0°C when the reaction is complete?

Additional Problems

68. What are the relative rates of effusion of N_2 and He?

*69. (a) What are the relative rates of effusion of methane, CH_4, and helium, He?

(b) If these two gases are simultaneously introduced into opposite ends of a 100. cm tube and allowed to diffuse toward each other, at what distance from the helium end will molecules of the two gases meet?

*70. A gas has a percent composition by mass of 85.7% carbon and 14.3% hydrogen. At STP the density of the gas is 2.50 g/L. What is the molecular formula of the gas?

*71. Assume that the reaction
$$2\ CO(g) + O_2(g) \longrightarrow 2\ CO_2(g)$$ goes to completion. When 10. mol of CO and 8.0 mol of O_2 react in a closed 10. L vessel,

(a) How many moles of CO, O_2, and CO_2 are present at the end of the reaction?

(b) What will be the total pressure in the flask at 0°C?

*72. 250 mL of O_2, measured at STP, were obtained by the decomposition of the $KClO_3$ in a 1.20 g mixture of KCl and $KClO_3$:

$$2\ KClO_3(s) \longrightarrow 2\ KCl(s) + 3\ O_2(g)$$

What is the percent by mass of $KClO_3$ in the mixture?

*73. Look at the apparatus below. When a small amount of water is squirted into the flask

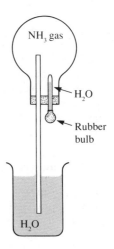

NH₃ gas

H₂O

Rubber bulb

H₂O

containing ammonia gas (by squeezing the bulb of the medicine dropper), water from the beaker fills the flask through the long glass tubing. Explain this phenomenon. (Remember that ammonia dissolves in water.)

*74. Determine the pressure of the gas in each of the figures below:

(a)   (b)

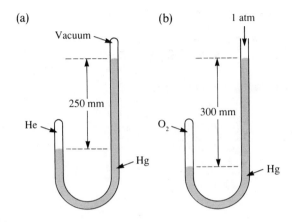

*75. Consider the arrangement of gases shown below. If the valve between the gases is opened and the temperature is held constant:
(a) Determine the pressure of each gas.
(b) Determine the total pressure in the system.

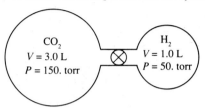

*76. Air has a mass of 1.29 g/L at STP. Calculate the density of air on Pikes Peak, where the pressure is 450 torr and the temperature is 17°C.

*77. A room is 16 ft × 12 ft × 12 ft. Would air enter or leave the room and how much, if the temperature in the room changed from 27°C to −3°C with the pressure remaining constant? Show evidence.

*78. A steel cylinder contained 50.0 L of oxygen gas under a pressure of 40.0 atm and a temperature of 25°C. What was the pressure in the cylinder during a storeroom fire that caused the temperature to rise 152°C? (Be careful!)

CHAPTER FOURTEEN

Water and the Properties of Liquids

Planet Earth, that magnificent blue sphere we enjoy viewing from the space shuttle, is spectacular. Over 75% of the earth is covered with water. We are born from it, drink it, bathe in it, cook with it, enjoy its beauty in waterfalls and rainbows, and stand in awe of the majesty of icebergs. Water supports and enhances life.

In chemistry water provides the medium for numerous reactions. The shape of the water molecule is the basis for hydrogen bonds. These bonds determine the unique properties and reactions of water. The tiny water molecule holds the answers to many of the mysteries of chemical reactions.

Chapter Preview

14.1   Liquids and Solids

In the last chapter, we found that a gas can be a substance containing particles that are far apart, in rapid random motion, and independent of each other. The Kinetic Molecular Theory, along with the ideal gas equation summarizes the behavior of most gases at relatively high temperatures and low pressures.

Solids are obviously very different from gases. A solid contains particles very close together, has a high density, compresses negligibly, and maintains its shape regardless of container. These characteristics indicate large attractive forces between particles. The model for solids is very different from the one for gases.

Liquids, on the other hand, lie somewhere between the extremes of gases and solids. A liquid contains particles that are close together, is essentially incompressible, and has a definite volume. These properties are very similar to solids. However, a liquid also takes the shape of its container, which is closer to the model of a gas.

Although liquids and solids show similar properties, they differ tremendously from gases. No simple mathematical relationship, like the ideal gas equation, works well for liquids or solids. Instead, these models are directly related to the forces of attraction between molecules. With these general statements in mind, let us consider some of the specific properties of liquids.

14.2   Evaporation

When beakers of water, ethyl ether, and ethyl alcohol are allowed to stand uncovered in an open room, the volumes of these liquids gradually decrease. The process by which this change takes place is called *evaporation*.

evaporation

Attractive forces exist between molecules in the liquid state. Not all of these molecules, however, have the same kinetic energy. Molecules that have greater than average kinetic energy can overcome the attractive forces and break away from the surface of the liquid to become a gas. **Evaporation** or **vaporization** is the escape of molecules from the liquid state to the gas or vapor state.

vaporization

In evaporation, molecules of higher than average kinetic energy escape from a liquid, leaving it cooler than it was before they escaped. For this reason, evaporation of perspiration is one way the human body cools itself and keeps its temperature constant. When volatile liquids such as ethyl chloride (C_2H_5Cl) are sprayed on the skin, they evaporate rapidly, cooling the area by removing heat. The numbing effect of the low temperature produced by evaporation of ethyl chloride allows it to be used as a local anesthetic for minor surgery.

sublimation

Solids such as iodine, camphor, naphthalene (moth balls), and, to a small extent, even ice will go directly from the solid to the gaseous state, bypassing the liquid state. This change is a form of evaporation and is called **sublimation**:

$$\text{liquid} \xrightarrow{\text{Evaporation}} \text{vapor}$$

$$\text{solid} \xrightarrow{\text{Sublimation}} \text{vapor}$$

14.3   Vapor Pressure

When a liquid vaporizes in a closed system as shown in Figure 14.1, part (b), some of the molecules in the vapor or gaseous state strike the surface and return to the liquid state by the process of *condensation*. The rate of condensation increases until it is equal to the rate of vaporization. At this point, the space above the liquid is said to be saturated with vapor, and an equilibrium, or steady state, exists between the liquid and the vapor. The equilibrium equation is

$$\text{liquid} \underset{\text{Condensation}}{\overset{\text{Vaporization}}{\rightleftharpoons}} \text{vapor}$$

This equilibrium is dynamic; both processes — vaporization and condensation — are taking place, even though one cannot see or measure a change. The number of molecules leaving the liquid in a given time interval is equal to the number of molecules returning to the liquid.

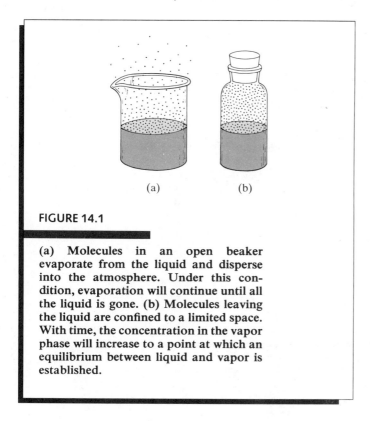

(a) (b)

FIGURE 14.1

(a) Molecules in an open beaker evaporate from the liquid and disperse into the atmosphere. Under this condition, evaporation will continue until all the liquid is gone. (b) Molecules leaving the liquid are confined to a limited space. With time, the concentration in the vapor phase will increase to a point at which an equilibrium between liquid and vapor is established.

vapor pressure

At equilibrium the molecules in the vapor exert a pressure like any other gas. The pressure exerted by a vapor in equilibrium with its liquid is known as the **vapor pressure** of the liquid. The vapor pressure may be thought of as an internal pressure, a measure of the "escaping" tendency of molecules to go from the liquid to the vapor state. The vapor pressure of a liquid is independent of the amount of liquid and vapor present, but it increases as the temperature rises (see Table 14.1). Figure 14.2 illustrates a liquid–vapor equilibrium and the measurement of vapor pressure.

When equal volumes of water, ethyl ether, and ethyl alcohol are placed in separate beakers and allowed to evaporate at the same temperature, we observe that the ether evaporates faster than the alcohol, which evaporates faster than the water. This order of evaporation is consistent with the fact that ether has a higher vapor pressure at any particular temperature than ethyl alcohol or water. One reason for this higher vapor pressure is that the attraction is less between ether molecules than between alcohol molecules or between water molecules. The vapor pressures of these three compounds at various temperatures are compared in Table 14.1.

volatile

Substances that evaporate readily are said to be **volatile**. A volatile liquid has a relatively high vapor pressure at room temperature. Ethyl ether is a very

TABLE 14.1

The Vapor Pressure of Water, Ethyl Alcohol, and Ethyl Ether at Various Temperatures

Temperature (°C)	Vapor pressure (torr)		
	Water	Ethyl alcohol	Ethyl ether[a]
0	4.6	12.2	185.3
10	9.2	23.6	291.7
20	17.5	43.9	442.2
30	31.8	78.8	647.3
40	55.3	135.3	921.3
50	92.5	222.2	1276.8
60	152.9	352.7	1729.0
70	233.7	542.5	2296.0
80	355.1	812.6	2993.6
90	525.8	1187.1	3841.0
100	760.0	1693.3	4859.4
110	1074.6	2361.3	6070.1

[a] Note that the vapor pressure of ethyl ether at temperatures of 40°C and higher exceeds standard pressure, 760 torr, which indicates that the substance has a low boiling point and therefore should be stored in a cool place in a tightly sealed container.

volatile liquid; water is not too volatile; mercury, which has a vapor pressure of 0.0012 torr at 20°C, is essentially a nonvolatile liquid. Most substances that are normally solids are nonvolatile (solids that sublime are exceptions).

14.4   Surface Tension

Have you ever observed water and mercury in the form of small drops? These liquids occur as drops due to *surface tension* of liquids. A droplet of liquid that is not falling or under the influence of gravity (as on the space shuttle) will form a sphere. Minimum surface area is found in the geometrical form of the sphere. The molecules within the liquid are attracted to the surrounding liquid molecules. However, at the surface of the liquid, the attraction is nearly all inward, pulling the surface into a spherical shape. The resistance of a liquid to an increase in its surface area is called the **surface tension** of the liquid. Substances with large attractive forces between molecules have high surface tensions. The effect of surface tension in water is illustrated by the phenomenon of floating a needle on

surface tension

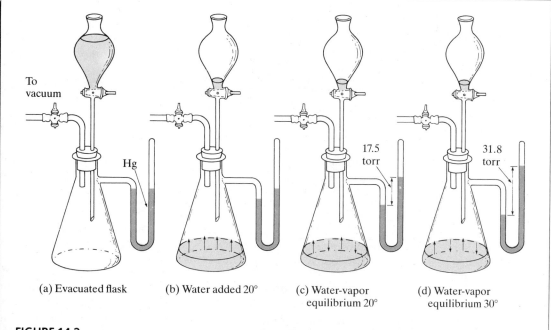

(a) Evacuated flask   (b) Water added 20°   (c) Water-vapor equilibrium 20°   (d) Water-vapor equilibrium 30°

FIGURE 14.2

Measurement of the vapor pressure of water at 20°C and 30°C. In flask (a) the system is evacuated. The mercury manometer attached to the flask shows equal pressure in both legs. In (b) water has been added to the flask and begins to evaporate, exerting pressure as indicated by the manometer. In (c), when equilibrium is established, the pressure inside the flask remains constant at 17.5 torr. In (d) the temperature is changed to 30°C, and equilibrium is reestablished with the vapor pressure at 31.8 torr.

the surface of still water. Other examples include the movement of the water strider insect across a calm pond, or the beading of water on a freshly waxed car.

capillary action

Liquids also exhibit a phenomenon known as **capillary action**, which is the spontaneous rising of a liquid in a narrow tube. This action results from the *cohesive forces* within the liquid, and the *adhesive forces* between the liquid and the walls of the container. If the forces between the liquid and the container are greater than those within the liquid itself, the liquid will climb the walls of the container. For example, take the California sequoia tree, which can be over 200 feet in height. Under atmospheric pressure, water will only rise 33 feet, but capillary action will cause water to rise from the roots to all parts of the tree.

The meniscus in liquids is further evidence of these cohesive and adhesive forces. When a liquid is placed in a glass cylinder the surface of the liquid shows a

meniscus

curve called the **meniscus** (see Figure 14.3). The concave shape of the meniscus

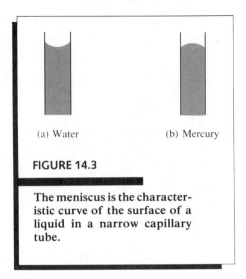

(a) Water   (b) Mercury

FIGURE 14.3

The meniscus is the characteristic curve of the surface of a liquid in a narrow capillary tube.

shows that the adhesive forces between the glass and the liquid (water) are stronger than the cohesive forces within the liquid. In a nonpolar substance such as mercury, the meniscus is convex, indicating that the cohesive forces within the mercury are greater than the adhesive forces within the glass walls.

14.5   Boiling Point

The boiling temperature of a liquid is associated with its vapor pressure. We have seen that the vapor pressure increases as the temperature increases. When the internal or vapor pressure of a liquid becomes equal to the external pressure, the liquid boils. (By external pressure we mean the pressure of the atmosphere above the liquid.) The boiling temperature of a pure liquid remains constant as long as the external pressure does not vary.

The boiling point (bp) of water is 100°C at 1 atm pressure. Table 14.1 shows that the vapor pressure of water at 100°C is 760 torr, a figure we have seen many times before. The significant fact here is that the boiling point is the temperature at which the vapor pressure of the water or other liquid is equal to standard, or atmospheric, pressure at sea level. These relationships lead to the following definition: The **boiling point** is the temperature at which the vapor pressure of a liquid is equal to the external pressure above the liquid.

boiling point

We can readily see that a liquid has an infinite number of boiling points. When we give the boiling point of a liquid, we should also state the pressure. When we express the boiling point without stating the pressure, we mean it to be the **standard** or **normal boiling point** at standard pressure (760 torr). Using

standard or normal boiling point

TABLE 14.2

Physical Properties of Ethyl Chloride, Ethyl Ether, Ethyl Alcohol, and Water

	Boiling point (°C)	Melting point (°C)	Heat of vaporization J/g (cal/g)	Heat of fusion J/g (cal/g)
Ethyl chloride	13	−139	387 (92.5)	—
Ethyl ether	34.6	−116	351 (83.9)	—
Ethyl alcohol	78.4	−112	855 (204.3)	104 (24.9)
Water	100.0	0	2259 (540)	335 (80)

Table 14.1 again, we see that the normal boiling point of ethyl ether is between 30°C and 40°C, and for ethyl alcohol it is between 70°C and 80°C, because, for each compound, 760 torr pressure lies within these stated temperature ranges. At the normal boiling point, 1 g of a liquid changing to a vapor (gas) absorbs an amount of energy equal to its heat of vaporization (see Table 14.2).

vapor pressure curves

The boiling point at various pressures may be evaluated by plotting the data of Table 14.2 on the graph in Figure 14.4, where temperature is plotted horizontally along the x axis and vapor pressure is plotted vertically along the y axis. The resulting curves are known as **vapor pressure curves**. Any point on these curves represents a vapor–liquid equilibrium at a particular temperature and pressure. We may find the boiling point at any pressure by tracing a horizontal line from the designated pressure to a point on the vapor pressure curve. From this point we draw a vertical line to obtain the boiling point on the temperature axis. Four such points are shown in Figure 14.4; they represent the normal boiling points of the four compounds at 760 torr pressure. By reversing this process, you can ascertain at what pressure a substance will boil at a specific temperature. The boiling point is one of the most commonly used physical properties for characterizing and identifying substances.

PRACTICE Use the graph in Figure 14.4 to determine the boiling points of ethyl chloride, ethyl ether, ethyl alcohol, and water at 600 torr.

Answers: 8.5°C, 28°C, 73°C, 93°C

PRACTICE The average atmospheric pressure in Denver is 0.83 atm. What is the boiling point of water in Denver?

Answer: approximately 93°C

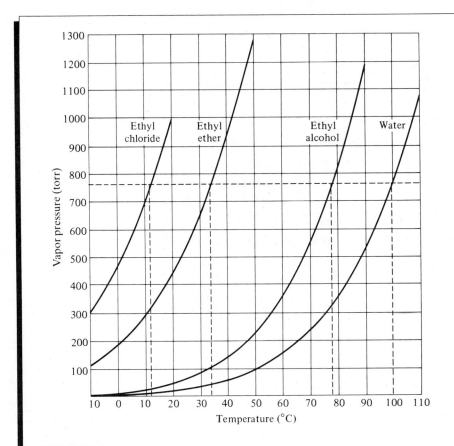

FIGURE 14.4

Vapor pressure–temperature curves for ethyl chloride, ethyl ether, ethyl alcohol, and water.

14.6   Freezing Point or Melting Point

As heat is removed from a liquid, the liquid becomes colder and colder, until a temperature is reached at which it begins to solidify. A liquid that is changing into a solid is said to be *freezing*, or *solidifying*. When a solid is heated continuously, a temperature is reached at which the solid begins to liquefy. A solid that is changing into a liquid is said to be *melting*. The temperature at which the solid phase of a substance is in equilibrium with its liquid phase is known as the

freezing or
melting point

freezing point or **melting point** of that substance. The equilibrium equation is

$$\text{solid} \underset{\text{Freezing}}{\overset{\text{Melting}}{\rightleftarrows}} \text{liquid}$$

When a solid is slowly and carefully heated so that a solid–liquid equilibrium is achieved and then maintained, the temperature will remain constant as long as both phases are present. The energy is used solely to change the solid to the liquid. The melting point is another physical property that is commonly used for characterizing substances.

The most common example of a solid–liquid equilibrium is ice and water. In a well-stirred system of ice and water, the temperature remains at 0°C as long as both phases are present. The melting point changes with pressure but only slightly unless the pressure change is very large.

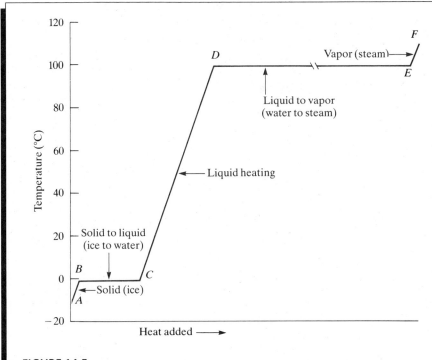

FIGURE 14.5

Heating curve—the absorption of heat by a substance from the solid state to the vapor state. Using water as an example, the interval *AB* represents the ice phase; *BC* interval, the melting of ice to water; *CD* interval, the elevation of the temperature of water from 0°C to 100°C; *DE* interval, the boiling of water to steam; and *EF* interval, the heating of steam.

14.7   Changes of State

The majority of solids undergo two changes of state upon heating. A solid changes to a liquid at its melting point, and a liquid changes to a gas at its boiling point. This warming process can be represented by a graph called a *heating curve* (Figure 14.5). This figure shows ice being heated at a constant rate. As energy flows into the ice, the vibrations within the crystal increase and the temperature rises ($A \longrightarrow B$). Eventually, the molecules begin to break free from the crystal and melting occurs ($B \longrightarrow C$). During the melting process all energy goes into breaking down the crystal structure; the temperature remains constant.

heat of fusion

The energy required to change 1 g of a solid at its melting point into a liquid is called the **heat of fusion**. When the solid has completely melted, the temperature once again rises ($C \longrightarrow D$); the energy input is increasing the molecular motion within the water. At 100°C, the water reaches its boiling point; the temperature remains constant while the added energy is used to vaporize the water to steam

heat of vaporization

($D \longrightarrow E$). The **heat of vaporization** is the energy required to change 1 g of liquid to vapor at its normal boiling point. The attractive forces between the liquid molecules are overcome during vaporization. Beyond this temperature, all the water exists as steam and is being further heated ($E \longrightarrow F$).

EXAMPLE 14.1   How many joules of energy are needed to change 10.0 g of ice at 0.°C to water at 20.°C?

Ice will absorb 335 J/g (heat of fusion) in going from a solid at 0°C to a liquid at 0°C. An additional 4.184 J/g are needed to raise the temperature of the water for each 1°C.

Joules needed to melt the ice:

$$10.0\ \cancel{g} \times \frac{335\ J}{1\ \cancel{g}} = 3.35 \times 10^3\ J\ (800\ cal)$$

Joules needed to heat the water from 0.°C to 20.°C:

$$10.0\ \cancel{g} \times \frac{4.184\ J}{1\ \cancel{g}\cancel{°C}} \times 20.\cancel{°C} = 837\ J\ (200\ cal)$$

Thus, 3350 J + 837 J = 4.2×10^3 J (1000 cal) are needed.

EXAMPLE 14.2   How many kilojoules of energy are needed to change 20.0 g of water at 20.°C to steam at 100.°C?

Kilojoules needed to heat the water from 20.°C to 100.°C:

$$20.0\ g \times \frac{4.184\ J}{1\ \cancel{g}\cancel{°C}} \times \frac{1\ kJ}{1000\ J} \times 80.\cancel{°C} = 7.0 \times 10^3\ kJ\ (1.6 \times 10^3\ kcal)$$

Kilojoules needed to change water at 100.°C to steam at 100.°C:

$$20.0 \, \cancel{g} \times \frac{2.26 \text{ kJ}}{1 \, \cancel{g}} = 4.52 \times 10^4 \text{ kJ} \ (1.08 \times 10^4 \text{ kcal})$$

Thus, 6.7×10^3 kJ $+ 4.52 \times 10^4$ kJ $= 5.2 \times 10^4$ kJ $(1.2 \times 10^3$ kcal) are needed.

PRACTICE   How many kilojoules of energy are required to change 50.0 g of ethyl alcohol at 60.0°C to vapor at 78.4°C? The specific heat of ethyl alcohol is 2.138 J/g°C. Answer: 44.7 kJ

14.8   Occurrence of Water

Water is our most common natural resource; it covers about 70% of the earth's surface. Not only is it found in the oceans and seas, in lakes, rivers, streams, and in glacial ice deposits, it is also always present in the atmosphere and in cloud formations.

About 97% of the earth's water is in the oceans. This *saline* water contains vast amounts of dissolved minerals. More than 70 elements have been detected in the mineral content of seawater. Only four of these—chlorine, sodium, magnesium, and bromine—are now commercially obtained from the sea. The world's *fresh* water comprises the other 3%, of which about two-thirds is locked up in polar ice caps and glaciers. The remaining fresh water is found in ground water, lakes, and the atmosphere.

Water is an essential constituent of all living matter. It is the most abundant compound in the human body, making up about 70% of total body mass. About 92% of blood plasma is water; about 80% of muscle tissue is water; and about 60% of a red blood cell is water. Water is more important than food in the sense that a person can survive much longer without food than without water.

14.9   Physical Properties of Water

Water is a colorless, odorless, tasteless liquid with a melting point of 0°C and a boiling point of 100°C at 1 atm pressure. The heat of fusion of water is 335 J/g (80 cal/g). The heat of vaporization of water is 2.26 kJ/g (540 cal/g). The values for water for both the heat of fusion and the heat of vaporization are high

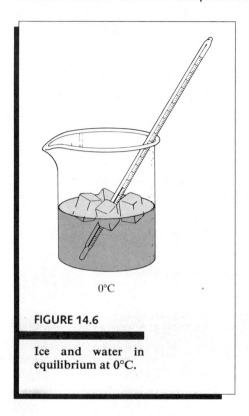

0°C

FIGURE 14.6

Ice and water in equilibrium at 0°C.

compared with those for other substances; these high values indicate that strong attractive forces are acting between the molecules.

Ice and water exist together in equilibrium at 0°C, as shown in Figure 14.6. When ice at 0°C melts, it absorbs 335 J/g (80 cal/g) in changing into a liquid; the temperature remains at 0°C. In order to refreeze the water we have to remove 335 J/g (80 cal/g) from the liquid at 0°C.

In Figure 14.7 both boiling water and steam are shown to have a temperature of 100°C. It takes 418 J (100 cal) to heat 1 g of water from 0°C to 100°C, but water at its boiling point absorbs 2.26 kJ/g (540 cal/g) in changing to steam. Although boiling water and steam are both at the same temperature, steam contains considerably more heat per gram and can cause more severe burns than hot water. In Table 14.3 the physical properties of water are tabulated and compared with those of other hydrogen compounds of Group VIA elements.

The maximum density of water is 1.000 g/mL at 4°C. Water has the unusual property of contracting in volume as it is cooled to 4°C and then expanding when cooled from 4°C to 0°C. Therefore 1 g of water occupies a volume greater than 1 mL at all temperatures except 4°C. Although most liquids contract in volume all the way down to the point at which they solidify, a large increase (about 9%) in

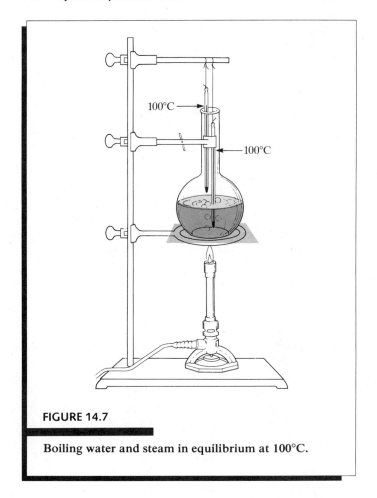

FIGURE 14.7

Boiling water and steam in equilibrium at 100°C.

TABLE 14.3

Physical Properties of Water and Other Hydrogen Compounds of Group VIA Elements

Formula	Color	Molar mass	Melting point (°C)	Boiling point, 1 atm (°C)	Heat of fusion J/g (cal/g)	Heat of vaporization J/g (cal/g)
H_2O	Colorless	18.0	0.00	100.0	335 (80.0)	2.26×10^3 (540)
H_2S	Colorless	34.1	−85.5	−60.3	69.9 (16.7)	548 (131)
H_2Se	Colorless	81.0	−65.7	−41.3	31 (7.4)	238 (57.0)
H_2Te	Colorless	129.6	−49	−2.2	—	179 (42.8)

volume occurs when water changes from a liquid at 0°C to a solid (ice) at 0°C. The density of ice at 0°C is 0.917 g/mL, which means that ice, being less dense than water, will float in water.

14.10   Structure of the Water Molecule

A single water molecule consists of two hydrogen atoms and one oxygen atom. Each H atom is attached to the O atom by a single covalent bond. This bond is formed by the overlap of the 1s orbital of hydrogen with an unpaired 2p orbital of oxygen. The average distance between the two nuclei is known as the *bond length*. The O—H bond length in water is 0.096 nm. The water molecule is nonlinear and has a V-shaped structure with an angle of about 105° between the two bonds (see Figure 14.8).

Oxygen is the second most electronegative element. As a result, the two covalent OH bonds in water are polar. If the three atoms in a water molecule were aligned in a linear structure such as H $\longmapsto$ O $\longleftarrow$ H, the two polar bonds would be acting in equal and opposite directions and the molecule would be nonpolar. However, water is a highly polar molecule. Therefore, it does not have a linear structure. When atoms are bonded together in a nonlinear fashion, the angle formed by the bonds is called the *bond angle*. In water the HOH bond angle is 105°. The two polar covalent bonds and the bent structure result in a partial negative charge on the oxygen atom and a partial positive charge on each hydrogen atom. The polar nature of water is responsible for many of its properties, including its behavior as a solvent.

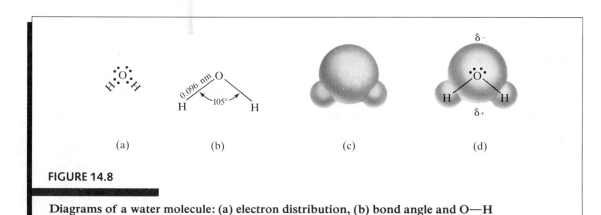

(a)   (b)   (c)   (d)

FIGURE 14.8

Diagrams of a water molecule: (a) electron distribution, (b) bond angle and O—H bond length, (c) molecular orbital structure, and (d) dipole representation.

14.11   The Hydrogen Bond

hydrogen bond

Table 14.3 compares the physical properties of H_2O, H_2S, H_2Se, and H_2Te. From this comparison it is apparent that four physical properties of water—melting point, boiling point, heat of fusion, and heat of vaporization—are extremely high and do not fit the trend relative to the molar masses of the four compounds. If the properties of water followed the progression shown by the other three compounds, we would expect the melting point of water to be below $-85°C$ and the boiling point to be below $-60°C$.

Why does water have these anomalous physical properties? The answer is that liquid water molecules are linked together by hydrogen bonds. A **hydrogen bond** is a chemical bond that is formed between polar molecules that contain hydrogen covalently bonded to a small, highly electronegative atom such as fluorine, oxygen, or nitrogen (F—H, O—H, N—H). The bond is actually the dipole–dipole attraction of polar molecules containing these three types of polar bonds.

> **Elements that have significant hydrogen bonding ability are F, O, and N.**

What is a hydrogen bond, or H-bond? Because a hydrogen atom has only one electron, it can form only one covalent bond. When it is attached to a strong electronegative atom such as oxygen, a hydrogen atom will also be attracted to an oxygen atom of another molecule, forming a bond (or bridge) between the two molecules. Water has two types of bonds: covalent bonds that exist between hydrogen and oxygen atoms within a molecule and hydrogen bonds that exist between hydrogen and oxygen atoms in different water molecules.

Hydrogen bonds are *intermolecular* bonds; that is, they are formed between atoms in different molecules. They are somewhat ionic in character because they are formed by electrostatic attraction. Hydrogen bonds are much weaker than the ionic or covalent bonds that unite atoms to atoms to form compounds. Despite their weakness, hydrogen bonds are of great chemical importance.

The oxygen atom in water can form two hydrogen bonds—one through each of the unbonded pairs of electrons. Figure 14.9 shows (a) two water molecules linked by a hydrogen bond and (b) six water molecules linked by hydrogen bonds. A dash (—) is used for the covalent bond and a dotted line (····) for the hydrogen bond. In water each molecule is linked to others through hydrogen bonds to form a three-dimensional aggregate of water molecules. This intermolecular hydrogen bonding effectively gives water the properties of a much larger, heavier molecule, explaining in part its relatively high melting point, boiling point, heat of fusion, and heat of vaporization. As water is heated and energy is absorbed, hydrogen

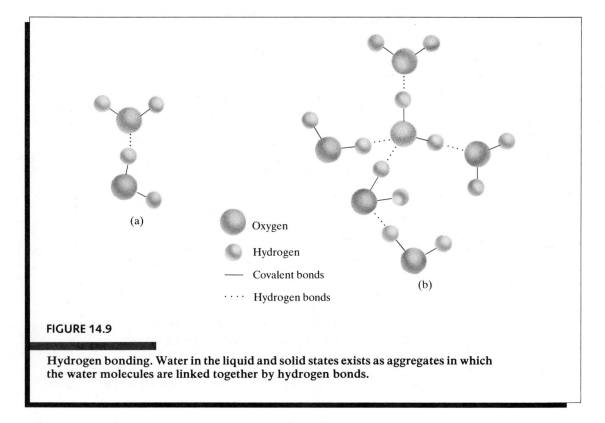

(a)

Oxygen

Hydrogen

—— Covalent bonds

· · · · Hydrogen bonds

(b)

FIGURE 14.9

Hydrogen bonding. Water in the liquid and solid states exists as aggregates in which the water molecules are linked together by hydrogen bonds.

bonds are continually being broken until at 100°C, with the absorption of an additional 2.26 kJ/g (540 cal/g), water separates into individual molecules, going into the gaseous state. Sulfur, selenium, and tellurium are not sufficiently electronegative for their hydrogen compounds to behave like water. As a result, H-bonding in H_2S, H_2Se, and H_2Te is only of small consequence (if any) to their physical properties. For example, the lack of hydrogen bonding is one reason why H_2S is a gas and not a liquid at room temperature.

Fluorine, the most electronegative element, forms the strongest hydrogen bonds. This bonding is strong enough to link hydrogen fluoride molecules together as *dimers*, H_2F_2, or as larger, $(HF)_n$, molecular units. The dimer structure may be represented in this way:

$$\begin{array}{c} \text{F} \\ \diagup \\ \text{H} \quad \text{..H} \\ \diagdown \quad \diagdown \\ \text{F} \quad \diagdown \\ \quad \text{H—bond} \end{array}$$

The existence of salts, such as KHF_2 and NH_4HF_2, verifies the hydrogen fluoride (bifluoride) structure, HF_2^- $(F—H \cdots F)^-$, where one H atom is bonded to two F atoms through one covalent bond and one hydrogen bond.

Hydrogen bonding can occur between two different atoms that are capable of forming H-bonds. Thus we may have an $O \cdots H{-}N$ or $O{-}H \cdots N$ linkage in which the hydrogen atom forming the H-bond is between an oxygen and a nitrogen atom. This form of the H-bond exists in certain types of protein molecules and many biologically active substances.

EXAMPLE 14.3   Would you expect hydrogen bonding to occur between molecules of the following substances?

(a)
$$
\begin{array}{ccc}
 & H & H \\
 & | & | \\
H{-} & C{-} & C{-}O{-}H \\
 & | & | \\
 & H & H
\end{array}
$$

(b)
$$
\begin{array}{ccc}
 & H & H \\
 & | & | \\
H{-} & C{-}O{-} & C{-}H \\
 & | & | \\
 & H & H
\end{array}
$$

Ethyl alcohol   Dimethyl ether

(a) Hydrogen bonding should occur in ethyl alcohol because one hydrogen atom is bonded to an oxygen atom.

$$
\begin{array}{cccccc}
H & H & & H & H & H \\
| & | & & | & | & | \\
H{-}C{-}C{-}O{-}H & \cdots & O{-}C{-}C{-}H \\
| & | & \uparrow & | & | \\
H & H & \text{H-bond} & H & H
\end{array}
$$

(b) There is no hydrogen bonding in dimethyl ether because all the hydrogen atoms are bonded only to carbon atoms.

Both ethyl alcohol and dimethyl ether have the same molar mass (46.0). Although both compounds have the same molecular formula, C_2H_6O, ethyl alcohol has a much higher boiling point (78.4°C) than dimethyl ether (−23.7°C), because of hydrogen bonding between the alcohol molecules.

PRACTICE   Would you expect hydrogen bonding to occur between molecules of the following substances?

(a)
$$
\begin{array}{ccc}
H & H & H \\
| & | & | \\
H{-}C{-}C{-}N{-}H \\
| & | \\
H & H
\end{array}
$$

(b)
$$
\begin{array}{ccc}
H & H & H \\
| & | & | \\
H{-}C{-}N{-}C{-}H \\
| & | \\
H & H
\end{array}
$$

(c)
$$
\begin{array}{cccc}
 & & H & \\
 & & | & \\
H & H & H{-}C{-}H \\
| & | & | \\
H{-}C{-}C{-} & & N \\
| & | & | \\
H & H & H{-}C{-}H \\
 & & | \\
 & & H
\end{array}
$$

Answers: (a) yes   (b) yes   (c) no

14.12 Formation of Water and Chemical Properties of Water

Water is very stable to heat; it decomposes to the extent of only about 1% at temperatures up to 2000°C. Pure water is a nonconductor of electricity. But when a small amount of sulfuric acid or sodium hydroxide is added, the solution is readily decomposed into hydrogen and oxygen by an electric current. Two volumes of hydrogen are produced for each volume of oxygen.

$$2\ H_2O(l) \xrightarrow[\text{H}_2\text{SO}_4 \text{ or NaOH}]{\text{Electrical energy}} 2\ H_2(g) + O_2(g)$$

Formation   Water is formed when hydrogen burns in air. Pure hydrogen burns very smoothly in air, but mixtures of hydrogen and air or oxygen explode when ignited. The reaction is strongly exothermic.

$$2\ H_2(g) + O_2(g) \longrightarrow 2\ H_2O(g) + 484\ \text{kJ (115.6 kcal)}$$

Water is produced by a variety of other reactions, especially by (1) acid–base neutralizations, (2) combustion of hydrogen-containing materials, and (3) metabolic oxidation in living cells:

1. $HCl(aq) + NaOH(aq) \longrightarrow NaCl(aq) + H_2O(l)$
2. $2\ C_2H_2(g) + 5\ O_2(g) \longrightarrow 4\ CO_2(g) + 2\ H_2O(g) + 1212\ \text{kJ (289.6 kcal)}$
    Acetylene

    $CH_4 + 2\ O_2 \longrightarrow CO_2 + 2\ H_2O + 803\ \text{kJ (192 kcal)}$
    Methane
3. $C_6H_{12}O_6 + 6\ O_2 \xrightarrow{\text{Enzymes}} 6\ CO_2 + 6\ H_2O + 2519\ \text{kJ (673 kcal)}$
    Glucose

The combustion of acetylene shown in (2) is strongly exothermic and is capable of producing very high temperatures. It is used in oxygen–acetylene torches to cut and weld steel and other metals. Methane is known as natural gas and is commonly used as fuel for heating and cooking. The reaction of glucose with oxygen shown in (3) is the reverse of photosynthesis. It is the overall reaction by which living cells obtain needed energy by metabolizing glucose to carbon dioxide and water.

Reactions of Water with Metals and Nonmetals   The reactions of metals with water at different temperatures show that these elements vary greatly in their reactivity. Metals such as sodium, potassium, and calcium react with cold water to produce hydrogen and a metal hydroxide. A small piece of sodium added to water melts from the heat produced by the reaction, forming a silvery metal ball, which rapidly flits back and forth on the surface of the water. Caution must be used when experimenting with this reaction, because the hydrogen produced is frequently ignited by the sparking of the sodium, and it will explode, spattering sodium. Potassium reacts even more vigorously than sodium. Calcium sinks in water and liberates a gentle stream of hydrogen. The equations for these reactions are

$$2\ Na(s) + 2\ H_2O(l) \longrightarrow H_2\uparrow + 2\ NaOH(aq)$$

$$2\ K(s) + 2\ H_2O(l) \longrightarrow H_2\uparrow + 2\ KOH(aq)$$

$$Ca(s) + 2\ H_2O(l) \longrightarrow H_2\uparrow + Ca(OH)_2(aq)$$

Zinc, aluminum, and iron do not react with cold water but will react with steam at high temperatures, forming hydrogen and a metallic oxide. The equations are

$$Zn(s) + H_2O(steam) \longrightarrow H_2\uparrow + ZnO(s)$$

$$2\ Al(s) + 3\ H_2O(steam) \longrightarrow 3\ H_2\uparrow + Al_2O_3(s)$$

$$3\ Fe(s) + 4\ H_2O(steam) \longrightarrow 4\ H_2\uparrow + Fe_3O_4(s)$$

Copper, silver, and mercury are examples of metals that do not react with cold water or steam to produce hydrogen. We conclude that sodium, potassium, and calcium are chemically more reactive than zinc, aluminum, and iron, which are more reactive than copper, silver, and mercury.

Certain nonmetals react with water under various conditions. For example, fluorine reacts violently with cold water, producing hydrogen fluoride and free oxygen. The reactions of chlorine and bromine are much milder, producing what is commonly known as "chlorine water" and "bromine water," respectively. Chlorine water contains HCl, HOCl, and dissolved Cl_2; the free chlorine gives it a yellow-green color. Bromine water contains HBr, HOBr, and dissolved Br_2; the free bromine gives it a reddish-brown color. Steam passed over hot coke (carbon) produces a mixture of carbon monoxide and hydrogen that is known as "water gas." Since water gas is combustible, it is useful as a fuel. It is also the starting material for the commercial production of several alcohols. The equations for these reactions are

$$2\ F_2(g) + 2\ H_2O(l) \longrightarrow 4\ HF(aq) + O_2(g)$$

$$Cl_2(g) + H_2O(l) \longrightarrow HCl(aq) + HOCl(aq)$$

$$Br_2(l) + H_2O(l) \longrightarrow HBr(aq) + HOBr(aq)$$

$$C(s) + H_2O(g) \xrightarrow{1000°C} CO(g) + H_2(g)$$

Reactions of Water with Metal and Nonmetal Oxides Metal oxides that react with water to form bases are known as **basic anhydrides**. Examples are

basic anhydride

$$CaO(s) + H_2O \longrightarrow Ca(OH)_2(aq)$$

Calcium hydroxide

$$Na_2O(s) + H_2O \longrightarrow 2\,NaOH(aq)$$
Sodium hydroxide

Certain metal oxides, such as CuO and Al_2O_3, do not form basic solutions because the oxides are insoluble in water.

acid anhydride

Nonmetal oxides that react with water to form acids are known as **acid anhydrides**. Examples are

$$CO_2(g) + H_2O(l) \rightleftharpoons H_2CO_3(aq)$$
Carbonic acid

$$SO_2(g) + H_2O(l) \rightleftharpoons H_2SO_3(aq)$$
Sulfurous acid

$$N_2O_5(s) + H_2O(l) \longrightarrow 2\,HNO_3(aq)$$
Nitric acid

The word *anhydrous* means "without water." An anhydride is a metal oxide or a nonmetal oxide derived from a base or an oxy-acid by the removal of water. To determine the formula of an anhydride, the elements of water, H_2O, are removed from an acid or base formula until all the hydrogen is removed. Sometimes more than one formula unit is needed to remove all the hydrogen as water. The formula of the anhydride then consists of the remaining metal or nonmetal and the remaining oxygen atoms. In calcium hydroxide, removal of water as indicated leaves CaO as the anhydride:

$$Ca\!\!\!\begin{array}{c} O\,H \\ OH \end{array} \xrightarrow{\Delta} CaO + H_2O$$

In sodium hydroxide, H_2O cannot be removed from one formula unit, so two formula units of $NaOH$ must be used, leaving Na_2O as the formula of the anhydride:

$$\begin{array}{c} NaO\,H \\ Na\,OH \end{array} \xrightarrow{\Delta} Na_2O + H_2O$$

The removal of water from sulfuric acid, H_2SO_4, gives the acid anhydride SO_3:

$$H_2SO_4 \xrightarrow{\Delta} SO_3 + H_2O$$

The foregoing are examples of typical reactions of water but are by no means a complete list of the known reactions of water.

14.13   Hydrates

hydrate

water of hydration

water of
crystallization

When certain salt solutions are allowed to evaporate, some water molecules
remain as part of the crystalline salt that is left after evaporation is complete.
Solids that contain water molecules as part of their crystalline structure are
known as **hydrates**. Water in a hydrate is known as **water of hydration**, or **water of
crystallization**.

Formulas for hydrates are expressed by first writing the usual anhydrous
(without water) formula for the compound and then adding a dot followed by the
number of water molecules present. An example is $BaCl_2 \cdot 2\ H_2O$. This formula
tells us that each formula unit of this salt contains one barium ion, two chloride
ions, and two water molecules. A crystal of the salt contains many of these units
in its crystalline lattice.

In naming hydrates, we first name the compound exclusive of the water and
then add the term *hydrate*, with the proper prefix representing the number of
water molecules in the formula. For example, $BaCl_2 \cdot 2\ H_2O$ is called *barium
chloride dihydrate*. Hydrates are true compounds and follow the Law of Definite
Composition. The molar mass of $BaCl_2 \cdot 2\ H_2O$ is 244.3 g/mol; it contains
56.20% barium, 29.06% chlorine, and 14.74% water.

Water molecules in hydrates are bonded by electrostatic forces between
polar water molecules and the positive or negative ions of the compound. These
forces are not as strong as covalent or ionic chemical bonds. As a result, water
of crystallization can be removed by moderate heating of the compound. A
partially dehydrated or completely anhydrous compound may result. When
$BaCl_2 \cdot 2\ H_2O$ is heated, it loses its water at about 100°C:

$$BaCl_2 \cdot 2\ H_2O \xrightarrow{100°C} BaCl_2 + 2\ H_2O\uparrow$$

When a solution of copper(II) sulfate ($CuSO_4$) is allowed to evaporate,
beautiful blue crystals containing 5 moles of water per mole of $CuSO_4$ are
formed. The formula for this hydrate is $CuSO_4 \cdot 5\ H_2O$; it is called copper(II)
sulfate pentahydrate, or cupric sulfate pentahydrate. When $CuSO_4 \cdot 5\ H_2O$ is
heated, water is lost, and a pale green-white powder, anhydrous $CuSO_4$, is
formed.

$$CuSO_4 \cdot 5\ H_2O \xrightarrow{250°C} CuSO_4 + 5\ H_2O\uparrow$$

When water is added to anhydrous copper(II) sulfate, the foregoing reaction is
reversed, and the salt turns blue again. Because of this outstanding color change,
anhydrous copper(II) sulfate has been used as an indicator to detect small
amounts of water. The formation of the hydrate is noticeably exothermic.

The formula for plaster of paris is $(CaSO_4)_2 \cdot H_2O$. When mixed with the
proper quantity of water, plaster of paris forms a dihydrate and sets to a hard
mass. It is, therefore, useful for making patterns for the reproduction of art

TABLE 14.4

Selected Hydrates

Hydrate	Name
$CaCl_2 \cdot 2\, H_2O$	Calcium chloride dihydrate
$Ba(OH)_2 \cdot 8\, H_2O$	Barium hydroxide octahydrate
$MgSO_4 \cdot 7\, H_2O$	Magnesium sulfate heptahydrate
$SnCl_2 \cdot 2\, H_2O$	Tin(II) chloride dihydrate
$CoCl_2 \cdot 6\, H_2O$	Cobalt(II) chloride hexahydrate
$Na_2CO_3 \cdot 10\, H_2O$	Sodium carbonate decahydrate
$(NH_4)_2C_2O_4 \cdot H_2O$	Ammonium oxalate monohydrate
$NaC_2H_3O_2 \cdot 3\, H_2O$	Sodium acetate trihydrate
$Na_2B_4O_7 \cdot 10\, H_2O$	Sodium tetraborate decahydrate
$Na_2S_2O_3 \cdot 5\, H_2O$	Sodium thiosulfate pentahydrate

objects, molds, and surgical casts. The chemical reaction is

$$(CaSO_4)_2 \cdot H_2O(s) + 3\, H_2O(l) \longrightarrow 2\, CaSO_4 \cdot 2\, H_2O(s)$$

The occurrence of hydrates is commonplace in salts. Table 14.4 lists a number of common hydrates.

14.14   Hygroscopic Substances: Deliquescence; Efflorescence

hygroscopic substances

Many anhydrous salts and other substances readily absorb water from the atmosphere. Such substances are said to be **hygroscopic**. This property can be observed in the following simple experiment: Spread a 10–20 g sample of anhydrous copper(II) sulfate on a watch glass and set it aside so that the salt is exposed to the air. Then determine the mass of the sample periodically for 24 hours, noting the increase in mass and the change in color. Water is absorbed from the atmosphere, forming the blue pentahydrate $CuSO_4 \cdot 5\, H_2O$.

deliquescence

Some compounds continue to absorb water beyond the hydrate stage to form solutions. A substance that absorbs water from the air until it forms a solution is said to be **deliquescent**. A few granules of anhydrous calcium chloride or pellets of sodium hydroxide exposed to the air will appear moist in a few minutes, and within an hour will absorb enough water to form a puddle of solution. Diphosphorus pentoxide (P_2O_5) picks up water so rapidly that its mass cannot be determined accurately except in an anhydrous atmosphere.

Compounds that absorb water are useful as drying agents (desiccants). Refrigeration systems must be kept dry with such agents or the moisture will freeze and clog the tiny orifices in the mechanism. Bags of drying agents are often enclosed in packages containing iron or steel parts to absorb moisture and prevent rusting. Anhydrous calcium chloride, magnesium sulfate, sodium sulfate, calcium sulfate, silica gel, and diphosphorus pentoxide are some of the compounds commonly used for drying liquids and gases that contain small amounts of moisture.

efflorescence

The process by which crystalline materials spontaneously lose water when exposed to the air is known as **efflorescence**. Glauber's salt ($Na_2SO_4 \cdot 10\,H_2O$), a transparent crystalline salt, loses water when exposed to the air. One can actually observe these well-defined, large crystals crumbling away as they lose water and form a white, noncrystalline-appearing powder. From our discussion of the decomposition of hydrates, we can predict that heat will increase the rate of efflorescence. The rate also depends on the concentration of moisture in the air. A dry atmosphere will allow the process to take place more rapidly.

14.15 Natural Waters

Natural fresh waters are not pure, but contain dissolved minerals, suspended matter, and sometimes harmful bacteria. The water supplies of large cities are usually drawn from rivers or lakes. Such water is generally unsafe to drink without treatment. To make such water potable (that is, safe to drink), it is treated by some or all of the following processes (See Figure 14.10).

1. **Screening** Removal of relatively large objects, such as trash, fish, and so on.
2. **Flocculation and sedimentation** Chemicals, usually lime (CaO) and alum (aluminum sulfate), are added to form a flocculent jellylike precipitate of aluminum hydroxide. This precipitate traps most of the fine suspended matter in the water and carries it to the bottom of the sedimentation basin.
3. **Sand filtration** Water is drawn from the top of the sedimentation basin and passed downward through fine sand filters. Nearly all the remaining suspended matter and bacteria are removed by the sand filters.
4. **Aeration** Water is drawn from the bottom of the sand filters and is aerated by spraying. The purpose of this process is to remove objectionable odors and tastes.
5. **Disinfection** In the final stage chlorine gas is injected into the water to kill harmful bacteria before the water is distributed to the public. Ozone is also used in some countries to disinfect water. In emergencies water may be disinfected by simply boiling for a few minutes.

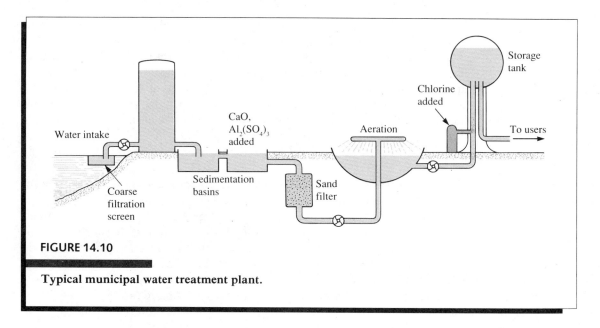

FIGURE 14.10

Typical municipal water treatment plant.

If the drinking water of children contains an optimum amount of fluoride ion, their teeth will be more resistant to decay. Therefore, in many communities NaF or Na_2SiF_6 is added to the water supply to bring the fluoride ion concentration up to the optimum level of about 1.0 ppm. Excessively high concentrations of fluoride ion can cause mottling of the teeth.

Water that contains dissolved calcium and magnesium salts is called *hard water*. One drawback of hard water is that ordinary soap does not lather well in it; the soap reacts with the calcium and magnesium ions to form an insoluble greasy scum. However, synthetic soaps, known as detergents or syndets, are available; they have excellent cleaning qualities and do not form precipitates with hard water. Hard water is also undesirable because it causes "boiler scale" to form on the walls of water heaters and steam boilers, which greatly reduces their efficiency.

Four techniques used to "soften" hard water are distillation, chemical precipitation, ion exchange, and demineralization. In distillation the water is boiled, and the steam thus formed is condensed to a liquid again, leaving the minerals behind in the distilling vessel. Figure 14.11 illustrates a simple laboratory distillation apparatus. Commercial stills are available that are capable of producing hundreds of liters of distilled water per hour.

Calcium and magnesium ions are precipitated from hard water by adding sodium carbonate and lime. Insoluble calcium carbonate and magnesium hydroxide are precipitated and are removed by filtration or sedimentation.

In the ion-exchange method, used in many households, hard water is effectively softened as it is passed through a bed or tank of zeolite. Zeolite is a complex sodium aluminum silicate. In this process sodium ions replace

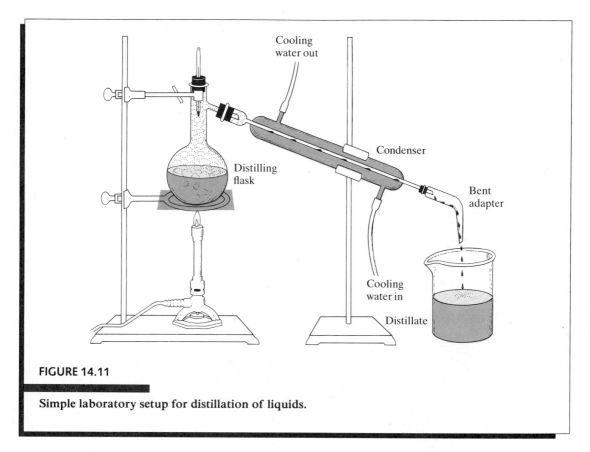

FIGURE 14.11

Simple laboratory setup for distillation of liquids.

objectionable calcium and magnesium ions, and the water is thereby softened:

$$Na_2Zeolite(s) + Ca^{2+}(aq) \longrightarrow CaZeolite(s) + 2\ Na^+(aq)$$

The zeolite is regenerated by back-flushing with concentrated sodium chloride solution, reversing the foregoing reaction.

The sodium ions that are present in water softened either by chemical precipitation or by the zeolite process are not objectionable to most users of soft water.

In demineralization both cations and anions are removed by a two-stage ion-exchange system. Special synthetic organic resins are used in the ion-exchange beds. In the first stage metal cations are replaced by hydrogen ions. In the second stage anions are replaced by hydroxide ions. The hydrogen and hydroxide ions react, and essentially pure, mineral-free water leaves the second stage (see Figure 14.12).

The oceans are an inexhaustible source of water; however, seawater contains about 3.5 lb of salts per 100 lb of water. This 35,000 ppm of dissolved salts makes seawater unfit for agricultural and domestic uses. Water that contains less than

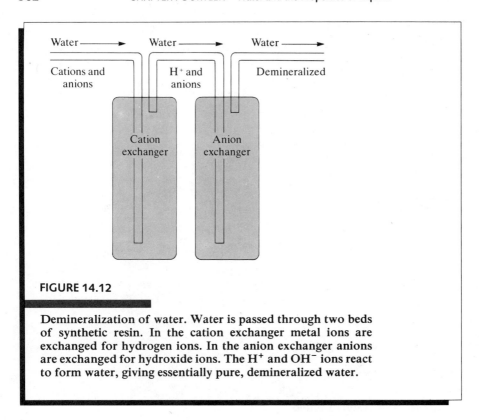

FIGURE 14.12

Demineralization of water. Water is passed through two beds of synthetic resin. In the cation exchanger metal ions are exchanged for hydrogen ions. In the anion exchanger anions are exchanged for hydroxide ions. The H^+ and OH^- ions react to form water, giving essentially pure, demineralized water.

1000 ppm of salts is considered reasonably good for drinking, and potable (safe to drink) water is already being obtained from the sea in many parts of the world. Continuous research is being done in an effort to make usable water from the oceans more abundant and economical.

14.16   Water Pollution

Polluted water was formerly thought of as water that was unclear, had a bad odor or taste, and contained disease-causing bacteria. However, such factors as increased population, industrial requirements for water, atmospheric pollution, toxic waste dumps, and use of pesticides have greatly modified the problem of water pollution.

Many of the newer pollutants are not removed or destroyed by the usual water-treatment processes. For example, among the 66 organic compounds

TABLE 14.5

Classification of Water Pollutants

Type of pollutant	Examples
Oxygen-demanding wastes	Decomposable organic wastes from domestic sewage and industrial wastes of plant and animal origin
Infectious agents	Bacteria, viruses, and other organisms from domestic sewage, animal wastes, and animal process wastes
Plant nutrients	Principally compounds of nitrogen and phosphorus
Organic chemicals	Large numbers of chemicals synthesized by industry, pesticides, chlorinated organic compounds
Other minerals and chemicals	Inorganic chemicals from industrial operations, mining, oil field operations, and agriculture
Radioactive substances	Waste products from mining and processing of radioactive materials, airborne radioactive fallout, increased use of radioactive materials in hospitals and research
Heat from industry	Large quantities of heated water returned to water bodies from power plants and manufacturing facilities after use for cooling
Sediment from land erosion	Solid matter washed into streams and oceans by erosion, rain, and water run off

found in the drinking water of a major city on the Mississippi River, 3 are labeled slightly toxic, 17 moderately toxic, 15 very toxic, 1 extremely toxic, and 1 supertoxic. Two are known carcinogens (cancer-producing agents), 11 are suspect, and 3 are metabolized to carcinogens. The United States Public Health Service classifies water pollutants under eight broad categories. These categories are shown in Table 14.5.

Many outbreaks of disease or poisoning, such as typhoid, dysentery, and cholera have been attributed directly to drinking water. Rivers and streams are an easy means for municipalities to dispose of their domestic and industrial waste products. Much of this water is used again by people downstream, and then discharged back into the water source. Then another community still farther downstream draws the same water and discharges its own wastes. Thus, along waterways such as the Mississippi and Delaware rivers, water is withdrawn and discharged many times. If this water is not properly treated, harmful pollutants will build up, causing epidemics of various diseases.

Hazardous waste products are unavoidable in the manufacture of many products that we use in everyday life. One common way to dispose of these wastes is to place them in toxic waste dumps. What has been found after many years of disposing of wastes in this manner is that toxic substances have seeped into the ground-water deposits. As a result many people have become ill, and water wells have been closed until satisfactory methods of detoxifying this water are found. This problem is serious, because one-half the United States population gets its drinking water from ground water. To clean up the thousands of industrial dumps and to find and implement new and safe methods of disposing of wastes will be very costly.

Mercury and its compounds have long been known to be highly toxic. Mercury gets into the body primarily in the foods we eat. Although it is not an essential mineral for the body, mercury accumulates in the blood, kidneys, liver, and brain tissues. Mercury in the brain causes serious damage to the central nervous system.

The sequence of events that have led to incidents of mercury poisoning is as follows: Mercury and its compounds are used in many industries and in agriculture, primarily as a fungicide in the treatment of seeds. One of the largest uses is in the electrochemical conversion of sodium chloride brines to chlorine and sodium hydroxide, as represented by this equation:

$$2\,NaCl + 2\,H_2O \xrightarrow{\text{Electrolysis}} Cl_2 + 2\,NaOH + H_2$$

Although no mercury is shown in the chemical equation, it is used in the process for electrical contact, and small amounts are discharged along with spent brine solutions. Thus considerable quantities of mercury, in low concentrations, have been discharged into lakes and other surface waters from the effluents of these manufacturing plants. The mercury compounds discharged into the water are converted by bacterial action and other organic compounds to methyl mercury, $(CH_3)_2Hg$, which then accumulates in the bodies of fish. Several major episodes of mercury poisoning that have occurred in the past years were the result of eating mercury-contaminated fish. The best way to control this contaminant is at the source, and much has been done since 1970 to eliminate the discharge of mercury in industrial wastes. In 1976 the Environmental Protection Agency banned the use of all mercury-containing insecticides and fungicides.

Many other major water pollutants have been recognized and steps have been taken to eliminate them. Three that pose serious problems are lead, detergents, and chlorine-containing organic compounds. Lead poisoning, for example, has been responsible for many deaths in past years. The toxic action of lead in the body is the inhibition of the enzyme necessary for the production of hemoglobin in the blood. The usual intake of lead into the body is through food. However, extraordinary amounts of lead can be ingested from water running through lead pipes and by using lead-containing ceramic containers for storage of food and beverages.

It has been clearly demonstrated that waterways rendered so polluted that the water is neither fit for human use nor able to sustain marine life can be successfully restored. However, keeping our lakes and rivers free from pollution is a very costly and complicated process.

Concepts in Review

1. List the common properties of liquids and solids. Explain how they are different from gases.

2. Explain the process of evaporation from the standpoint of kinetic energy.

3. Relate vapor pressure data or vapor pressure curves of different substances to their relative rates of evaporation and to their relative boiling points.

4. Explain the forces involved in surface tension of a liquid. Give two common examples.

5. Explain why a meniscus forms on the surface of liquids in a container.

6. Explain what is occurring throughout the heating curve for water.

7. Describe a water molecule with respect to the Lewis structure, bond angle, and polarity.

8. Make sketches showing hydrogen bonding (a) between water molecules, (b) between hydrogen fluoride molecules, and (c) between ammonia molecules.

9. Explain the effect of hydrogen bonding on the physical properties of water.

10. Determine whether a compound will or will not form hydrogen bonds.

11. Identify metal oxides as basic anhydrides and write balanced equations for their reactions with water.

12. Identify nonmetal oxides as acid anhydrides and write balanced equations for their reactions with water.

13. Deduce the formula of the acid anhydride or basic anhydride when given the formula of the corresponding acid or base.

14. Identify, name, and write equations for the complete dehydration of hydrates.

15. Outline the processes necessary to prepare potable water from a contaminated river source.

16. Describe how water may be softened by distillation, chemical precipitation, ion exchange, and demineralization.

17. Complete and balance equations for (a) the reactions of water with Na, K, and Ca; (b) the reaction of steam with Zn, Al, Fe, and C; and (c) the reaction of water with halogens.

Key Terms in Review

The terms listed here have all been defined within the chapter. You should review the definitions of each. Remember to use the glossary and the margin notations within the chapter to help you.

acid anhydride
basic anhydride
boiling point
capillary action
deliquescence
efflorescence
evaporation
freezing or melting point
heat of fusion
heat of vaporization
heating curve
hydrate

hydrogen bond
hygroscopic substances
meniscus
standard or normal boiling point
sublimation
surface tension
vapor pressure
vapor pressure curve
vaporization
volatile
water of crystallization
water of hydration

Exercises

An asterisk indicates a more challenging question or problem.

1. Compare the potential energy of the two states of water shown in Figure 14.6.
2. In what state (solid, liquid, or gas) would H_2S, H_2Se, and H_2Te be at 0°C? (See Table 14.3.)
3. The two thermometers in Figure 14.7 read 100°C. What is the pressure of the atmosphere?
4. Draw a diagram of a water molecule and point out the areas that are the negative and positive ends of the dipole.
5. If the water molecule were linear, with all three atoms in a straight line rather than in the shape of a V, as shown in Figure 14.8, what effect would this have on the physical properties of water?
6. Based on Table 14.4, how do we specify 1, 2, 3, 4, 5, 6, 7, and 8 molecules of water in the formulas of hydrates?
7. Would the distillation setup in Figure 14.11 be satisfactory for separating salt and water? Ethyl alcohol and water? Explain.

8. If the liquid in the flask in Figure 14.11 is ethyl alcohol and the atmospheric pressure is 543 torr, what temperature will show on the thermometer? (Use Figure 14.4.)
9. If water were placed in both containers in Figure 14.1, would both have the same vapor pressure at the same temperature? Explain.
10. In Figure 14.1, in which case, (a) or (b), will the atmosphere above the liquid reach a point of saturation?
11. Suppose that a solution of ethyl ether and ethyl alcohol were placed in the closed bottle in Figure 14.1. Use Figure 14.4 for information on the substances.
    (a) Would both substances be present in the vapor?
    (b) If the answer to part (a) is yes, which would have more molecules in the vapor?
12. In Figure 14.2, if 50% more water had been added in part (b), what equilibrium vapor pressure would have been observed in (c)?

13. At approximately what temperature would each of the substances listed in Table 14.2 boil when the pressure is 30 torr? (See Figure 14.4.)
14. Use the graph in Figure 14.4 to find the following:
    (a) The boiling point of water at 500 torr pressure
    (b) The normal boiling point of ethyl alcohol
    (c) The boiling point of ethyl ether at 0.50 atm
15. Consider Figure 14.5.
    (a) Why is line BC horizontal? What is happening in this interval?
    (b) What phases are present in the interval BC?
    (c) When heating is continued after point C, another horizontal line, DE, is reached at a higher temperature. What does this line represent?
16. List six physical properties of water.
17. What condition is necessary for water to have its maximum density? What is its maximum density?
18. Account for the fact that an ice–water mixture remains at 0°C until all the ice is melted, even though heat is applied to it.
19. Which contains less heat, ice at 0°C or water at 0°C? Explain.
20. Why does ice float in water? Would ice float in ethyl alcohol ($d = 0.789$ g/mL)? Explain.
21. If water molecules were linear instead of bent, would the heat of vaporization be higher or lower? Explain.
22. The heat of vaporization for ethyl ether is 351 J/g (83.9 cal/g) and that for ethyl alcohol is 855 J/g (204.3 cal/g). Which of these compounds has hydrogen bonding? Explain.
23. Would there be more or less H-bonding if water molecules were linear instead of bent? Explain.
24. Which would show hydrogen bonding, ammonia, NH_3, or methane, CH_4? Explain.
25. In which condition are there fewer hydrogen bonds between molecules: water at 40°C or water at 80°C?
26. Which compound, $H_2NCH_2CH_2NH_2$ or $CH_3CH_2CH_2NH_2$, would you expect to have the higher boiling point? Explain your answer. (Both compounds have similar molar masses.)
27. Explain why rubbing alcohol, warmed to body temperature, still feels cold when applied to your skin.

28. The vapor pressure at 20°C is given for the following compounds:

Methyl alcohol	96	torr
Acetic acid	11.7	torr
Benzene	74.7	torr
Bromine	173	torr
Water	17.5	torr
Carbon tetrachloride	91	torr
Mercury	0.0012	torr
Toluene	23	torr

   (a) Arrange these compounds in their order of increasing rate of evaporation.
   (b) Which substance listed would have the highest boiling point, and which would have the lowest?
29. Suggest a method whereby water could be made to boil at 50°C.
30. If a dish of water initially at 20°C is placed in a living room maintained at 20°C, the water temperature will fall below 20°C. Explain.
31. Explain why a higher temperature is obtained in a pressure cooker than in an ordinary cooking pot.
32. What is the relationship between vapor pressure and boiling point?
33. On the basis of the Kinetic-Molecular Theory, explain why vapor pressure increases with temperature.
34. Why does water have such a relatively high boiling point?
35. The boiling point of ammonia, NH_3, is -33.4°C and that of sulfur dioxide, SO_2, is -10.0°C. Which has the higher vapor pressure at -40°C?
36. Explain what is occurring physically when a substance is boiling.
37. Explain why HF (bp = 19.4°C) has a higher boiling point than HCl (bp = -85°C), whereas F_2 (bp = -188°C) has a lower boiling point than Cl_2 (bp = -34°C).
38. Can ice be colder than 0°C? Explain.
39. Why does a boiling liquid maintain a constant temperature when heat is continuously being added?
40. At what specific temperature will copper have a vapor pressure of 760 torr?
41. Why does a lake freeze from the top down?
42. What water temperature would you theoretically expect to find at the bottom of a very deep lake? Explain.

43. Write equations to show how the following metals react with water: aluminum, calcium, iron, sodium, zinc. State the conditions for each reaction.

44. Is the formation of hydrogen and oxygen from water an exothermic or an endothermic reaction? How do you know?

45. (a) What is an anhydride?
    (b) What type of compound will be an acid anhydride?
    (c) What type of compound will be a basic anhydride?

46. (a) Write the formulas for the anhydrides of the following acids:
    H_2SO_3, H_2SO_4, HNO_3, $HClO_4$, H_2CO_3, H_3PO_4
    (b) Write the formulas for the anhydrides of the following bases:
    NaOH, KOH, $Ba(OH)_2$, $Ca(OH)_2$, $Mg(OH)_2$

47. Complete and balance the following equations:
    (a) $Ba(OH)_2 \xrightarrow{\Delta}$
    (b) $CH_3OH + O_2 \longrightarrow$
    Methyl alcohol
    (c) $Rb + H_2O \longrightarrow$
    (d) $SnCl_2 \cdot 2 H_2O \xrightarrow{\Delta}$
    (e) $HNO_3 + NaOH \longrightarrow$
    (f) $Li_2O + H_2O \longrightarrow$
    (g) $KOH \xrightarrow{\Delta}$
    (h) $Ba + H_2O \longrightarrow$
    (i) $Cl_2 + H_2O \longrightarrow$
    (j) $SO_3 + H_2O \longrightarrow$
    (k) $H_2SO_3 + KOH \longrightarrow$
    (l) $CO_2 + H_2O \longrightarrow$

48. Name each of the following hydrates:
    (a) $BaBr_2 \cdot 2 H_2O$ (d) $MgNH_4PO_4 \cdot 6 H_2O$
    (b) $AlCl_3 \cdot 6 H_2O$ (e) $FeSO_4 \cdot 7 H_2O$
    (c) $FePO_4 \cdot 4 H_2O$ (f) $SnCl_4 \cdot 5 H_2O$

49. Explain how anhydrous copper(II) sulfate ($CuSO_4$) can act as an indicator for moisture.

50. Write formula for magnesium sulfate heptahydrate and disodium hydrogen phosphate dodecahydrate.

51. Distinguish between deionized water and:
    (a) Hard water   (c) Distilled water
    (b) Soft water

52. How can soap function to make soft water from hard water? What objections are there to using soap for this purpose?

53. What substance is commonly used to destroy bacteria in water?

54. What chemical, other than chlorine or chlorine compounds, can be used to disinfect water for domestic use?

55. Some organic pollutants in water can be oxidized by dissolved molecular oxygen. What harmful effect can result from this depletion of oxygen in the water?

56. Why should you not drink liquids that are stored in ceramic containers, especially unglazed ones?

57. Write the chemical equation showing how magnesium ions are removed by a zeolite water softener.

58. Write an equation to show how hard water containing calcium chloride ($CaCl_2$) is softened by using sodium carbonate (Na_2CO_3).

59. Which of the following statements are correct? Rewrite each incorrect statement to make it correct.
    (a) The process of a substance changing directly from a solid to a gas is called sublimation.
    (b) When water is decomposed, the volume ratio of H_2 to O_2 is 2:1, but the mass ratio of H_2 to O_2 is 1:8.
    (c) Hydrogen sulfide is a larger molecule than water.
    (d) The changing of ice into water is an exothermic process.
    (e) Water and hydrogen fluoride are both non-polar molecules.
    (f) Hydrogen bonding is stronger in H_2O than in H_2S because oxygen is more electronegative than sulfur.
    (g) $H_2O_2 \longrightarrow 2 H_2O + O_2$ represents a balanced equation for the decomposition of hydrogen peroxide.
    (h) Steam at 100°C can cause more severe burns than liquid water at 100°C.
    (i) The density of water is independent of temperature.
    (j) Liquid A boils at a lower temperature than liquid B. This fact indicates that liquid A has a lower vapor pressure than liquid B at any particular temperature.
    (k) Water boils at a higher temperature in the mountains than at sea level.
    (l) No matter how much heat you put under an open pot of pure water on a stove, you cannot heat the water above its boiling point.
    (m) The vapor pressure of a liquid at its boiling

point is equal to the prevailing atmospheric pressure.

(n) The normal boiling temperature of water is 273°C.

(o) The pressure exerted by a vapor in equilibrium with its liquid is known as the vapor pressure of the liquid.

(p) Sodium, potassium, and calcium each react with water to form hydrogen gas and a metal hydroxide.

(q) Calcium oxide reacts with water to form calcium hydroxide and hydrogen gas.

(r) Carbon dioxide is the hydride of carbonic acid.

(s) Water in a hydrate is known as water of hydration or water of crystallization.

(t) A substance that spontaneously loses its water of hydration when exposed to the air is said to be efflorescent.

(u) A substance that absorbs water from the air until it forms a solution is deliquescent.

(v) Distillation is effective for softening water because the minerals boil away, leaving soft water behind.

(w) The original source of mercury-contaminated fish is industrial pollution.

(x) Disposal of toxic industrial wastes in toxic waste dumps has been found to be a very satisfactory long-term solution to the problem of what to do with these wastes.

(y) The amount of heat needed to change 1 mole of ice at 0°C to a liquid at 0°C is 6.02 kJ (1.44 kcal).

(z) $BaCl_2 \cdot 2 H_2O$ has a higher percentage of water than does $CaCl_2 \cdot 2 H_2O$.

60. How many moles of compound are in 100. g of each of these hydrates?
   (a) $CoCl_2 \cdot 6 H_2O$ (b) $FeI_2 \cdot 4 H_2O$

61. How many moles of water can be obtained from 100. g of each of these hydrates?
   (a) $CoCl_2 \cdot 6 H_2O$ (b) $FeI_2 \cdot 4 H_2O$

62. When a person purchases epsom salts, $MgSO_4 \cdot 7 H_2O$, what percent of the compound is water?

63. Calculate the mass percent of water in the hydrate $Al_2(SO_4)_3 \cdot 18 H_2O$.

64. Sugar of lead, a hydrate of lead acetate, $Pb(C_2H_3O_2)_2$, contains 14.2% H_2O. What is the formula for the hydrate?

65. A 25.0 g sample of a hydrate of $FePO_4$ was heated until no more water was driven off. The mass of anhydrous sample is 16.9 g. What is the formula of the hydrate?

66. How many joules are needed to change 120. g of water at 20.°C to steam at 100.°C?

67. How many joules of energy must be removed from 126 g of water at 24°C to form ice at 0°C?

68. How many calories are required to change 225 g of ice at 0°C to steam at 100.°C?

69. The molar heat of vaporization is the number of joules required to change 1 mole of a substance from liquid to vapor at its boiling point. What is the molar heat of vaporization of water?

*70. Suppose 100. g of ice at 0°C are added to 300. g of water at 25°C. Is this sufficient ice to lower the temperature of the system to 0°C and still have ice remaining? Show evidence for your answer.

*71. If 75 g of $H_2O(s)$ at 0.0°C were added to 1.5 L of water at 75°C, what would be the final temperature of the mixture?

*72. The specific heat of zinc is 0.096 cal/g°C. Determine the energy required to raise the temperature of 250. g of zinc from room temperature to 150°C.

*73. If 9560 J of energy were absorbed by 500. g of ice at 0.0°C, what would be the final temperature?

*74. Suppose 150. g of ice at 0.0°C is added to 0.120 L of water at 45°C. If the mixture is stirred and allowed to cool to 0.0°C, how many grams of ice remain?

*75. Suppose 35.0 g of steam at 100.°C are added to 300. g of water at 25°C. Is there sufficient steam to heat all the water to 100.°C and still have steam remaining? Show evidence for your answer.

76. How many joules of energy would be liberated by condensing 50.0 mol of steam at 100.0°C and allowing the liquid to cool to 30.0°C?

77. How many kilojoules of energy are needed to convert 100. g of ice at −10.0°C to water at 20.0°C. (The specific heat of ice at −10.0°C is 2.01 J/g°C.)

78. What mass of water must be decomposed to produce 25.0 L of oxygen at STP?

79. Compare the volume occupied by 1.00 mol of liquid water at 0°C and 1.00 mol of water vapor at STP.

80. How many grams of water will react with each of the following?

(a) 1.00 mol K   (d) 1.00 mole SO_3

(b) 1.00 mol Ca   (e) 1.00 g MgO

(c) 1.00 g Na   (f) 1.00 g N_2O_5

*81. Suppose 1.00 mol of water evaporates in 1.00 day. How many water molecules, on the average, leave the liquid each second?

*82. A quantity of sulfuric acid is added to 100. mL of water. The final volume of the solution is 122 mL and it has a density of 1.26 g/mL. What mass of acid was added? Assume the density of the water is 1.00 g/mL.

83. A mixture of 80.0 mL of hydrogen and 60.0 mL of oxygen is ignited by a spark to form water.

(a) Does any gas remain unreacted? Which one, H_2 or O_2?

(b) What volume of which gas (if any) remains unreacted? (Assume the same conditions before and after the reaction.)

CHAPTER FIFTEEN

Solutions

Most of the substances we encounter in our daily lives are mixtures. Often they are homogeneous mixtures, which are called solutions. When you think of a solution, juices, blood plasma, shampoo, soft drinks, or wine may come to mind. These solutions all have water as a main component. However, many common items, such as air gasoline, and steel are also solutions. These do not contain water. What are the necessary components in a solution? Why do some substances mix while others do not? What effect does a dissolved substance have on the properties of the solution? The chemistry of solutions plays a significant role in our everyday lives.

Chapter Preview

15.1 Components of a Solution

solution

solute

solvent

The term **solution** is used in chemistry to describe a system in which one or more substances are homogeneously mixed or dissolved in another substance. A simple solution has two components, a solute and a solvent. The **solute** is the component that is dissolved or the least abundant component in the solution. The **solvent** is the dissolving agent or the most abundant component in the solution. For example, when salt is dissolved in water to form a solution, salt is the solute and water is the solvent. Complex solutions containing more than one solute and/or more than one solvent are common.

15.2 Types of Solutions

From the three states of matter—solid, liquid, and gas—it is possible to have nine different types of solutions: solid dissolved in solid, solid dissolved in liquid, solid dissolved in gas, liquid dissolved in liquid, and so on. Of these, the most common solutions are solid dissolved in liquid, liquid dissolved in liquid, gas dissolved in liquid, and gas dissolved in gas. Some common types of solutions are listed in Table 15.1.

15.3 General Properties of Solutions

A true solution is one in which the particles of dissolved solute are molecular or ionic in size, generally in the range of 0.1 to 1 nm (10^{-8} to 10^{-7} cm). The

TABLE 15.1

Common Types of Solutions

Phase of solution	Solute	Solvent	Example
Gas	Gas	Gas	Air
Liquid	Gas	Liquid	Soft drinks
Liquid	Liquid	Liquid	Antifreeze
Liquid	Solid	Liquid	Salt water
Solid	Gas	Solid	H_2 in Pt
Solid	Solid	Solid	Brass

properties of a true solution are as follows:

1. It is a homogeneous mixture of two or more components, solute and solvent.
2. It has a variable composition; that is, the ratio of solute to solvent may be varied.
3. The dissolved solute is molecular or ionic in size.
4. It may be either colored or colorless but is usually transparent.
5. The solute remains uniformly distributed throughout the solution and will not settle out with time.
6. The solute generally can be separated from the solvent by purely physical means (for example, by evaporation).

These properties are illustrated by water solutions of sugar and of potassium permanganate. Suppose that we prepare two sugar solutions, the first containing 10 g of sugar added to 100 mL of water and the second containing 20 g of sugar added to 100 mL of water. Each solution is stirred until all the solute dissolves, demonstrating that we can vary the composition of a solution. Every portion of the solution has the same sweet taste because the sugar molecules are uniformly distributed throughout. If confined so that no solvent is lost, the solution will taste and appear the same a week or a month later. The properties of the solution are unaltered after the solution is passed through filter paper. But by carefully evaporating the water, we can recover the sugar from the solution.

To observe the dissolving of potassium permanganate ($KMnO_4$) we affix a few crystals of $KMnO_4$ to paraffin wax or rubber cement at the end of a glass rod and submerge the entire rod, with the wax-permanganate end up, in a cylinder of water. Almost at once the beautiful purple color of dissolved permanganate ions (MnO_4^-) appears at the top of the rod and streams to the bottom of the cylinder as the crystals dissolve. The purple color at first is mostly at the bottom of the cylinder because potassium permanganate is denser than water. But after a while the purple color disperses until it is evenly distributed throughout the solution. This dispersal demonstrates that molecules and ions move about freely and spontaneously (diffuse) in a liquid or solution.

Once formed, a solution is permanent; the solute particles do not settle out. Solution permanency is explained in terms of the Kinetic-Molecular Theory (see Section 13.2). According to the KMT both the solute and solvent particles (molecules and/or ions) are in constant random thermal motion. This motion is energetic enough to prevent the solute particles from settling out under the influence of gravity. This same ceaseless, random thermal motion is also responsible for diffusion in liquids as well as in gases.

15.4   Solubility

solubility

The term **solubility** describes the amount of one substance (solute) that will dissolve in a specified amount of another substance (solvent) under stated conditions. For example, 36.0 g of sodium chloride (NaCl) will dissolve in 100 g of water at 20°C. We say, then, that the solubility of NaCl in water is 36.0 g per 100 g of water at 20°C.

Solubility is often used in a relative way. We say that a substance is very soluble, moderately soluble, slightly soluble, or insoluble. Although these terms do not accurately indicate how much solute will dissolve, they are frequently used to describe the solubility of a substance qualitatively.

Two other terms often used to describe solubility are miscible and immiscible. Liquids that are capable of mixing and forming a solution are

miscible

miscible; those that do not form solutions or are generally insoluble in each other are **immiscible**. Methyl alcohol and water are miscible in each other in all

immiscible

proportions. Carbon tetrachloride and water are immiscible, forming two separate layers when they are mixed. Miscible and immiscible systems are illustrated in Figure 15.1.

The general rules for the solubility of common salts and hydroxides are given in Table 15.2. The solubilities of over 200 compounds are given in the Solubility Table in Appendix IV. Solubility data for thousands of compounds can be found by consulting standard reference sources.*

The quantitative expression of the amount of dissolved solute in a particular

concentration of a
solution

quantity of solvent is known as the **concentration of a solution**. Several methods of expressing concentration will be described in Section 15.8.

---

* Two commonly used handbooks are *Lange's Handbook of Chemistry*, 13th ed. (New York: McGraw-Hill, 1985), and *Handbook of Chemistry and Physics*, 70th ed. (Cleveland: Chemical Rubber Co., 1989).

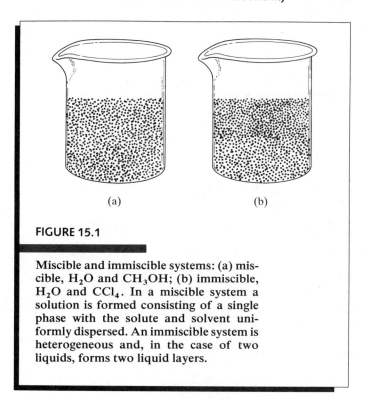

(a)   (b)

FIGURE 15.1

Miscible and immiscible systems: (a) miscible, H_2O and CH_3OH; (b) immiscible, H_2O and CCl_4. In a miscible system a solution is formed consisting of a single phase with the solute and solvent uniformly dispersed. An immiscible system is heterogeneous and, in the case of two liquids, forms two liquid layers.

15.5   Factors Related to Solubility

Predicting solubilities is complex and difficult. Many variables, such as size of ions, charge on ions, interaction between ions, interaction between solute and solvent, and temperature, bear upon the problem. Because of the factors involved, the general rules of solubility given in Table 15.2 have many exceptions. However, the rules are very useful, because they do apply to many of the more common compounds that we encounter in the study of chemistry. Keep in mind that these are rules, not laws, and are therefore subject to exceptions. Fortunately the solubility of a solute is relatively easy to determine experimentally. Factors related to solubility are discussed in the following paragraphs.

The Nature of the Solute and Solvent   The old adage "like dissolves like" has merit, in a general way. Polar or ionic substances tend to be more miscible, or

TABLE 15.2

General Solubility Rules for Common Salts and Hydroxides[a]

Class	Solubility in cold water[b]
Nitrates	Most nitrates are soluble.
Acetates	Most acetates are soluble.
Chlorides Bromides Iodides	Most chlorides, bromides, and iodides are soluble, except those of Ag, Hg(I), and Pb(II); $PbCl_2$ and $PbBr_2$ are slightly soluble in hot water.
Sulfates	Most sulfates are soluble except those of Ba, Sr and Pb; Ca and Ag sulfates are slightly soluble.
Carbonates Phosphates	Most carbonates and phosphates are insoluble except those of Na, K, and NH_4^+. Many bicarbonates and acid phosphates are soluble.
Hydroxides	Most hydroxides are insoluble except those of the alkali metals and NH_4OH; $Ba(OH)_2$ and $Ca(OH)_2$ are slightly soluble.
Sodium salts Potassium salts Ammonium salts	Most common salts of these ions are soluble.
Sulfides	Most sulfides are insoluble except those of the alkali metals, ammonium, and the alkaline earth metals (Ca, Mg, Ba).

[a] When we say a substance is soluble, we mean that the substance is reasonably soluble. All substances have some solubility in water, although the amount of solubility may be very small; the solubility of silver iodide, for example, is about 1×10^{-8} mol AgI/liter H_2O.
[b] These rules have exceptions.

soluble, with other polar substances. Nonpolar substances tend to be miscible with other nonpolar substances and less miscible with polar substances. Thus, mineral acids, bases, and salts, which are polar, tend to be much more soluble in water, which is polar, than in solvents such as ether, carbon tetrachloride, or benzene, which are essentially nonpolar. Sodium chloride, an ionic substance, is soluble in water, slightly soluble in ethyl alcohol (less polar than water), and insoluble in ether and benzene. Pentane (C_5H_{12}), a nonpolar substance, is only slightly soluble in water but is very soluble in benzene and ether.

At the molecular level the formation of a solution from two nonpolar substances, such as carbon tetrachloride and benzene, can be visualized as a process of simple mixing. The nonpolar molecules, having little tendency to either attract or repel one another, easily intermingle to form a homogeneous mixture.

Solution formation between polar substances is much more complex. For example, the process by which sodium chloride dissolves in water is illustrated in Figure 15.2. Water molecules are very polar and are attracted to other polar molecules or ions. When salt crystals (NaCl) are put into water, polar water

= Water   = Na^+ = Cl^-

FIGURE 15.2

Dissolution of sodium chloride in water. Polar water molecules are attracted to Na^+ and Cl^- ions in the salt crystal, weakening the attraction between the ions. As the attraction between the ions weakens, the ions move apart and become surrounded by water dipoles. The hydrated ions slowly diffuse away from the crystal to become dissolved in solution.

molecules become attracted to the sodium and chloride ions on the crystal surfaces and weaken the attraction between Na^+ and Cl^- ions. The positive end of the water dipole is attracted to the Cl^- ions, and the negative end of the water dipole to the Na^+ ions. The weakened attraction permits the ions to move apart, making room for more water dipoles. Thus, the surface ions are surrounded by water molecules, becoming hydrated ions, $Na^+(aq)$ and $Cl^-(aq)$, and slowly diffuse away from the crystals and dissolve in solution.

$$NaCl(crystal) \longrightarrow Na^+(aq) + Cl^-(aq)$$

Examination of the data in Table 15.3 reveals some of the complex questions relating to solubility. For example: Why are lithium halides, except for lithium fluoride (LiF), more soluble than sodium and potassium halides? Why are the solubilites of LiF and sodium fluoride (NaF) so low in comparison with those of the other salts? Why does not the solubility of LiF, NaF, and NaCl increase proportionately with temperature, as do the solubilities of the other salts? Sodium chloride is appreciably soluble in water but is insoluble in concentrated hydrochloric acid (HCl) solution. On the other hand, LiF and NaF are not very soluble in water but are quite soluble in hydrofluoric acid (HF) solution—why? These questions will not be answered directly here, but it is hoped that your curiosity will be aroused to the point that you will do some reading and research on the properties of solutions.

The Effect of Temperature on Solubility   Temperature has major effects on the solubility of most substances, and most solutes have a limited solubility in a specific solvent at a fixed temperature. For most solids dissolved in a liquid, an increase in temperature results in increased solubility (see Figure 15.3). However, no completely valid general rule governs the solubility of solids in liquids with change in temperature. Some solids increase in solubility only slightly with increasing temperature (see NaCl in Figure 15.3); other solids decrease in solubility with increasing temperature (see Li_2SO_4 in Figure 15.3).

On the other hand, the solubility of a gas in a liquid always decreases with increasing temperature (see HCl and SO_2 in Figure 15.3). The tiny bubbles that are formed when water is heated are due to the decreased solubility of air at higher temperatures. The decreased solubility of gases at higher temperatures is explained in terms of the KMT by assuming that, in order to dissolve, the gas molecules must form "bonds" of some sort with the molecules of the liquid. An increase in temperature decreases the solubility of the gas because it increases the kinetic energy (speed) of the gas molecules and thereby decreases their ability to form "bonds" with the liquid molecules.

The Effect of Pressure on Solubility   Small changes in pressure have little effect on the solubility of solids in liquids or liquids in liquids but have a marked effect

TABLE 15.3

Solubility of Alkali Metal Halides in Water

Salt	Solubility (g salt/100 g H_2O)	
	0°C	**100°C**
LiF	0.12	0.14 (at 35°C)
LiCl	67	127.5
LiBr	143	266
LiI	151	481
NaF	4	5
NaCl	35.7	39.8
NaBr	79.5	121
NaI	158.7	302
KF	92.3 (at 18°C)	Very soluble
KCl	27.6	57.6
KBr	53.5	104
KI	127.5	208

on the solubility of gases in liquids. The solubility of a gas in a liquid is directly proportional to the pressure of that gas above the solution. Thus, the amount of a gas that is dissolved in solution will double if the pressure of that gas over the solution is doubled. For example, carbonated beverages contain dissolved carbon dioxide at pressures greater than atmospheric pressure. When a bottle of carbonated soda is opened, the pressure is immediately reduced to the atmospheric pressure, and the excess dissolved carbon dioxide bubbles out of the solution.

15.6 Rate of Dissolving Solids

The rate at which a solid dissolves is governed by (1) the size of the solute particles, (2) the temperature, (3) the concentration of the solution, and (4) agitation or stirring.

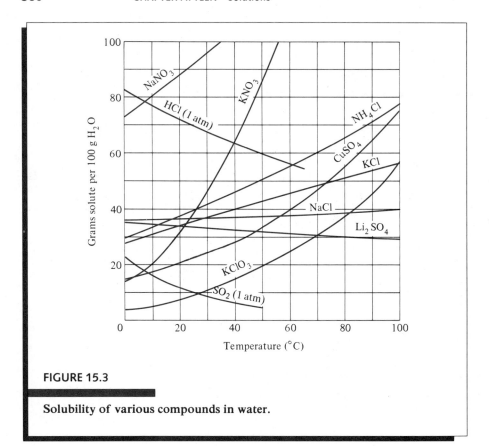

FIGURE 15.3

Solubility of various compounds in water.

Particle Size　A solid can dissolve only at the surface that is in contact with the solvent. Because the surface to volume ratio increases as size decreases, smaller crystals dissolve faster than large ones. For example, if a salt crystal 1 cm on a side (6 cm² surface area) is divided into 1000 cubes, each 0.1 cm on a side, the total surface of the smaller cubes is 60 cm²—a tenfold increase in surface area (see Figure 15.4).

Temperature　In most cases the rate of dissolving of a solid increases with temperature. This increase is due to kinetic effects: The solvent molecules move more rapidly at higher temperatures and strike the solid surfaces more often and harder, causing the rate of dissolving to increase.

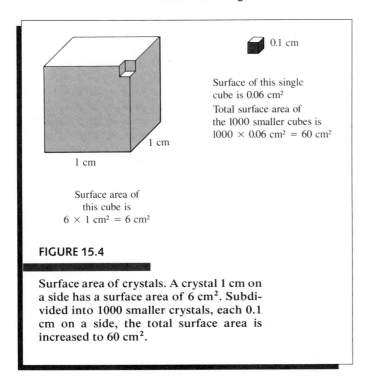

0.1 cm

Surface of this single
cube is 0.06 cm²
Total surface area of
the 1000 smaller cubes is
1000 × 0.06 cm² = 60 cm²

1 cm

1 cm

Surface area of
this cube is
6 × 1 cm² = 6 cm²

FIGURE 15.4

**Surface area of crystals. A crystal 1 cm on
a side has a surface area of 6 cm². Subdivided into 1000 smaller crystals, each 0.1
cm on a side, the total surface area is
increased to 60 cm².**

Concentration of the Solution   When the solute and solvent are first mixed, the
rate of dissolving is at its maximum. As the concentration of the solution
increases and the solution becomes more nearly saturated with the solute, the rate
of dissolving decreases greatly. The rate of dissolving is pictured graphically in
Figure 15.5. Note that about 17 g dissolve in the first 5 minute interval, but only
about 1 g dissolves in the fourth 5 minute interval. Although different solutes
show different rates, the rate of dissolving always becomes very slow as the
concentration approaches the saturation point.

Agitation or Stirring   The effect of agitation or stirring is kinetic. When a solid is
first put into water, the only solvent with which it comes in contact is in the
immediate vicinity. As the solid dissolves, the amount of dissolved solute around
the solid becomes more and more concentrated, and the rate of dissolving slows
down. If the mixture is not stirred, the dissolved solute diffuses very slowly
through the solution; weeks may pass before the solid is entirely dissolved.
Stirring distributes the dissolved solute rapidly through the solution, and more
solvent is brought into contact with the solid, causing it to dissolve more rapidly.

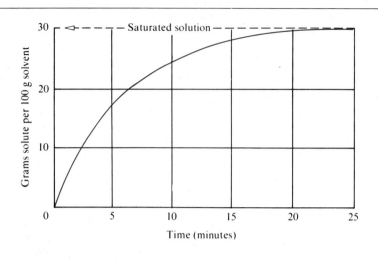

FIGURE 15.5

Rate of dissolution of a solid solute in a solvent. The rate is maximum at the beginning and decreases as the concentration approaches saturation.

15.7   Solutions: A Reaction Medium

Many solids must be put into solution in order to undergo appreciable chemical reaction. We can write the equation for the double-displacement reaction between sodium chloride and silver nitrate:

$$NaCl + AgNO_3 \longrightarrow AgCl + NaNO_3$$

But suppose we mix solid NaCl and solid $AgNO_3$ and look for a chemical change. If any reaction occurs, it is slow and virtually undetectable. In fact, the crystalline structures of NaCl and $AgNO_3$ are so different that we could separate them by tediously picking out each kind of crystal from the mixture. But if we dissolve the sodium chloride and silver nitrate separately in water and mix the two solutions, we observe the immediate formation of a white, curdy precipitate of silver chloride.

Molecules or ions must come into intimate contact or collide with one another in order to react. In the foregoing example, the two solids did not react because the ions were securely locked within their crystal structures. But when the sodium chloride and silver nitrate are dissolved, their crystal lattices are broken down and the ions become mobile. When the two solutions are mixed, the mobile Ag^+ and Cl^- ions come into contact and react to form insoluble AgCl, which precipitates out of solution. The soluble Na^+ and NO_3^- ions remain mobile in solution but form the crystalline salt $NaNO_3$ when the water is evaporated:

$$NaCl(aq) + AgNO_3(aq) \longrightarrow AgCl\downarrow + NaNO_3(aq)$$

$$(Na^+ + Cl^-) + (Ag^+ + NO_3^-) \xrightarrow{H_2O} AgCl\downarrow + Na^+ + NO_3^-$$

| Sodium chloride solution | Silver nitrate solution | Silver chloride | Sodium nitrate in solution |

The mixture of the two solutions provides a medium or space in which the Ag^+ and Cl^- ions can react. (See Chapter 16 for further discussion of ionic reactions.)

Solutions also function as diluents (diluting agents) in reactions in which the undiluted reactants would combine with each other too violently. Moreover, a solution of known concentration provides a convenient method for delivering specific amounts of reactants.

15.8   Concentration of Solutions

The concentration of a solution expresses the amount of solute dissolved in a given quantity of solvent or solution. Because reactions are often conducted in solution, it is important to understand the methods of expressing concentration and to know how to prepare solutions of particular concentrations. The concentration of a solution may be expressed qualitatively or quantitatively. Let's begin with a look at the qualitative methods of expressing concentration.

Dilute and Concentrated Solutions

When we say that a solution is *dilute* or *concentrated*, we are expressing, in a relative way, the amount of solute present. One gram of salt and 2 g of salt in solution are both dilute solutions when compared with the same volume of a solution containing 20 g of salt. Ordinary concentrated hydrochloric acid (HCl) contains 12 moles of HCl per liter of solution. In some laboratories the dilute acid is made by mixing equal volumes of water and the concentrated acid. In other laboratories the concentrated acid is diluted with two or three volumes of water, depending on its use. The term **dilute solution**, then, describes a solution that contains a relatively small amount of dissolved solute. Conversely, a **concentrated solution** contains a relatively large amount of dissolved solute.

dilute solution

concentrated solution

Saturated, Unsaturated, and Supersaturated Solutions

At a specific temperature there is a limit to the amount of solute that will dissolve in a given amount of solvent. When this limit is reached, the resulting solution is said to be *saturated*. For example, when we put 40.0 g of KCl into 100 g of H_2O at 20°C, we find that 34.0 g of KCl dissolve and 6.0 g of KCl remain undissolved. The solution formed is a saturated solution of KCl.

Two processes are occurring simultaneously in a saturated solution. The solid is dissolving into solution, and at the same time the dissolved solute is crystallizing out of solution. This may be expressed as

$$\text{solute (undissolved)} \rightleftharpoons \text{solute (dissolved)}$$

saturated solution

When these two opposing processes are occurring at the same rate, the amount of solute in solution is constant, and a condition of equilibrium is established between dissolved and undissolved solute. A **saturated solution** contains dissolved solute in equilibrium with undissolved solute.

It is important to state the temperature of a saturated solution, because a solution that is saturated at one temperature may not be saturated at another. If the temperature of a saturated solution is changed, the equilibrium is disturbed, and the amount of dissolved solute will change to reestablish equilibrium.

A saturated solution may be either dilute or concentrated, depending on the solubility of the solute. A saturated solution can be conveniently prepared by dissolving a little more than the saturated amount of solute at a temperature somewhat higher than room temperature. Then the amount of solute in solution will be in excess of its solubility at room temperature; and, when the solution cools, the excess solute will crystallize, leaving the solution saturated. (In this case, the solute must be more soluble at higher temperatures and must not form a supersaturated solution.) Examples expressing the solubility of saturated solutions at two different temperatures are given in Table 15.4.

unsaturated solution

An **unsaturated solution** contains less solute per unit of volume than does its corresponding saturated solution. In other words, additional solute can be dissolved in an unsaturated solution without altering any other conditions. Consider a solution made by adding 40 g of KCl to 100 g of H_2O at 20°C (see Table 15.4). The solution formed will be saturated and will contain about 6 g of undissolved salt, because the maximum amount of KCl that can dissolve in 100 g of H_2O at 20°C is 34 g. If the solution is now heated and maintained at 50°C, all the salt will dissolve and, in fact, even more can be dissolved. Thus the solution at 50°C is unsaturated.

supersaturated solution

In some circumstances, solutions can be prepared that contain more solute than that needed for a saturated solution at a particular temperature. Such solutions are said to be **supersaturated**. However, we must qualify this definition by noting that a supersaturated solution is unstable. Disturbances such as jarring, stirring, scratching the walls of the container, or dropping in a "seed" crystal cause the supersaturation to break. When a supersaturated solution is disturbed, the excess solute crystallizes out rapidly, returning the solution to a saturated state.

TABLE 15.4

Saturated Solutions at 20°C and 50°C

Solute	Solubility (g solute/100 g H_2O)	
	20°C	50°C
NaCl	36.0	37.0
KCl	34.0	42.6
$NaNO_3$	88.0	114.0
$KClO_3$	7.4	19.3
$AgNO_3$	222.0	455.0
$C_{12}H_{22}O_{11}$	203.9	260.4

Supersaturated solutions are not easy to prepare but may be made from certain substances by dissolving, in warm solvent, an amount of solute greater than that needed for a saturated solution at room temperature. The warm solution is then allowed to cool very slowly. With the proper solute and careful work, a supersaturated solution will result. Two substances commonly used to demonstrate this property are sodium thiosulfate pentahydrate, $Na_2S_2O_3 \cdot 5\,H_2O$, and sodium sulfate, Na_2SO_4 (from a saturated solution at 30°C).

EXAMPLE 15.1   Will a solution made by adding 2.5 g of $CuSO_4$ to 10 g of H_2O be saturated or unsaturated at 20°C?

To answer this question we first need to know the solubility of $CuSO_4$ at 20°C. From Figure 15.3 we see that the solubility of $CuSO_4$ at 20°C is about 21 g per 100 g of H_2O. This amount is equivalent to 2.1 g of $CuSO_4$ per 10 g of H_2O.

Since 2.5 g per 10 g of H_2O is greater than 2.1 g per 10 g of H_2O, the solution will be saturated and 0.4 g of $CuSO_4$ will be undissolved.

PRACTICE   Will a solution made by adding 9.0 g NH_4Cl to 20 g of H_2O be saturated or unsaturated at 50°C?

Answer: unsaturated

Each of these methods for expressing concentration is useful on a qualitative basis. But suppose we wish to compare two solutions with concentrations that are nearly the same. We would need to introduce quantities into our concentration system. A variety of concentration units are quantitative. Let's examine several of them.

Mass Percent Solution

This method expresses concentration as the percent of solute in a given mass of solution. It says that for a given mass of solution a certain percent of that mass is solute. Suppose that we take a bottle from the reagent shelf that reads "sodium hydroxide, NaOH, 10%." This statement means that for every 100 g of this solution, 10 g will be NaOH and 90 g will be water. (Note that this amount of solution is 100 g and not 100 mL.) We could also make this same concentration of solution by dissolving 2.0 g of NaOH in 18 g of water. Mass percent concentrations are most generally used for solids dissolved in liquids.

$$\text{mass percent} = \frac{\text{g solute}}{\text{g solute} + \text{g solvent}} \times 100 = \frac{\text{g solute}}{\text{g solution}} \times 100$$

As instrumentation advances in chemistry, our ability to measure the concentration of dilute solutions is increasing as well. Instead of mass percent, chemists now commonly use **parts per million (ppm).**

parts per million (ppm)

$$\text{parts per million} = \frac{\text{g solute}}{\text{g solute} + \text{g solvent}} \times 1,000,000$$

Currently, air and water contaminants, drugs in the human body, and pesticide residues are some substances measured in parts per million.

EXAMPLE 15.2   What is the mass percent of sodium hydroxide in a solution that is made by dissolving 8.00 g of NaOH in 50.0 g of H_2O?

grams of solute (NaOH) = 8.00 g

grams of solvent (H_2O) = 50.0 g

$$\frac{8.00 \text{ g NaOH}}{8.00 \text{ g NaOH} + 50.0 \text{ g } H_2O} \times 100 = 13.8\% \text{ NaOH solution}$$

EXAMPLE 15.3   What masses of potassium chloride (KCl) and water are needed to make 250. g of 5.00% solution?

The percent expresses the mass of the solute.

250 g = total mass of solution

5.00% of 250 g = $0.0500 \times 250.$ g = 12.5 g KCl (solute)

250. g − 12.5 g = 237.5 g H_2O

Dissolving 12.5 g of KCl in 237.5 g of H_2O gives a 5.00% KCl solution.

EXAMPLE 15.4 A 34.0% sulfuric acid solution has a density of 1.25 g/mL. How many grams of H_2SO_4 are contained in 1.00 L of this solution?

Since H_2SO_4 is the solute, we first solve the mass percent equation for grams of solute:

$$\text{mass percent} = \frac{\text{g solute}}{\text{g solution}} \times 100$$

$$\text{g solute} = \frac{\text{mass percent} \times \text{g solution}}{100}$$

The mass percent is given in the problem. We need to determine the grams of solution. The mass of the solution can be calculated from the density data.

Convert density (g/mL) to grams:

$$1.00 \text{ L} = 1000 \text{ mL}$$

$$\frac{1.25 \text{ g}}{\text{mL}} \times 1000 \text{ mL} = 1250 \text{ g} \quad (\text{g of solution})$$

Now we have all the figures to calculate the grams of solute.

$$\text{g solute} = \frac{34.0\% \times 1250 \text{ g}}{100\%} = 425 \text{ g } H_2SO_4$$

1.00 L of 34.0% H_2SO_4 solution contains 425 g of H_2SO_4.

PRACTICE What is the mass percent of Na_2SO_4 in a solution that is made by dissolving 25.0 g of Na_2SO_4 in 225.0 g of H_2O?

Answer: 10.0% Na_2SO_4 solution

The student should note that the concentration expressed as mass percent is independent of the formula of the solute.

Mass/Volume Percent (m/v)

This method expresses concentration as grams of solute per 100 mL of solution. With this system, a 10.0% (m/v) glucose solution is made by dissolving 10.0 g of glucose in water, diluting to 100 mL, and mixing. The 10.0% (m/v) solution could also be made by diluting 20.0 g to 200 mL, 50.0 g to 500 mL, and so on. Of course, any other appropriate dilution ratio may be used.

$$\text{mass/volume percent} = \frac{\text{g solute}}{\text{mL solution}} \times 100$$

Volume Percent

Solutions that are formulated from two liquids are often expressed as *volume percent* with respect to the solute. The volume percent is the volume of a liquid in 100 mL of solution. The label on a bottle of ordinary rubbing alcohol reads "isopropyl alcohol, 70% by volume." Such a solution could be made by mixing 70 mL of alcohol with water to make a total volume of 100 mL. We cannot use 30 mL of water because the two volumes are not necessarily additive.

$$\text{volume percent} = \frac{\text{volume of liquid in question}}{\text{total volume of solution}} \times 100$$

Volume percent is used to express the concentration of alcohol in beverages. Wines generally contain 12% alcohol by volume. This translates into 12 mL of alcohol in each 100 mL of wine. The beverage industry also uses the concentration unit of *proof* (twice the volume percent). Pure alcohol is 100%, therefore 200 proof. Scotch is 86 proof or 43% alcohol.

Molarity

Mass percent solutions do not equate or express the molar masses of the solute in solution. For example, 1000 g of 10% NaOH solution contain 100 g of NaOH; 1000 g 10% KOH solution contain 100 g of KOH. In terms of moles of NaOH and KOH, these solutions contain

$$\text{moles NaOH} = 100 \text{ g NaOH} \times \frac{1 \text{ mol NaOH}}{40.0 \text{ g NaOH}} = 2.50 \text{ mol NaOH}$$

$$\text{moles KOH} = 100 \text{ g KOH} \times \frac{1 \text{ mol KOH}}{56.1 \text{ g KOH}} = 1.78 \text{ mol KOH}$$

From these figures we see that the two 10% solutions do not contain the same number of moles of NaOH and KOH. Yet one mole of each of these bases will neutralize the same amount of acid. As a result we find that a 10% NaOH solution has more reactive alkali than a 10% KOH solution.

We need a method of expressing concentration that will easily indicate how many moles of solute are present per unit volume of solution. For this purpose the molar method of expressing concentration is used.

A 1 molar solution contains 1 mol, or 1 molar mass of solute per liter of solution. For example, to make a 1 molar solution of sodium hydroxide (NaOH), we dissolve 40.0 g of NaOH (1 mol) in water and dilute the solution with more water to a volume of 1 liter. The solution contains 1 mol of the solute in 1 L of solution and is said to be 1 molar (1 M) in concentration. Figure 15.6 illustrates the preparation of a 1 molar solution. Note that the volume of the solute and the solvent together is 1 liter.

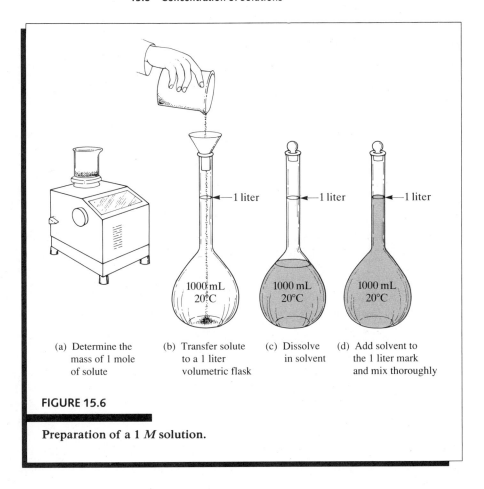

FIGURE 15.6

Preparation of a 1 *M* solution.

molarity

The concentration of a solution can, of course, be varied by using more or less solute or solvent; but in any case the **molarity** of a solution is the number of moles of solute per liter of solution. A capital *M* is the abbreviation for molarity. The units of molarity are moles per liter. The expression "2.0 *M* NaOH" means a 2.0 molar solution of NaOH (2.0 mol, or 80 g, of NaOH dissolved in 1 L of solution).

$$\text{molarity} = M = \frac{\text{number moles of solute}}{\text{liter of solution}} = \frac{\text{moles}}{\text{liter}}$$

Flasks that are calibrated to contain specific volumes at a particular temperature are used to prepare solutions of a desired concentration. These *volumetric flasks* have a calibration mark on the neck to indicate accurately the measured volume. Molarity is based on a specific volume of solution and therefore will vary slightly with temperature because volume varies with temperature (1000 mL of H_2O at 20°C = 1001 mL at 25°C).

Suppose we want to make 500 mL of 1 M solution. This solution can be prepared by determining the mass of 0.5 mol of the solute and diluting with water in a 500 mL (0.5 L) volumetric flask. The molarity will be

$$M = \frac{0.5 \text{ mol solute}}{0.5 \text{ L solution}} = 1 \text{ molar}$$

Thus you can see that it is not necessary to have a liter of solution to express molarity. All we need to know is the number of moles of dissolved solute and the volume of solution. Thus 0.001 mol of NaOH in 10 mL of solution is 0.1 M:

$$\frac{0.001 \text{ mol}}{10 \text{ mL}} \times \frac{1000 \text{ mL}}{1 \text{ L}} = 0.1 \text{ } M$$

When we stop to think that a balance is not calibrated in moles but in grams, we can incorporate grams into the molarity formula. We do so by using the relationship

$$\text{moles} = \frac{\text{grams of solute}}{\text{molar mass}}$$

Substituting this relationship into our expression for molarity, we get

$$M = \frac{\text{mol}}{\text{L}} = \frac{\text{g solute}}{\text{molar mass solute} \times \text{L solution}}$$

$$= \frac{\text{g}}{\text{molar mass} \times \text{L}}$$

We can now determine the mass of any amount of a solute that has a known formula, dilute it to any volume, and calculate the molarity of the solution using this formula.

The molarities of the concentrated acids commonly used in the laboratory are

HCl 12 M	HC$_2$H$_3$O$_2$ 17 M	HNO$_3$ 16 M	H$_2$SO$_4$ 18 M

EXAMPLE 15.5

What is the molarity of a solution containing 1.4 mol of acetic acid (HC$_2$H$_3$O$_2$) in 250 mL of solution?

Substitute the data, 1.4 mol and 250. mL (0.250 L), directly into the equation for molarity.

$$M = \frac{\text{mol}}{\text{L}} = \frac{1.4 \text{ mol}}{0.250 \text{ L}} = \frac{5.6 \text{ mol}}{\text{L}} = 5.6 \text{ } M \quad \text{(Answer)}$$

By the unit conversion method we note that the concentration given in the problem statement is 1.4 mol per 250. mL (mol/mL). Since molarity = mol/L, the needed conversion is

$$\frac{mol}{mL} \longrightarrow \frac{mol}{L} = M$$

$$\frac{1.40 \text{ mol}}{250 \text{ mL}} \times \frac{1000 \text{ mL}}{1 \text{ L}} = \frac{5.6 \text{ mol}}{1 \text{ L}} = 5.6 \ M \quad \text{(Answer)}$$

EXAMPLE 15.6 What is the molarity of a solution made by dissolving 2.00 g of potassium chlorate ($KClO_3$) in enough water to make 150. mL of solution?

This problem can be solved using the unit conversion method. The steps in the conversions must lead to units of moles/liter.

$$\frac{g \ KClO_3}{mL} \longrightarrow \frac{g \ KClO_3}{L} \longrightarrow \frac{mol \ KClO_3}{L} = M$$

The data are

$$g = 2.00 \text{ g} \qquad \text{molar mass } KClO_3 = 122.6 \text{ g/mol} \qquad \text{volume} = 150. \text{ mL}$$

$$\frac{2.00 \text{ g } KClO_3}{150. \text{ mL}} \times \frac{1000 \text{ mL}}{1 \text{ L}} \times \frac{1 \text{ mol } KClO_3}{122.6 \text{ g } KClO_3} = \frac{0.109 \text{ mol}}{1 \text{ L}} = 0.109 \ M$$

EXAMPLE 15.7 How many grams of potassium hydroxide are required to prepare 600. mL of 0.450 M KOH solution?

The conversion is

$$\text{milliliters} \longrightarrow \text{liters} \longrightarrow \text{moles} \longrightarrow \text{grams}$$

The data are

$$\text{volume} = 600. \text{ mL} \qquad M = \frac{0.450 \text{ mol}}{L} \qquad \text{molar mass } KOH = \frac{56.1 \text{ g } KOH}{mol}$$

The calculation is

$$600. \text{ mL} \times \frac{1 \text{ L}}{1000 \text{ mL}} \times \frac{0.450 \text{ mol}}{L} \times \frac{56.1 \text{ g } KOH}{mol} = 15.1 \text{ g } KOH$$

PRACTICE What is the molarity of a solution made by dissolving 7.50 g of magnesium nitrate, $Mg(NO_3)_2$, in enough water to make 25.0 mL of solution?

Answer: 2.02 M

EXAMPLE 15.8   How many milliliters of 2.00 M HCl will react with 28.0 g of NaOH?

Step 1 Write and balance the equation for the reaction:

$$\text{HCl}(aq) + \text{NaOH}(aq) \longrightarrow \text{NaCl}(aq) + \text{H}_2\text{O}(aq)$$

The equation states that 1 mol of HCl reacts with 1 mol of NaOH.

Step 2 Find the number of moles of NaOH in 28.0 g of NaOH:

$$\text{g NaOH} \longrightarrow \text{mol NaOH}$$

$$28.0 \text{ g NaOH} \times \frac{1 \text{ mol}}{40.0 \text{ g}} = 0.700 \text{ mol NaOH}$$

$$28.0 \text{ g NaOH} = 0.700 \text{ mol NaOH}$$

Step 3 Solve for moles and volume of HCl needed. From Steps 1 and 2 we see that 0.700 mol of HCl will react with 0.700 mol of NaOH, because the ratio of moles reacting is 1:1. We know that 2.00 M HCl contains 2.00 mol of HCl per liter; therefore, the volume that contains 0.700 mol of HCl will be less than 1 L.

$$\text{mol NaOH} \longrightarrow \text{mol HCl} \longrightarrow \text{L HCl} \longrightarrow \text{mL HCl}$$

$$0.700 \text{ mol NaOH} \times \frac{1 \text{ mol HCl}}{1 \text{ mol NaOH}} \times \frac{1 \text{ L HCl}}{2.00 \text{ mol HCl}} = 0.350 \text{ L HCl}$$

$$0.350 \text{ L HCl} \times \frac{1000 \text{ mL}}{1 \text{ L}} = 350 \text{ mL HCl}$$

Therefore, 350 mL of 2.00 M HCl contain 0.700 mol HCl and will react with 0.700 mol, or 28.0 g, of NaOH.

EXAMPLE 15.9   What volume of 0.250 M solution can be prepared from 16.0 g of potassium carbonate (K_2CO_3)?

We are starting with 16.0 g of K_2CO_3 and need to find the volume of 0.250 M solution that can be prepared from this K_2CO_3.

The conversion therefore is

$$\text{g K}_2\text{CO}_3 \longrightarrow \text{mol K}_2\text{CO}_3 \longrightarrow \text{L solution}$$

The data are

$$16.0 \text{ g K}_2\text{CO}_3 \qquad M = \frac{0.250 \text{ mol}}{1 \text{ L}} \qquad \text{molar mass K}_2\text{CO}_3 = \frac{138.2 \text{ g K}_2\text{CO}_3}{1 \text{ mol}}$$

$$16.0 \text{ g K}_2\text{CO}_3 \times \frac{1 \text{ mol K}_2\text{CO}_3}{138.2 \text{ g K}_2\text{CO}_3} \times \frac{1 \text{ L}}{0.250 \text{ mol K}_2\text{CO}_3} = 0.463 \text{ L (463 mL)}$$

Thus, a 0.250 M solution can be made by dissolving 16.0 g of K_2CO_3 in water and diluting to 463 mL.

EXAMPLE 15.10   Calculate the number of moles of nitric acid in 325 mL of 16 M HNO$_3$ solution.

Use the equation   moles = liters × M

Substitute the data given in the problem and solve:

$$\text{moles} = 0.325\,\cancel{L} \times \frac{16 \text{ mol HNO}_3}{1\,\cancel{L}} = 5.2 \text{ mol HNO}_3 \quad (\text{Answer})$$

PRACTICE   What volume of 0.035 M AgNO$_3$ can be made from 5.0 g of AgNO$_3$?
Answer: 0.84 L (840 mL)

Dilution Problems

Chemists often find it necessary to dilute solutions from one concentration to another by adding more solvent to the solution. If a solution is diluted by adding pure solvent, the volume of the solution increases, but the number of moles of solute in the solution remains the same. Thus, the moles/liter (molarity) of the solution decreases. It is important to read a problem carefully to distinguish between (1) how much solvent must be added to dilute a solution to a particular concentration and (2) to what volume a solution must be diluted to prepare a solution of a particular concentration.

EXAMPLE 15.11   Calculate the molarity of a sodium hydroxide solution that is prepared by mixing 100. mL of 0.20 M NaOH with 150. mL of water.
This problem is a dilution problem. If we double the volume of a solution by adding water, we cut the concentration in half. Therefore, the concentration of the above solution should be less than 0.10 M. In the dilution, the moles of NaOH remain constant; the molarity and volume change. The final volume is (100. mL + 150. mL) or 250. mL.
To solve this problem, (1) calculate the moles of NaOH in the original solution, and (2) divide the moles of NaOH by the final volume of the solution to obtain the new molarity.

Step 1   Calculate the moles of NaOH in the original solution.

$$M = \frac{\text{mol}}{\text{L}} \qquad \text{mol} = \text{L} \times M$$

$$0.100\,\cancel{L} \times \frac{0.20 \text{ mol NaOH}}{1\,\cancel{L}} = 0.020 \text{ mol NaOH}$$

Step 2   Solve for the new molarity, taking into account that the total volume of the solution after dilution is 250. mL (0.250 L).

$$M = \frac{0.020 \text{ mol NaOH}}{0.250 \text{ L}} = 0.080 \; M \text{ NaOH} \quad (\text{Answer})$$

Alternate Solution When the moles of solute in a solution before and after dilution are the same, then the moles before and after dilution may be set equal to each other:

$$mol_1 = mol_2$$

where mol_1 = moles before dilution, and mol_2 = moles after dilution. Then

$$mol_1 = L_1 \times M_1 \qquad mol_2 = L_2 \times M_2$$

$$L_1 \times M_1 = L_2 \times M_2$$

When both volumes are in the same units, a more general statement can be made:

$$V_1 \times M_1 = V_2 \times M_2$$

For this problem

$$V_1 = 100 \text{ mL} \qquad M_1 = 0.20 \ M$$
$$V_2 = 250 \text{ mL} \qquad M_2 = M_2 \ \text{(unknown)}$$

Then

$$100 \text{ mL} \times 0.20 \ M = 250 \text{ mL} \times M_2$$

Solving for M_2, we get

$$M_2 = \frac{100 \text{ mL} \times 0.20 \ M}{250 \text{ mL}} = 0.080 \ M \text{ NaOH}$$

PRACTICE   Calculate the molarity of a solution prepared by diluting 125 mL of 0.400 M $K_2Cr_2O_7$ with 875 mL of water.

Answer: $5.00 \times 10^{-2} \ M$

EXAMPLE 15.12   How many grams of silver chloride, AgCl, will be precipitated by adding sufficient silver nitrate, $AgNO_3$, to react with 1500. mL of 0.400 M $BaCl_2$ (barium chloride) solution?

$$2 \text{ AgNO}_3(aq) + \text{BaCl}_2(aq) \longrightarrow 2 \text{ AgCl}\downarrow + \text{Ba(NO}_3)_2(aq)$$
$$\qquad\qquad\qquad\qquad 1 \text{ mol} \qquad\qquad 2 \text{ mol}$$

This problem is a stoichiometry problem. The fact that $BaCl_2$ is in solution means that we need to consider the volume and concentration of the solution in order to know the number of moles of $BaCl_2$ reacting.

Step 1 Determine the number of moles of $BaCl_2$ in 1500. mL of 0.400 M solution:

$$M = \frac{mol}{L} \qquad mol = L \times M \qquad 1500 \text{ mL} = 1.500 \text{ L}$$

$$1.500 \, \cancel{L} \times \frac{0.400 \text{ mol BaCl}_2}{\cancel{L}} = 0.600 \text{ mol BaCl}_2$$

Step 2 Calculate the moles and grams of AgCl formed by the mole-ratio method:

$$\text{mol BaCl}_2 \longrightarrow \text{mol AgCl} \longrightarrow \text{g AgCl}$$

$$0.600 \, \cancel{\text{mol BaCl}_2} \times \frac{2 \, \cancel{\text{mol AgCl}}}{1 \, \cancel{\text{mol BaCl}_2}} \times \frac{143.3 \text{ g AgCl}}{\cancel{\text{mol AgCl}}} = 172 \text{ g AgCl}$$

PRACTICE How many grams of lead(II) iodide will be precipitated by adding sufficient $Pb(NO_3)_2$ to react with 750 mL of 0.250 M KI solution?

$$2 \text{ KI} + Pb(NO_3)_2 \longrightarrow PbI_2 \!\downarrow + 2 \text{ KNO}_3$$

Answer: 31 g

Normality

normality

Normality is another way of expressing the concentration of a solution. It is based on an alternate chemical unit of mass called the *equivalent mass*. The **normality** of a solution is the concentration expressed as the number of equivalent masses (equivalents, abbreviated equiv) of solute per liter of solution. A 1 normal (1 N) solution contains 1 equivalent mass of solute per liter of solution. Normality is widely used in analytical chemistry because it simplifies many of the calculations involving solution concentration.

$$\text{normality} = N = \frac{\text{number of equivalents of solute}}{1 \text{ liter of solution}} = \frac{\text{equivalents}}{\text{liter}}$$

where

$$\text{number of equivalents of solute} = \frac{\text{grams of solute}}{\text{equivalent mass of solute}}$$

Every substance may be assigned an equivalent mass. The equivalent mass may be equal either to the molar mass of the substance or to an integral fraction of the molar mass (that is, the molar mass divided by 2, 3, 4, and so on). To gain an understanding of the meaning of equivalent mass, let us start by considering these two reactions:

$$\text{HCl}(aq) + \text{NaOH}(aq) \longrightarrow \text{NaCl}(aq) + \text{H}_2\text{O}$$

1 mole	1 mol
(36.5 g)	(40.0 g)

TABLE 15.5

Comparison of Molar and Normal Solutions of HCl and H_2SO_4 reacting with NaOH

	Molar mass	Concentration	Volumes that react	Equivalent mass	Concentration	Volumes that react
HCl	36.5	1 M	1 liter	36.5	1 N	1 liter
NaOH	40.0	1 M	1 liter	40.0	1 N	1 liter
H_2SO_4	98.1	1 M	1 liter	49.0	1 N	1 liter
NaOH	40.0	1 M	2 liters	40.0	1 N	1 liter

$$H_2SO_4(aq) + 2\ NaOH(aq) \longrightarrow Na_2SO_4(aq) + 2\ H_2O$$

$$\underset{(98.1\ g)}{1\ mol} \qquad \underset{(80.0\ g)}{2\ mol}$$

We note first that 1 mol of hydrochloric acid (HCl) reacts with 1 mol of sodium hydroxide (NaOH) and 1 mol of sulfuric acid (H_2SO_4) reacts with 2 mol of NaOH. If we make 1 molar solutions of these substances, 1 L of 1 M HCl will react with 1 L of 1 M NaOH, and 1 L of 1 M H_2SO_4 will react with 2 L of 1 M NaOH. From this reaction, we can see that H_2SO_4 has twice the chemical capacity of HCl when reacting with NaOH. We can, however, adjust these acid solutions to be equivalent in reactivity by dissolving only 0.5 mol of H_2SO_4 per liter of solution. By doing so, we find that we are required to use 49.0 g of H_2SO_4 per liter (instead of 98.1 g of H_2SO_4 per liter) to make a solution that is equivalent to one made from 36.5 g of HCl per liter. These masses, 49.0 g of H_2SO_4 and 36.5 g of HCl, are chemically equivalent and are known as the equivalent masses of these substances, because each will react with the same amount of NaOH (40.0 g). The equivalent mass of HCl is equal to its molar mass, but that of H_2SO_4 is one-half its molar mass. Table 15.5 summarizes these relationships.

Thus, 1 L of solution containing 36.5 g of HCl would be 1 N, and 1 L of solution containing 49.0 g of H_2SO_4 would also be 1 N. A solution containing 98.1 g of H_2SO_4 (1 mol) per liter would be 2 N when reacting with NaOH in the given equation.

equivalent mass   The **equivalent mass** is the mass of a substance that will react with, combine with, contain, replace, or in any other way be equivalent to 1 mole of hydrogen atoms or hydrogen ions.

Normality and molarity can be interconverted in the following manner.

$$N = \frac{equiv}{L} \qquad M = \frac{mol}{L}$$

$$N = M \times \frac{equiv}{mol} = \frac{\cancel{mol}}{L} \times \frac{equiv}{\cancel{mol}} = \frac{equiv}{L}$$

$$M = N \times \frac{mol}{equiv} = \frac{\cancel{equiv}}{L} \times \frac{mol}{\cancel{equiv}} = \frac{mol}{L}$$

Thus a 2.0 N H_2SO_4 solution is 1.0 M.

One application of normality and equivalents is in acid–base neutralization reactions. An equivalent of an acid is that mass of the acid that will furnish 1 mol of H^+ ions. An equivalent of a base is that mass of base that will furnish 1 mol of OH^- ions. Using concentrations in normality, one equivalent of acid (A) will react with one equivalent of base (B).

$$N_A = \frac{equiv_A}{L_A} \quad \text{and} \quad N_B = \frac{equiv_B}{L_B}$$

$$equiv_A = L_A \times N_A \quad \text{and} \quad equiv_B = L_B \times N_B$$

Since $equiv_A = equiv_B$,

$$L_A \times N_A = L_B \times N_B$$

When both volumes are in the same units, we can write a more general equation:

$$V_A N_A = V_B N_B$$

which states that the volume of acid times the normality of the acid equals the volume of base times the normality of the base.

EXAMPLE 15.13

(a) What is the normality of an H_2SO_4 solution if 25.00 mL of the solution requires 22.48 mL of 0.2018 N NaOH for complete neutralization? (b) What is the molarity of the H_2SO_4 solution?

(a) Solve for N_A by substituting the data into

$$V_A N_A = V_B N_B$$

$$25.00 \text{ mL} \times N_A = 22.48 \text{ mL} \times 0.2018 \ N$$

$$N_A = \frac{22.48 \cancel{mL} \times 0.2018 \ N}{25.00 \cancel{mL}} = 0.1815 \ N \ H_2SO_4$$

(b) When H_2SO_4 is completely neutralized it furnishes 2 equivalents of H^+ ions per mole of H_2SO_4. The conversion from N to M is

$$\frac{equiv}{L} \longrightarrow \frac{mol}{L} \qquad H_2SO_4 = 0.1815 \ N$$

$$\frac{0.1815 \cancel{equiv}}{1 \text{ L}} \times \frac{1 \text{ mol}}{2 \cancel{equiv}} = 0.09075 \text{ mol/L}$$

The H_2SO_4 solution is 0.09075 M.

TABLE 15.6

Concentration Units for Solutions

Units	Symbol	Definition
Mass percent	%m/m	$\dfrac{\text{Mass solute}}{\text{Mass solution}} \times 100$
Parts per million	ppm	$\dfrac{\text{Mass solute}}{\text{Mass solution}} \times 1{,}000{,}000$
Mass/volume percent	%m/v	$\dfrac{\text{Mass solute}}{\text{mL solution}} \times 100$
Volume percent	%v/v	$\dfrac{\text{mL solute}}{\text{mL solution}} \times 100$
Molarity	M	$\dfrac{\text{Moles solute}}{\text{L solution}}$
Normality	N	$\dfrac{\text{Equivalents solute}}{\text{L solution}}$
Molality	m	$\dfrac{\text{Moles solute}}{\text{kg solvent}}$

PRACTICE   What is the normality of a NaOH solution if 50.0 mL of the solution requires 23.72 mL of 0.0250 N H_2SO_4? What is the molarity?

Answers: $1.19 \times 10^{-2}\ N$ $1.19 \times 10^{-2}\ M$

The equivalent mass of a substance may be variable; its value is dependent on the reaction that the substance is undergoing. Consider the reactions represented by these equations:

$$NaOH + H_2SO_4 \longrightarrow NaHSO_4 + H_2O$$
$$2\,NaOH + H_2SO_4 \longrightarrow Na_2SO_4 + 2\,H_2O$$

In the first reaction 1 mol of sulfuric acid furnishes 1 mol of hydrogen. Therefore the equivalent mass of sulfuric acid is the molar mass, namely 98.1 g. But in the second reaction 1 mol of H_2SO_4 furnishes 2 mol of hydrogen. Therefore, the equivalent mass of the sulfuric acid is one-half the molar mass, or 49.0 g. A summary of the quantitative concentration units is found in Table 15.6.

15.9   Colligative Properties of Solutions

Two solutions, one containing 1 mol (60.0 g) of urea (NH_2CONH_2) and the other containing 1 mol (342 g) of sucrose ($C_{12}H_{22}O_{11}$) in 1 kg of water, both have a freezing point of $-1.86°C$, not $0°C$ as for pure water. Urea and sucrose are distinctly different substances, yet they lower the freezing point of the water by the same amount. The only thing apparently common to these two solutions is that each contains 1 mol (6.022×10^{23} molecules) of solute and 1 kg of solvent. In fact, if we dissolve one mole of any nonionizable solute in 1 kg of water, the freezing point of the resulting solution will be $-1.86°C$.

These results lead us to conclude that the freezing point depression for a solution containing 6.022×10^{23} solute molecules (particles) and 1 kg of water is a constant, namely, $1.86°C$. Freezing point depression is a general property of solutions. Furthermore the amount by which the freezing point is depressed is the same for all solutions made with a given solvent; that is, each solvent shows a characteristic *freezing point depression constant*. Freezing point depression constants for several solvents are given in Table 15.7.

The solution formed by the addition of a nonvolatile solute to a solvent has a lower freezing point, a higher boiling point, and a lower vapor pressure than that of the pure solvent. All these effects are related and are known as colligative properties. The **colligative properties** are properties that depend only upon the number of solute particles in a solution and not on the nature of those particles. Freezing point depression, boiling point elevation, and vapor pressure lowering are colligative properties of solutions.

The colligative properties of a solution can be considered in terms of vapor pressure. The vapor pressure of a pure liquid depends on the tendency of

colligative
properties

TABLE 15.7

Freezing Point Depression and Boiling Point Elevation Constants of Selected Solvents

Solvent	Freezing point of pure solvent (°C)	Freezing point depression constant, K_f $\left(\dfrac{°C \ kg \ solvent}{mol \ solute}\right)$	Boiling point of pure solvent (°C)	Boiling point elevation constant, K_b $\left(\dfrac{°C \ kg \ solvent}{mol \ solute}\right)$
Water	0.00	1.86	100.0	0.512
Acetic acid	16.6	3.90	118.5	3.07
Benzene	5.5	5.1	80.1	2.53
Camphor	178	40	208.2	5.95

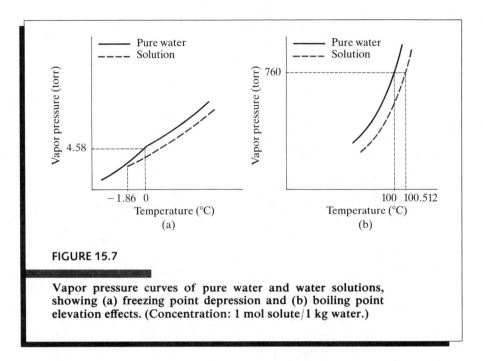

FIGURE 15.7

Vapor pressure curves of pure water and water solutions, showing (a) freezing point depression and (b) boiling point elevation effects. (Concentration: 1 mol solute/1 kg water.)

molecules to escape from its surface. Thus, if 10% of the molecules in a solution are nonvolatile solute molecules, the vapor pressure of the solution is 10% lower than that of the pure solvent. The vapor pressure is lower because the surface of the solution contains 10% nonvolatile molecules and 90% of the volatile solvent molecules. A liquid boils when its vapor pressure equals the pressure of the atmosphere. Thus, we can see that the solution just described as having a lower vapor pressure will have a higher boiling point than the pure solvent. The solution with a lowered vapor pressure does not boil until it has been heated above the boiling point of the solvent (see Figure 15.7). Each solvent has its own characteristic boiling point elevation constant (see Table 15.7). The boiling point elevation constant is based on a solution that contains 1 mole of solute particles per kilogram of solvent. For example, the boiling point elevation constant for a solution containing 1 mole of solute particles per kilogram of water is 0.512°C, which means that this water solution will boil at 100.512°C.

The freezing behavior of a solution can also be considered in terms of lowered vapor pressure. Figure 15.7 shows the vapor pressure relationships of ice, water, and a solution containing 1 mole of solute per kilogram of water. The freezing point of water is at the intersection of the water and ice vapor pressure curves—that is, at the point where water and ice have the same vapor pressure. Because the vapor pressure of water is lowered by the solute, the vapor pressure curve of the solution does not interesect the vapor pressure curve of ice until the solution has been cooled below the freezing point of pure water. Thus it is necessary to cool the solution below 0°C in order to freeze out ice.

The foregoing discussion dealing with freezing point depressions is restricted to *un-ionized* substances. The discussion of boiling point elevations is restricted to *nonvolatile* and un-ionized substances. The colligative properties of ionized substances (Electrolytes, Chapter 16) are not under consideration at this point.

Some practical applications involving colligative properties are (1) use of salt-ice mixtures to provide low freezing temperatures for homemade ice creams, (2) use of sodium chloride or calcium chloride to melt ice from streets, and (3) use of ethylene glycol and water mixtures as antifreeze in automobile radiators (ethylene glycol also raises the boiling point of radiator fluid and thus allows the engine to operate at a higher temperature).

Both the freezing point depression and the boiling point elevation are directly proportional to the number of moles of solute per kilogram of solvent. When we deal with the colligative properties of solutions, another concentration expression, *molality*, is used. The **molality** (*m*) of a solute is the number of moles of solute per kilogram of solvent:

molality

$$m = \frac{\text{mol solute}}{\text{kg solvent}}$$

Note that a lowercase *m* is used for molality concentrations and a capital *M* for molarity. The difference between molality and molarity is that molality refers to moles of solute *per kilogram of solvent*, whereas molarity refers to moles of solute *per liter of solution*. For un-ionized substances, the colligative properties of a solution are directly proportional to its molality.

Molality is independent of volume. It is a mass-to-mass relationship of solute to solvent and allows for experiments, such as freezing point depression and boiling point elevation, to be conducted at variable temperatures.

The following equations are used in calculations involving colligative properties and molality.

$$\Delta t_f = mK_f \qquad \Delta t_b = mK_b \qquad m = \frac{\text{mol solute}}{\text{kg solvent}}$$

m = molality; mol solute/kg solvent

Δt_f = freezing point depression; °C

Δt_b = boiling point elevation; °C

K_f = freezing point depression constant; °C kg solvent/mol solute

K_b = boiling point elevation constant; °C kg solvent/mol solute

---

EXAMPLE 15.14   What is the molality (*m*) of a solution prepared by dissolving 2.70 g of CH_3OH in 25.0 g of H_2O?

Since $\quad m = \dfrac{\text{mol solute}}{\text{kg solvent}} \quad$ the conversion is

$$\dfrac{2.70 \text{ g CH}_3\text{OH}}{25.0 \text{ g H}_2\text{O}} \longrightarrow \dfrac{\text{mol CH}_3\text{OH}}{25.0 \text{ g H}_2\text{O}} \longrightarrow \dfrac{\text{mol CH}_3\text{OH}}{1 \text{ kg H}_2\text{O}}$$

The molar mass of CH_3OH is $(12.0 + 4.0 + 16.0)$ or 32.0 g/mol.

$$\dfrac{2.70 \text{ g CH}_3\text{OH}}{25.0 \text{ g H}_2\text{O}} \times \dfrac{1 \text{ mol CH}_3\text{OH}}{32.0 \text{ g CH}_3\text{OH}} \times \dfrac{1000 \text{ g H}_2\text{O}}{1 \text{ kg H}_2\text{O}} = \dfrac{3.38 \text{ mol CH}_3\text{OH}}{1 \text{ kg H}_2\text{O}}$$

The molality is 3.38 m.

PRACTICE   What is the molality of a solution prepared by dissolving 150.0 g of $C_6H_{12}O_6$ in 600.0 g of H_2O?

Answer: 1.39 m

EXAMPLE 15.15   A solution is made by dissolving 100. g of ethylene glycol ($C_2H_6O_2$) in 200. g of water. What is the freezing point of this solution?

To calculate the freezing point of the solution, we first need to calculate Δt_f, the change in freezing point. Use the equation

$$\Delta t_f = m K_f = \dfrac{\text{mol solute}}{\text{kg solvent}} \times K_f$$

K_f (for water): $\quad \dfrac{1.86°\text{C kg solvent}}{\text{mol solute}} \quad$ (from Table 15.7)

mol solute: $\quad 100 \text{ g C}_2\text{H}_6\text{O}_2 \times \dfrac{1 \text{ mol C}_2\text{H}_6\text{O}_2}{62.0 \text{ g C}_2\text{H}_6\text{O}_2} = 1.61 \text{ mol C}_2\text{H}_6\text{O}_2$

kg solvent: $\quad 200. \text{ g H}_2\text{O} \times \dfrac{1 \text{ kg}}{1000 \text{ g}} = 0.200 \text{ kg H}_2\text{O}$

$$\Delta t_f = \dfrac{1.61 \text{ mol C}_2\text{H}_6\text{O}_2}{0.200 \text{ kg H}_2\text{O}} \times \dfrac{1.86°\text{C kg H}_2\text{O}}{1 \text{ mol C}_2\text{H}_6\text{O}_2} = 15.0°\text{C}$$

The freezing point depression, 15.0°C, must be subtracted from 0°C, the freezing point of the pure solvent.

$$\text{freezing point of solution} = \text{freezing point of solvent} - \Delta t_f$$
$$= 0.0°\text{C} - 15.0°\text{C} = -15°\text{C}$$

Therefore, the freezing point of the solution is $-15°\text{C}$.

The calculation can also be done using the equation

$$\Delta t_f = K_f \times \dfrac{\text{g solute}}{\text{molar mass solute}} \times \dfrac{1}{\text{kg solvent}}$$

EXAMPLE 15.16 A solution made by dissolving 4.71 g of a compound of unknown molar mass in 100.0 g of water has a freezing point of $-1.46°C$. What is the molar mass of the compound?
First substitute the data in $\Delta t_f = mK_f$ and solve for m.

$\Delta t_f = +1.46$ since the solvent, water, freezes at $0°C$.

$$K_f = \frac{1.86°C \text{ kg } H_2O}{\text{mol solute}}$$

$$1.46°C = mK_f = m \times \frac{1.86°C \text{ kg } H_2O}{\text{mol solute}}$$

$$m = \frac{1.46°C \times \text{mol solute}}{1.86°C \times \text{kg } H_2O} = \frac{0.785 \text{ mol solute}}{\text{kg } H_2O}$$

Now convert the data, 4.71 g solute/100.0 g H_2O, to g/mol.

$$\frac{4.71 \text{ g solute}}{100.0 \text{ g } H_2O} \times \frac{100 \text{ g } H_2O}{1 \text{ kg } H_2O} \times \frac{1 \text{ kg } H_2O}{0.785 \text{ mol solute}} = 60.0 \text{ g/mol}$$

The molar mass of the compound is 60.0 g/mol.

PRACTICE What is the freezing point of the solution in the previous practice problem? What is the boiling point?
Answers: freezing point $= -2.59°C$ boiling point $= 100.72°C$

Concepts in Review

1. Describe the types of solutions.

2. List the general properties of solutions.

3. Describe and illustrate the process by which an ionic substance dissolves in water.

4. Indicate the effects of temperature and pressure on the solubility of solids and gases in liquids.

5. Identify and explain the factors affecting the rate at which a solid dissolves in a liquid.

6. Use a solubility table or graph to determine whether a solution is saturated, unsaturated, or supersaturated at a given temperature.

7. Calculate the mass percent or volume percent for a solution.

8. Calculate the amount of solute in a given quantity of a solution when given the mass percent or volume percent of a solution.

9. Calculate the molarity of a solution from the volume and the mass, or moles, of solute.

10. Calculate the mass of a substance necessary to prepare a solution of specified volume and molarity.

11. Determine the resulting molarity in a typical dilution problem.

12. Apply stoichiometry to chemical reactions involving solutions.

13. Understand the concepts of equivalent mass and normality and use them in calculations.

14. Explain the effect of a solute on the vapor pressure of a solvent.

15. Explain the effect of a solute on boiling point and freezing point of a solution.

16. Calculate the boiling and freezing points of a solution from concentration data.

17. Calculate molality and molar mass of a solute from boiling or freezing point data.

Key Terms in Review

The terms listed here have all been defined within the chapter. Review the definitions of each. Use the glossary and the margin notations within the chapter as study aids.

colligative properties	normality
concentrated solution	parts per million
concentration of a solution	saturated solution
dilute solution	solubility
equivalent mass	solute
immiscible	solution
miscible	supersaturated solution
molality	solvent
molarity	unsaturated solution

Exercises

An asterisk indicates a more challenging question or problem.

1. Make a sketch indicating the orientation of water molecules (a) about a single sodium ion and (b) about a single chloride ion in solution.

2. Which of the substances listed below are reasonably soluble and which are insoluble in water?

(a) KOH	(f) PbI_2
(b) $NiCl_2$	(g) $MgCO_3$
(c) ZnS	(h) $CaCl_2$
(d) $AgC_2H_3O_2$	(i) $Fe(NO_3)_3$
(e) Na_2CrO_4	(j) $BaSO_4$

3. Estimate the number of grams of sodium fluoride that would dissolve in 100 g of water at 50°C. (See Table 15.3.)

4. What is the solubility at 25°C of each of the substances listed below? (See Figure 15.3.)
   (a) Potassium chloride (c) Potassium nitrate
   (b) Potassium chlorate

5. What is different in the solubility trend of the potassium halides compared with that of the lithium halides and the sodium halides? (See Table 15.3.)

6. What is the solubility, in grams of solute per 100 g of H_2O, of (a) $KClO_3$ at 60°C, (b) HCl at 20°C, (c) Li_2SO_4 at 80°C, and (d) KNO_3 at 0°C? (See Figure 15.3.)

7. Which substance, KNO_3 or NH_4Cl, shows the greater increase in solubility with increased temperature? (See Figure 15.3.)

8. Does a 2 molal solution in benzene or a 1 molal solution in camphor show the greater freezing point depression? (See Table 15.7)

9. What would be the total surface area if the 1 cm cube in Figure 15.4 were cut into cubes 0.01 cm on a side?

10. At which temperatures—10°C, 20°C, 30°C, 40°C, or 50°C—would you expect a solution made from 63 g of ammonium chloride and 150 g of water to be unsaturated? (See Figure 15.3.)

11. Explain why the rate of dissolving decreases as shown in Figure 15.5.

12. Would the volumetric flasks in Figure 15.6 be satisfactory for preparing normal solutions? Explain.

13. Name and distinguish between the two components of a solution.

14. Is it always apparent in a solution which component is the solute, for example, in a solution of a liquid in a liquid?

15. Explain why the solute does not settle out of a solution.

16. Is it possible to have one solid dissolved in another? Explain.

17. An aqueous solution of KCl is colorless, $KMnO_4$ is purple, and $K_2Cr_2O_7$ is orange. What color would you expect of an aqueous solution of $Na_2Cr_2O_7$? Explain.

18. Explain why carbon tetrachloride will dissolve benzene but will not dissolve sodium chloride.

19. Some drinks like tea are consumed either hot or cold, whereas others like Coca Cola are drunk only cold. Why?

20. Why is air considered to be a solution?

21. In which will a teaspoonful of sugar dissolve more rapidly, 200 mL of iced tea or 200 mL of hot coffee? Explain in terms of the KMT.

22. What is the effect of pressure on the solubility of gases in liquids? Solids in liquids?

23. Why do smaller particles dissolve faster than large ones?

24. In a saturated solution containing undissolved solute, solute is continuously dissolving, but the concentration of the solution remains unchanged. Explain.

25. Explain why there is no apparent reaction when crystals of $AgNO_3$ and NaCl are mixed, but a reaction is apparent immediately when solutions of $AgNO_3$ and NaCl are mixed.

26. What do we mean when we say that concentrated nitric acid (HNO_3) is 16 molar?

27. Will 1 liter of 1 molar NaCl contain more chloride ions than 0.5 liter of 1 molar $MgCl_2$? Explain.

28. Champagne is usually cooled in a refrigerator prior to opening. It is also opened very carefully. What would happen if a warm bottle of champagne is shaken and opened quickly and forcefully?

29. Explain how a supersaturated solution of $NaC_2H_3O_2$ can be prepared and proven to be supersaturated.

30. Which of the following statements are correct? Rewrite the incorrect statements to make them correct.
   (a) A solution is a homogeneous mixture.
   (b) It is possible for the same substance to be the solvent in one solution and the solute in another.
   (c) A solute can be removed from a solution by filtration.
   (d) Saturated solutions are always concentrated solutions.
   (e) If a solution of sugar in water is allowed to stand undisturbed for a long time, the sugar will gradually settle to the bottom of the container.
   (f) It is not possible to prepare an aqueous 1.0 M AgCl solution.
   (g) Gases are generally more soluble in hot water than in cold water.
   (h) It is impossible to prepare a two-phase liquid mixture from two liquids that are miscible with each other in all proportions.

(i) A solution that is 10% NaCl by mass always contains 10 g of NaCl.

(j) Small changes in pressure have little effect on the solubility of solids in liquids but a marked effect on the solubility of gases in liquids.

(k) How fast a solute dissolves depends mainly on the size of the solute particles, the temperature of the solvent, and the degree of agitation or stirring taking place.

(l) In order to have a 1 molar solution you must have 1 mol of solute dissolved in sufficient solvent to give 1 L of solution.

(m) Dissolving 1 mole of NaCl in 1 liter of water will give a 1 molar solution.

(n) One mole of solute in 1 L of solution has the same concentration as 0.1 mol of solute in 100 mL of solution.

(o) When 100 mL of 0.200 M HCl is diluted to 200 mL volume by the addition of water, the resulting solution is 0.100 M and contains one-half the number of moles of HCl as were in the original solution.

(p) Fifty milliliters of 0.1 M H_2SO_4 will neutralize the same volume of 0.1 M NaOH as 100 mL of 0.1 M HCl.

(q) Fifty milliliters of 0.1 N H_2SO_4 will neutralize the same volume of 0.1 M NaOH as 100 mL of 0.1 M HCl.

(r) The molarity of a solution will vary slightly with temperature.

(s) The equivalent mass of $Ca(OH)_2$ is one-half its molar mass.

(t) Gram for gram methyl alochol, CH_3OH, is more effective than ethyl alcohol, C_2H_5OH, in lowering the freezing point of water.

(u) An aqueous solution that freezes below 0°C will have a normal boiling point below 100°C.

(v) The colligative properties of a solution depend on the number of solute particles dissolved in solution.

31. What disadvantages are there in expressing the concentration of solutions as dilute or concentrated?

32. Explain how concentrated H_2SO_4 can be both 18 molar and 36 normal in concentration.

33. Describe how you would prepare 750 mL of 5 molar NaCl solution.

34. Arrange the following bases (in descending order) according to the volume of each that will react with 1 liter of 1 M HCl: (a) 1 M NaOH, (b) 1.5 M $Ca(OH)_2$, (c) 2 M KOH, and (d) 0.6 M $Ba(OH)_2$.

35. Explain in terms of vapor pressure why the boiling point of a solution containing a non-volatile solute is higher than that of the pure solvent.

36. Explain why the freezing point of a solution is lower than the freezing point of the pure solvent.

37. Which would be colder, a glass of water and crushed ice or a glass of Seven-Up and crushed ice? Explain.

38. When water and ice are mixed, the temperature of the mixture is 0°C. But, if methyl alcohol and ice are mixed, a temperature of −10°C is readily attained. Explain why the two mixtures show such different temperature behavior.

39. Which would be more effective in lowering the freezing point of 500. g of water?
(a) 100. g of sucrose ($C_{12}H_{22}O_{11}$) or 100. g of ethyl alcohol (C_2H_5OH)
(b) 100. g of sucrose or 20.0 g of ethyl alcohol
(c) 20.0 g of ethyl alcohol or 20.0 g of methyl alcohol (CH_3OH)

40. Is the molarity of a 5 molal aqueous solution of NaCl greater or less than 5 molar? Explain.

41. Express, in terms of its molarity, the normality of an H_2SO_4 solution used to titrate NaOH solutions. (Assume both hydrogens react.)

Percent Solutions

42. Calculate the mass percent of the following solutions.
(a) 25.0 g NaBr + 100. g H_2O
(b) 1.20 g K_2SO_4 + 10.0 g H_2O
(c) 40.0 g $Mg(NO_3)_2$ + 500. g H_2O

43. How many grams of a solution that is 12.5% by mass $AgNO_3$ would contain the following?
(a) 30.0 g of $AgNO_3$ (b) 0.400 mol of $AgNO_3$

44. Calculate the mass percent of the following solutions:
(a) 60.0 g NaCl + 200.0 g H_2O
(b) 0.25 mol $HC_2H_3O_2$ + 3.0 mol H_2O
(c) 1.0 molal solution of $C_6H_{12}O_6$ in water

45. How much solute is present in each of the following?
(a) 65 g of 5.0% KCl solution
(b) 250. g of 15.0% K_2CrO_4 solution

*(c) A solution that contains 100.0 g of water and is 6.0% by mass sodium bicarbonate, $NaHCO_3$.

46. What mass of 5.50% solution can be prepared from 25.0 g of KCl?

47. Physiological saline (NaCl) solutions used in intravenous injections have a concentration of 0.90% NaCl by mass.
    (a) How many grams of NaCl are needed to prepare 500. g of this solution?
    *(b) How much water must evaporate from this solution to give a solution that is 9.0% NaCl by mass?

48. A solution is made from 50.0 g of KNO_3 and 175 g of H_2O. How many grams of water must evaporate to give a saturated solution of KNO_3 in water at 20°C? (See Figure 15.3.)

49. Calculate the mass/volume percent of a solution made by dissolving
    (a) 22.0 g of CH_3OH dissolved in C_2H_5OH to make 100. mL of solution
    (b) 4.20 g of NaCl dissolved in H_2O to make 12.5 mL of solution

50. What is the volume percent of these solutions?
    (a) 10.0 mL of CH_3OH dissolved in water to a volume of 40.0 mL.
    (b) 2.0 mL of CCl_4 dissolved in benzene to a volume of 9.0 mL.

51. What volume of 70.% rubbing alcohol can you prepare if you have only 150 mL of pure isopropyl alcohol on hand?

52. At 20°C an aqueous solution of HNO_3 that is 35.0% HNO_3 by mass has a density of 1.21 g/mL.
    (a) How many grams of HNO_3 are present in 1.00 L of this solution?
    (b) What volume of this solution will contain 500. g of HNO_3?

Molarity Problems

53. Calculate the molarity of the following solutions:
    (a) 0.10 mol of solute in 250 mL of solution
    (b) 2.5 mol of NaCl in 0.650 L of solution.
    (c) 0.025 mol of HCl in 10. mL of solution
    (d) 0.35 mol $BaCl_2 \cdot 2 H_2O$ in 593 mL of solution

54. Calculate the molarity of the following solutions:

(a) 53.0 g of Na_2CrO_4 in 1.00 L of solution
(b) 260 g of $C_6H_{12}O_6$ in 800. mL of solution
(c) 1.50 g of $Al_2(SO_4)_3$ in 2.00 L of solution
(d) 0.0282 g of $Ca(NO_3)_2$ in 1.00 mL of solution

55. Calculate the number of moles of solute in each of the following solutions:
    (a) 40.0 L of 1.0 M LiCl
    (b) 25.0 mL of 3.00 M H_2SO_4
    (c) 349 mL of 0.0010 M NaOH
    (d) 5000. mL of 3.1 M $CoCl_2$

56. Calculate the grams of solute in each of the following solutions:
    (a) 150 L of 1.0 M NaCl
    (b) 0.035 L of 10.0 M HCl
    (c) 260 mL of 18 M H_2SO_4
    (d) 8.00 mL of 8.00 M $Na_2C_2O_4$

57. How many milliliters of 0.256 M KCl solution will contain the following?
    (a) 0.430 mol of KCl
    (b) 10.0 mol of KCl
    (c) 20.0 g of KCl
    *(d) 71.0 g of chloride ion, Cl^-

*58. What is the molarity of a nitric acid solution, if the solution is 35.0% HNO_3 by mass and has a density of 1.21 g/mL?

Dilution Problems

59. What will be the molarity of the resulting solutions made by mixing the following (Assume volumes are additive.)
    (a) 200. mL 12 M HCl + 200. mL H_2O
    (b) 60.0 mL 0.60 M $ZnSO_4$ + 500. mL H_2O
    (c) 100. mL 1.0 M HCl + 150 mL 2.0 M HCl

60. Calculate the volume of concentrated reagent required to prepare the diluted solutions indicated:
    (a) 12 M HCl to prepare 400. mL of 6.0 M HCl
    (b) 15 M NH_3 to prepare 50. mL of 6.0 M NH_3
    (c) 16 M HNO_3 to prepare 100. mL of 2.5 M HNO_3
    (d) 18 M H_2SO_4 to prepare 250 mL of 10.0 N H_2SO_4

61. To what volume must a solution of 80.0 g of H_2SO_4 in 500. mL of solution be diluted to give a 0.10 M solution?

62. What will be the molarity of each of the solutions made by mixing 250 mL of 0.75 M H_2SO_4 with (a) 150 mL of H_2O, (b) 250 mL of 0.70 M H_2SO_4, and (c) 400. mL of 2.50 M H_2SO_4?

63. How many milliliters of water must be added to 300. mL of 1.40 M HCl to make a solution that is 0.500 M HCl?

64. A 10.0 mL sample of 16 M HNO_3 solution is diluted to 500. mL. What is the molarity of the final solution?

65. Given a 5.00 M KOH solution, how would you prepare 250 mL of 0.625 M KOH?

Stoichiometry Problems

66. $BaCl_2(aq) + K_2CrO_4(aq) \longrightarrow$

$$BaCrO_4(s) + 2\ KCl(aq)$$

Using the above equation, calculate the following:
(a) The grams of $BaCrO_4$ that can be obtained from 100. mL of 0.300 M $BaCl_2$
(b) The volume of 1.0 M $BaCl_2$ solution needed to react with 50.0 mL of 0.300 M K_2CrO_4 solution

67. $3\ MgCl_2(aq) + 2\ Na_3PO_4(aq) \longrightarrow$

$$Mg_3(PO_4)_2(s) + 6\ NaCl(aq)$$

Using the above equation, calculate:
(a) The milliliters of 0.250 M Na_3PO_4 that will react with 50.0 mL of 0.250 M $MgCl_2$.
(b) The grams of $Mg_3(PO_4)_2$ that will be formed from 50.0 mL of 0.250 M $MgCl_2$.

68. (a) How many moles of hydrogen will be liberated from 200. mL of 3.00 M HCl reacting with an excess of magnesium? The equation is

$$Mg(s) + 2\ HCl(aq) \longrightarrow$$

$$MgCl_2(aq) + H_2(g)$$

(b) How many liters of hydrogen gas, H_2, measured at 27°C and 720 torr, will be obtained? (*Hint*: Use the ideal gas equation.]

*69. What is the molarity of an HCl solution, 150. mL of which, when treated with excess magnesium, liberates 3.50 L of H_2 gas measured at STP?

70. Given the balanced equation

$$6\ FeCl_2 + K_2Cr_2O_7 + 14\ HCl \longrightarrow$$

$$6\ FeCl_3 + 2\ CrCl_3 + 2\ KCl + 7\ H_2O$$

(a) How many moles of KCl will be produced from 2.0 mol of $FeCl_2$?
(b) How many moles of $CrCl_3$ will be produced from 1.0 mol of $FeCl_2$?
(c) How many moles of $FeCl_2$ will react with 0.050 mol of $K_2Cr_2O_7$?
(d) How many milliliters of 0.060 M $K_2Cr_2O_7$ will react with 0.025 mol of $FeCl_2$?
(e) How many milliliters of 6.0 M HCl will react with 15.0 mL of 6.0 M $FeCl_2$?

71. $2\ KMnO_4 + 16\ HCl \longrightarrow$

$$2\ MnCl_2 + 5\ Cl_2 + 8\ H_2O + 2\ KCl$$

Calculate the following using the above equation:
(a) The moles of Cl_2 produced from 0.50 mol of $KMnO_4$
(b) The moles of HCl required to react with 1.0 L of 2.0 M $KMnO_4$
(c) The milliliters of 6.0 M HCl required to react with 200. mL of 0.50 M $KMnO_4$
(d) The liters of Cl_2 gas at STP produced by the reaction of 75.0 mL of 6.0 M HCl

Equivalent Mass and Normality Problems

72. Calculate the equivalent mass of the acid and base in each of the following reactions:
(a) $HCl + NaOH \longrightarrow NaCl + H_2O$
(b) $2\ HCl + Ba(OH)_2 \longrightarrow BaCl_2 + 2\ H_2O$
(c) $H_2SO_4 + Ca(OH)_2 \longrightarrow CaSO_4 + 2\ H_2O$
(d) $H_2SO_4 + KOH \longrightarrow KHSO_4 + H_2O$
(e) $H_3PO_4 + 2\ LiOH \longrightarrow Li_2HPO_4 + 2\ H_2O$

73. What is the normality of the following solutions? Assume complete neutralization.
(a) 4.0 M HCl   (d) 1.85 M H_3PO_4
(b) 0.243 M HNO_3 (e) 0.250 M $HC_2H_3O_2$
(c) 3.0 M H_2SO_4

74. What is the normality of an H_2SO_4 solution if 36.26 mL are required to neutralize 2.50 g of $Ca(OH)_2$?

75. Which will be more effective in neutralizing stomach acid, HCl, a tablet containing 12.0 g of $Mg(OH)_2$ or a tablet containing 10.0 g of $Al(OH)_3$? Show evidence for your answer.

76. What volume of 0.2550 N NaOH is required to neutralize
(a) 20.22 mL of 0.1254 N HCl

(b) 14.86 mL of 0.1246 N H_2SO_4
(c) 18.00 mL of 0.1430 M H_2SO_4

Molality and Colligative Properties Problems

77. Calculate the molality of these solutions.
    (a) 14.0 g of CH_3OH in 100. g of H_2O
    (b) 2.50 mol of benzene (C_6H_6) in 250 g of CCl_4
    (c) 1.0 g of $C_6H_{12}O_6$ in 1.0 g of H_2O

78. Which would be more effective as an antifreeze in an automobile radiator? (a) 10 kg of methyl alcohol, CH_3OH, or 10 kg of ethyl alcohol, C_2H_5OH (b) 10 m solution of methyl alcohol or 10 m solution of ethyl alcohol?

79. Automobile battery acid is 38% H_2SO_4 and has a density of 1.29 g/mL. Calculate the molality and the molarity of this solution.

*80. A sugar solution made to feed humming birds contains 1.00 lb of sugar to 4.00 lb of water. Can this solution be put outside, without freezing, where the temperature falls to 20.0°F at night? Show evidence for your answer.

*81. What would be (a) the boiling point and (b) the molality of an aqueous sugar ($C_{12}H_{22}O_{11}$) solution that freezes at −5.4°C?

82. (a) What is the molality of a solution containing 100.0 g of ethylene glycol, $C_2H_6O_2$, in 150.0 g of water?
    (b) What is the boiling point of this solution?
    (c) What is the freezing point of this solution?

83. What is (a) the molality, (b) the freezing point, and (c) the boiling point of a solution containing 2.68 g of naphthalene, $C_{10}H_8$, in 38.4 g of benzene (C_6H_6)?

*84. The freezing point of a solution of 8.00 g of an unknown compound dissolved in 60.0 g of acetic acid is 13.2°C. Calculate the molar mass of the compound.

85. What is the molar mass of a compound if 4.80 g

of the compound dissolved in 22.0 g of H_2O gives a solution that freezes at −2.50°C?

86. A solution of 6.20 g of $C_2H_6O_2$ in water has a freezing point of −0.372°C. How many grams of H_2O are in the solution?

87. What (a) mass and (b) volume of ethylene glycol ($C_2H_6O_2$, density = 1.11 g/mL) should be added to 12.0 L of water in an automobile radiator to protect it from freezing at −20.0°C? (c) To what temperature Fahrenheit will the radiator be protected?

Additional Problems

88. How many grams of solution, 10.% NaOH by mass, are required to neutralize 150 mL of a 1.0 M HCl solution?

*89. How many grams of solution, 10.% NaOH by mass, are required to neutralize 250. g of a 1.0 molal solution of HCl?

*90. A sugar syrup solution contains 15.0% sugar, $C_{12}H_{22}O_{11}$, by mass, and has a density of 1.06 g/mL.
    (a) How many grams of sugar are in 1.0 L of this syrup?
    (b) What is the molarity of this solution?
    (c) What is the molality of this solution?

*91. A solution of 3.84 g of C_4H_2N (empirical formula) in 250. g of benzene depresses the freezing point of benzene 0.614°C. What is the molecular formula of the compound?

*92. Hydrochloric acid (HCl) is sold as a concentrated aqueous solution at 12.0 mol/L. If the density of the solution is 1.18 g/mL, determine the molality of the solution.

*93. How many grams of KNO_3 must be used to make 450 mL of a solution that is to contain 5.5 mg/mL of the potassium ion? Calculate the molarity of the solution.

Review Exercises for Chapters 13–15

CHAPTER THIRTEEN   The Gaseous State of Matter

True–False.   *Answer the following as either true or false.*

1. The average kinetic energy of molecules of a gas increases with increased temperature.
2. Pressure is defined as force per unit area.
3. One torr is equal to 760 mm Hg.
4. One mole of hydrogen gas and 1 mole of oxygen gas, each in a box of equal volume, and each at the same temperature, will exert the same pressure.
5. Boyle's law states that the volume of a gas is inversely proportional to the pressure.
6. Charles' law states that the pressure of a gas is directly proportional to the temperature.
7. One mole of any gas always occupies 22.4 liters.
8. Avogadro's law states that equal volumes of different gases at the same temperature and pressure contain the same masses of gas.
9. An ideal gas is one whose behavior is described exactly by the gas laws for all possible values of P, V, and T.
10. According to Avogadro, there are 6.022×10^{23} molecules of gas in a liter at STP.
11. When the temperature of a gas is decreased at fixed volume, the pressure decreases.
12. At constant temperature and pressure, N_2 will effuse more rapidly than O_2.
13. A barometer is an instrument used to measure the mass of a certain quantity of mercury.
14. The volume of a gas is dependent on the number of gas molecules, the pressure, the absolute temperature, and the gas constant, R.
15. Chlorofluorocarbons, which were used in aerosol spray cans and are used in refrigeration and air conditioners, escape to the stratosphere and react to destroy part of the ozone layer.
16. Ozone in the stratosphere aids ultraviolet radiation in reaching the earth's surface.
17. The formula for ozone is $3 O_2$.
18. The phenomenon in which an element can exist in two or more molecular or crystalline forms is known as allotropy.

Multiple Choice.   *Choose the correct answer to each of the following.*

1. Which of the following is not one of the principal assumptions of the Kinetic-Molecular Theory for an ideal gas?
   (a) All collisions of gaseous molecules are perfectly elastic.
   (b) A mole of any gas occupies 22.4 L at STP.
   (c) Gas molecules have no attraction for one another.
   (d) The average kinetic energy for molecules is the same for all gases at the same temperature
2. Which of the following is not equal to 1.00 atm pressure?
   (a) 760. cm Hg   (c) 760. mm Hg
   (b) 29.9 in. Hg   (d) 760. torr
3. If the pressure on 45 mL of gas is changed from 600. torr to 800. torr, the new volume will be:
   (a) 60 mL   (c) 0.045 L
   (b) 34mL   (d) 22.4 L
4. The volume of a gas is 300. mL at 740. torr and 25°C. If the pressure remains constant and the temperature is raised to 100.°C, the new volume will be:
   (a) 240. mL   (c) 376 mL
   (b) 1.20 L   (d) 75.0 mL
5. The volume of a dry gas is 4.00 L at 15°C and 745 torr. What volume will the gas occupy at 40°C and 700. torr?
   (a) 4.63 L   (b) 3.46 L   (c) 3.92 L   (d) 4.08 L
6. A sample of Cl_2 occupies 8.50 L at 80°C and 740. mm Hg. What volume will the Cl_2 occupy at STP?
   (a) 10.7 L   (b) 6.75 L   (c) 11.3 L   (d) 6.40 L
7. What volume will 8.00 g of O_2 occupy at 45°C and 2.00 atm?
   (a) 0.462 L   (b) 104 L   (c) 9.62 L   (d) 3.26 L

8. The density of NH_3 gas at STP is:
   (a) 0.759 g/mL   (c) 1.32 g/mL
   (b) 0.759 g/L   (d) 1.32 g/L
9. The ratio of the relative rate of effusion of methane, CH_4, to sulfur dioxide, SO_2, is:
   (a) $\frac{64}{16}$ (b) $\frac{16}{64}$ (c) $\frac{1}{4}$ (d) $\frac{2}{1}$
10. Measured at 65°C and 500. torr, the mass of 3.21 L of a gas is 3.5 g. The molar mass of this gas is:
    (a) 21 g/mole   (c) 24 g/mole
    (b) 46 g/mole   (d) 130 g/mole
11. Box A contains O_2 (molar mass = 32.0) at a pressure of 200 torr. Box B, which is identical to Box A in volume, contains twice as many molecules of CH_4 (molar mass = 16.0) as the molecules of O_2 in Box A. The temperatures of the gases are identical. The pressure in Box B is:
    (a) 100 torr (c) 400 torr
    (b) 200 torr (d) 800 torr
12. A 300. mL sample of oxygen, O_2, was collected over water at 23°C and 725 torr pressure. If the vapor pressure of water at 23°C is 21.0 torr, the volume of dry O_2 at STP would be:
    (a) 256 mL   (c) 341 mL
    (b) 351 mL   (d) 264 mL
13. A tank containing 0.01 mol of neon and 0.04 mol of helium shows a pressure of 1 atm. What is the partial pressure of neon in the tank?
    (a) 0.8 atm   (c) 0.2 atm
    (b) 0.01 atm   (d) 0.5 atm
14. How many liters of NO_2 (at STP) can be produced from 25.0 g of Cu reacting with concentrated nitric acid?

    $Cu(s) + 4\,HNO_3(aq) \longrightarrow$

    $\qquad Cu(NO_3)_2(aq) + 2\,H_2O(l) + 2\,NO_2\uparrow$

    (a) 4.41 L   (b) 8.82 L   (c) 17.6 L   (d) 44.8 L

15. How many liters of butane vapor are required to produce 2.0 L of CO_2 at STP?

    $2\,C_4H_{10}(g) + 13\,O_2(g) \longrightarrow$

    $\quad$ Butane $\qquad\qquad 8\,CO_2(g) + 10\,H_2O(g)$

    (a) 2.0 L   (b) 4.0 L   (c) 0.80 L   (d) 0.50 L
16. What volume of CO_2 (at STP) can be produced when 15.0 g of C_2H_6 and 50.0 g of O_2 are reacted?

    $2\,C_2H_6(g) + 7\,O_2(g) \longrightarrow 4\,CO_2(g) + 6\,H_2O(g)$

    (a) 20.0 L   (b) 22.4 L   (c) 35.0 L   (d) 5.6 L
17. Which of these gases has the highest density at STP?
    (a) N_2O (b) NO_2 (c) Cl_2 (d) SO_2
18. What is the density of CO_2 at 25°C and 0.954 atm pressure?
    (a) 1.72 g/L   (c) 0.985 g/L
    (b) 2.04 g/L   (d) 1.52 g/L
19. How many molecules are present in 0.025 mol of H_2 gas?
    (a) 1.5×10^{22} molecules
    (b) 3.37×10^{23} molecules
    (c) 2.40×10^{25} molecules
    (d) 1.50×10^{22} molecules
20. 5.60 L of a gas at STP has a mass of 13.0 g. What is the molar mass of the gas?
    (a) 33.2 g/mol   (c) 66.4 g/mol
    (b) 26.0 g/mol   (d) 52.0 g/mol

CHAPTER FOURTEEN   Water and the Properties of Liquids

True–False. *Answer the following as either true or false.*

1. Sodium, potassium, and calcium each react with cold water to form hydrogen gas and a metal hydroxide.

2. Calcium oxide reacts with water to form calcium hydroxide and hydrogen gas.
3. Water in a hydrate is known as water of hydration or water of crystallization.
4. Substances that can spontaneously lose their

water of hydration when exposed to air are said to be hygroscopic.

5. A substance that absorbs water from the air until it forms a solution is said to be deliquescent.

6. At pressures below 760 torr, water will boil above 100°C.

7. Evaporation is the escape of molecules from the liquid state to the vapor state.

8. Sublimation is the change from the vapor to the liquid state.

9. $CuSO_4 \cdot 5H_2O$ is named copper(II) sulfate pentahydrate.

10. The pressure exerted by a vapor in equilibrium with its liquid is known as the vapor pressure of the liquid.

11. The heat of vaporization of water is more than six times as large as its heat of fusion.

12. Of two liquids, the one having the higher vapor pressure at a given temperature will have the higher normal boiling point.

13. In treating city water supplies, fluorides are often added to prevent tooth deccay.

14. The molar heat of fusion of water is 335 J/g × 18.0 g/mol.

15. In treating city water supplies, chlorine is injected into the water to kill harmful bacteria before it is distributed to the public.

16. Hard water contains relatively large amounts of Ca^{2+} and Mg^{2+} ions.

17. Substances that evaporate readily are said to be volatile.

18. Metal oxides that react with water to form bases are known as basic anhydrides.

19. Nonmetal oxides that react with water to form acids are known as acid anhydrides.

20. SO_2 is the anhydride of H_2SO_4.

21. When P_2O_5 reacts with water, it would be expected to make the solution basic.

22. Hydrogen bonds are stronger than covalent bonds.

23. Substances that readily absorb water from the atmosphere are said to be hygroscopic.

24. Ice at 0°C is at the same temperature as water at 0°C, but ice contains less heat energy than the water.

25. The vapor pressure of a liquid depends on the temperature and the atmospheric pressure.

26. The bond angle between the atoms in a water molecule is about 90°.

27. Capillary action results from the adhesive forces within a liquid and the cohesive forces between the liquid and the walls of the container.

28. When mercury is placed in a glass tube, the meniscus will be convex.

Multiple Choice.   *Choose the correct answer to each of the following.*

1. The heat of fusion of water is:
   (a) 4.184 J/g   (c) 2.26 kJ/g
   (b) 335 J/g   (d) 2.26 kJ/mol

2. The heat of vaporization of water is:
   (a) 4.184 J/g   (c) 2.26 kJ/g
   (b) 335 J/g   (d) 2.26 kJ/mol

3. The specific heat of water is:
   (a) 4.184 J/g °C   (c) 2.26 kJ/g °C
   (b) 335 J/g °C   (d) 18 J/g °C

4. The density of water at 4°C is:
   (a) 1.0 g/mL   (c) 18.0 g/mL
   (b) 80 g/mL   (d) 14.7 lb/in.^{3}

5. SO_2 can be properly classified as a (n)
   (a) basic anhydride   (c) anhydrous salt
   (b) hydrate   (d) acid anhydride

6. When compared to H_2S, H_2Se, and H_2Te, water is found to have the highest boiling point because it:
   (a) has the lowest molar mass
   (b) is the smallest molecule
   (c) has the highest bonding
   (d) will form hydrogen bonds better than the others

7. In which of the following molecules will hydrogen bonding be important?
   (a) H—F   (c) H—Br

   (b) $\underset{\displaystyle H}{\overset{\displaystyle H}{S}}$ (d) $H-\overset{\displaystyle H}{\underset{\displaystyle H}{C}}-O-\overset{\displaystyle H}{\underset{\displaystyle H}{C}}-H$

8. Which of the following is an incorrect equation?
   (a) $H_2SO_4 + 2 NaOH \longrightarrow Na_2SO_4 + 2 H_2O$
   (b) $C_2H_6 + O_2 \longrightarrow 2 CO_2 + 3 H_2$
   (c) $2 H_2O \xrightarrow[H_2SO_4]{Electrolysis} 2 H_2 + O_2$
   (d) $Ca + 2 H_2O \longrightarrow H_2 + Ca(OH)_2$

9. Which of the following is an incorrect equation?
   (a) $C + H_2O(g) \xrightarrow{1000°C} CO(g) + H_2(g)$
   (b) $CaO + H_2O \longrightarrow Ca(OH)_2$

(c) $2 NO_2 + H_2O \longrightarrow 2 HNO_3$
(d) $Cl_2 + H_2O \longrightarrow HCl + HOCl$

10. How many kilojoules are required to change 85 g of water at 25°C to steam at 100.°C?
(a) 219 kJ (b) 27 kJ (c) 590 kJ (d) 192 kJ

11. A chunk of 0°C ice, mass 145 g, is dropped into 75 g of water at 62°C. The heat of fusion of water is 335 J/g. The result, after thermal equilibrium is attained, will be:
(a) 87 g of ice and 133 g of liquid water, all at 0°C
(b) 58 g of ice and 162 g of liquid water, all at 0°C
(c) 220 g of water at 7°C
(d) 220 g of water at 17°C

12. The formula for iron(II) sulfate heptahydrate is
(a) $Fe_2SO_4 \cdot 7 H_2O$ (c) $FeSO_4 \cdot 7 H_2O$
(b) $Fe(SO_4)_2 \cdot 6 H_2O$ (d) $Fe_2(SO_4)_3 \cdot 7 H_2O$

13. The process by which a solid changes directly to a vapor is called
(a) vaporization   (c) sublimation
(b) evaporation   (d) condensation

14. The anhydride of permanganic acid, $HMnO_4$, is
(a) MnO_3 (c) MnO_4
(b) Mn_2O_3 (d) Mn_2O_7

15. Hydrogen bonding
(a) occurs only between water molecules

(b) is stronger than covalent bonding
(c) can occur between NH_3 and H_2O
(d) results from strong attractive forces in ionic compounds

16. A liquid boils when
(a) the vapor pressure of the liquid equals the external pressure above the liquid
(b) the heat of vaporization exceeds the vapor pressure
(c) the vapor pressure equals one atmosphere
(d) the normal boiling temperature is reached

17. Consider two beakers, one containing 50 mL of liquid A and the other 50 mL of liquid B. The boiling point of A is 90°C and that of B is 72°C. Which of these statements is correct?
(a) A will evaporate faster than B.
(b) B will evaporate faster than A.
(c) Both A and B evaporate at the same rate.
(d) Insufficient data to answer the question.

18. 95.0 g of 0.0°C ice is added to exactly 100. g of water at 60.0°C. When the temperature of the mixture first reaches 0.0°C, the mass of ice still present is:
(a) 0.0 g   (c) 10.0 g
(b) 20.0 g   (d) 75.0 g

CHAPTER FIFTEEN   Solutions

True–False. *Answer the following as either true or false.*

1. Solubility describes the amount of one substance that will dissolve into another substance.
2. When solid NaCl dissolves in water, the sodium ions attract the positive end of the water dipole.
3. Bromine is more soluble in polar water than in nonpolar carbon tetrachloride because the polar water molecules help it form ions.
4. For most solids or gases dissolved in liquid, an increase in temperature results in an increase in solubility.
5. An increase in temperature almost always increases the rate at which a solid will dissolve in a liquid.

6. A supersaturated solution contains dissolved solute in equilibrium with undissolved solute.
7. A 1 molar solution contains 1 mole of solute per liter of solution.
8. When a solute is dissolved in a solvent, the freezing point of the solution will be higher than that of the pure solvent.
9. Freezing point depression, boiling point elevation, and vapor pressure lowering are colligative properties of solutions.
10. Liquids that mix with water in all proportions are usually ionic in solution or are polar substances.
11. Liquids that do not mix are said to be miscible.
12. As a general rule, sodium and potassium salts of the common ions are soluble in water.

13. The vapor pressure of the solvent in a solution is lower than the vapor pressure of the pure solvent.
14. Salt water has a higher boiling point than distilled water.
15. One definition of equivalent mass is the mass of a substance that will combine with, contain, replace, or in any other way be equivalent to 1 mol of hydrogen atoms.
16. One mole of H_2SO_4 will react with twice as much NaOH as 1 mol of HCl.
17. A solution containing a nonvolatile solute has a lower boiling point than the pure solvent as a result of having a lower vapor pressure than the pure solvent.
18. A sulfuric acid solution with a molarity of 4.0 M would have a normality of 2.0 N.
19. Large crystals dissolve faster than do smaller ones because they expose a larger surface area to the solvent.
20. A major use of a solvent is as a medium for chemical reactions.
21. Molar solutions are temperature dependent, because the volume of the solution can vary with the temperature but the number of moles of solute remains the same.
22. A 15% by mass solution contains 15 g of solute per 100 mL of solution.
23. The molarity of a solution is the number of moles of solute per liter of solution.
24. A 1.0 M HCl solution will also be 1.0 N HCl.
25. A solution containing 0.001 mol of solute in 1 mL of solvent is 1 M.

Multiple Choice.   *Choose the correct answer to each of the following.*

1. Which of the following is not a general property of solutions?
   (a) It is a homogeneous mixture of two or more substances.
   (b) It has a variable composition.
   (c) The dissolved solute breaks down to individual molecules.
   (d) The solution has the same chemical composition, the same chemical properties, and the same physical properties in every part.
2. If NaCl is soluble in water to the extent of 36.0 g NaCl/100 g H_2O at 20°C, then a solution at 20°C containing 45 g NaCl/150 g H_2O

would be:
(a) Dilute   (c) Supersaturated
(b) Saturated   (d) Unsaturated
3. If 5.00 g of NaCl are dissolved in 25.0 g of water, the percent of NaCl by mass is:
   (a) 16.7   (c) 0.20
   (b) 20.0   (d) No correct answer given
4. How many grams of 9.0% $AgNO_3$ solution will contain 5.3 g $AgNO_3$?
   (a) 47.7   (c) 59
   (b) 0.58   (d) No correct answer given
5. The molarity of a solution containing 2.5 mol of acetic acid, $HC_2H_3O_2$, in 400. mL of solution is:
   (a) 0.063 M (c) 0.103 M
   (b) 1.0 M (d) 6.3 M
6. What volume of 0.300 M KCl will contain 15.3 g of KCl?
   (a) 1.46 L   (c) 61.5 mL
   (b) 684 mL (d) 4.60 L
7. What mass of $BaCl_2$ will be required to prepare 200. mL of 0.150 M solution?
   (a) 0.750 g (b) 156 g (c) 6.25 g (d) 31.2 g
8. If 15 g of H_2SO_4 are dissolved in water to make 80. mL of solution, the normality is:
   (a) 0.19 N (b) 1.9 N (c) 6.1 N (d) 3.8 N

Problems 9–11 relate to the reaction

$$CaCO_3 + 2\ HCl \longrightarrow CaCl_2 + H_2O + CO_2$$

9. What volume of 6.0 M HCl will be needed to react with 0.350 mol of $CaCO_3$?
   (a) 42.0 mL   (c) 117 mL
   (b) 1.17 L   (d) 583 mL
10. If 400. mL of 2.0 M HCl react with excess $CaCO_3$, the volume of CO_2 produced, measured at STP, is:
    (a) 18 L   (b) 5.6 L   (c) 9.0 L   (d) 56 L
11. If 5.3 g of $CaCl_2$ were produced in the reaction, what was the molarity of the HCl used if 25 mL of it reacted with excess $CaCO_3$?
    (a) 3.8 M (b) 0.19 M (c) 0.38 M (d) 0.42 M
12. In the reaction $Ca + 2\ HCl \longrightarrow CaCl_2 + H_2$, the equivalent mass of Ca is:
    (a) 36.5 g   (b) 20.0 g   (c) 40.1 g   (d) 80.2 g
13. In the reaction

$$3\ HNO_3 + Cr(OH)_3 \longrightarrow Cr(NO_3)_3 + 3\ H_2O,$$

the equivalent mass of $Cr(OH)_3$ is:
(a) 34.3 g   (b) 52 g   (c) 103 g   (d) 309 g

14. If 20.0 g of the nonelectrolyte urea, $CO(NH_2)_2$, are dissolved in 25.0 g of of water, the freezing point of the solution will be:
    (a) $-2.48°C$ (c) $-24.8°C$
    (b) $-1.40°C$ (d) $-3.72°C$

15. When 256 g of a nonvolatile, nonelectrolyte unknown were dissolved in 500. g of H_2O, the freezing point was found to be $-2.79°C$. The molar mass of the unknown solute is:
    (a) 357   (b) 62.0   (c) 768   (d) 341

16. How many milliliters of 6.0 M H_2SO_4 must you use to prepare 500. mL of 0.20 M sulfuric acid solution?
    (a) 30   (b) 17   (c) 12   (d) 100

17. How many milliliters of water must be added to 200. mL of 1.40 M HCl to make a solution that is 0.500 M HCl?
    (a) 360 mL   (c) 140 mL
    (b) 560 mL   (d) 280 mL

18. Which procedure is most likely to increase the solubility of most solids in liquids?
    (a) Stirring
    (b) Pulverizing the solid
    (c) Heating the solution
    (d) Increasing the pressure

19. The addition of a crystal of $NaClO_3$ to a solution of $NaClO_3$ causes additional crystals to precipitate. The orginal solution was
    (a) Unsaturated   (c) Saturated
    (b) Dilute   (d) Supersaturated

20. Which of the following anions will not form a precipitate with silver ions, Ag^+?
    (a) Cl^- (b) NO_3^- (c) Br^- (d) CO_3^{2-}

21. Which of the following salts are considered to be soluble in water?
    (a) $BaSO_4$ (b) NH_4Cl (c) AgI (d) PbS

22. A solution of ethyl alcohol and benzene is 40% alcohol by volume. Which statement is correct?
    (a) The solution contains 40 mL of alcohol in 100 mL of solution.
    (b) The solution contains 60 mL of benzene in 100 mL of solution.
    (c) The solution contains 40 mL of alcohol in 100 g of solution.
    (d) The solution is made by dissolving 40 mL of alcohol in 60 mL of benzene.

23. The number of grams of solute needed to prepare 1.00 L of 3.00 N of $Ba(OH)_2$ is:
    (a) 171 g   (c) 514 g
    (b) 257 g   (d) 85.7 g

24. To what volume must 10.0 mL of 10.0 M HNO_3 be diluted to produce a 0.100 N solution?
    (a) 1.00 L   (c) 2.00 mL
    (b) 500 mL   (d) 1.50 L

CHAPTER SIXTEEN

Ionization: Acids, Bases, Salts

Many significant chemical reactions occur in aqueous solutions. Water is the medium for reactions that provide the foundation for successful life on this planet. Volcanos and hot springs yield acidic solutions formed from hydrochloric acid and sulfur dioxide. These acids mix with bases, such as calcium carbonate, dissolved from rocks and microscopic animals to form the salts of our oceans and the beauty of stalactites and stalagmites in limestone caves. Photosynthesis and respiration reactions cannot occur without an adequate balance between acids and bases. Even small deviations in this balance can be detrimental. In recent years, we have become very concerned about the effects of acid rain upon our environment. Consumers are rapidly becoming more informed about household products, such as shampoos, cleaning supplies, antacids, and paints, because of their effects upon our daily lives and on our planet.

Chapter Preview

16.1   Acids and Bases

The word *acid* is derived from the Latin *acidus*, meaning "sour" or "tart," and is also related to the Latin word *acetum*, meaning "vinegar." Vinegar has been known since antiquity as the product of the fermentation of wine and apple cider. The sour constituent of vinegar is acetic acid ($HC_2H_3O_2$).

Some of the characteristic properties commonly associated with acids are the following: Water solutions of acids taste sour and change the color of litmus, a vegetable dye, from blue to red. Water solutions of nearly all acids react with (1) metals such as zinc and magnesium to produce hydrogen gas, (2) bases to produce water and a salt, and (3) carbonates to produce carbon dioxide. These properties are due to the hydrogen ions, H^+, released by acids in a water solution.

Classically, a *base* is a substance capable of liberating hydroxide ions, OH^-, in water solution. Hydroxides of the alkali metals (Group IA) and alkaline earth metals (Group IIA), such as LiOH, NaOH, KOH, $Ca(OH)_2$, and $Ba(OH)_2$, are the most common inorganic bases. Water solutions of bases are called *alkaline solutions* or *basic solutions*. They have a bitter or caustic taste, a slippery, soapy feeling, the ability to change litmus from red to blue, and the ability to interact with acids to form a salt and water.

Several theories have been proposed to answer the question "What is an acid and a base?" One of the earliest, most significant of these theories was advanced in a doctoral thesis in 1884 by Svante Arrhenius (1859–1927), a Swedish scientist, who stated that an acid is a hydrogen-containing substance that dissociates to produce hydrogen ions, and that a base is a hydroxide-containing substance that dissociates to produce hydroxide ions in aqueous solutions. Arrhenius postulated that the hydrogen ions were produced by the dissociation of acids in water, and that the hydroxide ions were produced by the dissociation of bases in water:

$$HA \longrightarrow H^+ + A^-$$
Acid

$$MOH \longrightarrow M^+ + OH^-$$
Base

Thus, an acid solution contains an excess of hydrogen ions and a base an excess of hydroxide ions.

In 1923 the Brønsted–Lowry proton transfer theory was introduced by J. N. Brønsted (1897–1947), a Danish chemist, and T. M. Lowry (1847–1936), an English chemist. This theory states that an acid is a proton donor and a base is a proton acceptor.

Consider the reaction of hydrogen chloride gas with water to form hydrochloric acid:

$$HCl(g) + H_2O(l) \longrightarrow H_3O^+(aq) + Cl^- \ (aq) \tag{1}$$

In the course of the reaction, HCl donates, or gives up, a proton to form a Cl^- ion, and H_2O accepts a proton to form the H_3O^+ ion. Thus, HCl is an acid and H_2O is a base, according to the Brønsted–Lowry theory.

A hydrogen ion, H^+, is nothing more than a bare proton and does not exist by itself in an aqueous solution. In water a proton combines with a polar water molecule to form a hydrated hydrogen ion, H_3O^+ [that is, $H(H_2O)^+$], commonly called a **hydronium ion**. The proton is attracted to a polar water molecule, forming a coordinate-covalent bond with one of the two pairs of unshared electrons:

hydronium
ion

$$H^+ + H\!:\!\overset{\cdot\cdot}{\underset{H}{O}}\!: \longrightarrow \left[H\!:\!\overset{\cdot\cdot}{\underset{H}{O}}\!:\!H \right]^+$$

Hydronium ion

Note the electron structure of the hydronium ion. For simplicity of expression in equations, we often use H^+ instead of H_3O^+, with the explicit understanding that H^+ is always hydrated in solution.

Whereas the Arrhenius theory is restricted to aqueous solutions, the Brønsted–Lowry approach has applications in all media and has become the more important theory when the chemistry of substances in solutions other than water is studied. Ammonium chloride (NH_4Cl) is a salt, yet its water solution has an acidic reaction. From this test we must conclude that NH_4Cl has acidic properties. The Brønsted–Lowry explanation shows that the ammonium ion, NH_4^+, is a proton donor, and water is the proton acceptor:

$$NH_4^+ \rightleftharpoons NH_3 + H^+ \tag{2}$$
Acid   Base   Acid

$$NH_4^+ + H_2O \longrightarrow H_3O^+ + NH_3 \tag{3}$$
Acid   Base   Acid   Base

> **A Brønsted–Lowry acid is a proton (H^+) donor.**
> **A Brønsted–Lowry base is a proton (H^+) acceptor.**

The Brønsted–Lowry theory also applies to certain cases where no solution is involved. For example, in the reaction of hydrogen chloride and ammonia gases, HCl is the proton donor and NH_3 is the base. [Remember that (g) after a formula in equations stands for a gas.]

$$HCl(g) + NH_3(g) \longrightarrow NH_4^+ + Cl^- \qquad\qquad (4)$$

Acid   Base   Acid   Base

When a Brønsted–Lowry acid donates a proton, it forms the conjugate base of that acid. When a base accepts a proton, it forms the conjugate acid of that base.

In equations (1), (3), and (4) a conjugate acid and base are produced as products. The formulas of a conjugate acid–base pair differ by one proton (H^+). In equation (1) the conjugate acid–base pairs are $HCl-Cl^-$ and $H_3O^+-H_2O$. Cl^- is the conjugate base of HCl, and HCl is the conjugate acid of Cl^-. H_2O is the conjugate base of H_3O^+, and H_3O^+ is the conjugate acid of H_2O.

conjugate acid–base pair

$$HCl(g) + H_2O(l) \longrightarrow Cl^-(aq) + H_3O^+(aq)$$

conjugate acid–base pair

acid   base   base   acid

In equation (3) the conjugate acid–base pairs are $NH_4^+-NH_3$ and $H_3O^+-H_2O$; in equation (4) they are $HCl-Cl^-$ and $NH_4^+-NH_3$.

EXAMPLE 16.1   Write the formula for (a) the conjugate base of H_2O and of HNO_3, and (b) the conjugate acid of SO_4^{2-} and of $C_2H_3O_2^-$. The difference between an acid or a base and its conjugate is one proton, H^+.

(a) To write the conjugate base of an acid, remove one proton from the acid formula. Thus,

$$H_2O \xrightarrow{-H^+} OH^- \quad \text{(Conjugate base)}$$

$$HNO_3 \xrightarrow{-H^+} NO_3^- \quad \text{(Conjugate base)}$$

Note that, by removing an H^+, the conjugate base becomes more negative than the acid by one minus charge.

(b) To write the conjugate acid of a base, add one proton to the formula of the base. Thus,

$$SO_4^{2-} \xrightarrow{+H^+} HSO_4^- \quad \text{(Conjugate acid)}$$

$$C_2H_3O_2^- \xrightarrow{+H^+} HC_2H_3O_2 \quad \text{(Conjugate acid)}$$

In each case the conjugate acid becomes more positive than the base by one positive charge due to the addition of H^+.

PRACTICE Indicate the conjugate base for each of the following acids:
(a) H_2CO_3 (b) HNO_2 (c) $HC_2H_3O_2$
Answers: (a) HCO_3^- (b) NO_2^- (c) $C_2H_3O_2^-$

PRACTICE Indicate the conjugate acids for each of the following bases:
(a) HSO_4^- (b) Cl^- (c) OH^-
Answers: (a) H_2SO_4 (b) HCl (c) H_2O

A more general concept of acids and bases was introduced by Gilbert N. Lewis. The Lewis theory deals with the way in which a substance with an unshared pair of electrons reacts in an acid–base type of reaction. According to this theory a base is any substance that has an unshared pair of electrons (electron-pair donor), and an acid is any substance that will attach itself to or accept a pair of electrons. In the reaction

$$H^+ + \underset{\displaystyle \underset{H}{\overset{\displaystyle H}{}}}{:\!\ddot{N}\!:H} \longrightarrow H:\underset{\displaystyle \underset{H}{\overset{\displaystyle H}{}}}{\ddot{N}}\!:H^+$$

Acid   Base

H^+ is a Lewis acid and $:NH_3$ is a Lewis base. According to the Lewis theory, substances other than proton donors (for example, BF_3) behave as acids:

$$F:\underset{\displaystyle F}{\overset{\displaystyle F}{\ddot{B}}} + :\underset{\displaystyle H}{\overset{\displaystyle H}{\ddot{N}}}:H \longrightarrow F:\underset{\displaystyle F}{\overset{\displaystyle F}{\ddot{B}}}:\underset{\displaystyle H}{\overset{\displaystyle H}{\ddot{N}}}:H$$

Acid   Base

The Lewis and Brønsted–Lowry bases are identical because, to accept a proton, a base must have an unshared pair of electrons.

The three theories are summarized in Table 16.1. These theories explain how acid–base reactions occur. We will generally use the theory that best explains the reaction that is under consideration. Most of our examples will refer to aqueous solutions. It is important to realize that in an aqueous acidic solution the H^+ ion concentration is always greater than OH^- ion concentration. And, vice versa, in an aqueous basic solution the OH^- ion concentration is always greater than the H^+ ion concentration. When the H^+ and OH^- ion concentrations in a solution are equal, the solution is neutral; that is, it is neither acidic nor basic.

TABLE 16.1

Summary of Acid–Base Definitions

Theory	Acid	Base
Arrhenius	A hydrogen-containing substance that produces hydrogen ions in aqueous solution	A hydroxide-containing substance that produces hydroxide ions in aqueous solution
Brønsted–Lowry	A proton (H^+) donor	A proton (H^+) acceptor
Lewis	Any species that will bond to an unshared pair of electrons (electron-pair acceptor)	Any species that has an unshared pair of electrons (electron-pair donor)

16.2   Reactions of Acids

In aqueous solutions the H^+ or H_3O^+ ions are responsible for the characteristic reactions of acids. All the following reactions are in an aqueous medium.

Reaction with Metals   Acids react with metals that lie above hydrogen in the activity series of elements to produce hydrogen and a salt (see Section 18.5).

$$acid + metal \longrightarrow hydrogen + salt$$
$$2\ HCl(aq) + Ca(s) \longrightarrow H_2\uparrow + CaCl_2(aq)$$
$$H_2SO_4(aq) + Mg(s) \longrightarrow H_2\uparrow + MgSO_4(aq)$$
$$6\ HC_2H_3O_2(aq) + 2\ Al(s) \longrightarrow 3\ H_2\uparrow + 2\ Al(C_2H_3O_2)_3(aq)$$

Acids such as nitric acid (HNO_3) are oxidizing substances (see Chapter 18) and react with metals to produce water instead of hydrogen. For example,

$$3\ Zn(s) + 8\ HNO_3(dilute) \longrightarrow 3\ Zn(NO_3)_2(aq) + 2\ NO(g) + 4\ H_2O$$

Reaction with Bases   The interaction of an acid and a base is called a *neutralization reaction*. In aqueous solutions, the products of this reaction are a salt and water.

$$\text{acid} + \text{base} \longrightarrow \text{salt} + \text{water}$$

$$HBr(aq) + KOH(aq) \longrightarrow KBr(aq) + H_2O$$

$$2\ HNO_3(aq) + Ca(OH)_2(aq) \longrightarrow Ca(NO_3)_2(aq) + 2\ H_2O$$

$$2\ H_3PO_4(aq) + 3\ Ba(OH)_2(aq) \longrightarrow Ba_3(PO_4)_2\downarrow + 6\ H_2O$$

Reaction with Metal Oxides   This reaction is closely related to that of an acid with a base. With an aqueous acid, the products are a salt and water.

$$\text{acid} + \text{metal oxide} \longrightarrow \text{salt} + \text{water}$$

$$2\ HCl(aq) + Na_2O(s) \longrightarrow 2\ NaCl(aq) + H_2O$$

$$H_2SO_4(aq) + MgO(s) \longrightarrow MgSO_4(aq) + H_2O$$

$$6\ HCl(aq) + Fe_2O_3(s) \longrightarrow 2\ FeCl_3(aq) + 3\ H_2O$$

Reaction with Carbonates   Many acids react with carbonates to produce carbon dioxide, water, and a salt. Carbonic acid (H_2CO_3) is not the product, because it is unstable and decomposes into water and carbon dioxide.

$$\text{acid} + \text{carbonate} \longrightarrow \text{salt} + \text{water} + \text{carbon dioxide}$$

$$2\ HCl(aq) + Na_2CO_3(aq) \longrightarrow 2\ NaCl(aq) + H_2O + CO_2\uparrow$$

$$H_2SO_4(aq) + MgCO_3(s) \longrightarrow MgSO_4(aq) + H_2O + CO_2\uparrow$$

16.3   Reactions of Bases

The OH^- ions are responsible for the characteristic reactions of bases. All the following reactions are in an aqueous medium.

Reaction with Acids   Bases react with acids to produce a salt and water. See reaction of acids with bases in Section 16.2(b).

Amphoteric Hydroxides   Hydroxides of certain metals, such as zinc, aluminum, and chromium, are **amphoteric**; that is, they are capable of reacting as either an acid or a base. When treated with a strong acid, they behave like bases; when reacted with a strong base, they behave like acids.

amphoteric

$$Zn(OH)_2(s) + 2\ HCl(aq) \longrightarrow ZnCl_2(aq) + 2\ H_2O$$

$$Zn(OH)_2(s) + 2\ NaOH(aq) \longrightarrow Na_2Zn(OH)_4(aq)$$

Reaction of NaOH and KOH with Certain Metals   Some amphoteric metals react directly with the strong bases, sodium hydroxide and potassium hydroxide, to produce hydrogen.

$$\text{base + metal + water} \longrightarrow \text{salt + hydrogen}$$
$$2\,NaOH(aq) + Zn(s) + 2\,H_2O \longrightarrow Na_2Zn(OH)_4(aq) + H_2\uparrow$$
$$2\,KOH(aq) + 2\,Al(s) + 6\,H_2O \longrightarrow 2\,KAl(OH)_4(aq) + 3\,H_2\uparrow$$

Reaction with Salts   Bases will react with many salts in solution due to the formation of insoluble metal hydroxides.

$$\text{base + salt} \longrightarrow \text{metal hydroxide}\downarrow + \text{salt}$$
$$2\,NaOH(aq) + MnCl_2(aq) \longrightarrow Mn(OH)_2\downarrow + 2\,NaCl(aq)$$
$$3\,Ca(OH)_2(aq) + 2\,FeCl_3(aq) \longrightarrow 2\,Fe(OH)_3\downarrow + 3\,CaCl_2(aq)$$
$$2\,KOH(aq) + CuSO_4(aq) \longrightarrow Cu(OH)_2\downarrow + K_2SO_4(aq)$$

16.4   Salts

Salts are very abundant in nature. Most of the rocks and minerals of the earth's mantle are salts of one kind or another. Huge quantities of dissolved salts also exist in the oceans. Salts may be considered to be compounds that have been derived from acids and bases. They consist of positive metal or ammonium ions (H^+ excluded) combined with negative nonmetal ions (OH^- and O^{2-} excluded). The positive ion is the base counterpart and the nonmetal ion is the acid counterpart:

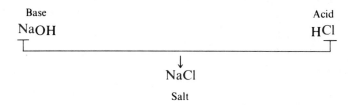

Salts are usually crystalline and have high melting and boiling points.

From a single acid such as hydrochloric acid (HCl), we can produce many chloride salts by replacing the hydrogen with metal ions (for example, NaCl, KCl, RbCl, $CaCl_2$, $NiCl_2$). Hence, the number of known salts greatly exceeds the number of known acids and bases. Salts are ionic compounds. If the hydrogen atoms of a binary acid are replaced by a nonmetal, the resulting compound has covalent bonding and is therefore not considered to be a salt (for example, PCl_3, S_2Cl_2, Cl_2O, NCl_3, ICl).

A review of Chapter 6 on the nomenclature of acids, bases, and salts may be
· beneficial at this point.

16.5   Electrolytes and Nonelectrolytes

Some of the most convincing evidence as to the nature of chemical bonding
within a substance is the ability (or lack of ability) of a water solution of the
substance to conduct electricity.

We can show that solutions of certain substances are conductors of
electricity by using a simple conductivity apparatus consisting of a pair of
electrodes connected to a voltage source through a light bulb and switch (see
Figure 16.1). If the medium between the electrodes is a conductor of electricity,
the light bulb will glow when the switch is closed. When chemically pure water is

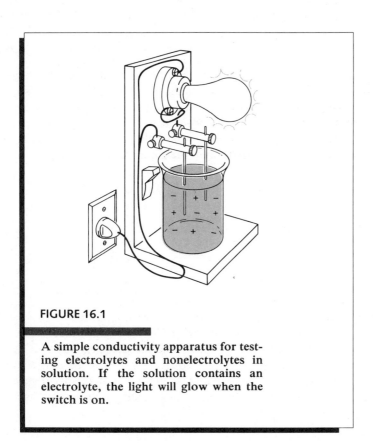

FIGURE 16.1

A simple conductivity apparatus for testing electrolytes and nonelectrolytes in
solution. If the solution contains an
electrolyte, the light will glow when the
switch is on.

TABLE 16.2

Representative Electrolytes and Nonelectrolytes

Electrolytes	Nonelectrolytes
H_2SO_4	$C_{12}H_{22}O_{11}$ (sugar)
HCl	C_2H_5OH (ethyl alcohol)
HNO_3	$C_2H_4(OH)_2$ (ethylene glycol)
NaOH	$C_3H_5(OH)_3$ (glycerol)
$HC_2H_3O_2$	CH_3OH (methyl alcohol)
NH_4OH	$CO(NH_2)_2$ (urea)
K_2SO_4	O_2
$NaNO_3$	H_2O

placed in the beaker and the switch is closed, the light does not glow, indicating that water is a virtual nonconductor. When we dissolve a small amount of sugar in the water and test the solution, the light still does not glow, showing that a sugar solution is also a nonconductor. But, when a small amount of salt, NaCl, is dissolved in water and this solution is tested, the light glows brightly. Thus, the salt solution conducts electricity. A fundamental difference exists between the chemical bonding of sugar and that of salt. Sugar is a covalently bonded (molecular) substance; salt is a substance with ionic bonds.

electrolyte

nonelectrolyte

Substances whose aqueous solutions are conductors of electricity are called **electrolytes**. Substances whose solutions are nonconductors are known as **nonelectrolytes**. The classes of compounds that are electrolytes are acids, bases, and salts. Solutions of certain oxides also are conductors because the oxides form an acid or a base when dissolved in water. One major difference between electrolytes and nonelectrolytes is that electrolytes are capable of producing ions in solution, whereas nonelectrolytes do not have this property. Solutions that contain a sufficient number of ions will conduct an electric current. Although pure water is essentially a nonconductor, many city water supplies contain enough dissolved ionic matter to cause the light to glow dimly when the water is tested in a conductivity apparatus. Table 16.2 lists some common electrolytes and nonelectrolytes.

Acids, bases, and salts are electrolytes.

16.6   Dissociation and Ionization of Electrolytes

Arrhenius received the 1903 Nobel Prize in chemistry for his work on electrolytes. He stated that a solution conducts electricity because the solute dissociates immediately upon dissolving into electrically charged particles called *ions*. The movement of these ions toward oppositely charged electrodes causes the solution to be a conductor. According to his theory, solutions that are relatively poor conductors contain electrolytes that are only partly dissociated. Arrhenius also believed that ions exist in solution whether or not an electric current is present. In other words, the electric current does not cause the formation of ions. Positive ions, attracted to the cathode, are cations; negative ions, attracted to the anode, are anions.

dissociation

We have seen that sodium chloride crystals consist of sodium and chloride ions held together by ionic bonds. **Dissociation** is the process by which the ions of a salt separate as the salt dissolves. When placed in water, the sodium and chloride ions are attracted by the polar water molecules, which surround each ion as it dissolves. In water, the salt dissociates, forming hydrated sodium and chloride ions (see Figure 16.2). The sodium and chloride ions in solution are bonded to a specific number of water dipoles and have less attraction for each other than they had in the crystalline state. The equation representing this dissociation is

$$NaCl(s) + (x + y)\, H_2O \longrightarrow Na^+(H_2O)_x + Cl^-(H_2O)_y$$

A simplified dissociation equation in which the water is omitted but understood to be present is

$$NaCl \longrightarrow Na^+ + Cl^-$$

It is important to remember that sodium chloride exists in an aqueous solution as hydrated ions and not as NaCl units, even though the formula NaCl or $Na^+ + Cl^-$ is often used in equations.

The chemical reactions of salts in solution are the reactions of their ions. For example, when sodium chloride and silver nitrate react and form a precipitate of silver chloride, only the Ag^+ and Cl^- ions participate in the reaction. The Na^+ and NO_3^- remain as ions in solution.

$$Ag^+ + Cl^- \longrightarrow AgCl\downarrow$$

In many cases the number of molecules of water associated with a particular ion is known. For example, the blue color of the copper(II) ion is due to the hydrated ion $Cu(H_2O)_4^{2+}$. The hydration of ions can be demonstrated in a striking way with cobalt(II) chloride. When cobalt(II) chloride hexahydrate is dissolved in water, a pink solution forms due to the $Co(H_2O)_6^{2+}$ ions. If concentrated hydrochloric acid is added to this pink solution, the color gradually changes to blue. If water is then added to the blue solution, the color changes to

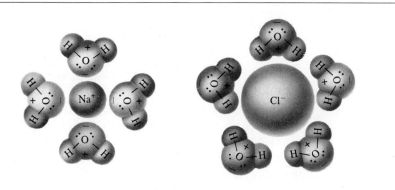

FIGURE 16.2

Hydrated sodium and chloride ions. When sodium chloride dissolves in water, each Na⁺ and Cl⁻ ion becomes surrounded by water molecules. The negative end of the water dipole is attracted to the Na⁺ ion, and the positive end is attracted to the Cl⁻ ion.

pink again. These color changes are due to the exchange of water molecules and chloride ions on the cobalt ion. The $CoCl_4^{2-}$ ion is blue. Thus, the hydration of the cobalt ion is a reversible or equilibrium reaction (see Chapter 17). The equilibrium equation representing these changes is

$$Co(H_2O)_6^{2+} + 4\ Cl^- \rightleftharpoons CoCl_4^{2-} + 6\ H_2O$$
Pink   Blue

ionization   **Ionization** is the formation of ions; it occurs as a result of chemical reaction of certain substances with water. Glacial acetic acid (100% $HC_2H_3O_2$) is a liquid that behaves as a nonelectrolyte when tested by the method described in Section 16.5. But a water solution of acetic acid conducts an electric current (as indicated by the dull-glowing light of the conductivity apparatus). The equation for the reaction with water, which forms hydronium and acetate ions, is

$$HC_2H_3O_2 + H_2O \rightleftharpoons H_3O^+ + C_2H_3O_2^-$$
Acid   Base   Acid   Base

or, in the simplified equation,

$$HC_2H_3O_2 \rightleftharpoons H^+ + C_2H_3O_2^-$$

In this ionization reaction, water serves not only as a solvent but also as a base according to the Brønsted–Lowry theory.

Hydrogen chloride is predominantly covalently bonded, but when dissolved in water it reacts to form hydronium and chloride ions:

$$HCl(g) + H_2O \longrightarrow H_3O^+ + Cl^-$$

When a hydrogen chloride solution is tested for conductivity, the light glows brilliantly, indicating many ions in the solution.

Ionization occurs in each of the above two reactions with water, producing ions in solution. The necessity for water in the ionization process can be demonstrated by dissolving hydrogen chloride in a nonpolar solvent such as benzene, and testing the solution for conductivity. The solution fails to conduct electricity, indicating that no ions are produced.

The terms *dissociation* and *ionization* are often used interchangeably to describe processes taking place in water. But, strictly speaking, the two are different. In the dissociation of a salt, the salt already exists as ions; when it dissolves in water, the ions separate, or dissociate, and increase in mobility. In the ionization process, ions are produced by the reaction of a compound with water.

16.7   Strong and Weak Electrolytes

strong electrolyte

weak electrolyte

Electrolytes are classified as strong or weak depending on the degree, or extent, of dissociation or ionization. **Strong electrolytes** are essentially 100% ionized in solution; **weak electrolytes** are much less ionized (based on comparing 0.1 M solutions). Most electrolytes are either strong or weak, with a few classified as moderately strong or weak. Most salts are strong electrolytes. Acids and bases that are strong electrolytes (highly ionized) are called *strong acids* and *strong bases*. Acids and bases that are weak electrolytes (slightly ionized) are called *weak acids* and *weak bases*.

For equivalent concentrations, solutions of strong electrolytes contain many more ions than do solutions of weak electrolytes. As a result, solutions of strong electrolytes are better conductors of electricity. Consider the two solutions, 1 M HCl and 1 M HC$_2$H$_3$O$_2$. Hydrochloric acid is almost 100% ionized; acetic acid is about 1% ionized. Thus HCl is a strong acid, and HC$_2$H$_3$O$_2$ is a weak acid. Hydrochloric acid has about 100 times as many hydronium ions in solution as acetic acid, making the HCl solution much more acidic.

One can distinguish between strong and weak electrolytes experimentally using the apparatus described in Section 16.5. A 1 M HCl solution causes the light to glow brilliantly, but a 1 M HC$_2$H$_3$O$_2$ solution causes only a dim glow. In a similar fashion the strong base sodium hydroxide (NaOH) may be distinguished from the weak base ammonium hydroxide (NH$_4$OH). The ionization of a weak electrolyte in water is represented by an equilibrium equation showing that both the un-ionized and ionized forms are present in solution. In the equilibrium equation of HC$_2$H$_3$O$_2$ and its ions, we say that the equilibrium lies "far to the

TABLE 16.3

Strong and Weak Electrolytes

Strong electrolytes	Weak electrolytes
Most soluble salts	$HC_2H_3O_2$ — *Ace*
H_2SO_4	H_2CO_3
HNO_3	HNO_2
HCl	H_2SO_3
HBr	H_2S
$HClO_4$	$H_2C_2O_4$
$NaOH$	H_3BO_3
KOH	$HClO$
$Ca(OH)_2$	NH_4OH
$Ba(OH)_2$	HF

left" because relatively few hydrogen and acetate ions are present in solution:

$$HC_2H_3O_2(aq) \rightleftharpoons H^+ + C_2H_3O_2^-$$

We have previously used a double arrow in an equation to represent reversible processes in the equilibrium between dissolved and undissolved solute in a saturated solution. A double arrow ($\rightleftharpoons$) is also used in the ionization equation of soluble weak electrolytes to indicate that the solution contains a considerable amount of the un-ionized compound in equilibrium with its ions in solution. (See Section 17.1 for a discussion of reversible reactions). A single arrow is used to indicate that the electrolyte is essentially all in the ionic form in the solution. For example, nitric acid is a strong acid; nitrous acid is a weak acid. Their ionization equations in water may be indicated as

$$HNO_3(aq) \xrightarrow{H_2O} H^+ + NO_3^-$$
$$HNO_2(aq) \xrightleftharpoons{H_2O} H^+ + NO_2^-$$

Practically all soluble salts; acids such as sulfuric, nitric, and hydrochloric acids; and bases such as sodium, potassium, calcium, and barium hydroxides are strong electrolytes. Weak electrolytes include numerous other acids and bases such as acetic acid, nitrous acid, carbonic acid, and ammonium hydroxide. The terms *strong acid*, *strong base*, *weak acid*, and *weak base* refer to whether an acid or base is a strong or weak electrolyte. A brief list of strong and weak electrolytes is given in Table 16.3.

Electrolytes yield two or more ions per formula unit upon dissociation, the actual number being dependent on the compound. Dissociation is complete or nearly complete for nearly all soluble salts and for certain other strong electrolytes such as those given in Table 16.3. The following are dissociation equations for several strong electrolytes. In all cases the ions are actually hydrated.

$$NaOH \xrightarrow{H_2O} Na^+ + OH^-$$ 2 ions in solution per formula unit

$$Na_2SO_4 \xrightarrow{H_2O} 2\,Na^+ + SO_4^{2-}$$ 3 ions in solution per formula unit

$$AlCl_3 \xrightarrow{H_2O} Al^{3+} + 3\,Cl^-$$ 4 ions in solution per formula unit

$$Fe_2(SO_4)_3 \xrightarrow{H_2O} 2\,Fe^{3+} + 3\,SO_4^{2-}$$ 5 ions in solution per formula unit

One mole of NaCl will give 1 mol of Na^+ ions and 1 mol of Cl^- ions in solution, assuming complete dissociation of the salt. One mole of $CaCl_2$ will give 1 mol of Ca^{2+} ions and 2 mol of Cl^- ions in solution.

$$NaCl \xrightarrow{H_2O} Na^+ + Cl^-$$

1 mol   1 mol   1 mol

$$CaCl_2 \xrightarrow{H_2O} Ca^{2+} + 2\,Cl^-$$

1 mol   1 mol   2 mol

EXAMPLE 16.2   What is the molarity of each ion in a solution of (a) 2.0 M NaCl, and (b) 0.40 M K_2SO_4? (Assume complete dissociation.)

(a) According to the dissociation equation,

$$NaCl \xrightarrow{H_2O} Na^+ + Cl^-$$

1 mol   1 mol   1 mol

the concentration of Na^+ is equal to that of NaCl (1 mol NaCl $\longrightarrow$ 1 mol Na^+) and the concentration of Cl^- is also equal to that of NaCl. Therefore, the concentrations of the ions in 2.0 M NaCl are 2.0 M Na^+ and 2.0 M Cl^-.

(b) According to the dissociation equation,

$$K_2SO_4 \xrightarrow{H_2O} 2\,K^+ + SO_4^{2-}$$

1 mol   2 mol   1 mol

The concentration of K^+ is twice that of K_2SO_4 and the concentration of SO_4^{2-} is equal to that of K_2SO_4. Therefore, the concentrations of the ions in 0.40 M K_2SO_4 are 0.80 M K^+ and 0.40 M SO_4^{2-}.

PRACTICE   What is the molarity of each ion in a solution of (a) 0.050 M of $MgCl_2$, and (b) 0.070 M of $AlCl_3$?

Answers: (a) 0.050 M Mg^{2+}, 0.1 M Cl^- (b) 0.070 M Al^{3+}, 0.21 M Cl^-

Colligative Properties of Electrolyte Solutions

We have learned that when 1 mol of sucrose, a nonelectrolyte, is dissolved in 1000 g of water, the solution freezes at $-1.86°C$. When 1 mol of NaCl is dissolved in 1000 g of water, the freezing point of the solution is not $-1.86°C$, as might be expected, but is closer to $-3.72°C$ (-1.86×2). The reason for the lower freezing point is that 1 mol of NaCl in solution produces 2 mol of particles ($2 \times 6.022 \times 10^{23}$ ions) in solution. Thus, the freezing point depression produced by 1 mol of NaCl is essentially equivalent to that produced by 2 mol of a nonelectrolyte. An electrolyte such as $CaCl_2$, which yields three ions in water, gives a freezing point depression of about three times that of a nonelectrolyte. These freezing point data provide additional evidence that electrolytes dissociate when dissolved in water. The other colligative properties are similarly affected by substances that yield ions in aqueous solutions.

16.8   Ionization of Water

The more we study chemistry, the more intriguing the little molecule of water becomes. Two equations commonly used to show how water ionizes are

$$H_2O + H_2O \rightleftharpoons H_3O^+ + OH^-$$

   Acid   Base   Acid   Base

and

$$H_2O \rightleftharpoons H^+ + OH^-$$

The first equation represents the Brønsted–Lowry concept, with water reacting as both an acid and a base, forming a hydronium ion and a hydroxide ion. The second equation is a simplified version, indicating that water ionizes to give a hydrogen and a hydroxide ion. Actually, the proton, H^+, is hydrated and exists as a hydronium ion. In either case equal molar amounts of acid and base are produced so that water is neutral, having neither H^+ nor OH^- ions in excess. The ionization of water at 25°C produces an H^+ ion concentration of 1.0×10^{-7} mole per liter and an OH^- ion concentration of 1.0×10^{-7} mole per liter. These concentrations are usually expressed as

$$[H^+] \text{ or } [H_3O^+] = 1.0 \times 10^{-7} \text{ mol/L}$$
$$[OH^-] = 1.0 \times 10^{-7} \text{ mol/L}$$

These figures mean that about two out of every billion water molecules are ionized. This amount of ionization, small as it is, is a significant factor in the behavior of water in many chemical reactions.

The square brackets, [ ], indicate that the concentration is in moles per liter. Thus $[H^+]$ means the concentration of H^+ in moles per liter.

16.9   Introduction to pH

The acidity of an aqueous solution depends on the concentration of hydrogen or hydronium ions. The acidity of solutions involved in a chemical reaction is often critically important, especially for biochemical reactions. The pH scale of acidity was devised to fill the need for a simple, convenient numerical way to state the acidity of a solution. Values on the pH scale are obtained by mathematical conversion of H^+ ion concentrations to pH by these expressions:

$$pH = \log \frac{1}{[H^+]} \quad \text{or} \quad pH = -\log[H^+]$$

pH

where $[H^+] = H^+$ or H_3O^+ ion concentration in moles per liter. The **pH** is defined as the logarithm (log) of the reciprocal of the H^+ or H_3O^+ ion concentration in moles per liter. The scale itself is based on the H^+ concentration in water at 25°C. At this temperature, water has an H^+ concentration of 1×10^{-7} mole per liter and is calculated to have a pH of 7.

$$pH = \log \frac{1}{[H^+]} = \log \frac{1}{[1 \times 10^{-7}]} = \log 1 \times 10^7 = 7$$

By an alternate and mathematically equivalent definition, pH is the *negative* logarithm of the H^+ or H_3O^+ concentration in moles per liter:

$$pH = -\log[H^+] = -\log[1 \times 10^{-7}] = -(-7) = 7$$

The pH of pure water at 25°C is 7 and is said to be neutral; that is, it is neither acidic nor basic, because the concentrations of H^+ and OH^- are equal. Solutions that contain more H^+ ions than OH^- ions have pH values less than 7, and solutions that contain less H^+ ions than OH^- ions have values greater than 7.

> **pH < 7.00**   **Acidic solution**
> **pH = 7.00**   **Neutral solution**
> **pH > 7.00**   **Basic solution**

When $[H^+] = 1 \times 10^{-5}$ mol/L, pH = 5   (acidic)
When $[H^+] = 1 \times 10^{-9}$ mol/L, pH = 9   (basic)

Instead of saying that the hydrogen ion concentration in the solution is 1×10^{-5} mole per liter, it is customary to say that the pH of the solution is 5. The smaller the pH value, the more acidic the solution (see Figure 16.3).

At a given molarity a strong acid is more acidic (has a higher H^+ concentration) than a weak acid. For example, in a 0.100 *M* concentration, the

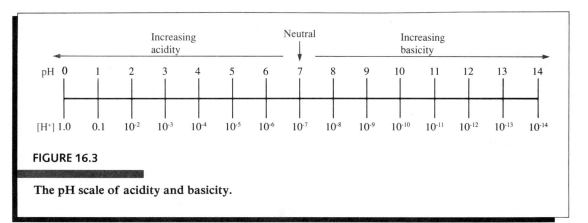

FIGURE 16.3

The pH scale of acidity and basicity.

pH of HCl is 1.00 and that of $HC_2H_3O_2$ is 2.87. As a weak acid is made more dilute, greater percentages of its molecules ionize, and the pH tends to approach that of a strong acid at comparable dilutions. This behavior is illustrated by the following comparative data for hydrochloric acid (100% ionized) and acetic acid.

HCl solution			$HC_2H_3O_2$ solution		
M	**pH**	**% ionized**	**M**	**pH**	**% ionized**
0.100	1.00	100	0.100	2.87	1.35
0.0100	2.00	100	0.0100	3.37	4.27
0.00100	3.00	100	0.00100	3.90	12.6

The pH scale, along with its interpretation, is given in Table 16.4, and Table 16.5 lists the pH of some common solutions. Note that a change of only 1 pH unit means a tenfold increase or decrease in H^+ ion concentration. For example, a solution with a pH of 3.0 is ten times more acidic than a solution with a pH of 4.0. A simplified method of determining pH from $[H^+]$ follows:

$$[H^+] = 1 \times 10^{-5}$$

When this number   pH = this number (5)
is exactly 1   pH = 5

$$[H^+] = 2 \times 10^{-5}$$

When this number   pH is between this number and
is between 1 and 10   next lower number (4 and 5)
                           pH = 4.7

TABLE 16.4		
The pH Scale for Expressing Acidity		
$[H^+]$ (mol/L)	pH	
1×10^{-14}	14	↑
1×10^{-13}	13	
11×10^{-12}	12	Increasing
1×10^{-11}	11	basicity
1×10^{-10}	10	
1×10^{-9}	9	
1×10^{-8}	8	
1×10^{-7}	7	Neutral
1×10^{-6}	6	
1×10^{-5}	5	
1×10^{-4}	4	Increasing
1×10^{-3}	3	acidity
1×10^{-2}	2	
1×10^{-1}	1	
1×10^{0}	0	↓

TABLE 16.5	
The pH of Some Common Solutions	
Solution	pH
Gastric juice	1.0
0.1 M HCl	1.0
Lemon juice	2.3
Vinegar	2.8
0.1 M $HC_2H_3O_2$	2.9
Orange juice	3.7
Tomato juice	4.1
Coffee, black	5.0
Urine	6.0
Milk	6.6
Pure water (25°C)	7.0
Blood	7.4
Household ammonia	11.0
1 M NaOH	14.0

logarithm

Calculation of the pH value corresponding to any H^+ ion concentration requires the use of logarithms. Logarithms are exponents. The **logarithm** (log) of a number is simply the power to which 10 must be raised to give that number. Thus the log of 100 is 2 ($100 = 10^2$), and the log of 1000 is 3 ($1000 = 10^3$). The log of 500 is 2.70, but you cannot easily determine this value without a scientific calculator. Determining a logarithm used to be a tedious task requiring a published standard log table, but with the availability of affordable calculators with log capability, the log table method is virtually obsolete.

Let us determine the pH of a solution with $[H^+] = 2 \times 10^{-5}$. The exponent (-5) indicates that the pH is between 4 and 5. Enter 2×10^{-5} into your calculator and press the log key. The number -4.69. will be displayed. The pH is then

$$pH = -\log[H^+] = -(-4.69\ldots) = 4.7$$

EXAMPLE 16.3

What is the pH of a solution with an $[H^+]$ of (a) 1.0×10^{-11} and (b) 6.0×10^{-4}?

(a) $[H^+] = 1.0 \times 10^{-11}$
$pH = -\log(1.0 \times 10^{-11})$
$pH = 11$

(b) $[H^+] = 6.0 \times 10^{-4}$
$\log 6.0 \times 10^{-4} = -3.22$
$pH = -\log[H^+]$
$pH = -(-3.22) = 3.22$

PRACTICE   What is the pH of a solution with $[H^+]$ of (a) 3.9×10^{-12}, (b) 1.3×10^{-3}, and (c) 3.72×10^{-6}?
Answers: (a) 11.41   (b) 2.89   (c) 5.43

The measurement and control of pH is extremely important in many fields of science and technology. The proper soil pH is necessary to grow certain types of plants successfully. The pH of certain foods is too acidic for some diets. Many biological processes are delicately controlled pH systems. The pH of human blood is regulated to very close tolerances through the uptake or release of H^+ by mineral ions such as HCO_3^-, HPO_4^{2-}, and $H_2PO_4^-$. Changes in the pH of the blood by as little as 0.4 pH unit result in death.

Compounds with colors that change at particular pH values are used as indicators in acid–base reactions. For example, phenolphthalein, an organic compound, is colorless in acid solution and changes to pink at a pH of 8.3. When a solution of sodium hydroxide is added to a hydrochloric acid solution containing phenolphthalein, the change in color (from colorless to pink) indicates that all the acid is neutralized. Commercially available pH test paper, such as shown in Figure 16.4, contains chemical indicators. The indicator in the paper takes on different colors when wetted with solutions of different pH. Thus the pH of a solution can be estimated by placing a drop on the test paper and comparing

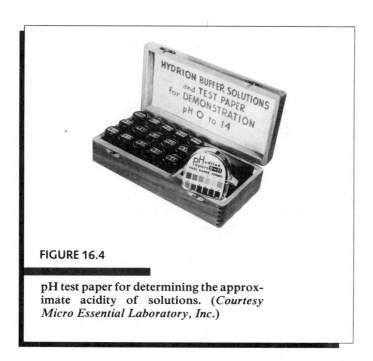

FIGURE 16.4

pH test paper for determining the approximate acidity of solutions. (*Courtesy Micro Essential Laboratory, Inc.*)

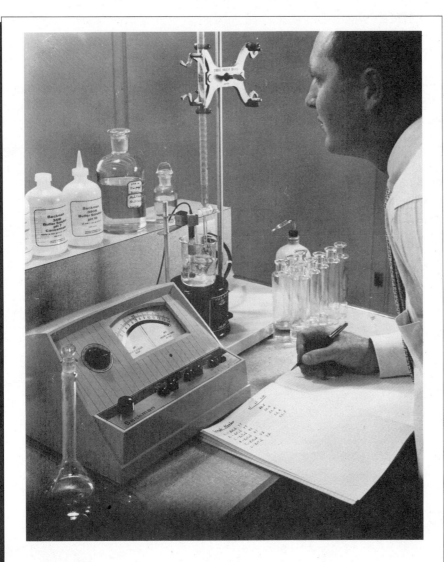

FIGURE 16.5

An electronic pH meter. Accurate measurements may be made by meters of this type. (*Courtesy Beckman Instruments, Inc. Zeromatic is a registered trademark.*)

the color of the test paper with a color chart calibrated at different pH values. Electronic pH meters of the type shown in Figure 16.5 are used for making rapid and precise pH determinations.

16.10 Neutralization

neutralization

The reaction of an acid and a base to form a salt and water is known as **neutralization**. We have seen this reaction before; but now, in the light of what we have learned about ions and ionization, let us reexamine the process of neutralization.

Consider the reaction that occurs when solutions of sodium hydroxide and hydrochloric acid are mixed. The ions present initially are Na^+ and OH^- from the base and H^+ and Cl^- from the acid. The products, sodium chloride and water, exist as Na^+ and Cl^- ions and H_2O molecules. A chemical equation representing this reaction is

$$HCl(aq) + NaOH(aq) \longrightarrow NaCl(aq) + H_2O \qquad (5)$$

This equation, however, does not show that HCl, NaOH, and NaCl exist as ions in solution. The following total ionic equation gives a better representation of the reaction:

$$(H^+ + Cl^-) + (Na^+ + OH^-) \longrightarrow Na^+ + Cl^- + H_2O \qquad (6)$$

spectator ion

Equation (6) shows that the Na^+ and Cl^- ions did not react. These ions are called **spectator ions** because they were present but did not take part in the reaction. The only reaction that occurred was that between the H^+ and OH^- ions. Therefore, the equation for the neutralization can be written as this net ionic equation:

$$\underset{\text{Acid}}{H^+} + \underset{\text{Base}}{OH^-} \longrightarrow \underset{\text{Water}}{H_2O} \qquad (7)$$

This simple net ionic equation (7) represents not only the reaction of sodium hydroxide and hydrochloric acid but also the reaction of any acid with any base in an aqueous solution. The driving force of a neutralization reaction is the ability of an H^+ ion and an OH^- ion to react and form a molecule of un-ionized water.

titration

The amount of acid, base, or other species in a sample may be determined by titration. **Titration** is the process of measuring the volume of one reagent that is required to react with a measured mass or volume of another reagent. Let us consider the titration of an acid with a base. A measured volume of acid of unknown concentration is placed in a flask, and a few drops of an indicator solution are added. Base solution of known concentration is slowly added from a buret to the acid until the indicator changes color (see Figure 16.6). The indicator selected is one that changes color when the stoichiometric quantity (according to the equation) of base has been added to the acid. At this point, known as the *end point of the titration*, the titration is complete, and the volume of base used to

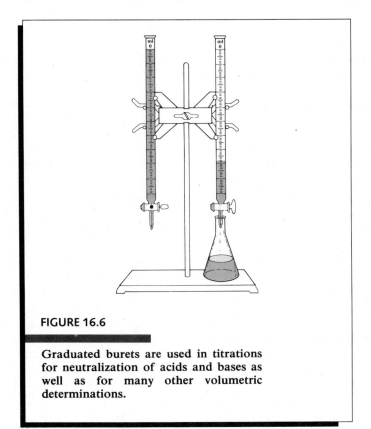

FIGURE 16.6

Graduated burets are used in titrations for neutralization of acids and bases as well as for many other volumetric determinations.

neutralize the acid is read from the buret. The concentration or amount of acid in solution can be calculated from the titration data and the chemical equation for the reaction. Illustrative problems follow.

EXAMPLE 16.4   Suppose that 42.00 mL of 0.150 M NaOH solution is required to titrate 50.00 mL of hydrochloric acid solution. What is the molarity of the acid solution?

The equation for the reaction is

$$NaOH(aq) + HCl(aq) \longrightarrow NaCl(aq) + H_2O(l)$$

In this neutralization NaOH and HCl react in a 1:1 mole ratio. Therefore, the moles of HCl in solution are equal to the moles of NaOH required to react with it. First we calculate the moles of NaOH used, and from this value we determine the moles of HCl.

Data: 42.00 mL of 0.150 M NaOH   50.00 mL of HCl
             Molarity of acid = M (unknown)

Moles of NaOH:

$$M = \text{mol/L} \qquad 42.00 \text{ mL} = 0.04200 \text{ L}$$

$$0.04200 \cancel{L} \times \frac{0.150 \text{ mol NaOH}}{1 \cancel{L}} = 0.00630 \text{ mol NaOH}$$

Since NaOH and HCl react in a 1:1 mole ratio, 0.00630 mol of HCl was present in the 50.00 mL of HCl solution. Therefore the molarity of the HCl is

$$M = \frac{\text{mol}}{\text{L}} = \frac{0.00630 \text{ mol HCl}}{0.05000 \text{ L}} = 0.126 \text{ } M \text{ HCl} \quad \text{(Answer)}$$

EXAMPLE 16.5   Suppose that 42.00 mL of 0.150 M NaOH solution is required to titrate 50.00 mL of sulfuric acid (H_2SO_4) solution. What is the molarity of the acid solution?
The equation for the reaction is

$$2 \text{ NaOH}(aq) + H_2SO_4(aq) \longrightarrow Na_2SO_4(aq) + 2 H_2O(l)$$

The same amount of base (0.00630 mol of NaOH) is used in this titration as in Example 16.4, but the mole ratio of acid to base in the reaction is 1:2. The moles of H_2SO_4 reacted can be calculated by using the mole-ratio method.

Data: 42.00 mL of 0.150 M NaOH = 0.00630 mol NaOH

$$0.00630 \cancel{\text{mol NaOH}} \times \frac{1 \text{ mol } H_2SO_4}{2 \cancel{\text{mol NaOH}}} = 0.00315 \text{ mol } H_2SO_4$$

Therefore 0.00315 mol of H_2SO_4 was present in 50.00 mL of H_2SO_4 solution. The molarity of the H_2SO_4 is

$$M = \frac{\text{mol}}{\text{L}} = \frac{0.00315 \text{ mol } H_2SO_4}{0.05000} = 0.0630 \text{ } M \text{ } H_2SO_4 \quad \text{(Answer)}$$

EXAMPLE 16.6   A 25.00 mL sample of H_2SO_4 solution required 14.26 mL of 0.2240 N NaOH for complete neutralization. What is the normality and the molarity of the sulfuric acid?
The equation for the reaction is

$$2 \text{ NaOH}(aq) + H_2SO_4(aq) \longrightarrow Na_2SO_4(aq) + 2 H_2O(l)$$

The normality of the acid can be calculated from

$$V_A N_A = V_B N_B$$

Substitute the data in the problem and solve for N_A.

$$25.00 \text{ mL} \times N_A = 14.26 \text{ mL} \times 0.2240 \text{ } N$$

$$N_A = \frac{14.26 \text{ mL} \times 0.2240 \text{ } N}{25.00 \text{ mL}} = 0.1278 \text{ } N \text{ } H_2SO_4$$

The normality of the acid is 0.1278 N.

Because H_2SO_4 furnishes 2 equivalents of H^+ per mole, the conversion to molarity is

$$\frac{\text{equiv.}}{L} \times \frac{\text{mol}}{\text{equiv.}}$$

$$\frac{0.1278 \text{ equiv. } H_2SO_4}{1 \text{ L}} \times \frac{1 \text{ mol } H_2SO_4}{2 \text{ equiv. } H_2SO_4} = 0.06390 \text{ mol/L}$$

The H_2SO_4 solution is 0.06390 M.

PRACTICE   A 50.00-mL sample of HCl required 24.81 mL of 0.1250 M NaOH for neutralization. What is the molarity of the acid?

Answer: 0.06203 M HCl

16.11   Writing Ionic Equations

In Section 16.10 we wrote the reaction of hydrochloric acid and sodium hydroxide in three different equations. Equation (5) was the un-ionized equation; equation (6) was the total ionic equation; and equation (7) was the net ionic equation. In the **un-ionized equation**, compounds are written in their molecular, or normal, formula expressions. In the **total ionic equation**, compounds are written to show the form in which they are predominantly present: strong electrolytes as ions in solution; and nonelectrolytes, weak electrolytes, precipitates, and gases in their molecular (or un-ionized) forms. In the **net ionic equation**, only those molecules or ions that have changed are included in the equation; ions or molecules that do not change (spectators) are omitted.

un-ionized
equation

total ionic
equation

net ionic
equation

Up to this point, when balancing an equation we have been concerned only with the atoms of the individual elements. Because ions are electrically charged, ionic equations often end up with a net electrical charge. A balanced equation must have the same net charge on each side, whether that charge is positive, negative, or zero. Therefore, when balancing an ionic equation, we must make sure that both the same number of each kind of atom and the same net electrical charge are present on each side.

Following is a list of rules for writing ionic equations:

1. Strong electrolytes in solution are written in their ionic form.
2. Weak electrolytes are written in their molecular (un-ionized) from.
3. Nonelectrolytes are written in their molecular form.
4. Insoluble substances, precipitates, and gases are written in their molecular forms.
5. The net ionic equation should include only those substances that have

undergone a chemical change. Spectator ions are omitted from the net ionic equation.

6. Equations must be balanced, both in atoms and in electrical charge.

Study the examples below. Note that all reactions are in solution.

EXAMPLE 16.7

$HNO_3(aq) + KOH(aq) \longrightarrow KNO_3(aq) + H_2O$ Un-ionized equation

$(H^+ + NO_3^-) + (K^+ + OH^-) \longrightarrow$

$\qquad\qquad\qquad (K^+ + NO_3^-) + H_2O$ Total ionic equation

$H^+ + OH^- \longrightarrow H_2O$ Net ionic equation

HNO_3, KOH, and KNO_3 are soluble, strong electrolytes. K^+ and NO_3^- are spectator ions, have not changed, and are not included in the net ionic equation. Water is a nonelectrolyte and is written in the molecular form.

EXAMPLE 16.8

$2\,AgNO_3(aq) + BaCl_2(aq) \longrightarrow 2\,AgCl\!\downarrow + Ba(NO_3)_2(aq)$

$(2\,Ag^+ + 2\,NO_3^-) + (Ba^{2+} + 2\,Cl^-) \longrightarrow 2\,AgCl\!\downarrow + (Ba^{2+} + 2\,NO_3^-)$

$Ag^+ + Cl^- \qquad AgCl\!\downarrow$ Net ionic equation

Although silver chloride (AgCl) is an ionic salt, it is written in the un-ionized form on the right side of the ionic equations because most of the Ag^+ and Cl^- ions are no longer in solution but have formed a precipitate of AgCl. Ba^{2+} and NO_3^- are spectator ions.

EXAMPLE 16.9

$Na_2CO_3(aq) + H_2SO_4(aq) \longrightarrow Na_2SO_4(aq) + H_2O + CO_2\!\uparrow$

$(2\,Na^+ + CO_3^{2-}) + (2\,H^+ + SO_4^{2-}) \longrightarrow (2\,Na^+ + SO_4^{2-}) + H_2O + CO_2\!\uparrow$

$CO_3^{2-} + 2\,H^+ \longrightarrow H_2O + CO_2\!\uparrow$ Net ionic equation

Carbon dioxide (CO_2) is a gas and evolves from the solution; Na^+ and SO_4^{2-} are spectator ions.

EXAMPLE 16.10

$HC_2H_3O_2(aq) + NaOH(aq) \longrightarrow NaC_2H_3O_2(aq) + H_2O$

$HC_2H_3O_2 + (Na^+ + OH^-) \longrightarrow (Na^+ + C_2H_3O_2^-) + H_2O$

$HC_2H_3O_2 + OH^- \longrightarrow C_2H_3O_2^- + H_2O$ Net ionic equation

Acetic acid ($HC_2H_3O_2$), a weak acid, is written in the molecular form, but sodium acetate ($NaC_2H_3O_2$), a soluble salt, is written in the ionic form. The Na^+ ion is the only spectator ion in this reaction. Both sides of the net ionic equation have a -1 electrical charge.

EXAMPLE 16.11

$Mg(s) + 2\,HCl(aq) \longrightarrow MgCl_2(aq) + H_2\!\uparrow$

$Mg + (2\,H^+ + 2\,Cl^-) \longrightarrow (Mg^{2+} + 2\,Cl^-) + H_2\!\uparrow$

$Mg + 2\,H^+ \longrightarrow Mg^{2+} + H_2\!\uparrow$ Net ionic equation

The net electrical charge on both sides of the equation is $+2$.

EXAMPLE 16.12

$$H_2SO_4(aq) + Ba(OH)_2(aq) \longrightarrow BaSO_4\downarrow + 2\,H_2O$$

$$(2\,H^+ + SO_4^{2-}) + (Ba^{2+} + 2\,OH^-) \longrightarrow BaSO_4\downarrow + 2\,H_2O$$

$$2\,H^+ + SO_4^{2-} + Ba^{2+} + 2\,OH^- \longrightarrow BaSO_4\downarrow + 2\,H_2O \quad \text{Net ionic equation}$$

Barium sulfate ($BaSO_4$) is a highly insoluble salt. If we conduct this reaction using the conductivity apparatus described in Section 16.5, the light glows brightly at first but goes out when the reaction is complete, because almost no ions are left in solution. The $BaSO_4$ precipitates out of solution, and water is a nonconductor of electricity.

Concepts in Review

1. State the general characteristics of acids and bases.

2. Define an acid and base in terms of the Arrhenius, Brønsted-Lowry, and Lewis theories.

3. Identify acid-base conjugate pairs in a reaction.

4. When given the reactants, complete and balance equations for the reactions of acids with bases, metals, metal oxides, and carbonates.

5. When given the reactants, complete and balance equations of the reaction of an amphoteric hydroxide with either a strong acid or a strong base.

6. Write balanced equations for the reaction of sodium hydroxide or potassium hydroxide with zinc and with aluminum.

7. Classify common compounds as electrolytes or nonelectrolytes.

8. Distinguish between strong and weak electrolytes.

9. Understand the process of dissociation and ionization.

10. Write equations for the dissociation and ionization of acids, bases, and salts in water.

11. Describe and write equations for the ionization of water.

12. Understand pH as an expression of hydrogen ion concentration or hydronium ion concentration.

13. Given pH as an integer, indicate the H^+ molarity, and vice versa.

14. Use a calculator to estimate pH values from corresponding H^+ molarities.

15. Understand the process of acid-base neutralization.

16. Calculate the molarity, normality, or volume of an acid or base solution from appropriate titration data.

17. Write un-ionized, total ionic, and net ionic equations for neutralization equations.

Key Terms in Review

The terms listed here have all been defined within the chapter. Review the definitions of each. Use the glossary and the margin notations within the chapter as study aids.

amphoteric	nonelectrolyte
dissociation	pH
electrolyte	spectator ion
hydronium ion	strong electrolyte
ionization	titration
logarithm	total ionic equation
net ionic equation	un-ionized equation
neutralization	weak electrolyte

Exercises

An asterisk indicates a more challenging question or problem.

1. Since a hydrogen ion and a proton are identical, what differences exist between the Arrhenius and Brønsted–Lowry definitions of an acid? (See Table 16.1.)

2. According to Figure 16.1, what type of substance must be in solution in order for the bulb to light?

3. Which of the following classes of compounds are electrolytes: acids, alcohols, bases, salts? (See Table 16.2.)

4. What two differences are apparent in the arrangement of water molecules about the hydrated ions as depicted in Figure 16.2?

5. The pH of a solution with a hydrogen ion concentration of 0.003 M is between what two whole numbers? (See Table 16.4.)

6. Which is the more acidic, tomato juice or blood? (See Table 16.5.)

7. Using each of the three acid–base theories (Arrhenius, Brønsted–Lowry, and Lewis), define an acid and a base.

8. For each of the acid–base theories referred to in Exercise 7, write an equation illustrating the neutralization of an acid with a base.

9. Identify the conjugate acid–base pairs in the following equations:
   (a) $HCl + NH_3 \longrightarrow NH_4^+ + Cl^-$
   (b) $HCO_3^- + OH^- \rightleftharpoons CO_3^{2-} + H_2O$
   (c) $HCO_3^- + H_3O^+ \rightleftharpoons H_2CO_3 + H_2O$
   (d) $HC_2H_3O_2 + H_2O \rightleftharpoons H_3O^+ + C_2H_3O_2^-$
   (e) $HC_2H_3O_2 + H_2SO_4 \rightleftharpoons$
        $H_2C_2H_3O_2^+ + HSO_4^-$
   (f) The two-step ionization of sulfuric acid,
        $H_2SO_4 + H_2O \longrightarrow H_3O^+ + HSO_4^-$
        $HSO_4^- + H_2O \rightleftharpoons H_3O^+ + SO_4^{2-}$
   (g) $HClO_4 + H_2O \longrightarrow H_3O^+ + ClO_4^-$
   (h) $CH_3O^- + H_3O^+ \longrightarrow CH_3OH + H_2O$

10. Write the Lewis structure for (a) bromide ion, (b) hydroxide ion, and (c) cyanide ion. Why are these ions considered to be bases according to the Brønsted–Lowry and Lewis acid–base theories?

11. Complete and balance the following equations:
    (a) $Mg(s) + HCl(aq) \longrightarrow$
    (b) $BaO(s) + HBr(aq) \longrightarrow$
    (c) $Al(s) + H_2SO_4(aq) \longrightarrow$
    (d) $Na_2CO_3(aq) + HCl(aq) \longrightarrow$
    (e) $Fe_2O_3(s) + HBr(aq) \longrightarrow$
    (f) $Ca(OH)_2(aq) + H_2CO_3(aq) \longrightarrow$
    (g) $NaOH(aq) + HBr(aq) \longrightarrow$

(h) $KOH(aq) + HCl(aq) \longrightarrow$
(i) $Ca(OH)_2(aq) + HI(aq) \longrightarrow$
(j) $Al(OH)_3(s) + HBr(aq) \longrightarrow$
(k) $Na_2O(s) + HClO_4(aq) \longrightarrow$
(l) $LiOH(aq) + FeCl_3(aq) \longrightarrow$
(m) $NH_4OH(aq) + FeCl_2(aq) \longrightarrow$

12. Into what three classes of compounds do electrolytes generally fall?

13. Which of the following compounds are electrolytes? Consider each substance to be mixed with water.
    (a) HCl
    (b) CO_2
    (c) $CaCl_2$
    (d) $C_{12}H_{22}O_{11}$ (sugar)
    (e) C_3H_7OH (rubbing alcohol)
    (f) CCl_4 (insoluble)
    (g) $NaHCO_3$ (baking soda)
    (h) N_2 (insoluble gas)
    (i) $AgNO_3$
    (j) $HCOOH$ (formic acid)
    (k) $RbOH$
    (l) K_2CrO_4

14. Name each compound listed in Table 16.3.

15. A solution of HCl in water conducts an electric current, but a solution of HCl in benzene does not. Explain this behavior in terms of ionization and chemical bonding.

16. How do salts exist in their crystalline structure? What occurs when they are dissolved in water?

17. An aqueous methyl alcohol, CH_3OH, solution does not conduct an electric current, but a solution of sodium hydroxide, NaOH, does. What does this information tell us about the OH group in the alcohol?

18. Why does molten sodium chloride conduct electricity?

19. Explain the difference between dissociation of ionic compounds and ionization of molecular compounds.

20. Distinguish between strong and weak electrolytes.

21. Explain why ions are hydrated in aqueous solutions.

22. Indicate, by simple equations, how the following substances dissociate or ionize in water:
    (a) $Cu(NO_3)_2$
    (b) $HC_2H_3O_2$
    (c) HNO_2
    (d) $LiOH$

(e) NH_4Br
(f) K_2SO_4
(g) $NaClO_3$
(h) K_3PO_4

23. What is the main distinction between water solutions of strong and weak electrolytes?

24. What are the relative concentrations of $H^+(aq)$ and $OH^-(aq)$ in (a) a neutral solution, (b) an acid solution, and (c) a basic solution?

25. Write the net ionic equation for the reaction of an acid with a base in an aqueous solution.

26. The solubility of hydrogen chloride gas in water, a polar solvent, is much greater than its solubility in benzene, a nonpolar solvent. How can you account for this difference?

27. Pure water, containing both acid and base ions, is neutral. Why?

28. Which of the following statements are correct? Rewrite each incorrect statement to make it correct.
    (a) The Arrhenius theory of acids and bases is restricted to aqueous solutions.
    (b) The Brønsted–Lowry theory of acids and bases is restricted to solutions other than aqueous solutions.
    (c) All substances that are acids according to the Brønsted–Lowry theory will also be acids by the Lewis theory.
    (d) All substances that are acids according to the Lewis theory will also be acids by the Brønsted–Lowry theory.
    (e) An electron-pair donor is a Lewis acid.
    (f) All Arrhenius acid–base neutralization reactions can be represented by a single net ionic equation.
    (g) When an ionic compound dissolves in water, the ions separate; this process is called ionization.
    (h) In the autoionization of water

    $$2\,H_2O \rightleftharpoons H_3O^+ + OH^-$$

    the H_3O^+ and the OH^- constitute a conjugate acid–base pair.
    (i) In the reaction in part (h) H_2O is both the acid and the base.
    (j) Most common Na, K, and NH_4^+ salts are soluble in water.
    (k) A solution of pH 3 is 100 times more acidic than a solution of pH 5.

(l) In general, ionic substances when placed in water will give a solution capable of conducting an electric current.

(m) The terms *dissociation* and *ionization* are synonymous.

(n) A solution of $Mg(NO_3)_2$ will produce three ions per formula unit in solution.

(o) The terms *strong acid, strong base, weak acid,* and *weak base* refer to whether an acid or base solution is concentrated or dilute.

(p) pH is defined as the negative logarithm of the molar concentration of H^+ ions (or H_3O^+ ions).

(q) All reactions may be represented by net ionic equations.

(r) One mole of $CaCl_2$ contains more anions than cations.

(s) It is possible to boil seawater at a lower temperature than that required to boil pure water (both at the same pressure).

(t) It is possible to have a neutral aqueous solution whose pH is not 7.

Review problems

29. Calculate the molarity of the ions present in each of the following salt solutions. Assume each salt to be 100% dissociated.
(a) 0.015 M NaCl
(b) 4.25 M NaKSO$_4$
(c) 0.75 M ZnBr$_2$
(d) 1.65 M Al$_2$(SO$_4$)$_3$
(e) 0.20 M CaCl$_2$
(f) 22.0 g KI in 500 mL of solution
(g) 900 g (NH$_4$)$_2$SO$_4$ in 20.0 L of solution
(h) 0.0120 g Mg(ClO$_3$)$_2$ in 1.00 mL of solution

30. In Exercise 29, how many grams of each ion would be present in 100 mL of each solution?

31. What is the concentration of Ca^{2+} ions in a solution of CaI$_2$ having an I^- ion concentration of 0.520 M?

32. What is the molar concentration of all ions present in a solution prepared by mixing the following?
(a) 30.0 mL of 1.0 M NaCl and 40.0 mL of 1.0 M NaCl
(b) 30.0 mL of 1.0 M HCl and 30.0 mL of 1.0 M NaOH
(c) 100.0 mL of 2.0 M KCl and 100.0 mL of 1.0 M CaCl$_2$

*(d) 100.0 mL of 0.40 M KOH and 100.0 mL of 0.80 M HCl
(e) 35.0 mL of 0.20 M Ba(OH)$_2$ and 35.0 mL of 0.20 M H$_2$SO$_4$
(f) 1.00 L of 1.0 M AgNO$_3$ and 500 mL of 2.0 M NaCl.
(Neglect the concentration of H^+ and OH^- from water. Also, assume volumes of solutions are additive.)

33. How many milliliters of 0.40 M HCl can be made by diluting 100. mL of 12 M HCl with water?

34. Given the data for the following six titrations, calculate the molarity of the HCl in titrations (a), (b), and (c), and the molarity of the NaOH in titrations (d), (e), and (f).

	mL HCl	Molarity HCl	mL NaOH	Molarity NaOH
(a)	40.13	M HCl	37.70	0.728
(b)	19.00	M HCl	33.66	0.306
(c)	27.25	M HCl	18.00	0.555
(d)	37.19	0.126	31.91	M NaOH
(e)	48.04	0.482	24.02	M NaOH
(f)	13.13	1.425	39.39	M NaOH

35. If 29.26 mL of 0.430 M HCl neutralizes 20.40 mL of Ba(OH)$_2$ solution, what is the molarity of the Ba(OH)$_2$ solution? The reaction is

$$Ba(OH)_2(aq) + 2 HCl(aq) \longrightarrow BaCl_2(aq) + 2 H_2O$$

36. A 1 molal solution of acetic acid in water freezes at a lower temperature than a 1 molal solution of ethyl alcohol, C_2H_5OH, in water. Explain.

37. At the same cost per pound, which alcohol, CH_3OH or C_2H_5OH, would be more economical to purchase as an antifreeze for your car?

38. How does a hydronium ion differ from a hydrogen ion?

39. Arrange, in decreasing order of freezing points, 1 molal aqueous solutions of HCl, $HC_2H_3O_2$, $C_{12}H_{22}O_{11}$ (sucrose), and $CaCl_2$. (List the one with the highest freezing point first.)

40. At 100°C the H^+ concentration in water is about 1×10^{-6} mol/L, about 10 times that of water at 25°C. At which of these temperatures is (a) the pH of water the greater, (b) the hydrogen ion (hydronium ion) concentration the higher, and (c) the water neutral?

41. What is the relative difference in H^+ concentration in solutions that differ by 1 pH unit?

42. Rewrite the following unbalanced equations, changing them into balanced net ionic equations. All reactions are in water solution.
(a) $K_2SO_4(aq) + Ba(NO_3)_2(aq) \longrightarrow$
$$KNO_3(aq) + BaSO_4(s)$$
(b) $CaCO_3(s) + HCl(aq) \longrightarrow$
$$CaCl_2(aq) + CO_2(g) + H_2O$$
(c) $Mg(s) + HC_2H_3O_2(aq) \longrightarrow$
$$Mg(C_2H_3O_2)_2(aq) + H_2(g)$$
(d) $H_2S(g) + CdCl_2(aq) \longrightarrow CdS(s) + HCl(aq)$
(e) $Zn(s) + H_2SO_4(aq) \longrightarrow$
$$ZnSO_4(aq) + H_2(g)$$
(f) $AlCl_3(aq) + Na_3PO_4(aq) \longrightarrow$
$$AlPO_4(s) + NaCl(aq)$$

43. In each of the following pairs which solution is more acidic? (All are water solutions.)
(a) 1 molar HCl or 1 molar H_2SO_4?
(b) 1 molar HCl or 1 molar $HC_2H_3O_2$?
(c) 1 molar HCl or 2 molar HCl?
(d) 1 normal H_2SO_4 or 1 molar H_2SO_4?

44. What volume (in milliliters) of 0.245 M HCl will neutralize (a) 50.0 mL of 0.100 M $Ca(OH)_2$ and (b) 10.0 g of $Al(OH)_3$? The equations are
(a) $2 HCl(aq) + Ca(OH)_2(aq) \longrightarrow$
$$CaCl_2(aq) + 2 H_2O$$
(b) $3 HCl(aq) + Al(OH)_3(s) \longrightarrow$
$$AlCl_3(aq) + 3 H_2O$$

45. A sample of pure sodium carbonate with a mass of 0.452 g was dissolved in water and neutralized with 42.4 mL of hydrochloric acid. Calculate the molarity of the acid:

$$Na_2CO_3(aq) + 2 HCl(aq) \longrightarrow$$
$$2 NaCl(aq) + CO_2(g) + H_2O$$

46. What volume (mL) of 0.1234 M HCl is needed to neutralize 2.00 g $Ca(OH)_2$?

47. How many grams of KOH are required to neutralize 50.00 mL of 0.240 M HNO_3?

*48. A 0.200 g sample of impure NaOH requires 18.25 mL of 0.2406 M HCl for neutralization.

What is the mass percent of NaOH in the sample?

*49. A batch of sodium hydroxide was found to contain sodium chloride as an impurity. To determine the amount of impurity, a 1.00 g sample was analyzed and found to require 49.90 mL of 0.466 M HCl for neutralization. What is the percent of NaCl in the sample?

*50. What volume of H_2 gas, measured at 27°C and 700 torr pressure, can be obtained by reacting 5.00 g of zinc metal with (a) 100 mL of 0.350 M HCl and (b) 200 mL of 0.350 M HCl? The equation is

$$Zn(s) + 2 HCl(aq) \longrightarrow ZnCl_2(aq) + H_2(g)$$

51. Calculate the pH of solutions having the following H^+ ion concentrations:
(a) 0.01 M (d) 1×10^{-7} M
(b) 1.0 M (e) 0.50 M
(c) 6.5×10^{-9} M (f) 0.00010 M

52. Calculate the pH of the following:
(a) Orange juice, 3.7×10^{-4} M H^+
(b) Vinegar, 2.8×10^{-3} M H^+
(c) Black coffee, 5.0×10^{-5} M H^+
(d) Limewater, 3.4×10^{-11} M H^+

53. Two drops (0.1 mL) of 1.0 M HCl are added to water to make 1.0 L of solution. What is the pH of this solution if the HCl is 100% ionized?

54. What volume of concentrated (18.0 M) sulfuric acid must be used to prepare 50.0 L of 5.00 M solution?
Three (3.0) grams of NaOH are added to 500 mL of 0.10 M HCl. Will the resulting solution be acidic or basic? Show evidence for your answer.

56. A 10.00 mL sample of base solution requires 28.92 mL of 0.1240 N H_2SO_4 for neutralization. What is the normality of the base?
How many milliliters of 0.325 N HNO_3 are required to neutralize 32.8 mL of 0.225 N NaOH?

58. How many milliliters of 0.325 N H_2SO_4 are required to neutralize 32.8 mL of 0.225 N NaOH?

59. What is the normality and the molarity of a 25.00 mL sample of H_3PO_4 solution that requires 22.68 mL of 0.5000 N NaOH for complete neutralization?

*60. A 50. mL of an unknown acid solution is titrated to neutrality with 20. mL of NaOH solution, which has a normality of 0.40 equiv/L. What is the normality of the unknown acid solution?

61. How many milliliters of 0.10 N NaOH is required to neutralize 60. mL of 0.20 N H_2SO_4?

*__62.__ A 25 mL solution of H_2SO_4 requires 40. mL of 0.20 NaOH for neutralization.

   (a) What is the normality of the sulfuric acid solution?

   (b) How many grams of sulfuric acid are contained in the 25 mL?

*__63.__ A solution of 40.0 mL of HCl is neutralized by 20.0 mL of NaOH solution. The resulting neutral solution is evaporated to dryness, and the residue is found to have a mass of 0.117 g. Calculate the normality of HCl and NaOH solutions.

*__64.__ The equivalent masses of organic acids are often determined by titration with a standard solution of a base. Determine the equivalent mass of benzoic acid, if 0.305 g of it require 25 mL of 0.10 N NaOH for neutralization.

*__65.__ Determine the equivalent mass of succinic acid, if 0.738 g is required to neutralize 125 mL of 0.10 N base.

CHAPTER SEVENTEEN

Chemical Equilibrium

Thus far, we have considered chemical change to proceed from reactants to products. Does that mean that reactions stop? No. A solute dissolves until the solution becomes saturated. Once a solid remains undissolved in a container, the system appears to be at rest. The human body is a marvelous chemical factory, yet from day to day it appears to be quite the same. For example, the blood remains at a constant pH, even though all sorts of chemical reactions are taking place. A terrarium can be watered and sealed for long periods of time with no ill effects. An antacid is advertised to absorb excess stomach acid and *not* change the pH of the stomach. In all of these cases, reactions are proceeding, even though visible signs of chemical change are absent. When the system is at equilibrium, chemical reactions are dynamic. As the Frenchman Alphonse Karr so succinctly stated in 1849, "The more things change the more they stay the same."

Chapter Preview

17.1   Reversible Reactions

In the preceding chapters we have treated chemical reactions mainly as reactants going to products. However, many reactions do not go to completion. Some reactions do not go to completion because they are reversible; that is, when the products are formed, they react to produce the starting reactants.

We have encountered reversible systems before. One is the vaporization of a liquid by heating and its subsequent condensation by cooling:

liquid + heat $\longrightarrow$ vapor

vapor + cooling $\longrightarrow$ liquid

The interconversion of nitrogen dioxide (NO_2) and dinitrogen tetroxide (N_2O_4) offers visible evidence of the reversibility of a reaction. NO_2 is a reddish-brown gas that changes, with cooling, to N_2O_4, a yellow liquid that boils at 21.2°C, and then to a colorless solid (N_2O_4) that melts at -11.2°C. The reaction is reversible by heating N_2O_4.

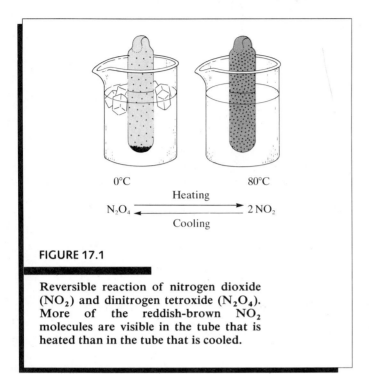

$$N_2O_4 \xleftarrow{\text{Heating}} 2\,NO_2$$
$$\text{Cooling}$$

0°C   80°C

FIGURE 17.1

Reversible reaction of nitrogen dioxide (NO₂) and dinitrogen tetroxide (N₂O₄). More of the reddish-brown NO₂ molecules are visible in the tube that is heated than in the tube that is cooled.

$$2\,NO_2(g) \xrightarrow{\text{Cooling}} N_2O_4(l)$$
$$N_2O_4(l) \xrightarrow{\text{Heating}} 2\,NO_2(g)$$

These two reactions may be represented by a single equation with a double arrow, $\rightleftarrows$, to indicate that the reactions are taking place in both directions at the same time.

$$2\,NO_2(g) \rightleftarrows N_2O_4(l)$$

This reversible reaction can be demonstrated by sealing samples of NO₂ in two tubes and placing one tube in warm water and the other in ice water (see Figure 17.1). Heating promotes disorder or randomness in a system, so we would expect more NO₂, a gas, to be present at higher temperatures.

reversible chemical reaction

A **reversible chemical reaction** is one in which the products formed react to produce the original reactants. Both the forward and reverse reactions occur simultaneously. The forward reaction is called *the reaction to the right*, and the reverse reaction is called *the reaction to the left*. A double arrow is used in the equation to indicate that the reaction is reversible.

17.2 Rates of Reaction

chemical
kinetics

Every reaction has a rate, or speed, at which it proceeds. Some are fast and some are extremely slow. The study of reaction rates and reaction mechanisms is known as **chemical kinetics**.

The rate of a reaction is variable and depends on the concentration of the reacting species, the temperature, the presence or absence of catalytic agents, and the nature of the reactants. Consider the hypothetical reaction

$$A + B \longrightarrow C + D \quad \text{(forward reaction)}$$
$$C + D \longrightarrow A + B \quad \text{(reverse reaction)}$$

in which a collision between A and B is necessary for a reaction to occur. The rate at which A and B react depends on the concentration or the number of A and B molecules present; it will be fastest, for a fixed set of conditions, when they are first mixed. As the reaction proceeds, the number of A and B molecules available for reaction decreases, and the rate of reaction slows down. If the reaction is reversible, the speed of the reverse reaction is zero at first and gradually increases as the concentrations of C and D increase. As the number of A and B molecules decreases, the forward rate slows down because A and B cannot find one another as often in order to accomplish a reaction. To counteract this diminishing rate of reaction, an excess of one reagent is often used to keep the reaction from becoming impractically slow. Collisions between molecules may be likened to the space games at the video arcades. When many objects are on the screen, collisions occur frequently; but if only a few objects are present, collisions can usually be avoided.

17.3 Chemical Equilibrium

equilibrium

Any system at **equilibrium** represents a dynamic state in which two or more opposing processes are taking place at the same time and at the same rate. A chemical equilibrium is a dynamic system in which two or more opposing chemical reactions are going on at the same time and at the same rate. When the rate of the forward reaction is exactly equal to the rate of the reverse reaction, a

chemical
equilibrium

condition of **chemical equilibrium** exists (see Figure 17.2). The concentrations of the products and the reactants are not changing, and the system appears to be at a standstill because the products are reacting at the same rate at which they are being formed.

> **Chemical equilibrium:**
> **rate of forward reaction = rate of reverse reaction**

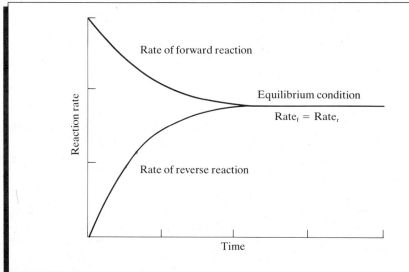

FIGURE 17.2

The graph illustrates that the rates of the forward and reverse reactions become equal at some point in time. The forward reaction rate decreases as a result of decreasing amounts of reactants. The reverse reaction rate starts at zero and increases as the amount of product increases. When the two rates become equal, a state of chemical equilibrium has been reached.

A saturated salt solution is in a condition of equilibrium:

$$NaCl(s) \rightleftharpoons Na^+(aq) + Cl^-(aq)$$

At equilibrium, salt crystals are continuously dissolving, and Na^+ and Cl^- ions are continuously crystallizing. Both processes are occurring at the same rate.

The ionization of weak electrolytes is another common chemical equilibrium system:

$$HC_2H_3O_2(aq) + H_2O(l) \rightleftharpoons H_3O^+(aq) + C_2H_3O_2^-(aq)$$

In this reaction, the equilibrium is established in a 1 M solution when the forward reaction has gone about 1%; that is, when only 1% of the acetic acid molecules in solution have ionized. Therefore, only a relatively few ions are present, and the acid behaves as a weak electrolyte. In any acid–base equilibrium system the position of equilibrium is toward the weaker conjugate acid and base. In the ionization of acetic acid, $HC_2H_3O_2$ is a weaker acid than H_3O^+, and H_2O is a weaker base than $C_2H_3O_2^-$.

The reactions represented by

$$H_2(g) + I_2(g) \rightleftharpoons 2\ HI(g)$$

provide another example of a chemical equilibrium. Theoretically, 1.00 mol of hydrogen should react with 1.00 mol of iodine to yield 2.00 mol of hydrogen iodide. Actually, when 1.00 mol of H_2 and 1.00 mol of I_2 are reacted at 700 K, only 1.58 mol of HI are present when equilibrium is attained. Since 1.58 is 79% of the theoretical yield of 2.00 mol of HI, the forward reaction is only 79% complete at equilibrium. The equilibrium mixture will also contain 0.21 mol each of unreacted H_2 and I_2 (1.00 mol − 0.79 mol = 0.21 mol).

$$H_2 + I_2 \xrightarrow{700\ K} 2\ HI$$
1.00 mol   1.00 mol   2.00 mol

(This equation would represent the condition if the reaction were 100% complete; 2.00 mol of HI would be formed and no H_2 and I_2 would be left unreacted.)

$$H_2 + I_2 \xrightleftharpoons{700\ K} 2\ HI$$
0.21 mol   0.21 mol   1.58 mol

(This equation represents the actual equilibrium attained starting with 1.00 mol each of H_2 and I_2. It shows that the forward reaction is only 79% complete.)

17.4   Principle of Le Chatelier

principle of
Le Chatelier

In 1888 the French chemist Henri Le Chatelier (1850–1936) set forth a simple, far-reaching generalization on the behavior of equilibrium systems. This generalization, known as the **principle of Le Chatelier**, states: If a stress is applied to a system in equilibrium, the system will respond in such a way as to relieve that stress and restore equilibrium under a new set of conditions. The application of Le Chatelier's principle helps us to predict the effect of changing conditions in chemical reactions. We will examine the effect of changes in concentration, temperature, and pressure.

17.5   Effect of Concentration on Reaction Rate and Equilibrium

The way in which the rate of a chemical reaction depends on the concentration of the reactants must be determined experimentally. Many simple, one-step reactions result from a collision between two molecules or ions. The rate of such one-step reactions can be altered by changing the concentration of the reactants or products. An increase in concentration of the reactants provides more individual reacting species for collisions and results in an increase in the rate of reaction.

An equilibrium is disturbed when the concentration of one or more of its components is changed. As a result the concentration of all the species will change, and a new equilibrium mixture will be established. Consider the hypothetical equilibrium represented by the equation

$$A + B \rightleftharpoons C + D$$

where A and B react in one step to form C and D. When the concentration of B is increased, the following occurs:

1. The rate of the reaction to the right (forward) increases. This rate is proportional to the concentration of A times the concentration of B.
2. The rate to the right becomes greater than the rate to the left.
3. Reactants A and B are used faster than they are produced; C and D are produced faster than they are used.
4. After a period of time, rates to the right and left become equal, and the system is again in equilibrium.
5. In the new equilibrium the concentration of A is less and the concentrations of B, C, and D are greater than in the original equilibrium.

Conclusion: The equilibrium has shifted to the right.

Applying this change in concentration to the equilibrium mixture of 1.00 mol of hydrogen and 1.00 mol of iodine from Section 17.3, we find that, when an additional 0.20 mol of I_2 is added, the yield of HI (based on H_2) is 85% (1.70 mol) instead of 79%. A comparison of the two systems, after the new equilibrium mixture is reached, follows.

Original equilibrium	New equilibrium
1.00 mol H_2 + 1.00 mol I_2	1.00 mol H_2 + 1.20 mol I_2
Yield: 79% HI	Yield: 85% HI (based on H_2)
Equilibrium mixture contains:	Equilibrium mixture contains:
1.58 mol HI	1.70 mol HI
0.21 mol H_2	0.15 mol H_2
0.21 mol I_2	0.35 mol I_2

Analyzing this new system, we see that, when the 0.20 mol I_2 was added, the equilibrium shifted to the right in order to counteract the increase in I_2 concentration. Some of the H_2 reacted with added I_2 and produced more HI, until an equilibrium mixture was established again. When I_2 was added, the concentration of I_2 increased, the concentration of H_2 decreased, and the concentration of HI increased. What do you think would be the effects of adding (a) more H_2 or (b) more HI?

The equation

$$Fe^{3+}(aq) + SCN^-(aq) \rightleftharpoons Fe(SCN)^{2+}(aq)$$

Pale yellow   Colorless   Red

represents an equilibrium that is used in certain analytical procedures as an indicator because of the readily visible, intense red color of the complex $Fe(SCN)^{2+}$ ion. A very dilute solution of iron(III), Fe^{3+}, and thiocyanate, SCN^-, is light red. When the concentration of either Fe^{3+} or SCN^- is increased, the equilibrium shift to the right is observed by an increase in the intensity of the color, resulting from the formation of additional $Fe(SCN)^{2+}$.

If either Fe^{3+} or SCN^- is removed from solution, the equilibrium will shift to the left, and the solution will become lighter in color. When Ag^+ is added to the solution, a white precipitate of silver thiocyanate (AgSCN) is formed, thus removing SCN^- ion from the equilibrium.

$$Ag^+(aq) + SCN^-(aq) \longrightarrow AgSCN\downarrow$$

The system accordingly responds to counteract the change in SCN^- concentration by shifting the equilibrium to the left. This shift is evident by a decrease in the intensity of the red color due to a decreased concentration of $Fe(SCN)^{2+}$.

Let us now consider the effect of changing the concentrations in the equilibrium mixture of chlorine water. The equilibrium equation is

$$Cl_2(aq) + 2\ H_2O \rightleftharpoons HOCl(aq) + H_3O^+(aq) + Cl^-(aq)$$

The variation in concentrations and the equilibrium shifts are tabulated in the following table. An X in the second or third column indicates the reagent that is increased or decreased. The fourth column indicates the direction of the equilibrium shift.

| Reagent | Concentration | | Equilibrium shift |
	Increase	Decrease	
Cl_2	—	X	Left
H_2O	X	—	Right
HOCl	X	—	Left
H_3O^+	—	X	Right
Cl^-	X	—	Left

Consider the equilibrium in a 0.100 M acetic acid solution:

$$HC_2H_3O_2 + H_2O \rightleftharpoons H_3O^+ + C_2H_3O_2^-$$

In this solution the concentration of the hydronium ion (H_3O^+), which is a measure of the acidity, is 1.34×10^{-3} mol/L, corresponding to a pH of 2.87. What will happen to the acidity when 0.100 mol of sodium acetate ($NaC_2H_3O_2$) is added to 1 L of 0.100 M $HC_2H_3O_2$? When $NaC_2H_3O_2$ dissolves, it dissociates into sodium ions (Na^+) and acetate ions ($C_2H_3O_2^-$). The acetate ion from the salt is a common ion to the acetic acid equilibrium system and increases the total acetate ion concentration in the solution. As a result the equilibrium shifts to the left, decreasing the hydronium ion concentration and lowering the acidity of the solution. Evidence of this decrease in acidity is shown by the fact that the pH of a solution that is 0.100 M in $HC_2H_3O_2$ and 0.100 M in $NaC_2H_3O_2$ is 4.74. The pH of several different solutions of $HC_2H_3O_2$ and $NaC_2H_3O_2$ is shown in the table that follows. Each time the acetate ion is increased, the pH increases, indicating a further shift in the equilibrium toward un-ionized acetic acid.

Solution	pH
1 L 0.100 M $HC_2H_3O_2$	2.87
1 L 0.100 M $HC_2H_3O_2$ + 0.100 mol $NaC_2H_3O_2$	4.74
1 L 0.100 M $HC_2H_3O_2$ + 0.200 mol $NaC_2H_3O_2$	5.05
1 L 0.100 M $HC_2H_3O_2$ + 0.300 mol $NaC_2H_3O_2$	5.23

In summary, we can say that, when the concentration of a reagent on the left side of an equation is increased, the equilibrium shifts to the right. When the concentration of a reagent on the right side of an equation is increased, the equilibrium shifts to the left. In accordance with Le Chatelier's principle the equilibrium always shifts in the direction that tends to reduce the concentration of the added reactant.

PRACTICE Aqueous chromate ion (CrO_4^{2-}) exists in equilibrium with aqueous dichromate ion ($Cr_2O_7^{2-}$) in an acidic solution. What effect will (a) increasing the dichromate ion and (b) adding HCl have on the equilibrium?

$$2\,CrO_4^{2-} + 2\,H^+ \Longrightarrow Cr_2O_7^{2-} + H_2O$$

Answers: (a) Equilibrium shifts left   (b) Equilibrium shifts right

17.6   Effect of Pressure on Reaction Rate and Equilibrium

Changes in pressure significantly affect the reaction rate only when one or more of the reactants or products is a gas. In these cases the effect of increasing the pressure of the reacting gases is equivalent to increasing their concentrations. In the reaction

$$CaCO_3(s) \overset{\Delta}{\rightleftharpoons} CaO(s) + CO_2(g)$$

calcium carbonate decomposes into calcium oxide and carbon dioxide when heated above 825°C. Increasing the pressure of the equilibrium system by adding CO_2 or by decreasing the volume, speeds up the reverse reaction and causes the equilibrium to shift to the left. The increased pressure gives the same effect as that caused by increasing the concentration of CO_2, the only gaseous substance in the reaction.

When the pressure on a gas is increased, its volume is decreased. In a system composed entirely of gases, an increase in pressure will cause the reaction and the equilibrium to shift to the side that contains the smaller volume or smaller number of moles. This shift occurs because the increase in pressure is partially relieved by the system's shifting its equilibrium toward the side in which the substances occupy the smaller volume.

Prior to World War I, Fritz Haber (1868–1934) in Germany invented the first major process for the fixation of nitrogen. In the Haber process nitrogen and hydrogen are reacted together in the presence of a catalyst at moderately high temperature and pressure to produce ammonia. The catalyst consists of iron and iron oxide with small amounts of potassium and aluminum oxides. For this process, Haber received the Nobel Prize in chemistry in 1918.

$$N_2(g) \; + \; 3\,H_2(g) \rightleftharpoons 2\,NH_3(g) + 92.5 \; kJ(22.1 \; kcal) \qquad (at \; 25°C)$$

1 mol	3 mol	2 mol
1 volume	3 volumes	2 volumes

The left side of the equation in the Haber process represents four volumes of gas combining to give two volumes of gas on the right side of the equation. An increase in the total pressure on the system shifts the equilibrium to the right. This increase in pressure results in a higher concentration of both reactants and products. The equilibrium shifts to the right when the pressure is increased, because fewer moles of NH_3 than moles of N_2 and H_2 are in the equilibrium reaction.

Ideal conditions for the Haber process are 200°C and 1000 atm pressure. However, at 200°C the rate of reaction is very slow, and at 1000 atm extraordinarily heavy equipment is required. As a compromise the reaction is run at 400–600°C and 200–350 atm pressure, which gives a reasonable yield at a reasonable rate. The effect of pressure on the yield of ammonia at one particular temperature is shown in Table 17.1.

When the total number of gaseous molecules on both sides of an equation is the same, a change in pressure does not cause an equilibrium shift. The following reaction is an example.

$$N_2(g) \; + \; O_2(g) \; \rightleftharpoons \; 2\,NO(g)$$

1 mol	1 mol	2 mol
1 volume	1 volume	2 volumes
6.022×10^{23} molecules	6.022×10^{23} molecules	$2 \times 6.022 \times 10^{23}$ molecules

TABLE 17.1

The Effect of Pressure on the Conversion of H_2 and N_2 to NH_3 at 450°C[a]

Pressure (atm)	Yield of NH_3(%)	Pressure (atm)	Yield of NH_3(%)
10	2.04	300	35.5
30	5.8	600	53.4
50	9.17	1000	69.4
100	16.4		

[a] The starting ratio of H_2 to N_2 is 3 moles to 1 mole.

When the pressure on this system is increased, the rate of both the forward and the reverse reactions will increase because of the higher concentrations of N_2, O_2, and NO. But the equilibrium will not shift, because the increase in concentration of molecules is the same on both sides of the equation and the decrease in volume is the same on both sides of the equation.

EXAMPLE 17.1   What effect would an increase in pressure have on the position of equilibrium in the following reactions?

(a) $2 SO_2(g) + O_2(g) \rightleftharpoons 2 SO_3(g)$
(b) $H_2(g) + Cl_2(g) \rightleftharpoons 2 HCl(g)$
(c) $N_2O_4(l) \rightleftharpoons 2 NO_2(g)$

(a) The equilibrium will shift to the right because the substance on the right has a smaller volume than those on the left.
(b) The equilibrium position will be unaffected because the volumes (or moles) of gases on both sides of the equation are the same.
(c) The equilibrium will shift to the left because $N_2O_4(l)$ occupies a much smaller volume than does $2 NO_2(g)$.

PRACTICE   What effect would an increase in pressure have on the position of the equilibrium in the following reactions?
(a) $2 NO(s) + Cl_2(g) \rightleftharpoons 2 NOCl(g)$
(b) $COBr_2(g) \rightleftharpoons CO(g) + Br_2(g)$
Answers: (a) Equilibrium shifts right   (b) Equilibrium shifts left

17.7 Effect of Temperature on Reaction Rate and Equilibrium

An increase in temperature generally increases the rate of reaction. Molecules at elevated temperatures are more energetic and have more kinetic energy; thus, their collisions are more likely to result in a reaction. However, we cannot assume that the rate of a desired reaction will increase indefinitely as the temperature is raised. High temperatures may cause the destruction or decomposition of the reactants and products or may initiate reactions other than the one desired. For example, when calcium oxalate (CaC_2O_4) is heated to 500°C, it decomposes into calcium carbonate and carbon monoxide:

$$CaC_2O_4(s) \xrightarrow{500°C} CaCO_3(s) + CO(g)$$

If calcium oxalate is heated to 850°C, the products are calcium oxide, carbon monoxide, and carbon dioxide:

$$CaC_2O_4(s) \xrightarrow{850°C} CaO(s) + CO(g) + CO_2(g)$$

When heat is applied to a system in equilibrium, the reaction that absorbs heat is favored. When the process, as written, is endothermic, the forward reaction is increased. When the reaction is exothermic, the reverse reaction is favored. In this sense heat may be treated as a reactant in endothermic reactions or as a product in exothermic reactions. Therefore, temperature is analogous to concentration when applying Le Chatelier's principle to heat effects on a chemical reaction.

Hot coke (C) is a very reactive element. In the reaction

$$C(s) + CO_2(g) + heat \rightleftharpoons 2\,CO(g)$$

very little, if any, CO is formed at room temperature. At 1000°C the equilibrium mixture contains about an equal number of moles of CO and CO_2. At higher temperatures the equilibrium shifts to the right, increasing the yield of CO. The reaction is endothermic, and, as can be seen, the equilibrium is shifted to the right at higher temperatures.

Phosphorus trichloride reacts with dry chlorine gas to form phosphorus pentachloride. The reaction is exothermic:

$$PCl_3(l) + Cl_2(g) \rightleftharpoons PCl_5(s) + 88 \text{ kJ}(21 \text{ kcal})$$

Heat must continuously be removed during the reaction to obtain a good yield of the product. According to the principle of Le Chatelier, heat will cause the product, PCl_5, to decompose, re-forming PCl_3 and Cl_2. The equilibrium mixture at 200°C contains 52% PCl_5, and at 300°C it contains 3% PCl_5, verifying that heat causes the equilibrium to shift to the left.

When the temperature of a system is raised, the rate of reaction increases because of increased kinetic energy and more frequent collisions of the reacting species. In a reversible reaction the rate of both the forward and the reverse reactions is increased by an increase in temperature; however, the reaction that absorbs heat increases to a greater extent, and the equilibrium shifts to favor that reaction. The following example illustrates these effects.

EXAMPLE 17.2

$$4\,HCl(g) + O_2(g) \rightleftharpoons 2\,H_2O(g) + 2\,Cl_2(g) + 95.4\ kJ\ (28.4\ kcal) \tag{1}$$

$$H_2(g) + Cl_2(g) \rightleftharpoons 2\,HCl(g) + 185\ kJ\ (44.2\ kcal) \tag{2}$$

$$CH_4(g) + 2\,O_2(g) \rightleftharpoons CO_2(g) + 2\,H_2O(g) + 890\ kJ\ (212.8\ kcal) \tag{3}$$

$$N_2O_4(l) + 58.6\ kJ\ (14\ kcal) \rightleftharpoons 2\,NO_2(g) \tag{4}$$

$$2\,CO_2(g) + 566\ kJ\ (135.2\ kcal) \rightleftharpoons 2\,CO(g) + O_2(g) \tag{5}$$

$$H_2(g) + I_2(g) + 51.9\ kJ\ (12.4\ kcal) \rightleftharpoons 2\,HI(g) \tag{6}$$

Reactions (1), (2), and (3) are exothermic; an increase in temperature will cause the equilibrium to shift to the left. Reactions (4), (5), and (6) are endothermic; an increase in temperature will cause the equilibrium to shift to the right.

PRACTICE   What effect would an increase in temperature have on the position of the equilibrium in the following reactions?

(a) $2\,SO_2(g) + O_2(g) \rightleftharpoons 2\,SO_3(g) + 198\ kJ$
(b) $H_2(g) + CO_2(g) + 41\ kJ \rightleftharpoons H_2O(g) + CO(g)$

Answers: (a) Equilibrium shifts left   (b) Equilibrium shifts right   .

17.8   Effect of Catalysts on Reaction Rate and Equilibrium

catalyst

A **catalyst** is a substance that influences the rate of a reaction and can be recovered essentially unchanged at the end of the reaction. A catalyst does not shift the equilibrium of a reaction; it affects only the speed at which the equilibrium is reached. If a catalyst does not affect the equilibrium, then it follows that it must affect the rate of both the forward and the reverse reactions equally.

The reaction between phosphorus trichloride (PCl_3) and sulfur is highly exothermic, but it is so slow that very little product, thiophosphoryl chloride ($PSCl_3$), is obtained, even after prolonged heating. When a catalyst, such as aluminum chloride ($AlCl_3$), is added, the reaction is complete in a few seconds:

$$PCl_3(l) + S(s) \xrightarrow{AlCl_3} PSCl_3(l)$$

In the laboratory preparation of oxygen, manganese dioxide is used as a catalyst to increase the rates of decomposition of both potassium chlorate and hydrogen peroxide.

$$2 \text{ KClO}_3(s) \xrightarrow[\Delta]{\text{MnO}_2} 2 \text{ KCl}(s) + 3 \text{ O}_2(g)$$

$$2 \text{ H}_2\text{O}_2(aq) \xrightarrow{\text{MnO}_2} 2 \text{ H}_2\text{O}(l) + \text{O}_2(g)$$

Catalysts are extremely important to industrial chemistry. Hundreds of chemical reactions that are otherwise too slow to be of practical value have been put to commercial use once a suitable catalyst was found. But in the area of biochemistry catalysts are of supreme importance because nearly all chemical reactions in all forms of life are completely dependent on biochemical catalysts known as *enzymes*.

17.9 Equilibrium Constants

In a reversible chemical reaction at equilibrium, the concentrations of the reactants and products are constant; that is, they are not changing. The rates of the forward and reverse reactions are constant, and an equilibrium constant expression can be written relating the products to the reactants. For the general reaction

$$a \text{ A} + b \text{ B} \rightleftharpoons c \text{ C} + d \text{ D}$$

at constant temperature, the following equilibrium constant expression can be written:

$$K_{eq} = \frac{[\text{C}]^c [\text{D}]^d}{[\text{A}]^a [\text{B}]^b}$$

equilibrium constant, K_{eq}

where K_{eq} is constant at a particular temperature and is known as the **equilibrium constant**. The quantities in brackets are the concentrations of each substance in moles per liter. The superscript letters a, b, c, and d are the coefficients of the substances in the balanced equation. The units for K_{eq} are not the same for every equilibrium reaction; however, the units are generally omitted. Observe that the concentration of each substance is raised to a power that is the same as the substance's numerical coefficient in the balanced equation. The convention is to place the concentrations of the products (the substances on the right side of the equation as written) in the numerator and the concentrations of the reactants in the denominator.

EXAMPLE 17.3   Write equilibrium constant expressions for

> (a) $3 H_2(g) + N_2(g) \rightleftharpoons 2 NH_3(g)$
> (b) $CO(g) + 2 H_2(g) \rightleftharpoons CH_3OH(g)$

> (a) The only product, NH_3, has a coefficient of 2. Therefore the numerator of the equilibrium constant will be $[NH_3]^2$. Two reactants are present, H_2 with a coefficient of 3 and N_2 with a coefficient of 1. Therefore the denominator will be $[H_2]^3[N_2]$. The equilibrium constant expression is

$$K_{eq} = \frac{[NH_3]^2}{[H_2]^3[N_2]}$$

> (b) For this equation the numerator is $[CH_3OH]$ and the denominator is $[CO][H_2]^2$. The equilibrium constant expression is

$$K_{eq} = \frac{[CH_3OH]}{[CO][H_2]^2}$$

PRACTICE   Write equilibrium constant expressions for

(a) $2 N_2O_5(g) \rightleftharpoons 4 NO_2(g) + O_2(g)$
(b) $4 NH_3(g) + 3 O_2(g) \rightleftharpoons 2 N_2(g) + 6 H_2O(g)$

Answers: (a) $K_{eq} = \dfrac{[NO_2]^4[O_2]}{[N_2O_5]^2}$ (b) $K_{eq} = \dfrac{[N_2]^2[H_2O]^6}{[NH_3]^4[O_2]^3}$

The magnitude of an equilibrium constant indicates the extent to which the forward and reverse reactions take place. When K_{eq} is greater than 1, the amount of the products at equilibrium is greater than the amount of the reactants. When K_{eq} is less than 1, the amount of reactants at equilibrium is greater than the amount of the products. A very large value for K_{eq} indicates that the forward reaction goes essentially to completion. A very small K_{eq} means that the reverse reaction goes nearly to completion and that the equilibrium is far to the left (toward the reactants). Two examples follow:

$$H_2(g) + I_2(g) \rightleftharpoons 2 HI(g) \qquad K_{eq} = 54.8 \text{ at } 425°C$$

This K_{eq} indicates that considerably more product than reactants is present at equilibrium.

$$COCl_2(g) \rightleftharpoons CO(g) + Cl_2(g) \qquad K_{eq} = 7.6 \times 10^{-4} \text{ at } 400°C$$

This K_{eq} indicates that $COCl_2$ is stable and that very little decomposition to CO and Cl_2 occurs at 400°C. The equilibrium is far to the left.

When the molar concentrations of all the species in an equilibrium reaction are known, the K_{eq} can be calculated by substituting the concentrations into the equilibrium constant expression.

EXAMPLE 17.4   Calculate the K_{eq} for the following reaction based on concentrations of: $PCl_5 = 0.030$ mol/L; $PCl_3 = 0.97$ mol/L; $Cl_2 = 0.97$ mol/L at 300°C.

$$PCl_5(g) \rightleftharpoons PCl_3(g) + Cl_2(g)$$

First write the K_{eq} expression; then substitute the respective concentrations into this equation and solve:

$$K_{eq} = \frac{[PCl_3][Cl_2]}{[PCl_5]} = \frac{[0.97][0.97]}{[0.030]} = 31$$

This K_{eq} is considered to be a fairly large value, indicating that at 300°C the decomposition of PCl_5 proceeds far to the right.

PRACTICE   Calculate the K_{eq} for the following reaction. Is the forward or the reverse reaction favored?

$$2\,NO(g) + O_2(g) \rightleftharpoons 2\,NO_2(g)$$

when $[NO] = 0.050$ M, $[O_2] = 0.75$ M, $[NO_2] = 0.25$ M
Answer: $K_{eq} = 33$; The forward reaction is favored

17.10   Ionization Constants

acid ionization
constant, K_a

As a first application of an equilibrium constant, let us consider the constant for acetic acid in solution. Because it is a weak acid, an equilibrium is established between molecular $HC_2H_3O_2$ and its ions in solution. The constant is called the **acid ionization constant**, K_a, a special type of equilibrium constant. The concentration of water in the solution is large compared to the other concentrations and does not change appreciably, so we may use the following simplified equation to set up the constant:

$$HC_2H_3O_2 \rightleftharpoons H^+ + C_2H_3O_2^-$$

The ionization constant expression is the concentration of the products divided by the concentration of the reactants:

$$K_a = \frac{[H^+][C_2H_3O_2^-]}{[HC_2H_3O_2]}$$

It states that the ionization constant, K_a, is equal to the product of the hydrogen ion $[H^+]$ concentration and the acetate ion $[C_2H_3O_2^-]$ concentration divided by the concentration of the un-ionized acetic acid $[HC_2H_3O_2]$.

At 25°C a 0.1 M $HC_2H_3O_2$ solution is 1.34% ionized and has a hydrogen ion concentration of 1.34×10^{-3} mol/L. From this information we can calculate the ionization constant for acetic acid.

A 0.10 M solution initially contains 0.10 mol of acetic acid per liter. Of this 0.10 mol, only 1.34%, or 1.34×10^{-3} mol, is ionized, which gives an H^+ ion concentration of 1.34×10^{-3} mol/L. Because each molecule of acid that ionizes yields one H^+ and one $C_2H_3O_2^-$, the concentration of $C_2H_3O_2^-$ ions is also 1.34×10^{-3} mol/L. This ionization leaves $0.10 - 0.00134 = 0.09866$ mol/L of un-ionized acetic acid.

	Initial concentration (mol/L)	Equilibrium concentration (mol/L)
$[HC_2H_3O_2]$	0.10	0.09866
$[H^+]$	0	0.00134
$[C_2H_3O_2^-]$	0	0.00134

Substituting these concentrations in the equilibrium expression, we obtain the value for K_a:

$$K_a = \frac{[H^+][C_2H_3O_2^-]}{[HC_2H_3O_2]} = \frac{[1.34 \times 10^{-3}][1.34 \times 10^{-3}]}{[0.09866]} = 1.8 \times 10^{-5}$$

The K_a for acetic acid, 1.8×10^{-5}, is small and indicates that the position of the equilibrium is far toward the un-ionized acetic acid. In fact, a 0.10 M acetic acid solution is 98.66% un-ionized.

Once the K_a for acetic acid is established, it can be used to describe other systems containing H^+, $C_2H_3O_2^-$, and $HC_2H_3O_2$ in equilibrium at 25°C. The ionization constants for several other weak acids are listed in Table 17.2.

EXAMPLE 17.5   What is the H^+ ion concentration in a 0.50 M $HC_2H_3O_2$ solution? The ionization constant, K_a, for $HC_2H_3O_2$ is 1.8×10^{-5}.

To solve this problem, first write the equilibrium equation and the K_a expression:

$$HC_2H_3O_2 \rightleftharpoons H^+ + C_2H_3O_2^- \qquad K_a = \frac{[H^+][C_2H_3O_2^-]}{[HC_2H_3O_2]} = 1.8 \times 10^{-5}$$

We know that the initial concentration of $HC_2H_3O_2$ is 0.50 M. We also know from the ionization equation that one $C_2H_3O_2^-$ is produced for every H^+ produced; that is, the $[H^+]$ and the $[C_2H_3O_2^-]$ are equal. To solve, let $Y = [H^+]$, which also equals

TABLE 17.2

Ionization Constants (K_a) of Weak Acids at 25°C

Acid	Formula	K_a
Acetic	$HC_2H_3O_2$	1.8×10^{-5}
Benzoic	$HC_7H_5O_2$	6.3×10^{-5}
Carbolic (phenol)	HC_6H_5O	1.3×10^{-10}
Cyanic	$HCNO$	2.0×10^{-4}
Formic	$HCHO_2$	1.8×10^{-4}
Hydrocyanic	HCN	4.0×10^{-10}
Hypochlorous	$HClO$	3.5×10^{-8}
Nitrous	HNO_2	4.5×10^{-4}
Hydrofluoric	HF	6.5×10^{-4}

the $[C_2H_3O_2^-]$. The un-ionized $[HC_2H_3O_2]$ remaining will then be $0.50 - Y$, the starting concentration minus the amount that ionized.

$$[H^+] = [C_2H_3O_2^-] = Y \qquad [HC_2H_3O_2] = 0.50 - Y$$

Substituting these values into the K_a expression, we obtain

$$K_a = \frac{(Y)(Y)}{0.50 - Y} = \frac{Y^2}{0.50 - Y} = 1.8 \times 10^{-5}$$

An exact solution of this equation for Y requires the use of a mathematical equation known as the quadratic equation. However, an approximate solution is obtained if we assume that Y is small and can be neglected compared with 0.50. Then $0.50 - Y$ will be equal to approximately 0.50. The equation now becomes

$$\frac{Y^2}{0.50} = 1.8 \times 10^{-5}$$

$$Y^2 = 0.50 \times 1.8 \times 10^{-5} = 0.90 \times 10^{-5} = 9.0 \times 10^{-6}$$

Taking the square root of both sides of the equation, we obtain

$$Y = \sqrt{9.0 \times 10^{-6}} = 3.0 \times 10^{-3} \text{ mol/L}$$

Thus, $[H^+]$ is approximately 3.0×10^{-3} mol/L in a $0.50\,M$ $HC_2H_3O_2$ solution. The exact solution to this problem, using the quadratic equation, gives a value of 2.99×10^{-3} mol/L for $[H^+]$, showing that we were justified in neglecting Y compared with 0.50.

PRACTICE   Calculate the hydrogen ion concentration in (a) 0.100 M hydrocyanic acid (HCN) solution and (b) 0.0250 M carbolic acid (HC_6H_5O) solution.

Answers: (a) 6.23×10^{-6} (b) 1.80×10^{-6}

EXAMPLE 17.6   Calculate the percent ionization in a 0.50 M $HC_2H_3O_2$ solution.

The percent ionization of a weak acid, $HA(aq) \rightleftharpoons H^+ + A^-$, is found by dividing the concentration of the H^+ or A^- ions at equilibrium by the initial concentration of HA. For acetic acid

$$\frac{\text{concentration of } [H^+] \text{ or } [C_2H_3O_2^-]}{\text{initial concentration of } [HC_2H_3O_2]} \times 100 = \text{percent ionized}$$

To solve this problem we first need to calculate $[H^+]$. This calculation has already been done in Example 17.5 for a 0.50 M solution.

$[H^+] = 3.0 \times 10^{-3}$ mol/L in a 0.50 M solution   (from Example 17.5)

This $[H^+]$ represents a fractional amount of the initial 0.50 M $HC_2H_3O_2$. Therefore

$$\frac{3.0 \times 10^{-3} \text{ mol/L}}{0.50 \text{ mol/L}} \times 100 = 0.60\% \text{ ionized}$$

A 0.50 M $HC_2H_3O_2$ solution is 0.60% ionized.

PRACTICE   Calculate the percent ionization for the solution in the previous practice problem.

Answers: (a) $6.23 \times 10^{-3}\%$ ionized   (b) $7.20 \times 10^{-3}\%$ ionized

17.11   Ion Product Constant for Water

We have seen that water ionizes to a slight degree. This ionization is represented by these equilibrium equations:

$$H_2O + H_2O \rightleftharpoons H_3O^+ + OH^- \tag{7}$$

$$H_2O \rightleftharpoons H^+ + OH^- \tag{8}$$

Equation (7) is the more accurate representation of the equilibrium, since free protons (H^+) do not exist in water. Equation (8) is a simplified and often-used representation of the water equilibrium. The actual concentration of H^+ produced in pure water is very minute and amounts to only 1.00×10^{-7} mol/L

TABLE 17.3

Relationship of H⁺ and OH⁻ Concentrations in Water Solutions

$[H^+]$	$[OH^-]$	K_w	pH	pOH
1.00×10^{-2}	1.00×10^{-12}	1.00×10^{-14}	2.00	12.0
1.00×10^{-4}	1.00×10^{-10}	1.00×10^{-14}	4.00	10.0
2.00×10^{-6}	5.00×10^{-9}	1.00×10^{-14}	5.70	8.30
1.00×10^{-7}	1.00×10^{-7}	1.00×10^{-14}	7.00	7.00
1.00×10^{-9}	1.00×10^{-5}	1.00×10^{-14}	9.00	5.00

at 25°C. In pure water

$$[H^+] = [OH^-] = 1.00 \times 10^{-7} \text{ mol/L}$$

since both ions are produced in equal molar amounts, as shown in equation (8).

ion product constant for water, K_w

The $H_2O \rightleftharpoons H^+ + OH^-$ equilibrium exists in water and in all water solutions. A special equilibrium constant called the **ion product constant for water**, K_w, applies to this equilibrium. The constant K_w is defined as the product of the H^+ ion concentration and the OH^- ion concentration, each in moles per liter:

$$K_w = [H^+][OH^-]$$

The numerical value of K_w is 1.00×10^{-14}, since for pure water at 25°C

$$K_w = [H^+][OH^-] = [1.00 \times 10^{-7}][1.00 \times 10^{-7}] = 1.00 \times 10^{-14}$$

The value of K_w for all water solutions at 25°C is the constant 1.00×10^{-14}. It is important to realize that, as the concentration of one of these ions, H^+ or OH^-, increases, the other decreases. However, the product of $[H^+]$ and $[OH^-]$ always equals the constant 1.00×10^{-14}. This relationship can be seen in the examples shown in Table 17.3. If the concentration of one ion is known, the concentration of the other can be calculated from the K_w expression.

$$K_w = [H^+][OH^-] \qquad [H^+] = \frac{K_w}{[OH^-]} \qquad [OH^-] = \frac{K_w}{[H^+]}$$

EXAMPLE 17.7   What is the concentration of (a) H^+ and (b) OH^- in a 0.001 M HCl solution? Assume that HCl is 100% ionized.

(a) Since all the HCl is ionized, H^+ = 0.001 mol/L.

$$HCl \longrightarrow H^+ + Cl^-$$

0.00 M 0.001 M 0.001 M

$$[H^+] = 1 \times 10^{-3} \text{ mol/L}   \text{(Answer)}$$

(b) To calculate the $[OH^-]$ in this solution, use the following equation and substitute the values for K_w and $[H^+]$:

$$[OH^-] = \frac{K_w}{[H^+]}$$

$$[OH^-] = \frac{1.00 \times 10^{-14}}{1 \times 10^{-3}} = 1 \times 10^{-11} \text{ mol/L}   \text{(Answer)}$$

PRACTICE   Determine the $[H^+]$ and $[OH^-]$ in
(a) 5.0×10^{-5} M HNO_3 (b) 2.0×10^{-6} M KOH
Answers: (a) $[H^+] = 5.0 \times 10^{-5}$ $[OH^-] = 2.0 \times 10^{-10}$
              (b) $[H^+] = 5.0 \times 10^{-9}$ $[OH^-] = 2.0 \times 10^{-6}$

EXAMPLE 17.8   What is the pH of a 0.010 M NaOH solution? Assume that NaOH is 100% ionized.
Since all the NaOH is ionized, OH^- = 0.010 mol/L or 1.0×10^{-2} mol/L.

$$NaOH \longrightarrow Na^+ + OH^-$$

0.00 M 0.010 M 0.010 M

To find the pH of the solution we first calculate the H^+ concentration. Use the following equation and substitute the values for K_w and $[OH^-]$.

$$[H^+] = \frac{K_w}{[OH^-]} = \frac{1.00 \times 10^{-14}}{1.0 \times 10^{-2}} = 1.0 \times 10^{-12} \text{ mol/L}$$

$$pH = -\log[H^+] = -\log 1 \times 10^{-12} = 12   \text{(Answer)}$$

PRACTICE   Determine the pH for the two solutions in the previous practice problem.
Answers: (a) 4.30   (b) 8.30

Just as pH is used to express the acidity of a solution, pOH is used to express the basicity of an aqueous solution. The pOH is related to the OH^- ion concentration in the same way that the pH is related to the H^+ ion concentration:

$$pOH = \log \frac{1}{[OH^-]} \quad \text{or} \quad pOH = -\log[OH^-]$$

Thus, a solution in which $[OH^-] = 1.0 \times 10^{-2}$, as in EXAMPLE 17.8, will have pOH = 2.0.

In pure water, where $[H^+] = 1 \times 10^{-7}$ and $[OH^-] = 1 \times 10^{-7}$, the pH is 7, and the pOH is 7. The sum of the pH and pOH is always 14.

$$pH + pOH = 14$$

In Example 17.8 the pH can also be found by first calculating the pOH (2) from the OH^- ion concentration and then subtracting from 14.

$$pH = 14 - pOH = 14 - 2 = 12$$

17.12 Solubility Product Constant

solubility
product
constant, K_{sp}

The **solubility product constant**, abbreviated K_{sp}, is another application of the equilibrium constant. It is the equilibrium constant of a slightly soluble salt. The following example illustrates how K_{sp} is evaluated.

The solubility of silver chloride (AgCl) in water is 1.3×10^{-5} mol/L at 25°C. The equation for the equilibrium between AgCl and its ions in solution is

$$AgCl(s) \rightleftharpoons Ag^+ + Cl^-$$

The equilibrium constant expression is

$$K_{eq} = \frac{[Ag^+][Cl^-]}{[AgCl(s)]}$$

The amount of solid AgCl does not affect the equilibrium system provided that some is present. In other words, the concentration of solid silver chloride is constant whether 1 mg or 10 g of the salt are present. Therefore, the product obtained by multiplying the two constants K_{eq} and $[AgCl(s)]$ is also a constant. This constant is the solubility product constant, K_{sp}.

$$K_{eq} \times [AgCl(s)] = [Ag^+][Cl^-] = K_{sp}$$
$$K_{sp} = [Ag^+][Cl^-]$$

The K_{sp} is equal to the product of the Ag^+ ion and the Cl^- ion concentrations, each in moles per liter. When 1.3×10^{-5} mol/L of AgCl dissolves, it produces 1.3×10^{-5} mol/L each of Ag^+ and Cl^-. From these concentrations the K_{sp} can be evaluated.

$$[Ag^+] = 1.3 \times 10^{-5} \text{ mol/L} \qquad [Cl^-] = 1.3 \times 10^{-5} \text{ mol/L}$$
$$K_{sp} = [Ag^+][Cl^-] = [1.3 \times 10^{-5}][1.3 \times 10^{-5}] = 1.7 \times 10^{-10}$$

TABLE 17.4

Solubility Product Constants (K_{sp}) at 25°C

Compound	K_{sp}
AgCl	1.7×10^{-10}
AgBr	5×10^{-13}
AgI	8.5×10^{-17}
$AgC_2H_3O_2$	2×10^{-3}
Ag_2CrO_4	1.9×10^{-12}
$BaCrO_4$	8.5×10^{-11}
$BaSO_4$	1.5×10^{-9}
CaF_2	3.9×10^{-11}
CuS	9×10^{-45}
$Fe(OH)_3$	6×10^{-38}
PbS	7×10^{-29}
$PbSO_4$	1.3×10^{-8}
$Mn(OH)_2$	2.0×10^{-13}

Once the K_{sp} value for AgCl is established, it can be used to describe other systems containing Ag^+ and Cl^-.

The K_{sp} expression does not have a denominator. It consists only of the concentrations (mol/L) of the ions in solution. As in other equilibria expressions, each of these concentrations is raised to a power that is the same number as its coefficient in the balanced equation. The equilibrium equations and the K_{sp} expressions for several other substances follow.

$$AgBr(s) \rightleftharpoons Ag^+ + Br^- \qquad K_{sp} = [Ag^+][Br^-]$$
$$BaSO_4(s) \rightleftharpoons Ba^{2+} + SO_4^{2-} \qquad K_{sp} = [Ba^{2+}][SO_4^{2-}]$$
$$Ag_2CrO_4(s) \rightleftharpoons 2\,Ag^+ + CrO_4^{2-} \qquad K_{sp} = [Ag^+]^2[CrO_4^{2-}]$$
$$CuS(s) \rightleftharpoons Cu^{2+} + S^{2-} \qquad K_{sp} = [Cu^{2+}][S^{2-}]$$
$$Mn(OH)_2(s) \rightleftharpoons Mn^{2+} + 2\,OH^- \qquad K_{sp} = [Mn^{2+}][OH^-]^2$$
$$Fe(OH)_3(s) \rightleftharpoons Fe^{3+} + 3\,OH^- \qquad K_{sp} = [Fe^{3+}][OH^-]^3$$

Table 17.4 lists K_{sp} values for these and several other substances.

When the product of the molar concentration of the ions in solution, each raised to its proper power, is greater than the K_{sp} for that substance, precipitation should occur. If the ion product is less than the K_{sp} value, no precipitation will occur.

EXAMPLE 17.9   Write K_{sp} expressions for AgI and PbI$_2$, both of which are slightly soluble salts.
First write the equilibrium equations:

$$AgI(s) \rightleftharpoons Ag^+ + I^-$$
$$PbI_2(s) \rightleftharpoons Pb^{2+} + 2\,I^-$$

Since the concentration of the solid crystals is constant, the K_{sp} equals the product of the molar concentrations of the ions in solution. In the case of PbI$_2$, the $[I^-]$ must be squared.

$$K_{sp} = [Ag^+][I^-]$$
$$K_{sp} = [Pb^{2+}][I^-]^2$$

PRACTICE   Write the K_{sp} expression for
(a) Cr(OH)$_3$ (b) Cu$_3$(PO$_4$)$_2$
Answers: (a) $K_{sp} = [Cr^{3+}][OH^-]^3$ (b) $K_{sp} = [Cu^{2+}]^3[PO_4^{3-}]^2$

EXAMPLE 17.10   The K_{sp} value for lead sulfate is 1.3×10^{-8}. Calculate the solubility of PbSO$_4$ in grams per liter.
First write the equilibrium equation and the K_{sp} expression:

$$PbSO_4 \rightleftharpoons Pb^{2+} + SO_4^{2-}$$
$$K_{sp} = [Pb^{2+}][SO_4^{2-}] = 1.3 \times 10^{-8}$$

Since the lead sulfate that is in solution is completely dissociated, the concentration of $[Pb^{2+}]$ or $[SO_4^{2-}]$ is equal to the solubility of PbSO$_4$ in moles per liter.
Let

$$Y = [Pb^{2+}] = [SO_4^{2-}]$$

Substitute Y into the K_{sp} equation and solve.

$$[Pb^{2+}][SO_4^{2-}] = [Y][Y] = 1.3 \times 10^{-8}$$
$$Y^2 = 1.3 \times 10^{-8}$$
$$Y = 1.1 \times 10^{-4} \text{ mol/L}$$

The solubility of PbSO$_4$, therefore, is 1.1×10^{-4} mol/L. Now convert mol/L to g/L:

1 mol of PbSO$_4$ has a mass of (207.2 g + 32.1 g + 64.0 g) or 303.3 g

$$\frac{1.1 \times 10^{-4} \text{ mol}}{L} \times \frac{303.3 \text{ g}}{\text{mol}} = 3.3 \times 10^{-2} \text{ g/L}$$

The solubility of PbSO$_4$ is 3.3×10^{-2} g/L.

PRACTICE   The K_{sp} value for CuS is 9.0×10^{-45}. Calculate the solubility in grams per liter.

Answer: 9.1×10^{-21} g/L

An ion added to a solution that already contains that ion is called a *common ion*. When a common ion is added to an equilibrium solution of a weak electrolyte or a slightly soluble salt, the equilibrium shifts according to Le Chatelier's principle. For example, when silver nitrate, $AgNO_3$, is added to a saturated solution of silver chloride, AgCl ($AgCl(s) \rightleftharpoons Ag^+ + Cl^-$), the equilibrium shifts to the left due to the increase in the Ag^+ concentration. As a result, the Cl^- concentration and the solubility of AgCl decreases. AgCl and $AgNO_3$ have the common ion Ag^+. A shift in the equilibrium position upon addition of an ion already contained in the solution is known as the **common ion effect**.

common ion
effect

EXAMPLE 17.11   Silver nitrate, $AgNO_3$, is added to a saturated AgCl solution until the Ag^+ concentration is 0.10 M. What will be the Cl^- concentration remaining in solution?

This problem is an example of the common ion effect. The addition of $AgNO_3$ puts more Ag^+ in solution; the Ag^+ combines with Cl^- and causes the equilibrium to shift to the left, reducing the Cl^- concentration in solution.

We use the K_{sp} to calculate the Cl^- ion concentration remaining in solution. The K_{sp} is constant at a particular temperature and remains the same no matter how we change the concentration of the species involved.

$$K_{sp} = [Ag^+][Cl^-] = 1.7 \times 10^{-10} \qquad [Ag^+] = 0.10 \text{ mol/L}$$

We then substitute the concentration of Ag^+ ion into the K_{sp} expression and calculate the Cl^- concentration.

$$[0.10][Cl^-] = 1.7 \times 10^{-10}$$

$$[Cl^-] = \frac{1.7 \times 10^{-10}}{0.10} = 1.7 \times 10^{-9} \text{ mol/L}$$

This calculation shows a 10,000-fold reduction of Cl^- ions in solution. It illustrates that Cl^- ions may be quantitatively removed from solution with an excess of Ag^+ ions.

PRACTICE   Sodium sulfate (Na_2SO_4) is added to a saturated solution of $BaSO_4$ until the concentration of sulfate ion is 2.0×10^{-2} M. What will be the concentration of the Ba^{2+} ions remaining in solution?

Answer: 7.5×10^{-8} mol/L

TABLE 17.5

Ionic Composition of Salts and the Nature of the Aqueous Solutions They Form

Type of salt	Nature of aqueous solution	Examples
Weak base–strong acid	Acidic	NH_4Cl, NH_4NO_3
Strong base–weak acid	Basic	$NaC_2H_3O_2$, K_2CO_3
Weak base–weak acid	Depends on the salt	$NH_4C_2H_3O_2$, NH_4NO_2
Strong base–strong acid	Neutral	$NaCl$, KBr

17.13 Hydrolysis

hydrolysis

Hydrolysis is the term used for the general reaction in which a water molecule is split. For example, the net ionic hydrolysis reaction for a sodium acetate solution is

$$C_2H_3O_2^- + H_2O(l) \rightleftharpoons HC_2H_3O_2(aq) + OH^-$$

In the reaction above the water molecule is split, with the H^+ combining with $C_2H_3O_2^-$ to give the weak acid $HC_2H_3O_2$ and the OH^- going into solution, making the solution more basic.

Salts that contain an ion of a weak acid or a weak base undergo hydrolysis. For example, a 0.10 M NH_4Cl solution has a pH of 5.1, and a 0.10 M NaCN solution has a pH of 11.1. The hydrolysis reactions that cause these solutions to become acidic or basic are

$$NH_4^+ + H_2O(l) \rightleftharpoons NH_4OH(aq) + H^+$$
$$CN^- + H_2O(l) \rightleftharpoons HCN(aq) + OH^-$$

The ions of a salt derived from a strong acid and a strong base, such as NaCl, do not undergo hydrolysis and thus form neutral solutions. Table 17.5 lists the ionic composition of various salts and the nature of the aqueous solutions that they form.

PRACTICE Indicate whether each of the following salts would produce acidic, basic, or neutral aqueous solution.
(a) KCN   (b) $NaNO_3$ (c) NH_4Br
Answers: (a) basic   (b) neutral   (c) acidic

17.14   Buffer Solutions: The Control of pH

The control of pH within narrow limits is critically important in many chemical applications and vitally important in many biological systems. For example, human blood must be maintained between pH 7.35 and 7.45 for the efficient transport of oxygen from the lungs to the cells. This narrow pH range is maintained by buffer systems in the blood.

buffer solution

A **buffer solution** resists changes in pH when diluted or when small amounts of acid or base are added. Two common types of buffer solutions are (1) a weak acid mixed with a salt of that weak acid and (2) a weak base mixed with a salt of that weak base.

The action of a buffer system can be understood by considering a solution of acetic acid and sodium acetate. The weak acid, $HC_2H_3O_2$, is mostly un-ionized and is in equilibrium with its ions in solution. The salt, $NaC_2H_3O_2$, is completely ionized.

$$HC_2H_3O_2(aq) \rightleftharpoons H^+(aq) + C_2H_3O_2^-(aq)$$
$$NaC_2H_3O_2(aq) \longrightarrow Na^+(aq) + C_2H_3O_2^-(aq)$$

Because the salt is completely ionized, the solution contains a much higher concentration of acetate ions than would be present if only acetic acid were in solution. The acetate ion represses the ionization of acetic acid and also reacts with water, causing the solution to have a higher pH (be more basic) than an acetic acid solution (see Section 17.5). Thus, a 0.1 M acetic acid solution has a pH of 2.87, but a solution that is 0.1 M in acetic acid and 0.1 M in sodium acetate has a pH of 4.74. This difference in pH is the result of the common ion effect.

A buffer solution has a built-in mechanism that counteracts the effect of adding acid or base. Consider the effect of adding HCl or NaOH to an acetic acid–sodium acetate buffer. When a small amount of HCl is added, the acetate ions of the buffer combine with the H^+ ions from HCl to form un-ionized acetic acid, thus neutralizing the added acid and maintaining the approximate pH of the solution. When NaOH is added, the OH^- ions react with acetic acid to neutralize the added base and thus maintain the approximate pH. The equations for these reactions are

$$H^+ + C_2H_3O_2^- \rightleftharpoons HC_2H_3O_2(aq)$$
$$OH^- + HC_2H_3O_2(aq) \rightleftharpoons H_2O(l) + C_2H_3O_2^-$$

Data comparing the changes in pH caused by adding HCl and NaOH to pure water and to an acetic acid–sodium acetate buffer solution are shown in Table 17.6.

The human body has a number of buffer systems. One of these, the bicarbonate–carbonic acid buffer, $HCO_3^- - H_2CO_3$, maintains the blood plasma at a pH of 7.4. The phosphate system, $HPO_4^{2-} - H_2PO_4^-$, is an important buffer in the red blood cells as well as in other places in the body.

TABLE 17.6

Changes in pH Caused by the Addition of HCl and NaOH

Solution	pH	Change in pH
H_2O (1000 mL)	7	—
$\quad$ H_2O + 0.010 mol HCl	2	5
$\quad$ H_2O + 0.010 mol NaOH	12	5
Buffer solution (1000 mL)		
$\quad$ 0.10 M $HC_2H_3O_2$ + 0.10 M $NaC_2H_3O_2$	4.74	—
$\quad$ Buffer + 0.010 mol HCl	4.66	0.08
$\quad$ Buffer + 0.010 mol NaOH	4.83	0.09

17.15   Mechanism of Reactions

mechanism of a reaction

How a reaction occurs—that is, the manner in which it proceeds—is known as the **mechanism of the reaction**. The mechanism is the path, or route, the atoms and molecules take to arrive at the products. Our aim here is not to study the mechanisms themselves but to show that chemical reactions occur by specific routes.

When hydrogen and iodine are mixed at room temperature, we observe no appreciable reaction. In this case, the reaction takes place as a result of collisions between H_2 and I_2 molecules, but at room temperature the collisions do not result in reaction because the molecules lack sufficient energy to react. We say that an energy barrier to reaction exists. If heat is added, the kinetic energy of the molecules increases. When molecules of H_2 and I_2 with sufficient energy collide,

activated complex

activation energy

an intermediate product, known as the **activated complex**, is formed. The amount of energy needed to form the activated complex is known as the **activation energy**. The activated complex, H_2I_2, is in a metastable form and has an energy level higher than that of the reactants or the product. It can decompose to form either the reactants or the product. Three steps constitute the mechanism of the reaction: (1) collision of an H_2 and an I_2 molecule; (2) formation of the activated complex, H_2I_2; and (3) decomposition to the product, HI. The various steps in the formation of HI are shown in Figure 17.3. Figure 17.4 illustrates the energy relationships in this reaction.

The reaction of hydrogen and chlorine proceeds by a different mechanism. When H_2 and Cl_2 are mixed and kept in the dark, essentially no product is formed. But, if the mixture is exposed to sunlight or ultraviolet radiation, it reacts

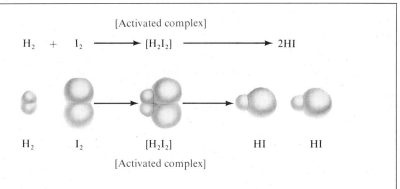

[Activated complex]

$$H_2 \; + \; I_2 \longrightarrow [H_2I_2] \longrightarrow 2HI$$

H₂   I₂   [H₂I₂]   HI   HI

[Activated complex]

FIGURE 17.3

Mechanism of the reaction between hydrogen and iodine. H₂ and I₂ molecules of sufficient energy unite, forming the intermediate activated complex that decomposes to the product, hydrogen iodide.

very rapidly. The overall reaction is

$$H_2(g) + Cl_2(g) \longrightarrow 2\,HCl(g)$$

free radical

This reaction proceeds by what is known as a *free radical mechanism*. A **free radical** is a neutral atom or group of atoms containing one or more unpaired electrons. Both atomic chlorine ($:\ddot{C}l\cdot$) and atomic hydrogen ($H\cdot$) have an unpaired electron and are free radicals. The reaction occurs in three steps:

Step 1 *Initiation*

$$:\ddot{C}l:\ddot{C}l: \; + \; hv \longrightarrow \; :\ddot{C}l\cdot + \; :\ddot{C}l\cdot$$

Chlorine free radicals

In this step a chlorine molecule absorbs energy in the form of a photon, *hv*, of light or ultraviolet radiation. The energized chlorine molecule then splits into two chlorine free radicals.

Step 2 *Propagation*

$$:\ddot{C}l\cdot + \; H:H \longrightarrow HCl + \quad H\cdot$$

Hydrogen
free radical

$$H\cdot + :\ddot{C}l:\ddot{C}l: \longrightarrow HCl + :\ddot{C}l\cdot$$

This step begins when a chlorine free radical reacts with a hydrogen molecule to produce a molecule of hydrogen chloride and a

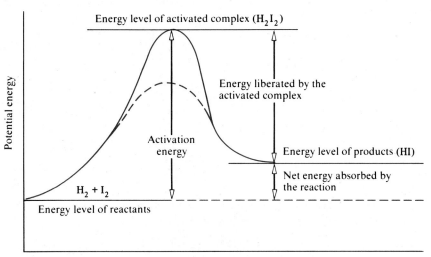

FIGURE 17.4

Relative energy diagram for the reaction between hydrogen and iodine.

$$H_2 + I_2 \longrightarrow [H_2I_2] \longrightarrow 2\,HI$$

Energy equal to the activation energy is put into the system to form the activated complex, H_2I_2. When this complex decomposes, it liberates energy, forming the product. In this case, the product is at a higher energy level than the reactants, indicating that the reaction is endothermic and that energy is absorbed during the reaction. The dotted line represents the effect that a catalyst would have on the reaction. The catalyst lowers the activation energy, thereby increasing the rate of the reaction.

hydrogen free radical. The hydrogen radical then reacts with another chlorine molecule to form hydrogen chloride and another chlorine free radical. This chlorine radical can repeat the process by reacting with another hydrogen molecule, and the reaction continues to propagate itself in this manner until one or both of the reactants are used up. Almost all of the product is formed in this step.

Step 3 *Termination*

$$:\overset{\cdot\cdot}{\underset{\cdot\cdot}{Cl}}\cdot + :\overset{\cdot\cdot}{\underset{\cdot\cdot}{Cl}}\cdot \longrightarrow Cl_2$$

$$H\cdot + H\cdot \longrightarrow H_2$$

$$H\cdot + :\overset{\cdot\cdot}{\underset{\cdot\cdot}{Cl}}\cdot \longrightarrow HCl$$

Hydrogen and chlorine free radicals can react in any of the three ways shown. Unless further activation occurs, the formation of hydrogen chloride will terminate when the radicals form molecules. In an exothermic reaction such as that between hydrogen and chlorine, usually enough heat and light energy is available to maintain the supply of free radicals, and the reaction will continue until at least one reactant is exhausted.

Concepts in Review

1. Describe a reversible reaction.

2. Explain why the rate of the forward reaction decreases and the rate of the reverse reaction increases as a chemical reaction approaches equilibrium.

3. State and understand the qualitative effect of Le Chatlier's principle.

4. Predict how the rate of a chemical reaction is affected by (a) changes in the concentration of reactants, (b) changes in pressure of gaseous reactants, (c) changes in temperature, and (d) the presence of a catalyst.

5. Write the equilibrium constant expression for a chemical reaction from a balanced chemical equation.

6. Explain the meaning of the numerical constant, K_{eq}, when given the concentration of the reactants and products in equilibrium.

7. Calculate the concentration of one substance in an equilibrium when given the equilibrium constant and the concentrations of all the other substances.

8. Calculate the equilibrium constant, K_{eq}, when given the concentration of reactants and products in equilibrium.

9. Calculate the ionization constant for a weak acid from appropriate data.

10. Calculate the concentrations of all the chemical species in a solution of a weak acid when given the percent ionization or the ionization constant.

11. Compare the relative strengths of acids by using their ionization constants.

12. Use the ion product constant for water, K_w, to calculate $[H^+]$, $[OH^-]$, pH, and pOH when given any one of these quantities.

13. Calculate the solubility product constant, K_{sp}, of a slightly soluble salt when given its solubility, or vice versa.

14. Compare relative solubilities of salts if solubility products are known.

15. Discuss the common ion effect on a system at equilibrium.

16. Explain hydrolysis and why some salts form acidic or basic aqueous solutions.

17. Explain how a buffer solution is able to counteract the addition of small amounts of either H^+ or OH^- ions.

18. Draw the relative energy diagram of a reaction including activation energy, exothermic or endothermic reaction, and the effect of a catalyst.

Key Terms in Review

The terms listed here have all been defined within the chapter. Review the definitions of each, and use the glossary and margin notations within the chapter as study aids

acid ionization constant, K_a equilibrium constant, K_{eq}
activated complex   free radical
activation energy   hydrolysis
buffer solution   ionization constants
catalyst   ion product constant for water, K_w
chemical equilibrium   mechanism for a reaction
chemical kinetics   principle of Le Chatlier
common ion effect   reversible chemical reaction
equilibrium   solubility product constant, K_{sp}

Exercises

An asterisk indicates a more challenging question or problem.

1. How would you expect the two tubes in Figure 17.1 to appear if both are at 25°C?
2. Is the reaction $N_2O_4 \rightleftharpoons 2\ NO_2$ (Figure 17.1) exothermic or endothermic?
3. At equilibrium how do the forward and reverse reaction rates compare? (See Figure 17.2.)
4. Would the reaction of 30 mol of H_2 and 10 mol of N_2 produce a greater yield of NH_3 if carried out in a 1 L or a 2 L vessel? (See Table 17.1.)
5. Of the acids listed in Table 17.2, which ones are stronger than acetic acid and which are weaker?
6. For each of the solutions in Table 17.3, what is the sum of the pH plus the pOH? What would be the pOH of a solution whose pH was −1?
7. Tabulate the relative order of molar solubilities of AgCl, AgBr, AgI, $AgC_2H_3O_2$, $PbSO_4$, $BaSO_4$, $BaCrO_4$, and PbS. (Use Table 17.4.) List the most soluble first.
8. Which compound in each of the following pairs has the greater molar solubility? (See Table 17.4.)

(a) $Mn(OH)_2$ or Ag_2CrO_4
(b) $BaCrO_4$ or Ag_2CrO_4
9. Using Table 17.6, explain how the acetic acid–sodium acetate buffer system maintains its pH when 0.010 mol of HCl is added to 1 L of the buffer solution.
10. How would Figure 17.4 be altered if the reaction were exothermic?
11. Express the following reversible systems in equation form:
(a) A mixture of ice and liquid water at 10°C
(b) Liquid water and vapor at 100°C in a pressure cooker
(c) Crystals of Na_2SO_4 in a saturated aqueous solution of Na_2SO_4
(d) A closed system containing boiling sulfur dioxide, SO_2
12. Explain why a precipitate of NaCl forms when hydrogen chloride gas is passed into a saturated aqueous solution of NaCl.
13. Why does the rate of a reaction usually increase when the concentration of one of the reactants is increased?

14. Consider the following system at equilibrium:

$$4 NH_3(g) + 3 O_2(g) \rightleftharpoons$$
$$2 N_2(g) + 6 H_2O(g) + 1531 \text{ kJ}$$

(a) Is the reaction exothermic or endothermic?

(b) If the system's state of equilibrium is disturbed by the addition of O_2, in which direction, left or right, must reaction occur to reestablish equilibrium? After the new equilibrium has been established, how will the final molar concentrations of NH_3, O_2, N_2, and H_2O compare (increase or decrease) with their concentrations before the addition of the O_2?

(c) If the system's state of equilibrium is disturbed by the addition of heat, in which direction will reaction occur, left or right, to reestablish equilibrium?

15. Consider the following system at equilibrium:

$$N_2(g) + 3 H_2(g) \rightleftharpoons 2 NH_3(g) + 92.5 \text{ kJ}$$

Complete the following table. Indicate changes in moles by entering I, D, N, or ? in the table. (I = increase, D = decrease, N = no change, ? = insufficient information to determine.)

Change or stress imposed on the system at equilibrium	Direction of reaction, left or right, to reestablish equilibrium	Change in number of moles		
		N_2	H_2	NH_3
(a) Add N_2				
(b) Remove H_2				
(c) Decrease volume of reaction vessel				
(d) Increase volume of reaction vessel				
(e) Increase temperature				
(f) Add catalyst				
(g) Add both H_2 and NH_3				

16. If pure hydrogen iodide, HI, is placed in a vessel at 700 K, will it decompose? Explain.

17. For each of the equations that follow, tell in which direction, left or right, the equilibrium will shift when the following changes are made: The temperature is increased; the pressure is increased by decreasing the volume of the reaction vessel; a catalyst is added.

(a) $3 O_2(g) + 271 \text{ kJ} \rightleftharpoons 2 O_3(g)$

(b) $CH_4(g) + Cl_2(g) \rightleftharpoons$
$$CH_3Cl(g) + HCl(g) + 110 \text{ kJ}$$

(c) $2 NO(g) + 2 H_2(g) \rightleftharpoons$
$$N_2(g) + 2 H_2O(g) + 665 \text{ kJ}$$

(d) $2 SO_3(g) + 197 \text{ kJ} \rightleftharpoons 2 SO_2(g) + O_2(g)$

(e) $4 NH_3(g) + 3 O_2(g) \rightleftharpoons$
$$2 N_2(g) + 6 H_2O(g) + 1531 \text{ kJ}$$

18. Explain why an increase in temperature causes the rate of reaction to increase.

19. Give a word description of how equilibrium is reached when the substances A and B are first mixed and react as

$$A + B \rightleftharpoons C + D$$

20. Utilizing Le Chatelier's principle, indicate the shift (if any) that would occur to

$$C_2H_6(g) + \text{heat} \rightleftharpoons C_2H_4(g) + H_2(g)$$

(a) If the concentration of hydrogen gas is decreased

(b) If the temperature is lowered

(c) If a catalyst is added

(d) If C_2H_6 is removed from the system

(e) If the volume of the container is increased

21. With dilution, aqueous solutions of acetic acid, $HC_2H_3O_2$, show increased ionization. For example, a 1.0 M solution of acetic acid is 0.42% ionized, whereas a 0.10 M solution is 1.34% ionized. Explain this behavior using the ionization equation and equilibrium principles.

22. A 1.0 M solution of acetic acid ionizes less and has a higher concentration of H^+ ions than a 0.10 M acetic acid solution. Explain this behavior. (See Exercise 21 for data.)

23. Write the equilibrium constant expression for each of the following reactions:

(a) $4 HCl(g) + O_2(g) \rightleftharpoons 2 Cl_2(g) + 2 H_2O(g)$

(b) $N_2(g) + 3 H_2(g) \rightleftharpoons 2 NH_3(g)$

(c) $PCl_5(g) \rightleftharpoons PCl_3(g) + Cl_2(g)$

(d) $HClO_2(aq) \rightleftharpoons H^+(aq) + ClO_2^-(aq)$

Chemistry Is Everywhere

The mention of the word "chemistry" brings to mind test tubes, equations, and people in white lab coats. We rarely stop to think about the many ways in which chemistry surrounds us. Most items we use or come in contact with at home, work, school, or in our leisure time require some aspect of chemistry, either during manufacturing or while being used.

A compact disc is made by stamping pits into a substance coated with aluminum and acrylic. As it is scanned, it detects the contrast between the dark pits and the reflective flat areas.
▼

The brilliant flashes of color seen in a fireworks display are produced by compounds such as sodium salts (yellow), strontium salts (red), copper sulfate (blue), and metallic magnesium (white).
▼

Optical fibers made of ultra-pure glass (SiO_2) conduct light the same way as metal wires conduct electricity.
▼

▲
In this colorful sport, heated air expands, thus decreasing in density, which causes the balloons to fill and rise.

Petrochemicals

Petroleum is a thick, flammable liquid composed of a complex mixture of hydrocarbon compounds. During the refining process, it is separated into different fractions, based on boiling points of the various types of molecules. Natural gas, gasoline, fuel oils, lubricants, plastics, and drugs are some of the familiar products derived from the petroleum industry.

Petroleum can be found in surface springs and pools but is usually obtained deep beneath the earth's surface. ▶

▲ After petroleum is collected, it is transported to a refinery by pipeline or tanker.

◀ At the refinery, the hydrocarbons are separated. Natural gas consists of hydrocarbon molecules containing one to four carbons. Hydrocarbon molecules in gasoline contain five to twelve carbons.

Nuclear Technology

Since the dawn of the nuclear age, the energy produced by splitting atoms has been harnessed and used productively. The process of fission begins when the nucleus of a radioactive element is struck in a particular way by a neutron. The nucleus splits into new elements and is accompanied by a release of subatomic particles and a large amount of energy.

◀ A technician carefully handles the radioactive fuel rods, which contain enriched uranium (U-235).

Radionuclide technology can be used to determine the age of geological structures and ancient artifacts.
▼

▲
A luminous glow is emitted from a vial containing a mixture of radium bromide ($RaBr_2$) and zinc sulfide (ZnS).

◀ A nuclear power plant uses heat produced by the fission of enriched uranium (U-235) fuel in a nuclear reactor to turn turbines and generate electricity.

Medicine and Pharmacy

Developing and manufacturing medical products, such as pharmaceuticals, diagnostic tools, and prosthetics demands much human labor from an array of chemistry-oriented disciplines. Chemical reactions are used to synthesize or isolate molecules, and extensive chemical testing is performed before new drugs and medical products can be pronounced safe and effective.

Researchers are attempting to isolate new drugs that are chemically similar to AZT, but produce fewer side effects.
▼

Computer-simulated models of molecules allow researchers to see three-dimensional views of possible new drugs.
▼

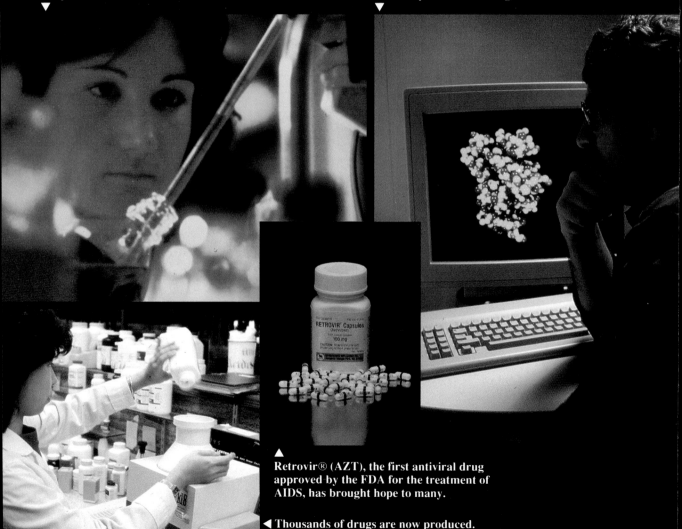

▲
Retrovir® (AZT), the first antiviral drug approved by the FDA for the treatment of AIDS, has brought hope to many.

◄ Thousands of drugs are now produced. Pharmacists must be aware of drug interactions, dosages, and side effects.

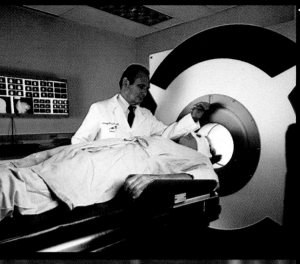

◀ The use of radioactive forms of nitrogen or carbon in positive emission tomography (PET) scans allows visualization of the inside of a human body without the use of invasive techniques.

This PET scan image reveals the areas of the brain utilized during sensory activities.
▼

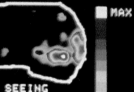

▲
This artificial leg is a single unit with an inner metal bar that allows the ankle and foot to simulate almost natural motion.

▲
Artificial limbs have come a long way since the wooden peg. This prosthesis is made from a carbon graphite composite.

Agriculture

The major use of chemistry in agriculture is for the creation of fertilizers and pesticides. Agricultural chemists are constantly striving to create new products that are ecologically compatible.

Biochemists develop species of plants that are pest and drought resistant. ▶

Large-scale applications of pesticides and fertilizers control pests and diseases while increasing crop yields.
▼

▲
An expansive assortment of garden chemicals are available for home use.

Treating vegetables with radiation from ▶ cobalt-60 eliminates microorganisms and gives a longer shelf life.

Paper Making

Have you ever wondered how paper is made? Chemistry plays a major role in this industry. Millions of trees must be grown to satisfy the great demand for paper, causing the paper industry to be diligent in replenishing the forests for future use.

Trees that are ready for harvest are cut and transported to the logging mill where they are preserved by spraying with pest-deterring chemicals.
▼

Paper fibers are examined by a technician. These fibers can be manipulated to create different textures and strengths.
▼

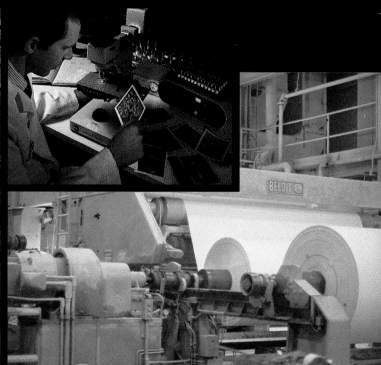

▲
Water is squeezed out of the pulp and is run through steam-heated cylinders, which form the chemical bonds that hold the paper together.

◄ The logs are cut into chips, mashed into pulp, and treated with compounds, such as $NaHSO_3$ and $NaOH$. The pulp is then heated in a digester to dissolve the noncellulose components of the wood.

Consumer Chemistry

We have touched upon a small number of industries that use chemistry in the development and manufacture of their products. An inspection around a typical household will reveal many other consumer goods that, somewhere between their discovery and their arrival in the marketplace, depend on a chemical process.

The pigment titanium dioxide (TiO_2) has ▶ great hiding powers. It is used in enamels, inks, shoe whitener, ceramics, and plastics.

Passing electricity through a tube of neon gas causes it to glow with an orange-red color. Changing the color of the outer tube allows a wide variety of hues.
▼

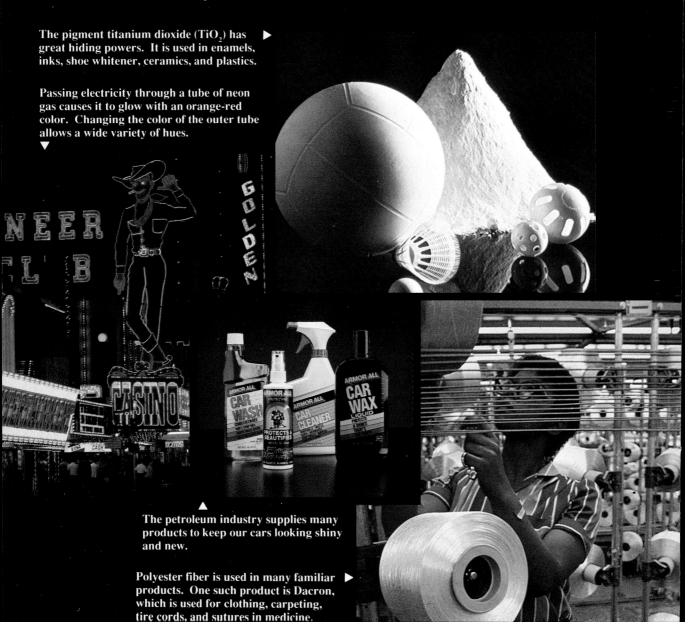

▲
The petroleum industry supplies many products to keep our cars looking shiny and new.

Polyester fiber is used in many familiar ▶ products. One such product is Dacron, which is used for clothing, carpeting, tire cords, and sutures in medicine.

(e) $NH_4OH(aq) \rightleftharpoons NH_4^+(aq) + OH^-(aq)$
(f) $4\,NH_3(g) + 5\,O_2(g) \rightleftharpoons$
$$4\,NO(g) + 6\,H_2O(g)$$

24. What would cause two separate samples of pure water to have slightly different pH values?

25. Why are the pH and pOH equal in pure water?

26. What effect will increasing the H^+ ion concentration of a solution have on (a) pH, (b) pOH, (c) $[OH^-]$, and (d) K_w?

27. Write the solubility product expression (K_{sp}) for each of the following substances:
    (a) CuS   (e) $Fe(OH)_3$
    (b) $BaSO_4$ (f) Sb_2S_5
    (c) $PbBr_2$ (g) CaF_2
    (d) Ag_3AsO_4 (h) $Ba_3(PO_4)_2$

28. Explain why silver acetate, $AgC_2H_3O_2$, is more soluble in nitric acid than in water. [*Hint:* Write the equilibrium equation first and then consider the effect of the acid on the acetate ion.] What would happen if hydrochloric acid, HCl, were used in place of nitric acid?

29. Decide whether each of the following salts forms an acidic, a basic, or a neutral aqueous solution.
    (a) KCl   (d) $(NH_4)_2SO_4$ (g) $NaNO_2$
    (b) Na_2CO_3 (e) $Ca(CN)_2$ (h) NaF
    (c) K_2SO_4 (f) $BaBr_2$

30. Dissolution of sodium acetate, $NaC_2H_3O_2$, in pure water gives a basic solution. Why? [*Hint:* A small amount of $HC_2H_3O_2$ is formed.]

31. Write hydrolysis equations for aqueous solutions of these salts.
    (a) KNO_2 (c) NH_4NO_3
    (b) $Mg(C_2H_3O_2)_2$ (d) Na_2SO_3

32. Write hydrolysis equations for the following ions.
    (a) HCO_3^- (b) NH_4^+ (c) OCl^- (d) ClO_2^-

33. Describe why the pH of a buffer solution remains almost constant when a small amount of acid or base is added to it.

34. One of the important pH-regulating systems in the blood consists of a carbonic acid–sodium bicarbonate buffer.

$$H_2CO_3(aq) \rightleftharpoons H^+(aq) + HCO_3^-(aq)$$

$$NaHCO_3(aq) \longrightarrow Na^+(aq) + HCO_3^-(aq)$$

Explain how this buffer resists changes in pH when (a) excess acid and (b) excess base get into the bloodstream.

35. Which of the following statements are correct?

Rewrite the incorrect statements to make them correct.

(a) In a reaction at equilibrium the concentrations of reactants and products are equal.

(b) A catalyst increases the concentrations of products present at equilibrium.

(c) Enzymes are the catalysts in living systems.

(d) A catalyst lowers the activation energy of a reaction by equal amounts for both the forward and the reverse reactions.

(e) If an increase in temperature causes an increase in the concentration of products present at equilibrium, the reaction is exothermic.

(f) The magnitude of an equilibrium constant is independent of the reaction temperature.

(g) A large equilibrium constant for a reaction indicates that the reaction, at equilibrium, favors products over reactants.

(h) The amount of product obtained at equilibrium is proportional to how fast equilibrium is attained.

(i) The study of reaction rates and reaction mechanisms is known as chemical kinetics.

(j) For the reaction

$$CaCO_3(s) \overset{\Delta}{\rightleftharpoons} CaO(s) + CO_2(g)$$

increasing the pressure of CO_2 present at equilibrium will cause the reaction to shift left.

(k) The larger the value of the equilibrium constant, the greater the proportion of products present at equilibrium.

(l) At chemical equilibrium the rate of the reverse reaction is equal to the rate of the forward reaction.

Statements (m)–(s) pertain to the equilibrium system.

$$2\,NO(g) + O_2(g) \rightleftharpoons 2\,NO_2(g) + heat$$

(m) The reaction as shown is endothermic.

(n) Increasing the temperature will cause the equilibrium to shift left.

(o) Increasing the temperature will increase the magnitude of the equilibrium constant, K_{eq}.

(p) Decreasing the volume of the reaction vessel will shift the equilibrium to the right and decrease the concentrations of the reactants.

(q) Removal of some of the O_2 will cause an increase in the concentration of NO.

(r) High temperatures and pressures favor increased yields of NO_2.

(s) The equilibrium constant expression for the reaction is

$$K_{eq} = \frac{[NO_2]^2}{[NO]^2[O_2]}$$

(t) A solution with an H^+ ion concentration of 1×10^{-5} mol/L has a pOH of 9.

(u) An aqueous solution that has an OH^- ion concentration of 1×10^{-4} mol/L has an H^+ ion concentration of 1×10^{-10} mol/L.

(v) $K_w = [H^+][OH^-] = 1.00 \times 10^{-14}$, and pH + pOH = 14.

(w) As solid $BaSO_4$ is added to a saturated solution of $BaSO_4$, the magnitude of its K_{sp} increases.

(x) A solution of pOH 10 is basic.

(y) The pH increases as the $[H^+]$ increases.

(z) The pH of 0.050 M $Ca(OH)_2$ is 13.

Equilibrium Constants

36. What is the maximum number of moles of HI that can be obtained from a reaction mixture containing 2.30 mol of I_2 and 2.10 mol of H_2?

37. (a) How many moles of hydrogen iodide, HI, will be produced when 2.00 mol of H_2 and 2.00 mol of I_2 are reacted at 700 K? (Reaction is 79% complete.)

(b) Addition of 0.27 mol of I_2 to the system increases the yield of HI to 85%. How many moles of H_2, I_2, and HI are now present?

(c) From the data in part (a), calculate K_{eq} for the reaction at 700 K.

*38. 6.00 g of hydrogen, H_2, and 200 g of iodine, I_2, are reacted at 500 K. After equilibrium is reached, analysis shows that the flask contains 64.0 g of HI. How many moles of H_2, I_2, and HI are present in this equilibrium mixture?

39. What is the equilibrium constant of the reaction shown if a 20 L flask contains 0.10 mol of PCl_3, 1.50 mol of Cl_2, and 0.22 mol of PCl_5?

$$PCl_3(g) + Cl_2(g) \rightleftharpoons PCl_5$$

40. If the rate of a reaction doubles for every 10° rise in temperature, how much faster will the reaction go at 100°C than at 30°C?

Ionization Constants

41. Calculate the ionization constant for each of the following monoprotic acids. Each acid ionizes as follows: $HA \rightleftharpoons H^+ + A^-$.

Acid	Acid concentration	$[H^+]$
Hypochlorous, HOCl	0.10 M	5.9×10^{-5} mol/L
Propanoic, $HC_3H_5O_2$	0.15 M	1.4×10^{-3} mol/L
Hydrocyanic, HCN	0.20 M	8.9×10^{-6} mol/L

42. Calculate (a) the H^+ ion concentration, (b) the pH, and (c) the percent ionization of a 0.25 M solution of $HC_2H_3O_2$. ($K_a = 1.8 \times 10^{-5}$.)

43. A 1.0 M solution of a weak acid, HA, is 0.52% ionized. Calculate the ionization constant, K_a, for the acid.

44. A 0.15 M solution of a weak acid, HA, has a pH of 5. Calculate the ionization constant, K_a, for the acid.

45. Calculate the percent ionization and pH of solutions of $HC_2H_3O_2$ having the following molarities: (a) 1.0 M, (b) 0.10 M, and (c) 0.010 M. ($K_a = 1.8 \times 10^{-5}$.)

46. A 0.37 M solution of a weak acid, HA, has a pH of 3.7. What is the K_a for this acid?

47. A 0.23 M solution of a weak acid, HA, has a pH of 2.89. What is the K_a for this acid?

48. A common laboratory reagent is 6.0 M HCl. Calculate the $[H^+]$, $[OH^-]$, pH, and pOH of this solution.

49. Calculate the pH and the pOH of the following solutions:

(a) 0.00010 M HCl

(b) 0.010 M NaOH

(c) 0.0025 M NaOH

(d) 0.10 M HClO ($K_a = 3.5 \times 10^{-8}$)

*(e) Saturated $Fe(OH)_2$ solution ($K_{sp} = 8.0 \times 10^{-16}$)

50. Calculate the $[OH^-]$ in each of these solutions:

(a) $[H^+] = 1.0 \times 10^{-4}$

(b) $[H^+] = 2.8 \times 10^{-6}$

(c) $[H^+] = 4.0 \times 10^{-9}$

51. Calculate the $[H^+]$ in each of these solutions:
    (a) $[OH^-] = 6.0 \times 10^{-7}$
    (b) $[OH^-] = 1 \times 10^{-8}$
    (c) $[OH^-] = 4.5 \times 10^{-6}$

Solubility Product Constant

52. Given the following solubility data, calculate the solubility product constant for each substance.
    (a) $BaSO_4$, 3.9×10^{-5} mol/L
    (b) Ag_2CrO_4, 7.8×10^{-5} mol/L
    (c) ZnS, 3.5×10^{-12} mol/L
    (d) $Pb(IO_3)_2$, 4.0×10^{-5} mol/L
    (e) Bi_2S_3, 4.9×10^{-15} mol/L
    (f) $AgCl$, 0.0019 g/L
    (g) $CaSO_4$, 0.67 g/L
    (h) $Zn(OH)_2$, 2.33×10^{-4} g/L
    (i) Ag_3PO_4, 6.73×10^{-3} g/L

53. Calculate the molar solubility for each of the following substances:
    (a) $BaCO_3$, $K_{sp} = 2.0 \times 10^{-9}$
    (b) $AlPO_4$, $K_{sp} = 5.8 \times 10^{-19}$
    (c) Ag_2SO_4, $K_{sp} = 1.5 \times 10^{-5}$
    (d) $Mg(OH)_2$, $K_{sp} = 7.1 \times 10^{-12}$

54. Calculate, for each of the substances in Question 18, the solubility in grams per 100 mL of water.

55. The K_{sp} of CaF_2 is 3.9×10^{-11}. Calculate (a) the molar concentrations of Ca^{2+} and F^- in a saturated solution, and (b) the grams of CaF_2 that will dissolve in 500 mL of water.

*56. The following pairs of solutions are mixed. Show by calculation whether or not a precipitate will form.
    (a) 100. mL of 0.010 M Na_2SO_4 and 100 mL of 0.001 M $Pb(NO_3)_2$
    (b) 50.0 mL of 1.0×10^{-4} M $AgNO_3$ and 100. mL of 1.0×10^{-4} M $NaCl$
    (c) 1.0 g $Ca(NO_3)_2$ in 150 mL H_2O and 250 mL of 0.01 M $NaOH$
    $K_{sp}\ PbSO_4 = 1.3 \times 10^{-8}$
    $K_{sp}\ AgCl = 1.7 \times 10^{-10}$
    $K_{sp}\ Ca(OH)_2 = 1.3 \times 10^{-6}$

57. $BaCl_2$ is added to a saturated $BaSO_4$ solution until the Ba^{2+} concentration is 0.050 M. (a) What concentration of SO_4^{2-} remains in solution? (b) How many grams of $BaSO_4$ remain dissolved in 100. mL of the solution? ($K_{sp} = 1.5 \times 10^{-9}$ for $BaSO_4$.)

58. The concentration of a solution is 0.10 M Ba^{2+} and 0.10 M Sr^{2+}. Which sulfate, $BaSO_4$ or $SrSO_4$, will precipitate first when a dilute solution of H_2SO_4 is added dropwise to the solution? Show evidence for your answer. ($K_{sp} = 1.5 \times 10^{-9}$ for $BaSO_4$ and $K_{sp} = 3.5 \times 10^{-7}$ for $SrSO_4$.)

59. How many moles of $AgBr$ will dissolve in 1.0 L of (a) 0.10 M $NaBr$ and (b) 0.10 M $MgBr_2$? ($K_{sp} = 5.0 \times 10^{-13}$ for $AgBr$.)

60. The K_{sp} for $PbCl_2$ is 2.0×10^{-5}. Will a precipitate form when 0.050 mol of $Pb(NO_3)_2$ and 0.010 mol of $NaCl$ are dissolved in 1.0 L H_2O? Show evidence for your answer.

61. Calculate the K_{eq} for the reaction $SO_2(g) + O_2(g) \rightleftharpoons SO_3(g)$ when the equilibrium concentrations of the gases are at 530°C. $SO_3 = 11.0$ M, $SO_2 = 4.20$ M, and $O_2 = 0.6 \times 10^{-3}$ M.

62. If it takes 0.048 g of BaF_2 to saturate 15.0 mL of water, what is the K_{sp} of BaF_2?

63. The K_{eq} for the formation of ammonia gas from its elements is 4.0. If the equilibrium concentrations of nitrogen gas and hydrogen gas are both 2.0 M, what is the equilibrium concentration of the ammonia gas?

64. The K_{sp} of $SrSO_4$ is 7.6×10^{-7}. Should precipitation occur when 25.0 mL of 1.0×10^{-3} M of $SrCl_2$ solution is mixed with 15.0 mL of 2.0×10^{-3} M Na_2SO_4? Prove.

*65. The solubility of Hg_2I_2 in H_2O is 3.04×10^{-7} g/L. The reaction $Hg_2I_2 \rightleftharpoons Hg_2^{2+} + 2\,I^-$ represents equilibrium. Calculate the K_{sp}.

66. Calculate the H^+ ion concentration and the pH of buffer solutions that are 0.20 M in $HC_2H_3O_2$ and contain sufficient sodium acetate to make the $C_2H_3O_2^-$ ion concentration equal to (a) 0.10 M and (b) 0.20 M. ($K_a\ HC_2H_3O_2 = 1.8 \times 10^{-5}$.)

67. (a) When 1.0 mL of 1.0 M HCl is added to 50 mL of 1.0 M $NaCl$, the H^+ ion concentration changes from 1×10^{-7} M to 2.0×10^{-2} M.
    (b) When 1.0 mL of 1.0 M HCl is added to 50. mL of a buffer solution that is 1.0 M in $HC_2H_3O_2$ and 1.0 M in $NaC_2H_3O_2$, the H^+ ion concentration changes from 1.8×10^{-5} M to 1.9×10^{-5} M.
    Calculate the initial pH and the pH change in each solution (log 1.8 = 0.26; log 1.9 = 0.28; log 2.0 = 0.30).

CHAPTER EIGHTEEN

Oxidation–Reduction

The variety of chemical reactions that occur in our daily lives is amazing. Our society seems to run on batteries—in calculators, cars, toys, lights, thermostats, radios, televisions, and more. We polish sterling silver, paint iron railings, and galvanize nails to combat corrosion. Jewelry and computer chips are electroplated with very thin coatings of gold or silver. Clothes are bleached, and photographs are developed in solutions using chemical reactions that involve electron transfer. Tests for glucose in urine, or alcohol in the breath show vivid color changes. Plants turn energy into chemical compounds through a series of reactions called the *electron transport chain*. All of these reactions involve the transfer of electrons between substances in a chemical process called *oxidation-reduction*.

Chapter Preview

18.1   Oxidation Number

The oxidation number of an atom (sometimes called its oxidation state) can be considered to represent the number of electrons lost, gained, or unequally shared by the atom. Oxidation numbers can be zero, positive, or negative. When the oxidation number of an atom is zero, the atom has the same number of electrons assigned to it as there are in the free neutral atom. When the oxidation number is positive, the atom has fewer electrons assigned to it than there are in the neutral atom. When the oxidation number is negative, the atom has more electrons assigned to it than there are in the neutral atom.

The oxidation number of an atom that has lost or gained electrons to form an ion is the same as the plus or minus charge of the ion. In the ionic compound NaCl the oxidation numbers are clearly established to be $+1$ for the Na^+ ion and -1 for the Cl^- ion. The Na^+ ion has one less electron than the neutral Na atom; and the Cl^- ion has one more electron than the neutral Cl atom. In $MgCl_2$ two electrons have transferred from the Mg atom to the Cl atoms; thus, the oxidation number of Mg is $+2$.

In covalently bonded substances, where electrons are shared between two atoms, oxidation numbers are assigned by a somewhat arbitrary system based on relative electronegativities. For symmetrical covalent molecules, such as H_2 and Cl_2, each atom is assigned an oxidation number of zero because the bonding pair of electrons is shared equally between two like atoms, neither of which is more electronegative than the other.

$$H:H \qquad :\ddot{C}l:\ddot{C}l:$$

When the covalent bond is between two unlike atoms, the bonding electrons are shared unequally because the more electronegative element has a greater attraction for them. In this case the oxidation numbers are determined by assigning both electrons to the more electronegative element.

Thus in compounds with covalent bonds, such as NH_3 and H_2O,

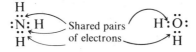

485

TABLE 18.1

Arbitrary Rules for Assigning Oxidation Numbers

---

1. All elements in their free state (uncombined with other elements) have an oxidation number of zero (for example, Na, Cu, Mg, H_2, O_2, Cl_2, N_2).
2. H is $+1$, except in metal hydrides, where it is -1 (for example, NaH, CaH_2).
3. O is -2, except in peroxides, where it is -1, and in OF_2, where it is $+2$.
4. The metallic element in an ionic compound has a positive oxidation number.
5. In covalent compounds the negative oxidation number is assigned to the most electronegative atom.
6. The algebraic sum of the oxidation numbers of the elements in a compound is zero.
7. The algebraic sum of the oxidation numbers of the elements in a polyatomic ion is equal to the charge of the ion.

---

the pairs of electrons are unequally shared between the atoms and are attracted toward the more electronegative elements, N and O. This unequal sharing causes the N and O atoms to be relatively negative with respect to the H atoms. At the same time it causes the H atoms to be relatively positive with respect to the N and O atoms. In H_2O both pairs of shared electrons are assigned to the O atom, giving it two electrons more than the neutral O atom. At the same time, each H atom is assigned one electron less than the neutral H atom. Therefore, the O atom is assigned an oxidation number of -2, and each H atom is assigned an oxidation number of $+1$. In NH_3 the three pairs of shared electrons are assigned to the N atom, giving it three electrons more than the neutral N atom. At the same time, each H atom has one electron less than the neutral atom. Therefore, the N atom is assigned an oxidation number of -3, and each H atom is assigned an oxidation number of $+1$.

The assignment of correct oxidation numbers to elements is essential for balancing oxidation–reduction equations. Review Section 6.2 regarding oxidation numbers, oxidation number tables, and the determination of oxidation numbers from formulas. Table 12.3 lists relative electronegativities of the elements. Rules for assigning oxidation numbers are given in Section 6.2 and are summarized in Table 18.1. Examples showing oxidation numbers in compounds and ions are given in Table 18.2.

EXAMPLE 18.1   Determine the oxidation number of each element in (a) KNO_3 and (b) SO_4^{2-}.

(a) K is a Group IA metal; therefore it has an oxidation number of $+1$. The oxidation number of each O atom is -2 (Table 18.1, Rule 3). Using these values and the fact that the sum of the oxidation numbers of all the atoms in a compound is zero, we can determine the oxidation number of N.

TABLE 18.2

Oxidation Numbers of Atoms in Selected Compounds

Ion or compound	Oxidation number	
H_2O	H, +1;	O, −2
SO_2	S, +4;	O, −2
CH_4	C, −4;	H, +1
CO_2	C, +4;	O, −2
$KMnO_4$	K, +1;	Mn, +7; O, −2
Na_3PO_4	Na, +1;	P, +5; O, −2
$Al_2(SO_4)_3$	Al, +3;	S, +6; O, −2
NO	N, +2;	O, −2
BCl_3	B, +3;	Cl, −1
SO_4^{2-}	S, +6;	O, −2
NO_3^-	N, +5;	O, −2
CO_3^{2-}	C, +4;	O, −2

$$KNO_3$$

$$+1 + N + 3(-2) = 0$$

$$N = +6 - 1 = +5$$

The oxidation numbers are K, +1; N, +5; O, −2.

(b) SO_4^{2-} is an ion; therefore, the sum of oxidation numbers of the S and the O atoms must be −2, the charge of the ion. The oxidation number of each O atom is −2 (Table 18.1, Rule 3). Then

$$SO_4^{2-}$$

$$S + 4(-2) = -2$$

$$S = -2 + 8 = +6$$

The oxidation numbers are S, +6; O, −2.

PRACTICE Determine the oxidation number of each element in the following compounds:

(a) $BeCl_2$ (b) HClO (c) H_2O_2 (d) NH_4^+ (e) BrO_3^- (f) CrO_4^{2-}

Answers: (a) Be = +2; Cl = −1 (d) N = −3; H = +1
(b) H = +1; Cl = +1(O = −2) (e) Br = +5; O = −2
(c) H = +1; O = −1 (f) Cr = +6; O = −2

18.2   Oxidation–Reduction

oxidation–
reduction

redox

oxidation

reduction

Oxidation–reduction, also known as **redox**, is a chemical process in which the oxidation number of an element is changed. The process may involve the complete transfer of electrons to form ionic bonds or only a partial transfer or shift of electrons to form covalent bonds.

Oxidation occurs whenever the oxidation number of an element increases as a result of losing electrons. Conversely, **reduction** occurs whenever the oxidation number of an element decreases as a result of gaining electrons. For example, a change in oxidation number from $+2$ to $+3$ or from -1 to 0 is oxidation; a change from $+5$ to $+2$ or from -2 to -4 is reduction (see Figure 18.1). Oxidation and reduction occur simultaneously in a chemical reaction; one cannot take place without the other.

Many combination, decomposition, and single-displacement reactions involve oxidation–reduction. Let us examine the combustion of hydrogen and oxygen from this point of view:

$$2\,H_2 + O_2 \longrightarrow 2\,H_2O$$

Both reactants, hydrogen and oxygen, are elements in the free state and have an oxidation number of zero. In the product, water, hydrogen has been oxidized to $+1$ and oxygen reduced to -2. The substance that causes an increase in the oxidation state of another substance is called an **oxidizing agent**. The substance that causes a decrease in the oxidation state of another substance is called a **reducing agent**. In this reaction the oxidizing agent is free oxygen, and the reducing agent is free hydrogen. In the reaction

oxidizing agent

reducing agent

$$Zn(s) + H_2SO_4(aq) \longrightarrow ZnSO_4(aq) + H_2\uparrow$$

metallic zinc is oxidized, and hydrogen ions are reduced. Zinc is the reducing agent, and hydrogen ions, the oxidizing agent. Electrons are transferred from the zinc metal to the hydrogen ions. The reaction is better expressed as

$$Zn^0 + 2\,H^+ + SO_4^{2-} \longrightarrow Zn^{2+} + SO_4^{2-} + H_2^0\uparrow$$

Oxidation:	**Increase in oxidation number**
	Loss of electrons
Reduction:	**Decrease in oxidation number**
	Gain of electrons

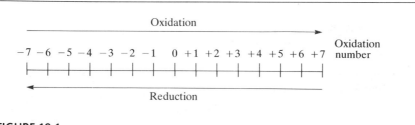

FIGURE 18.1

Oxidation and reduction. Oxidation results in an increase in the oxidation number, and reduction results in a decrease in the oxidation number.

The oxidizing agent is reduced and gains electrons. The reducing agent is oxidized and loses electrons. The transfer of electrons is characteristic of all redox reactions.

18.3 Balancing Oxidation–Reduction Equations

Many simple redox equations can be balanced readily by inspection, or trial and error.

$$Na + Cl_2 \longrightarrow NaCl \qquad \text{(unbalanced)}$$
$$2\,Na + Cl_2 \longrightarrow 2\,NaCl \quad \text{(balanced)}$$

Balancing this equation is certainly not complicated. But as we study more complex reactions and equations, such as

$$P + HNO_3 + H_2O \longrightarrow NO + H_3PO_4 \qquad \text{(unbalanced)}$$
$$3\,P + 5\,HNO_3 + 2\,H_2O \longrightarrow 5\,NO + 3\,H_3PO_4 \quad \text{(balanced)}$$

the trial-and-error method of finding the proper numbers to balance the equation would take an unnecessarily long time.

One systematic method for balancing oxidation–reduction equations is based on the transfer of electrons between the oxidizing and reducing agents. Consider the first equation again.

$$Na^0 + Cl_2^0 \longrightarrow Na^+Cl^- \quad \text{(unbalanced)}$$

In this reaction sodium metal loses one electron per atom when it changes to a sodium ion. At the same time chlorine gains one electron per atom. Because chlorine is diatomic, two electrons per molecule are needed to form a chloride ion from each atom. These electrons are furnished by two sodium atoms. Stepwise, the reaction may be written as two half-reactions, the oxidation half-reaction and the reduction half-reaction:

Oxidation half-reaction $\qquad$ $2\,Na^0 \longrightarrow 2\,Na^+ + 2\,e^-$
Reduction half-reaction $\qquad$ $\dfrac{Cl_2^0 + 2\,e^- \longrightarrow 2\,Cl^-}{2\,Na^0 + Cl_2^0 \longrightarrow 2\,Na^+Cl^-}$

When the two half-reactions, each containing the same number of electrons, are added together algebraically, the electrons cancel out. In this reaction there are no excess electrons; the two electrons lost by the two sodium atoms are utilized by chlorine. In all redox reactions the loss of electrons by the reducing agent must equal the gain of electrons by the oxidizing agent. Sodium is oxidized; chlorine is reduced. Chlorine is the oxidizing agent; sodium is the reducing agent.

The following examples illustrate a systematic method of balancing more complicated redox equations by the change-in-oxidation-number method.

EXAMPLE 18.2   Balance the equation

$$Sn + HNO_3 \longrightarrow SnO_2 + NO_2 + H_2O \quad \text{(unbalanced)}$$

Step 1 Assign oxidation numbers to each element to identify the elements that are being oxidized and those that are being reduced. Write the oxidation numbers below each element in order to avoid confusing them with the charge on an ion.

$$Sn + H\,N\,O_3 \longrightarrow SnO_2 + NO_2 + H_2O$$
$$\;0 \quad\; +1\,+5\,-2 \qquad\; +4\,-2 \quad +4\,-2 \quad +1\,-2$$

Note that the oxidation numbers of Sn and N have changed.

Step 2 Now write two new equations, using only the elements that change in oxidation number. Then add electrons to bring the equations into electrical balance. One equation represents the oxidation step; the other represents the reduction step. The oxidation step produces electrons; the reduction step uses electrons.

oxidation $\qquad$ $Sn^0 \longrightarrow Sn^{4+} + 4\,e^-$ $\quad$ (Sn^0 loses 4 electrons)
reduction $\qquad$ $N^{5+} + 1\,e^- \longrightarrow N^{4+}$ $\quad$ (N^{5+} gains 1 electron)

Step 3 Now multiply the two equations by the smallest integral numbers that will make the loss of electrons by the oxidation step equal to the number of electrons gained in the reduction step. In this reaction the oxidation step is multiplied by 1 and the reduction step by 4. The equations become

oxidation $\qquad$ $Sn^0 \longrightarrow Sn^{4+} + 4\,e^-$ $\qquad$ (Sn^0 loses 4 electrons)
reduction $\qquad$ $4\,N^{5+} + 4\,e^- \longrightarrow 4\,N^{4+}$ $\quad$ (4 N^{5+} gain 4 electrons)

We have now established the ratio of the oxidizing to the reducing agent as being four atoms of N to one atom of Sn.

Step 4 Now transfer the coefficient that appears in front of each substance in the balanced oxidation–reduction equations to the corresponding substance in the original equation. We need to use 1 Sn, 1 SnO_2, 4 HNO_3, and 4 NO_2:

$$Sn + 4\,HNO_3 \longrightarrow SnO_2 + 4\,NO_2 + H_2O \quad \text{(unbalanced)}$$

Step 5 In the usual manner, balance the remaining elements that are not oxidized or reduced to give the final balanced equation:

$$Sn + 4\,HNO_3 \longrightarrow SnO_2 + 4\,NO_2 + 2\,H_2O \quad \text{(balanced)}$$

In balancing the final elements, we must not change the ratio of the elements that were oxidized and reduced. We should make a final check to ensure that both sides of the equation have the same number of atoms of each element. The final balanced equation contains 1 atom of Sn, 4 atoms of N, 4 atoms of H, and 12 atoms of O on each side.

Because each new equation may present a slightly different problem and because proficiency in balancing equations requires practice, we will work through a few more problems.

EXAMPLE 18.3 Balance the equation

$$I_2 + Cl_2 + H_2O \longrightarrow HIO_3 + HCl \quad \text{(unbalanced)}$$

Step 1 Assign oxidation numbers:

$$\underset{0 \quad\quad 0 \quad\quad +1\,-2}{I_2 + Cl_2 + H_2O} \longrightarrow \underset{+1\,+5\,-2 \quad\quad +1\,-1}{H\ I\ O_3 +\ HCl}$$

The oxidation numbers of I_2 and Cl_2 have changed, I_2 from 0 to $+5$, and Cl_2 from 0 to -1.

Step 2 Write oxidation and reduction steps. Balance the number of atoms and then balance the electrical charge using electrons.

oxidation $\quad I_2 \longrightarrow 2\,I^{5+} + 10\,e^- \quad$ (I_2 loses 10 electrons)

reduction $\quad Cl_2 + 2\,e^- \longrightarrow 2\,Cl^- \quad$ (Cl_2 gains 2 electrons)

Step 3 Adjust loss and gain of electrons so that they are equal. Multiply the oxidation step by 1 and the reduction step by 5.

oxidation $\quad I_2 \longrightarrow 2\,I^{5+} + 10\,e^- \quad\quad$ (I_2 loses 10 electrons)

reduction $\quad 5\,Cl_2 + 10\,e^- \longrightarrow 10\,Cl^- \quad$ ($5\,Cl_2$ gain 10 electrons)

Step 4 Transfer the coefficients from the balanced redox equations into the original equation. We need to use $1 \, I_2$, $2 \, HIO_3$, $5 \, Cl_2$, and $10 \, HCl$.

$$I_2 + 5 \, Cl_2 + H_2O \longrightarrow 2 \, HIO_3 + 10 \, HCl \quad \text{(unbalanced)}$$

Step 5 Balance the remaining elements, H and O:

$$I_2 + 5 \, Cl_2 + 6 \, H_2O \longrightarrow 2 \, HIO_3 + 10 \, HCl \quad \text{(balanced)}$$

Check: The final balanced equation contains 2 atoms of I, 10 atoms of Cl, 12 atoms of H, and 6 atoms of O on each side.

EXAMPLE 18.4   Balance the equation

$$K_2Cr_2O_7 + FeCl_2 + HCl \longrightarrow CrCl_3 + KCl + FeCl_3 + H_2O \quad \text{(unbalanced)}$$

Step 1 Assign oxidation numbers (Cr and Fe have changed):

$$K_2Cr_2O_7 + FeCl_2 + HCl \longrightarrow CrCl_3 + KCl + FeCl_3 + H_2O$$
$$\underset{+1 \ +6 \ -2}{} \quad \underset{+2 \ -1}{} \quad \underset{+1 \ -1}{} \quad \underset{+3 \ -1}{} \quad \underset{+1 \ -1}{} \quad \underset{+3 \ -1}{} \quad \underset{+1 \ -2}{}$$

Step 2 Write the oxidation and reduction steps. Balance the number of atoms and then balance the electrical charge using electrons.

oxidation $Fe^{2+} \longrightarrow Fe^{3+} + 1 \, e^-$ (Fe^{2+} loses 1 electron)

reduction $2 \, Cr^{6+} + 6 \, e^- \longrightarrow 2 \, Cr^{3+}$ ($2 \, Cr^{6+}$ gain 6 electrons)

Step 3 Balance the loss and gain of electrons. Multiply the oxidation step by 6 and the reduction step by 1 to equalize the transfer of electrons.

oxidation $6 \, Fe^{2+} \longrightarrow 6 \, Fe^{3+} + 6 \, e^-$ ($6 \, Fe^{2+}$ lose 6 electrons)

reduction $2 \, Cr^{6+} + 6 \, e^- \longrightarrow 2 \, Cr^{3+}$ ($2 \, Cr^{6+}$ gain 6 electrons)

Step 4 Transfer the coefficients from the balanced redox equations into the original equation. (Note that one formula unit of $K_2Cr_2O_7$ contains two Cr atoms.) We need to use $1 \, K_2Cr_2O_7$, $2 \, CrCl_3$, $6 \, FeCl_2$, and $6 \, FeCl_3$.

$$K_2Cr_2O_7 + 6 \, FeCl_2 + HCl \longrightarrow$$
$$2 \, CrCl_3 + KCl + 6 \, FeCl_3 + H_2O \quad \text{(unbalanced)}$$

Step 5 Balance the remaining elements in this order: K, Cl, H, O.

$$K_2Cr_2O_7 + 6 \, FeCl_2 + 14 \, HCl \longrightarrow$$
$$2 \, CrCl_3 + 2 \, KCl + 6 \, FeCl_3 + 7 \, H_2O \quad \text{(balanced)}$$

Check: The final balanced equation contains 2 K atoms, 2 Cr atoms, 7 O atoms, 6 Fe atoms, 26 Cl atoms, and 14 H atoms on each side.

PRACTICE   Balance the following equations using the half-reaction method:
(a) $HNO_3 + S \longrightarrow NO_2 + H_2SO_4 + H_2O$
(b) $CrCl_3 + MnO_2 + H_2O \longrightarrow MnCl_2 + H_2CrO_4$
(c) $KMnO_4 + HCl + H_2S \longrightarrow KCl + MnCl_2 + S + H_2O$
Answers: The coefficients for the balanced equations from left to right are
(a) 6, 1, 6, 1, 2   (b) 2, 3, 2, 3, 2   (c) 2, 6, 5, 2, 2, 5, 8

18.4   Balancing Ionic Redox Equations

The main difference between balancing ionic and balancing molecular redox equations is in the handling of ions. In addition to having the same number of each kind of element on both sides of the final equation, the net charges must also be equal. In assigning oxidation numbers we must be careful to consider the charge on the ions. In many respects, balancing ionic equations is much simpler than balancing molecular equations.

Several methods can be used to balance ionic redox equations. These methods include, with slight modification, the oxidation-number method just shown for molecular equations. But the most popular is probably the ion–electron method, which is explained in the following paragraphs.

The ion–electron method uses ionic charges and electrons to balance ionic redox equations. Oxidation numbers are not formally used, but it is necessary to determine what is being oxidized and what is being reduced. The method is as follows:

1. Write the two half-reactions that contain the elements being oxidized and reduced.
2. Balance the elements other than oxygen and hydrogen.
3. Balance oxygen and hydrogen: (a) acidic solutions; (b) basic solutions.
    (a) For reactions that occur in acidic solution, use H^+ and H_2O to balance oxygen and hydrogen. For each oxygen needed, use one H_2O. Then add H^+ as needed to balance the hydrogen atoms.
    (b) Balancing equations that occur in alkaline solutions is a bit more complicated. For reactions that occur in alkaline solutions, first balance as though the reaction were in an acidic solution, using Steps 1, 2, and 3(a). Then add as many OH^- ions to each side of the equation as there are H^+ ions in the equation. Now combine the H^+ and OH^- ions into water (for example, $4\,H^+$ and $4\,OH^-$ give $4\,H_2O$). Rewrite the equation, canceling equal numbers of water molecules that appear on opposite sides of the equation. Now proceed to Step 4 for electrical balance.
4. Add electrons (e^-) to each half-reaction to bring them into electrical balance.

5. Since the loss and gain of electrons must be equal, multiply each half-reaction by the appropriate number to make the number of electrons the same in each half-reaction.

6. Add the two half-reactions together, canceling electrons and any other identical substances that appear on opposite sides of the equation.

EXAMPLE 18.5   Balance this equation using the ion–electron method:

$$MnO_4^- + S^{2-} \longrightarrow Mn^{2+} + S^0 \quad \text{(acidic solution)}$$

Step 1   Write two half-reactions, one containing the element being oxidized and the other, the element being reduced. (Use the entire molecule or ion.)

oxidation $S^{2-} \longrightarrow S^0$

reduction $MnO_4^- \longrightarrow Mn^{2+}$

Step 2   Balance elements other than oxygen and hydrogen (accomplished in Step 1 in this example — 1 S and 1 Mn on each side).

Step 3   Balance O and H. The oxidation requires neither O nor H, but the reduction equation needs 4 H_2O on the right and 8 H^+ on the left [Step 3(a)].

$$S^{2-} \longrightarrow S^0$$
$$8\,H^+ + MnO_4^- \longrightarrow Mn^{2+} + 4\,H_2O$$

Step 4   Balance electrically with electrons.

$S^{2-} \longrightarrow S^0 + 2\,e^-$ (net charge $= -2$ on each side)

$5\,e^- + 8\,H^+ + MnO_4^- \longrightarrow Mn^{2+} + 4\,H_2O$ (net charge $= +2$ on each side)

Step 5   Equalize loss and gain of electrons. In this case multiply the oxidation equation by 5 and the reduction equation by 2.

$$5\,S^{2-} \longrightarrow 5\,S^0 + 10\,e^-$$
$$10\,e^- + 16\,H^+ + 2\,MnO_4^- \longrightarrow 2\,Mn^{2+} + 8\,H_2O$$

Step 6   Add the two half-reactions together, cancelling the 10 e^- from each side, to obtain the balanced equation.

$$5\,S^{2-} \longrightarrow 5\,S^0 + \cancel{10\,e^-}$$
$$\underline{\cancel{10\,e^-} + 16\,H^+ + 2\,MnO_4^- \longrightarrow 2\,Mn^{2+} + 8\,H_2O}$$
$$16\,H^+ + 2\,MnO_4^- + 5\,S^{2-} \longrightarrow 2\,Mn^{2+} + 5\,S^0 + 8\,H_2O \quad \text{(balanced)}$$

Check: Both sides of the equation have a charge of $+4$ and contain the same number of atoms of each element.

EXAMPLE 18.6   Balance the following equation.

$$CrO_4^{2-} + Fe(OH)_2 \longrightarrow Cr(OH)_3 + Fe(OH)_3 \quad \text{(basic solution)}$$

Step 1 Write the two half-reactions.

oxidation $Fe(OH)_2 \longrightarrow Fe(OH)_3$

reduction $CrO_4^{2-} \longrightarrow Cr(OH)_3$

Step 2 Balance elements other than H and O (accomplished in Step 1).

Step 3 Balance O and H as though the solution were acidic (Step 3a). Use H_2O and H^+ To balance O and H in the oxidation equation, add 1 H_2O on the left and 1 H^+ on the right side of the equation.

$$Fe(OH)_2 + H_2O \longrightarrow Fe(OH)_3 + H^+$$

Add 1 OH^- to each side.

$$Fe(OH)_2 + H_2O + OH^- \longrightarrow Fe(OH)_3 + H^+ + OH^-$$

Combine H^+ and OH^- as H_2O and rewrite, cancelling H_2O on each side [Step 3(b)].

$$Fe(OH)_2 + \cancel{H_2O} + OH^- \longrightarrow Fe(OH)_3 + \cancel{H_2O}$$

$$\boxed{Fe(OH)_2 + OH^- \longrightarrow Fe(OH)_3}$$

To balance O and H in the reduction equation, add 1 H_2O on the right and 5 H^+ on the left.

$$CrO_4^{2-} + 5\,H^+ \longrightarrow Cr(OH)_3 + H_2O$$

Add 5 OH^- to each side.

$$CrO_4^{2-} + 5\,H^+ + 5\,OH^- \longrightarrow Cr(OH)_3 + H_2O + 5\,OH^-$$

Combine $5\,H^+ + 5\,OH^- \longrightarrow 5\,H_2O$

$$CrO_4^{2-} + 5\,H_2O \longrightarrow Cr(OH)_3 + H_2O + 5\,OH^-$$

Rewrite. cancelling 1 H_2O from each side.

$$\boxed{CrO_4^{2-} + 4\,H_2O \longrightarrow Cr(OH)_3 + 5\,OH^-}$$

Step 4 Balance electrically with electrons.

$Fe(OH)_2 + OH^- \longrightarrow Fe(OH)_3 + e^-$ (balanced oxidation equation)

$CrO_4^{2-} + 4\,H_2O + 3\,e^- \longrightarrow Cr(OH)_3 + 5\,OH^-$ (balanced reduction equation)

Step 5 Equalize the loss and gain of electrons. Multiply the oxidation reaction by 3.

$$3\,Fe(OH)_2 + 3\,OH^- \longrightarrow 3\,Fe(OH)_3 + 3\,e^-$$

$$CrO_4^{2-} + 4\,H_2O + 3\,e^- \longrightarrow Cr(OH)_3 + 5\,OH^-$$

Step 6   Add the two half-reactions together, cancelling the 3 e$^-$ and 3 OH$^-$ from each side of the equation.

$$3\ Fe(OH)_2 + 3\ OH^- \longrightarrow 3\ Fe(OH)_3 + 3\ e^-$$
$$CrO_4^{2-} + 4\ H_2O + 3\ e^- \longrightarrow Cr(OH)_3 + 5\ OH^-$$
$$\overline{CrO_4^{2-} + 3\ Fe(OH)_2 + 4\ H_2O \longrightarrow Cr(OH)_3 + 3\ Fe(OH)_3 + 2\ OH^-} \quad \text{(balanced)}$$

Check: Each side of the equation has a charge of -2 and contains the same number of atoms of each element.

PRACTICE   Balance the following equations using the ion electron method:
(a) $I^- + NO_2^- \longrightarrow I_2 + NO$ (acid solution)
(b) $Cl_2 + IO_3^- \longrightarrow IO_4^- + Cl^-$ (basic solution)
(c) $AuCl_4^- + Sn^{2+} \longrightarrow Sn^{4+} + AuCl + Cl^-$

Answers: (a) $4H^+ + 2I^- + 2NO_2^- \longrightarrow I_2 + 2NO + 2H_2O$
          (b) $2OH^- + Cl_2 + IO_3^- \longrightarrow IO_4^- + H_2O + 2Cl^-$
          (c) $AuCl_4^- + Sn^{2+} \longrightarrow Sn^{4+} + AuCl + 3Cl^-$

Ionic equations can also be balanced by using the electron-transfer method shown in Section 18.3. Steps 1, 2, 3, and 4 are the same. In Step 5 the electrical charges are balanced with H^+ in acidic solutions and with OH^- in basic solutions. H_2O is then used to complete the balancing of the equation. Let us use the same equation as in Example 18.6 as an example of this method.

EXAMPLE 18.7   Balance the following equation using the electron-transfer method.

$$CrO_4^{2-} + Fe(OH)_2 \longrightarrow Cr(OH)_3 + Fe(OH)_3 \quad \text{(basic solution)}$$

Steps 1 and 2   Assign oxidation numbers and balance the charges with electrons.

reduction $Cr^{6+} + 3\ e^- \longrightarrow Cr^{3+}$ (Cr^{6+} gains 3 e$^-$)
oxidation $Fe^{2+} \longrightarrow Fe^{3+} + e^-$ (Fe^{2+} loses 1 e$^-$)

Step 3   Equalize the loss and gain of electrons. Multiply the oxidation step by 3.

$3\ Fe^{2+} \longrightarrow 3\ Fe^{3+} + 3\ e^-$ (3 Fe^{2+} lose 3 e$^-$)
$Cr^{6+} + 3\ e^- \longrightarrow Cr^{3+}$ (Cr^{6+} gains 3 e$^-$)

Step 4   Transfer coefficients back to the original equation.

$$CrO_4^{2-} + 3\ Fe(OH)_2 \longrightarrow Cr(OH)_3 + 3\ Fe(OH)_3$$

Step 5   Balance electrically. Because the solution is basic, use OH^- to balance charges. The charge on the left side is -2, and on the right side is 0. Add 2 OH^- ions to the right side of the equation.

$$CrO_4^{2-} + 3\ Fe(OH)_2 \longrightarrow Cr(OH)_3 + 3\ Fe(OH)_3 + 2\ OH^-$$

Adding 4 H_2O to the left side balances the equation.

$$CrO_4^{2-} + 3\ Fe(OH)_2 + 4\ H_2O \longrightarrow Cr(OH)_3 + 3\ Fe(OH)_3 + 2\ OH^-\ \ \ \text{(balanced)}$$

Check: Each side of the equation has a charge of -2 and contains the same number of atoms of each element.

PRACTICE Balance each of the following equations using the electron transfer method:
(a) $Zn \longrightarrow Zn(OH)_4^{2-} + H_2$ (basic solution)
(b) $H_2O_2 + Sn^{2+} \longrightarrow Sn^{4+}$ (acid solution)
(c) $Cu + Cu^{2+} \longrightarrow Cu_2O$ (basic solution)
Answers: (a) $Zn + H_2O + 2\ OH^- \longrightarrow Zn(OH)_4^{2-} + H_2$
         (b) $H_2O_2 + Sn^{2+} + 2\ H^+ \longrightarrow Sn^{4+} + 2\ H_2O$
         (c) $Cu + Cu^{2+} + 2\ OH \longrightarrow Cu_2O + H_2O$

18.5 Activity Series of Metals

Knowledge of the relative chemical reactivities of the elements helps us predict the course of many chemical reactions.

Calcium reacts with cold water to produce hydrogen, and magnesium reacts with steam to produce hydrogen. Therefore, calcium is considered to be a more reactive metal than magnesium.

$$Ca(s) + 2\ H_2O(l) \longrightarrow Ca(OH)_2(aq) + H_2(g)$$
$$Mg(s) + H_2O(g) \longrightarrow MgO(s) + H_2(g)$$
Steam

The difference in their activity is attributed to the fact that calcium loses its two valence electrons more easily than does magnesium and is therefore more reactive and/or more readily oxidized than magnesium.

When a strip of copper is placed in a solution of silver nitrate ($AgNO_3$), free silver begins to plate out on the copper. After the reaction has continued for some time, we can observe a blue color in the solution, indicating the presence of copper(II) ions. If a strip of silver is placed in a solution of copper(II) nitrate, $Cu(NO_3)_2$, no reaction is visible. The equations are

$$Cu^0 + 2\ AgNO_3(aq) \longrightarrow 2\ Ag^0 + Cu(NO_3)_2(aq)$$

$Cu^0 + 2\ Ag^+ \longrightarrow 2\ Ag^0 + Cu^{2+}$ net ionic equation
$Cu^0 \longrightarrow Cu^{2+} + 2\ e^-$ oxidation of Cu^0
$Ag^+ + e^- \longrightarrow Ag^0$ reduction of Ag^+
$Ag^0 + Cu(NO_3)_2(aq) \longrightarrow$ no reaction

TABLE 18.3

Activity Series of Metals

Ease of oxidation ↑

$$K \longrightarrow K^+ + e^-$$
$$Ba \longrightarrow Ba^{2+} + 2e^-$$
$$Ca \longrightarrow Ca^{2+} + 2e^-$$
$$Na \longrightarrow Na^+ + e^-$$
$$Mg \longrightarrow Mg^{2+} + 2e^-$$
$$Al \longrightarrow Al^{3+} + 3e^-$$
$$Zn \longrightarrow Zn^{2+} + 2e^-$$
$$Cr \longrightarrow Cr^{3+} + 3e^-$$
$$Fe \longrightarrow Fe^{2+} + 2e^-$$
$$Ni \longrightarrow Ni^{2+} + 2e^-$$
$$Sn \longrightarrow Sn^{2+} + 2e^-$$
$$Pb \longrightarrow Pb^{2+} + 2e^-$$
$$\mathbf{H_2 \longrightarrow 2H^+ + 2e^-}$$
$$Cu \longrightarrow Cu^{2+} + 2e^-$$
$$As \longrightarrow As^{3+} + 3e^-$$
$$Ag \longrightarrow Ag^+ + e^-$$
$$Hg \longrightarrow Hg^{2+} + 2e^-$$
$$Au \longrightarrow Au^{3+} + 3e^-$$

In the reaction between Cu and $AgNO_3$, electrons are transferred from Cu^0 atoms to Ag^+ ions in solution. Copper has a greater tendency than silver to lose electrons, so an electrochemical force is exerted upon silver ions to accept electrons from copper atoms. When a Ag^+ ion accepts an electron, it is reduced to a Ag^0 atom and is no longer soluble in solution. At the same time, Cu^0 is oxidized and goes into solution as Cu^{2+} ions. From this reaction we can conclude that copper is more reactive than silver.

Metals such as sodium, magnesium, zinc, and iron, which react with solutions of acids to liberate hydrogen, are more reactive than hydrogen. Metals such as copper, silver, and mercury, which do not react with solutions of acids to liberate hydrogen, are less reactive than hydrogen. By studying a series of reactions such as those given above, we can list metals according to their chemical activity, placing the most active at the top and the least active at the bottom. This list is called the **activity series of metals**. Table 18.3 shows some of the common metals in the series. The arrangement corresponds to the ease with which the elements listed are oxidized or lose electrons, with the most easily oxidizable element listed first. More extensive tables are available in chemistry reference books.

activity series
of metals

The general principles governing the arrangement and use of the activity series are as follows:

1. The reactivity of the metals listed decreases from top to bottom.
2. A free metal can displace the ion of a second metal from solution provided that the free metal is above the second metal in the Activity Series.
3. Free metals above hydrogen react with nonoxidizing acids in solution to liberate hydrogen gas.
4. Free metals below hydrogen do not liberate hydrogen from acids.
5. Conditions such as temperature and concentration may affect the relative position of some of these elements.

Two examples of the application of the activity series are given in the following problems.

EXAMPLE 18.8 Will zinc metal react with dilute sulfuric acid?

From Table 18.3 we see that zinc is above hydrogen; therefore zinc atoms will lose electrons more readily than hydrogen atoms. Hence, zinc atoms will reduce hydrogen ions from the acid to form hydrogen gas and zinc ions. In fact, these reagents are commonly used for the laboratory preparation of hydrogen. The equation is

$$Zn(s) + H_2SO_4(aq) \longrightarrow ZnSO_4(aq) + H_2(g)$$

$$Zn + 2\,H^+ \longrightarrow Zn^{2+} + H_2(g) \qquad \text{net ionic equation}$$

EXAMPLE 18.9 Will a reaction occur when copper metal is placed in an iron(II) sulfate solution?

No, copper lies below iron in the series, loses electrons less easily than iron, and therefore will not displace iron(II) ions from solution. In fact, the reverse is true. When an iron nail is dipped into a copper(II) sulfate solution, it becomes coated with free copper. The equations are

$$Cu(s) + FeSO_4(aq) \longrightarrow \text{no reaction}$$

$$Fe(s) + CuSO_4(aq) \longrightarrow FeSO_4(aq) + Cu\downarrow$$

From Table 18.3 we may abstract the following pair in their relative position to each other.

$$Fe \longrightarrow Fe^{2+} + 2\,e^-$$

$$Cu \longrightarrow Cu^{2+} + 2\,e^-$$

According to the second principle listed above on the use of the activity series, we can predict that free iron will react with copper(II) ions in solution to form free copper metal and iron(II) ions in solution.

$$Fe(s) + Cu^{2+}(aq) \longrightarrow Fe^{2+}(aq) + Cu(s) \quad \text{net ionic equation}$$

PRACTICE   Indicate whether the following reactions will occur:
(a) Sodium metal is placed in dilute hydrochloric acid.
(b) A piece of lead is placed in magnesium nitrate solution.
(c) Mercury is placed in a solution of silver nitrate.

Answers: (a) yes   (b) no   (c) no

18.6   Electrolytic and Voltaic Cells

electrolysis

electrolytic cell

The process in which electrical energy is used to bring about chemical change is known as **electrolysis**. An **electrolytic cell** uses electrical energy to produce a nonspontaneous chemical reaction. The use of electrical energy has many applications in the chemical industry—for example, in the production of sodium, sodium hydroxide, chlorine, fluorine, magnesium, aluminum, and pure hydrogen and oxygen, and in the purification and electroplating of metals.

What happens when an electric current is passed through a solution? Let us consider a hydrochloric acid solution in a simple electrolytic cell, as shown in Figure 18.2. The cell consists of a source of direct current (a battery) connected to two electrodes that are immersed in a solution of hydrochloric acid. The negative

cathode

anode

electrode is called the **cathode** because cations are attracted to it. The positive electrode is called the **anode** because anions are attracted to it. The cathode is attached to the negative pole and the anode to the positive pole of the battery. The battery supplies electrons to the cathode.

When the electric circuit is completed, positive hydronium ions (H_3O^+) migrate to the cathode, where they pick up electrons and evolve hydrogen gas. At the same time the negative chloride ions (Cl^-) migrate to the anode, where they lose electrons and evolve chlorine gas.

Reaction at the cathode	$H_3O^+ + 1\,e^- \longrightarrow H^0 + H_2O$
(reduction)	$H^0 + H^0 \longrightarrow H_2(g)$
Reaction at the anode	$Cl^- \longrightarrow Cl^0 + 1\,e^-$
(oxidation)	$Cl^0 + Cl^0 \longrightarrow Cl_2(g)$
Net reaction	$2\,HCl(aq) \xrightarrow{\text{Electrolysis}} H_2(g) + Cl_2(g)$

Note that oxidation–reduction has taken place. Chloride ions lose electrons (are oxidized) at the anode, and hydronium ions gain electrons (are reduced) at the cathode.

> **Oxidation always occurs at the anode and reduction at the cathode.**

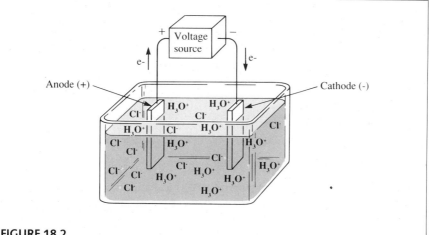

FIGURE 18.2

Electrolysis. During the electrolysis of a hydrochloric acid solution, positive hydronium ions are attracted to the cathode, where they gain electrons and form hydrogen gas. Chloride ions migrate to the anode, where they lose electrons and form chlorine gas. The equation for this process is
$$2\ HCl(aq) \longrightarrow H_2(g) + Cl_2(g).$$

When concentrated sodium chloride solutions (brines) are electrolyzed, the products are sodium hydroxide, hydrogen, and chlorine. The overall reaction is

$$2\ Na^+ + 2\ Cl^- + 2\ H_2O \xrightarrow{\text{Electrolysis}} 2\ Na^+ + 2\ OH^- + H_2(g) + Cl_2(g)$$

The net ionic equation is

$$2\ Cl^-(aq) + 2\ H_2O(l) \longrightarrow 2\ OH^-(aq) + H_2(g) + Cl_2(g)$$

During the electrolysis, Na^+ ions move toward the cathode and Cl^- ions move toward the anode. The anode reaction is similar to that of hydrochloric acid; chlorine is liberated.

$$2\ Cl^-(aq) \longrightarrow Cl_2(g) + 2\ e^-$$

Even though Na^+ ions are attracted by the cathode, the facts show that hydrogen is liberated there. No evidence of metallic sodium is found, but the area around the cathode tests alkaline from accumulated OH^- ions. The reaction at the cathode is

$$2\ H_2O(l) + 2\ e^- \longrightarrow H_2(g) + 2\ OH^-(aq)$$

If the electrolysis is allowed to continue until all the chloride is reacted, the solution remaining will contain only sodium hydroxide, which on evaporation yields solid NaOH. Large tonnages of sodium hydroxide and chlorine are made by this process.

When molten sodium chloride (without water) is subjected to electrolysis, metallic sodium and chlorine gas are formed:

$$2\,Na^+(l) + 2\,Cl^-(l) \xrightarrow{\text{Electrolysis}} 2\,Na(l) + Cl_2(g)$$

An important electrochemical application is the electroplating of metals. Electroplating is the art of covering a surface or an object with a thin adherent electrodeposited metal coating. Electroplating is done for protection of the surface of the base metal or for a purely decorative effect. The layer deposited is surprisingly thin, varying from as little as 5×10^{-5} cm to 2×10^{-3} cm, depending on the metal and the intended use. The object to be plated is set up as the cathode and is immersed in a solution containing ions of the metal to be plated. When an electric current passes through the solution, metal ions that migrate to the cathode are reduced, depositing on the object as the free metal. In most cases the metal deposited on the object is replaced in the solution by using an anode of the same metal. The following equations show the chemical changes in the electroplating of nickel:

Reaction at the cathode $Ni^{2+}(aq) + 2\,e^- \longrightarrow Ni(s)$ Ni plated out on an object

Reaction at the anode $Ni(s) \longrightarrow Ni^{2+}(aq) + 2\,e^-$ Ni replenished in solution

Metals commonly used in commercial electroplating are copper, nickel, zinc, lead, cadmium, chromium, tin, gold, and silver.

In the electrolytic cell shown in Figure 18.2, electrical energy from the voltage source is used to bring about nonspontaneous redox reactions. The hydrogen and chlorine produced have more potential energy than was present in the hydrochloric acid before electrolysis.

Conversely, some spontaneous redox reactions can be made to supply useful amounts of electrical energy. When a piece of zinc is put in a copper(II) sulfate solution, the zinc quickly becomes coated with metallic copper. We expect this coating to happen because zinc is above copper in the activity series; copper(II) ions are therefore reduced by zinc atoms:

$$Zn^0(s) + Cu^{2+}(aq) \longrightarrow Zn^{2+}(aq) + Cu^0(s)$$

This reaction is clearly a spontaneous redox reaction, but simply dipping a zinc rod into a copper(II) sulfate solution will not produce useful electric current. However, when we carry out this reaction in the cell shown in Figure 18.3, an electric current is produced. The cell consists of a piece of zinc immersed in a zinc sulfate solution and connected by a wire through a voltmeter to a piece of copper immersed in copper(II) sulfate solution. The two solutions are connected by a salt bridge. Such a cell produces an electric current and a potential of about 1.1 volts

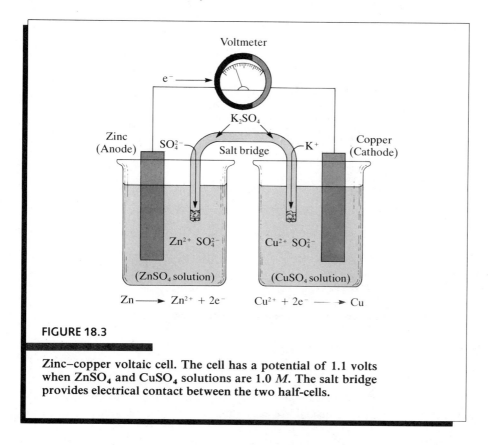

FIGURE 18.3

Zinc–copper voltaic cell. The cell has a potential of 1.1 volts when ZnSO$_4$ and CuSO$_4$ solutions are 1.0 M. The salt bridge provides electrical contact between the two half-cells.

when both solutions are 1.0 M in concentration. A cell that produces electric current from a spontaneous chemical reaction is called a **voltaic cell**. A voltaic cell is also known as a *galvanic cell*.

voltaic cell

The driving force responsible for the electric current in the zinc–copper cell originates in the great tendency of zinc atoms to lose electrons relative to the tendency of copper(II) ions to gain electrons. In the cell shown in Figure 18.3 zinc atoms lose electrons and are converted to zinc ions at the zinc electrode surface; the electrons flow through the wire (external circuit) to the copper electrode. Here copper(II) ions pick up electrons and are reduced to copper atoms, which plate out on the copper electrode. Sulfate ions flow from the CuSO$_4$ solution via the salt bridge into the ZnSO$_4$ solution (internal circuit) to complete the circuit. The equations for the reactions of this cell are

anode	$Zn^0(s) \longrightarrow Zn^{2+}(aq) + 2\,e^-$	(oxidation)
cathode	$Cu^{2+}(aq) + 2\,e^- \longrightarrow Cu^0(s)$	(reduction)
net ionic	$Zn^0(s) + Cu^{2+}(aq) \longrightarrow Zn^{2+}(aq) + Cu^0(s)$	
overall	$Zn(s) + CuSO_4(aq) \longrightarrow Cu(s) + ZnSO_4(aq)$	

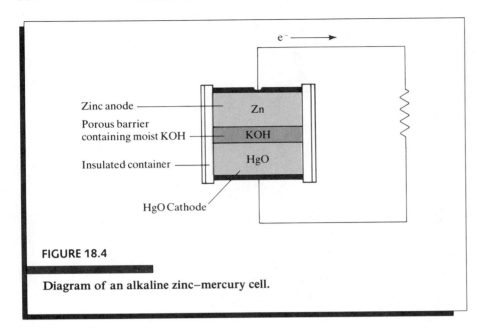

FIGURE 18.4

Diagram of an alkaline zinc–mercury cell.

The redox reaction, the movement of electrons in the metallic or external part of the circuit, and the movement of ions in the solution or internal part of the circuit of the copper–zinc cell are very similar to the actions that occur in the electrolytic cell of Figure 18.2. The only important difference is that the reactions of the zinc–copper cell are spontaneous. This spontaneity is the crucial difference between all voltaic and electrolytic cells.

> **Voltaic cells use chemical reactions to produce electrical energy, and electrolytic cells use electrical energy to produce chemical reactions.**

Although the zinc–copper voltaic cell is no longer used commercially, it was used to energize the first transcontinental telegraph lines. Such cells were the direct ancestors of the many different kinds of "dry" cells that operate portable radio and television sets, automatic cameras, tape recorders, and so on.

One such "dry" cell, the alkaline zinc–mercury cell, is shown diagrammatically in Figure 18.4. The reactions occurring in this cell are

anode	$Zn^0 + 2\,OH^- \longrightarrow ZnO + H_2O + 2\,e^-$	(oxidation)
cathode	$HgO + H_2O + 2\,e^- \longrightarrow Hg^0 + 2\,OH^-$	(reduction)
net ionic	$Zn^0 + Hg^{2+} \longrightarrow Zn^{2+} + Hg^0$	
overall	$Zn^0 + HgO \longrightarrow ZnO + Hg^0$	

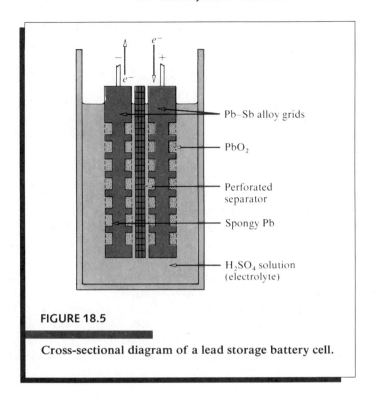

FIGURE 18.5

Cross-sectional diagram of a lead storage battery cell.

To offset the relatively high initial cost, this cell (a) provides current at a very steady potential of about 1.5 volts, (b) has an exceptionally long service life—that is, high energy output to weight ratio, (c) is completely self-contained, and (d) can be stored for relatively long periods of time when not in use.

An automobile storage battery is an energy reservoir. The charged battery acts as a voltaic cell and through chemical reactions furnishes electrical energy to operate the starter, lights, radio, and so on. When the engine is running, a generator, or alternator, produces and forces an electric current through the battery and, by electrolytic chemical action, restores it to the charged condition.

The cell unit consists of a lead plate filled with spongy lead and a lead dioxide plate, both immersed in dilute sulfuric acid solution, which serves as the electrolyte (see Figure 18.5). When the cell is discharging, or acting as a voltaic cell, these reactions occur:

Pb plate (anode)	$Pb^0 \longrightarrow Pb^{2+} + 2\,e^-$	(oxidation)
PbO_2 plate (cathode)	$PbO_2 + 4\,H^+ + 2\,e^- \longrightarrow Pb^{2+} + 2\,H_2O$	(reduction)
Net ionic redox reaction	$Pb^0 + PbO_2 + 4\,H^+ \longrightarrow 2\,Pb^{2+} + 2\,H_2O$	
Precipitation reaction on plates	$Pb^{2+} + SO_4^{2-} \longrightarrow PbSO_4(s)$	

Because lead(II) sulfate is insoluble, the Pb^{2+} ions combine with SO_4^{2-} ions to form a coating of $PbSO_4$ on each plate. The overall chemical reaction of the cell is

$$Pb(s) + PbO_2(s) + 2\ H_2SO_4(aq) \xrightarrow[\text{cycle}]{\text{Discharge}} 2\ PbSO_4(s) + 2\ H_2O(l)$$

The cell can be recharged by reversing the chemical reaction. This reversal is accomplished by forcing an electric current through the cell in the opposite direction. Lead sulfate and water are reconverted to lead, lead dioxide, and sulfuric acid:

$$2\ PbSO_4(s) + 2\ H_2O(l) \xrightarrow[\text{cycle}]{\text{Charge}} Pb(s) + PbO_2(s) + 2\ H_2SO_4(aq)$$

The electrolyte in a lead storage battery is a 38% by mass sulfuric acid solution having a density of 1.29 g/mL. As the battery is discharged, sulfuric acid is removed, thereby decreasing the density of the electrolyte solution. The state of charge or discharge of the battery can be estimated by measuring the density (or specific gravity) of the electrolyte solution with a hydrometer. When the density has dropped to about 1.05 g/mL, the battery needs recharging.

In a commercial battery, each cell consists of a series of cell units of alternating lead–lead dioxide plates separated and supported by wood, glass wool, or fiberglass. The energy storage capacity of a single cell is limited, and its electrical potential is only about 2 volts. Therefore, a bank of six cells is connected in series to provide the 12 volt output of the usual automobile battery.

Concepts in Review

1. Assign oxidation numbers to all the elements in a given compound or ion.
2. Determine which element is being oxidized and which element is being reduced in an oxidation–reduction reaction.
3. Identify the oxidizing agent and the reducing agent in an oxidation–reduction reaction.
4. Balance oxidation–reduction equations in molecular and ionic forms.
5. Outline the general principles concerning the activity series of the metals.
6. Use the activity series to determine whether a proposed single-displacement reaction will occur.
7. Distinguish between an electrolytic and a voltaic cell.
8. Draw a voltaic cell that will produce electric current from an oxidation–reduction reaction involving two metals and their salts.
9. Identify the anode reaction and the cathode reaction in a given electrolytic or voltaic cell.

Key Terms in Review

The terms listed here have all been defined within the chapter. Review the definitions of each, and use the glossary and margin notations within the chapter as study aids.

activity series of metals
anode
Cathode
electrolysis
electrolytic cell
oxidation

oxidation–reduction
oxidizing agent
redox
reducing agent
reduction
voltaic cell

Exercises

An asterisk indicates a more challenging question or problem.

1. In the equation

$$I_2 + 5 Cl_2 + 6 H_2O \longrightarrow 2 HIO_3 + 10 HCl$$

   (a) Has iodine been oxidized or has it been reduced ?
   (b) Has chlorine been oxidized or has it been reduced? (See Figure 18.1.)
2. Based on Table 18.3, which element of each pair is more active?
   (a) Ag or Al   (b) Na or Ba   (c) Ni or Cu
3. Based on Table 18.3, will the following combinations react in aqueous solution?
   (a) $Zn + Cu^{2+}$ (e) $Ba + FeCl_2$
   (b) $Ag + H^+$ (f) $Pb + NaCl$
   (c) $Sn + Ag^+$ (g) $Ni + Hg(NO_3)_2$
   (d) $As + Mg^{2+}$ (h) $Al + CuSO_4$
4. The reaction between powdered aluminum and iron(III) oxide (in the Thermite process), producing molten iron, is very exothermic.
   (a) Write the equation for the chemical reaction that occurs.
   (b) Explain in terms of Table 18.3 why a reaction occurs.
   (c) Would you expect a reaction between powdered iron and aluminum oxide?
   (d) Would you expect a reaction between powdered aluminum and chromium(III) oxide?

5. Write equations for the chemical reaction of each of the following metals with dilute solutions of (a) hydrochloric acid and (b) sulfuric acid: aluminum, chromium, gold, iron, copper, magnesium, mercury, and zinc. If a reaction will not occur, write "no reaction" as the product. (See Table 18.3.)
6. A $NiCl_2$ solution is placed in the apparatus shown in Figure 18.2, instead of the HCl solution shown. Write equations for:
   (a) The anode reaction
   (b) The cathode reaction
   (c) The net electrochemical reaction
7. What is the major distinction between the reactions occurring in Figure 18.2 and those in Figure 18.3?
8. In the cell shown in Figure 18.3:
   (a) What would be the effect of removing the voltmeter and connecting the wires shown coming to the voltmeter?
   (b) What would be the effect of removing the salt bridge?
9. What is the oxidation number of the underlined element in each compound?
   (a) $\underline{N}aCl$ (e) $H_2\underline{S}O_3$ (i) $\underline{N}H_3$
   (b) $Fe\underline{Cl}_3$ (f) $\underline{N}H_4Cl$ (j) $KClO_3$
   (c) $\underline{Pb}O_2$ (g) $K\underline{Mn}O_4$ (k) $K_2\underline{Cr}O_4$
   (d) $Na\underline{N}O_3$ (h) $\underline{I}_2$ (l) $K_2\underline{Cr}_2O_7$

10. What is the oxidation number of the underlined elements?
    (a) $\underline{S}^{2-}$ (d) $\underline{Mn}O_4^-$ (g) $As\underline{O}_3^{3-}$
    (b) $\underline{N}O_2^-$ (e) $\underline{Bi}^{3+}$ (h) $Fe(\underline{O}H)_3$
    (c) $Na_2\underline{O}_2$ (f) $\underline{O}_2$ (i) $\underline{I}O_3^-$

11. In the following half-reactions, which element is changing oxidation state? Is the half-reaction an oxidation or a reduction? Supply the proper number of electrons to the proper side to balance each equation.
    (a) $Zn^{2+} \longrightarrow Zn$
    (b) $2\,Br^- \longrightarrow Br_2$
    (c) $MnO_4^- + 8\,H^+ \longrightarrow Mn^{2+} + 4\,H_2O$
    (d) $Ni \longrightarrow Ni^{2+}$
    (e) $SO_3^{2-} + H_2O \longrightarrow SO_4^{2-} + 2\,H^+$
    (f) $NO_3^- + 4\,H^+ \longrightarrow NO + 2\,H_2O$
    (g) $S_2O_4^{2-} + 2\,H_2O \longrightarrow 2\,SO_3^{2-} + 4\,H^+$
    (h) $Fe^{2+} \longrightarrow Fe^{3+}$

12. In the following unbalanced equations,
    (a) Identify the element that is oxidized and the element that is reduced.
    (b) Identify the oxidizing agent and the reducing agent.
        (1) $Cr + HCl \longrightarrow CrCl_3 + H_2$
        (2) $SO_4^{2-} + I^- + H^+ \longrightarrow H_2S + I_2 + H_2O$
        (3) $AsH_3 + Ag^+ + H_2O \longrightarrow$
            $\qquad\qquad H_3AsO_4 + Ag + H^+$
        (4) $Cl_2 + NaBr \longrightarrow NaCl + Br_2$

13. Balance these equations by the change-in-oxidation-number method.
    (a) $Zn + S \longrightarrow ZnS$
    (b) $AgNO_3 + Pb \longrightarrow Pb(NO_3)_2 + Ag$
    (c) $Fe_2O_3 + CO \longrightarrow Fe + CO_2$
    (d) $H_2S + HNO_3 \longrightarrow S + NO + H_2O$
    (e) $MnO_2 + HBr \longrightarrow MnBr_2 + Br_2 + H_2O$
    (f) $Cl_2 + KOH \longrightarrow KCl + KClO_3 + H_2O$
    (g) $Ag + HNO_3 \longrightarrow AgNO_3 + NO + H_2O$
    (h) $CuO + NH_3 \longrightarrow N_2 + Cu + H_2O$
    (i) $PbO_2 + Sb + NaOH \longrightarrow$
        $\qquad\qquad PbO + NaSbO_2 + H_2O$
    (j) $H_2O_2 + KMnO_4 + H_2SO_4 \longrightarrow$
        $\qquad\qquad O_2 + MnSO_4 + K_2SO_4 + H_2O$

14. Balance the following ionic redox equations using the ion–electron method. All reactions occur in acidic solution.
    (a) $Zn + NO_3^- \longrightarrow Zn^{2+} + NH_4^+$
    (b) $NO_3^- + S \longrightarrow NO_2 + SO_4^{2-}$
    (c) $PH_3 + I_2 \longrightarrow H_3PO_2 + I^-$
    (d) $Cu + NO_3^- \longrightarrow Cu^{2+} + NO$
    (e) $ClO_3^- + I^- \longrightarrow I_2 + Cl^-$
    (f) $Cr_2O_7^{2-} + Fe^{2+} \longrightarrow Cr^{3+} + Fe^{3+}$

    (g) $MnO_4^- + SO_2 \longrightarrow Mn^{2+} + SO_4^{2-}$
    (h) $H_3AsO_3 + MnO_4^- \longrightarrow H_3AsO_4 + Mn^{2+}$
    *(i) $ClO_3^- + Cl^- \longrightarrow Cl_2$
    *(j) $Cr_2O_7^{2-} + H_3AsO_3 \longrightarrow Cr^{3+} + H_3AsO_4$

15. Balance the following ionic redox equations using the ion–electron method. All reactions occur in basic solutions.
    (a) $Cl_2 + IO_3^- \longrightarrow Cl^- + IO_4^-$
    (b) $MnO_4^- + ClO_2^- \longrightarrow MnO_2 + ClO_4^-$
    (c) $Se \longrightarrow Se^{2-} + SeO_3^{2-}$
    (d) $MnO_4^- + SO_3^{2-} \longrightarrow MnO_2 + SO_4^{2-}$
    (e) $ClO_2(g) + SbO_2^- \longrightarrow ClO_2^- + Sb(OH)_6^-$
    *(f) $Fe_3O_4 + MnO_4^- \longrightarrow Fe_2O_3 + MnO_2$
    *(g) $BrO^- + Cr(OH)_4^- \longrightarrow Br^- + CrO_4^{2-}$
    *(h) $P_4 \longrightarrow HPO_3^{2-} + PH_3$
    *(i) $Al + OH^- \longrightarrow Al(OH)_4^- + H_2$
    (j) $Al + NO_3^- \longrightarrow NH_3 + Al(OH)_4^-$

16. Why are oxidation and reduction said to be complementary processes?

17. When molten $CaBr_2$ is electrolyzed, calcium metal and bromine are produced. Write equations for the two half-reactions that occur at the electrodes. Label the anode half-reaction and the cathode half-reaction.

18. Why is direct current used instead of alternating current in the electroplating of metals?

19. The chemical reactions taking place during discharge in a lead storage battery are

    $$Pb + SO_4^{2-} \longrightarrow PbSO_4$$
    $$PbO_2 + SO_4^{2-} + 4\,H^+ \longrightarrow PbSO_4 + 2\,H_2O$$

    (a) Complete each half-reaction by supplying electrons.
    (b) Which reaction is oxidation and which is reduction?
    (c) Which reaction occurs at the anode of the battery?

20. What property of lead dioxide and lead(II) sulfate makes it unnecessary to have salt bridges in the cells of a lead storage battery?

21. Explain why the density of the electrolyte in a lead storage battery decreases during the discharge cycle.

22. In one type of alkaline cell used to power devices such as portable radios, Hg^{2+} ions are reduced to metallic mercury when the cell is being discharged. Does this reduction occur at the anode or the cathode? Explain.

23. Differentiate between an electrolytic cell and a voltaic cell.

24. Why is a porous barrier or a salt bridge necessary in some voltaic cells?
25. Use the following redox equation to indicate:

$$KMnO_4 + HCl \longrightarrow$$
$$KCl + MnCl_2 + H_2O + Cl_2$$

    (a) The oxidizing agent
    (b) The reducing agent
    (c) The number of electrons that are transferred per mole of oxidizing agent
26. Which of the following statements are correct? Rewrite the incorrect statements to make them correct.
    (a) An atom of an element in the uncombined state has an oxidation number of zero.
    (b) The oxidation number of molybdenum in Na_2MoO_4 is $+4$.
    (c) The oxidation number of an ion is the same as the electrical charge on the ion.
    (d) The process in which an atom or an ion loses electrons is called reduction.
    (e) The reaction $Fe^{3+} + e^- \longrightarrow Fe^{2+}$ is a reduction reaction.
    (f) In the reaction

        $2\,Al + 3\,CuCl_2 \longrightarrow 2\,AlCl_3 + 3\,Cu$

        aluminum is the oxidizing agent.
    (g) In a redox reaction the oxidizing agent is reduced, and the reducing agent is oxidized.
    (h) $Cu^0 \longrightarrow Cu^{2+}$ is a balanced oxidation half-reaction.
    (i) In the electrolysis of sodium chloride brine (solution), Cl_2 gas is formed at the cathode, and hydroxide ions are formed at the anode.
    (j) In any cell, electrolytic or voltaic, reduction takes place at the cathode, and oxidation occurs at the anode.
    (k) In the Zn–Cu voltaic cell, the reaction at the anode is $Zn \longrightarrow Zn^{2+} + 2\,e^-$.

The statements in (l) through (o) pertain to this activity series:

    Ba Mg Zn Fe H Cu Ag

    (l) The reaction $Zn + MgCl_2 \rightarrow Mg + ZnCl_2$ is a spontaneous reaction.
    (m) Barium is more active than copper.
    (n) Silver metal will react with acids to liberate hydrogen gas.
    (o) Iron is a better reducing agent than zinc.
    (p) Oxidation and reduction occur simultaneously in a chemical reaction; one cannot take place without the other.

    (q) A free metal can displace from solution the ions of a metal that lies below the free metal in the Activity Series.
    (r) In electroplating, the piece to be electroplated with a metal is attached to the cathode.
    (s) In an automobile lead storage battery, the density of the sulfuric acid solution decreases as the battery discharges.
    (t) In an electrolytic cell, chemical energy is used to produce electrical energy.
27. How many moles of NO gas will be formed by the reaction of 25.0 g of silver with nitric acid? [See the equation given in Exercise 13(g).]
28. What volume of chlorine gas, measured at STP, is required to react with excess KOH to form 0.300 mol of $KClO_3$? [See the equation given in Exercise 13(f).]
29. What mass of $KMnO_4$ would be needed to react with 100 mL of H_2O_2 solution ($d = 1.031$ g/mL, 9.0% H_2O_2 by mass)? [See the equation given in Exercise 13(j).]
*30. What volume of $0.200\ M$ $K_2Cr_2O_7$ will be required to oxidize 5.00 g of H_3AsO_3? [See the equation given in Exercise 14(j).]
*31. What volume of $0.200\ M$ $K_2Cr_2O_7$ will be required to oxidize the Fe^{2+} ion in 60.0 mL of $0.200\ M$ $FeSO_4$ solution? [See the equation given in Exercise 14(f).]
32. A sample of crude potassium iodide was analyzed using the following reaction (not balanced):

$$I^- + SO_4^{2-} \longrightarrow I_2 + H_2S \quad \text{(acid solution)}$$

    If a 4.00 g sample of crude KI produced 2.79 g of iodine, what is the percent purity of the KI?
33. What mass of copper is formed when 35.0 L of ammonia gas, measured at STP, reacts with copper(II) oxide? [See the equation given in Exercise 13(h).]
*34. What volume of NO gas, measured at 28°C and 744 torr, will be formed by the reaction of 0.500 mol of Ag reacting with excess nitric acid? [See the equation given in Exercise 13(g).]
35. How many moles of H_2 can be produced from 100. g of Al according to the following reaction?

$$Al + OH^- \longrightarrow$$
$$Al(OH)_4^- + H_2 \quad \text{(basic solution)}$$

Review Exercises for Chapters 16–18

CHAPTER SIXTEEN   Ionization: Acids, Bases, Salts

True–False. *Answer the following as either true or false.*

1. Arrhenius defined a base as a hydroxide-containing substance that dissociates in water to produce hydroxide ions.
2. The Brønsted–Lowry theory defined an acid as a proton donor.
3. The Lewis theory defined an acid as an electron-pair donor.
4. In the reaction

$$HCl + NH_3 \longrightarrow NH_4^+ + Cl^-$$

according to the Brønsted–Lowry theory, the conjugate base of HCl is NH_4^+.

5. In the reaction

$$HNO_3 + H_2O \longrightarrow H_3O^+ + NO_3^-$$

the conjugate base of HNO_3 is NO_3^-.

6. When an acid reacts with a carbonate, the products are a salt, water, and carbon dioxide.
7. When electrolytes dissociate, they break into individual molecules.
8. The equation for the ionization of acetic acid is:

$$HC_2H_3O_2 + H_2O \rightleftharpoons H_3O^+ + C_2H_3O_2^-$$

9. In water, zinc nitrate dissociates as shown by the equation

$$Zn(NO_3)_2 \xrightarrow{H_2O} Zn^{2+} + 2\ NO_3^-$$

10. The equation for the dissociation of sodium carbonate in water is

$$Na_2CO_3 \xrightarrow{H_2O} Na_2^{2+} + CO_3^{2-}$$

11. When 1 mol of $CaCl_2$ dissolves in water, it will give 1 mol of Ca^{2+} and 2 mol of Cl^- ions in solution.

12. A solution with $[H^+] = 1.0 \times 10^{-9}$ M has a pH of -9.0.
13. A solution with $[H^+] = 7.8 \times 10^{-2}$ M has a pH between 1 and 2.
14. The reaction of an acid and a base to form a salt and water is known as neutralization.
15. In neutralizations between a strong acid and a strong base, the net ionic equation is

$$H^+ + OH^- \longrightarrow H_2O$$

16. For the aqueous reaction of NaOH with $FeCl_3$, the net ionic equation is

$$Fe^{3+} + 3\ OH^- \longrightarrow Fe(OH)_3(s)$$

17. A 1.0 M $HC_2H_3O_2$ solution will freeze at a lower temperature than a 1.0 M solution of KBr.
18. In writing net ionic equations, weak electrolytes are written in their un-ionized form.
19. Acids, bases, and salts are electrolytes.
20. Positive ions are called spectator ions.
21. Negative ions are called anions.
22. Titration is the process of measuring the volume of one reagent required to react with a measured mass or volume of another reagent.
23. A pH meter is an instrument used to measure the acidity of a solution.
24. As the acidity of a solution increases, the pH decreases.
25. When 50.0 mL of 0.20 M NaOH and 100. mL of 0.10 M HCl are mixed, the resulting solution will have a pH of 7.
26. The conjugate acid of HSO_4^- is SO_4^{2-}.
27. H_3O^+ is called a hydronium ion and NH_4^+ is called a nitronium ion.
28. A 0.1 M HCl and a 0.1 M $HC_2H_3O_2$ solution will have the same pH.
29. HNO_2, $HC_2H_3O_2$, HClO, and HBr are weak acids.
30. When water ionizes at 100°C it produces more H^+ ions than OH^- ions.

Multiple Choice. *Choose the correct answer to each of the following.*

1. When the reaction $Al + HCl \longrightarrow$ is completed and balanced, a term appearing in the balanced equation is:
(a) $3 HCl$ (b) $AlCl_2$ (c) $3 H_2$ (d) $4 Al$

2. When the reaction $CaO + HNO_3 \longrightarrow$ is completed and balanced, a term appearing in the balanced equation is:
(a) H_2 (b) $2 H_2$ (c) $2 CaNO_3$ (d) H_2O

3. When the reaction $H_3PO_4 + KOH \longrightarrow$ is completed and balanced, a term appearing in the balanced equation is:
(a) H_3PO_4 (c) KPO_4
(b) $6 H_2O$ (d) $3 KOH$

4. When the reaction $HCl + Cr_2(CO_3)_3 \longrightarrow$ is completed and balanced, a term appearing in the balanced equation is:
(a) Cr_2Cl (b) $3 HCl$ (c) $3 CO_2$ (d) H_2O

5. Which of the following is not a salt?
(a) $K_2Cr_2O_7$ (c) $Ca(OH)_2$
(b) $NaHCO_3$ (d) $Na_2C_2O_4$

6. Which of the following is not an acid?
(a) H_3PO_4 (b) H_2S (c) H_2SO_4 (d) NH_3

7. Which of the following is a weak electrolyte?
(a) NH_4OH (c) K_3PO_4
(b) $Ni(NO_3)_2$ (d) $NaBr$

8. Which of the following is a nonelectrolyte?
(a) $HC_2H_3O_2$ (c) $KMnO_4$
(b) $MgSO_4$ (d) CCl_4

9. Which of the following is a strong electrolyte?
(a) H_2CO_3 (c) NH_4OH
(b) HNO_3 (d) H_3BO_3

10. Which of the following is a weak electrolyte?
(a) $NaOH$ (c) $HC_2H_3O_2$
(b) $NaCl$ (d) H_2SO_4

11. A solution has a concentration of H^+ of $3.4 \times 10^{-5} M$. The pH is:
(a) 4.47 (b) 5.53 (c) 3.53 (d) 5.47

12. A solution with a pH of 5.85 has an H^+ concentration of:
(a) $7.1 \times 10^{-5} M$ (c) $3.8 \times 10^{-4} M$
(b) $7.1 \times 10^{-6} M$ (d) $1.4 \times 10^{-6} M$

13. 16.55 mL of 0.844 M NaOH is required to titrate 10.00 mL of a hydrochloric acid solution. The molarity of the acid solution is:
(a) 0.700 M (c) 1.40 M
(b) 0.510 M (d) 0.255 M

14. What volume of 0.462 M NaOH is required to neutralize 20.00 mL of 0.391 M HNO_3?
(a) 23.6 mL (c) 9.03 mL
(b) 16.9 mL (d) 11.8 mL

15. 25.00 mL of H_2SO_4 solution required 18.92 mL of 0.1024 N NaOH for complete neutralization. The normality of the acid is
(a) 0.1550 N (c) 0.07750 N
(b) 0.1353 N (d) 0.06765 N

16. Dilute hydrochloric acid is a typical acid, as shown by its:
(a) color (b) odor (c) solubility (d) taste

17. What is the pH of a 0.00015 M HCl solution?
(a) 4.0 (c) between 3 and 4
(b) 2.82 (d) No correct answer given

18. The chloride ion concentration in 300. mL of 0.10 M $AlCl_3$ is
(a) 0.30 M (c) 0.030 M
(b) 0.10 M (d) 0.90 M

19. The amount of $BaSO_4$ that will precipitate when 100. mL of 0.10 M $BaCl_2$ and 100. mL of 0.10 M Na_2SO_4 are mixed is:
(a) 0.010 mol (c) 23 g
(b) 0.10 mol (d) No correct answer given

20. The freezing point of a 0.50 molal NaCl aqueous solution will be about:
(a) $-1.86°C$ (c) $-2.79°C$
(b) $-0.93°C$ (d) No correct answer given

CHAPTER SEVENTEEN Chemical Equilibrium

True–False. *Answer the following as either true or false.*

1. A reversible reaction is one in which the products formed in a chemical reaction are reacting to produce the original reactants.

2. The study of reaction rates is known as chemical kinetics.

3. When the rate of the forward reaction is exactly equal to the rate of the reverse reaction, a condition of chemical equilibrium exists.

4. A statement of Le Chatelier's principle is that, if a stress is applied to a system in equilibrium, the system will behave in such a way as to relieve that stress and restore equilibrium but under a new set of conditions.

5. A catalyst will shift the point of equilibrium of a reaction but will not alter the reaction rates.

6. The reaction $CaCO_3(s) \longrightarrow CaO(s) + CO_2(g)$ will proceed to the right better in a closed container where the pressure of the CO_2 can build up than in an open container where the CO_2 can escape.

7. When heat is applied to a system in equilibrium, the reaction that absorbs the heat is favored.

8. The ionization constant expression for the ionization of the weak acid HCN is
$K_a = [H^+][CN^-]$.

9. If K_a for acetic acid is 1.8×10^{-5}, and K_a for nitrous acid is 4.5×10^{-4}, then, at equal concentrations, acetic acid is a stronger acid.

10. If the K_{sp} for AgI is 1.6×10^{-16}, and the K_{sp} for CuS is 8×10^{-45}, then AgI is more soluble than CuS.

11. The amount of energy needed to form an activated complex is known as the activation energy of a reaction.

12. A catalyst can lower the activation energy, thus increasing the speed of a reaction.

13. A chemical reaction at equilibrium will have different equilibrium constants at different temperatures.

14. Generally, as the concentrations of the reactants increase in a chemical reaction, the speed of the reaction decreases.

15. When the temperature of an exothermic reaction is increased, the forward reaction is favored.

16. The ion product constant for water at 25°C is 1×10^{-14}.

17. A solution of pOH 12 has an H^+ concentration of 0.010 mol per liter.

18. A solution of pOH 3 will turn blue litmus to red.

19. KNO_2 dissolved in water will give a solution with a pH less than 7.

20. A solution made from NaCl and HCl will act as a buffer solution.

21. A solution made from 100 mL of 0.1 M NaOH and 100 mL of 0.2 M $HC_2H_3O_2$ will act as a buffer solution.

22. KCN will hydrolyze to give an alkaline solution.

23. The equilibrium constant expression for

$$CH_4(g) + 2\,O_2(g) \longrightarrow CO_2(g) + 2\,H_2O(g)$$

is $\dfrac{[CO_2][H_2O]}{[CH_4][O_2]}$

24. An increase in pressure for a system composed entirely of gases will cause the reaction to shift to the side that contains the larger number of moles.

25. A catalyst will increase the speed of the forward and reverse reactions equally.

26. When K_a is large ($\gg 1$), the concentration of the reactants at equilibrium is greater than the concentration of the products.

27. The K_{sp} expression for $Fe(OH)_3$:

$$Fe(OH)_3(s) \rightleftharpoons Fe^{3+} + 3(OH)^-$$

is $K_{sp} = \dfrac{[Fe^{3+}][OH^-]^3}{[Fe(OH)_3]}$

28. If $[H^+]$ is known, $[OH^-]$ can be calculated from the expression for K_w.

Multiple Choice. *Choose the correct answer to each of the following.*

1. The equation
$HC_2H_3O_2 + H_2O \rightleftharpoons H_3O^+ + C_2H_3O_2^-$
implies that
   (a) If you start with 1.0 mol of $HC_2H_3O_2$, 1.0 mol of H_3O^+ and 1.0 mole of $C_2H_3O_2^-$ will be produced.
   (b) An equilibrium exists between the forward reaction and the reverse reaction.
   (c) At equilibrium, equal molar amounts of all four substances will exist.
   (d) The reaction proceeds all the way to the products, then reverses, going all the way back to the reactants.

2. If the reaction $A + B \rightleftharpoons C + D$ is initially at equilibrium, and then more A is added, which of the following is not true?
   (a) More collisions of A and B will occur, thus the rate of the forward reaction will be increased.
   (b) The equilibrium will shift toward the right.
   (c) The moles of B will be increased.
   (d) The moles of D will be increased.

3. What will be the H^+ concentration in a 1.0 M HCN solution? ($K_a = 4.0 \times 10^{-10}$)
   (a) 2.0×10^{-5} M (c) 4.0×10^{-10} M
   (b) 1.0 M (d) 2.0×10^{-10} M

4. What is the percent ionization of HCN in Exercise 3?
   (a) 100%   (c) 2.0×10^{-3} %
   (b) 2.0×10^{-8} %   (d) 4.0×10^{-8} %

5. If $[H^+] = 1 \times 10^{-5}$ M, which of the following is not true?
   (a) pH $= 5$ (c) $[OH^-] = 1 \times 10^{-5}$ M
   (b) pOH $= 9$ (d) The solution is acidic.

6. If $[H^+] = 2.0 \times 10^{-4}$ M, then $[OH^-]$ will be:
   (a) 5.0×10^{-9} M (c) 2.0×10^{-4} M
   (b) 3.70   (d) 5.0×10^{-11} M

7. The solubility product of $PbCrO_4$ is 2.8×10^{-13}. The solubility of $PbCrO_4$ is:
   (a) 5.3×10^{-7} M (c) 7.8×10^{-14} M
   (b) 2.8×10^{-13} M (d) 1.0 M

8. The solubility of AgBr is 6.3×10^{-7} M. The value of the solubility product is:
   (a) 6.3×10^{-7} (c) 4.0×10^{-48}
   (b) 4.0×10^{-13} (d) 4.0×10^{-15}

9. Which of the following solutions would be the best buffer solution?
   (a) 0.10 M $HC_2H_3O_2$ + 0.10 M $NaC_2H_3O_2$
   (b) 0.10 M HCl
   (c) 0.10 M HCl + 0.10 M NaCl
   (d) Pure water

10. For the reaction $H_2(g) + I_2(g) \rightleftharpoons 2\,HI(g)$, at 700 K, $K_{eq} = 56.6$. If an equilibrium mixture at 700 K was found to contain 0.55 M HI and 0.21 M H_2, the I_2 concentration must be:
    (a) 0.046 M (c) 22 M
    (b) 0.025 M (d) 0.21 M

11. The equilibrium constant for the reaction $2\,A + B \rightleftharpoons 3\,C + D$ is
    (a) $\dfrac{[C]^3[D]}{[A]^2[B]}$ (c) $\dfrac{[3C][D]}{[2A][B]}$
    (b) $\dfrac{[2A][B]}{[3C][D]}$ (d) $\dfrac{[A]^2[B]}{[C]^3[D]}$

12. In the equilibrium represented by

    $$N_2(g) + O_2(g) \rightleftharpoons 2\,NO_2(g)$$

    as the pressure is increased, the amount of NO_2 formed:
    (a) Increases   (b) Decreases
    (c) Remains the same
    (d) Increases and decreases irregularly

13. Which factor will not increase the concentration of ammonia as represented by the following equation?

    $$3\,H_2(g) + N_2(g) \rightleftharpoons 2\,NH_3(g) + 92.5\ kJ$$

    (a) Increasing the temperature
    (b) Increasing the concentration of N_2
    (c) Increasing the concentration of H_2
    (d) Increasing the pressure

14. If $HCl(g)$ is added to a saturated solution of AgCl, the concentration of Ag^+ in solution:
    (a) Increases
    (b) Decreases
    (c) Remains the same
    (d) Increases and decreases irregularly

15. The solubility of $CaCO_3$ at 20°C is 0.013 g/L. What is the K_{sp} for $CaCO_3$?
    (a) 1.3×10^{-8} (c) 1.7×10^{-8}
    (b) 1.3×10^{-4} (d) 1.7×10^{-4}

16. The K_{sp} for $BaCrO_4$ is 8.5×10^{-11}. What is the solubility of $BaCrO_4$ in grams per liter?
    (a) 9.2×10^{-6} (c) 2.3×10^{-3}
    (b) 0.073   (d) 8.5×10^{-11}

17. What will be the $[Ba^{2+}]$ when 0.010 mol of Na_2CrO_4 is added to 1.0 L of saturated $BaCrO_4$ solution? See Exercise 16 for K_{sp}.
    (a) 8.5×10^{-11} M (c) 9.2×10^{-6} M
    (b) 8.5×10^{-9} M (d) 9.2×10^{-4} M

18. Which would occur if a small amount of sodium acetate crystals, $NaC_2H_3O_2$, were added to 100 mL of 0.1 M $HC_2H_3O_2$ at constant temperature?
    (a) The number of acetate ions in the solution would decrease.
    (b) The number of acetic acid molecules would decrease.
    (c) The number of sodium ions in solution would decrease.
    (d) The H^+ concentration in the solution would decrease.

19. If the temperature is decreased for the endothermic reaction:

    $$A + B \rightleftharpoons C + D$$

    which of the following is true?
    (a) The concentration of A will increase.
    (b) No change will occur.
    (c) The concentration of B will decrease.
    (d) The concentration of D will increase.

CHAPTER EIGHTEEN   Oxidation–Reduction

True–False.   *Answer the following as either true or false.*

1. In oxidation, the oxidation number of an element increases in a positive direction as a result of gaining electrons.
2. The oxidation number of chlorine in Cl_2 is -1.
3. Oxidation and reduction occur simultaneously in a chemical reaction; one cannot take place without the other.
4. The negative electrode is called the cathode.
5. The cathode is the electrode at which oxidation takes place.
6. The change in the oxidation number of an element from -2 to 0 is reduction.
7. A free metal can displace from solution the ions of a metal that lies below the free metal in the Activity Series.
8. Metallic zinc will react with hydrochloric acid.
9. In electroplating, the piece to be electroplated with a metal is attached to the cathode.
10. In a lead storage battery, $PbSO_4$ is produced at both electrodes in the discharging cycle.
11. As a lead storage battery discharges, the electrolyte becomes less dense.
12. The algebraic sum of the oxidation numbers of all the atoms in $K_2Cr_2O_7$ is zero.
13. A reducing agent will always decrease in oxidation number.
14. Potassium is a better reducing agent than sodium.
15. In the reaction $2\ Cl^- \longrightarrow Cl_2$, each chloride ion loses one electron.
16. $2\ Ag(s) + 2\ HCl(aq) \longrightarrow 2\ AgCl(s) + H_2(g)$
17. In a voltaic cell, reduction occurs at the cathode.
18. The oxidation number of P in $Mg_2P_2O_7$ is $+7$.
19. $CrO_4^{2-} + 4\ H_2O + 2\ e^- \longrightarrow$
$$Cr(OH)_3 + 5\ OH^-$$
is a balanced reduction half-reaction.
20. In an electrolytic cell, electrical energy is used to bring about a chemical reaction.

Multiple Choice.   *Choose the correct answer to each of the following.*

1. In K_2SO_4, the oxidation number of sulfur is:
   (a) $+2$ (b) $+4$ (c) $+6$ (d) -2

2. In $Ba(NO_3)_2$, the oxidation number of N is:
   (a) $+5$ (b) -3 (c) $+4$ (d) -1
3. In the reaction $H_2S + 4\ Br_2 + 4\ H_2O \longrightarrow H_2SO_4 + 8\ HBr$, the oxidizing agent is:
   (a) H_2S (b) Br_2 (c) H_2O (d) H_2SO_4
4. In the reaction

$$VO_3^- + Fe^{2+} + 4\ H^+ \longrightarrow$$
$$VO^{2+} + Fe^{3+} + 2\ H_2O$$

the element reduced is:
   (a) V   (b) Fe   (c) O   (d) H

Questions 5, 6, and 7 pertain to the activity series.

K   Ca   Mg   Al   Zn   Fe   H   Cu   Ag

5. Which of the following pairs will not react in water solution?
   (a) Zn, $CuSO_4$ (c) Fe, $AgNO_3$
   (b) Cu, $Al_2(SO_4)_3$ (d) Mg, $Al_2(SO_4)_3$
6. Which element is the most easily oxidized?
   (a) K   (b) Mg   (c) Zn   (d) Cu
7. Which element will reduce Cu^{2+} to Cu but will not reduce Zn^{2+} to Zn?
   (a) Fe   (b) Ca   (c) Ag   (d) Mg
8. In the electrolysis of fused (molten) $CaCl_2$, the product at the negative electrode is:
   (a) Ca^{2+} (b) Cl^- (c) Cl_2 (d) Ca
9. In its reactions, a free element from Group IIA in the periodic table is most likely to:
   (a) Be oxidized   (c) Be unreactive
   (b) Be reduced   (d) Gain electrons
10. In the partially balanced redox equation

$$3\ Cu + HNO_3 \longrightarrow$$
$$3\ Cu(NO_3)_2 + 2\ NO + H_2O$$

the coefficient needed to balance H_2O is:
   (a) 8   (b) 6   (c) 4   (d) 2
11. Which reaction does not involve oxidation–reduction?
   (a) Burning sodium in chlorine
   (b) Chemical union of Fe and S
   (c) Decomposition of $KClO_3$
   (d) Neutralization of NaOH with H_2SO_4

12. How many moles of Fe^{2+} can be oxidized to Fe^{3+} by 2.50 mol of Cl_2 according to the following equation?

$$Fe^{2+} + Cl_2 \longrightarrow Fe^{3+} + Cl^-$$

    (a) 2.50 mol   (c) 1.00 mol
    (b) 5.00 mol   (d) 22.4 mol

13. How many grams of sulfur can be produced from 100 mL of 6.00 M HNO_3?

$$HNO_3 + H_2S \longrightarrow S + NO + H_2O$$

    (a) 28.9 g   (b) 19.3 g   (c) 32.1 g   (d) 289 g

14. Which of the following ions can be reduced by H_2?
    (a) Hg^{2+} (b) Sn^{2+} (c) Zn^{2+} (d) K^+

15. Which of the following is *not* true of a zinc-mercury cell?
    (a) It provides current at a steady potential.
    (b) It has a short service life.
    (c) It is self-contained.
    (d) It can be stored for long periods of time.

Balancing Oxidation-Reduction Equations. *Balance each of the following equations.*

1. $P + HNO_3 \longrightarrow HPO_3 + NO + H_2O$
2. $MnSO_4 + PbO_2 + H_2SO_4 \longrightarrow$
   $$HMnO_4 + PbSO_4 + H_2O$$
3. $Cr_2O_7^{2-} + Cl^- \longrightarrow Cr^{3+} + Cl_2$
   (acidic solution)
4. $MnO_4^- + AsO_3^{3-} \longrightarrow Mn^{2+} + AsO_4^{3-}$
   (acidic solution)
5. $S^{2-} + Cl_2 \longrightarrow SO_4^{2-} + Cl^-$ (basic solution)
6. $Zn + NO_3^- \longrightarrow Zn(OH)_4^{2-} + NH_3$
   (basic solution)
7. $KOH + Cl_2 \longrightarrow HCl + KClO + H_2O$
8. $As + ClO_3^- \longrightarrow H_3AsO_3 + HClO$
   (acidic solution)
9. $MnO_4^- + Cl^- \longrightarrow Mn^{2+} + Cl_2$
   (acidic solution)
10. $H_2O_2 + Cl_2O_7 \longrightarrow ClO_2^- + O_2$
    (basic solution)

CHAPTER NINETEEN

Nuclear Chemistry

An unusual telephone conversation originated from the University of Chicago in December of 1942. The caller, British physicist Arthur Compton, said, "Jim, the Italian navigator has just landed in the new world." The reference was to the great physicist Enrico Fermi; the landing was the first controlled atomic chain reaction; and the new world was the dawning of the nuclear age.

A new world indeed! Locked within the nucleus of an atom is a source of tremendous energy. Its power can be devastating, as seen in the effects of nuclear weapons or accidents, such as Three Mile Island or Chernobyl. Nuclear energy can be harnassed to perform such useful tasks as generating power, treating cancer, and preservating food. Isotopes are used in medicine to diagnose illness and detect minute quantities of drugs or hormones. Researchers are using radioactive tracers to the sequence of human genome. The applications of nuclear chemistry are important in medicine, industry, art, and research. Its impact both threatens and enhances our lives and our future.

Chapter Preview

19.1   Discovery of Radioactivity

One of the most important steps leading to the discovery of radioactivity was made in 1895 by Wilhelm Konrad Roentgen (1845–1923). Roentgen discovered X rays when he observed that a vacuum discharge tube, enclosed in a thin, black cardboard box, caused a nearby piece of paper coated with barium platinocyanide to glow with a brilliant phosphorescence. From this and other experiments he concluded that certain rays, which he called X rays, were emitted from the discharge tube, penetrated the box, and caused the salt to glow. Roentgen also showed that X rays could penetrate other bodies and affect photographic plates. This observation led to the development of X-ray photography.

Shortly after this discovery Antoine Henri Becquerel (1852–1908) attempted to show a relationship between X rays and the phosphorescence of uranium salts. In one of his experiments he wrapped a photographic plate in black paper, placed a sample of uranium salt on it, and exposed it to sunlight. The developed photographic plate showed that rays emitted from the salt had penetrated the paper. Later Becquerel prepared to repeat the experiment, but, because the sunlight was intermittent, he placed the entire setup in a drawer. Several days later he developed the photographic plate, expecting to find it only slightly affected. He was amazed to observe an intense image on the plate. He repeated the experiment in total darkness and obtained the same results, proving that the uranium salt emitted rays that affected the photographic plate without its being exposed to sunlight. In this way Becquerel discovered radioactivity. The name radioactivity was given to this phenomenon two years later (in 1898) by Marie Curie. **Radioactivity** is the spontaneous emission of particles and/or rays from the nucleus of an atom. Elements having this property are said to be radioactive. Becquerel later showed that the rays coming from uranium were able to ionize air and were also capable of penetrating thin sheets of metal.

In 1898 Marie Sklodowska Curie (1867–1934) and her husband Pierre Curie (1859–1906) turned their research interests to radioactivity. In a short time the Curies discovered two new elements, polonium and radium, both of which are radioactive. To confirm their work on radium they processed 1 ton of pitchblende residue ore to obtain 0.1 g of pure radium chloride, which they used to make further studies on the properties of radium and to determine its atomic mass.

In 1899 Ernest Rutherford began to investigate the nature of the rays emitted from uranium. He found two rays, which he called *alpha* and *beta rays*. Soon he realized that uranium, while emitting these rays, was changing into another element. By 1912, over 30 radioactive isotopes were known, and many more are known today. The *gamma ray*, a third ray emitted from radioactive materials and similar to an X ray, was discovered by Paul Villard in 1900. After the description of the nuclear atom by Rutherford, the phenomenon of radioactivity was attributed to reactions taking place in the nuclei of atoms.

The symbolism and notation described for isotopes in Chapter 5 is very useful in nuclear chemistry and is briefly reviewed here.

$$^A_Z X \quad \begin{array}{l} \longleftarrow \text{Mass number} \\ \longleftarrow \text{Symbol of element} \\ \longleftarrow \text{Atomic number} \end{array}$$

For example, $^{238}_{92}$ represents a uranium isotope with an atomic number of 92 and a mass number of 238. This isotope is also designated as U-238 or uranium-238 and contains 92 protons and 146 neutrons. The protons and neutrons collectively are known as **nucleons**. The mass number is the total number of nucleons in the nucleus. Table 19.1 shows the iostopic notations for several particles associated with nuclear chemistry.

When we speak of isotopes we generally mean atoms of the same element with different masses, such as $^{16}_8 O$, $^{17}_8 O$, $^{18}_8 O$. In nuclear chemistry we use the term **nuclide** to mean any isotope of any atom. Thus $^{16}_8 O$ and $^{235}_{92} U$ are referred to as nuclides. Nuclides that spontaneously emit radiation are referred to as *radionuclides*.

TABLE 19.1

Symbols in Isotopic Notation for Several Particles (and Small Isotopes) Associated with Nuclear Chemistry

Particle	Symbol	Atomic number Z	Approximate atomic mass A
Neutron	$_0^1n$	0	1
Proton	$_1^1H$	1	1
Beta particle (electron)	$_{-1}^0e$	-1	0
Positron (positive electron)	$_{+1}^0e$	1	0
Alpha particle (helium nucleus)	$_2^4He$	2	4
Deuteron (heavy hydrogen nucleus)	$_1^2H$	1	2

19.2 Natural Radioactivity

radioactive decay

Radioactive elements continuously undergo **radioactive decay**, or disintegration, to form different elements. The chemical properties of an element are associated with its electronic structure, but radioactivity is a property of the nucleus. Therefore, neither ordinary changes of temperature and pressure nor the chemical or physical state of an element has any effect on its radioactivity.

The principal emissions from the nuclei of radionuclides are known as alpha rays (or particles), beta rays (or particles), and gamma rays. Upon losing an alpha or beta particle, the radioactive element changes into a different element. This process will be explained in detail later.

half-life

Each radioactive nuclide disintegrates at a specific and constant rate, which is expressed in units of half-life. The **half-life** ($t_{1/2}$) is the time required for one-half of a specific amount of a radioactive nuclide to disintegrate. The half-lives of the elements range from a fraction of a second to billions of years. For example, $_{92}^{238}U$ has a half-life of 4.5×10^9 years, $_{88}^{226}Ra$ has a half-life of 1620 years, and $_6^{15}C$ has a half-life of 2.4 seconds. To illustrate, if we start today with 1.0 g of $_{88}^{226}Ra$, we will have 0.50 g of $_{88}^{226}Ra$ remaining at the end of 1620 years; at the end of another 1620 years, 0.25 g will remain; and so on.

$$1.0 \text{ g } _{88}^{226}Ra \xrightarrow[\text{1620 years}]{t_{1/2}} 0.50 \text{ g } _{88}^{226}Ra \xrightarrow[\text{1620 years}]{t_{1/2}} 0.25 \text{ g } _{88}^{226}Ra$$

The half-lives of the various radioisotopes of the same element are different from one another. Half-lives for some isotopes of radium, carbon, and uranium are listed in Table 19.2. A radioactive decay curve is illustrated in Figure 19.1.

TABLE 19.2

Half-Lives for Radium, Carbon, and Uranium Isotopes

Isotope	Half-life	Isotope	Half-life
Ra-223	11.7 days	C-14	5668 years
Ra-224	3.64 days	C-15	2.4 seconds
Ra-225	14.8 days	U-235	7.1×10^8 years
Ra-226	1620 years	U-238	4.5×10^9 years
Ra-228	6.7 years		

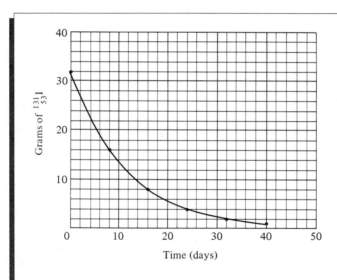

FIGURE 19.1

Radioactive decay curve for $^{131}_{53}I$, which has a half-life of 8 days:

$$^{131}_{53}I \longrightarrow {}^{131}_{54}Xe + {}^{0}_{-1}e$$

EXAMPLE 19.1   The half-life of $^{131}_{53}I$ is 8 days. How much ^{131}I will be left of a 32-g sample after five half-lives?

This problem can be solved by using the graph in Figure 19.1. To find the number of grams of ^{131}I left after one half-life, trace a perpendicular line from 8 days on the x axis to the line on the graph. Now trace a horizontal line from this point on the plotted line to the y axis and read the corresponding grams of ^{131}I. This process continues for each half-life, adding 8 days to the previous value on the x axis.

Half-lives	0	1	2	3	4	5
Number of days		8	16	24	32	40
Amount remaining	32 g	16 g	8 g	4 g	2 g	1 g

Starting with 32 g, 1 g of ^{131}I will be left after five half-lives (40 days).

EXAMPLE 19.2   In how many half-lives will 10.0 g of a radioactive nuclide decay to less than 10% of its original value?

Ten percent of the original amount is 1.0 g. After the first half-life, half of the original material remains and half has decayed. After the second half-life, one fourth of the original material remains, which is one-half of the starting amount at the end of the first half-life. This progression continues, reducing the quantity remaining by half for each half-life that passes.

Half-lives	0	1	2	3	4
Percent remaining	100%	50%	25%	12.5%	6.25%
Amount remaining	10.0 g	5.00 g	2.50 g	1.25 g	0.625 g

Therefore, the amount remaining will be less than 10% sometime between the third and the fourth half-lives.

PRACTICE   The half-life of $^{14}_{6}C$ is 5668 years. How much $^{14}_{6}C$ will remain after six half-lives in a sample that initially contains 25.0 g?
Answer: 0.391 g

Nuclides are said to be either *stable* (nonradioactive) or *unstable* (radio-active). All elements that have atomic numbers greater than 83 (bismuth) are naturally radioactive, although some of the nuclides have extremely long half-lives. Some of the naturally occurring nuclides of elements 81, 82, and 83 are radioactive, and some are stable. Only a few naturally occurring elements that have atomic numbers less than 81 are radioactive. However, no stable isotopes of element 43 (technetium) or of element 61 (promethium) are known.

Radioactivity is believed to be a result of an unstable ratio of neutrons to protons in the nucleus. Stable nuclides of elements up to about atomic number 20 generally have about a 1:1 neutron to proton ratio. In elements above number 20 the neutron to proton ratio in the stable nuclides gradually increases to about 1.5:1 in element number 83 (bismuth). When the neutron to proton ratio is too high or too low, alpha, beta, or other particles are emitted to achieve a more stable nucleus.

19.3   Properties of Alpha Particles, Beta Particles, and Gamma Rays

The classical experiment proving that alpha and beta particles are oppositely charged was performed by Marie Curie (see Figure 19.2). She placed a radio-active source in a hole in a lead block and positioned two poles of a strong electromagnet so that the radiations that were given off passed between them. The paths of three different kinds of radiation were detected by means of a photographic plate placed some distance beyond the electromagnet. The lighter beta particles were strongly deflected toward the positive pole of the electromagnet; the heavier alpha particles were less strongly deflected and in the opposite direction. The uncharged gamma rays were not affected by the electromagnet and struck the photographic plates after traveling along a path straight out of the lead block.

alpha particle

Alpha Particle   An **alpha particle** consists of two protons and two neutrons, has a mass of about 4 amu, a charge of $+2$, and is considered to be a doubly charged helium atom. It is usually given one of the following symbols: α, He^{2+}, or ^4_2He. When an alpha particle is emitted from the nucleus, a different element is formed. The atomic number of the new element is 2 less and the mass is 4 amu less than that of the starting element

Loss of an alpha particle from the nucleus results in
   loss of 4 in the mass number (A)
   loss of 2 in the atomic number (Z)

For example, when $^{238}_{92}U$ loses an alpha particle, $^{234}_{90}Th$ is formed, because two neutrons and two protons are lost from the uranium nucleus. This disintegration may be written as an equation:

$$^{238}_{92}U \longrightarrow {}^{234}_{90}Th + \alpha \qquad \text{or} \qquad {}^{238}_{92}U \longrightarrow {}^{234}_{90}Th + {}^4_2He$$

For the loss of an alpha particle from $^{226}_{88}Ra$, the equation is

$$^{226}_{88}Ra \longrightarrow {}^{222}_{86}Rn + {}^4_2He \qquad \text{or} \qquad {}^{226}_{88}Ra \longrightarrow {}^{222}_{86}Rn + \alpha$$

A nuclear equation, like a chemical equation, consists of reactants and products and must be balanced. To have a balanced nuclear equation the sum of the mass numbers (superscripts) on both sides of the equation must be equal, and

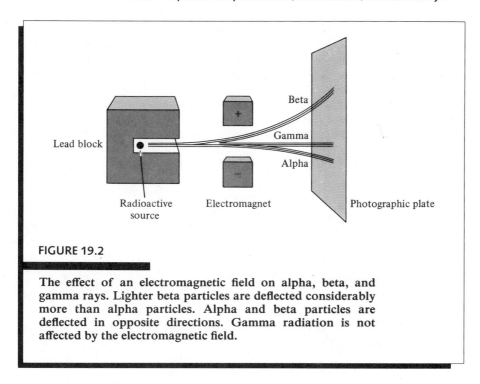

FIGURE 19.2

The effect of an electromagnetic field on alpha, beta, and gamma rays. Lighter beta particles are deflected considerably more than alpha particles. Alpha and beta particles are deflected in opposite directions. Gamma radiation is not affected by the electromagnetic field.

the sum of the atomic numbers (subscripts) on both sides of the equation must be equal.

Sum of mass numbers equals 226

$$^{226}_{88}\text{Ra} \longrightarrow \text{}^{222}_{86}\text{Rn} + \text{}^{4}_{2}\text{He}$$

Sum of atomic numbers equals 88

What new nuclide will be formed when $^{230}_{90}\text{Th}$ loses an alpha particle? This loss is equivalent to two protons and two neutrons. The new nuclide will have a mass of $(230 - 4)$ or 226 amu and will contain $(90 - 2)$ or 88 protons, so its atomic number is 88. Locate the corresponding element on the periodic chart. It is $^{226}_{88}\text{Ra}$ or radium-226.

beta particle

Beta Particle   The **beta particle** is identical in mass and charge to an electron; its charge is -1. Both a beta particle and a proton are produced by the decomposition of a neutron.

$$^{1}_{0}\text{n} \longrightarrow \text{}^{1}_{1}\text{p} + \text{}^{0}_{-1}\text{e}$$

The beta particle leaves, and the proton remains in the nucleus. When an atom loses a beta particle from its nucleus, a different element is formed that has essentially the same mass but an atomic number that is 1 greater than that of the starting element. The beta particle is written as β or $_{-1}^{0}e$.

Loss of a beta particle from the nucleus results in
no change in the mass number (A)
increase of 1 in the atomic number (Z)

Examples of equations in which a beta particle is lost are

$$_{90}^{234}\text{Th} \longrightarrow {}_{91}^{234}\text{Pa} + \beta$$
$$_{91}^{234}\text{Pa} \longrightarrow {}_{92}^{234}\text{U} + {}_{-1}^{0}e$$
$$_{82}^{210}\text{Pb} \longrightarrow {}_{83}^{210}\text{Bi} + \beta$$

gamma ray

Gamma Ray   **Gamma rays** are photons of energy. A gamma ray is similar to an X ray, but is more energetic. They have no electrical charge and no measurable mass. Gamma rays emanate from the nucleus in many radioactive changes along with either alpha or beta particles. The designation for a gamma ray is γ. Gamma radiation does not result in a change of atomic number or the mass of an element.

Loss of a gamma ray from the nucleus results in
no change in mass number or atomic number

EXAMPLE 19.3

(a) Write an equation for the loss of an alpha particle from the nuclide $_{78}^{194}\text{Pt}$.
(b) What nuclide is formed when $_{88}^{228}\text{Ra}$ loses a beta particle from its nucleus?

(a) Loss of an alpha particle, $_{2}^{4}\text{He}$, means the loss of two neutrons and two protons. This change results in a decrease of 4 in the mass number and a decrease of 2 in the atomic number.

Mass of new nuclide: $A - 4$ or $194 - 4 = 190$

Atomic number of new nuclide: $Z - 2$ or $78 - 2 = 76$

Looking up element number 76 on the periodic table, we find it to be osmium, Os. The equation, then, is

$$_{78}^{194}\text{Pt} \longrightarrow {}_{76}^{190}\text{Os} + {}_{2}^{4}\text{He}$$

(b) The loss of a beta particle from a $_{88}^{228}\text{Ra}$ nucleus means a gain of 1 in the atomic number with no essential change in mass.
The new nuclide will have an atomic number of $(Z + 1)$ or 89, which is actinium, Ac. The nuclide formed is $_{89}^{228}\text{Ac}$.

$$_{88}^{228}\text{Ra} \longrightarrow {}_{89}^{228}\text{Ac} + {}_{-1}^{0}e$$

EXAMPLE 19.4   What nuclide will be formed when $^{214}_{82}Pb$ successively emits β, β, and α particles from its nucleus? Write successive equations showing these changes.

The changes brought about in the three-steps outlined are as follows:

β loss: Increase of 1 in the atomic number; no change in mass

β loss: Increase of 1 in the atomic number; no change in mass

α loss: Decrease of 2 in the atomic number; decrease of 4 in the mass

The equations are

$$^{214}_{82}Pb \xrightarrow{-\beta} {}^{214}_{83}X \xrightarrow{-\beta} {}^{214}_{84}X \xrightarrow{-\alpha} {}^{210}_{82}X$$

where X stands for the new nuclide formed. Looking up each of these elements by their atomic numbers, we rewrite the equations

$$^{214}_{82}Pb \xrightarrow{-\beta} {}^{214}_{83}Bi \xrightarrow{-\beta} {}^{214}_{84}Po \xrightarrow{-\alpha} {}^{210}_{82}Pb$$

PRACTICE   What nuclide will be formed when $^{230}_{90}Th$ emits an alpha particle?

Answer: $^{226}_{88}Ra$

The ability of radioactive rays to pass through various objects is in proportion to the speed at which they leave the nucleus. Gamma rays travel at the velocity of light (186,000 miles per second) and are capable of penetrating several inches of lead. The velocities of beta particles are variable, the fastest being about nine-tenths the velocity of light. Alpha particles have velocities less than one-tenth the velocity of light. Figure 19.3 illustrates the relative penetrating power of these rays. A few sheets of paper will stop alpha particles; a thin sheet of aluminum will stop both alpha and beta particles; and a 5 cm block of lead will reduce, but not completely stop, gamma radiation. In fact it is difficult to stop all gamma radiation. Table 19.3 summarizes the properties of alpha, beta, and gamma radiation.

19.4 Radioactive Disintegration Series

The naturally occurring radioactive elements with a higher atomic number than lead (Pb) fall into three orderly disintegration series. Each series proceeds from one element to the next by the loss of either an alpha or a beta particle, finally ending in a nonradioactive nuclide. The uranium series starts with $^{238}_{92}U$ and ends with $^{206}_{82}Pb$. The thorium series starts with $^{232}_{90}Th$ and ends with $^{208}_{82}Pb$. The actinium series starts with $^{235}_{92}U$ and ends with $^{207}_{82}Pb$. A fourth series, the

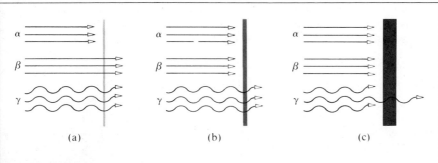

FIGURE 19.3

Relative penetrating ability of alpha, beta, and gamma radiation. (a) Thin sheet of paper; (b) thin sheet of aluminum; (c) 5 cm lead block.

TABLE 19.3

Characteristics of Nuclear Radiation

Radiation	Symbol	Mass (amu)	Electrical charge	Velocity	Composition	Ionizing power
Alpha	α ^4_2He	4	+2	Variable, less than 10% the speed of light	Identical to He^{2+}	High
Beta	β $^0_{-1}e$	$\dfrac{1}{1837}$	−1	Variable, up to 90% the speed of light	Identical to an electron	Moderate
Gamma	γ	0	0	Speed of light	Photons or electromagnetic waves of energy	Almost none

neptunium series, starts with the synthetic element $^{241}_{94}Pu$ and ends with the stable bismuth nuclide $^{209}_{83}Bi$. The uranium series is shown in Figure 19.4; gamma radiation, which accompanies alpha and beta radiation, is not shown in the figure.

By using such a series and the half-lives of its members, scientists have been able to approximate the age of certain geologic deposits. This approximation was done by comparing the amount of $^{238}_{92}U$ with the amount of $^{206}_{82}Pb$ and other nuclides in the series that are present in a particular geologic formation. Rocks

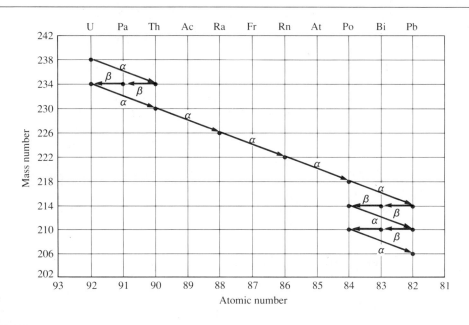

FIGURE 19.4

The uranium disintegration series. $^{238}_{92}U$ decays by a series of alpha (α) and beta (β) emissions to the stable nuclide $^{206}_{82}Pb$.

found in Canada and Finland have been calculated to be about 3.0×10^9 (3 billion) years old. Some meteorites have been determined to be 4.5×10^9 years old.

PRACTICE   What nuclides are formed when $^{238}_{92}U$ undergoes the following decays?
(a) an alpha particle and a beta particle
(b) 3 alpha particles and 2 beta particles
Answers: (a) $^{234}_{91}Pa$ (b) $^{226}_{88}Ra$

19.5   Transmutation of Elements

transmutation

Transmutation is the conversion of one element into another by either natural or artificial means. Transmutation occurs spontaneously in natural radioactive disintegrations. Alchemists tried for centuries to convert lead and mercury into gold by artificial means. But transmutation by artifical means was not achieved until 1919, when Ernest Rutherford succeeded in bombarding the nuclei of nitrogen atoms with alpha particles and produced oxygen nuclides and protons.

The nuclear equation for this transmutation can be written as

$$^{14}_{7}N + \alpha \longrightarrow \, ^{17}_{8}O + \, ^{1}_{1}H \quad \text{or} \quad ^{14}_{7}N + \, ^{4}_{2}He \longrightarrow \, ^{17}_{8}O + \, ^{1}_{1}H$$

It is believed that the alpha particle enters the nitrogen nucleus, forming $^{18}_{9}F$ as an intermediate, which then decomposes into the products

Rutherford's experiments opened the door to nuclear transmutations of all kinds. Atoms were bombarded by alpha particles, neutrons, protons, deuterons ($^{2}_{1}H$), electrons, and so forth. Massive instruments were developed for accelerating these particles to very high speeds and energies to aid their penetration of the nucleus. Some of these instruments are the famous cyclotron, developed by E. O. Lawrence at the University of California; the Van de Graaf electrostatic generator; the betatron; and the electron and proton synchrotrons. With these instruments many nuclear transmutations became possible. Equations for a few of these follows.

$$^{7}_{3}Li + \, ^{1}_{1}H \longrightarrow 2\, ^{4}_{2}He$$

$$^{40}_{18}Ar + \, ^{1}_{1}H \longrightarrow \, ^{40}_{19}K + \, ^{1}_{0}n$$

$$^{23}_{11}Na + \, ^{1}_{1}H \longrightarrow \, ^{23}_{12}Mg + \, ^{1}_{0}n$$

$$^{114}_{48}Cd + \, ^{2}_{1}H \longrightarrow \, ^{115}_{48}Cd + \, ^{1}_{1}H$$

$$^{2}_{1}H + \, ^{2}_{1}H \longrightarrow \, ^{3}_{1}H + \, ^{1}_{1}H$$

$$^{209}_{83}Bi + \, ^{2}_{1}H \longrightarrow \, ^{210}_{84}Po + \, ^{1}_{0}n$$

$$^{16}_{8}O + \, ^{1}_{0}n \longrightarrow \, ^{13}_{6}C + \, ^{4}_{2}He$$

$$^{238}_{92}U + \, ^{12}_{6}C \longrightarrow \, ^{244}_{98}Cf + 6\, ^{1}_{0}n$$

A shorthand notation for nuclear bombardment reactions includes, in order, the nuclide being bombarded, followed by the bombarding particle, a comma, and the particle released all in parentheses, and finally the nuclide produced. The shorthand notation for the last two equations above are

$$^{16}_{8}O(n, \alpha)^{13}_{6}C \quad ^{238}_{92}U(^{12}_{6}C, 6n)^{244}_{98}Cf$$

Note that the reactants are to the left and products to the right of the comma.

PRACTICE   Use shorthand notation to represent the following nuclear equation:

$$^{209}_{83}Bi + \, ^{2}_{1}H \longrightarrow \, ^{210}_{84}Po + \, ^{1}_{0}n$$

Answer: $^{209}_{83}Bi(^{2}_{1}H, \, ^{1}_{0}n)^{210}_{84}Po$

19.6   Artificial Radioactivity

Irene Joliot-Curie, a daughter of Pierre and Marie Curie, and her husband Frederic Joliot-Curie observed that when aluminum-27 was bombarded with alpha particles, neutrons and positrons (positive electrons) were emitted as part of the products. When the source of alpha particles was removed, neutrons ceased to be produced, but positrons continued to be emitted. This obervation suggested that the neutrons and positrons were coming from two separate reactions. It also indicated that one of the products of the first reaction was radioactive. After further investigation they discovered that, when aluminum-27 is bombarded with alpha particles, phosphorus-30 and neutrons are produced. Phosphorus-30 is radioactive, has a half-life of 2.5 minutes, and decays to silicon-30 with the emission of a positron. The equations for these reactions follow.

$$^{27}_{13}\text{Al} + {}^{4}_{2}\text{He} \longrightarrow {}^{30}_{15}\text{P} + {}^{1}_{0}\text{n}$$
$$^{30}_{15}\text{P} \longrightarrow {}^{30}_{14}\text{Si} + {}^{0}_{+1}\text{e}$$

**artificial
radioactivity**

**induced
radioactivity**

The radioactivity of nuclides produced in this manner is known as **artificial radioactivity** or **induced radioactivity**. Artificial radionuclides behave like natural radioactive elements in two ways: they disintegrate in a definite fashion and they have a specific half-life. The Joliot-Curies received the Nobel Prize in chemistry in 1935 for the discovery of artificial, or induced, radioactivity.

19.7 Measurement of Radioactivity

Radiation from radioactive sources is so energetic that it is called *ionizing radiation*. When it strikes an atom or a molecule, one or more electrons are knocked off, and an ion is created. One of the common instruments used to detect and measure radioactivity, the Geiger counter, depends on this fact. The instrument consists of a Geiger–Müller detecting tube and a counting device. The detector tube is a pair of oppositely charged electrodes in an argon gas–filled chamber fitted with a thin window. When radiation, such as a beta particle, passes through the window into the tube, some argon is ionized, and a momentary pulse of current (discharge) flows between the electrodes. These current pulses are electronically amplified in the counter and appear as signals in the form of audible clicks, flashing lights, meter defections, or numerical readouts (Figure 19.5a).

The amount of radiation that an individual encounters can be measured by a film badge (Figure 19.5(b)). A piece of photographic film in a light proof holder is worn in areas where radiation might be encountered. The silver grains in the film will darken when exposed to radiation. The badges are processed after a predetermined time interval to determine the amount of radiation the wearer has been exposed to.

A scintillation counter is used to measure radioactivity for biomedical application. A scintillator is composed of molecules that emit light when exposed to ionizing radiation. A light sensitive detector counts the flashes and converts them to a numerical readout (Figure 19.5(c)).

The *curie* is the unit used to express the amount of radioactivity produced by

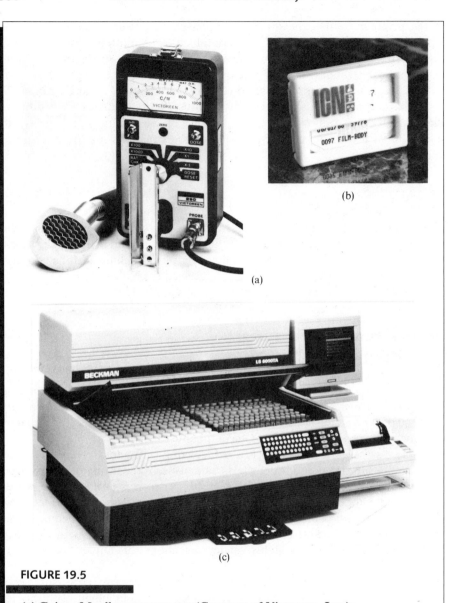

(a)

(b)

(c)

FIGURE 19.5

(a) Geiger-Mueller survey meter (Courtesy of Victoreen, Inc.);
(b) Film badge to measure radioactivity (Courtesy of ICN
Biomedicals, Inc.); (c) Liquid Scintillation counter (Courtesy
of Beckman Instruments, Inc.)

TABLE 19.4

Radiation Units

Curie (Ci)	A unit of radioactivity indicating the rate of decay of a radioactive substance. 1 Ci = 3.7 × 10^{10} disintegrations/sec/g Ra
Roentgen (R)	A unit of exposure of gamma or X radiation based on the quantity of ionization produced in air 1 R = 2.1 × 10^9 ions/cm^3
Rad	A unit of absorbed dose of radiation indicating the energy absorbed from any ionizing radiation. 1 rad = 0.01 J/kg matter
Rem	A unit of radiation dose equivalent. 1 rem is equal to the dose in rads multiplied by a factor dependent on the particular type of radiation. 1 rem = 1 rad × factor
Gray (Gy) (SI unit)	Energy absorbed by tissue. 1 Gy = 1 J/kg tissue (1 Gy = 100 rad)

curie

an element. One **curie** is defined as the quantity of radioactive material giving 3.7 × 10^{10} disintegrations per second. The basis for this figure is pure radium, which has an activity of 1 curie per gram. Because the curie is such a large quantity, the millicurie and microcurie, representing one-thousandth and one-millionth of a curie, respectively, are more practical and more commonly used.

The curie only measures radioactivity emitted by a radionuclide. Different units are required to measure exposure to radiation. The Roentgen (R) is the unit that quantitates exposure to gamma or X rays. One Roentgen is defined as the amount of radiation required to produce 2.1 × 10^9 ions per cm^3 of dry air. The rad (*r*adiation *a*bsorbed *d*ose) is defined as the amount of radiation that provides 0.01 J of energy per kilogram of matter. The amount of radiation absorbed will change depending on the type of matter. The Roentgen and the rad are numerically similar. One Roentgen of gamma radiation provides 0.92 rad in bone tissue.

Neither the rad nor the roentgen indicates the biological damage caused by radiation. One rad of alpha particles has the ability to cause ten times more damage than one rad of gamma rays or beta particles. Another unit, rem (*r*oentgen *e*quivalent to *m*an) takes into account the degree of biological effect caused by the type of radiation exposure. One rem is equal to the dose in rads multiplied by a factor specific to the form of radiation. The factor is 10 for alpha particles and 1 for both beta particles and gamma rays. Units of radiation are summarized in Table 19.4.

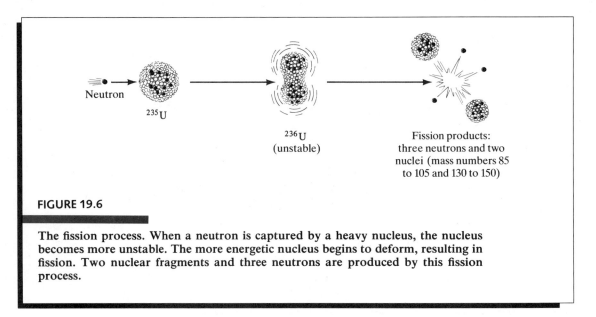

FIGURE 19.6

The fission process. When a neutron is captured by a heavy nucleus, the nucleus becomes more unstable. The more energetic nucleus begins to deform, resulting in fission. Two nuclear fragments and three neutrons are produced by this fission process.

19.8   Nuclear Fission

nuclear fission

In **nuclear fission** a heavy nuclide splits into two or more intermediate-sized fragments when struck in a particular way by a neutron. The fragments are called *fission products*. As the atom splits, it releases energy and two or three neutrons, each of which can cause another nuclear fission. The first instance of nuclear fission was reported in January 1939 by the German scientists Otto Hahn and F. Strassmann. Detecting isotopes of barium, krypton, cerium, and lanthanum after bombarding uranium with neutrons, the scientists were led to believe that the uranium nucleus had been split.

Characteristics of nuclear fission are

1. Upon absorption of a neutron, a heavy nuclide splits into two or more smaller nuclides (fission products).
2. The mass of the nuclides formed range from about 70 to 160 amu.
3. Two or more neutrons are produced from the fission of each atom.
4. Large quantities of energy are produced as a result of the conversion of a small amount of mass into energy.
5. Most nuclides produced are radioactive and continue to decay until they reach a stable nucleus.

One suggested process by which this fission takes place is illustrated in Figure 19.6. When a heavy nucleus captures a neutron, the energy increase may be sufficient to cause deformation of the nucleus until the mass finally splits into two fragments, releasing energy and usually two or more neutrons.

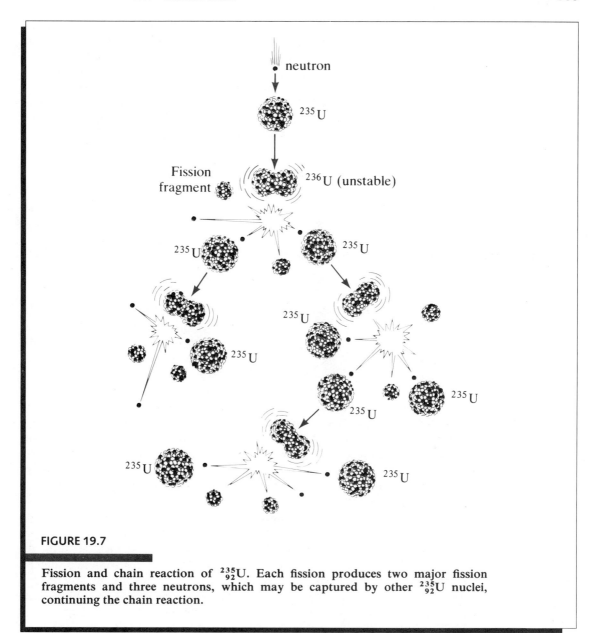

FIGURE 19.7

Fission and chain reaction of $^{235}_{92}$U. Each fission produces two major fission fragments and three neutrons, which may be captured by other $^{235}_{92}$U nuclei, continuing the chain reaction.

In a typical fission reaction, a $^{235}_{92}$U nucleus captures a neutron and forms unstable $^{236}_{92}$U. This $^{236}_{92}$U nucleus undergoes fission, quickly disintegrating into two fragments, such as $^{139}_{56}$Ba and $^{94}_{36}$Kr, and three neutrons. The three neutrons in turn may be captured by three other $^{235}_{92}$U atoms, each of which undergoes fission, producing nine neutrons, and so on. A reaction of this kind, in which the

chain reaction

critical mass

products cause the reaction to continue or magnify, is known as a **chain reaction**. For a chain reaction to continue, enough fissionable material must be present so that each atomic fission causes, on the average, at least one additional fission. The minimum quantity of an element needed to support a self-sustaining chain reaction is called the **critical mass**. Since energy is released in each atomic fission, chain reactions constitute a possible source of a steady supply of energy. A chain reaction is illustrated in Figure 19.7. Two of the many possible ways in which $^{235}_{92}U$ may fission are shown by the following equations.

$$^{235}_{92}U + {}^{1}_{0}n \longrightarrow {}^{139}_{56}Ba + {}^{94}_{36}Kr + 3\,{}^{1}_{0}n$$

$$^{235}_{92}U + {}^{1}_{0}n \longrightarrow {}^{144}_{54}Xe + {}^{90}_{38}Sr + 2\,{}^{1}_{0}n$$

19.9   Nuclear Power

Nearly all electricity for commercial use is produced by machines consisting of a turbine linked by a drive shaft to an electrical generator. The energy required to run the turbine can be supplied by falling water, as in hydroelectric power plants, or by steam generated by heat from fuel, as in thermal power plants. Thermal power plants burn fossil fuel—coal, oil, or natural gas.

The world's demand for energy, largely from fossil fuels, has continued to grow at an ever-increasing rate for about 250 years. Even at present rates of consumption, the estimated world supply of fossil fuels is sufficient for only a few centuries. Although the United States has large coal and oil shale deposits, it is currently importing about 40% of its oil supply. We clearly need to develop alternative energy sources. At present uranium is the most productive alternative energy source, and about 12% of the electrical energy used in the United States is generated in power plants using uranium fuel.

A nuclear power plant is a thermal power plant in which heat is produced by a nuclear reactor instead of by combustion of fossil fuel. The major components of a nuclear reactor are (1) an arrangement of nuclear fuel, called the reactor core; (2) a control system, which regulates the rate of fission and thereby the rate of heat generation; and (3) a cooling system, which removes the heat from the reactor and also keeps the core at the proper temperature. One type of reactor uses metal slugs containing uranium enriched from the normal 0.7% U-235 to about 3% U-235. The self-sustaining fission reaction is moderated, or controlled, by adjustable control rods containing substances that slow down and capture some of the neutrons produced. Ordinary water, heavy water, and molten sodium are typical coolants used. Energy obtained from nuclear reactions in the form of heat is used in the production of steam to drive turbines for generating electricity. (See Figure 19.8.)

Two events that demonstrate the potential dangers of nuclear power are the accidents at Three Mile Island, Pennsylvania (1979) and Chernobyl, U.S.S.R.

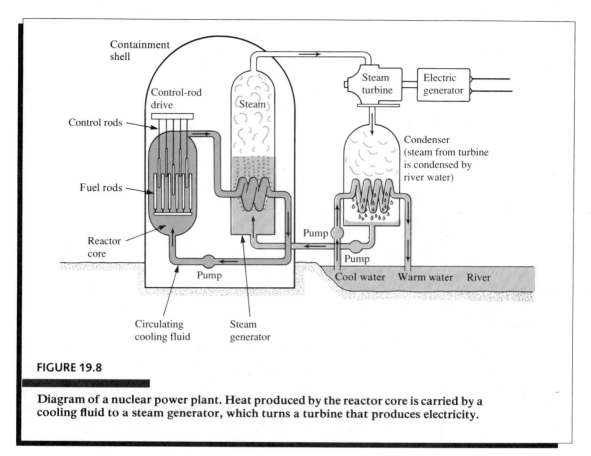

FIGURE 19.8

Diagram of a nuclear power plant. Heat produced by the reactor core is carried by a cooling fluid to a steam generator, which turns a turbine that produces electricity.

(1986). Both of these nuclear accidents resulted from the loss of coolant to the reactor core. The reactors at Three Mile Island are covered by concrete containment buildings and therefore released a relatively small amount of radioactive material into the atmosphere. Because the Soviet Union does not require containment structures on nuclear power plants, the Chernobyl accident resulted in 31 deaths and the resettlement of 135,000 people. The release of large quantities of I-131, Cs-134 and Cs-137 may cause long-term health problems in that exposed population.

Another major disadvantage of nuclear-fueled power plants is that they produce highly radioactive waste products, some of which have half-lives of thousands of years. As yet, no agreement has been reached on how to safely dispose of these dangerous wastes.

In the United States reactors designed for commercial power production use uranium oxide, U_3O_8, that is enriched with the relatively scarce fissionable U-235 isotope. Because the supply of U-235 is limited, a new type of reactor known as the *breeder reactor* has been developed. Breeder reactors are designed to produce additional fissionable material at the same time that the fission reaction is occurring. In a breeder reactor, excess neutrons convert non-fissionable isotopes, such as U-238 or Th-232, to the fissionable isotopes, Pu-239 or U-233.

$$^{238}_{92}\text{U} + {}^{1}_{0}\text{n} \longrightarrow {}^{239}_{92}\text{U} \xrightarrow{-\beta} {}^{239}_{93}\text{Np} \xrightarrow{-\beta} {}^{239}_{94}\text{Pu}$$

$$^{232}_{90}\text{Th} + {}^{1}_{0}\text{n} \longrightarrow {}^{233}_{90}\text{Th} \xrightarrow{-\beta} {}^{233}_{91}\text{Pa} \xrightarrow{-\beta} {}^{233}_{92}\text{U}$$

These transmutations make it possible to greatly extend the supply of fuel for nuclear reactors. No breeder reactors are presently in commercial operation in the United States, but several of them are being operated in Europe and Great Britain.

19.10   The Atomic Bomb

The atomic bomb is a fission bomb; it operates on the principle of a very fast chain reaction that releases a tremendous amount of energy. An atomic bomb and a nuclear reactor both depend on self-sustaining nuclear fission chain reactions. The essential difference is that in a bomb the fission is "wild," or uncontrolled, whereas in a nuclear reactor the fission is moderated and carefully controlled. A minimum critical mass of fissionable material is needed for a bomb, or a major explosion will not occur. When a quantity smaller than the critical mass is used, too many neutrons formed in the fission step escape without combining with another nucleus, and a chain reaction does not occur. Therefore, the fissionable material of an atomic bomb must be stored as two or more subcritical masses and brought together to form the critical mass at the desired time of explosion. The temperature developed in an atomic bomb is believed to be about 10 million degrees Celsius.

The nuclides used in atomic bombs are U-235 and Pu-239. Uranium deposits contain about 0.7% of the U-235 isotope, the remainder being U-238. Uranium-238 does not undergo fission except with very-high-energy neutrons. It was discovered, however, that U-238 captures a low-energy neutron without undergoing fission and that the product, U-239 changes to Pu-239 (plutonium) by a beta decay process. Plutonium-239 readily undergoes fission upon capture of a neutron and is therefore useful for nuclear weapons. The equations for the nuclear transformations are

$$^{238}_{92}\text{U} + {}^{1}_{0}\text{n} \longrightarrow {}^{239}_{92}\text{U} \underset{\searrow\beta}{\longrightarrow} {}^{239}_{93}\text{Np} \underset{\searrow\beta}{\longrightarrow} {}^{239}_{94}\text{Pu}$$

The hazards of an atomic bomb explosion include not only shock waves from the explosive pressure and tremendous heat, but also intense radiation in the form of alpha, beta, gamma, and ultraviolet rays. Gamma rays and X rays can penetrate deeply into the body, causing burns, sterilization, and mutation of the genes, which may have an adverse effect on future generations. Both radioactive fission products and unfissioned material are present after the explosion. If the bomb explodes near the ground, many tons of dust are lifted into the air. Radioactive material adhering to this dust, known as *fallout*, is spread by air currents over wide areas of the land and constitutes a lingering source of radiation hazard.

Today nuclear war is probably the most awesome threat facing civilization. Only two rather primitive fission-type atom bombs were used to destroy the Japanese cities of Hiroshima and Nagasaki and bring World War II to an early end. Now tens of thousands of nuclear weapons are in existence, and hundreds more are being manufactured each year. Many of these weapons are at least several hundred times more powerful than the Hiroshima and Nagasaki bombs. The United States and the Soviet Union, the so-called superpowers, have made considerable progress toward reaching an effective agreement on the control and reduction of nuclear armaments. The threat of nuclear war is increased by the fact that the number of nations possessing nuclear weapons is steadily increasing.

19.11   Nuclear Fusion

nuclear fusion

The process of uniting the nuclei of two light elements to form one heavier nucleus is known as **nuclear fusion**. Such reactions can be used for producing energy, because the mass of the two individual nuclei that are fused into a single nucleus is greater than the mass of the nucleus formed by their fusion. The mass differential is liberated in the form of energy. Fusion reactions are responsible for the tremendous energy output of the sun. Thus, aside from relatively small amounts from nuclear fission and radioactivity, fusion reactions are the ultimate source of our energy, even the energy from fossil fuels. They are also responsible for the devastating power of the thermonuclear, or hydrogen, bomb.

Fusion reactions require temperatures on the order of tens of millions of degrees for initiation. Such temperatures are present in the sun but have been produced only momentarily on earth. For example, the hydrogen, or fusion, bomb is triggered by the temperature of an exploding fission bomb.

Two typical fusion reactions are

$$\overset{3}{_1}\text{H} \; + \; \overset{2}{_1}\text{H} \; \longrightarrow \; \overset{4}{_2}\text{He} + \overset{1}{_0}\text{n} + \text{energy}$$

Tritium   Deuterium

$$\underset{\substack{3.01495 \\ \text{amu}}}{\overset{3}{_1}\text{H}} \; + \; \underset{\substack{1.00782 \\ \text{amu}}}{\overset{1}{_1}\text{H}} \; \longrightarrow \; \underset{\substack{4.00260 \\ \text{amu}}}{\overset{4}{_2}\text{He}} \; + \text{energy}$$

The total mass of the reactants in the second equation is 4.02277 amu, which is 0.02017 amu greater than the mass of the product. This difference in mass is manifested in the great amount of energy liberated.

During the past 40 to 45 years, a large amount of research on controlled nuclear fusion reactions has been done in the United States and in other countries, especially the Soviet Union. So far, the goal of controlled nuclear fusion has not been attained, although the required ignition temperature has been reached in several devices. Evidence to date leads us to believe that it is possible to develop a practical fusion power reactor. Fusion power, if it can be developed, will be far superior to fission power because: (1) Virtually infinite amounts of energy are to be had from fusion. Uranium supplies for fission power are limited, but heavy hydrogen, or deuterium (the most likely fusion fuel), is relatively abundant. It is estimated that the deuterium in a cubic mile of seawater is an energy resource (as fusion fuel) greater than the petroleum reserves of the entire world. (2) From an environmental viewpoint, fusion power is much "cleaner" than fission power because fusion reactions, in contrast to uranium and plutonium fission reactions, do not produce large amounts of long-lived and dangerously radioactive isotopes.

19.12   Mass-Energy Relationship in Nuclear Reactions

Because large amounts of energy are released in nuclear reactions, significant amounts of mass are converted to energy. We stated earlier that the amount of mass converted to energy in chemical changes is considered insignificant. In fission reactions about 0.1% of the mass is converted into energy. In fusion reactions as much as 0.5% of the mass may be changed into energy. The Einstein equation, $E = mc^2$, can be used to calculate the energy liberated, or available, when the mass loss is known. For example, in the reaction

$$\overset{7}{_3}\text{Li} \ + \ \overset{1}{_1}\text{H} \ \longrightarrow \ \overset{4}{_2}\text{He} \ + \ \overset{4}{_2}\text{He} \ + \text{energy}$$

$$\text{7.0160 g} \qquad \text{1.0078 g} \qquad \text{4.0026 g} \qquad \text{4.0026 g}$$

the mass difference between the reactants and the products (8.0238 g − 8.0052 g) is 0.0186 g. The energy equivalent to this amount of mass is 1.67×10^{12} J (4.0×10^{11} cal). By comparison this amount is more than 4 million times greater than the 3.9×10^5 J (9.4×10^4 cal) of energy obtained from the complete combustion of 12.0 g (1 mol) of carbon.

The mass of a nucleus is actually less than the sum of the masses of the protons and neutrons that make up that nucleus. The difference between the mass of the protons and the neutrons in a nucleus and the mass of the nucleus is known **mass defect** as the **mass defect**. The energy equivalent to this difference in mass is known as the **nuclear binding** **nuclear binding energy**. This energy is the amount that would be required to break **energy** up a nucleus into its individual protons and neutrons. The higher the binding energy, the more stable is the nucleus. Elements of intermediate atomic masses have high binding energies. For example, iron (element number 26) has a very high binding energy and therefore has a very stable nucleus. Just as electrons attain less energetic and more stable arrangements through ordinary chemical reactions, neutrons and protons attain less energetic and more stable arrangements through nuclear fission or fusion reactions. Thus, when uranium undergoes fission, the products have less mass (and greater binding energy) than the original uranium. In like manner, when hydrogen and lithium fuse to form helium, the helium has less mass (and greater binding energy) than the hydrogen and lithium. It is this conversion of mass to energy that accounts for the very large amounts of energy associated with both nuclear fission and fusion reactions.

19.13   Transuranium Elements

transuranium
elements

The elements following uranium on the periodic chart and having atomic numbers greater than 92 are known as the **transuranium elements**. All of them are synthetic radioactive elements; none of them occur naturally.

The first transuranium element, number 93, was discovered in 1939 by Edwin M. McMillan (b. 1907) at the University of California while he was investigating the fission of uranium. He named it neptunium for the planet Neptune. In 1941 element 94, plutonium, was identified as a beta decay product of neptunium:

$$^{238}_{93}Np \longrightarrow {}^{238}_{94}Pu + {}^{0}_{-1}e$$
$$^{239}_{93}Np \longrightarrow {}^{239}_{94}Pu + {}^{0}_{-1}e$$

Plutonium is one of the most important fissionable elements known today.

Since 1964 the discoveries of six new transuranium elements, numbers 104–109, have been announced. All of these elements were produced in minute quantities by high energy particle accelerators. In the case of element 109, the discovery was based on the detection of a single atom of $^{266}_{109}X$, which existed for about five milliseconds! The reported synthesis was achieved by bombarding a $^{209}_{83}Bi$ target with highly accelerated iron-58 nuclei.

$$^{209}_{83}Bi + {}^{58}_{26}Fe \longrightarrow {}^{266}_{109}X + {}^{1}_{0}n$$

Groups of scientists in the United States and the Soviet Union involved in high energy particle research to produce new elements disagree on the names for elements 104, 105 and 106. Consequently the International Union of Pure and Applied Chemistry (IUPAC) has suggested a systematic method for naming all elements beyond 103. The method combines letters from the Greek and Latin words for the numbers plus the ending *ium*. For example, *un* for 1, *nil* for 0, and *quad* for 4 plus the ending *ium* forms unnilquadium, element 104. Thus unnilpentium is 105 and unnilhexium is 106. Table 19.5 lists all the presently known transuranium elements.

TABLE 19.5

Transuranium Elements

Element	Symbol	Atomic number	Discovery date
Neptunium	Np	93	1939
Plutonium	Pu	94	1941
Americium	Am	95	1944
Curium	Cm	96	1944
Berkelium	Bk	97	1949
Californium	Cf	98	1950
Einsteinium	Es	99	1953
Fermium	Fm	100	1953
Mendelevium	Md	101	1955
Nobelium	No	102	1957
Lawrencium	Lr	103	1961
Unnilquadium	Unq	104	1964
Unnilpentium	Unp	105	1970
Unnilhexium	Unh	106	1974
Unnilseptium	Uns	107	1981
Unniloctium	Uno	108	1988
Unnilennium	Une	109	1982

19.14 Biological Effects of Radiation

Radiation that has enough energy to dislocate bonding electrons and create ions when passing through matter is classified as *ionizing radiation*. Alpha, beta, and gamma rays, along with X rays, fall into this classification. Ionizing radiation can damage or kill living cells. This damage is particularly devastating when it occurs in the cell nuclei and affects molecules involved in cell reproduction. The overall effects of radiation on living organisms fall into these general categories: (1) acute, or short-term, effects; (2) long-term effects, and (3) genetic effects.

Acute Radiation Damage High levels of radiation, especially of gamma or X rays, produce nausea, vomiting, and diarrhea. The effect has been likened to a sunburn throughout the body. If the dosage is high enough, death will occur in a few days. The damaging effects of radiation appear to be centered in the nuclei of the cells, and cells that are undergoing rapid cell division are most susceptible to damage. It is for this reason that cancers are often treated with gamma radiation from a Co-60 source. Cancerous cells are multiplying rapidly and are destroyed by a level of radiation that does not seriously damage normal cells.

Long-Term Radiation Damage   Protracted exposure to low levels of any form of ionizing radiation can weaken the organism and lead to the onset of malignant tumors, even after fairly long time delays. The largest exposure to man-made sources of radiation is from X rays. Evidence suggests that a number of early workers in radioactivity and X-ray technology may have had their lives shortened by long-term radiation damage.

A number of women who had been employed in the early 1920s to paint luminous numbers on watch dials died some years later from the effects of radiation. These women had ingested radium by using their lips to point the brushes used on the job. Radium was retained in their bodies and, as an alpha emitter with a half-life of about 1620 years, continued to inflict radiation damage.

Strontium-90 isotopes are present in the fallout from atmospheric testing of nuclear weapons. Strontium is in the same periodic-table group as calcium, and its chemical behavior is similar to that of calcium. Hence, when foods contaminated with Sr-90 are eaten, Sr-90 ions are laid down in the bone tissue along with ordinary calcium ions. Strontium-90 is a beta emitter with a half-life of 28 years. Blood cells that are manufactured in bone marrow are affected by the radiation from Sr-90. Hence, there is concern that Sr-90 accumulation in the environment may cause an increase in the incidence of leukemia and bone cancers. Fortunately, the United States and the Soviet Union agreed to stop atmospheric testing of nuclear weapons several years ago; however, some countries are still doing testing in the atmosphere.

Genetic Effects   All the information needed to create an individual of a particular species, be it a bacterial cell or a human being, is contained within the nucleus of a cell. This genetic information is encoded in the structure of DNA (deoxyribonucleic acid) molecules, which make up genes. The DNA molecules form precise duplicates of themselves when cells divide, thus passing genetic information from one generation to the next. Radiation can damage DNA molecules. If the damage is not severe enough to prevent the individual from reproducing, a mutation (a heritable variation in the offspring) may result. Most mutation-induced traits are undesirable. Unfortunately, if the bearer of the altered genes survives to reproduce, these traits are passed along to succeeding generations. In other words, the genetic effects of increased radiation exposure are found in future generations, not in the present generation.

Because radioactive rays are hazardous to health and living tissue, special precautions must be taken in designing laboratories and nuclear reactors, in disposing of waste materials, and in monitoring the radiation exposure of people working in this field. For example, personnel working in areas of hazardous radiation wear film badges or pocket dosimeters to provide them with an accurate indication of cumulative radiation exposure.

19.15   Applications of Nuclear Chemistry

To date, the largest uses of radioactive materials have been for making weapons and for the generation of electricity in nuclear power plants. Aside from these major uses, radionuclides have innumerable applications. They are used extensively in chemical, physical, biological, and medical research. Radionuclides now serve in a wide variety of almost routine technological applications in medicine and various branches of industry, including chemical, petroleum, and metallurgical processing. A major advantage of using radionuclides in research is that they continuously emit radiations that can be detected and measured. A few of these applications are briefly described here.

Agriculture   Agricultural research scientists use gamma radiation from Co-60 or other sources to develop disease-resistant and highly productive grains and other crops. The seeds are exposed to gamma radiation to induce mutations. The most healthy and vigorous plants grown from the irradiated seed are then selected and propagated to obtain new and improved varieties for commercial use. Preservation of foodstuffs by radiation is another beneficial application. Food is exposed to either gamma radiation or a beam of beta particles supplied by Co-60 or Cs-137. Microorganisms that can cause food spoilage are destroyed, but the temperature of the food is raised only slightly. The food does not become radioactive as a result of this process, but the shelf-life is extended significantly.

Isotopic Tracers   Compounds containing a radionuclide are described as being *labeled* or *tagged*. These compounds undergo their normal chemical reactions, but their location can be detected because of their radioactivity. When such compounds are given to a plant or an animal, the movement of the nuclide can be traced through the organism by the use of a Geiger counter or other detecting device.

One important use of the tracer technique was in determining the source of molecular oxygen produced by the process of photosynthesis. The net equation for the photosynthetic process is

$$6\,CO_2 + 6\,H_2O \longrightarrow C_6H_{12}O_6 + 6O_2$$

Plants were given water that was tagged with radioactive O-18. The carbon dioxide had stable O-16. The oxygen produced by the plant contained only atoms of O-18, so it was determined that molecular oxygen was derived only from water.

Some other examples of biological research in which tracer techniques have been employed are (1) in determining the rate of phosphate uptake by plants, using radiophosphorus; (2) the utilization of carbon dioxide in photosynthesis, using radioactive carbon; (3) the accumulation of iodine in the thyroid gland, using radioactive iodine; and (4) the absorption of iron by the hemoglobin of the blood, using radioactive iron. In chemistry, the uses are unlimited. The study of reaction mechanisms, the measurement of the rates of chemical reactions, and the determination of physical constants are just a few of the areas of application.

Age Dating An interesting outgrowth of the use of radionuclide techniques is *radiocarbon dating*. The method is based on the decay rate of C-14 and was devised by the American chemist W. F. Libby, who received the Nobel Prize in chemistry in 1960 for this work. The principle of radiocarbon dating is as follows: Carbon dioxide in the atmosphere contains a fixed ratio of radioactive C-14 to ordinary C-12, because C-14 is produced at a steady rate in the atmosphere by bombardment of N-14 by neutrons from cosmic ray sources.

$$^{14}_{7}N + ^{1}_{0}n \longrightarrow ^{14}_{6}C + ^{1}_{1}H$$

Plants that consume carbon dioxide during photosynthesis and animals that eat the plants contain the same proportion of C-14 to C-12 as long as they are alive. When an organism dies, the amount of C-12 remains fixed, but the C-14 content diminishes according to its half-life (5668 years). By comparing the ratio of C-14 to C-12 in an object to the same ratio in living plants, one can estimate the age of the object being evaluated. In 5668 years, one-half the radiocarbon initially present will have undergone decomposition. In 11,336 years, one-fourth of the original C-14 will be left. The age of fossil material, archaeological specimens, and old wood can be determined by this method. The age of specimens from ancient Egyptian tombs calculated by radiocarbon dating correlates closely with the chronological age established by Egyptologists. Charcoal samples obtained at Darrington Walls, a wood-henge in Great Britain, were determined to be about 4000 years old. Radiocarbon dating instruments currently in use enable researchers to date specimens back as far as 70,000 years. This technique was used recently to estimate the age of the shroud of Turin.

Radioactive decay has been used to date samples other than those containing carbon. For example, the age of rock formations containing uranium has been approximated by determining the ratio of U-238 to Pb-206. Lead-206 is the last isotope formed in the U-238 disintegration series. Thus, a geologic deposit containing a 1:1 ratio of U-238 to Pb-206 would correspond to a time lapse of one half-life of U-238, which is 4.5×10^9 years, assuming that all the lead came from the decay of U-238. The age of moon rocks returned to earth by the Apollo missions were calculated by similar techniques.

Insect Control Radioactivity has been used to control and, in some areas, to eliminate the screw-worm fly. The larvae of this obnoxious insect pest burrow into wounds in livestock. The female fly, like a queen bee, mates only once. When large numbers of gamma ray–sterilized male flies are released at the proper time in an area infested with screw-worm flies, the majority of the females mate with sterile males. As a consequence, the flies fail to reproduce sufficiently to maintain their numbers. This technique has also been used to eradicate the Mediterranean fruit fly in some areas.

Diagnosis   Radioactive traces are commonly used in medical diagnosis. Because radiation must be detected outside of the body, radionuclides that emit gamma rays are usually chosen. The radionuclide must also be effective at a low concentration and have a short half-life to reduce the possibility of damage to the patient.

Radioactive iodine (I-131) is used to determine thyroid function, where the body concentrates iodine. In this process, a small amount of radioactive potassium or sodium iodide is ingested. A detector is focused on the thyroid gland and measures the amount of iodine in the gland. This picture can be compared to that of a normal thyroid to detect any differences.

Doctors can examine the heart's pumping performance and check for evidence of obstruction in the coronary arteries by *nuclear scanning*. The radionuclide Tl-201, when rejected into the bloodstream, lodges in healthy heart muscle. Thallium-201 emits gamma radiation, which is detected by a special imaging device called a *scintillation camera*. The data obtained are simultaneously translated into pictures by a computer. With this technique doctors can observe whether heart tissue has died after a heart attack and whether blood is flowing freely through the coronary passages.

One of the newest applications of nuclear chemistry is the use of positron emission tomography (PET) in the measurement of dynamic processes in the body, such as oxygen use or blood flow. In this application a compound is made that contains a positron-emitting nuclide such as C-11, O-15, or N-13. The compound is injected into the body, and the patient is placed in an instrument that detects the positron emissions. A computer produces a three-dimensional image of the area.

PET scans have been used to locate the areas of the brain involved with epileptic seizures. The brain uses glucose almost exclusively for energy. Glucose that has been tagged with C-11 is injected, and an image of the brain is produced. Diseased areas that use glucose at a rate different than normal tissue can then be visualized.

Radiation and Chemotherapy   For many years radium has been used in the treatment of cancer. Co-60 and Cs-137 are now extensively used for radiation therapy. The effectiveness of this therapy is dependent on the fact that rapidly growing or dividing malignant cells are more susceptible to radiation damage than normal cells. Cobalt-60 emits both beta particles and gamma rays. The radiation is focused on the area where the tumor is located, but it is very difficult to limit exposure only to malignant cells. Many patients suffer from radiation sickness following this type of therapy.

Iodine-131 can be used for the treatment of hyperthyroidism. The therapeutic dose is larger than that used for diagnosis. The thyroid gland selectively concentrates the I-131. The section of the gland that is hyperactive will be exposed to a large dose of the isotope and be specifically destroyed. First Lady Barbara Bush went through this procedure in 1989.

Concepts in Review

1. Outline the historical development of nuclear chemistry, including the major contributions of Henri Becquerel, Marie Curie, Ernest Rutherford, Irene Joliet-Curie, Otto Hahn, and Fritz Strassmann.

2. Write balanced nuclear chemical equations using isotopic notation.

3. Determine the amount of radionuclide remaining after a given period of time when the starting amount and half-life are given.

4. List the characteristics that distinguish alpha, beta, and gamma rays from the standpoint of mass, charge, relative velocities, and penetrating power.

5. Describe the effect of a magnetic field on alpha particles, beta particles, and gamma rays.

6. Describe a radioactive disintegration series, and predict which isotope would be formed by the loss of specified numbers of alpha and beta particles from a given radionuclide.

7. Discuss the transmutation of elements.

8. Indicate the methods used for the detection of radiation.

9. Distinguish between radioactive disintegration and nuclear fission reactions.

10. Explain how the fission of U-235 can lead to a chain reaction and why a critical mass is necessary.

11. Explain how the energy from nuclear fission is converted to electrical energy.

12. Explain the difference between fission reactions in a nuclear reactor and those of an atomic bomb.

13. Explain what is meant by the term *nuclear fusion* and why a massive effort to develop controlled nuclear fusion is in progress.

14. Indicate the significance of mass defect and nuclear binding energy.

15. Indicate the major effects of radiation on living organisms.

16. Explain how the age of objects can be determined using radioactivity.

17. Indicate several current uses for radioactive tracers.

Key Terms in Review

The terms listed here have all been defined within the chapter. Review the definitions of each, and use the glossary and margin notations within the chapter as study aids.

alpha particle   nuclear binding energy
artificial radioactivity   nuclear fission
beta particle   nuclear fusion
chain reaction   nucleon
critical mass   nuclide
curie   radioactive decay
gamma ray   radioactivity
half-life   transmutation
induced radioactivity   transuranium elements
mass defect

Exercises

An asterisk indicates a more challenging question or problem.

1. To afford protection from radiation injury, which kind of radiation requires (a) the most shielding? (b) the least shielding?
2. Why is an alpha particle deflected less than a beta particle in passing through an electromagnetic field?
3. Name three pairs of nuclides that might be obtained by fissioning three U-235 atoms.
4. Identify the following people and their associations with the early history of radioactivity.
   (a) Antoine Henri Becquerel
   (b) Marie and Pierre Curie
   (c) Wilhelm Roentgen
   (d) Ernest Rutherford
   (e) Paul Villard
5. Why is the radioactivity of an element unaffected by the usual factors that affect the rate of chemical reactions, such as ordinary changes of temperature and concentration?
6. Distinguish between the terms *isotope* and *nuclide*.

7. Indicate the number of protons, neutrons, and nucleons in each of the following nuclei.
   (a) $^{35}_{17}Cl$ (b) $^{226}_{88}Ra$ (c) $^{235}_{92}U$ (d) $^{82}_{35}Br$
8. Explain the phenomenon of half-life for a radionuclide.
9. The half-life of Pu-244 is 76 million years. If the age of the earth is about 5 billion years, discuss the feasibility of this nuclide's being found as a naturally occurring element.
10. Tell how alpha, beta, and gamma radiation are distinguished from the standpoint of:
    (a) Charge
    (b) Relative mass
    (c) Nature of particle or ray
    (d) Relative penetrating power
11. How are the mass and the atomic number of a nucleus affected by the loss of the following:
    (a) An alpha particle
    (b) A beta particle
12. Distinguish between natural and artificial radioactivity.
13. What is a radioactive disintegration series?
14. Briefly discuss the transmutation of elements.

15. Write nuclear equations for the alpha decay of:
    (a) $^{218}_{85}At$ (b) $^{221}_{87}Fr$ (c) $^{192}_{78}Pt$ (d) $^{210}_{84}Po$

16. Write nuclear equations for the beta decay of:
    (a) $^{14}_{6}C$ (b) $^{137}_{55}Cs$ (c) $^{239}_{93}Np$ (d) $^{90}_{38}Sr$

17. Stable Pb-208 is formed from Th-232 in the thorium disintegration series by successive α, β, β, α, α, α, α, β, β, α particle emissions. Write the symbol (including mass and atomic number) for each nuclide formed in this series.

18. The nuclide Np-237 loses a total of seven alpha particles and four beta particles. What nuclide remains after these losses?

19. Bismuth-211 decays by alpha emission to give a nuclide that in turn decays by beta emission to yield a stable nuclide. Show these two steps with nuclear equations.

20. Write nuclear equations for the following:
    (a) Conversion of $^{13}_{6}C$ to $^{14}_{6}C$
    (b) Conversion of $^{30}_{15}P$ to $^{30}_{14}Si$

21. Complete and balance the following nuclear equations by supplying the missing particles.
    (a) $^{27}_{13}Al + ^{4}_{2}He \longrightarrow ^{30}_{15}P + ?$
    (b) $^{27}_{14}Si \longrightarrow ^{0}_{+1}e + ?$
    (c) $? + ^{2}_{1}H \quad ^{13}_{7}N + ^{1}_{0}n$
    (d) $? \longrightarrow ^{82}_{36}Kr + ^{0}_{-1}e$
    (e) $^{66}_{29}Cu \longrightarrow ^{66}_{30}Zn + ?$
    (f) $^{0}_{-1}e + ? \longrightarrow ^{7}_{3}Li$
    (g) $^{27}_{13}Al + ^{4}_{2}He \longrightarrow ^{30}_{14}Si + ?$
    (h) $^{85}_{37}Rb + ? \longrightarrow ^{82}_{35}Br + ^{4}_{2}He$
    (i) $^{214}_{83}Bi \longrightarrow ^{4}_{2}He + ?$

22. Write shorthand notations for the following nuclear equations.
    (a) $^{27}_{13}Al + ^{4}_{2}He \longrightarrow ^{30}_{15}P + ^{1}_{0}n$
    (b) $^{14}_{7}N + ^{4}_{2}He \longrightarrow ^{17}_{8}O + ^{1}_{1}H$
    (c) $^{59}_{27}Co + ^{1}_{0}n \longrightarrow ^{60}_{27}Co + \gamma$
    (d) $^{249}_{98}Cf + ^{12}_{6}C \longrightarrow ^{257}_{104}Unq + 4\,^{1}_{0}n$

23. Write nuclear equations from these shorthand notations.
    (a) $^{23}_{11}Na(p,n)^{23}_{12}Mg$
    (b) $^{209}_{83}Bi(d,n)^{210}_{84}P_{0}$ (d = deuteron, $^{2}_{1}H$)
    (c) $^{238}_{92}U(^{16}_{8}O, 5\,n)^{249}_{100}Fm$
    (d) $^{9}_{4}Be(\alpha,n)^{12}_{6}C$

24. Describe the use of a Geiger counter in measuring radioactivity.

25. What does each of these units represent?
    (a) Curie   (c) Rad or Gray
    (b) Roentgen   (d) Rem

26. What was the contribution to nuclear physics of Otto Hahn and Fritz Strassmann?

27. What is a breeder reactor? Explain how it accomplishes the "breeding."

28. What is the essential difference between the nuclear reactions in a nuclear reactor and those in an atomic bomb?

29. Why must a certain minimum amount of fissionable material be present before a self-supporting chain reaction can occur?

30. Are the terms *atomic bomb* and *hydrogen bomb* synonymous? If not, what is the major distinction between them?

31. What is mass defect and nuclear binding energy?

32. Explain why radioactive rays are classified as ionizing radiation.

33. Give a brief description of the biological hazards associated with radioactivity.

34. Strontium-90 has been found to occur in radioactive fallout. Why is there so much concern about this radionuclide being found in cow's milk? (Half-life of Sr-90 is 28 years.)

35. What is a radioactive tracer? How is it used?

36. Describe the radiocarbon method for dating archaeological artifacts.

37. How might radioactivity be used to locate a leak in an underground pipe?

38. Anthropologists have found bones whose age suggests that the human line may have emerged in Africa as much as 4 million years ago. If wood or charcoal were found with such bones, would C-14 dating be useful in dating the bones? Explain.

39. Strontium-90 has a half-life of 28 years. If a 1.00 mg sample were stored for 112 years, what weight of Sr-90 would remain?

40. If radium costs $50,000 a gram, how much will 0.0100 g of $^{226}RaCl_2$ cost if the price is based only on the radium content?

41. Strontium-90 has a half-life of 28 years. If a sample was tested in 1980 and found to be emitting 240 counts/minute, in what year would the same sample be found to be emitting 30 counts/minute? How much of the original Sr-90 would be left?

42. An archaeological specimen was analyzed and found to be emitting only 25% as much carbon-14 radiation per gram of carbon as newly cut wood. How old is this specimen?

43. Barium-141 is a beta emitter. What is the half-life if a 16.0 g sample of the nuclide decays to 0.500 g in 90 minutes?

*44. Calculate (a) the mass defect and (b) the binding energy of $_3^7Li$. Mass data: $_3^7Li = 7.0160$ g; n = 1.0087 g; p = 1.0073 g; $e^- = 0.00055$ g; 1.0 g = 9.0×10^{13} J (from $E = mc^2$).

*45. For the fission reaction

$$_{92}^{235}U + _0^1n \longrightarrow _{38}^{94}Sr + _{54}^{139}Xe + 3\,_0^1n + energy$$

Mass data: U-235 = 235.0439; Sr-94 = 93.9154; Xe-139 = 138.9179 amu; n = 1.0087 amu; 1.0 g = 9.0×10^{13} J. Calculate:
   (a) The energy released in joules for a single event (one uranium atom splitting)
   (b) The energy released in joules per mole of uranium splitting
   (c) The percentage of the mass that is lost in the reaction

*46. For the fusion reaction

$$_1^1H + _1^2H \longrightarrow _2^3He + energy$$

Mass data: $_1^1H = 1.00782$; $_1^2H = 2.01410$; $_2^3He = 3.01603$; 1.0 g = 9.0×10^{13} J. Calculate:
   (a) The energy released in joules per mole of He-3 formed
   (b) The percentage of the mass that is lost in the reaction

*47. In the disintegration series $_{92}^{235}U \longrightarrow _{82}^{207}Pb$, how many alpha and beta particles are emitted?

48. List three devices used for radiation detection and explain their operation.

49. The half-life of I-123 is 13 hours. If 10 mg is given to a patient, how much remains of I-123 after 3 days and 6 hours?

50. Clearly distinguish between fission and fusion. Give an example of each.

51. Which of the following statements are correct? Rewrite the incorrect statements to make them correct.
   (a) Radioactivity was dicovered by Marie Curie.
   (b) An atom of $_{28}^{59}Ni$ has 59 neutrons.
   (c) The loss of a beta particle by an atom of $_{33}^{75}As$ forms an atom of increased atomic number.
   (d) The emission of an alpha particle from the nucleus of an atom lowers its atomic number by 4 and lowers its atomic mass by 2.
   (e) Emission of gamma radiation from the nucleus of an atom leaves both the atomic number and atomic mass unchanged.
   (f) Relatively few naturally occurring radioactive nuclides have atomic numbers below 81.
   (g) The longer the half-life of a radionuclide, the more slowly it decays.
   (h) The beta ray has the greatest penetrating power of all the rays emitted from the nucleus of an atom.
   (i) Radioactivity is due to an unstable ratio of neutrons to nucleons in an atom.
   (j) The half-life of different nuclides can vary from a fraction of a second to millions of years.
   (k) The symbol $_{+1}^0e$ is used to indicate a positron, which is a positively charged particle with the mass of an electron.
   (l) The disintegration of $_{88}^{226}Ra$ into $_{83}^{214}Po$ involves the loss of three alpha particles and two beta particles.
   (m) If 1.0 g of a radionuclide has a half-life of 7.2 days, the half-life of 0.50 g of that nuclide is 3.6 days.
   (n) If the mass of a radionuclide is reduced by radioactive decay from 12 g to 0.75 g in 22 hours, the half-life of the isotope is 5.5 hours.
   (o) A very high temperature is required to initiate nuclear fusion reactions.
   (p) Radiocarbon dating of archaeological artifacts is based on an increase in the ratio of C-14 to C-12 in the object.
   (q) Cancers are often treated by radiation from Co-60, which destroys the rapidly growing cancer cells.
   (r) High levels of radiation produce nausea, vomiting, and diarrhea.
   (s) Carbon-14, used in radiocarbon dating, is produced in living matter by the beta decay of N-14.
   (t) Radioactive tracers are small amounts of radioactive nuclides of selected elements whose progress in chemical or biological processes can be followed using a radiation detection instrument.

CHAPTER TWENTY

Introduction to Organic Chemistry

Many substances throughout nature incorporate silicon or carbon within their structures. Silicon is the staple of the geologist—it combines with oxygen in a variety of ways to produce silica and a family of compounds known as the silicates. These compounds form the chemical foundation of most types of sand, rocks, and soil. In the living world, carbon provides the basis for millions of compounds in combination with hydrogen, oxygen, nitrogen, and sulfur.

The petroleum industry and the myriad of polymer products that surround us in our homes and offices are two of the many areas incorporated within the field of carbon chemistry. Synthetic fibers found in clothing and carpeting, and plastics including containers, compact discs, computer terminals, and pens, have emerged from carbon compounds. Cold remedies, cleaning products, nutrients for space travel, convenience foods, and countless drugs, both legal and illegal, have all come about from understanding the endless variety of carbon compounds.

Chapter Preview

20.1   Organic Chemistry: History and Scope

During the late 18th and the early 19th centuries chemists were baffled by the fact that compounds obtained from animal and vegetable sources defied the established rules for inorganic compounds—namely, that compound formation is due to a simple attraction between positively and negatively charged elements. In their experience with inorganic chemistry only one, or at most a few, compounds were composed of any group of only two or three elements. However, they observed that a group of only four elements—carbon, hydrogen, oxygen, and nitrogen—gave rise to a large number of different compounds that often were remarkably stable.

Because no organic compounds had been synthesized from inorganic substances and because they had no other explanation for the complexities of organic compounds, chemists believed that organic compounds were formed by some "vital force." The **vital force theory** held that organic substances could originate only from living material. In 1828 Friedrich Wöhler (1800–1882), a German chemist, did a simple experiment that eventually proved to be the death blow to this theory. In attempting to prepare ammonium cyanate (NH_4CNO) by heating cyanic acid (HCNO) and ammonia, Wöhler obtained a white crystalline substance that he identified as urea ($H_2N—CO—NH_2$). Wöhler knew that urea is an authentic organic substance because it is a product of metabolism that had been isolated from urine. Although Wöhler's discovery was not immediately and generally recognized, the vital force theory was overthrown by this simple experiment because *one* organic compound had been made from nonliving materials.

After the work of Wöhler, it was apparent that no vital force other than skill and knowledge was needed to make organic chemicals in the laboratory and that inorganic as well as organic substances could be used as raw materials. Today, **organic chemistry** designates the branch of chemistry that deals with carbon compounds but does not imply that these compounds must originate from some form of life. A few special kinds of carbon compounds (for example, carbon oxides, metal carbides, and metal carbonates) are excluded from the organic classification because their chemistry is more conveniently related to that of inorganic substances.

The field of organic chemistry is vast, for it includes not only the composition of all living organisms but also of a great many other materials that we use daily. Examples of organic materials are foodstuffs (fats, proteins, carbohydrates); fuels; fabrics (cotton, wool, rayon, nylon); wood and paper products; paints and varnishes; plastics; dyes; soaps and detergents; cosmetics; medicinals; rubber products; and explosives.

The sources of organic compounds are carbon-containing raw materials—petroleum and natural gas, coal, carbohydrates, fats, and oils. In the United States we produce about 250 billion pounds of organic chemicals per year from these sources, which amounts to more than 1100 pounds per year for every man, woman, and child in the United States. About 90% of this 250 billion pounds comes from petroleum and natural gas. Because the amount of petroleum and natural gas in the world is finite, we will, sometime in the future, have to rely on the other sources to make the voluminous amount of organic chemicals that we use. Fortunately we know how to synthesize many organic compounds from sources other than petroleum, but at much greater expense than from petroleum and natural gas.

An immense and ever-growing number of organic chemical compounds have been prepared and are described in the chemical literature. The number is estimated to be more than 7 million. The elements that make up most of the organic compounds are carbon, hydrogen, oxygen, nitrogen, the halogens

(margin notes)
vital force theory

organic chemistry

(Cl, Br, I), and sulfur. The number of organic compounds that can exist is theoretically unlimited. Several thousand different organic chemicals are available commercially.

20.2   The Need for Classification of Organic Compounds

functional group

It is impossible for anyone to study the properties of each of the millions of known organic compounds. Hence, organic compounds with similar structural features are grouped into classes. The members of each class of compounds contain a characteristic atom or group of atoms called a **functional group**. This functional group identifies a compound as belonging to a particular class. Each class of compounds exhibits similar chemical properties as a result of having a common functional group. For example, the hydroxyl group (—OH) is the functional group of the class of compounds called *alcohols*.

Even within a single class such a large number of possible compounds exists that it is virtually impossible to study each individual compound. The properties of a few members of a class of compounds are studied, and the information obtained is used to predict the behavior of other members of the class. For example, ethyl alcohol, a two-carbon alcohol, can be oxidized to a two-carbon acid. On the basis of this information, we can predict and prove that hexyl alcohol, a six-carbon alcohol, can be oxidized to a six-carbon acid because the oxidizable structures of the two alcohols are identical. Some of the classes of organic compounds we will study are hydrocarbons, alcohols, aldehydes, ketones, ethers, carboxylic acids, and esters. See Table 20.1 for the functional groups of these classes of compounds.

20.3   The Carbon Atom: Tetrahedral Structure

The carbon atom is central to all organic compounds. The atomic number of carbon is 6, and its electron structure is $1s^2 2s^2 2p^2$. Two stable isotopes of carbon exist, C-12 and C-13. In addition carbon has several radioactive isotopes; C-14 is the most widely known of these because of its use in radiocarbon dating. Having four electrons in its outer shell, carbon has oxidation numbers ranging from $+4$ to -4 and forms predominantly covalent bonds. Carbon occurs as the free element in diamond, graphite, coal, coke, carbon black, charcoal, and lampblack.

A carbon atom usually forms four covalent bonds. The most common geometric arrangement of these bonds is tetrahedral (see Figure 20.1). In this structure the four covalent bonds are not planar about the carbon atom but are directed toward the corners of a regular tetrahedron. (A tetrahedron is a solid figure with four sides.) The angle between these tetrahedral bonds is 109.5°.

TABLE 20.1

Classes of Organic Compounds

Class	General formula	Structure of functional group	Sample structural formula	Name IUPAC	Name Common
Alkane	$R^a H$	$R-H$	CH_4 CH_3CH_3	Methane Ethane	Methane Ethane
Alkene	$R-CH=CH_2$	$C=C$	$CH_2=CH_2$ $CH_3CH=CH_2$	Ethene Propene	Ethylene Propylene
Alkyne	$R-C\equiv C-H$	$-C\equiv C-$	$CH\equiv CH$ $CH_3C\equiv CH$	Ethyne Propyne	Acetylene Methyl acetylene
Alkyl halide	RX	$-X$ $X=F, Cl,$ Br, I	CH_3Cl CH_3CH_2Cl	Chloromethane Chloroethane	Methyl chloride Ethyl chloride
Alcohol	ROH	$-OH$	CH_3OH CH_3CH_2OH	Methanol Ethanol	Methyl alcohol Ethyl alcohol
Ether	$R-O-R$	$R-O-R$	CH_3-O-CH_3 $CH_3CH_2-O-CH_2CH_3$	Methoxymethane Ethoxyethane	Dimethyl ether Diethyl ether
Aldehyde	$R-\overset{H}{\underset{}{C}}=O$	$-\overset{H}{\underset{}{C}}=O$	$H-\overset{H}{\underset{}{C}}=O$ $CH_3-\overset{}{\underset{H}{C}}=O$	Methanal Ethanal	Formaldehyde Acetaldehyde
Ketone	$R-\overset{}{\underset{O}{C}}-R$	$R-\overset{}{\underset{O}{C}}-R$	$CH_3-\overset{}{\underset{O}{C}}-CH_3$ $CH_3-\overset{}{\underset{O}{C}}-CH_2CH_3$	Propanone 2-Butanone	Acetone Methyl ethyl ketone
Carboxylic acid	$R-C\overset{O}{\underset{OH}{}}$	$-C\overset{O}{\underset{OH}{}}$	$HCOOH$ CH_3COOH	Methanoic acid Ethanoic acid	Formic acid Acetic acid
Ester	$R-C\overset{O}{\underset{OR}{}}$	$-C\overset{O}{\underset{OR}{}}$	$HCOOCH_3$ CH_3COOCH_3	Methyl methanoate Methyl ethanoate	Methyl formate Methyl acetate

^{a} The letter "R" is used to indicate any of the many possible alkyl groups.

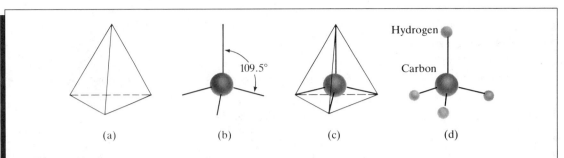

(a)   (b)   (c)   (d)

FIGURE 20.1

Tetrahedral structure of carbon. (a) A regular tetrahedron: (b) a carbon atom with tetrahedral bonds; (c) a carbon atom within a regular tetrahedron; (d) a methane molecule, CH_4.

20.4   Carbon–Carbon Bonds

With four outer-shell electrons, the carbon atom ($\cdot \ddot{C} \cdot$), following the octet rule, forms four single covalent bonds by sharing electrons with other atoms. The structures of methane and carbon tetrachloride illustrate this point:

$$H : \overset{\displaystyle H}{\underset{\displaystyle H}{\ddot{C}}} : H \qquad H - \overset{\displaystyle H}{\underset{\displaystyle H}{C}} - H \qquad Cl : \overset{\displaystyle Cl}{\underset{\displaystyle Cl}{\ddot{C}}} : Cl \qquad Cl - \overset{\displaystyle Cl}{\underset{\displaystyle Cl}{C}} - Cl$$

Methane   Carbon tetrachloride

Actually, these compounds have a tetrahedral shape with bond angles of 109.5°, but, for convenience of writing the bonds are drawn at right angles (see Figure 20.1). In methane each bond is formed by the sharing of electrons between a carbon and a hydrogen atom.

The formation of carbon–carbon bonds is due to the ability of a carbon atom to share electrons with other carbon atoms. One, two, or three pairs of electrons can be shared between two carbon atoms, forming a single, double, or triple bond, respectively:

$$\cdot \dot{C} : \dot{C} \cdot \qquad \cdot \dot{C} :: \dot{C} \cdot \qquad \cdot C ::: C \cdot$$

$$\cdot \dot{C} - \dot{C} \cdot \qquad \cdot \dot{C} = \dot{C} \cdot \qquad \cdot C \equiv C \cdot$$

Single bond   Double bond   Triple bond

Each dash above represents a covalent bond. Carbon, more than any other element, has the ability to form chains of covalently bonded atoms. This bonding ability is the main reason for the large number of organic compounds. Three examples are shown below. It is easy to see how, because of this bonding ability, long chains of carbon atoms can be formed by the linking of one carbon to another through covalent bonds.

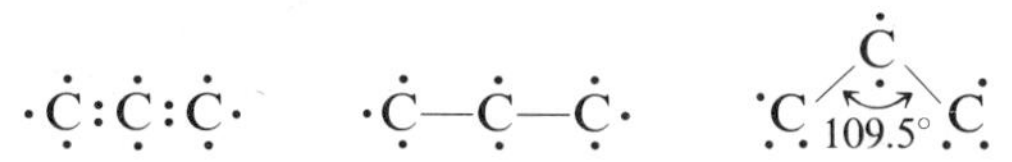

Three carbon atoms bonded by single bonds

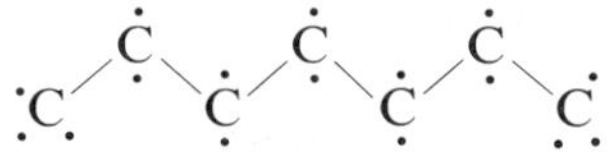

Seven-carbon chain

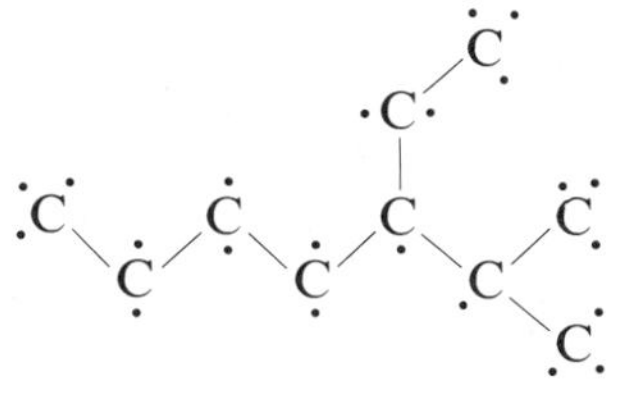

Ten carbon atoms bonded together

20.5 Hydrocarbons

hydrocarbons

Hydrocarbons are compounds that are composed entirely of carbon and hydrogen atoms bonded to each other by covalent bonds. Several classes of hydrocarbons are known. These classes include the alkanes, alkenes, alkynes, and aromatic hydrocarbons.

Benzene, C_6H_6, is an *aromatic* hydrocarbon. It contains carbon atoms in a special ring structure. Compounds containing this benzene ring system are known as *aromatic compounds*. We use the term *aliphatic* to indicate open carbon chain compounds. Thus the open chain alkanes, alkenes, and alkynes are called *aliphatic hydrocarbons*.

Fossil fuels—natural gas, petroleum, and coal—are the principal sources of hydrocarbons. Natural gas is primarily methane with small amounts of ethane, propane, and butane. Petroleum is a mixture of hydrocarbons from which gasoline, kerosene, fuel oil, lubricating oil, paraffin wax, and petrolatum (themselves mixtures of hydrocarbons) are separated. Coal tar, a volatile by-product of the process of making coke from coal for use in the steel industry, is the source of many valuable chemicals, including the aromatic hydrocarbons benzene, toluene, and naphthalene.

20.6   Saturated Hydrocarbons: Alkanes

alkanes

The **alkanes**, also known as *paraffins* or *saturated hydrocarbons,* are straight- or branched-chain hydrocarbons with only single covalent bonds between the carbon atoms. We shall study the alkanes in some detail because many other classes of organic compounds can be considered as derivatives of these substances. For example, it is necessary to learn the names of the first ten members of the alkane series, because these names are used as a basis for naming other classes of compounds.

Methane (CH_4) is the first member of the alkane series. Members having two, three, and four carbon atoms are ethane, propane, and butane, respectively. The names of the first four alkanes are of common or trivial origin and must be memorized; but the names beginning with the fifth member, pentane, are derived from Greek numbers and are relatively easy to recall. The names and formulas of the first ten members of the series are given in Table 20.2.

Successive compounds in the alkane series differ from each other in composition by one carbon and two hydrogen atoms. When each member of a series differs from the next member by a CH_2 group, the series is called a **homologous series**. The members of a homologous series are similar in structure but have a regular difference in formula. All common classes of organic compounds exist in homologous series. A homologous series can be represented by a general formula. For all open-chain alkanes the general formula is C_nH_{2n+2},

homologous series

TABLE 20.2

Names, Formulas, and Physical Properties of Straight-Chain Alkanes

Name	Molecular formula C_nH_{2n+2}	Condensed structural formula	Boiling point (°C)	Melting point (°C)
Methane	CH_4	CH_4	-161	-183
Ethane	C_2H_6	CH_3CH_3	-88	-172
Propane	C_3H_8	$CH_3CH_2CH_3$	-45	-187
Butane	C_4H_{10}	$CH_3CH_2CH_2CH_3$	-0.5	-138
Pentane	C_5H_{12}	$CH_3CH_2CH_2CH_2CH_3$	36	-130
Hexane	C_6H_{14}	$CH_3CH_2CH_2CH_2CH_2CH_3$	69	-95
Heptane	C_7H_{16}	$CH_3CH_2CH_2CH_2CH_2CH_2CH_3$	98	-90
Octane	C_8H_{18}	$CH_3CH_2CH_2CH_2CH_2CH_2CH_2CH_3$	125	-57
Nonane	C_9H_{20}	$CH_3CH_2CH_2CH_2CH_2CH_2CH_2CH_2CH_3$	151	-54
Decane	$C_{10}H_{22}$	$CH_3CH_2CH_2CH_2CH_2CH_2CH_2CH_2CH_2CH_3$	174	-30

where n corresponds to the number of carbon atoms in the molecule. The formula of any specific alkane is easily determined from this general formula. Thus, for pentane, $n = 5$ and $2n + 2 = 12$, so the formula is C_5H_{12}. For hexadecane, a 16-carbon alkane, the formula is $C_{16}H_{34}$.

20.7   Structural Formulas and Isomerism

The properties of an organic substance are dependent on its molecular structure. By structure we mean the way in which the atoms are bonded together within the molecule. The majority of organic compounds are made from relatively few elements—carbon, hydrogen, oxygen, nitrogen, and the halogens. In these compounds carbon is tetravalent, hydrogen is monovalent, oxygen is divalent, nitrogen is trivalent, and the halogens are monovalent. The valence bonds or points of attachment may be represented in structural formulas by a corresponding number of dashes attached to each atom:

$$-\overset{\displaystyle |}{\underset{\displaystyle |}{C}}- \qquad H- \qquad -O- \qquad -\overset{\displaystyle |}{N}- \qquad Cl- \qquad Br- \qquad I- \qquad F-$$

Thus carbon will have four bonds to each atom, nitrogen three bonds, oxygen two bonds, and hydrogen and the halogens one bond to each atom.

In an alkane each carbon atom is joined to four other atoms by four single covalent bonds. These bonds are separated by angles of 109.5° (the angles correspond to those formed by lines drawn from the center of a regular tetrahedron to its corners). Alkane molecules contain only carbon–carbon and carbon–hydrogen bonds and are essentially nonpolar. Because of this low polarity, alkane molecules have very little intermolecular attraction and therefore relatively low boiling points compared with other organic compounds of similar molar mass.

Without the use of models or perspective drawings, the three-dimensional character of atoms and molecules is difficult to portray accurately. Concepts of structure can be conveyed to some extent by Lewis-dot diagrams of structural formulas. But it is more convenient to use conventional structural formulas representing electron pairs by single short lines. Methane and ethane are shown below.

Methane   Ethane

To write the correct structural formula for propane (C_3H_8), the next member of the alkane series, we must determine how to place each atom in the molecule.

An alkane contains only single bonds, and carbon is tetravalent. Therefore, each carbon atom must be bonded to four other atoms by either C—C or C—H bonds. Hydrogen is univalent and therefore must be bonded to only one carbon atom by a C—H bond, since C—H—C bonds do not occur, and an H—H bond would simply represent a hydrogen molecule. Applying this information, we find that the only possible structure for propane is

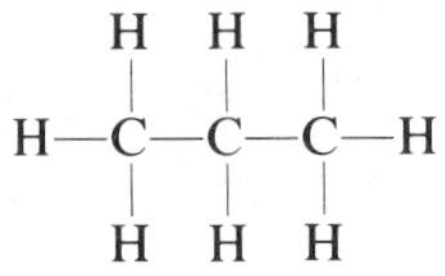

$$
\begin{array}{c}
\text{H} \quad \text{H} \quad \text{H} \\
| \quad\; | \quad\; | \\
\text{H—C—C—C—H} \\
| \quad\; | \quad\; | \\
\text{H} \quad \text{H} \quad \text{H}
\end{array}
$$

Propane

However, it is possible to write two structural formulas corresponding to the molecular formula C_4H_{10} (butane):

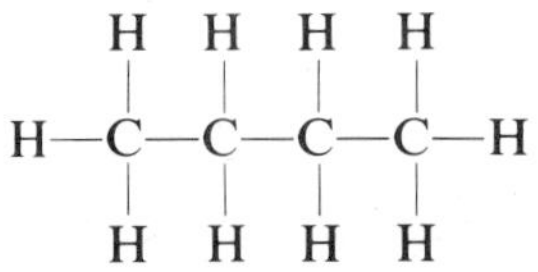

$$
\begin{array}{c}
\text{H} \quad \text{H} \quad \text{H} \quad \text{H} \\
| \quad\; | \quad\; | \quad\; | \\
\text{H—C—C—C—C—H} \\
| \quad\; | \quad\; | \quad\; | \\
\text{H} \quad \text{H} \quad \text{H} \quad \text{H}
\end{array}
\qquad \text{and} \qquad
\begin{array}{c}
\text{H} \\
| \\
\text{H} \backslash \text{C} / \text{H} \\
\text{H} \quad\quad\quad\;\; \text{H} \\
| \quad\quad\quad\;\; | \\
\text{H—C—C—C—H} \\
| \quad\; | \quad\; | \\
\text{H} \quad \text{H} \quad \text{H}
\end{array}
$$

Normal butane   Isobutane

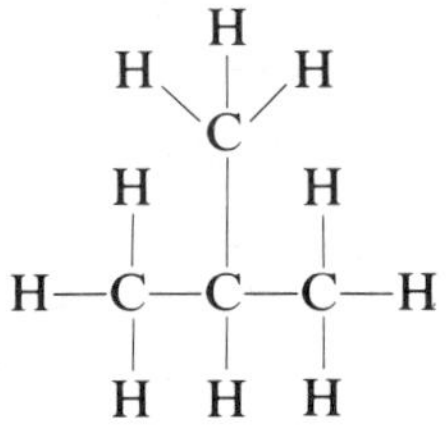

Two C_4H_{10} compounds with the structural formulas shown above actually exist. The butane with the unbranched carbon chain is called *normal butane* (abbreviated *n*-butane); it boils at 0.5°C and melts at −138.3°C. The branched-chain butane is called *isobutane*; it boils at −11.7°C and melts at −159.5°C. These differences in physical properties are sufficient to establish that the two compounds, though they have the same molecular formula, are different substances. Models illustrating the structural arrangement of the atoms in methane, ethane, propane, butane, and isobutane are shown in Figure 20.2.

isomerism
isomers
　　　This phenomenon of two or more compounds having the same molecular formula but different structural arrangements of their atoms is called **isomerism**. The various individual compounds are called **isomers**. For example, there are 2 isomers of butane, C_4H_{10}. Isomerism is very common among organic compounds and is one reason for the large number of known compounds. There are 3 isomers of pentane, 5 isomers of hexane, 9 isomers of heptane, 18 isomers of octane, 35 isomers of nonane, and 75 isomers of decane. The phenomenon of isomerism is a very compelling reason for the use of structural formulas.

> **Isomers are compounds that have the same molecular formula but different structural formulas.**

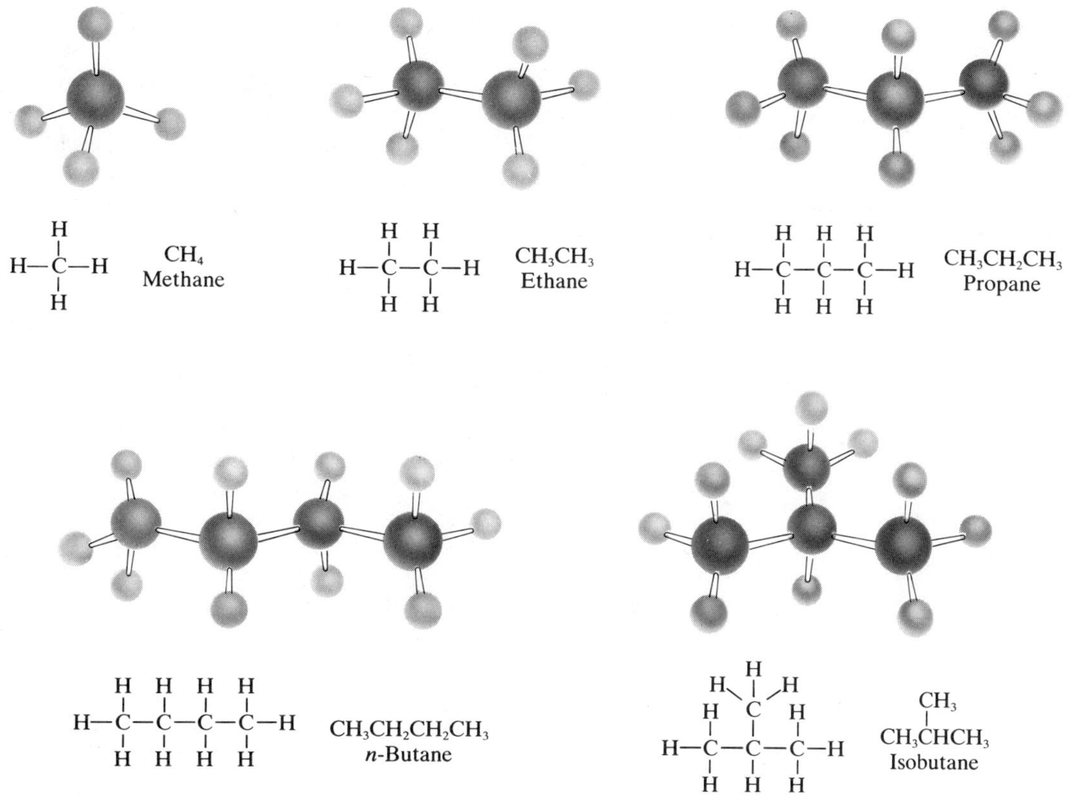

$$
\begin{array}{c}
\text{H} \\
| \\
\text{H}-\text{C}-\text{H} \\
| \\
\text{H}
\end{array}
\qquad
\begin{array}{l}
\text{CH}_4 \\
\text{Methane}
\end{array}
$$

$$
\begin{array}{c}
\text{H} \ \ \text{H} \\
| \ \ \ | \\
\text{H}-\text{C}-\text{C}-\text{H} \\
| \ \ \ | \\
\text{H} \ \ \text{H}
\end{array}
\qquad
\begin{array}{l}
\text{CH}_3\text{CH}_3 \\
\text{Ethane}
\end{array}
$$

$$
\begin{array}{c}
\text{H} \ \ \text{H} \ \ \text{H} \\
| \ \ \ | \ \ \ | \\
\text{H}-\text{C}-\text{C}-\text{C}-\text{H} \\
| \ \ \ | \ \ \ | \\
\text{H} \ \ \text{H} \ \ \text{H}
\end{array}
\qquad
\begin{array}{l}
\text{CH}_3\text{CH}_2\text{CH}_3 \\
\text{Propane}
\end{array}
$$

$$
\begin{array}{c}
\text{H} \ \ \text{H} \ \ \text{H} \ \ \text{H} \\
| \ \ \ | \ \ \ | \ \ \ | \\
\text{H}-\text{C}-\text{C}-\text{C}-\text{C}-\text{H} \\
| \ \ \ | \ \ \ | \ \ \ | \\
\text{H} \ \ \text{H} \ \ \text{H} \ \ \text{H}
\end{array}
\qquad
\begin{array}{l}
\text{CH}_3\text{CH}_2\text{CH}_2\text{CH}_3 \\
\textit{n}\text{-Butane}
\end{array}
$$

$$
\begin{array}{c}
\text{H} \\
\text{H} \diagdown \ | \diagup \text{H} \\
\text{H} \ \ \ \text{C} \ \ \ \text{H} \\
| \ \ \ | \ \ \ | \\
\text{H}-\text{C}-\text{C}-\text{C}-\text{H} \\
| \ \ \ | \ \ \ | \\
\text{H} \ \ \text{H} \ \ \text{H}
\end{array}
\qquad
\begin{array}{l}
\text{CH}_3 \\
\text{CH}_3\text{CHCH}_3 \\
\text{Isobutane}
\end{array}
$$

FIGURE 20.2

Ball-and-stick models illustrating structural formulas of methane, ethane, propane, butane, and isobutane. Condensed structural formulas are shown above the names.

To save time and space in writing, condensed structural formulas are often used. In the condensed structural formulas, atoms and groups that are attached to a carbon atom are generally written to the right of that carbon atom. For example, the condensed structural formula for pentane is $CH_3CH_2CH_2CH_2CH_3$ or $CH_3(CH_2)_3CH_3$. Some condensed structural formulas are shown in Figure 20.2.

Let us interpret the condensed structural formula for propane:

$$
\overset{1}{C}H_3\overset{2}{C}H_2\overset{3}{C}H_3
$$

Carbon number 1 has three hydrogen atoms attached to it and is bonded to carbon number 2, which has two hydrogen atoms on it and is bonded to carbon number 3, which has three hydrogen atoms bonded to it.

EXAMPLE 20.1   Pentane, C_5H_{12}, has three isomers. Write structural formulas and condensed structural formulas for these isomers.

In a problem of this kind it is best to start by writing the carbon skeleton of the compound containing the longest continuous carbon chain. In this case it is five carbon atoms. Now complete the structure by attaching hydrogen atoms around each carbon atom so that each carbon atom has four bonds attached to it. The carbon atoms at each end of the chain need three hydrogen atoms. The three inner carbon atoms each need two hydrogen atoms to give them four bonds.

$$C-C-C-C-C \qquad \begin{matrix} & H & H & H & H & H \\ & | & | & | & | & | \\ H-&C-&C-&C-&C-&C&-H \\ & | & | & | & | & | \\ & H & H & H & H & H \end{matrix} \qquad CH_3CH_2CH_2CH_2CH_3$$

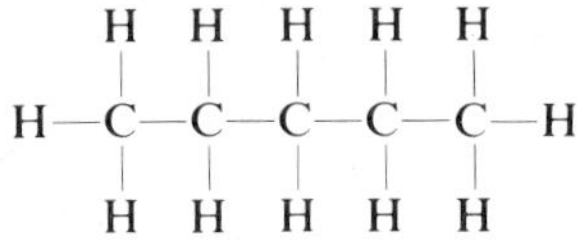

For the next isomer, start by writing a four-carbon chain and attach the fifth carbon atom to either of the middle carbon atoms; do not use the end ones.

$$\begin{matrix} & C & & & & C \\ & | & & & & | \\ C-&C-&C-C & \qquad & C-&C-&C-C \end{matrix} \qquad \text{Both of these structures represent the same compound.}$$

Now add the 12 hydrogen atoms to complete the structure:

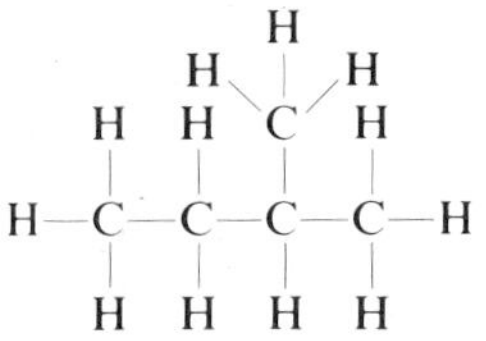

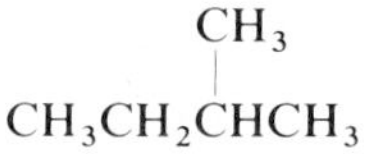

$$\begin{matrix} & & & H & & \\ & & H & | & H & \\ & H & H & C & H & \\ & | & | & | & | & \\ H-&C-&C-&C-&C&-H \\ & | & | & | & | & \\ & H & H & H & H & \end{matrix} \qquad CH_3CH_2CHCH_3 \quad \text{or} \quad CH_3CH_2CH(CH_3)_2$$

with CH_3 above the third carbon.

For the third isomer, write a three-carbon chain, attach the other two carbon atoms to the central carbon atom, and complete the structure by adding the 12 hydrogen atoms:

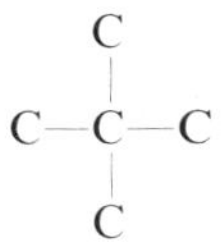

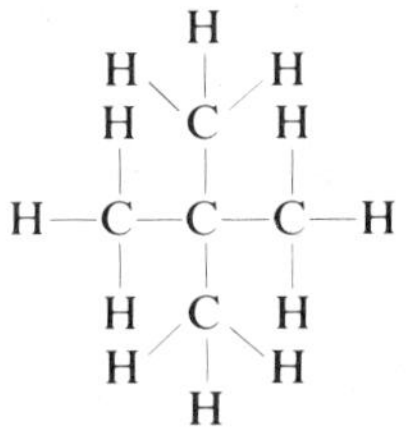

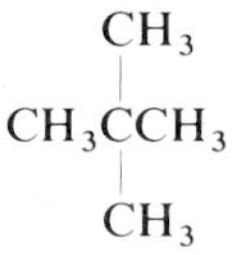

$$\begin{matrix} & C & & & & H & & \\ & | & & & H & | & H & \\ C-&C-&C & \qquad & H & C & H & \\ & | & & & | & | & | & \\ & C & & H-&C-&C-&C&-H \\ & & & & | & | & | & \\ & & & & H & C & H & \\ & & & & H & | & H & \\ & & & & & H & & \end{matrix} \qquad CH_3CCH_3 \quad \text{or} \quad C(CH_3)_4$$

with CH_3 above and CH_3 below the central carbon.

20.8   Naming Organic Compounds: Alkanes

In the early development of organic chemistry each new compound was given a name, usually by the person who had isolated or synthesized it. Names were not systematic but often did carry some information—usually about the origin of the

TABLE 20.3

Names and Formulas of Selected Alkyl Groups

Formula	Name	Formula	Name
$CH_3—$	Methyl	$\begin{array}{c} CH_3 \\ | \\ CH_3CH— \end{array}$	Isopropyl
$CH_3CH_2—$	Ethyl		
$CH_3CH_2CH_2—$	Propyl		
$CH_3CH_2CH_2CH_2—$	Butyl	$\begin{array}{c} CH_3 \\ | \\ CH_3CHCH_2— \end{array}$	Isobutyl
$CH_3(CH_2)_3CH_2—$	Pentyl		
$CH_3(CH_2)_4CH_2—$	Hexyl	$\begin{array}{c} CH_3 \\ | \\ CH_3CH_2CH— \end{array}$	sec-Butyl (secondary butyl)
$CH_3(CH_2)_5CH_2—$	Heptyl		
$CH_3(CH_2)_6CH_2—$	Octyl	$\begin{array}{c} CH_3 \\ | \\ CH_3C— \\ | \\ CH_3 \end{array}$	tert-Butyl (tertiary butyl)
$CH_3(CH_2)_7CH_2—$	Nonyl		
$CH_3(CH_2)_8CH_2—$	Decyl		

substance. Wood alcohol (methanol), for example, was so named because it was obtained by destructive distillation or pyrolysis of wood. Methane, formed during underwater decomposition of vegetable matter in marshes, was originally called *marsh gas*. A single compound was often known by several names. For example, the active ingredient in alcoholic beverages has been called *alcohol, ethyl alcohol, methyl carbinol, grain alcohol, spirit*, and *ethanol*.

Beginning with a meeting in Geneva in 1892, an international system for naming compounds was developed. In its present form the method recommended by the International Union of Pure and Applied Chemistry is systematic, generally unambiguous, and internationally accepted. It is called the IUPAC System. Despite the existence of the official IUPAC System a great many well-established common, or trivial, names and abbreviations (such as TNT and DDT) are used because of their brevity and/or convenience. So it is necessary to have a knowledge of both the IUPAC System and many common names.

alkyl group

In order to name organic compounds systematically, you must be able to recognize certain common alkyl groups. **Alkyl groups** have the general formula C_nH_{2n+1} (one less hydrogen atom than the corresponding alkane). The name of the group is formed from the name of the corresponding alkane by simply dropping *ane* and substituting a *yl* ending. The names and formulas of selected alkyl groups up to and including four carbon atoms are given in Table 20.3. The letter "R" is often used in formulas to mean any of the many possible alkyl groups.

$R = C_nH_{2n+1}$ (any alkyl group)

The following relatively simple rules are all that are needed to name a great many alkanes according to the IUPAC System. In later sections these rules will be extended to cover other classes of compounds, but advanced texts or references must be consulted for the complete system. The IUPAC rules for naming alkanes are

1. Select the longest continuous chain of carbon atoms as the parent compound, and consider all alkyl groups, attached to it as branch chains or substituents that have replaced hydrogen atoms of the parent hydrocarbon. If two chains of equal length are found, use the chain that has the larger number of substituents attached to it. The name of the alkane consists of the name of the parent compound prefixed by the names of the alkyl groups attached to it.

2. Number the carbon atoms in the parent carbon chain starting from the end closest to the first carbon atom that has an alkyl or other group substituted for a hydrogen atom. If the first substituent from each end is on the same-numbered carbon, go to the next substituent to determine which end of the chain to start numbering.

3. Name each alkyl group and designate its position on the parent carbon chain by a number (for example, 2-methyl means a methyl group attached to C-2).

4. When the same alkyl group branch chain occurs more than once, indicate this repetition by a prefix (*di, tri, tetra*, and so forth) written in front of the alkyl group name (for example, *dimethyl* indicates two methyl groups). The numbers indicating the positions of these alkyl groups are separated by a comma and followed by a hyphen and are placed in front of the name (for example, 2,3-dimethyl).

5. When several different alkyl groups are attached to the parent compound, list them in alphabetical order; for example, ethyl before methyl in 3-ethyl-4-methyloctane. Prefixes are not included in alphabetical ordering (ethyl comes before dimethyl).

The compound shown below is commonly called isopentane. Consider naming it by the IUPAC System:

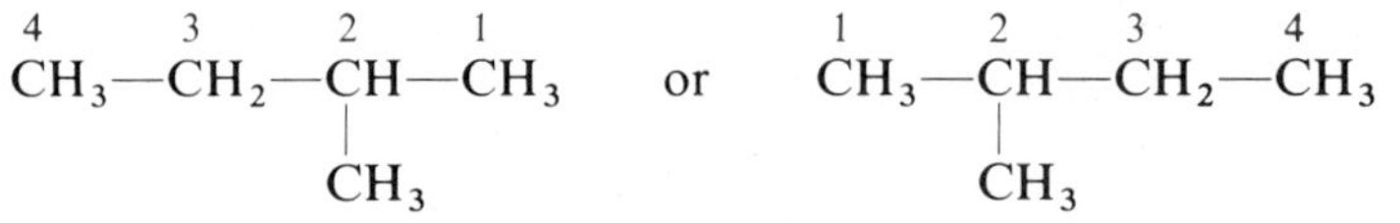

$$\overset{4}{CH_3}-\overset{3}{CH_2}-\overset{2}{\underset{|}{CH}}-\overset{1}{CH_3} \quad \text{or} \quad \overset{1}{CH_3}-\overset{2}{\underset{|}{CH}}-\overset{3}{CH_2}-\overset{4}{CH_3}$$
$$\qquad\qquad CH_3 \qquad\qquad\qquad\qquad CH_3$$

2-Methylbutane
(Isopentane)

The longest continuous chain contains four carbon atoms. Therefore, the parent compound is butane and the compound is named as a butane. The methyl group (CH_3—) attached to carbon number 2 is named as a prefix to butane, the "2-" indicating the point of attachment of the methyl group on the butane chain.

How would we write the structural formula for 2-methylpentane? An analysis of its name gives us this information. The parent compound, pentane, contains five carbons. Write and number the five-carbon skeleton of pentane. Put a methyl group on C-2 ("2-methyl" in the name gives this information). Add hydrogens to give each carbon four bonds.

$$\underset{5}{C}-\underset{4}{C}-\underset{3}{C}-\underset{2}{C}-\underset{1}{C} \qquad \underset{5}{C}-\underset{4}{C}-\underset{3}{C}-\underset{2}{\underset{|}{C}}-\underset{1}{C} \qquad CH_3-CH_2-CH_2-\underset{|}{CH}-CH_3$$
$$\qquad\qquad\qquad\qquad CH_3 \qquad\qquad\qquad\qquad\qquad CH_3$$

2-Methylpentane

Should this compound be called 4-methylpentane? No, because the IUPAC System specifically states that the parent carbon chain shall be numbered starting from the end nearest to the branch chain.

It is very important to understand that the sequence of atoms and groups, not the way the sequence is written, determines the name of a compound. Each of the following formulas represents 2-methylpentane:

$$\underset{1}{CH_3}-\underset{2}{\underset{|}{CH}}-\underset{3}{CH_2}-\underset{4}{CH_2}-\underset{5}{CH_3} \qquad\qquad \underset{5}{CH_3}-\underset{4}{CH_2}-\underset{3}{CH_2}-\overset{CH_3}{\underset{2}{\underset{|}{CH}}}-\underset{1}{CH_3}$$
$$\qquad CH_3$$

$$\underset{1}{CH_3}-\overset{CH_3}{\underset{2}{\underset{|}{CH}}}-\underset{3}{CH_2} \qquad\qquad\qquad \overset{CH_3}{\underset{2}{\underset{|}{CH}}}$$
$$\qquad\qquad\qquad \underset{4}{\underset{|}{CH_2}}-\underset{5}{CH_3} \qquad\qquad \underset{1}{CH_3}\diagdown\;\underset{3}{CH_2}\;\diagdown\underset{5}{CH_3}$$

The following formulas and names demonstrate other aspects of the official nomenclature system:

$$\underset{4}{CH_3}-\underset{3}{\underset{|}{CH}}-\underset{2}{\underset{|}{CH}}-\underset{1}{CH_3} \qquad\qquad \underset{4}{CH_3}-\underset{3}{CH_2}-\overset{CH_3}{\underset{2}{\underset{|}{C}}}-\underset{1}{CH_3}$$
$$\qquad CH_3 \;\; CH_3 \qquad\qquad\qquad\qquad\qquad CH_3$$

2,3-Dimethylbutane   2,2-Dimethylbutane

In 2,3-dimethylbutane, the longest carbon chain is four, indicating butane; "dimethyl" indicates two methyl groups; "2,3-" means that one CH_3 is on carbon 2 and one is on carbon 3. In 2,2-dimethylbutane both methyl groups are on the same carbon atom; both numbers are required.

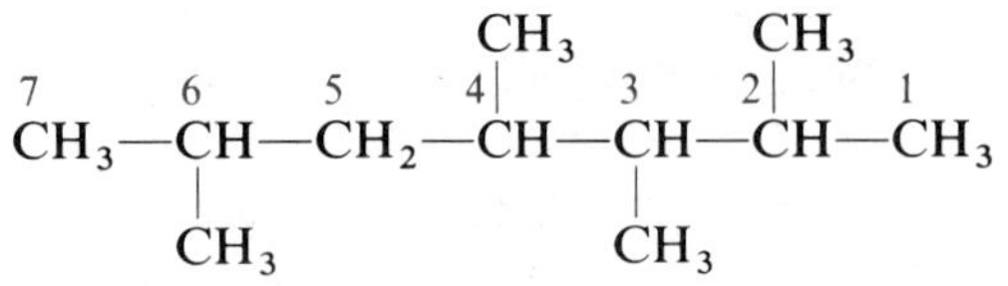

$$\overset{7}{CH_3}-\overset{6}{CH}-\overset{5}{CH_2}-\overset{4}{CH}-\overset{3}{CH}-\overset{2}{CH}-\overset{1}{CH_3}$$

with CH_3 groups at positions 4, 3, 2 and CH_3 groups below at 6 and 3.

2,3,4,6-Tetramethylheptane   (not 2,4,5,6-)

(The molecule is numbered from right to left.)

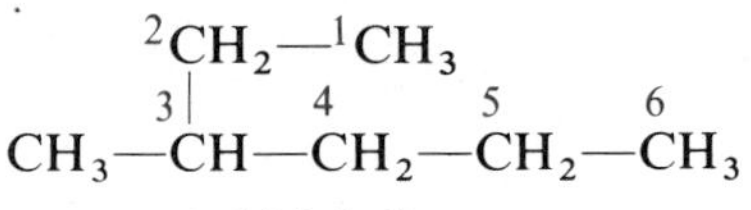

$$\overset{2}{CH_2}-\overset{1}{CH_3}$$
$$\overset{3}{CH_3}-\overset{}{CH}-\overset{4}{CH_2}-\overset{5}{CH_2}-\overset{6}{CH_3}$$

3-Methylhexane

The longest continuous chain in 3-methylhexane has six carbons.

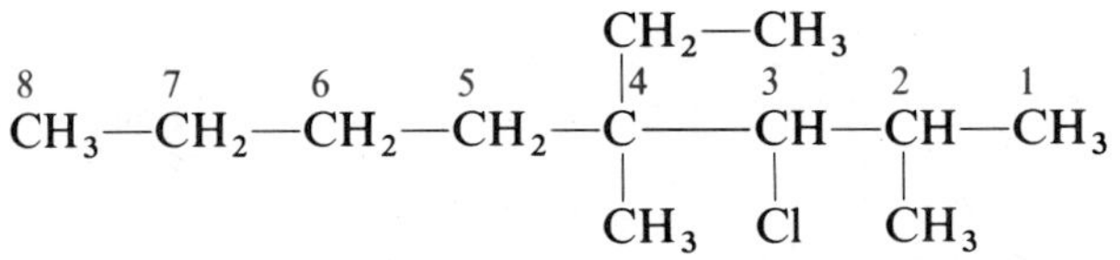

$$\overset{}{CH_2}-\overset{}{CH_3}$$
$$\overset{8}{CH_3}-\overset{7}{CH_2}-\overset{6}{CH_2}-\overset{5}{CH_2}-\overset{4}{C}-\overset{3}{CH}-\overset{2}{CH}-\overset{1}{CH_3}$$
with CH_3, Cl, CH_3 below

3-Chloro-4-ethyl-2,4-dimethyloctane

The longest carbon chain is eight. The groups that are attached or substituted for H on the octane chain are named in alphabetical order.

EXAMPLE 20.2   Write the formulas for (a) 3-ethylpentane and (b) 2,2,4-trimethylpentane

(a) The name *pentane* indicates a five-carbon chain. Write five connecting carbon atoms and number them. Attach an ethyl group, CH_3CH_2—, to C-3. Now add hydrogen atoms to give each carbon atom four bonds. C-1 and C-5 each need three H atoms; C-2 and C-4 each need two H atoms; and C-3 needs one H atom.

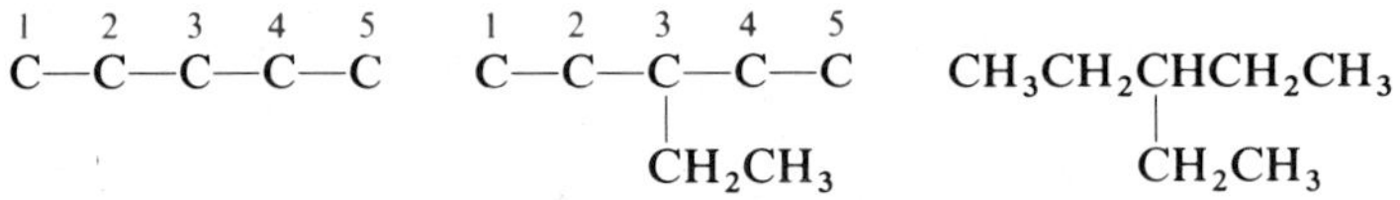

$$\overset{1}{C}-\overset{2}{C}-\overset{3}{C}-\overset{4}{C}-\overset{5}{C} \qquad \overset{1}{C}-\overset{2}{C}-\overset{3}{C}-\overset{4}{C}-\overset{5}{C} \qquad CH_3CH_2CHCH_2CH_3$$
with CH_2CH_3 below C-3, and CH_2CH_3 below

3-Ethylpentane

(b) Pentane indicates a five-carbon chain. Write five connecting carbon atoms and number them. There are three methyl groups (CH_3—) in the compound (trimethyl), two attached to C-2 and one attached to C-4. Attach these three methyl groups to their respective carbon atoms. Now add H atoms to give each carbon atom four bonds. Carbons 1 and 5 each need three H atoms; C-2 does not need any H atoms; C-3 needs two H atoms; and C-4 needs one H atom. The formula is complete.

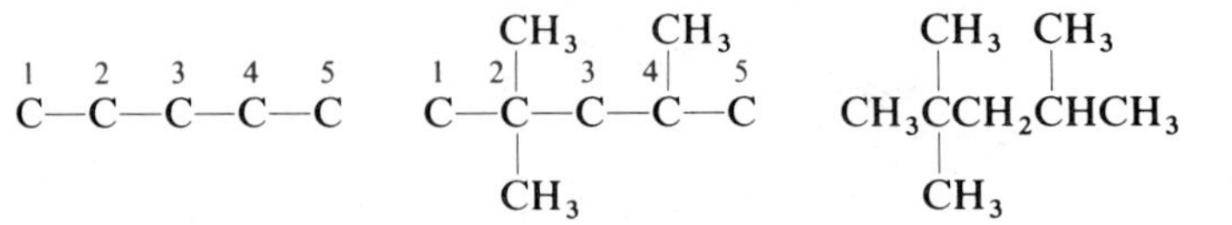

$$\overset{1}{C}-\overset{2}{C}-\overset{3}{C}-\overset{4}{C}-\overset{5}{C} \qquad \overset{1}{C}-\overset{2}{C}-\overset{3}{C}-\overset{4}{C}-\overset{5}{C} \qquad CH_3CCH_2CHCH_3$$
with CH_3 at C-2 and C-4 above, CH_3 below C-2, and CH_3, CH_3 above and CH_3 below

2,2,4-Trimethylpentane

EXAMPLE 20.3 Name the following compounds:

(a) $CH_3CH_2CH_2CH_2CHCH_3$ (b) $CH_3CH_2CH_2CHCH_2CHCH_3$

                CH_3 CH_3CH_2 CH_2CH_3

(a) The longest continuous carbon chain contains six carbon atoms (Rule 1). Thus the parent name of the compound is hexane. Number the carbon chain from right to left so that the methyl group attached to C-2 is given the lowest possible number (Rule 2). With a methyl group on carbon 2 the name of the compound is 2-methylhexane (Rule 3).

(b) The longest continuous carbon chain contains eight carbon atoms:

$$
\begin{array}{c}
\quad\quad\quad\quad C-C \\
8\;\;\; 7\;\;\; 6\;\;\; |5\;\; 4\;\;\; 3 \\
C-C-C-C-C-C-C \\
\quad\quad\quad\quad\quad\quad |2\;\;\; 1 \\
\quad\quad\quad\quad\quad\quad C-C
\end{array}
$$

Thus the parent name is octane. As the chain is numbered, a methyl group is on C-3 and an ethyl group is on C-5. Thus the name of the compound is 5-ethyl-3-methyloctane. Note that ethyl is named before methyl (alphabetical order) (Rule 5).

PRACTICE Name the following alkanes.

                CH_3 CH_3

(a) $CH_3CH{-}C{-}CH_2CH_2CH_3$ (b) $CH_3CHCHCH_2CH_3$

          CH_3 CH_3 CH_2CHCH_3

                                            CH_2CH_3

Answers: (a) 2,3,3-trimethylhexane (b) 3-ethyl-2,5-dimethylheptane

20.9 Reactions of Alkanes

One single type of reaction of alkanes has inspired people to explore equatorial jungles, endure the heat and sandstorms of the deserts of Africa and the Middle East, mush across the frozen Arctic, and drill holes in the earth more than 30,000 feet deep! These strenuous and expensive activities have been undertaken because oil contains alkanes (as well as other hydrocarbons) that undergo combustion with oxygen to evolve large amounts of heat energy. Methane, for example, reacts with oxygen:

$$CH_4(g) + 2\,O_2(g) \longrightarrow CO_2(g) + 2\,H_2O(g) + 802.5 \text{ kJ (191.8 kcal)}$$

The thermal energy can be converted to mechanical and electrical energy. Combustion reactions overshadow all other reactions of alkanes in economic importance. But combustion reactions are not usually of great interest to organic chemists, because carbon dioxide and water are the only chemical products of complete combustion.

Aside from their combustibility, alkanes are sluggish and limited in reactivity. But with proper activation, such as high temperature and/or catalysts, alkanes can be made to react in a variety of ways. We will consider only one type of reaction—halogenation.

halogenation   **Halogenation** is the substitution of a halogen atom for a hydrogen atom. Halogenation is a general term. When a specific halogen, such as chlorine, is used, the reaction is called chlorination.

$$RH + X_2 \xrightarrow[\text{light}]{\text{uv}} RX + HX \qquad (X = Cl \quad \text{or} \quad Br)$$

$$CH_3CH_3 + Cl_2 \xrightarrow[\text{light}]{\text{uv}} CH_3CH_2Cl + HCl$$
$$\text{Chloroethane}$$

This reaction yields alkyl halides (RX), which are useful as intermediates for the manufacture of other substances.

A well-known reaction of methane and chlorine is shown by the equation

$$CH_4 + Cl_2 \xrightarrow[\text{light}]{\text{uv}} \quad CH_3Cl \quad + \quad HCl$$
$$\text{Chloromethane}$$
$$\text{(Methyl chloride)}$$

The reaction of methane and chlorine will give a mixture of mono-, di-, tri-, and tetrasubstituted chloromethanes.

$$CH_4 \xrightarrow{Cl_2} CH_3Cl \xrightarrow{Cl_2} CH_2Cl_2 \xrightarrow{Cl_2} CHCl_3 \xrightarrow{Cl_2} CCl_4 + 4\,HCl$$

However, if an excess of chlorine is used, the reaction can be controlled to give all tetrachloromethane (carbon tetrachloride). On the other hand, if a large ratio of methane to chlorine is used, the product will be predominantly chloromethane (methyl chloride). Table 20.4 shows the formulas and names for all the chloromethanes. The names for the other halogen-substituted methanes follow the same pattern as for the chloromethanes; for example, CH_3Br is bromomethane, or methyl bromide, and CHI_3 is triiodomethane, or iodoform.

Chloromethane is a monosubstitution product of methane. The term *monosubstitution* refers to the fact that one hydrogen atom in an organic molecule is replaced by another atom or a group of atoms. In hydrocarbons, for example, when we substitute one chlorine atom for a hydrogen atom, the new compound is a monosubstitution (*monochloro* substitution) product. In a like manner we can have *di-*, *tri-*, *tetra-*, and so on, substitution products. For example, there are two *dichloro* substituted ethanes.

$$CH_3\text{—}CHCl_2 \qquad CH_2Cl\text{—}CH_2Cl$$
$$\text{1,1-Dichloroethane} \qquad \text{1,2-Dichloroethane}$$

TABLE 20.4

Chlorination Products of Methane

Formula	IUPAC name	Common name
CH_3Cl	Chloromethane	Methyl chloride
CH_2Cl_2	Dichloromethane	Methylene chloride
$CHCl_3$	Trichloromethane	Chloroform
CCl_4	Tetrachloromethane	Carbon tetrachloride

This kind of chlorination (or bromination) is general with alkanes. Ethane has nine different chlorination products. See if you can write the structural formulas for all of them.

When propane is chlorinated, two isomeric monosubstitution products are obtained, because a hydrogen atom may be replaced on either the first or second carbon.

$$CH_3CH_2CH_3 + Cl_2 \xrightarrow[25°]{Light} CH_3CH_2CH_2Cl + CH_3CHClCH_3 + HCl$$

1-Chloropropane 2-Chloropropane
(<i>n</i>-Propyl chloride) (Isopropyl chloride)

alkyl halide

The letter **X** is commonly used to indicate a halogen atom. The formula **RX** indicates a halogen atom attached to an alkyl group and represents the class of compounds known as the **alkyl halides.** When $R = CH_3$, CH_3X can be CH_3F, CH_3Cl, CH_3Br, or CH_3I.

Alkyl halides are named systematically in the same general way as alkanes. Halogen atoms are identified as *fluoro, chloro, bromo,* or *iodo* and are named as substituents like side-chain alkyl groups. Study these examples:

CH_3—$CHCl$—CH_2—CH_3 CH_2Cl—$CHBr$—CH_3 CH_2Cl—CH_2—CH_2Cl

2-Chlorobutane   2-Bromo-1-chloropropane   1,3-Dichloropropane

EXAMPLE 20.4

How many monochloro substitution products can be obtained from pentane?
First write the formula for pentane:

$$\overset{5}{CH_3}\overset{4}{CH_2}\overset{3}{CH_2}\overset{2}{CH_2}\overset{1}{CH_3}$$

Now rewrite the formula five times substituting a Cl atom for an H atom on each C atom:

I $CH_3CH_2CH_2CH_2\overset{1}{CH_2}Cl$ Cl on carbon 1

II $CH_3CH_2CH_2\overset{2}{CH}ClCH_3$ Cl on carbon 2

III $CH_3CH_2CHClCH_2CH_3$ Cl on carbon 3

IV $CH_3CHClCH_2CH_2CH_3$ Cl on carbon 4

V $CH_2ClCH_2CH_2CH_2CH_3$ Cl on carbon 5

Compound IV is identical to II and compound V is identical to I, because, if we number the carbon chains in IV and V from left to right, we find the Cl atoms substituted on C-1 and C-2. By naming the compounds we find that both I and V are 1-chloropentane and that II and IV are both 2-chloropentane. Thus, pentane has three monochloro substitution products—compounds I, II, and III, 1-chloropentane, 2-chloropentane, and 3-chloropentane.

PRACTICE   Name the monochloro substitution products that can be obtained from 2-methylbutane.

Answers: 1-chloro-2-methylbutane; 2-chloro-3-methylbutane
2-chloro-2-methylbutane; 1-chloro-3-methylbutane

20.10   Alkenes and Alkynes

The bulk of the organic compounds manufactured in the United States are derived from seven starting materials: ethylene, propylene, benzene, butylene, toluene, xylene, and methane. Of the seven substances listed, all are unsaturated hydrocarbons except methane. Ethylene alone is the base for almost half the petrochemicals made in the country. Over 31 billion pounds of ethylene are produced annually, making it the fifth-ranking chemical in production volume in the United States.

alkene

alkyne

Both **alkenes** and **alkynes** are classified as unsaturated hydrocarbons. They are said to be unsaturated because, unlike alkanes, their molecules do not contain the maximum possible number of hydrogen atoms. Alkenes have two less hydrogen atoms and alkynes have four less hydrogen atoms than alkanes with a comparable number of carbon atoms. Alkenes (also known as *olefins*) contain at least one double bond between adjacent carbon atoms. Alkynes (also known as *acetylenes*) contain at least one triple bond between adjacent carbon atoms.

> **Alkenes contain a carbon–carbon double bond.**
> **Alkynes contain a carbon–carbon triple bond.**

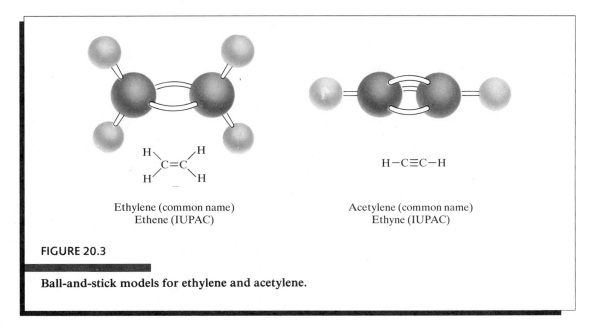

Ethylene (common name)
Ethene (IUPAC)

Acetylene (common name)
Ethyne (IUPAC)

FIGURE 20.3

Ball-and-stick models for ethylene and acetylene.

The simplest alkene is ethylene (or ethene), $CH_2{=}CH_2$, and the simplest alkyne is acetylene (or ethyne), $CH{\equiv}CH$ (Figure 20.3). Ethylene and acetylene are the first members of homologous series in which the formulas of successive members differ by increments of CH_2 (for example, $CH_2{=}CH_2$, $CH_3CH{=}CH_2$, and $CH_3CH_2CH{=}CH_2$). Huge quantities of alkenes are made by cracking and dehydrogenating alkanes during the processing of crude oils. These alkenes are used to manufacture motor fuels, polymers, and petrochemicals. Alkene molecules, like those of alkanes, have very little polarity. Hence, the physical properties of alkenes are very similar to those of the corresponding saturated hydrocarbons.

General formula for alkenes: C_nH_{2n}
General formula for alkynes: C_nH_{2n-2}

Table 20.5 gives the names and formulas for several alkenes and alkynes.

TABLE 20.5

Names and Formulas for Several Alkenes and Alkynes

Formula	IUPAC name	Common name
$CH_2{=}CH_2$	Ethene	Ethylene
$CH_3CH{=}CH_2$	Propene	Propylene
$CH_3CH_2CH{=}CH_2$	1-Butene	Butylene
$CH_3CH{=}CHCH_3$	2-Butene	
$CH_3C{=}CH_2$ $\quad\mid$ $\quad CH_3$	2-Methylpropene	Isobutylene
$CH{\equiv}CH$	Ethyne	Acetylene
$CH_3C{\equiv}CH$	Propyne	Methyl acetylene
$CH_3CH_2C{\equiv}CH$	1-Butyne	Ethyl acetylene
$CH_3C{\equiv}CCH_3$	2-Butyne	Dimethyl acetylene

20.11   Naming Alkenes and Alkynes

The names of alkenes and alkynes are derived from the names of corresponding alkanes. To name an alkene (or alkyne) by the IUPAC System:

1. Select the longest carbon–carbon chain that contains the double or triple bond.
2. Name this parent compound as you would an alkane but change the *ane* ending to *ene* for an alkene or to *yne* for an alkyne; for example, propane is changed to propene or propyne.

$$CH_3CH_2CH_3 \qquad CH_3CH{=}CH_2 \qquad CH_3C{\equiv}CH$$

Propane   Propene   Propyne

3. Number the carbon chain of the parent compound starting with the end nearer to the double or triple bond. Use the smaller of the two numbers on the double- or triple-bonded carbon atoms to indicate the position of the double or triple bond. Place this number in front of the alkene or alkyne name; for example, 2-butene means that the carbon–carbon double bond is between C-2 and C-3.
4. Side chains and other groups are treated as in naming alkanes. Name the substituent group, and designate its position on the parent chain with a number.

Study the following examples of named alkenes and alkynes:

$$\overset{4}{C}H_3\overset{3}{C}H_2\overset{2}{C}H=\overset{1}{C}H_2$$

1-Butene

$$\overset{1}{C}H_3\overset{2}{C}H=\overset{3}{C}H\overset{4}{C}H_3$$

2-Butene

$$\overset{6}{C}H_3\overset{5}{C}H_2\overset{4}{C}H_2$$
$$\overset{3}{\diagdown}\overset{2}{C}H\overset{2}{C}H=\overset{1}{C}H_2$$
$$CH_3CH_2CH_2$$

3-Propyl-1-hexene

$$\overset{4}{C}H_3\overset{3}{C}H_2\overset{2}{C}\equiv\overset{1}{C}H$$

1-Butyne

$$\overset{1}{C}H_3-\overset{2}{C}\equiv\overset{3}{C}-\overset{4}{\underset{\underset{CH_3}{|}}{C}H}-\overset{5}{C}H-\overset{6}{C}H_3$$
(with CH₃ above C-4)

4,5-Dimethyl-2-hexyne

To write a structural formula from a systematic name, the naming process is reversed. For example, how would we write the structural formula for 4-methyl-2-pentene? The name indicates: (1) five carbons in the longest chain, (2) a double bond between C-2 and C-3, and (3) a methyl group C-4.

Write five carbon atoms in a row. Place a double bond between C-2 and C-3, and place a methyl group on C-4. Now add hydrogen atoms to give each carbon atom four bonds. Carbons 1 and 5 each need three H atoms; C-2, C-3, and C-4 each need one H atom.

$$\overset{1}{C}-\overset{2}{C}=\overset{3}{C}-\overset{4}{\underset{\underset{CH_3}{|}}{C}}-\overset{5}{C}$$

Carbon skeleton

$$CH_3CH=CHCHCH_3$$
$$\underset{CH_3}{|}$$

4-Methyl-2-pentene

EXAMPLE 20.5

Write structural formulas for (a) 7-methyl-2-octene and (b) 3-hexyne.

(a) Octene, like octane, indicates an eight-carbon chain. The chain contains a double bond between C-2 and C-3 and a methyl group on C-7. Write eight carbon atoms in a row, place a double bond between C-2 and C-3, and place a methyl group on C-7. Now add hydrogen atoms to give each carbon atom four bonds.

$$\overset{1}{C}-\overset{2}{C}=\overset{3}{C}-\overset{4}{C}-\overset{5}{C}-\overset{6}{C}-\overset{7}{\underset{\underset{CH_3}{|}}{C}}-\overset{8}{C}$$

$$CH_3CH=CHCH_2CH_2CH_2CHCH_3$$
$$\underset{CH_3}{|}$$

7-Methyl-2-octene

(b) The stem *hex* indicates a six-carbon chain; the suffix *yne* indicates a carbon–carbon triple bond; the number 3 locates the triple bond between C-3 and C-4. Write six carbon atoms in a row and place a triple bond between C-3 and C-4.

Now add hydrogen atoms to give each carbon atom four bonds. Carbons 3 and 4 do not need any H atoms.

$$\overset{1}{C}-\overset{2}{C}-\overset{3}{C}\equiv\overset{4}{C}-\overset{5}{C}-\overset{6}{C} \qquad CH_3CH_2C\equiv CCH_2CH_3$$

3-Hexyne

PRACTICE   Name the following compounds.

(a) $CH_3CHCH{=}CHCHCH_3$ (b) $CH_3C\equiv CCHCH_2CH_2CH_2CH_3$
         | | |
       CH_3 CH_2CH_3 CH_2CH_3

Answers: (a) 2,5-dimethyl-3-heptene   (b) 4-ethyl-2-octyne

20.12   Reactions of Alkenes

The alkenes are much more reactive than the corresponding alkanes. This greater reactivity is due to the carbon–carbon double bonds. Once again we will limit our discussion to one type of reaction, an addition reaction.

addition reaction

Addition   In organic chemistry a reaction in which two substances join together to produce one compound is called an **addition reaction**. Addition at the carbon–carbon double bond is the most common reaction of alkenes. Hydrogen, halogens (Cl_2 or Br_2), hydrogen halides, and water are some of the reagents that can be added to unsaturated hydrocarbons. Ethylene (ethene), for example, reacts in this fashion:

$$CH_2{=}CH_2 + H_2 \xrightarrow[\text{1 atm}]{\text{Pt, 25°C}} CH_3-CH_3$$

Ethylene   Ethane

$$CH_2{=}CH_2 + Br-Br \longrightarrow CH_2Br-CH_2Br$$

1,2-Dibromoethane

Visible evidence of the Br_2 addition is the disappearance of the reddish brown color of bromine as it reacts.

$$CH_2{=}CH_2 + HCl \longrightarrow CH_3CH_2Cl$$

Chloroethane
(Ethyl chloride)

$$CH_2{=}CH_2 + HOH \xrightarrow{\text{H}^+} CH_3CH_2OH$$

Ethanol
(Ethyl alcohol)

(The H^+ indicates that the reaction is carried out under acidic conditions.)

Note that the double bond is broken and the unsaturated alkene molecules become saturated by an addition reaction.

The preceding examples dealt with ethylene, but reactions of this kind can be made to occur with almost any molecule that contains a carbon–carbon double bond.

20.13 Acetylene

Acetylene (ethyne), $HC\equiv CH$, is the simplest alkyne and is an industrial chemical of prime importance. It can be prepared from relatively inexpensive raw materials—lime and coke—or from methane. Lime and coke are heated to form calcium carbide, which is then reacted with water to form acetylene:

$$\underset{\substack{\text{Calcium oxide} \\ \text{(Lime)}}}{CaO} + \underset{\substack{\text{Carbon} \\ \text{(Coke)}}}{3\,C} \xrightarrow[\text{Electric furnace}]{2500°C} \underset{\text{Calcium carbide}}{CaC_2} + CO$$

$$CaC_2 + 2\,H_2O \longrightarrow \underset{\text{Acetylene}}{CH\equiv CH} + Ca(OH)_2$$

Until about 1940, nearly all acetylene was made from calcium carbide. Then cheaper methods of production, using methane as the raw material, were devised. When methane is subjected to high temperatures for a very short time (0.01–0.1 second) and the products are cooled rapidly, a reasonably high yield of acetylene is obtained. The equation for the reaction, which is endothermic, is

$$\underset{\text{Methane}}{2\,CH_4} \xrightarrow{1500°C} \underset{\text{Acetylene}}{CH\equiv CH} + 3\,H_2 - 400\text{ kJ }(-95.5\text{ kcal})$$

Special precautions must be taken when handling acetylene. Like all hydrocarbon gases, it forms explosive mixtures with air or oxygen. In addition to this hazard, acetylene is unstable at room temperature. When highly compressed or liquefied, acetylene may decompose violently (explode) either spontaneously or from a slight shock:

$$CH\equiv CH \longrightarrow H_2 + 2\,C + 227\text{ kJ }(54.3\text{ kcal})$$

To eliminate the danger of explosion, acetylene is dissolved under pressure in acetone and is packed in cylinders that contain a porous inert material.

Acetylene is used mainly (1) as fuel for oxyacetylene cutting and welding torches and (2) as an intermediate in the manufacture of other substances. Both uses are dependent on the great reactivity of acetylene. Acetylene and oxygen mixtures produce flame temperatures of about 2800°C.

20.14   Aromatic Hydrocarbons: Structure

aromatic
compound

Benzene and all substances that have structures and chemical properties resembling benzene are classified as **aromatic compounds**. The word *aromatic* originally referred to the rather pleasant odor possessed by many of these substances, but this meaning has been dropped. Benzene, the parent substance of the aromatic hydrocarbons, was first isolated by Michael Faraday in 1825; its correct molecular formula, C_6H_6, was established a few years later. The establishment of a reasonable structural formula that would account for the properties of benzene was a very difficult problem for chemists in the mid–19th century.

Finally, in 1865, August Kekulé proposed that the carbon atoms in a benzene molecule are arranged in a six-membered ring with one hydrogen atom bonded to each carbon atom and with three carbon–carbon double bonds:

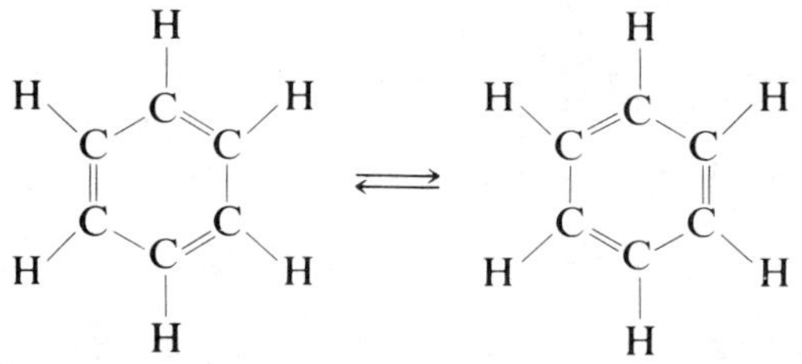

Kekulé's concepts are a landmark in the history of chemistry. They are the basis of the best representation of the benzene molecule devised in the 19th century, and they mark the beginning of our understanding of structure in aromatic compounds.

Kekulé's formulas have one serious shortcoming. They represent benzene and related substances as highly unsaturated compounds. Yet benzene does not readily undergo addition reactions like a typical alkene. For example, benzene does not decolorize bromine solutions rapidly. Instead, the chemical behavior of benzene resembles that of an alkane. Its typical reactions are the substitution type, wherein a hydrogen atom is replaced by some other group; for example,

$$C_6H_6 + Cl_2 \xrightarrow{\text{Fe}} C_6H_5Cl + HCl$$

Modern theory suggests that the benzene molecule is a hybrid of the two Kekulé structures shown above.

For convenience, the present-day chemist usually writes the structure of benzene as one or the other of these abbreviated forms:

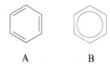

A   B

In both representations, it is understood that there is a carbon atom and a hydrogen atom at each corner of the hexagon. The classical Kekulé structure is represented by A; the modern molecular orbital structure is represented by B. Hexagonal structures, like A or B above, are used in representing the structural formulas of benzene derivatives—that is, substances in which one or more hydrogen atoms in the ring have been replaced by other atoms or groups. Chlorobenzene (C_6H_5Cl), for example, is written in this fashion:

Chlorobenzene, C_6H_5Cl

This notation indicates that the chlorine atom has replaced a hydrogen atom and is bonded directly to a carbon atom in the ring. Thus, the correct formula for chlorobenzene is C_6H_5Cl, not C_6H_6Cl.

20.15 Naming Aromatic Compounds

A substituted benzene is derived by replacing one or more hydrogen atoms of benzene by another atom or group of atoms. Thus, a monosubstituted benzene has the formula C_6H_5G, where G is the group replacing a hydrogen atom. Because all the hydrogen atoms in benzene are equivalent, it does not matter at which corner of the ring a monosubstituted group is written.

Monosubstituted Benzenes Some monosubstituted benzenes are named by adding the name of the substituent group as a prefix to the word *benzene*. The name is written as one word. Several examples follow:

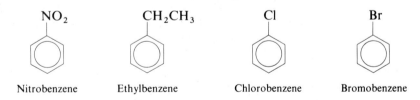

| Nitrobenzene | Ethylbenzene | Chlorobenzene | Bromobenzene |

Certain monosubstituted benzenes have special names. These are used as parent names for further substituted compounds, so they should be learned.

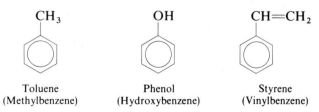

| Toluene (Methylbenzene) | Phenol (Hydroxybenzene) | Styrene (Vinylbenzene) |

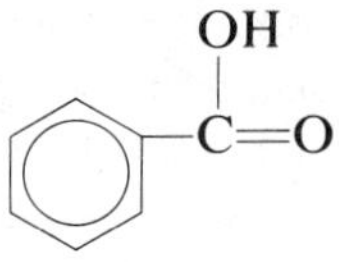

Benzoic acid
(Benzene carboxylic acid)

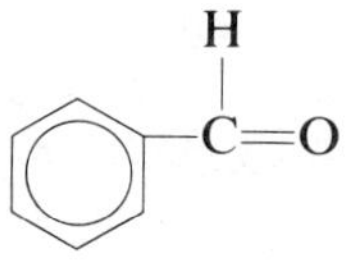

Benzaldehyde
(Benzene carboxaldehyde)

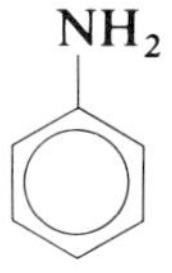

Aniline

The C_6H_5— group is known as the phenyl group (pronounced *fen-ill*), and the name *phenyl* is used for compounds that cannot easily be named as benzene derivatives. For example, the following compounds are named as derivatives of alkanes:

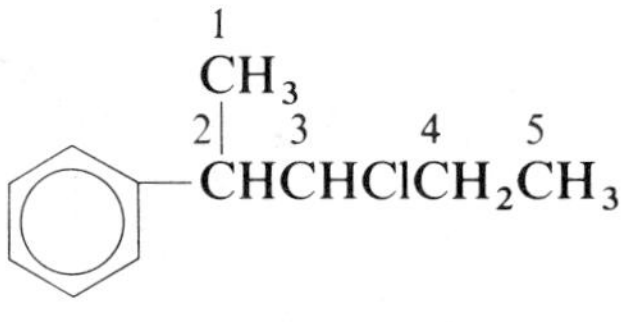

3-Chloro-2-phenylpentane

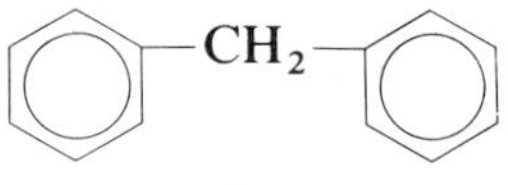

Diphenylmethane

Disubstituted Benzenes When two substituent groups replace two hydrogen atoms in a benzene molecule, three isomers are possible. The prefixes *ortho-*, *meta-*, and *para-* (abbreviated *o-*, *m-*, and *p-*) are used to name these disubstituted benzenes. In the ortho-compound the substituents are located on adjacent carbon atoms. In the meta-compound they are one carbon apart. And in the para-compound the substituents are located on opposite sides of the ring. When the two groups are different, name them alphabetically followed by the word *benzene*.

The dichlorobenzenes, $C_6H_4Cl_2$, illustrate this method of naming. Note that the three isomers have different physical properties, indicating that they are truly different substances. Note that the para isomer is a solid and the two others are liquids at room temperature.

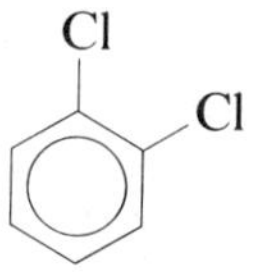

ortho-Dichlorobenzene
(1,2-Dichlorobenzene)
mp −17.2°C, bp 180.4°C

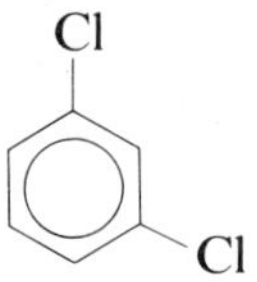

meta-Dichlorobenzene
(1,3-Dichlorobenzene)
mp −24.8°C, bp 172°C

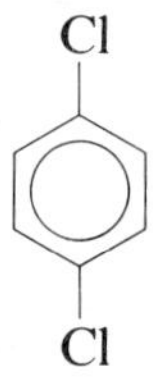

para-Dichlorobenzene
(1,4-Dichlorobenzene)
mp 53.1°C, bp 174.4°C

The dimethylbenzenes have the special name *xylene*. Also shown is one of the bromochlorobenzenes.

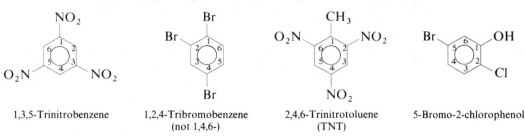

ortho-Xylene   meta-Xylene   para-Xylene   p-Bromochlorobenzene

When one of the substituents corresponds to a monosubstituted benzene that has a special name, the disubstituted compound is named as a derivative of that parent compound. In the following examples the parent compounds are phenol, aniline, toluene, and benzoic acid.

o-Nitrophenol   p-Bromoaniline   m-Nitrotoluene   p-Aminobenzoic acid

Tri- and Polysubstituted Benzenes   When a benzene ring has more than two substituents, the carbon atoms in the ring are numbered. Numbering starts at one of the substituted groups and may go either clockwise or counterclockwise, but must be done in the direction that gives the lowest possible numbers to the substituent groups. When the compound is named as a derivative of one of the special parent compounds, the substituent of the parent compound is considered to be on carbon 1 of the ring (the CH_3 group is on C-1 in 2,4,6-tribromotoluene). The following examples illustrate this system:

1,3,5-Trinitrobenzene   1,2,4-Tribromobenzene (not 1,4,6-)   2,4,6-Trinitrotoluene (TNT)   5-Bromo-2-chlorophenol

EXAMPLE 20.6   Write formulas and names for all the possible isomers of (a) chloronitrobenzene $[C_6H_4Cl(NO_2)]$ and (b) tribromobenzene $(C_6H_3Br_3)$.

(a) The name and formula indicate a chloro group (Cl) and a nitro group (NO_2) attached to a benzene ring. There are six positions in which to place these two groups. They can be ortho-, meta-, or para- to each other.

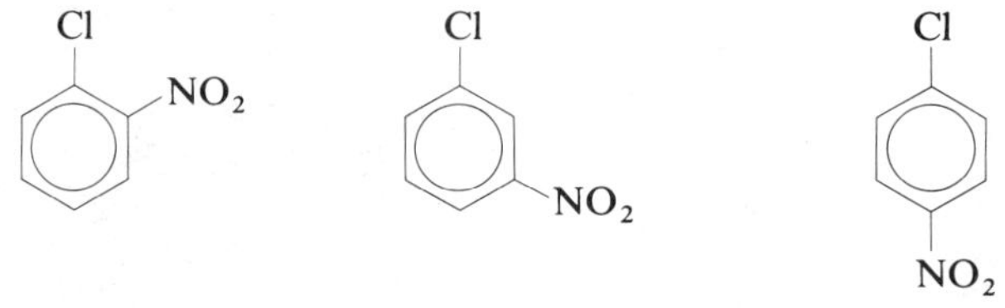

o-Chloronitrobenzene   m-Chloronitrobenzene   p-Chloronitrobenzene

(b) For tribromobenzene start by placing the three bromo groups in the 1, 2, and 3 positions; then the 1, 2, and 4 positions; and so on until all the possible isomers are formed. Name each isomer to check that no duplicate formulas have been written.

1,2,3-Tribromobenzene   1,2,4-Tribromobenzene   1,3,5-Tribromobenzene

Tribromobenzene has only three isomers. However, if one Br is replaced by a Cl, six isomers are possible.

PRACTICE   Name the following compounds.

(a)   CH$_2$CH$_3$ (b)   OH   (c)   NH$_2$

NO$_2$

Cl   Br   Br

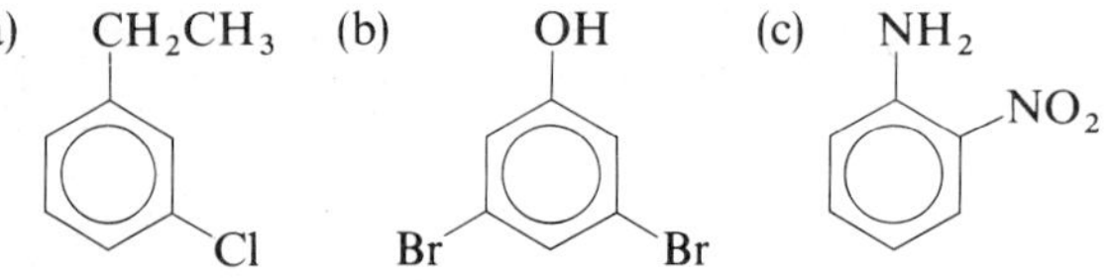

Answers: (a) *m*-chloroethylbenzene (b) 3,5-dibromophenol (c) *o*-nitroaniline

Polycyclic Aromatic Compounds

polycyclic or fused aromatic ring system

There are many other aromatic ring systems besides benzene. Their structures consist of two or more rings in which two carbon atoms are common to two rings. These compounds are known as **polycyclic or fused aromatic ring systems**. Three of the most common hydrocarbons in this category are naphthalene, anthracene, and phenanthrene. As you can see in the following structures, one hydrogen is attached to each carbon atom except at the carbons that are common to two rings.

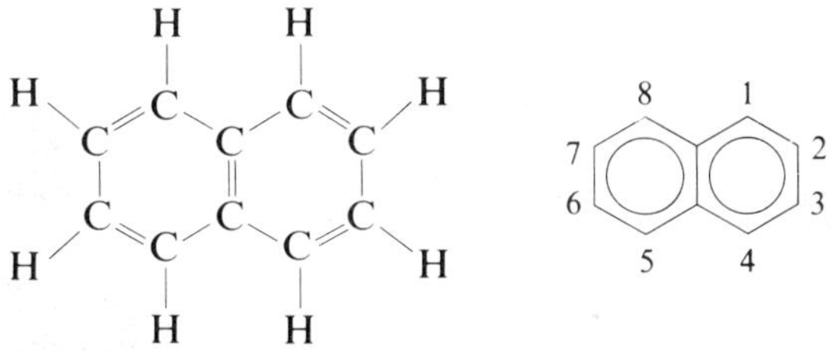

Naphthalene, $C_{10}H_8$

Anthracene, $C_{14}H_{10}$ 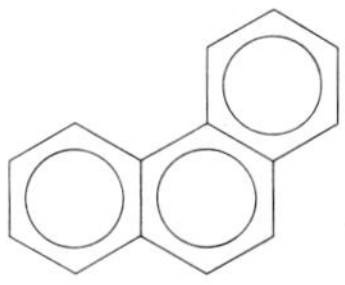   Phenanthrene, $C_{14}H_{10}$

All three of these substances may be obtained from coal tar. Naphthalene is known as mothballs and has been used as a moth repellent for many years.

20.16   Alcohols

alcohol

Alcohols are organic compounds whose molecules contain a hydroxyl (—OH) group bonded to a saturated carbon atom. Thus, if we substitute an —OH for an H in CH_4, we get CH_3OH, methyl alcohol. The functional group of the alcohols is —OH. The general formula for alcohols is ROH, where R is an alkyl or a substituted alkyl group.

Alcohols differ from metal hydroxides in that they do not dissociate or ionize in water. The —OH group is attached to the carbon atom by a covalent bond and not by an ionic bond. The alcohols form a homologous series, with methanol, CH_3OH, the first member of the series. Models illustrating the structural arrangements of the atoms in methanol and ethanol are shown in Figure 20.4.

primary alcohol

secondary alcohol

tertiary alcohol

Alcohols are classified as **primary** (1°), **secondary** (2°), or **tertiary** (3°), depending on whether the carbon atom to which the —OH group is attached is bonded to one, two, or three other carbon atoms, respectively. Generalized formulas for 1°, 2°, and 3° alcohols follow:

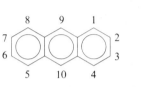

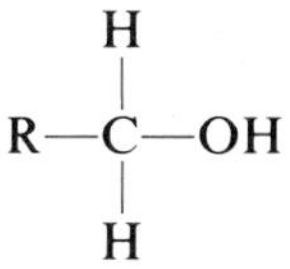

Primary alcohol 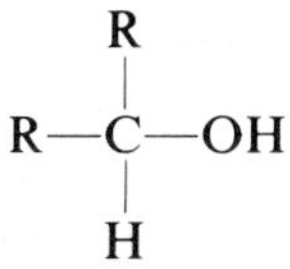   Secondary alcohol 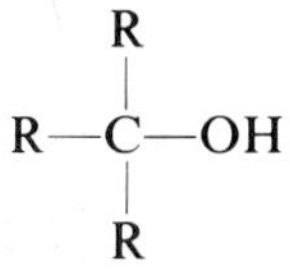   Tertiary alcohol

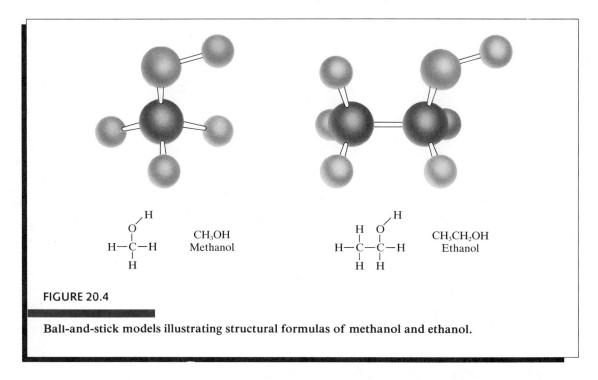

FIGURE 20.4

Ball-and-stick models illustrating structural formulas of methanol and ethanol.

Formulas of specific examples of these classes of alcohols are shown in Table 20.6. Methanol (CH_3OH) is grouped with the primary alcohols.

Molecular structures with more than one —OH group attached to a single carbon atom are generally not stable. But an alcohol molecule can contain two or more —OH groups if each —OH is attached to a different carbon atom. Accordingly, alcohols are also classified as monohydroxy, dihydroxy, trihydroxy, and so on, on the basis of the number of hydroxyl groups per molecule. **Polyhydroxy alcohols** are a general term for alcohols that have more than one —OH group per molecule.

polyhydroxy
alcohol

Methanol When wood is heated to a high temperature in an atmosphere lacking oxygen, methanol (wood alcohol) and other products are formed and driven off. The process is called *destructive distillation*, and until about 1925 nearly all methanol was obtained in this way. In the early 1920s the synthesis of methanol by high-pressure catalytic hydrogenation of carbon monoxide was developed in Germany. The reaction is

$$CO + 2\,H_2 \xrightarrow[\text{300–400°C, 200 atm}]{\text{ZnO–Cr}_2\text{O}_3} CH_3OH$$

Nearly all methanol is now manufactured by this method.

TABLE 20.6

Naming and Classification of Alcohols

Class	Formula	IUPAC name	Common name[a]	Boiling point, °C
Primary	CH_3OH	Methanol	Methyl alcohol	65.0
Primary	CH_3CH_2OH	Ethanol	Ethyl alcohol	78.5
Primary	$CH_3CH_2CH_2OH$	1-Propanol	*n*-Propyl alcohol	97.4
Primary	$CH_3CH_2CH_2CH_2OH$	1-Butanol	*n*-Butyl alcohol	118
Primary	$CH_3(CH_2)_3CH_2OH$	1-Pentanol	*n*-Amyl or *n*-pentyl alcohol	138
Primary	$CH_3(CH_2)_6CH_2OH$	1-Octanol	*n*-Octyl alcohol	195
Primary	$CH_3\underset{\underset{CH_3}{\mid}}{C}HCH_2OH$	2-Methyl-1-propanol	Isobutyl alcohol	108
Secondary	$CH_3\underset{\underset{OH}{\mid}}{C}HCH_3$	2-Propanol	Isopropyl alcohol	82.5
Secondary	$CH_3CH_2\underset{\underset{OH}{\mid}}{C}HCH_3$	2-Butanol	*sec*-Butyl alcohol	91.5
Tertiary	$CH_3\!-\!\underset{\underset{CH_3}{\mid}}{\overset{\overset{CH_3}{\mid}}{C}}\!-\!OH$	2-Methyl-2-propanol	*t*-Butyl alcohol	82.9
Dihydroxy	$HOCH_2CH_2OH$	1,2-Ethanediol	Ethylene glycol	197
Trihydroxy	$HOCH_2\underset{\underset{OH}{\mid}}{C}HCH_2OH$	1,2,3-Propanetriol	Glycerol or glycerine	290

[a] The abbreviations *n*, *sec*, and *t* stand for normal, secondary, and tertiary, respectively.

Methanol is a volatile (bp 65°C), highly flammable liquid. It is poisonous and capable of causing blindness or death if taken internally. Exposure to methanol vapors for even short periods of time is dangerous. Nevertheless over 8 billion lb (3.6×10^9 kg) annually are manufactured and used for

1. Conversion to formaldehyde (methanal) primarily for use in the manufacture of polymers
2. Manufacture of other chemicals, especially various kinds of esters
3. Denaturing ethyl alcohol (rendering it unfit as a beverage)
4. An industrial solvent
5. An inexpensive and temporary antifreeze for radiators (it is not a satisfactory permanent antifreeze, because its boiling point is lower than that of water)

The experimental use of 5% to 30% methanol or ethanol in gasoline (gasohol) has shown promising results in reducing the amount of air pollutants emitted in automobile exhausts. Another benefit of using methanol in gasoline is that methanol can be made from nonpetroleum sources. The most economical nonpetroleum source of carbon monoxide for making methanol is coal. In addition to coal, burnable materials such as wood, agricultural wastes, and sewage sludge also are potential sources of methanol.

Ethanol   Ethanol is without doubt the earliest and most widely known alcohol. It is or has been known by a variety of names such as ethyl alcohol, "alcohol," grain alcohol, methyl carbinol, spirit, and *aqua vitae*. The preparation of ethanol by fermentation is recorded in the Old Testament. Huge quantities of this substance are still prepared by fermentation. Starch and sugar are the raw materials. Starch is first converted to sugar by enzyme- or acid-catalyzed hydrolysis. (An enzyme is a biological catalyst.) Conversion of simple sugars to ethanol is accomplished by the yeast enzyme zymase:

$$C_6H_{12}O_6 \xrightarrow{\text{Zymase}} 2\ CH_3CH_2OH + 2\ CO_2$$

Glucose   Ethanol

For legal use in beverages ethanol is made by fermentation; but a large part of the alcohol for industrial uses [about 1.0 billion lb (4.5×10^8 kg) annually] is made from petroleum-derived ethylene. Ethylene is passed into an aqueous acid solution to form ethanol.

$$CH_2{=}CH_2 + H_2O \xrightarrow{H^+} CH_3CH_2OH$$

Some of the economically significant uses of ethanol are the following:

1. Intermediate in the manufacture of other chemicals such as acetaldehyde, acetic acid, ethyl acetate, and diethyl ether.
2. Solvent for many organic substances.
3. Compounding ingredient for pharmaceuticals, perfumes, flavorings, and so on.
4. Essential ingredient of alcoholic beverages.

Ethanol acts physiologically as a food, as a drug, and as a poison. It is a food in the limited sense that the body is able to metabolize small amounts of it to carbon dioxide and water with the production of energy. As a drug ethanol is often mistakenly considered to be a stimulant, but it is in fact a depressant. In moderate quantities ethanol causes drowsiness and depresses brain functions so that activities requiring skill and judgment (such as automobile driving) are impaired. In larger quantities ethanol causes nausea, vomiting, impaired perception, and incoordination. If a very large amount is consumed, unconsciousness and ultimately death may occur.

Authorities maintain that the effects of ethanol on automobile drivers are a factor in about half of all fatal traffic accidents in the United States. The gravity of this problem can be grasped when you realize that traffic accidents are responsible for more than 50,000 deaths each year in the United States.

Ethanol for industrial use is often denatured (rendered unfit for drinking). Denaturing is done by adding small amounts of methanol and other denaturants that are extremely difficult to remove. Denaturing is required by the federal government to protect the beverage alcohol tax source. Special tax-free use permits are issued to scientific and industrial users who require pure ethanol for nonbeverage uses.

Other widely used alcohols are: (1) isopropyl alcohol (2-propanol), the principal ingredient in rubbing alcohol formulations; (2) ethylene glycol, which is the main compound in permanent-type antifreezes, is used in the manufacture of Dacron polyester synthetic fibers and is widely used in the paint industry; (3) glycerol, also known as glycerine, which is a syrupy, sweet-tasting liquid. Its major uses are in the manufacture of polymers and explosives, as an emollient in cosmetics, as a humectant in tobacco, and as a sweetener.

20.17   Naming Alcohols

If you know how to name alkanes, it is easy to name alcohols by the IUPAC System. Unfortunately, several of the alcohols are generally known by common or nonsystematic names, so it is often necessary to know more than one name for a given alcohol. The common name is usually formed from the name of the alkyl group that is attached to the —OH group, followed by the word *alcohol*. See examples given below and in Table 20.6. To name an alochol by the IUPAC System:

1. Select the longest continuous chain of carbon atoms containing the hydroxyl group.
2. Number the carbon atoms in this chain so that the one bearing the —OH group has the lowest possible number.
3. Form the parent alcohol name by replacing the final *e* of the corresponding alkane name by *ol*. When isomers are possible, locate the position of the —OH by placing the number (hyphenated) of the carbon atom to which the —OH is bonded immediately before the parent alcohol name.
4. Name each alkyl side chain (or other group) and designate its position by number.

Study the application of this naming system to these examples and to those shown in Table 20.6.

$$\overset{3}{C}H_3 - \overset{2}{C}H_2 - \overset{1}{C}H_2OH$$

1-Propanol
(*n*-Propyl alcohol)

$$\overset{1}{C}H_3 - \overset{2}{C}H - \overset{3}{C}H_3$$
$$| \\ OH$$

2-Propanol
(Isopropyl alcohol)

$$
\begin{array}{cccc}
\overset{4}{C}H_3\!-\!\overset{3}{C}H\!-\!\overset{2}{C}H_2\!-\!\overset{1}{C}H_2OH & & \overset{2}{H}OCH_2\!-\!\overset{1}{C}H_2OH \\
\quad\quad | & & \\
\quad\quad CH_3 & &
\end{array}
$$

1,2-Ethanediol
(Ethylene glycol)

3-Methyl-1-butanol
(Isoamyl alcohol or
isopentyl alcohol)

EXAMPLE 20.7   Name this alcohol by the IUPAC method.

$$
CH_3CH_2CHCH_2CHCH_3
$$
$$
\qquad\quad | \qquad\quad |
$$
$$
\qquad\quad CH_3 \quad\; OH
$$

Step 1 The longest continuous carbon chain containing the —OH group has six carbon atoms.

Step 2 This carbon chain is numbered from right to left so that the —OH group has the smallest possible number. In this case, the —OH is on C-2.

$$
\overset{6}{C}\!-\!\overset{5}{C}\!-\!\overset{4}{C}\!-\!\overset{3}{C}\!-\!\overset{2}{C}\!-\!\overset{1}{C}
$$
$$
\qquad\qquad | \qquad\; |
$$
$$
\qquad\qquad CH_3 \quad\; OH
$$

Step 3 The name of the six-carbon alkane is hexane. Replace the final *e* in hexane by *ol*, forming the name hexanol. Since the —OH is on C-2, place a 2- before hexanol to give the parent alcohol name 2-hexanol.

Step 4 A methyl group (—CH₃) is located on carbon 4. Therefore the full name of the compound is 4-methyl-2-hexanol.

EXAMPLE 20.8   Write the structural formula of 3,3-dimethyl-2-hexanol.

Step 1 The 2-hexanol refers to a six-carbon chain with an —OH group on C-2. Write the skeleton structure with an —OH on C-2. Now place the two methyl groups (3,3-dimethyl) on C-3.

$$
\overset{1}{C}\!-\!\overset{2}{C}\!-\!\overset{3}{C}\!-\!\overset{4}{C}\!-\!\overset{5}{C}\!-\!\overset{6}{C} \qquad\qquad \overset{1}{C}\!-\!\overset{2}{C}\!-\!\overset{3}{C}\!-\!\overset{4}{C}\!-\!\overset{5}{C}\!-\!\overset{6}{C}
$$

$$
\qquad | \qquad\qquad\qquad\qquad\qquad\qquad | \quad |
$$
$$
\qquad OH \qquad\qquad\qquad\qquad\qquad HO \;\; CH_3
$$

with CH₃ on C-3 of the right structure.

Step 2 Finally, add H atoms to give each carbon atom four bonds:

$$
\qquad\qquad\qquad CH_3
$$
$$
\qquad\qquad\qquad |
$$
$$
CH_3CH\!-\!C\!-\!CH_2CH_2CH_3
$$
$$
\qquad\; | \qquad |
$$
$$
\qquad\; OH \quad CH_3
$$

3,3-Dimethyl-2-hexanol

PRACTICE Classify and name the following alcohols.

(a) $CH_3CH_2CHCH_2CH_2CHCH_3$ (b) $CH_3\overset{\overset{\displaystyle OH}{|}}{C}CH_2CHCH_3$
        $\underset{OH}{|}$ $\underset{CH_3}{|}$ $\underset{CH_3}{|}$ $\underset{CH_3}{|}$

Answers: (a) Secondary; 6-methyl-3-heptanol. (b) Tertiary; 2,4-dimethyl-2-pentanol.

20.18 Reactions of Alcohols

We will consider only two of the many chemical reactions of alcohols—oxidation and reaction with active metals.

Oxidation Alcohols may be oxidized to aldehydes, ketones, and carboxylic acids. Primary alcohols yield aldehydes, which on further oxidation yield carboxylic acids; secondary alcohols are oxidized to ketones; tertiary alcohols resist oxidation.

Primary alcohol $\xrightarrow{\text{Oxidation}}$ Aldehyde $\xrightarrow{\text{Oxidation}}$ Carboxylic acid

Secondary alcohol $\xrightarrow{\text{Oxidation}}$ Ketone

Tertiary alcohol $\xrightarrow{\text{Oxidation}}$ No reaction

Because our main interest is in the changes that occur in the functional groups, we often use abbreviated equations to convey this information. A balanced ionic equation is also shown.

$$CH_3CH_2OH \xrightarrow[\substack{H^+,\,K_2Cr_2O_7,\\ \text{or } KMnO_4}]{H_2O,\,\Delta} CH_3\overset{\overset{\displaystyle H}{|}}{C}{=}O \xrightarrow[\text{or } KMnO_4]{H^+,\,K_2Cr_2O_7,} CH_3COOH$$

Ethanol   Ethanal   Acetic acid
                               (an aldehyde)   (a carboxylic acid)

$$CH_3\underset{\underset{OH}{|}}{C}HCH_3 \xrightarrow[\substack{H^+,\,K_2Cr_2O_7,\\ \text{or } KMnO_4}]{H_2O,\,\Delta} CH_3\underset{\underset{O}{\|}}{C}CH_3$$

2-Propanol   Propanone (Acetone)
                                 (a ketone)

$$3\,CH_3CH_2OH + Cr_2O_7^{2-} + 8\,H^+ \xrightarrow{\Delta} 3\,CH_3\overset{\overset{\displaystyle H}{|}}{C}{=}O + 2\,Cr^{3+} + 7\,H_2O$$

Reaction with Active Metals Although alcohols are less acidic than water, they react with active metals such as sodium and potassium to replace the hydrogen atom of the alcohol —OH group. The products formed are a salt and hydrogen gas. The salt is an ionic compound and is called an alkoxide (CH_3CH_2ONa is sodium ethoxide). The equation is

$$2\,ROH + 2\,Na \longrightarrow 2\,RONa + H_2(g)$$
Alkoxide

$$2\,CH_3CH_2OH + 2\,Na \longrightarrow 2\,CH_3CH_2ONa + H_2(g)$$
Ethanol   Sodium ethoxide

The order of reactivity of alcohols with sodium is primary > secondary > tertiary. Methanol and ethanol react vigorously with sodium, but not as vigorously as sodium and water. The reactivity of an alcohol also decreases as its molar mass increases, because the —OH group becomes a relatively smaller and less significant part of the molecule.

20.19   Ethers

ether

Ethers have the general formula ROR′. The two groups, R and R′, may be derived from saturated, unsaturated, or aromatic hydrocarbons and, for a given ether, may be alike or different. Table 20.7 shows structural formulas and names for some of the different kinds of ethers.

Saturated ethers have little chemical reactivity, but, because they readily dissolve a great many organic substances, ethers are often used as solvents in both laboratory and manufacturing operations.

Alcohols (ROH) and ethers (ROR′) are isomeric, having the same molecular formula but different structural formulas. For example, the molecular formula for ethanol and dimethyl ether is C_2H_6O but the structural formulas are

$$CH_3CH_2OH \qquad CH_3—O—CH_3$$
Ethanol   Dimethyl ether

These two molecules are extremely different in both physical and chemical properties. Ethanol boils at 78.3°C, and dimethyl ether boils at −23.7°C. Ethanol is capable of intermolecular hydrogen bonding and therefore has a much higher boiling point. It also has a greater solubility in water than dimethyl ether. Ethanol reacts with metallic sodium to give hydrogen and sodium ethylate and can also be oxidized by dichromate and permanganate solutions. Dimethyl ether is non-reactive with all these reagents.

Naming Ethers   Individual ethers, like alcohols, may be known by several names. The ether having the formula $CH_3CH_2—O—CH_2CH_3$ and formerly

TABLE 20.7

Names and Structural Formulas of Ethers

Name[a]	Formula	Boiling point (°C)
Dimethyl ether (Methoxymethane)	$CH_3—O—CH_3$	−24
Methyl ethyl ether (Methoxyethane)	$CH_3CH_2—O—CH_3$	8
Diethyl ether (Ethoxyethane)	$CH_3CH_2—O—CH_2CH_3$	35
Ethyl isopropyl ether (2-Ethoxypropane)	$CH_3CH_2—O—\underset{\underset{CH_3}{\mid}}{C}HCH_3$	54
Divinyl ether	$CH_2{=}CH—O—CH{=}CH_2$	39
Anisole (Methoxybenzene)	⬡—OCH_3	154

[a] The IUPAC name is in parentheses when given.

widely used as an anesthetic is called diethyl ether, ethyl ether, ethoxyethane, or simply ether. Common names of ethers are formed from the names of the groups attached to the oxygen atom followed by the word *ether*.

$$CH_3—\boxed{O}—CH_3 \qquad CH_3—\boxed{O}—CH_2CH_3$$

Methyl   Ether   Methyl   Methyl   Ether   Ethyl

Dimethyl ether   Methyl ethyl ether

In the IUPAC System, ethers are named as alkoxy (RO—) derivatives of the alkane corresponding to the longest carbon–carbon chain in the molecule. To name an ether by this system,

1. Select the longest carbon–carbon chain and label it with the name of the corresponding alkane.
2. Change the *yl* ending of the other hydrocarbon group to *oxy* to obtain the alkoxy group name. For example, CH_3O— is called *methoxy*.
3. Combine the two names from Steps 1 and 2, giving the alkoxy name first, to form the ether name.

$$CH_3—O—CH_2CH_3$$

This is the longest carbon–carbon chain, so call it *ethane*.

This is the other radical; modify its name to *methoxy* and combine with ethane to obtain the name of the ether, *methoxyethane*.

Thus,

$CH_3CH_2—O—CH_2CH_3$ is ethoxyethane

$CH_3CH_2CH_2—O—CH_2CH_2CH_2CH_3$ is *n*-propoxybutane

Preparation of Ethers A general method for preparing ethers is by reacting sodium alkoxides with alkyl halides (Williamson synthesis).

$$RONa \quad + \quad R'X \quad \longrightarrow ROR' + \quad NaX$$

Sodium alkoxide   Alkyl halide   Ether   Sodium halide

The Williamson synthesis is especially useful in the preparation of mixed ethers (where $R \neq R'$) and aromatic ethers.

$$CH_3CH_2ONa + CH_3Br \longrightarrow CH_3CH_2—O—CH_3 + NaBr$$

Sodium ethoxide   Bromo-methane   Methoxyethane (Methyl ethyl ether)

20.20   Aldehydes and Ketones

carbonyl group

aldehyde

ketone

The aldehydes and ketones are closely related classes of compounds. Their structures contain the **carbonyl group**, $>C=O$, a carbon double-bonded to oxygen. **Aldehydes** have at least one hydrogen atom bonded to the carbonyl group, whereas **ketones** have two alkyl or aryl (aromatic, Ar) groups bonded to the carbonyl group.

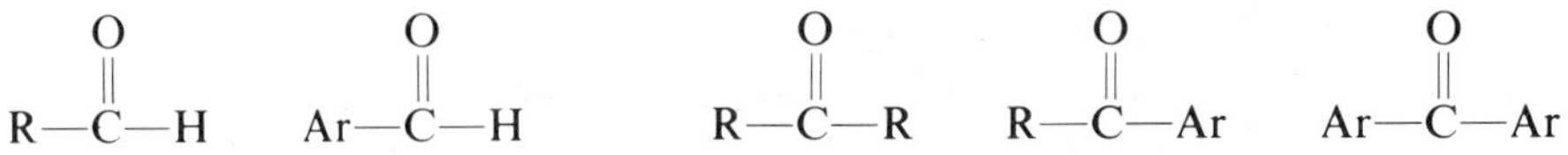

$$\underset{\text{Aldehydes}}{\overset{\displaystyle O}{R—\overset{\|}{C}—H} \quad \overset{\displaystyle O}{Ar—\overset{\|}{C}—H}} \qquad \underset{\text{Ketones}}{\overset{\displaystyle O}{R—\overset{\|}{C}—R} \quad \overset{\displaystyle O}{R—\overset{\|}{C}—Ar} \quad \overset{\displaystyle O}{Ar—\overset{\|}{C}—Ar}}$$

In a linear expression the aldehyde group is often written as CHO or CH=O. For example,

$$CH_3CHO \quad \text{is equivalent to} \quad CH_3\overset{\displaystyle O}{\overset{\|}{C}}—H$$

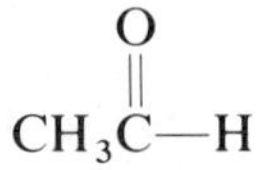

In the linear expression of a ketone, the carbonyl group is written as CO; for example,

$$CH_3COCH_3 \quad \text{is equivalent to} \quad CH_3\overset{\displaystyle O}{\overset{\displaystyle \|}{C}}CH_3$$

Formaldehyde is the most widely used aldehyde. It is a poisonous, irritating gas that is very soluble in water. It is marketed as a 40% aqueous solution called *formalin*. Since formaldehyde is a powerful germicide, it is used in embalming and to preserve biological specimens. Formaldehyde is also used for disinfecting dwellings, ships, and storage houses; for destroying flies; for tanning hides; and as a fungicide for plants and vegetables. But by far the largest use of this chemical is in the manufacture of polymers. About 2.1 billion pounds (9.6×10^8 kg) of formaldehyde are manufactured annually in the United States. Formaldehyde vapors are intensely irritating to the mucous membranes. Ingestion may cause severe abdominal pains, leading to coma and death.

Acetone and methyl ethyl ketone are widely used organic solvents. Acetone, in particular, is used in very large quantities for this purpose. U.S. production of acetone is about 1.9 billion pounds annually. It is used as a solvent in the manufacture of drugs, chemicals, and explosives; for removal of paints, varnishes, and fingernail polish; and as a solvent in the plastics industry. Methyl ethyl ketone (MEK) is also widely used as a solvent, especially for lacquers.

20.21 Naming Aldehydes and Ketones

Aldehydes The IUPAC names of aliphatic aldehydes are obtained by dropping the final *e* and adding *al* to the name of the parent hydrocarbon (that is, the longest carbon–carbon chain carrying the —CHO group). The aldehyde carbon is always at the end of the carbon chain, is understood to be carbon number 1, and does not need to be numbered. The first member of the homologous series, $H_2C\!=\!O$, is methanal. The name *methanal* is derived from the hydrocarbon methane, which contains one carbon atom. The second member of the series is ethanal; the third member of the series is propanal; and so on.

$$CH_4 \qquad H-\overset{\displaystyle O}{\overset{\displaystyle \|}{C}}-H \qquad CH_3CH_3 \qquad CH_3\overset{\displaystyle O}{\overset{\displaystyle \|}{C}}-H$$

Methane   Methanal   Ethane   Ethanal
(from methan*e* + *al*)   (from ethan*e* + *al*)

The longest carbon chain containing the aldehyde group is the parent compound. Other groups attached to this chain are numbered and named as we have done previously. For example,

$$
\begin{array}{c}
\overset{\displaystyle CH_3}{|} \qquad \overset{\displaystyle O}{\|} \\
CH_3CH_2CHCH_2CH_2C\!-\!H
\end{array}
$$

4-Methylhexanal

Common names for some aldehydes are widely used. The common names for the aliphatic aldehydes are derived from the common names of the carboxylic acids (see Table 20.8 p. 594). The *ic acid* or *oic acid* ending of the acid name is dropped and is replaced with the suffix *aldehyde*. Thus the name of the one-carbon acid, formic acid, becomes formaldehyde for the one-carbon aldehyde.

$$
\overset{O}{\overset{\|}{H\!-\!C\!-\!OH}} \qquad \overset{O}{\overset{\|}{H\!-\!C\!-\!H}} \qquad \overset{O}{\overset{\|}{CH_3C\!-\!OH}} \qquad \overset{O}{\overset{\|}{CH_3C\!-\!H}}
$$

Formic acid   Formaldehyde   Acetic acid   Acetaldehyde

The IUPAC name of a ketone is derived from the name of the alkane corresponding to the longest carbon chain that contains the ketone carbonyl group. The parent name is formed by changing the *e* ending of the alkane to *one*. If the chain is longer than four carbons, it is numbered so that the carbonyl carbon has the smallest number possible; and this number is prefixed to the name of the ketone. Other groups bonded to the parent chain are named and numbered as previously indicated for hydrocarbons and alcohols. See the following examples:

$$
\overset{O}{\overset{\|}{CH_3\!-\!C\!-\!CH_3}} \qquad \overset{O}{\overset{\;\;\;2\|\;\;1}{\underset{5\;\;\;4\;\;\;3}{CH_3CH_2CH_2\!-\!C\!-\!CH_3}}} \qquad \overset{O}{\overset{3\|\;\;4\;\;5\;\;6}{\underset{1\;\;\;2}{CH_3CH_2\!-\!C\!-\!CHCH_2CH_3}}}
$$
$$
\underset{CH_3}{|}
$$

Propanone   2-Pentanone   4-Methyl-3-hexanone

Note that in 4-methyl-3-hexanone the carbon chain is numbered from left to right to give the ketone group the lowest possible number.

An alternate non-IUPAC method commonly used to name simple ketones is to list the names of the alkyl or aromatic groups attached to the carbonyl carbon together with the word *ketone*.

$$
\overset{O}{\overset{\|}{CH_3\!-\!C\!-\!CH_2CH_3}} \qquad\qquad \overset{O\quad\; CH_3}{\overset{\|\qquad\;|}{CH_3\!-\!C\!-\!CHCH_3}}
$$
$$
\overset{\uparrow\qquad\;\;\uparrow\qquad\;\;\uparrow}{\text{Methyl}\;\;\;\text{Ketone}\;\;\;\text{Ethyl}} \qquad\qquad \overset{\uparrow\qquad\;\;\uparrow\qquad\;\;\uparrow}{\text{Methyl}\;\;\;\text{Ketone}\;\;\;\text{Isopropyl}}
$$

(Methyl ethyl ketone)   (Methyl isopropyl ketone)

Two of the most widely used ketones have special common names: propanone is called acetone, and butanone is known as methyl ethyl ketone, or MEK.

Aromatic aldehydes are commonly named after the corresponding acids; for example, benzaldehyde from benzoic acid and *p*-tolualdehyde from *p*-toluic acid. Aromatic ketones are named in a similar fashion to aliphatic ketones and often have special names as well.

Benzaldehyde   *p*-Tolualdehyde   Methyl phenyl ketone
(Acetophenone, 1-phenylethanone)

EXAMPLE 20.9   Write the formulas and the names for the straight-chain five- and six-carbon aldehydes.

The IUPAC names are based on the five- and six-carbon alkanes. Drop the *e* of the alkane name and add the suffix *al*. Pentane (C_5) becomes pentanal and hexane (C_6) becomes hexanal. The common names are derived from valeric acid and caproic acid, respectively.

$$CH_3CH_2CH_2CH_2\overset{\overset{O}{\|}}{C}-H \qquad CH_3CH_2CH_2CH_2CH_2\overset{\overset{O}{\|}}{C}-H$$

Pentanal (Valeraldehyde)   Hexanal (Caproaldehyde)

EXAMPLE 20.10   Give two names for each of the following ketones:

$$\text{(a) } CH_3CH_2\overset{\overset{O}{\|}}{C}CH_2\overset{\overset{CH_3}{|}}{C}HCH_3 \qquad \text{(b) } CH_3CH_2CH_2\overset{\overset{O}{\|}}{C}-$$

(a) The parent carbon chain that contains the carbonyl group has six carbons. Number this chain from the end nearest to the carbonyl group. The ketone group is on C-3, and a methyl group is on C-5. The six-carbon alkane is hexane. Drop the *e* from hexane and add *one* to give the parent name hexanone. Prefix the name hexanone with 3- to locate the ketone group and with 5-methyl- to locate the methyl group. The name is 5-methyl-3-hexanone. The common name is ethyl isobutyl ketone since the C=O has an ethyl group and an isobutyl group bonded to it.

(b) The longest aliphatic chain has four carbons. The parent ketone name is butanone, derived by dropping the *e* of butane and adding *one*. The butanone has a phenyl group attached to carbon 1. The IUPAC name is therefore 1-phenyl-1-butanone. The common name for this compound is phenyl *n*-propyl ketone, since the C=O group has a phenyl and an *n*-propyl group bonded to it.

PRACTICE   Write the IUPAC names for the following molecules.

(a)
$$\begin{array}{c} O \\ \parallel \\ H-C-CHCH_2CH_2 \\ \qquad | \qquad | \\ \qquad CH_3 \quad CH_2-CH_3 \end{array}$$

(b)
$$\begin{array}{c} CH_3 \quad CH_3 \\ | \qquad | \\ CH_3CH_2CH_2CHCH_2CHCCH_3 \\ \qquad\qquad\qquad\qquad | \\ \qquad\qquad\qquad\qquad O \end{array}$$

Answers: (a) 2-methylhexanal   (b) 3,5-dimethyl-2-octanone

20.22   Reactions of Aldehydes and Ketones

Oxidation   We have seen (Section 20.18) that aldehydes are easily oxidized to carboxylic acids. Ketones resist oxidation except under drastic conditions where carbon–carbon bonds are broken and a variety of products are formed.

Since aldehydes are easily oxidized they can readily be distinguished from ketones. The Tollens' or silver-mirror test for aldehydes is based on the ability of silver ions to oxidize aldehydes. The Ag^+ ions are thereby reduced to metallic silver. In practice, a little of the suspected aldehyde is added to a solution of silver nitrate and ammonia in a clean test tube. The appearance of a silver mirror on the inner wall of the tube is a positive test for the aldehyde group. The abbreviated equation is

$$\begin{array}{c} H \\ | \\ RC{=}O \end{array} + 2\ Ag^+ \xrightarrow[H_2O]{NH_3} \begin{array}{c} O \\ \parallel \\ RC-O^- \end{array} NH_4^+ + 2\ Ag\downarrow \quad \text{(general reaction)}$$

$$\begin{array}{c} H \\ | \\ CH_3C{=}O \end{array} + 2\ Ag^+ \xrightarrow[H_2O]{NH_3} CH_3COO^-\ NH_4^+ + 2\ Ag\downarrow$$

Fehling's and Benedict's solutions contain Cu^{2+} ions in an alkaline medium. In the Fehling's and Benedict's tests, the aldehyde group is oxidized to a carboxylic acid by Cu^{2+} ions. The blue Cu^{2+} ions are reduced and form brick-red copper(I) oxide (Cu_2O), which precipitates during the reaction. These tests can be used for detecting carbohydrates that have an available aldehyde group. The abbreviated equation is

$$\begin{array}{c} H \\ | \\ RC{=}O \\ \end{array} + Cu^{2+} \xrightarrow[H_2O]{NaOH} RCOO^-\ Na^+ + \underset{\text{Brick-red}}{Cu_2O}\downarrow \quad \text{(general reaction)}$$

$\underset{\text{Blue}}{\phantom{RC{=}O + }}$

Ketones do not give a positive test with Tollens', Fehling's, or Benedict's solutions.

Reduction Aldehydes and ketones are easily reduced to alcohols, either by elemental hydrogen in the presence of a catalyst or by chemical reducing agents such as lithium aluminum hydride ($LiAlH_4$) or sodium borohydride ($NaBH_4$). Aldehydes yield primary alcohols; ketones yield secondary alcohols:

$$R-\overset{\overset{\displaystyle H}{|}}{C}=O \xrightarrow[\Delta]{H_2/Ni} RCH_2OH \qquad\qquad R-\overset{\overset{\displaystyle \;}{\underset{\underset{\displaystyle O}{\|}}{C}}}{}-R \xrightarrow[\Delta]{H_2/Ni} R-\overset{\overset{\displaystyle \;}{\underset{\underset{\displaystyle OH}{|}}{CH}}}{}-R$$

Aldehyde   1° alcohol   Ketone   2° alcohol

(general reaction)   (general reaction)

The carbonyl group undergoes a great variety of additional reactions. Although there are differences, aldehydes and ketones undergo many similar reactions. However, ketones are generally less reactive than aldehydes.

20.23 Carboxylic Acids and Esters

carboxylic acid

carboxyl group

Organic acids, known as **carboxylic acids**, are characterized by the functional group called a **carboxyl group**. The carboxyl group is represented in the following ways:

$$-\overset{\overset{\displaystyle O}{\|}}{C}-OH \qquad or \qquad -COOH \qquad or \qquad -CO_2H$$

Aliphatic carboxylic acids form a homologous series. The carboxyl group is always at the end of a carbon chain, and the C atom in this group is understood to be carbon number 1 in naming the compound.

To name a carboxylic acid by the IUPAC System, first identify the longest carbon chain including the carboxyl group. Then form the acid name by dropping the *e* from the corresponding parent hydrocarbon name and adding *oic acid*. Thus, the names corresponding to the one-, two-, and three-carbon acids are methanoic acid, ethanoic acid, and propanoic acid. These names are, of course, derived from methane, ethane, and propane.

CH_4	Methane	HCOOH	Methanoic acid
CH_3CH_3	Ethane	CH_3COOH	Ethanoic acid
$CH_3CH_2CH_3$	Propane	CH_3CH_2COOH	Propanoic acid

The IUPAC method is neither the only nor the most generally used method of naming acids. Organic acids are usually known by common names. Methanoic, ethanoic, and propanoic acids are commonly called formic, acetic, and propionic acids, respectively. These names usually refer to a natural source of

TABLE 20.8

Formulas and Names of Saturated Aliphatic Carboxylic Acids

Formula	IUPAC name	Common name
HCOOH	Methanoic acid	Formic acid
CH_3COOH	Ethanoic acid	Acetic acid
CH_3CH_2COOH	Propanoic acid	Propionic acid
$CH_3(CH_2)_2COOH$	Butanoic acid	Butyric acid
$CH_3(CH_2)_3COOH$	Pentanoic acid	Valeric acid
$CH_3(CH_2)_4COOH$	Hexanoic acid	Caproic acid
$CH_3(CH_2)_6COOH$	Octanoic acid	Caprylic acid
$CH_3(CH_2)_8COOH$	Decanoic acid	Capric acid
$CH_3(CH_2)_{10}COOH$	Dodecanoic acid	Lauric acid
$CH_3(CH_2)_{12}COOH$	Tetradecanoic acid	Myristic acid
$CH_3(CH_2)_{14}COOH$	Hexadecanoic acid	Palmitic acid
$CH_3(CH_2)_{16}COOH$	Octadecanoic acid	Stearic acid
$CH_3(CH_2)_{18}COOH$	Eicosanoic acid	Arachidic acid

the acid and are not really systematic. Formic acid was named from the Latin word *formica*, meaning "ant." This acid contributes to the stinging sensation of ant bites. Acetic acid is found in vinegar and is so named from the Latin word for vinegar. The name of butyric acid is derived from the Latin term for butter, since it is a constituent of butterfat. Many of the carboxylic acids, principally those having even numbers of carbon atoms ranging from 4 to about 20, exist in combined form in plant and animal fats. These acids are called *saturated fatty acids*. Table 20.8 lists the IUPAC and common names of the more important saturated aliphatic acids.

The simplest aromatic acid is benzoic acid. Ortho-hydroxybenzoic acid is known as salicylic acid, the basis for many salicylate drugs such as aspirin. There are three methylbenzoic acids, known as *o*-, *m*-, and *p*-toluic acids.

Benzoic acid   Salicylic acid   Acetylsalicylic acid   *p*-Toluic acid
                      (*o*-Hydroxybenzoic acid)   (Aspirin)

One of the many known methods of preparing carboxylic acids is by the oxidation of primary alcohols. The following two equations illustrate this method:

$$CH_3CH_2CH_2CH_2OH \xrightarrow[\Delta,\ H^+]{Cr_2O_7^{2-}} CH_3CH_2CH_2COOH$$

1-Butanol   Butyric acid (Butanoic acid)

$$\text{Benzyl alcohol} \quad \xrightarrow[\Delta,\ H^+]{Cr_2O_7^{2-}} \quad \text{Benzoic acid}$$

The carboxylic acids are weak acids, ionizing in water only to a small degree. They react with active metals to produce hydrogen and a salt, and with bases to form salts:

$$2\ CH_3COOH(aq) + 2\ Na(s) \longrightarrow 2\ CH_3COONa(aq) + H_2\uparrow$$

Sodium acetate

$$CH_3COOH(aq) + NaOH(aq) \longrightarrow CH_3COONa(aq) + H_2O$$

Acids are reduced to the corresponding primary alcohol by reaction with lithium aluminum hydride, $LiAlH_4$. This reaction is a source of long-carbon-chain alcohols.

$$CH_3(CH_2)_{14}COOH \xrightarrow{LiAlH_4} CH_3(CH_2)_{14}CH_2OH$$

Palmitic acid   1-Hexadecanol (Cetyl alcohol)

ester

Carboxylic acids react with alcohols in an acidic medium to form esters. **Esters** have the general formula RCOOR′ where R′ can be an aliphatic or an aromatic group. The functional group of the ester is —COOR′.

$$RC\overset{O}{\underset{OR'}{\diagdown}} \qquad -C\overset{O}{\underset{OR'}{\diagdown}} \qquad \text{or} \qquad -COOR'$$

Ester   Functional group of an ester

The reaction of acetic acid and ethyl alcohol is shown as an example. In addition to the ester, a molecule of water is formed as a product. The method is called *esterification*.

$$CH_3\overset{O}{\overset{\|}{C}}\!-\!OH + H\!-\!O\!-\!CH_2CH_3 \overset{H^+}{\rightleftharpoons} CH_3\overset{O}{\overset{\|}{C}}\!-\!OCH_2CH_3 + H_2O$$

Acetic acid   Ethyl alcohol   Ethyl accetate
(Ethanoic acid)   (Ethanol)   (Ethyl ethanoate)

Esters are alcohol derivatives of carboxylic acids. They are named in much the same way as salts. The alcohol part (R′ in OR′) is named first, followed by the name of the acid modified to end in *ate*. The *ic* ending of the organic acid name is

TABLE 20.9

Odors and Flavors of Selected Esters

Formula	IUPAC name	Common name	Odor or flavor
$CH_3COOCH_2CH_2CHCH_3$ 　　　　　　　　$\|$ 　　　　　　　CH_3	Isopentyl ethanoate	Isoamyl acetate	Banana, pear
$CH_3CH_2CH_2COOCH_2CH_3$	Ethyl butanoate	Ethyl butyrate	Pineapple
$HCOOCH_2CHCH_3$ 　　　　　$\|$ 　　　　CH_3	Isobutyl methanoate	Isobutyl formate	Raspberry
$CH_3COOCH_2(CH_2)_6CH_3$	Octyl ethanoate	n-Octyl acetate	Orange
⬡—$COOCH_3$ 　—OH	Methyl-2-hydroxy benzoate	Methyl salicylate	Wintergreen

replaced by the ending *ate*. Thus, in the IUPAC System, ethanoic acid becomes ethanoate. In the common names, acetic acid becomes acetate. To name an ester it is necessary to recognize the portion of the ester molecule that comes from the acid and the portion that comes from the alcohol. In the general formula for an ester, the RC=O comes from the acid and the R'O comes from the alcohol:

$$
\underset{\substack{\text{Acid} \qquad \text{Alcohol}}}{R-\overset{\overset{\textstyle O}{\|}}{C}-O-R'}
\qquad
\underset{\substack{\text{Acetic} \quad \text{Methyl}\\ \text{acid} \quad\;\; \text{alcohol}}}{CH_3\overset{\overset{\textstyle O}{\|}}{C}-OCH_3}
$$

Methyl acetate
(Methyl ethanoate)

Esters occur naturally in many varieties of plant life. Many of them have pleasant, fragrant, fruity odors and are used as flavoring and scenting agents. Esters are insoluble in water, but soluble in alcohol. Table 20.9 gives a list of selected esters.

20.24   Polymers — Macromolecules

There exist in nature some very large molecules (macromolecules) containing tens of thousands of atoms. Some of these, such as starch, glycogen, cellulose, proteins, and DNA, have molar masses in the millions and are central to many of our life processes. Man-made macromolecules touch every phase of modern living. It is hard to imagine a world today without polymers. Textiles for clothing, carpeting, and draperies; shoes; toys; automobile parts; construction materials; synthetic rubber; chemical equipment; medical supplies; cooking utensils; synthetic leather; recreational equipment—the list could go on and on. All these and a host of others that we consider to be essential in our daily life are wholly or partly man-made polymers. Most of these polymers were unknown 60 to 70 years ago. The vast majority of these polymeric materials are based on petroleum. Since petroleum is nonreplaceable, our dependence on polymers is another good reason for not squandering our limited world supply of petroleum.

polymerization

polymer

monomer

copolymer

The process of forming very large, high-molar-mass molecules from smaller units is called **polymerization**. The large molecule, or unit, is called the **polymer** and the small unit, the **monomer**. Polymers containing more than one kind of monomer are called **copolymers**. The term *polymer* is derived from the Greek word *polumerēs*, meaning "having many parts." Ethylene is a monomer and polyethylene is a polymer. Because of their large size, polymers are often called *macromolecules*. Some synthetic polymers are called *plastics*. The word *plastic* means "capable of being molded, or pliable." Although all polymers are not pliable and capable of being remolded, the word *plastics* has gained general use and has come to mean any of a variety of polymeric substances.

Polyethylene is an example of a synthetic polymer. Ethylene, derived from petroleum, is made to react with itself to form polyethylene (or polythene).

Polyethylene is a long-chain hydrocarbon made from many ethylene units:

$$n\, CH_2{=}CH_2 \longrightarrow -CH_2CH_2CH_2[CH_2CH_2]_nCH_2CH_2CH_2-$$
Ethylene

A typical polyethylene molecule is made up of about 2,500–25,000 ethylene molecules joined in a continuous structure. Over 12 billion pounds of polyethylene are produced annually in the United States. Its uses are as varied as any single substance known and include chemical equipment, packaging material, electrical insulation, films, industrial protective clothing, and toys.

Ethylene derivatives, in which one or more hydrogen atoms have been replaced by other atoms or groups, can also be polymerized. Many of our commercial synthetic polymers are made from such modified ethylene monomers. For example, $CH_2{=}CH-$ is a vinyl group and $CH_2{=}CHCl$ is vinyl chloride. Vinyl chloride can be polymerized to polyvinylchloride (PVC), a widely used polymer. The names, structures, and some uses for several of these polymers are given in Table 20.10.

TABLE 20.10

Polymers Derived from Modified Ethylene Monomers

Monomer	Polymer	Uses
$CH_2{=}CH_2$ Ethylene	$-(CH_2{-}CH_2)_n$ Polyethylene	Packing material, molded articles, containers, toys
$CH_2{=}CH$ $\vert$ CH_3 Propylene	$-(CH_2{-}CH$ $\vert$ $CH_3)_n$ Polypropylene	Textile fibers, molded articles, lightweight ropes, autoclavable biological equipment
$CH_2{=}C{\diagup}^{CH_3}_{\diagdown CH_3}$ Isobutylene	$-(CH_2{-}C{\diagup}^{CH_3}_{\diagdown CH_3})_n$ Polyisobutylene	Pressure-sensitive adhesives, butyl rubber (contains some isoprene as copolymer)
$CH_2{=}CH$ $\vert$ Cl Vinyl chloride	$-(CH_2{-}CH$ $\vert$ $Cl)_n$ Polyvinyl chloride (PVC)	Phonograph records, garden hoses, pipes, molded articles, floor tile, electrical insulation, vinyl leather
$CH_2{=}CCl_2$ Vinylidene chloride	$-(CH_2{-}CCl_2)_n$ Saran	Food packaging, textile fibers, pipes, tubing (contains some vinyl chloride as copolymer)
$CH_2{=}CH$ $\vert$ CN Acrylonitrile	$-(CH_2{-}CH$ $\vert$ $CN)_n$ Orlon, Acrilan	Textile fibers

(*continued*) ▶

Concepts in Review

1. Describe the tetrahedral nature of the carbon atom.
2. Identify the different types of bonding between the carbon atoms.
3. Explain what is meant by a homologous series.
4. Describe the phenomenon of isomerism.
5. Write the names and formulas for the first ten normal alkanes.
6. Write the names and structural formulas for the common alkyl groups (C_nH_{2n+1}).

TABLE 20.10 (*continued*)

Monomer	Polymer	Uses
$CF_2{=}CF_2$ Tetrafluoroethylene	$-\!(CF_2{-}CF_2)_{\overline{n}}$ Teflon	Gaskets, valves, insulation, heat-resistant and chemically resistant coatings, linings for pots and pans
$CH_2{=}CH$ (with phenyl ring) Styrene	$-\!(CH_2{-}CH)_n$ (with phenyl ring) Polystyrene	Molded articles, styrofoam, insulation, toys, disposable food containers
$CH_2{=}CH$ $\quad OC{-}CH_3$ $\qquad \|\!\|$ $\qquad O$ Vinyl acetate	$-\!(CH_2{-}CH)_n$ $\quad OC{-}CH_3$ $\qquad \|\!\|$ $\qquad O$ Polyvinyl acetate	Adhesives, paint, and varnish, starting material for polyvinyl alcohol
$CH_2{=}C{-}CH_3$ $\quad C{-}O{-}CH_3$ $\quad \|\!\|$ $\quad O$ Methylmethacrylate	$-\!(CH_2{-}C(CH_3))_n$ $\quad C{-}OCH_3$ $\quad \|\!\|$ $\quad O$ Lucite, Plexiglas (acrylic resins)	Contact lenses, clear sheets for windows and optical uses, molded articles, automobile finishes

7. Write common or IUPAC names for each class of compounds discussed in this chapter.

8. When given a name, write the structural formulas for compounds from each classification in this chapter.

9. Write structural formals for the isomers of a compound when given a name or molecular formula.

10. Write equations for the combustion and halogenation of alkanes.

11. Write the names and formulas for all the monohalogen substitution isomers of a given alkane.

12. List the major classes of organic compounds and their functional groups.

13. Distinguish by structure an alkane, an alkene, and an alkyne.

14. Write equations for addition reactions of alkenes.

15. Describe the nature of benzene and how its properties differ from open-chain unsaturated hydrocarbons.

16. Recognize and write formulas for benzene compounds and the common polycyclic aromatic ring compounds.

17. Recognize and identify primary, secondary, and tertiary alcohols.

18. Write equations for the following reactions of alcohols: (a) oxidation to form aldehydes or ketones and (b) reaction with an active metal, such as sodium.

19. Be familiar with the methods of preparing methanol and ethanol and the general properties and uses of these two alcohols.

20. Write equations for the preparation of ethers by the Williamson method.

21. Select the proper alcohol to prepare an aldehyde or a ketone by oxidation.

22. Discuss Tollens', Fehling's, and Benedict's tests, including reagents used, evidence of a positive test, and the equations for the reactions that occur in positive tests.

23. Write equations for the reduction of aldehydes and ketones to form alcohols.

24. Write equations for (a) the oxidation of an alcohol to a carboxylic acid, (b) the reaction of carboxylic acids with sodium metal, (c) the reaction of carboxylic acids with a base, and (d) the reduction of carboxylic acids to alcohols.

25. Write the equations for the preparation of esters from carboxylic acids and alcohols.

26. Identify the alcohol and the carboxylic acid that would be needed to prepare a given ester.

27. Understand the general makeup of polymers (macromolecules).

28. Write structural formulas for polymers derived from modified ethylene monomers when given the monomer, or vice versa.

Key Terms in Review

The terms listed here have all been defined within the chapter. Review the definitions of each, and use the glossary and margin notation within the Chapter as study aids.

addition reaction
alcohol
aldehyde
alkane
alkene
alkyl group
alkyl halide
alkyne
aromatic compound
carbonyl group

carboxyl group
carboxylic acid
copolymer
ester
ether
functional group
halogenation
homologous series
hydrocarbon
isomerism

isomers   polymer
ketone   polymerization
monomer   primary alcohol
organic chemistry   secondary alcohol
polycyclic or fused   tertiary alcohol
  aromatic ring system   vital force theory
polyhydroxy alcohol

Exercises

1. What bonding characteristic of carbon is primarily responsible for the existence of so many organic compounds?
2. What is the most common geometric arrangement of covalent carbon bonds? Illustrate this structure.
3. In addition to single bonds, what other types of bonds can carbon atoms form? Give examples.
4. Draw a Lewis structure for:
   (a) A single carbon atom
   (b) Molecules of methane, ethylene (ethene), and acetylene (ethyne)
5. Write the names and draw the structural formulas for the first ten normal alkanes.
6. Write the names and draw the structural formulas for all possible alkyl groups (C_nH_{2n+1}) containing from one to four carbon atoms.
7. Name the following normal alkyl groups:
   (a) C_5H_{11}—   (c) C_8H_{17}—
   (b) C_7H_{15}—   (d) $C_{10}H_{21}$—
8. What are the three principal sources of hydrocarbons?
9. Which one of the following compounds belongs to a different homologous series than the others?
   (a) C_2H_4 (b) CH_4 (c) C_6H_{14} (d) C_5H_{12}
10. Which word does not belong with the others?
    (a) Alkane   (c) Saturated   (e) Ethylene
    (b) Paraffin (d) Ethane   (f) Pentane
11. Write condensed structural formulas for:
    (a) The five isomers of hexane
    (b) The nine isomers of heptane
12. Give common and IUPAC names for the following:
    (a) CH_3CH_2Cl (d) $CH_3CH_2CH_2Cl$
    (b) $CH_3CHClCH_3$ (e) $(CH_3)_3CCl$
    (c) $(CH_3)_2CHCH_2Cl$ (f) $CH_3CHClCH_2CH_3$

13. Give IUPAC names for the following compounds:
    (a) $CH_3CHCH_2CH\!-\!CH_2CH_2CH_3$
        $\quad\;\; |\qquad\quad |$
        $\quad CH_3\quad C_2CH_3$

    (b) $CH_3CHCH_2CHCH_2\!-\!CHCH_2CH_3$
        $\quad\;\; |\qquad\quad |\qquad\quad |$
        $\quad CH_3\quad CH_2CH_3\;\; CH_2CH_3$

    (c) $CH_3\!-\!CH\!-\!CH_2CH_2CHCH_3$
        $\qquad\quad |\qquad\qquad\qquad |$
        $\qquad\;\; CH_2\qquad\qquad\; CH_2$
        $\qquad\quad |\qquad\qquad\qquad |$
        $\qquad\;\; CH_3\qquad\qquad\; CH_3$

    (d) $CH_3CH_2CH_2CHCH_2CH_3$
        $\qquad\qquad\qquad |$
        $\qquad\quad CH_3CHCH_3$

14. Draw structural formulas of the following compounds:
    (a) 2,4-Dimethylpentane
    (b) 2,2-Dimethylpentane
    (c) 3-Isopropyloctane
    (d) 4-Ethyl-2-methylhexane
    (e) 4-t-Butylheptane
    (f) 4-Ethyl-7-isopropyl-2,4,8-trimethyldecane
15. One name in each of the following pairs is incorrect. Draw structures corresponding to each name and indicate which name is incorrect.
    (a) 2-Methylbutane and 3-methylbutane
    (b) 2-Ethylbutane and 3-methylpentane
    (c) 2-Dimethylbutane and 2,2-dimethylbutane
    (d) 2,4-Dimethylhexane and 2-ethyl-4-methylpentane
16. What is the single most important reaction of alkanes?

17. A hydrocarbon sample of formula C_4H_{10} was brominated, and four different monobromo compounds of formula C_4H_9Br were isolated. Was the sample a pure compound or a mixture of compounds? Explain your answer.

18. Draw structural formulas for all the isomers of
(a) CH_3Br (d) C_3H_7Br (g) C_3H_6BrCl
(b) CH_2Cl_2 (e) C_4H_9I (h) $C_4H_8Cl_2$
(c) C_2H_5Cl (f) $C_3H_6Cl_2$

19. Using alkanes and any other necessary inorganic reagents, write reactions showing the formation of:
(a) CH_3Cl
(b) $CHCl_3$
(c) $CH_3CH_2CH_2Br$
(d) $CH_3CHBrCH_3$

20. Draw structures containing two carbon atoms for each of the following classes of compounds:
(a) Alkene   (e) Ether
(b) Alkyne   (f) Aldehyde
(c) Alkyl halide (g) Carboxylic acid
(d) Alcohol   (h) Ester

21. Give the common name and the IUPAC name for each of the compounds in Exercise 20.

22. Which compounds in Exercise 20 are isomers?

23. Draw structural formulas for the following:
(a) Chloromethane
(b) Vinyl chloride
(c) Chloroform
(d) Hexachloroethane
(e) Iodoethyne
(f) 1-Bromo-4-methyl-2-hexene-5-yne
(g) 1,1-Dibromoethene
(h) 1,2-Dibromoethene

24. Draw the structure for vinyl acetylene, which has the formula C_4H_4 and contains one double bond and one triple bond.

25. An open chain hydrocarbon has the formula C_6H_8. What possible combinations of carbon-carbon double bonds and/or carbon-carbon triple bonds can be in this compound.

26. Draw structural formulas for the following:
(a) 2,5-Dimethyl-3-hexene
(b) 2-Ethyl-3-methyl-1-pentene
(c) 4-Methyl-2-pentene
(d) 1,2-Diphenylethene
(e) 3-Penten-1-yne
(f) 3-Phenyl-1-butyne
(g) Vinyl bromide

27. Name the following compounds:

(a) $CH_3CH{=}CCH_2CH_2CH_3$
      |
      CH_3

(b) $CH_3C{=\!=\!=}C{-}CH_3$
        | |
      CH_3 CH_3

(c) $CH_3CH_2CHCH{=}CH_2$
           |
           CH
        H_3C CH_3

(d) $CH_3CH_2CH{=}CCH_2CH_3$
                 |
                 CH_3

(e)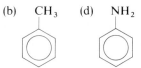

28. Draw the structural formulas and write the IUPAC names for all the isomers of (a) pentyne and (b) hexyne.

29. Complete the following reactions and name the products:
(a) $CH_2{=}CHCH_3 + Br_2 \longrightarrow$
(b) $CH_2{=}CH_2 + HBr \longrightarrow$

(c) $CH_3CH{=}CHCH_3 + H_2 \xrightarrow[\text{1 atm}]{\text{Pt, 25°C}}$

(d) $CH_2{=}CH_2 + H_2O \xrightarrow{H^+}$

(e) $CH{\equiv}CH + 2\,Br_2 \longrightarrow$

30. Write equations showing two methods for the preparation of acetylene.

31. Name the following aromatic compounds:

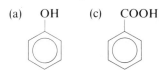

(a)   OH   (c)   COOH

(b) CH_3 (d) NH_2

(e)

(i)

(f)

(j)

(g)

(k)

(h)

(l)

32. Draw structural formulas for
(a) Benzene   (f) Phenol
(b) Toluene   (g) *o*-Bromochlorobenzene
(c) *p*-Xylene   (h) 1,3-Dichloro-5-nitrobenzene
(d) Styrene   (i) *m*-Dinitrobenzene
(e) Aniline

33. Draw structural formulas for
(a) Ethylbenzene
(b) Benzoic acid
(c) 1,3,5-Tribromobenzene
(d) Naphthalene
(e) Anthracene
(f) *tert*-Butylbenzene
(g) 1,1-Diphenylethane

34. Draw structural formulas and write the IUPAC names for all the isomers of:
(a) Trichlorobenzene ($C_6H_3Cl_3$)
(b) Dichlorobromobenzene ($C_6H_3Cl_2Br$)
(c) The benzene derivatives of formula C_8H_{10}

35. Write the IUPAC name for each of these alcohols. Classify each as primary, secondary, or tertiary alcohols.

(a) $CH_3CH_2\underset{\underset{\displaystyle OH}{|}}{C}H—CH_3$

(b) $CH_3\underset{\underset{\displaystyle CH_3}{|}}{C}H—CH_2CH_2\underset{\underset{\displaystyle CH_3}{|}}{C}H—OH$

(c) $CH_3\underset{\underset{\displaystyle CH_2CH_3}{|}}{C}H—\overset{\overset{\displaystyle OH}{|}}{C}HCH_2CH_2\underset{\underset{\displaystyle CH_2CH_3}{|}}{C}H_2CH_3$

(d) $CH_2\underset{\underset{\displaystyle OH}{|}}{C}HCHCH_2\underset{\underset{\displaystyle CH—CH_3}{|}}{C}H_2CH_3$ $\underset{\underset{\displaystyle CH_3}{}}{}$

36. Eight open-chain isomeric alcohols have the formula $C_5H_{11}OH$.
(a) Draw the structural formula and write the IUPAC name for each of these alcohols.
(b) Indicate which of these isomers are primary, secondary, and tertiary alcohols.

37. Draw structural formulas for the following:
(a) 2-Pentanol
(b) Isopropyl alcohol
(c) 2,2-Dimethyl-1-heptanol
(d) 1,3-Propanediol
(e) Glycerol

38. Alcohols can be made by the reaction of alkyl halides with aqueous sodium hydroxide:

$$RX + NaOH(aq) \longrightarrow ROH + NaX$$

What alkyl bromide should be used to prepare (a) isopropyl alcohol, (b) 3-methyl-1-butanol, and (c) 1,3-propanediol by this method? Write the equation for each reaction.

39. Complete the following equations and name the organic product formed in each reaction.

(a) $CH_3CH_2CH_2OH \xrightarrow[\Delta]{H^+, K_2Cr_2O_7}$

(b) $CH_3CH(OH)CH_3 \xrightarrow[\Delta]{H^+, K_2Cr_2O_7}$

(c) $(CH_3)_2C(OH)CH_2CH_3 \xrightarrow[\Delta]{H^+, K_2Cr_2O_7}$

(d) $CH_3CH_2CH_2OH + Na \longrightarrow$

40. Write equations to show how each of the transformations can be accomplished. Some conversions may require more than one step.

(a) $CH_3CH\!=\!CHCH_3 \longrightarrow$
$\qquad\qquad CH_3CH_2CH(OH)CH_3$

(b) $CH_3CH\!=\!CHCH_3 \longrightarrow$
$\qquad\qquad CH_3CH_2CHBrCH_3$

(c) $CH_2\!=\!CH_2 \longrightarrow CH_3\overset{\text{H}}{\underset{}{C}}\!=\!O$

(d) $CH_3CH_2CH(OH)CH_3 \longrightarrow$
$\qquad\qquad CH_3CH_2\underset{\overset{\|}{O}}{C}CH_3$

41. What is the molar mass of myricyl alcohol, an open-chain saturated alcohol containing 30 carbon atoms? Myricyl alcohol is present in beeswax as an ester.

42. Why is ethylene glycol (1,2-ethanediol) superior to methyl alcohol (methanol) as an antifreeze for automobile radiators?

43. Alcohols are considered to be toxic to the human body, with ethanol, the alcohol in alcoholic beverages, the least toxic one. What are the hazards of ingesting (a) methanol, (b) ethanol?

44. Write the name and structure of the ether that is isomeric with (a) 1-propanol, (b) ethanol, and (c) isopropyl alcohol.

45. Give the names and structures of all ethers having the molecular formula $C_5H_{12}O$.

46. What possible combinations of RONa and RCl might be used in preparing each of the following ethers by the Williamson synthesis?

(a) $CH_3CH_2OCH_3$

(b) $CH_3CH_2OCH_2CH_3$

(c) CH_3OCH_3

47. Draw structural formulas for propanal and propanone. From these formulas, do you think that aldehydes and ketones are isomeric with each other? Show evidence and substantiate your answer by testing with a four-carbon aldehyde and ketone.

48. Name each of these aldehydes:

(a) $CH_2\!=\!O$ (two names)

(b) $CH_3CH_2CH_2\underset{\overset{\|}{O}}{C}\!-\!H$ (two names)

(c) $CH_3\underset{\underset{CH_3}{|}}{CH}CH_2\overset{\overset{O}{\|}}{C}\!-\!H$ (one name)

(d) $C_6H_5\overset{\overset{O}{\|}}{C}\!-\!H$ (one name)

(e) $O\!=\!\overset{\overset{H}{|}}{C}CH_2CH_2\overset{\overset{O}{\|}}{C}\!-\!H$ (one name)

(f) (chlorobenzaldehyde structure) $\overset{\overset{O}{\|}}{C}\!-\!H$ (one name)

(g) $CH_3\underset{\underset{OH}{|}}{CH}CH_2\overset{\overset{O}{\|}}{C}\!-\!H$ (one name)

49. Name each of these ketones:

(a) CH_3COCH_3 (three names)

(b) $CH_3CH_2COCH_3$ (two names)

(c) $C_6H_5\overset{\overset{O}{\|}}{C}\!-\!CH_2CH_3$ (two names)

(d) $CH_3\overset{\overset{O}{\|}}{C}\!-\!\underset{\underset{CH_3}{|}}{\overset{\overset{CH_3}{|}}{C}}CH_3$ (two names)

(e) $CH_3\overset{\overset{O}{\|}}{C}CH_2CH_2\overset{\overset{O}{\|}}{C}CH_3$ (one name)

(f) $CH_3\underset{\underset{OH}{|}}{\overset{\overset{CH_3}{|}}{C}}\!-\!CH_2\overset{\overset{O}{\|}}{C}CH_3$ (one name)

(g) $C_6H_5\!-\!CH_2\overset{\overset{O}{\|}}{C}CH_3$ (one name)

50. Ketones are prepared by oxidation of secondary alcohols. What alcohol should be used to prepare each of the following?
  (a) Dimethyl ketone
  (b) Methyl ethyl ketone
  (c) Ethyl isopropyl ketone
  (d) 1-Phenyl-3-hexanone
51. (a) What functional group is present in a compound that gives a positive Tollens' test?
  (b) What is the visible evidence for a positive Tollens' test?
  (c) Write an equation showing the reaction involved in a positive Tollens' test.
52. Give a chemical test or reaction by which it is possible to distinguish between the compounds in each pair:
  (a) $CH_3CH_2OCH_2CH_3$ and CH_3OH
  (b) CH_3CHO and CH_3CCH_3
$$\overset{||}{O}$$
  (c) $CH_3(CH_2)_5CH\!=\!CH_2$ and
   $CH_3(CH_2)_6CH_3$
  (d) $CH_3(CH_2)_3CH_2OH$ and CH_3CH_2COOH
  (e) $CH_2\!=\!CHOCH\!=\!CH_2$ and
    Divinyl ether
   $CH_3CH_2OCH_2CH_3$
    Diethyl ether
53. Give the common and the IUPAC names for the first five straight-chain carboxylic acids.
54. Draw the structural formulas for lauric, myristic, palmitic, and stearic acids.
55. Draw the structural formulas and write the IUPAC names for all the isomers of hexanoic acid, $CH_3(CH_2)_4COOH$.
56. Name each of the following acids:
  (a) $CH_3CHBrCOOH$
  (b) $CH_2\!=\!CHCH_2COOH$
  (c) $CH_3CHCOOH$
     |
     CH_2CH_3

  (d)

  (e)

57. Draw structural formulas for the following esters:
  (a) Ethyl formate
  (b) Methyl ethanoate
  (c) Isopropyl propanoate
  (d) n-Nonyl acetate
  (e) Ethyl benzoate
  (f) Methyl salicylate
  (g) Vinyl butanoate
58. Write IUPAC names for the following esters:
  (a) $CH_3COCH_2CH_3$ (c) CH_3CO—
     $\overset{||}{O}$ $\overset{||}{O}$

  (b) $HCOCH(CH_3)_2$ (d)
     $\overset{||}{O}$

59. Write equations for the preparation of the following esters:
  (a) Ethyl formate
  (b) Methyl propanoate
  (c) n-Propyl benzoate
60. Complete the following equations:
  (a) $CH_3COOH + NaOH \longrightarrow$
  (b) $CH_3CHCOOH + NH_3 \longrightarrow$
     |
     OH
  (c) $CH_3COOH + K \longrightarrow$
  (d)

61. Which of the following statements are correct? Rewrite each incorrect statement to make it correct.

(a) In the alkane homologous series the formula of each member differs from the preceding member by CH_3.

(b) Carbon atoms can form single, double, and triple bonds with other carbon atoms.

(c) The IUPAC name for
$CH_3CH_2CH_2CHClCH_3$
is 4-chloropentane.

(d) Propane is an isomer of propene.

(e) Isopropyl alcohol is an isomer of both methyl ethyl ether and methyl ethyl ketone.

(f) Depending on which hydrogen atom is removed, it is possible to get two alkyl groups, t-butyl and isobutyl, from isobutane.

(g) Isobutane, methylpropane, and 1,1-dimethylethane are all correct names for the same compound.

(h) Compounds belonging to the same homologous series will generally exhibit similar chemical properties.

(i) A single monosubstituted product will result from the chlorination of butane.

(j) There are eight carbon atoms in a molecule of 3,3,4-trimethylpentane.

(k) The compound C_6H_{10} can have in its structure two carbon–carbon double bonds or one carbon–carbon triple bond.

(l) The addition of HCl to ethylene and monochlorination of ethane yield the same product, chloroethane.

(m) Dimethyl ketone, acetone, and propanone all have the same molecular formula but different structural formulas.

(n) Methyl alcohol is a very poisonous substance that can lead to blindness when ingested.

(o) The number of isomers of dichlorobenzene, trichlorobenzene, and tetrachlorobenzene is the same.

(p) $(CH_3)_3CCH_2OH$ is a primary alcohol.

(q) Phenanthrene is isomeric with anthracene.

(r) Polymerization is the process of forming very large high-molar mass molecules from smaller units.

(s) Ethanal may be distinguished from propanal by use of Tollens' reagent.

(t) The monomer of Teflon is tetrafluoroethylene, $CF_2{=}CF_2$.

(u) The chemical characteristics of CH_3OH and NaOH are similar.

(v) Ethers, because of their relative unreactivity, are often used as solvents in organic reactions.

(w) Both aldehydes and ketones are easily oxidized.

(x) Esters are derivatives of alcohols and carboxylic acids.

(y) Acetic acid is the most common and structurally the simplest carboxylic acid.

(z) Methyl propanoate and propyl methanoate are isomers.

62. Ethylene and its derivatives are the most common monomers for polymers. Write formulas for the following ethylene-based polymers:

(a) Polyethylene

(b) Polyvinyl chloride

(c) Polystyrene

(d) Saran

(e) Polyacrylonitrile (Orlon or Acrilan)

(f) Teflon

63. (a) Draw a structural formula showing the polymer that can be formed from each of the following monomers (show four units): (a) propylene, (2) 1-butene, (3) 2-butene.

(b) How many ethylene units are in a polyethylene molecule that has a molar mass of 35,000?

CHAPTER TWENTY-ONE

Introduction to Biochemistry

Biochemistry is the study of the molecular basis of life. The chemical bases for some of the fundamental processes in biology such as the conversion of sunlight and carbon dioxide into a food source for other living organisms are well understood. Now we are beginning to pinpoint the location of specific genes. We can use DNA in much the same manner as a fingerprint to identify the presence of a suspect at a crime scene.

There are also common molecular patterns that underscore the diversity of life. These principles resulted in the production of essential substances, such as insulin for diabetics, using bacteria as the manufacturing machine. Other substances, such as the human growth factor, are being genetically engineered as well.

Biochemistry is also making an impact in the fields of medicine and nutrition. For example, the severity of a heart attack can be determined by the levels of certain enzymes in the blood. The role of cholesterol, fats, and trace elements in our diet are current topics of great interest in biochemistry laboratories.

607

Chapter Preview

21.1   Chemistry in Living Organisms

The study of life has long fascinated people—and it is probably the most intriguing of all scientific studies, although the answer to the question, "What is life?" still eludes us.

The chemical substances present in all living organisms—from microbes to humans—range in complexity from water and simple salts to DNA (deoxyribonucleic acid) molecules containing tens of thousands of atoms. Four of the chemical elements, hydrogen, carbon, nitrogen, and oxygen, make up approximately 95% of the mass of living matter. Small amounts of sulfur, phosphorus, calcium, sodium, chlorine, magnesium, and iron, together with trace amounts of many other elements such as copper, manganese, zinc, cobalt, and iodine, are also found in living organisms. The human body consists of about 60% water with some tissues having a water content as high as 80%.

biochemistry
Biochemistry is the branch of chemistry that is concerned with the chemical reactions occurring in living organisms. Its scope includes such processes as growth, respiration, digestion, metabolism, and reproduction.

The four major classes of biomolecules upon which all life depends are carbohydrates, lipids, proteins, and nucleic acids. Each kind of living organism has an amazing ability to select and synthesize a large portion of the many complicated molecules needed for its existence. In fact, the processes carried out in a living organism can be likened to those of a highly automated, smoothly running "chemical factory." But unlike a chemical factory, a living organism is able to expand (grow), repair damage (if not too severe), and, finally, reproduce itself.

Of necessity, we are limited here to brief consideration of only a few important aspects of biochemistry. These include carbohydrates, lipids, proteins, nucleic acids, and enzymes.

21.2 Carbohydrates

carbohydrates

Chemically, **carbohydrates** are polyhydroxy aldehydes or polyhydroxy ketones or substances that yield these compounds when hydrolyzed. The name *carbohydrate* was given to this class of compounds many years ago by French scientists who called them *hydrates de carbone* because their empirical formulas approximated $(C \cdot H_2O)_n$. However, the hydrogen and oxygen do not actually exist as water or in hydrate form as we have seen in compounds such as $BaCl_2 \cdot 2 H_2O$. Empirical formulas used to represent carbohydrates are $C_x(H_2O)_y$ and $(CH_2O)_n$.

Carbohydrates, also known as saccharides, occur naturally in plants and are one of the three principal classes of animal food. The other two classes of foods are lipids (fats) and proteins. Plants are able to synthesize carbohydrates by the photosynthetic process. Animals are incapable of this synthesis and are dependent on the plant kingdom for their source of carbohydrates. Animals are capable, however, of converting plant carbohydrate into glycogen (animal carbohydrate) and storing this glycogen throughout the body as a reserve energy source. The amount of energy available from carbohydrates is about 17 J/g (4 kcal/g).

Carbohydrates exist as sugars, starches, and cellulose. The simplest of these are the sugars. The names of the sugars end in *ose* (for example, glucose, sucrose, maltose). Carbohydrates are classified as monosaccharides, disaccharides, oligosaccharides, and polysaccharides according to the number of monosaccharide units linked together to form a molecule.

Monosaccharides

monosaccharides

Monosaccharides are carbohydrates that cannot be hydrolyzed to simpler carbohydrate units. They are often called simple sugars, the most common of which is glucose. Monosaccharides containing three, four, five, and six carbon atoms are called *trioses*, *tetroses*, *pentoses*, and *hexoses*, respectively. Monosaccharides that contain an aldehyde ($-\overset{\overset{\displaystyle H}{|}}{C}=O$) group on one carbon atom and a hydroxy ($-OH$) group on each of the other carbon atoms are called *aldoses*.

Ketoses are monosaccharides that contain a ketone ($\overset{\overset{\displaystyle O}{\|}}{\diagup C \diagdown}$) group on one carbon atom and a hydroxyl ($-OH$) group on each of the other carbons. Fructose, also known as levulose, is a ketohexose. Fructose is the sweetest of the sugars, followed by sucrose, and then glucose. Structural formulas of several monosaccharides are as follows:

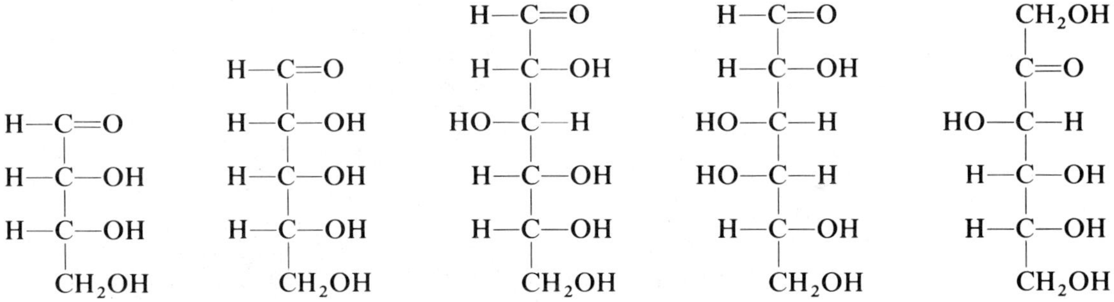

$$
\begin{array}{ccccc}
 & & H-C=O & H-C=O & CH_2OH \\
 & H-C=O & H-C-OH & H-C-OH & C=O \\
H-C=O & H-C-OH & HO-C-H & HO-C-H & HO-C-H \\
H-C-OH & H-C-OH & H-C-OH & HO-C-H & H-C-OH \\
H-C-OH & H-C-OH & H-C-OH & H-C-OH & H-C-OH \\
CH_2OH & CH_2OH & CH_2OH & CH_2OH & CH_2OH \\
\text{Erythrose} & \text{Ribose} & \text{Glucose} & \text{Galactose} & \text{Fructose} \\
\text{(an aldotetrose)} & \text{(an aldopentose)} & \text{(an aldohexose)} & \text{(an aldohexose)} & \text{(a ketohexose)}
\end{array}
$$

It is possible to write structural formulas for many monosaccharides, but only a limited number are of biological importance. Sixteen different isomeric aldohexoses of formula $C_6H_{12}O_6$ are known; glucose and galactose are the most important of these. Most sugars exist predominantly in a cyclic structure, forming a five- or six-membered ring. For example, glucose exists as a ring in which C-1 is bonded through an oxygen atom to C-5.

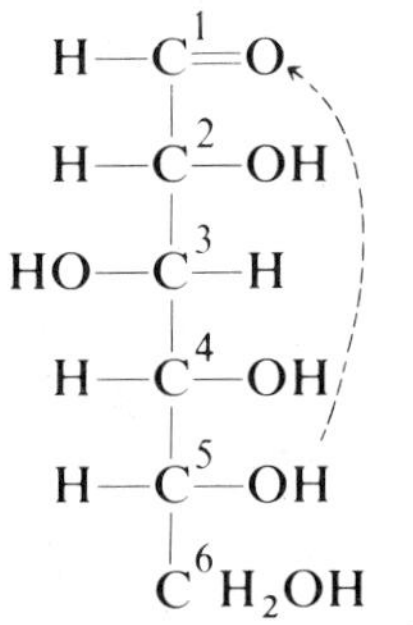

Open-chain form of glucose

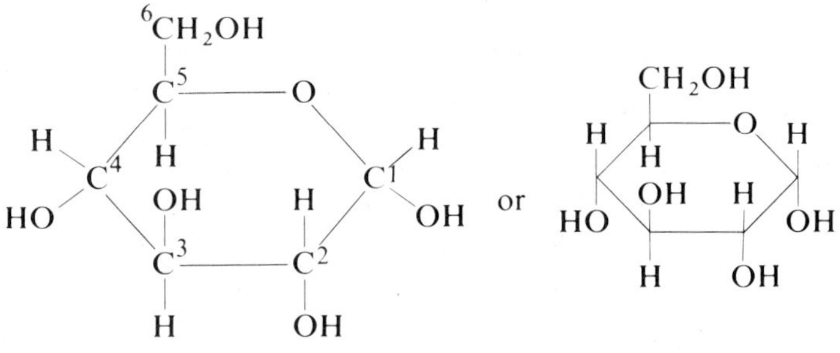

Cyclic form of glucose

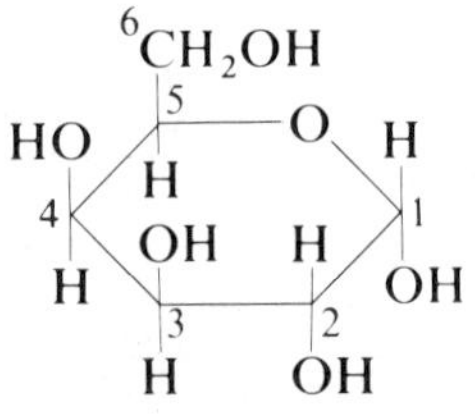

Cyclic form of galactose

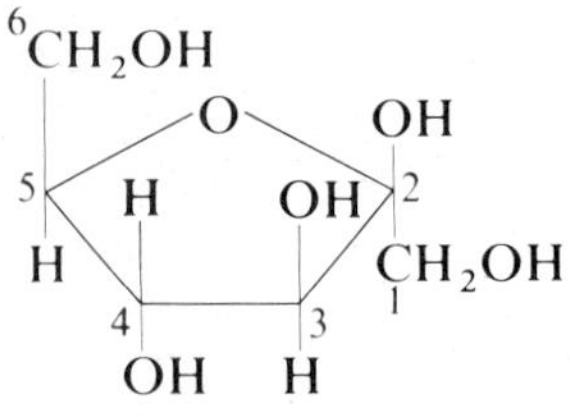

Cyclic form of fructose

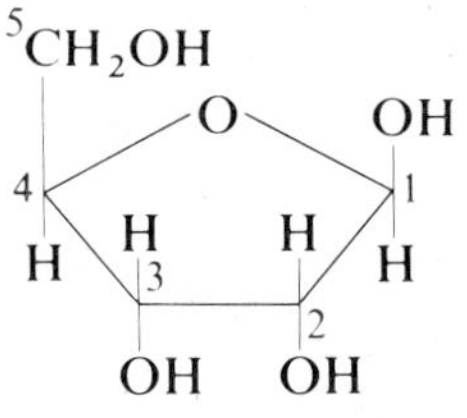

Cyclic form of ribose

TABLE 21.1

Relative Sweetness of Sugars

Fructose	100	Galactose	19
Sucrose	58	Lactose	9.2
Glucose	43	Invert sugar	75
Maltose	19		

The properties of four important monosaccharides follow.

Glucose Glucose ($C_6H_{12}O_6$) is the most important of the monosaccharides. It is an aldohexose and is found in the free state in plants and animal tissue. Glucose is commonly known as *dextrose* or *grape sugar*. It is a component of the disaccharides sucrose, maltose, and lactose and is also the monomer of the polysaccharides starch, cellulose, and glycogen. Among the common sugars, glucose is of intermediate sweetness (see Table 21.1).

Glucose is the key sugar of the body and is carried by the bloodstream to all body parts. The concentration of glucose in the blood is normally 80 to 100 mg per 100 mL of blood. Because glucose is the most abundant carbohydrate in the blood, it is also sometimes known as *blood sugar*. Glucose requires no digestion and therefore may be given intravenously to patients who cannot take food by mouth. Glucose is found in the urine of those who have diabetes mellitus (sugar diabetes). The condition in which glucose is excreted in the urine is called glycosuria.

Galactose Galactose ($C_6H_{12}O_6$) is also an aldohexose and occurs along with glucose in the disaccharide lactose and in many oligo- and polysaccharides, such as pectin, gums, and mucilages. Galactose is an isomer of glucose, differing only in the spatial arrangement of the —H and —OH groups around C-4. Galactose is synthesized in the mammary glands to make the lactose of milk. Galactose is less than half as sweet as glucose.

A severe inherited disease, called galactosemia, is the inability of infants to metabolize galactose. The galactose concentration increases markedly in the blood and also appears in the urine. Galactosemia causes vomiting, diarrhea, enlargement of the liver, and often mental retardation. If not recognized within a few days after birth, it can lead to death. If diagnosis is made early and lactose is excluded from the diet, the symptoms disappear and normal growth may be resumed.

Fructose   Fructose ($C_6H_{12}O_6$), also known as levulose, is a ketohexose and occurs in fruit juices, honey, and (along with glucose) as a constituent of the disaccharide sucrose. Fructose is the major constituent of the polysaccharide inulin, a starchlike substance present in many plants such as dahlia tubers, chicory roots, and Jerusalem artichokes. Fructose is the sweetest of all the sugars, being about twice as sweet as glucose. This sweetness accounts for the sweet taste of honey; the enzyme invertase present in bees splits sucrose into glucose and fructose. Fructose is metabolized directly but is also readily converted to glucose in the liver.

Ribose   Ribose ($C_5H_{10}O_5$) is an aldopentose and is present in adenosine triphosphate (ATP), one of the chemical energy carriers in the body. Ribose and one of its derivatives, deoxyribose, are also important components of the nucleic acids DNA and RNA, the genetic information carriers in the body.

Disaccharides

disaccharides

Disaccharides are carbohydrates whose molecules yield two molecules of the same or of different monosaccharides when hydrolyzed. The three disaccharides that are especially important from a biological viewpoint are sucrose, lactose, and maltose. Sucrose ($C_{12}H_{22}O_{11}$), which is commonly known as *table sugar*, is found in the free state throughout the plant kingdom. Sugar cane contains 15–20% sucrose, and sugar beets contain 10–17%. Maple syrup and sorghum are also good sources of sucrose.

Lactose ($C_{12}H_{22}O_{11}$), also known as *milk sugar*, is found free in nature mainly in the milk of mammals. Human milk contains about 6.7% lactose, and cow's milk contains about 4.5% of this sugar.

Maltose ($C_{12}H_{22}O_{11}$) is found in sprouting grain but occurs much less commonly (in nature) than either sucrose or lactose. Maltose is prepared commercially by the partial hydrolysis of starch, catalyzed either by enzymes or by dilute acids.

Disaccharides are not used directly in the body but are first hydrolyzed to monosaccharides. Disaccharides yield two monosaccharide molecules when hydrolyzed in the laboratory at elevated temperatures in the presence of hydrogen ions (acids) as catalyts. In biological systems, enzymes (biochemical catalysts) carry out the reaction. A different enzyme is required for the hydrolysis of each of the three disaccharides:

$$\text{Sucrose} + \text{Water} \xrightarrow{\text{H}^+ \text{ or Sucrase}} \text{Glucose} + \text{Fructose}$$

$$\text{Lactose} + \text{Water} \xrightarrow{\text{H}^+ \text{ or Lactase}} \text{Galactose} + \text{Glucose}$$

$$\text{Maltose} + \text{Water} \xrightarrow{\text{H}^+ \text{ or Maltase}} \text{Glucose} + \text{Glucose}$$

The structure of a disaccharide is derived from two monosaccharide molecules by the elimination of a water molecule between them. In maltose, for example, the two monosaccharides are glucose. The water molecule is eliminated between the OH group on C-1 of one glucose unit and the OH group on C-4 of the other glucose unit. Thus the two glucose units are joined at C-1 and C-4. Sucrose consists of a glucose unit and a fructose unit linked together through an oxygen atom from C-1 on glucose to C-2 on fructose. In lactose, the linkage is from C-1 of galactose through an oxygen atom to C-4 of glucose. The structures of maltose and sucrose follow.

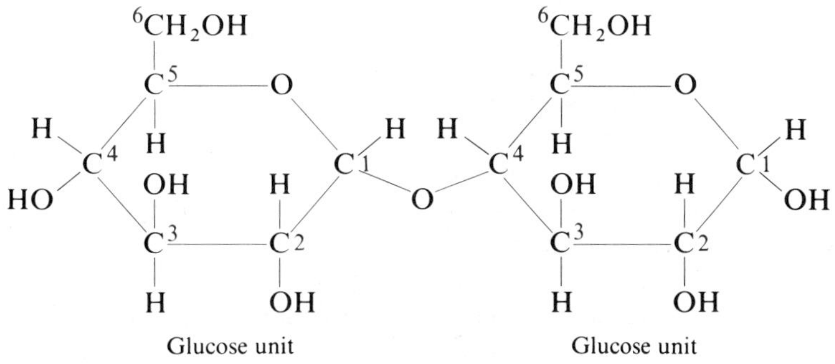

Maltose, a disaccharide

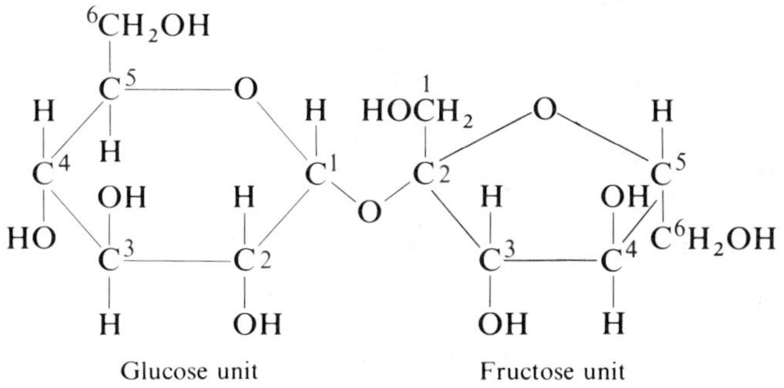

Sucrose, a disaccharide

Polysaccharides

polysaccharides

Polysaccharides are also called *complex carbohydrates* and can be hydrolyzed to a large number of monosaccharide units. The molar masses of polysaccharides range up to 1 million or more. Three of the most important polysaccharides are starch, glycogen, and cellulose.

Starch is a polymer of glucose. It is found mainly in the seeds, roots, and tubers of plants. Corn, wheat, potatoes, rice, and cassava are the chief sources of starch. The principal use of starch is for food.

Glycogen is the reserve carbohydrate of the animal kingdom. It is often called animal starch. Glycogen is formed in the body by polymerization of glucose and is stored especially in the liver and in muscle tissue. Glycogen also occurs in some insects and lower plants including fungi and yeasts.

Cellulose, like starch and glycogen, is also a polymer of glucose. It differs from starch and glycogen in the manner in which the cyclic glucose units are linked together to form chains. Cellulose is the most abundant organic substance found in nature. It is the chief structural component of plants and wood. Cotton fibers are almost pure cellulose, and wood, after removal of moisture, consists of about 50% cellulose. Cellulose is an important substance in the textile and paper industries. It is also used to make rayon fibers, photographic film, guncotton, celluloid, and cellophane. Humans cannot utilize cellulose as food because they lack the necessary enzymes to hydrolyze it to glucose. Thus, cellulose is an important source of bulk in the diet.

The digestion or metabolism of carbohydrates is a very complex biochemical process. It starts in the mouth where the enzyme amylase in the saliva begins the hydrolysis of starch to maltose and temporarily stops in the stomach where the hydrochloric acid present deactivates the enzyme. Digestion continues again in the intestines where the hydrochloric acid is neutralized and pancreatic enzymes complete the hydrolysis to maltose. The enzyme maltase then catalyzes the digestion of maltose to glucose. Other specific enzymes in the intestines convert sucrose and lactose to monosaccharides.

$$\text{Starch} \xrightarrow{\text{Amylase}} \text{Dextrins} \xrightarrow{\text{Amylase}} \text{Maltose} \xrightarrow{\text{Maltase}} \text{Glucose}$$

Glucose is absorbed through the intestinal walls into the bloodstream where it is transported to the cells to be used for energy. Excess glucose is rapidly removed by the liver and muscle tissue, where it is polymerized and stored as glycogen. As the body calls for it, glycogen is converted back to glucose, which is ultimately oxidized to carbon dioxide and water with the release of energy. This energy is used by the body for maintenance, growth, and other normal functions.

21.3   Lipids

lipids

fats

oils

triacylglycerols or triglycerides

Lipids are a group of oily, greasy organic substances found in living organisms that are water insoluble but soluble in fat solvents such as diethyl ether, benzene, and chloroform. Unlike carbohydrates lipids share no common chemical structure.

The most abundant lipids are the fats and oils, which make up one of the three important classes of foods.

Fats and **oils** are esters of glycerol and predominantly long-chain fatty acids. Fats and oils are also called **triacylglycerols** or **triglycerides**, since each molecule is derived from one molecule of glycerol and three molecules of fatty acid:

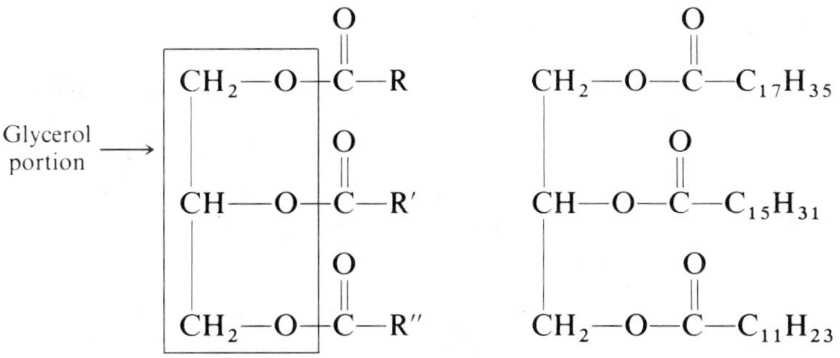

General formula
for a triacylglycerol

Typical triacylglycerol
containing three different
fatty acids

The formulas of triacylglycerol molecules vary for the following reasons:

1. The length of the fatty acid chain may vary from 4 to 20 carbons, but the number of carbon atoms in the chain is nearly always even.
2. Each fatty acid may be saturated, or it may be unsaturated and contain one, two, or three carbon–carbon double bonds.
3. An individual triacylglycerol may, and frequently does, contain three different fatty acids.

The most abundant saturated fatty acids in fats and oils are lauric, myristic, palmitic, and stearic acids (see Table 20.8). The most abundant unsaturated acids in fats and oils contain 18 carbon atoms and have one, two, or three carbon–carbon double bonds, Their formulas are

$$CH_3(CH_2)_7CH{=}CH(CH_2)_7COOH$$
Oleic acid

$$CH_3(CH_2)_4CH{=}CHCH_2CH{=}CH(CH_2)_7COOH$$
Linoleic acid

$$CH_3CH_2CH{=}CHCH_2CH{=}CHCH_2CH{=}CH(CH_2)_7COOH$$
Linolenic acid

Other significant unsaturated fatty acids are palmitoleic acid (with 16 carbons) and arachidonic acid (with 20 carbons).

$$CH_3(CH_2)_5CH{=}CH(CH_2)_7COOH \qquad CH_3(CH_2)_4(CH{=}CHCH_2)_4CH_2CH_2COOH$$
Palmitoleic acid   Arachidonic acid

Three unsaturated fatty acids—linoleic, linolenic, and arachidonic—are essential for animal nutrition and must be supplied in the diet. Diets lacking these fatty acids lead to impaired growth and reproduction, and skin disorders such as

TABLE 21.2

Fatty Acid Composition of Selected Fats and Oils

Fat or oil	Fatty acid (%)				
	Myristic acid	Palmitic acid	Stearic acid	Oleic acid	Linoleic acid
Animal fat					
Butter[a]	7–10	23–26	10–13	30–40	4–5
Lard	1–2	28–30	12–18	41–48	6–7
Tallow	3–6	24–32	14–32	35–48	2–4
Vegetable oil					
Olive	0–1	5–15	1–4	49–84	4–12
Peanut	—	6–9	2–6	50–70	13–26
Corn	0–2	7–11	3–4	43–49	34–42
Cottonseed	0–2	19–24	1–2	23–33	40–48
Soybean	0–2	6–10	2–4	21–29	50–59
Linseed[b]	—	4–7	2–5	9–38	3–43

[a] Butyric acid, 3–4%
[b] Linolenic acid, 25–58%

eczema and dermatitis. Fats are not required in our diet except as a source of these three fatty acids.

The major physical difference between fats and oils is that fats are solid and oils are liquid at room temperature. Since the glycerol part of the structure is the same for a fat and an oil, the difference must be due to the fatty acid end of the molecule. Fats contain a higher proportion of saturated fatty acids, whereas oils contain higher amounts of unsaturated fatty acids. The term *polyunsaturated* has been popularized in recent years; it means that the molecules of a particular product each contain several double bonds.

Fats and oils are obtained from natural sources. In general, fats come from animal sources and oils from vegetable sources. Thus, lard is obtained from hogs and tallow from cattle and sheep. Olive, cottonseed, corn, soybean, linseed, and other oils are obtained from the fruit or seed of their respective vegetable sources. Table 21.2 shows the major constituents of several fats and oils.

Solid fats are preferable to oils for the manufacture of soaps and for use as certain food products. Hydrogenation of oils to make them solid is carried out on a large commercial scale. In this process hydrogen, bubbled through hot oil containing a finely dispersed nickel catalyst, adds to the carbon–carbon double bonds of the oil to saturate the double bonds and form fats. In practice, only some of the double bonds are allowed to become saturated. The product that is marketed as solid "shortening" (Crisco or Spry for example) is used for cooking and baking. Oils and fats are also partially hydrogenated to improve their keeping qualities. Rancidity in fats and oils results from air oxidation at points of unsaturation, producing low molar mass aldehydes and acids of disagreeable odor and flavor.

Fats are an important energy source for humans and normally account for about 25–50% of caloric intake. When oxidized to carbon dioxide and water, fats supply about 39 kJ (9.4 kcal) energy per gram, which is more than twice the amount obtained from carbohydrates and proteins.

Fats are digested in the small intestine where they are first emulsified by the bile salts and then hydrolyzed to di- and monoglycerides, fatty acids, and glycerol. The fatty acids pass through the intestinal wall and are coated with a protein to increase solubility in the blood. They are transported to various parts of the body where they are broken down in a series of enzyme-catalyzed reactions for the production of energy. Part of the hydrolyzed fat is converted back into fat in the adipose tissue and stored as a reserve source of energy. These fat deposits also function to insulate against loss of heat as well as to protect vital organs against mechanical injury.

Soap is made by hydrolyzing fats or oils with caustic soda (aqueous NaOH). This hydrolysis process, called *saponification*, requires 3 moles of NaOH per mole of fat:

$$
\begin{array}{c}
CH_2-O-C-R \\
\quad\quad\| \\
\quad\quad O \\
CH-O-C-R' + 3\,NaOH \longrightarrow \\
\quad\quad\| \\
\quad\quad O \\
CH_2-O-C-R'' \\
\quad\quad\| \\
\quad\quad O
\end{array}
\qquad
\begin{array}{c}
CH_2-OH \\
| \\
CH-OH \\
| \\
CH_2-OH \\
\text{Glycerol}
\end{array}
\quad + \quad
\begin{array}{c}
RCOONa \\
R'COONa \\
R''COONa \\
\text{Soap}
\end{array}
$$

A fat

The most common soaps are the sodium salts of long-chain fatty acids, such as sodium stearate, $C_{17}H_{35}COONa$; sodium palmitate, $C_{15}H_{31}COONa$; and sodium oleate, $C_{17}H_{33}COONa$.

Other principal classes of lipids, besides fats and oils, are phospholipids, glycolipids, and steroids (see Figure 21.1). The phospholipids are found in all animal and vegetable cells and are abundant in the brain, the spinal cord, egg yolk, and liver. Glycolipids (cerebrosides) contain a long-chain alcohol called *sphingosine*. They contain no glycerol but do contain a monosaccharide (usually galactose). Glycolipids are found in many different tissues but, as the name *cerebroside* indicates, occur in large quantities in brain tissue.

Steroids all have a four-fused-carbocyclic-ring system (as in cholesterol) with various side groups attached to the rings. Cholesterol is the most abundant steroid in the body. It occurs in the brain, the spinal column, and nervous tissue, and it is the principal constituent of gallstones. The body synthesizes about 1 g of cholesterol per day, whereas about 0.3 g per day is ingested in the average diet. The major sources of cholesterol in the diet are meat, liver, and egg yolk.

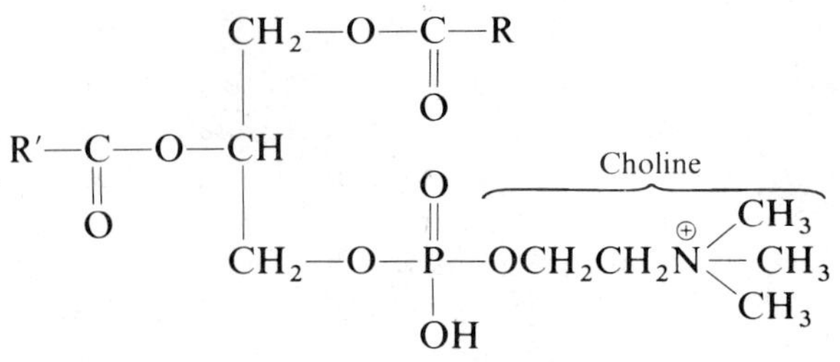

A lecithin (a phospholipid)

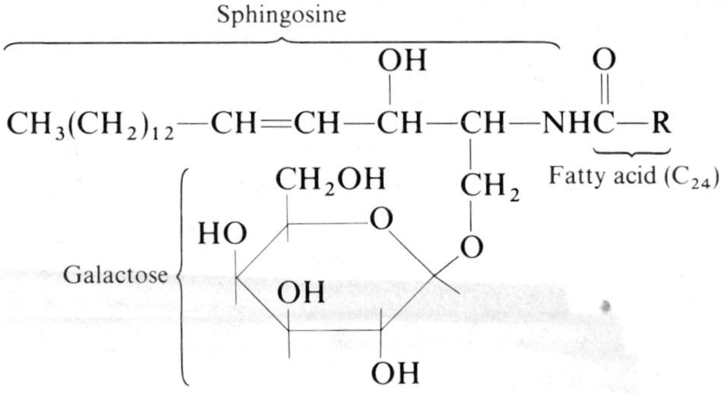

A cerebroside (a glycolipid)

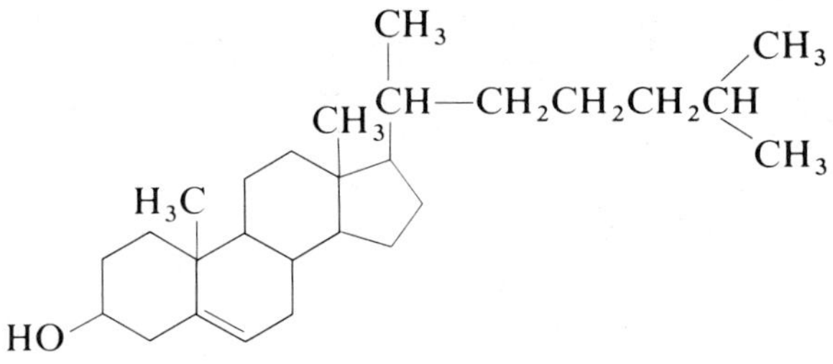

Cholesterol (a steroid)

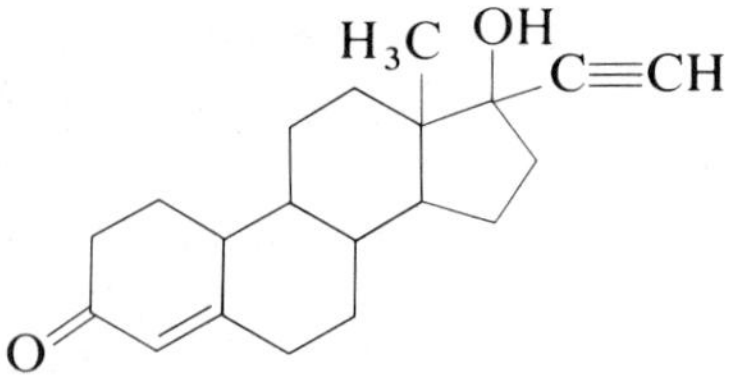

Norlutin
(a birth control pill)

FIGURE 21.1

Formulas for a phospholipid, a glycolipid, and steroids.

The cholesterol level in the blood generally rises with a person's age and body weight. In recent years, a high blood-level cholesterol has been associated with atherosclerosis (hardening of the arteries), which results in reduced flow of blood and high blood pressure. Cholesterol is needed by the body to synthesize other steroids, some of which regulate male and female sexual characteristics. Many of the synthetic birth control pills such as *norlutin* are modified steroids that interfere with the normal conception cycle in the female.

21.4 Amino Acids and Proteins

Proteins are the third important class of foodstuffs. Some common foods with high (over 10%) protein content are gelatin, fish, beans, nuts, cheese, eggs, poultry, and meat. These foods are the kinds that are most desired and needed and yet the least available to the undernourished people of the world. Proteins are present in all body tissue. They can form structural elements such as hair, fingernails, wool, and silk. Proteins also function as enzymes that regulate the countless chemical reactions taking place in every living organism. About 15% of the human body weight is protein. Chemically, proteins are polymers of amino acids with high molar masses, ranging up to more than 50 million.

amino acids

Amino acids are carboxylic acids that contain an amino ($-NH_2$) group attached to C-2 (the alpha carbon), and thus are called α *amino acids*. They also contain another variable group, R. The R group represents any of the various groups that make up the specific amino acids. For example, when R is H—, the amino acid is glycine; when R is CH_3—, the amino acid is alanine; when R is $CH_3SCH_2CH_2$—, the amino acid is methionine.

An α amino acid

Some amino acids have two amino groups and some contain two acid groups. All naturally occurring ones have an amino group in the alpha (α) position to the carboxyl group; they are called α-amino acids. The alpha position is the carbon atom adjacent to the carboxyl group. The beta (β) position is the next adjacent carbon, the gamma (γ) position the next carbon, and so on.

$$\overset{\gamma}{C}H_3\overset{\beta}{C}H_2\overset{\alpha}{C}HCOOH$$
$$|$$
$$NH_2$$

α-Aminobutyric acid
(an alpha (α) amino acid)

TABLE 21.3

Common Amino Acids Derived from Proteins

Name	Abbreviation	Formula (R groups are in black)
Alanine	Ala	$CH_3CHCOOH$ $\mid$ NH_2
Arginine	Arg	NH_2—C—NH—$CH_2CH_2CH_2CHCOOH$ $\parallel$ $\mid$ NH NH_2
Asparagine	Asn	NH_2C—$CH_2CHCOOH$ $\parallel$ $\mid$ O NH_2
Aspartic acid	Asp	$HOOCCH_2CHCOOH$ $\mid$ NH_2
Cysteine	Cys	$HSCH_2CHCOOH$ $\mid$ NH_2
Glutamic acid	Glu	$HOOCCH_2CH_2CHCOOH$ $\mid$ NH_2
Glutamine	Gln	$NH_2CCH_2CH_2CHCOOH$ $\parallel$ $\mid$ O NH_2
Glycine	Gly	$HCHCOOH$ $\mid$ NH_2
Histidine	His	N——CH HC C—$CH_2CHCOOH$ N NH_2 H
Isoleucine[a]	Ile	CH_3CH_2CH—$CHCOOH$ $\mid$ $\mid$ CH_3 NH_2
Leucine[a]	Leu	$(CH_3)_2CHCH_2CHCOOH$ $\mid$ NH_2

There are approximately 200 different known amino acids in nature. Some are found in only one particular species of plant or animal, others in only a few life forms. But 20 of these amino acids are found in almost all proteins. Furthermore, these same 20 amino acids are used by all forms of life in the synthesis of

TABLE 21.3 (*continued*)

Name	Abbreviation	Formula (R groups are in black)
Lysine[a]	Lys	$NH_2CH_2CH_2CH_2CH_2CHCOOH$ $\qquad\qquad\qquad\quad NH_2$
Methionine[a]	Met	$CH_3SCH_2CH_2CHCOOH$ $\qquad\qquad\qquad NH_2$
Phenylalanine[a]	Phe	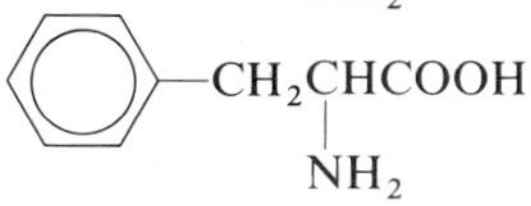 $-CH_2CHCOOH$ $\qquad\quad NH_2$
Proline	Pro	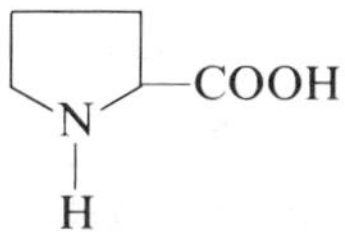 $-COOH$ (ring structure with N, H)
Serine	Ser	$HOCH_2CHCOOH$ $\qquad\quad NH_2$
Threonine[a]	Thr	$CH_3CH-CHCOOH$ $\qquad\; OH\quad NH_2$
Tryptophan[a]	Trp	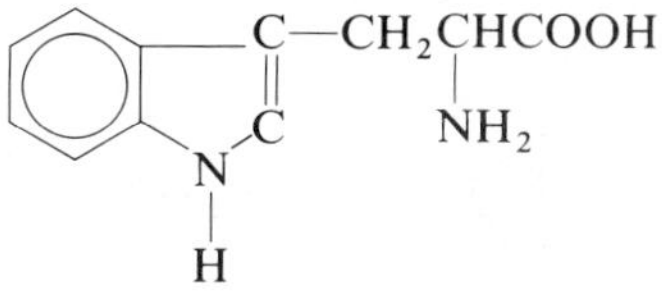 $C-CH_2CHCOOH$ $C\qquad\quad NH_2$ (indole ring, N, H)
Tyrosine	Tyr	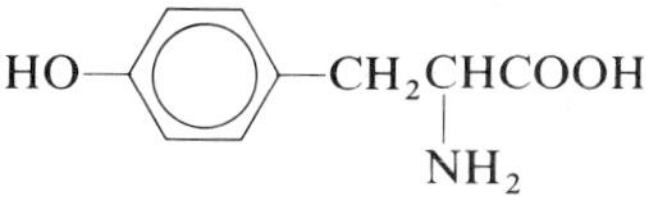 $HO-$ (ring) $-CH_2CHCOOH$ $\qquad\qquad\qquad NH_2$
Valine[a]	Val	$(CH_3)_2CHCHCOOH$ $\qquad\qquad\quad NH_2$

[a] Amino acids essential in human nutrition

proteins. The names, formulas, and abbreviations of these 20 amino acids are given in Table 21.3. Eight of these are considered essential amino acids, since the human body is not capable of synthesizing them. Therefore, they must be supplied in our diets if we are to enjoy normal health.

proteins

Proteins are polymeric substances that yield primarily amino acids on hydrolysis. The bond connecting the amino acids in a protein is commonly called

peptide linkage

a **peptide linkage** or peptide bond. If we combine two glycine molecules with the elimination of a water molecule between the amino group of one and the carboxyl group of the second glycine, we form a compound containing the amide structure and the peptide linkage. The compound containing the two amino acid groups is called a *dipeptide*.

Amide structure   Peptide linkage

$$CH_2C\overset{O}{\diagup}_{OH} + H-N-H \longrightarrow CH_2C\overset{O}{\diagup}_{N-H} - CH_2COOH + H_2O$$

$$\underset{NH_2}{|} \qquad\qquad \underset{NH_2}{|}$$

Glycine   Glycine   Glycylglycine (Gly-Gly)
(a dipeptide)

The product formed from two glycine molecules is called *glycylglycine* (abbreviated Gly-Gly). Note that the molecule still has a free amino group at one end and a free carboxyl group at the other end. The formation of glycylglycine may be considered to be the first step in the synthesis of a protein, since each end of the molecule is capable of joining to another amino acid. We can thus visualize the formation of a protein by joining a great many amino acids in this fashion. Another example, showing a tripeptide (three amino acids linked together), follows. This compound contains two peptide linkages.

$$HO-\bigcirc-CH_2CHCOOH + CH_3CHCOOH + CH_2COOH$$

$$\underset{NH_2}{|} \qquad\qquad \underset{NH_2}{|} \qquad \underset{NH_2}{|}$$

Tyrosine   Alanine   Glycine

$$HO-\bigcirc-CH_2CHC\overset{O}{\diagup}_{NH} - \overset{CH_3}{\underset{|}{CH}}-C\overset{O}{\diagup}_{NH} - CH_2COOH$$

$$\underset{NH_2}{|}$$

Tyrosylalanylglycine (a tripeptide) (Tyr-Ala-Gly)

There are five other tripeptide combinations of these three amino acids using only one unit of each amino acid. Peptides containing up to about 40–50 amino acid units in a chain are **polypeptides**. Still longer chains of amino acids are proteins.

polypeptides

The amino acid units in a peptide are called *amino acid residues* or simply residues. (They no longer are amino acids because they have lost an H atom from their amino groups and an OH from their carboxyl groups.) In linear peptides one end of the chain has a free amino group and the other end a free carboxyl group. The amino group end is called the *N-terminal residue* and the other end the *C-terminal residue*.

$$\overset{\overset{\displaystyle 1 \quad 2 \quad 3 \quad 4 \quad 5 \quad 6 \quad 7}{}}{\text{Ala-Pro-Tyr-Met-Gly-Lys-Gly}}$$

$\nearrow$ *N*-terminal residue

$\nwarrow$ *C*-terminal residue

The sequence of amino acids in a chain is numbered starting with the *N*-terminal residue, which is usually written to the left with the *C*-terminal residue at the right.

Peptides are named as acyl derivatives of the *C*-terminal amino acid with the *C*-terminal unit keeping its complete name. The *ine* ending of all but the *C*-terminal amino acids is changed to *yl*, and these are listed in the order in which they appear, starting with the *N*-terminal amino acid.

$$CH_3CHC \overset{O}{\overset{\|}{}} — NHCHC \overset{O}{\overset{\|}{}} — NHCH_2COOH$$

$$\underset{NH_2}{|} \qquad \underset{CH_2}{|}$$

Alanyl   Tyrosyl   Glycine

Ala-Tyr-Gly

Thus Ala-Tyr-Gly is called *alanyltyrosylglycine*. The name of Arg-Gln-His-Ala is arginylglutamylhistidylalanine.

A number of small, naturally occurring polypeptides have significant biochemical functions. The amino acid sequence of two of these, oxytocin and vasopressin, are shown in Figure 21.2. Oxytocin controls uterine contractions during labor in childbirth and also causes contraction of the smooth muscles of the mammary gland, resulting in milk excretion. Vasopressin in high concentration raises the blood pressure and has been used in surgical shock treatment for this purpose. Vasopressin is also an antidiuretic, regulating the excretion of fluid by the kidneys. The absence of vasopressin leads to diabetes insipidus. This condition, which is characterized by excretion of up to 30 liters of urine per day, may be controlled by administration of vasopressin or its derivatives. Oxytocin and vasopressin are similar nonapeptides, differing only at positions 3 and 8.

Determining the sequence of the amino acids in even one protein molecule was a formidable task. The amino acid sequence of beef insulin was announced in 1955 by the British biochemist Frederick Sanger (b. 1918). This structure determination required several years of effort by a team under Sanger's direction.

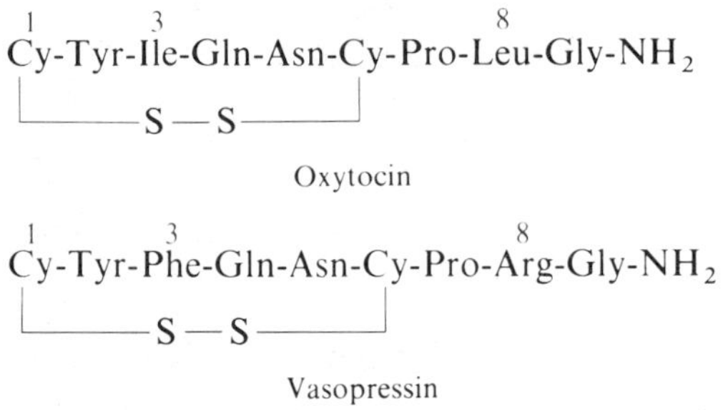

Cy-Tyr-Ile-Gln-Asn-Cy-Pro-Leu-Gly-NH$_2$
$\overset{1}{}$ $\overset{3}{}$ $\overset{8}{}$
└────S—S────┘

Oxytocin

Cy-Tyr-Phe-Gln-Asn-Cy-Pro-Arg-Gly-NH$_2$
└────S—S────┘

Vasopressin

FIGURE 21.2

Amino acid sequences of oxytocin and vasopressin. The difference in only two amino acids in these two compounds results in very different physiological activity. The C-terminal amino acid has an amide structure instead of the free COOH (indicated as -Gly-NH$_2$).

He was awarded the 1958 Nobel Prize in chemistry for this work. Recently, automated amino acid sequencers have been developed that can determine the sequence of an average-sized protein in a few days. Beef insulin consists of 51 amino acid units in two polypeptide chains. The two chains are connected by disulfide linkages (—S—S—) of two cysteine residues at two different sites. The structure is shown in Figure 21.3. Insulins from other animals, including humans, differ slightly by one, two, or three amino acid residues.

Protein digestion takes place in the stomach and the small intestine. Here digestive enzymes hydrolyze proteins to smaller peptides and amino acids, which pass through the walls of the intestines, are absorbed by the blood, and are transported to the liver and other tissues of the body. The body does not store free amino acids. They are used in many ways: (1) to synthesize proteins to replace and repair body tissues; (2) to synthesize other nitrogen-containing substances, such a certain hormones and heme; (3) to synthesize nucleic acids; and (4) to make enzymes that control the synthesis of other necessary products, such as carbohydrates and fats. Proteins are catabolized (degraded) to carbon dioxide, water, and urea. Urea, containing the protein nitrogen, is eliminated from the body in the urine.

Carbohydrates and fats are used primarily to supply heat and energy to the body. Proteins, on the other hand, are used mainly to repair and replace worn-out tissue. Tissue proteins are continuously being broken down and resynthesized.

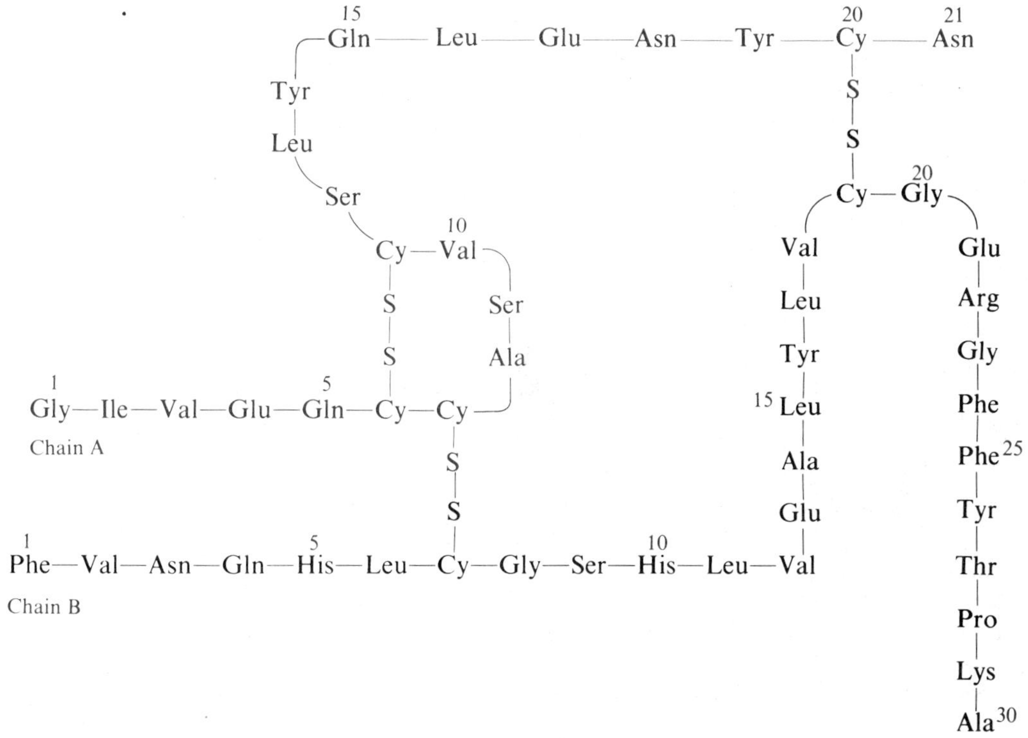

FIGURE 21.3

Amino acid sequence of beef insulin.

Therefore, protein must be continually supplied to the body in the diet. It is nothing short of amazing how the organism picks out the desired amino acids from the bloodstream and puts them together in proper order to synthesize a needed protein. The synthesis of proteins is controlled by nucleic acids.

21.5 Nucleic Acids

Explaining how hereditary material duplicates itself was one of the most baffling problems of biology. For many years biologists attempted in vain to solve this problem and also to find an answer to the question "Why are the offspring of a

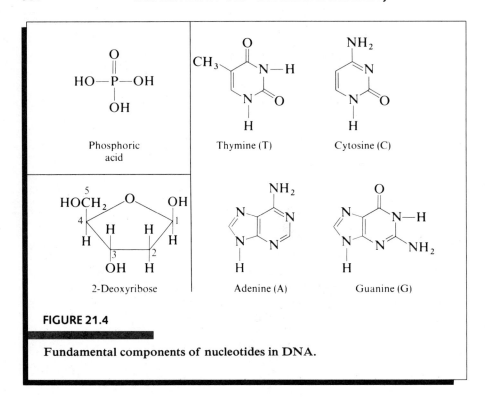

FIGURE 21.4

Fundamental components of nucleotides in DNA.

given species undeniably of that species?" Many thought the chemical basis for heredity lay in the structure of the proteins. But no one was able to provide sufficient evidence as to how protein could reproduce itself. The answer to the hereditary problem was finally found in the structure of the nucleic acids.

The unit structure of all living things is the cell. Suspended in the nucleus of cells are chromosomes, which consist largely of proteins and nucleic acids. The nucleic acids and the proteins are intimately associated into complexes called nucleoproteins. The two types of nucleic acids are those that contain the sugar deoxyribose and those that contain the sugar ribose. Accordingly, they are called deoxyribonucleic acid (DNA) and ribonucleic acid (RNA). DNA was discovered in 1869 by Swiss physiologist Friedrich Miescher (1844–1895), who extracted it from the nuclei of cells.

DNA   **DNA** is a polymeric substance made up of thousands of units called nucleotides. The fundamental components of the nucleotides in DNA are phosphoric acid, deoxyribose (a pentose sugar), and the four nitrogen-containing bases, adenine, thymine, guanine, and cytosine (abbreviated as A, T, G, and C). phosphoric acid, deoxyribose (a pentose sugar), and the four nitrogen-containing in the body from glucose; and the four nitrogen bases are made in the body from amino acids. The formulas for these compounds are given in Figure 21.4.

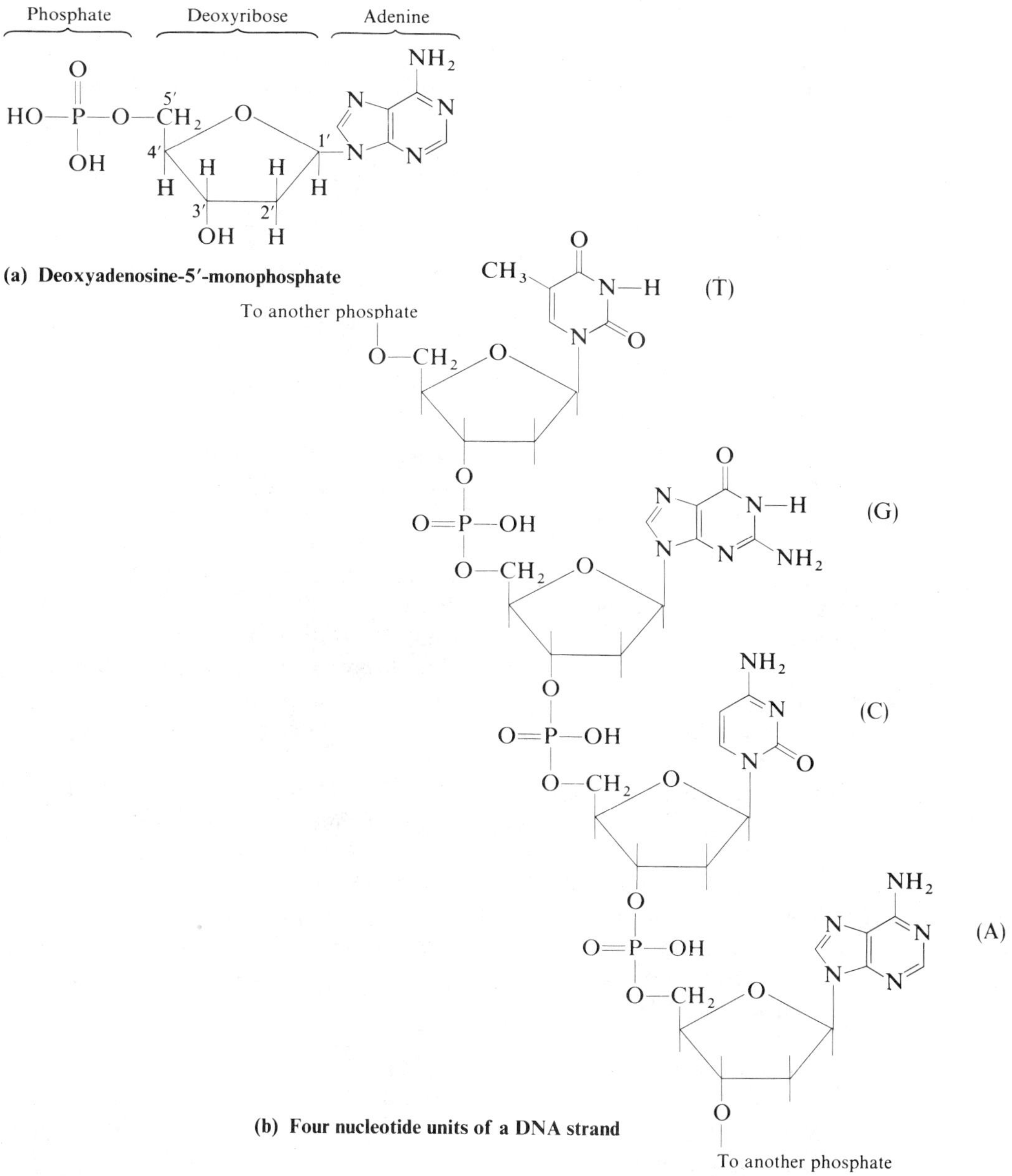

(a) Deoxyadenosine-5′-monophosphate

(b) Four nucleotide units of a DNA strand

FIGURE 21.5

(a) A single nucleotide, adenine deoxyribonucleotide (deoxyadenosine-5′-monophosphate). (b) A segment of one strand of deoxyribonucleic acid (DNA) showing four nucleotides, including those of adenine (A), cytosine (C), guanine (G), and thymine (T). The names of the last three nucleotides are, respectively: cytosine deoxyribonucleic acid (deoxycytidine-5′-monophosphate); guanine deoxyribonucleic acid (deoxyguanosine-5′-monophosphate); and thymine deoxyribonucleic acid (deoxythymidine-5′monophosphate).

nucleotide

A **nucleotide** in DNA consists of one of the four bases linked to a deoxyribose sugar, which in turn is linked to a phosphate group. Each nucleotide has the following sequence:

The structures for a single nucleotide and a segment of a polynucleotide (DNA) are shown in Figure 21.5.

Deoxyribonucleic acid (DNA) is a polymeric substance containing thousands of nucleotides. The order in which the four nucleotides (Figure 21.5) occur differs in different DNA molecules, and it is this order that determines the specificity of each DNA molecule.

In 1953, the American biologist James D. Watson (b. 1928) and the British physicist Francis H. C. Crick (b. 1916) announced their now-famous double-stranded helix structure for DNA. This concept was a milestone in the history of biology, and in 1962 Watson and Crick, along with Maurice H. F. Wilkin (b. 1916), who did the brilliant X-ray diffraction studies on DNA, were awarded the Nobel Prize in medicine and physiology.

The structure of DNA, according to Watson and Crick, consists of two polymeric strands of nucleotides in the form of a double helix, with both nucleotide strands coiled around the same axis (see Figure 21.6). Along each strand are alternate phosphate and deoxyribose units with one of the four bases adenine, guanine, cytosine, or thymine attached to deoxyribose as a side group. The double helix is held together by hydrogen bonds extending from the base on one strand of the double helix to a complementary base on the other strand. Furthermore, Watson and Crick ascertained that adenine was always hydrogen bonded to thymine, and guanine was always hydrogen bonded to cytosine. Previous analytical work by others, substantiating this concept of complementary bases, showed that the molar ratio of adenine to thymine in DNA was approximately 1 to 1 and the molar ratio of guanine to cytosine was also approximately 1 to 1.

The structure of DNA has been likened to a ladder that has been twisted into a double helix, with the rungs of the ladder kept perpendicular to the twisted railings. The phosphate and deoxyribose units alternate along the two railings of the ladder and two nitrogen bases form each rung of the ladder. The DNA structure is illustrated in Figure 21.6.

For any individual of any species, the sequence of base pairs and the length of the nucleotide chains in DNA molecules contain the coded messages that determine all of the characteristics of that individual. In this sense the DNA molecule is a template that stores information for recall as needed. DNA contains the genetic code of life, which is passed on from one generation to another.

RNA

RNA is a polymer of nucleotides but differs from DNA in that (1) it is single stranded, (2) it contains the pentose sugar ribose instead of deoxyribose, and (3) it contains uracil instead of thymine as one of its four nitrogen bases.

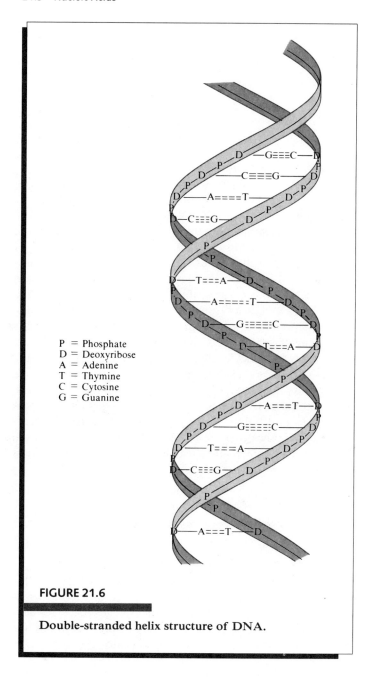

P = Phosphate
D = Deoxyribose
A = Adenine
T = Thymine
C = Cytosine
G = Guanine

FIGURE 21.6

Double-stranded helix structure of DNA.

Uracil   Thymine   Ribose

transcription

Transcription is the process by which DNA directs the synthesis of RNA. The nucleotide sequence of only one strand of DNA is transcribed into a single strand of RNA. This transcription occurs in a complementary fashion. Where there is a guanine base in DNA, a cytosine base will occur in RNA. Cytosine is transcribed to guanine, thymine to adenine, and adenine to uracil.

The main function of RNA is to direct the synthesis of proteins. RNA is produced in the cell nucleus but performs its function outside of the nucleus. Three kinds of RNA are produced directly from DNA: messenger RNA (*m*RNA), transfer RNA (*t*RNA), and ribosomal RNA (*r*RNA). Messenger RNA contains bases in the exact order transcribed from a strip of the master code in DNA. The base sequence on the *m*RNA in turn establishes the sequence of amino acids that are put together to make a specific protein. The function of the relatively small transfer RNA molecules is to bring specific amino acids to the site of protein synthesis. There is at least one different *t*RNA for each amino acid. The actual site of protein synthesis is a ribosome, which is composed of *r*RNA and protein. The function of the *r*RNA is not completely understood. However, the ribosome is believed to move along the *m*RNA chain and to aid in the polymerization of amino acids in the order prescribed by the base sequence of the *m*RNA chain. The flow of genetic information usually is in one direction, from DNA to RNA to proteins (Figure 21.7).

21.6   DNA and Genetics

Heredity is the process by which the physical and mental characteristics of parents are transferred to their offspring. In order for this process to occur it is necessary for the material responsible for genetic transfer to be able to make exact copies of itself. The design for replication is built into the DNA structure of Watson and Crick, first by the nature of its double helical structure and second by the complementary nature of its nitrogen bases, where adenine will bond only to thymine and guanine only to cytosine. The DNA double helix unwinds, or "unzips," into two separate strands at the hydrogen bonds between the bases. Each strand then serves as a template combining only with the proper free nucleotides to produce two identical replicas of itself. This replication of DNA occurs in the cell just before the cell divides, thereby giving each daughter cell

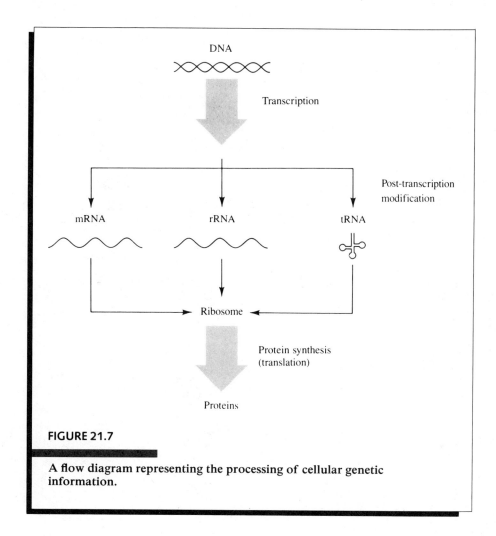

FIGURE 21.7

A flow diagram representing the processing of cellular genetic information.

the full genetic code of the cell that it came from. This process is illustrated in Figure 21.8.

DNA, as we have indicated before, is an integral part of the chromosomes. Each species carries a specific number of chromosomes in the nucleus of each of its cells. The number of chromosomes varies with different species. Humans have 23 pairs, or 46 chromosomes. Chromosomes are long, threadlike bodies composed of nucleic acids and proteins that contain the basic units of heredity called genes. **Genes** are segments of the DNA chain that contain the codes for the formation of polypeptides and RNAs. Hundreds of genes can exist along a DNA chain.

genes

In ordinary cell division, known as **mitosis**, each DNA molecule forms a duplicate by uncoiling to single strands. Each strand then assembles the

mitosis

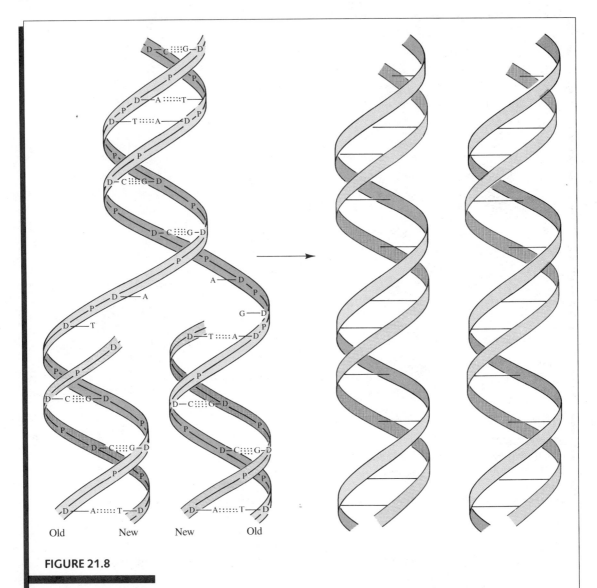

Old   New   New   Old

FIGURE 21.8

Method of replication of DNA. The two helices unwind, separating at the point of the hydrogen bonds. Each strand then serves as a template, recombining with the proper nucleotides to duplicate itself as a double-stranded helix.

complementary portion from available free nucleotides to form a duplicate of the original DNA molecule. After cell division is completed, each daughter cell contains the genetic material that correspond exactly to those that were present in the original cell before division.

meiosis

However, in almost all higher forms of life, reproduction takes place by union of the male sperm with the female egg. Cell splitting to form the sperm cell and the egg cell occurs by a different and more complicated process called *meiosis*. In **meiosis**, the genetic material is divided so that each daughter cell receives one chromosome from each pair. After meiosis, the egg cell and the sperm cell each carry only half of the chromosomes from its original cell. Between them they form a new cell that once again contains the correct number of chromosomes and all of the hereditary characteristics of the species. Thus, the offspring derives half of its genetic characteristics from the father and half from the mother.

Nature is not 100% perfect. Occasionally, DNA replication is not perfect or a section of the DNA molecule is damaged by X rays, radioactive rays, or drugs, and a mutant organism is produced. In the disease of sickle-cell anemia, a large proportion of the red blood cells form into sickle shapes instead of the usual globular shape. This irregularity limits the ability of the blood to carry oxygen and causes the person to be weak and unable to fight infection, leading to early death. Sickle-cell anemia is due to one misplaced amino acid in the structure of hemoglobin. The sickle-cell-producing hemoglobin has a valine residue where a glutamic acid residue should be located. Sickle-cell anemia is an inherited disease indicating a fault in the DNA coding that is transmitted from parent to child. Many biological disorders and ailments have been traced directly to a deficiency in the genetic information of DNA.

21.7 Enzymes

enzymes

Enzymes are the catalysts of biochemical reactions. All enzymes are proteins and they catalyze nearly all of the myriad reactions that occur in living cells. Uncatalyzed reactions that may require hours of boiling in the presence of a strong acid or a strong base can occur in a fraction of a second in the presence of the proper enzyme at room temperature and nearly neutral pH. This process is all the more remarkable when we realize that enzymes do not actually cause chemical reactions. They act as catalysts by greatly lowering the activation energy of specific biochemical reactions. The lowered activation energy permits these reactions to proceed at high speed at body temperature.

Louis Pasteur (1822–1895) was one of the first scientists to study enzyme-catalyzed reactions. He believed that living yeasts or bacteria were required for these reactions, which he called fermentations—for example, the conversion of glucose to alcohol by yeasts. In 1897, Eduard Büchner (1860–1917) made a cell-free filtrate that contained enzymes prepared by grinding yeast cells with very fine sand. The enzymes in this filtrate converted glucose to alcohol, thus proving that the presence of living cells was not required for enzyme activity. For this work Büchner received the Nobel Prize in chemistry in 1907.

Each organism contains thousands of enzymes. Some enzymes are simple proteins consisting only of amino acid units. Others are conjugated and consist of a protein part, or *apoenzyme*, and a nonprotein part, or *coenzyme*. Both parts are essential, and a functioning enzyme consisting of both the protein and nonprotein parts is called a *holoenzyme*.

Apoenzyme + Coenzyme = Holoenzyme

Often the coenzyme is a vitamin, and the same coenzyme may be associated with many different enzymes.

For some enzymes an inorganic component, such as a metal ion (for example, Ca^{2+}, Mg^{2+}, or Zn^{2+}) is required. This inorganic component is an *activator*. From the standpoint of function, an activator is analogous to a coenzyme, but inorganic components are not called coenzymes.

Another remarkable property of enzymes is their specificity of reaction; that is, a certain enzyme will catalyze the reaction of a specific type of substance. For example, the enzyme maltase catalyzes the reaction of maltose and water to form glucose. Maltase has no effect on the other two common disaccharides, sucrose and lactose. Each of these sugars requires a specific enzyme; sucrase to hydrolyze sucrose; lactase to hydrolyze lactose. (See the hydrolysis equations in Section 21.2.)

substrate

The substance acted on by an enzyme is called the **substrate**. Sucrose is the substrate of the enzyme sucrase. Enzymes have been named by adding the suffix *ase* to the root of the substrate name. Note the derivations of maltase, sucrase, and lactase from maltose, sucrose, and lactose. Many enzymes, especially digestive enzymes, have trivial names such as pepsin, rennin, trypsin, and so on. These names have no systematic significance.

Enzymes act according to the following general sequence. Enzyme (E) and substrate (S) combine to form an enzyme–substrate intermediate (E–S). This intermediate decomposes to give the product (P) and regenerate the enzyme:

$$E + S \rightleftharpoons E\text{–}S \longrightarrow E + P$$

For the hydrolysis of maltose the sequence is

Maltase + Maltose   Maltase–Maltose
    E   S   E–S

Maltase–Maltose + H_2O Maltase + 2 Glucose
    E–S   E   P

Enzyme specificity is believed to be due to the particular shape of a small part of the enzyme, its active site, which exactly fits a complementary-shaped part of the substrate (see Figure 21.9). This interaction is analogous to a lock and key; the substrate is the lock and the enzyme, the key. Just as a key will open only the lock

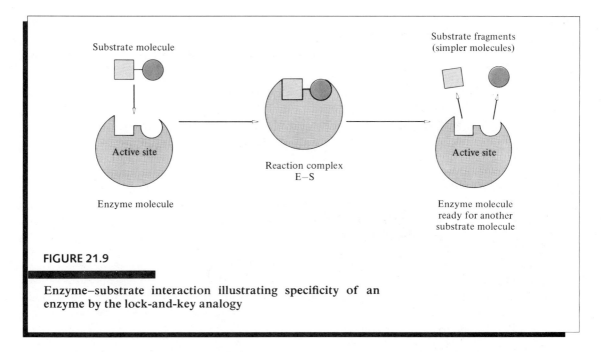

FIGURE 21.9

Enzyme–substrate interaction illustrating specificity of an enzyme by the lock-and-key analogy

it fits, the enzyme will act only on a molecule that fits its particular shape. When the substrate and the enzyme come together, they form a substrate–enzyme complex unit. The substrate, activated by the enzyme in the complex, reacts to form the products, regenerating the enzyme.

A more recently suggested model of the enzyme–substrate catalytic site is known as the "induced fit" model. In this model, the enzyme site of attachment to the substrate is flexible, with the substrate inducing a change in the enzyme shape to fit the shape of the substrate. This theory allows for the possibility that in some cases the enzyme might wrap itself around the substrate and so form the correct shape of lock and key. Thus, the enzyme does not need to have an exact preformed catalytic site to match the substrate.

Concepts in Review

1. Classify carbohydrates as mono-, di-, or polysaccharides.

2. Draw structural formulas in the open-chain and cyclic forms for glucose, fructose, galactose, and ribose.

3. Draw structural formulas for maltose and sucrose.

4. State the properties of and the general occurrence of glucose, galactose, fructose, and ribose.

5. Understand the manner in which monosaccharides are linked together in maltose, lactose, and sucrose.

6. Write equations for the enzyme-catalyzed hydrolysis of maltose, lactose, and sucrose.

7. Give the monosaccharide composition of maltose, lactose, sucrose, starch, cellulose, and glycogen.

8. Discuss the similarities and differences between starch and cellulose.

9. Discuss, in simple terms, the metabolism of carbohydrates in the human body.

10. Rate the relative sweetness of the common mono- and disaccharides.

11. Give the general formula for fats and oils.

12. Write the names and formulas of the fatty acids that most commonly occur in fats and oils.

13. State which fatty acids are essential to human diets.

14. Write the structure of a triacylglycerol when given the fatty acid composition.

15. Write an equation for the saponification of a fat or an oil with caustic soda (NaOH).

16. Tell how a fat differs from an oil.

17. Explain the "hydrogenation" of vegetable oils, and the purposes for this hardening.

18. Draw the structural formula for cholesterol and the structural feature that is common to all steroids.

19. Distinguish among these three lipids: fats, phospholipids, and glycolipids.

20. List five foods that are major sources of proteins.

21. Explain the meaning of α-amino acids and the significance of these compounds in naturally occurring protein material.

22. Show the structural formula of a di-, tri-, or polypeptide that will be formed by combining amino acids.

23. Describe the functions of and the metabolic fate of amino acids and proteins.

24. Name and write the formulas of the six fundamental components of DNA.

25. Write the structure for a segment of a polynucleotide that contains up to four nucleotides.

26. Explain the three structural differences between DNA and RNA.

27. Describe the double-helix structure of DNA according to Watson and Crick.

28. Explain the concept of complementary bases and how it relates to DNA.

29. Discuss the role of DNA in genetics.

30. Distinguish between cell division in mitosis and meiosis.

31. Discuss the role of enzymes in the body and the theory of how they function.

32. Tell what is meant by the specificity of an enzyme.

Key Terms in Review

The terms listed here have all been defined within the chapter. Review the definitions of each, and use the glossary and margin notations within the chapter as study aids.

amino acids	nucleotide
biochemistry	oils
carbohydrates	peptide linkage
disaccharides	polypeptides
DNA	polysaccharides
enzymes	proteins
fats	RNA
genes	substrate
lipids	transcription
meiosis	triacylglycerols
mitosis	(triglycerides)
monosaccharides	

Exercises

1. Of the sugars tabulated in Table 21.1, which is the sweetest disaccharide? Which is the sweetest monosaccharide? (Invert sugar is a mixture of glucose and fructose, so it should not be considered.)

2. According to Table 21.2, are the fatty acids in vegetable oils more saturated or unsaturated than those in animal fats? Explain your answer.

3. Which of the common amino acids in Table 21.3 have more than one carboxyl group? Which have more than one amino group?

4. How many disulfide linkages are there in each molecule of beef insulin? (See Figure 21.3.)

5. In the four nucleotide units of DNA shown in Figure 21.5, which of these components are part of the backbone chain, and which are off to the side: the nitrogen bases, the deoxyribose, and the phosphoric acid?

6. In the double-stranded helix structure of DNA (Figure 21.6), which nitrogen bases are always hydrogen bonded to each of the following: cytosine, thymine, adenine, and guanine?

7. Life is dependent on four major classes of biomolecules. What are they?

8. Indicate the three types of carbohydrates. Which is the simplest?

9. What is an aldose, an aldotetrose, a ketose, a ketohexose? Give an example of each.

10. Classify each of the following as a monosaccharide, disaccharide, or polysaccharide: glucose, sucrose, maltose, fructose, cellulose, lactose, glycogen, galactose, starch, and ribose.

11. Draw structural formulas in the open-chain form for ribose, glucose, fructose, and galactose.

12. Draw structural formulas in the cyclic form for ribose, glucose, fructose, and galactose.

13. State the properties and the sources of ribose, glucose, fructose, and galactose.

14. The molecular formula for lactic acid is $C_3H_6O_3$, and its structural formula is $CH_3CH(OH)COOH$. Is this compound a carbohydrate? Explain.

15. What is the monosaccharide composition of (a) sucrose, (b) maltose, (c) lactose, (d) starch, (e) cellulose, and (f) glycogen?

16. Draw structural formulas in the cyclic form for sucrose and maltose.

17. If the most common monosaccharides have the formula $C_6H_{12}O_6$, why do the resulting disaccharides have the formula $C_{12}H_{22}O_{11}$, rather than $C_{12}H_{24}O_{12}$?

18. Write equations, using structural formulas in the cyclic form, for the hydrolysis of (a) sucrose, and (b) maltose. What enzymes catalyze these reactions?

19. Discuss the similarities and differences between starch and cellulose.

20. In what form is carbohydrate stored in the body?

21. Discuss, in simple terms, the metabolism of carbohydrates in the human body.

22. State the natural sources of sucrose, maltose, lactose, and starch.

23. Invert sugar, obtained by the hydrolysis of sucrose to an equal-molar mixture of fructose and glucose, is commonly used as a sweetener in commercial food preparations. Why is invert sugar sweeter than the original sucrose?

24. What properties of molecules cause them to be classified as lipids?

25. Write structural formulas for glycerol, stearic acid, palmitic acid, oleic acid, and linoleic acid.

26. Distinguish both chemically and physically between a fat and a vegetable oil.

27. What is a triacylglycerol? Give an example.

28. Write the structure for tristearin, a fat in which all the fatty acid units are stearic acid.

29. Write the structure of a triacylglycerol that contains one unit each of linoleic, stearic, and oleic acids. How many other formulas are possible in which the triacylglycerol contains one unit each of these acids?

30. Write equations for the saponification of (a) tripalmitin and (b) the triacylglycerol of Exercise 29. Name the product(s) that is a soap.

31. How can vegetable oils be solidified? What is the advantage of solidifying these oils?

32. What functions do fats have in the human body?

33. Which fatty acids are essential to human diets?

34. Draw the structural formula of cholesterol.

35. Draw the ring structure that is common to all steroids.

36. List six foods that are major sources of proteins.

37. What functional groups are present in amino acids?

38. Why are the amino acids of proteins called α-amino acids?

39. Write out the full structure for the two possible dipeptides containing glycine and phenylalanine.

40. Write structures for (a) glycylglycine, (b) glycylglycylalanine, and (c) leucylmethionylglycylserine.

41. Using amino acid abbreviations, write all the possible tripeptides containing one unit each of glycine, phenylalanine, and leucine.

42. What are essential amino acids? Write the names of the amino acids that are essential to humans.

43. When proteins are eaten by a human, what are the metabolic fates of the protein material?

44. Why should protein be continually included in a balanced diet?

45. Write structural formulas for the compounds that make up DNA.

46. (a) What are the three units that make up a nucleotide?
    (b) List the components of the four types of nucleotides found in DNA.
    (c) Write the structure and name of one of these nucleotides.

47. Briefly describe the structure of DNA as proposed by Watson and Crick.

48. What is the role of hydrogen bonding in the structure of DNA?

49. Explain the concept of complementary bases and how it relates to DNA.

50. A segment of a DNA strand has a base sequence of C-G-A-T-T-G-C-A. What is the base sequence of the other complementary strand of the double helix?

51. Explain the replication process of DNA.

52. Briefly discuss the relationship of DNA to genetics.

53. What are the three differences between DNA and RNA in terms of structure?

54. Distinguish between cell division in mitosis and in meiosis.

55. What are enzymes and what is their role in the body?

56. What is meant by specificity of an enzyme?

57. In the polypeptide Tyr-Gly-His-Phe-Val, identify the *N*-terminal and the *C*-terminal residues.

58. How are polypeptides numbered? Number the polypeptide in Exercise 57.

59. Which of the following statements are correct? Rewrite each incorrect statement to make it correct.

   (a) Biochemistry is the branch of chemistry that is concerned with the chemical reactions occurring in living organisms.

   (b) Carbohydrates are polyhydroxy aldehydes or polyhydroxy ketones or compounds that will yield them when hydrolyzed.

   (c) The body stores glucose as glycogen until it is needed.

   (d) The ultimate use of carbohydrates in the body is oxidation to carbon dioxide and water and the utilization of the energy released.

   (e) A disaccharide molecule consists of two monosaccharide units linked together, minus a water molecule.

   (f) The most common lipids are fats and oils.

   (g) Oils have a higher percentage of saturated fatty acids than fats.

   (h) Fats are digested in the stomach and converted to monosaccharides.

   (i) Oleic, linoleic, and linolenic acids are the three essential fatty acids required by humans.

   (j) Proteins are high molar mass polymers of amino acids.

   (k) The bond connecting the amino acids in a protein is commonly called a peptide linkage.

   (l) The double helix of DNA is tied together by hydrogen bonds between the deoxyribose and the phosphoric acid units.

   (m) When a cell divides, the DNA double helix unwinds and each helix serves as a template combining only with the proper free nucleotides to produce two identical replicas of itself.

   (n) The amino acid residues in a polypeptide chain are numbered beginning with the *C*-terminal amino acid.

   (o) A major difference between RNA and DNA is that RNA contains the pentose ribose and DNA contains deoxyribose.

   (p) A nucleotide is made up of components called nucleic acids.

   (q) The backbone of each strand of DNA is made up of alternating molecules of deoxyribose and nitrogen bases.

   (r) The molecular difference between ribose and deoxyribose is an oxygen atom on carbon 2.

   (s) Genes are contained in the nucleotide sequence of DNA.

   (t) The substrate acted on by an enzyme is called a coenzyme.

   (u) DNA is a polymer made from nucleotides.

   (v) The molar ratio of adenine to thymine and guanine to cytosine is about 1:1 in DNA.

   (w) The main function of RNA is to see that each cell contains 46 chromosomes.

Review Exercises for Chapters 19–21

CHAPTER NINETEEN Nuclear Chemistry

True–False. *Answer the following as either true or false.*

1. Radioactivity was first discovered by Antoine Henri Becquerel.
2. A gamma ray has more penetrating ability than either an alpha or a beta particle.
3. A thin sheet of paper will normally block beta radiation.
4. In an electromagnetic field, an alpha particle undergoes greater deflection than does a beta particle.
5. The half-life is the time required for one-half of a specified amount of a radioactive nuclide to disintegrate.
6. When a nucleus emits an alpha particle or a beta particle, it always changes into a nuclide of a different element.
7. A beta particle consists of two protons and two neutrons.
8. When an atom loses a beta particle from its nucleus, a different element is formed, having essentially the same mass and an atomic number one greater than the starting element.
9. A radioactive disintegration series shows the succession of alpha and beta emissions by which naturally occurring radioactive elements decay to reach stability.
10. The fission of U-235 can become a chain reaction because of all the energy liberated.
11. The energy from nuclear fission can be harnessed to produce steam, which can drive turbines and produce electricity.
12. The process of uniting the nuclei of two light elements to form one heavier nucleus is known as nuclear fusion.
13. The high energy released from nuclear fusion gives other nuclei enough kinetic energy to sustain a chain reaction.
14. In radiocarbon dating, the ratio of C-14 to C-12 gives data relative to the age of the object being dated.
15. Cancers are often treated by gamma radiation from Co-60, which destroys the rapidly growing cancer cells.
16. Protracted exposure to low levels of any form of ionizing radiation can weaken the body and lead to the onset of malignant tumors.
17. An atom of $^{235}_{92}U$ has 327 nucleons.
18. A positron has the mass of an electron and the electrical charge of a proton.
19. Transmutation is the changing of a nuclide of one element to a nuclide of another element.
20. Rem is the unit of radiation that relates to the biological effect of absorbed radioactive radiation.
21. $^{232}_{92}U$ represents a uranium isotope with 92 protons and 146 neutrons.
22. A geiger counter measures ionizing radiation.
23. The Curie is the unit of radiation that measures the rate of decay of a radionuclide.

Multiple Choice. *Choose the correct answer to each of the following.*

1. If $^{238}_{92}U$ loses an alpha particle, the resulting nuclide is:
   (a) $^{237}_{92}U$ (b) $^{234}_{90}Th$ (c) $^{238}_{93}Np$ (d) $^{210}_{83}Bi$
2. If $^{210}_{82}Pb$ loses a beta particle, the resulting nuclide is:
   (a) $^{209}_{83}Bi$ (b) $^{210}_{81}Ti$ (c) $^{206}_{80}Hg$ (d) $^{210}_{83}Bi$
3. In the equation $^{209}_{83}Bi + ? \longrightarrow ^{210}_{84}Po + ^{1}_{0}n$, the missing bombarding particle would be:
   (a) $^{2}_{1}H$ (b) $^{1}_{0}n$ (c) $^{4}_{2}He$ (d) $_{-1}^{0}e$
4. Which of the following is not a characteristic of nuclear fission?
   (a) Upon absorption of a proton, a heavy nucleus splits into two or more smaller nuclei.
   (b) Two or more neutrons are produced from the fission of each atom.
   (c) Large quantities of energy are produced.
   (d) All nuclei formed are radioactive, giving off beta and gamma radiation.

5. The half-life of Sn-121 is 10 days. If you started with 40 g of this isotope, how much would you have left 30 days later?
   (a) 10 g   (b) None   (c) 15 g   (d) 5 g
6. $^{241}_{94}$Pu successively emits β, α, α, β, α, α. At that point, the nuclide has become:
   (a) $^{225}_{94}$Pu   (b) $^{225}_{88}$Ra   (c) $^{207}_{84}$Po   (d) $^{219}_{84}$Po
7. Calculate the nuclear binding energy of $^{56}_{26}$Fe.
   Mass data: $^{56}_{26}$Fe = 55.9349 g/mol;
   n = 1.0087 g/mol; p = 1.0073 g/mol;
   e^- = 0.00055 g/mol; 1.0 g = 9.0×10^{13} J
   (a) 4.8×10^{13} J/mol   (c) 0.5302 g/mol
   (b) 56.4651 g/mol   (d) 4.9×10^{15} J/mol
8. The radioactivity ray with the greatest penetrating ability is:
   (a) Alpha   (c) Gamma
   (b) Beta   (d) Proton
9. In a nuclear reaction:
   (a) Mass is lost
   (b) Mass is gained
   (c) Mass is converted into energy
   (d) Energy is converted into mass
10. As the temperature of a radionuclide increases, its half-life:
   (a) Increases   (c) Remains the same
   (b) Decreases   (d) Fluctuates
11. The nuclide that has the longest half-life is:
   (a) $^{238}_{92}$U   (b) $^{210}_{82}$Pb   (c) $^{234}_{90}$Th   (d) $^{222}_{88}$Ra

12. Which of the following is not a unit of radiation?
   (a) Curie   (b) Roentgen   (c) Rod   (d) Rem
13. When $^{235}_{92}$U is bombarded by a neutron, the atom can fission into:
   (a) $^{124}_{53}$I + $^{109}_{47}$Ag + 2 $^{1}_{0}$n
   (b) $^{123}_{50}$Sn + $^{110}_{42}$Mo + 2 $^{1}_{0}$n
   (c) $^{134}_{56}$Ba + $^{128}_{36}$Xe + 2 $^{1}_{0}$n
   (d) $^{90}_{38}$Sr + $^{143}_{58}$Ce + 2 $^{1}_{0}$n
14. In the nuclear equation

$$^{45}_{21}\text{Sc} + ^{1}_{0}\text{n} \longrightarrow X + ^{1}_{1}\text{H}$$

   the nuclide X that is formed is:
   (a) $^{45}_{22}$Ti   (b) $^{45}_{20}$Ca   (c) $^{46}_{22}$Ti   (d) $^{45}_{20}$K
15. What type of radiation is a very energetic form of photon?
   (a) Alpha   (c) Gamma
   (b) Beta   (d) Positron
16. When $^{239}_{92}$U decays to $^{239}_{93}$Np, what particle is emitted?
   (a) Positron   (c) Alpha particle
   (b) Neutron   (d) Beta particle
17. The roentgen is the unit of radiation that measures:
   (a) An absorbed dose of radiation
   (b) Exposure to X-rays
   (c) The dose from a different type of radiation
   (d) The rate of decay of a radioactive substance

CHAPTER TWENTY   Introduction to Organic Chemistry

True–False. *Answer the following as either true or false.*

1. Hydrocarbons are compounds composed entirely of carbon and hydrogen atoms bonded to each other with covalent bonds.
2. Alkanes, alkenes, and alkynes are all hydrocarbons.
3. Alkenes and alkynes are isomers.
4. C_2H_2 and C_4H_8 belong to the same class of compounds.
5. The IUPAC name for

$$CH_3-CH-CH-CH_2-CH_3$$
$$\quad\quad\;\; | \quad\; |$$
$$\quad\quad CH_3 \; CH_3$$

   is 3,4-dimethylpentane.

6. If Cl_2 is added to $CH_3CH{=}CH_2$, the product is $CH_3CH_2CH_2Cl$.
7. Chlorination of $CH_3CH_2CH_2CH_3$ produces three monochloro substitution products.
8. When each member of a series of compounds differs from the next higher member by a CH_2 group, the series is called a homologous series.
9. The IUPAC name for $CH_3CHClCH_3$ is chloropropane.
10. The carbonyl group is $\overset{\displaystyle O}{\overset{\|}{-C}}-OH$.
11. Formaldehyde is $H-\overset{\displaystyle O}{\overset{\|}{C}}-H$.
12. The name of $(CH_3)_3CBr$ is *tert*-butyl bromide or 2-bromo-2-methylpropane.

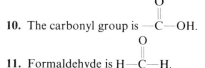

13. 2-Methylpentane and 2,3-dimethylbutane are isomers.
14. Dimethyl ketone and propanal have the same molar mass.
15. A compound of formula $C_6H_{12}O$ cannot be a carboxylic acid.
16. Acetylene is the common name for ethyne.
17. Although ethyl alcohol is used in beverages, it is classified physiologically as a depressant and a poison.
18. Dichlorobenzene, $C_6H_4Cl_2$, has the same number of isomers as bromochlorobenzene, C_6H_4BrCl.
19. Ethanal may be distinguished from propanal by reaction with Tollens' reagent.
20. Another name for stearic acid is octadecanoic acid.
21. 2-butanone and butanal are isomers.

Multiple Choice. *Choose the correct answer to each of the following.*

1. Which of the following is not a correct name for the alkane shown with it?
   (a) C_2H_6, ethane   (c) C_7H_{16}, heptane
   (b) C_5H_{12}, propane   (d) $C_{10}H_{22}$, decane
2. The structural formula of *o*-xylene is:

(a)   (c)

(b)   (d)

3. The ester $CH_3C\!-\!O\!-\!CHCH_3$ can be made
from which alcohol and carboxylic acid?

(a) $CH_3C\!-\!OH$, $CH_3CH_2CH_2OH$

(b) $CH_3C\!-\!OH$, $HO\!-\!CHCH_2OH$

(c) $H\!-\!C\!-\!OH$, CH_3CHCH_3

(d) $CH_3C\!-\!OH$, CH_3CHCH_3

4. The product of the reaction
$$CH_3CH\!=\!CH_2 + H_2O + H^+ \longrightarrow \quad \text{is a(n)}$$
   (a) alcohol   (c) alkyne
   (b) aldehyde (d) carboxylic acid
5. The number of isomers of butyl alcohol, C_4H_9OH, is
   (a) 2 (b) 3 (c) 4 (d) 6
6. The correct name for

$$CH_3\!-\!CH\!-\!CH\!-\!CH\!-\!CH\!=\!CH_2 \quad \text{is:}$$

   (a) Isobutane
   (b) 2,4-Methyl-3-propyl-5-hexene
   (c) 3,5-Dimethyl-4-isopropyl-1-hexene
   (d) 4,4-Diisopropylhexene
7. When CH_3CH_2Br reacts with CH_3CH_2ONa, the main organic product is:
   (a) $CH_3C\!-\!O\!-\!CH_2CH_3$ (c) $CH_3C\!=\!O$
   (b) CH_3COOH (d) $CH_3CH_2\!-\!O\!-\!CH_2CH_3$
8. Which of the following would give a positive Tollens' test?
   (a) CH_3CCH_3 (c) CH_3CHCH_3
   (b) $CH_3C\!=\!O$ (d) $CH_3C\!-\!OH$
9. Which of the following acids is named incorrectly?
   (a) $CH_3CH_2CH_2COOH$, butyric acid
   (b) $HCOOH$, formic acid
   (c) CH_3CH_2COOH, propic acid
   (d) CH_3COOH, acetic acid
10. With acid as a catalyst, ethanol and formic acid will react to form:
    (a) $CH_3C\!-\!NH_2$ (c) $CH_3C\!-\!O\!-\!CH_3$
    (b) $H\!-\!C\!-\!O\!-\!CH_2CH_3$ (d) $CH_3C\!-\!O\!-\!CCH_3$

11. The number of isomers of C_6H_{14} is:
    (a) 3   (b) 5   (c) 6   (d) 8
12. An open-chain hydrocarbon of formula C_6H_8 can have in its formula
    (a) one carbon–carbon double bond
    (b) two carbon–carbon double bonds
    (c) one carbon–carbon triple bond
    (d) one carbon–carbon double bond and one carbon–carbon triple bond
13. The reaction $CH_2{=}CH_2 + Br_2 \longrightarrow$ CH_2BrCH_2Br represents:
    (a) Dehalogenation   (c) Addition
    (b) Substitution   (d) Dehydration
14. Which alcohol will give diethyl ketone when oxidized?
    (a) 1-Pentanol   (c) 2-Pentanol
    (b) 2-Butanol   (d) 3-Pentanol
15. The general formula for a ketone is:
    (a) RCHO   (c) RCOOR
    (b) ROR   (d) R_2CO
16. Which of the following cannot be an aromatic compound?
    (a) C_6H_5OH (c) C_6H_{14}
    (b) C_6H_6 (d) $C_6H_5CH_3$
17. What is the correct name for

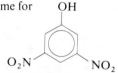

    (a) *m*-Dinitrophenol   (c) 3,5-Dinitrophenol
    (b) 2,4-Dinitrophenol   (d) 1,3-Dinitrophenol
18. Which of the following pairs are not isomers?
    (a) CH_3OCH_2Cl and CH_2ClCH_2OH
    (b) CH_3CH_2CHO and $CH_3OCH_2CH_3$
    (c) $CH_3OCH_2OCH_3$ and $CH_3CH(OH)CH_2OH$
    (d) $C_6H_4(CH_3)_2$ and $C_6H_5CH_2CH_3$

19. The reaction of $CH_3CH{=}CHCH_3 + HBr$ produces
    (a) $CH_3CH_2CH_2CH_3 + Br_2$
    (b) $CH_3CHBrCHBrCH_3 + H_2$
    (c) $CH_3CHBrCH_2CH_3$
    (d) $CH_3CH_2CH_2CH_2Br$
20. Polyvinyl chloride is a polymer of
    (a) $CH_2{=}CCl_2$
    (b) $CF_2{=}CF_2$
    (c) $CH_2{=}CHCl$
    (d) $C_6H_5CH{=}CHCl$
21. The correct name for

$$\underset{H}{\overset{\overset{O}{\underset{|}{\|}}}{CH_3C}}{-}O{-}\underset{\underset{H}{|}}{\overset{\overset{CH_3}{|}}{C}}{-}CH_3$$

    is
    (a) acetyl-2-propanoate
    (b) propyl acetate
    (c) ethyl isopropylate
    (d) isopropyl ethanoate
22. When $CH_3CH_2\overset{\overset{OH}{|}}{CH}{-}CH_3$ reacts with $KMnO_4$, the main organic product is:
    (a) $CH_3CH_2CH_2CH_3$ (c) $CH_3CH_2CH_2\overset{\overset{O}{\|}}{CH}$
    (b) $CH_3CH_2\underset{\underset{O}{\|}}{C}CH_3$ (d) no reaction
23. Teflon is a polymer of
    (a) $CH_2{=}CCl_2$ (c) $CH_2{=}CHCl$
    (b) $CF_2{=}CF_2$ (d) $C_6H_5CH{=}CHCl$

CHAPTER TWENTY-ONE   Introduction to Biochemistry

True–False. *Answer the following as either true or false.*

1. Carbohydrates are polyhydroxy alcohols or polyhydroxy ketones, or compounds that will yield them when hydrolyzed.
2. The body stores glucose as glycogen until it is needed.

3. The ultimate use of carbohydrates in the body is oxidation to carbon dioxide and the utilization of the energy released.
4. The structure of a disaccharide molecule consists of two monosaccharide units linked together, minus a water molecule.
5. Polysaccharides are macromolecules made up of many monosaccharide units linked together.

6. Maltose is a disaccharide composed of two glucose units.
7. Starch and cellulose are both polysaccharides, but cellulose consists only of glucose units whereas starch is alternating glucose and fructose units.
8. Triacylglycerols are also called triglycerides.
9. Fats and oils are esters of glycerol and the higher molar-mass fatty acids.
10. Oils have a higher percentage of unsaturated fatty acids than fats.
11. Fats are digested in the stomach and converted to glucose.
12. Fats are burned in the body or stored as fats for future use.
13. Palmitic acid contains one carbon–carbon double bond.
14. Vegetable oils are hydrogenated by reacting them with sodium hydroxide.
15. When a fat is saponified, the products are glycerol and soap.
16. Proteins are high molar-mass polymers of nucleotides.
17. Amino acids are organic compounds containing at least two functional groups: an amino group and a carboxyl group.

18. $H_2N{-}\overset{\displaystyle O}{\overset{\|}{C}}{-}OH$ is an alpha amino acid.
19. The bond connecting the amino acids in a protein is commonly called a peptide linkage.
20. The double helix of DNA is held together by hydrogen bonds between the complementary base units.
21. For any individual of any species, the sequence of base combinations and the length of the nucleotide chains in DNA molecules contain the coded messages that determine all the characteristics of the individual.
22. The flow of genetic information is in one direction, from DNA to RNA to proteins.
23. Enzymes are protein molecules that act as catalysts by greatly lowering the activation energy of specific biochemical reactions.
24. In the polypeptide Ala-Glu-Asn-Arg-Gly-Gly, the *N*-terminal amino acid is alanine.
25. Ala-Glu-Asn-Arg-Gly-Gly has five peptide bonds and is, therefore, a pentapeptide.
26. RNA is transcribed from DNA and controls the synthesis of proteins.

Multiple Choice. *Choose the correct answer to each of the following.*

1. Sugars are members of a group of compounds with the general name:
   (a) Carbohydrates (c) Proteins
   (b) Lipids   (d) Steroids
2. The products formed when maltose is hydrolyzed are:
   (a) Glucose and fructose
   (b) Glucose and galactose
   (c) Glucose and glucose
   (d) Galactose and fructose
3. Which is not true about starch?
   (a) It is a polysaccharide.
   (b) It is hydrolyzed to maltose.
   (c) It is composed of glucose units.
   (d) It is not digestible by humans.
4. Lactose is:
   (a) A monosaccharide
   (b) A disaccharide composed of galactose and glucose
   (c) A disaccharide composed of two glucose units
   (d) A decomposition product of starch
5. Which is not true about glucose?
   (a) It is a monosaccharide.
   (b) It is a component of sucrose, maltose, lactose, starch, glycogen, and cellulose.
   (c) It is a ketohexose.
   (d) It is the main source of energy for the body.
6. The sweetest of the common sugars is:
   (a) Fructose (c) Glucose
   (b) Sucrose   (d) Maltose
7. Which of the following is not true?
   (a) Lactose is sweeter than sucrose.
   (b) Glycogen is known as animal starch.
   (c) Cellulose is the most abundant organic substance in nature.
   (d) Lactose is a disaccharide known as milk sugar.
8. Which is not formed in the saponification of a fat?
   (a) Glycerol
   (b) Amino acids
   (c) Soap
   (d) A metal salt of a long-chain fatty acid
9. Which is not an essential fatty acid?
   (a) Oleic acid   (c) Linolenic acid
   (b) Linoleic acid (d) Arachidonic acid

Review Exercises for Chapters 19–21

10. Which of the following lipids does not contain a glycerol unit as part of its structure?
    (a) A fat   (c) A glycolipid
    (b) A phospholipid (d) An oil

11. An alpha amino acid always contains:
    (a) An amino group on the carbon atom adjacent to the carboxyl group
    (b) A carboxyl group at each end of the molecule
    (c) Two amino groups
    (d) Alternating amino and carboxyl groups

12. Which of the following amino acids contains sulfur?
    (a) Alanine   (c) Cysteine
    (b) Histidine (d) Glycine

13. A compound containing ten amino acid molecules linked together is called a:
    (a) Protein   (c) Deca-amino acid
    (b) Polypeptide   (d) Nucleotide

14. The major end-product(s) of protein nitrogen metabolism in humans is (are):
    (a) Amino acids   (c) Urea
    (b) Ammonium salts   (d) Dipeptides

15. Which of the following is not a correct statement about DNA and RNA?
    (a) DNA contains deoxyribose, whereas RNA contains ribose.
    (b) Both DNA and RNA are polymers made up of nucleotides.
    (c) DNA directs the synthesis of proteins and RNA contains the genetic code of life.
    (d) DNA exists as a double helix, whereas RNA exists as a single helix.

16. Which of the following bases is found in RNA, but not in DNA?
    (a) Thymine (c) Guanine
    (b) Adenine (d) Uracil

17. Complementary base pairs in DNA are linked through the formation of:
    (a) Phosphate ester bonds
    (b) Peptide linkages
    (c) Hydrogen bonds
    (d) Ionic bonds

18. In a DNA double helix, hydrogen bonding occurs between:
    (a) Adenine and thymine
    (b) Thymine and guanine
    (c) Adenine and uracil
    (d) Cytosine and thymine

19. Which of the following scientists did not receive the Nobel Prize for the structure of DNA?
    (a) Crick   (c) Sanger
    (b) Watson   (d) Wilkins

20. The substance acted on by an enzyme is called a(n):
    (a) Catalyst   (c) Coenzyme
    (b) Apoenzyme   (d) Substrate

21. The process during which a cell splits to form a sperm or an egg cell is called:
    (a) Mitosis   (c) Translation
    (b) Meiosis   (d) Transcription

APPENDIX I

Mathematical Review

1. Multiplication   Multiplication is a process of adding any given number or quantity to itself a certain number of times. Thus, 4 times 2 means 4 added two times, or 2 added together four times, to give the product 8. Various ways of expressing multiplication are

$$ab \qquad a \times b \qquad a \cdot b \qquad a(b) \qquad (a)(b)$$

All mean a times b, or a multiplied by b, or b times a.

When $a = 16$ and $b = 24$,   we have $16 \times 24 = 384$.

The expression $°F = (1.8 \times °C) + 32$ means that we are to multiply 1.8 times $°C$ and add 32 to the product. When $°C$ equal 50,

$$°F = (1.8 \times 50) + 32 = 90 + 32 = 122°F$$

The result of multiplying two or more numbers together is known as the *product*.

2. Division The word *division* has several meanings. As a mathematical expression, it is the process of finding how many times one number or quantity is contained in another. Various ways of expressing division are:

$$a \div b \qquad \frac{a}{b} \qquad a/b$$

All mean *a* divided by *b*.

When $a = 15$ and $b = 3$, $\frac{15}{3} = 5$.

The number above the line is called the *numerator*; the number below the line is the *denominator*. Both the horizontal and the slanted (/) division signs also mean "per." For example, in the expression for density, the mass per unit volume:

$$\text{density} = \text{mass/volume} = \frac{\text{mass}}{\text{volume}} = \text{g/mL}$$

The diagonal line still refers to a division of grams by the number of milliliters occupied by that mass. The result of dividing one number into another is called the *quotient*.

3. Fractions and Decimals A fraction is an expression of division, showing that the numerator is divided by the denominator. A *proper fraction* is one in which the numerator is smaller than the denominator. In an *improper fraction*, the numerator is the larger number. A decimal or a decimal fraction is a proper fraction in which the denominator is some power of 10. The decimal fraction is determined by carrying out the division of the proper fraction. Examples of proper fractions and their decimal fraction equivalents are shown in the following table.

Proper fraction		Decimal fraction		Proper fraction
$\frac{1}{8}$	=	0.125	=	$\frac{125}{1000}$
$\frac{1}{10}$	=	0.1	=	$\frac{1}{10}$
$\frac{3}{4}$	=	0.75	=	$\frac{75}{100}$
$\frac{1}{100}$	=	0.01	=	$\frac{1}{100}$
$\frac{1}{4}$	=	0.25	=	$\frac{25}{100}$

4. Addition of Numbers with Decimals To add numbers with decimals we use the same procedure as that used when adding whole numbers, but we always line up the decimal points in the same column. For example, add $8.21 + 143.1 + 0.325$

```
    8.21
+ 143.1
+   0.325
---------
  151.635
```

When adding numbers that express units of measurement, we must be certain that the numbers added together represent the same units. For example, what is the total length of three pieces of glass tubing: 10.0 cm, 125 mm, and 8.4 cm? If we simply add the numbers, we obtain a value of 143.4, but we are not certain what the unit of measurement is. To add these lengths correctly, first change 125 mm to 12.5 cm. Now all the lengths are expressed in the same units and can be added.

```
10.0 cm
12.5 cm
 8.4 cm
--------
30.9 cm
```

5. Subtraction of Numbers with Decimals To subtract numbers containing decimals, we use the same procedure as for subtracting whole numbers, but we always line up the decimal points in the same column. For example, subtract 20.60 from 182.49.

```
  182.49
-  20.60
--------
  161.89
```

6. Multiplication of Numbers with Decimals To multiply two or more numbers together that contain decimals, we first multiply as if they were whole numbers. Then, to locate the decimal point in the product, we add together the number of digits to the right of the decimal in all the numbers multiplied together. The product should have this same number of digits to the right of the decimal point.

Multiply 2.05×2.05 (total of four digits to the right of the decimal):

```
   2.05
× 2.05
------
   1025
  4100
------
 4.2025     (four digits to the right of the decimal)
```

Here are more examples:

$14.25 \times 6.01 \times 0.75 = 64.231875$ (six digits to the right of the decimal)

$39.26 \times 60 = 2355.60$ (two digits to the right of the decimal)

[*Note*: When at least one of the numbers that is multiplied is a measurement, the answer must be adjusted to contain the correct number of significant figures. (See Section 2.5 on significant figures.)]

7. Division of Numbers with Decimals   To divide numbers containing decimals, we first relocate the decimal points of the numerator and denominator by moving them to the right as many places as needed to make the denominator a whole number. (Move the decimal of both the numerator and the denominator the same amount and in the same direction.) For example,

$$\frac{136.94}{4.1} = \frac{1369.4}{41}$$

The decimal point adjustment in this example is equivalent to multiplying both numerator and denominator by 10. Now we carry out the division normally, locating the decimal point immediately above its position in the dividend.

$$\frac{0.441}{26.25} = \frac{44.1}{2625}$$

$$
\begin{array}{r}
33.4 \\
41\overline{)1369.4} \\
123 \\
\hline
139 \\
123 \\
\hline
164 \\
164 \\
\hline
\end{array}
\qquad
\begin{array}{r}
0.0168 \\
2625\overline{)44.1000} \\
2625 \\
\hline
17850 \\
15750 \\
\hline
21000 \\
21000 \\
\hline
\end{array}
$$

[*Note*: When at least one of the numbers in the division is a measurement, the answer must be adjusted to contain the correct number of significant figures. (See Section 2.5 on significant figures.)]

The foregoing examples are merely guides to the principles used in performing the various mathematical operations illustrated. There are, no doubt, shortcuts and other methods, and the student will discover these with experience. Every student of chemistry should learn to use a scientific electronic calculator for solving mathematical problems. The use of a calculator will save many hours of doing tedious longhand calculations. After solving a problem, the student should check for errors and evaluate the answer to see if it is logical and consistent with the data given.

8. Algebraic Equations   Many mathematical problems that are encountered in chemistry fall into the following algebraic forms. Solutions to these problems are simplified by first isolating the desired term on one side of the equation.

This rearrangement is accomplished by treating both sides of the equation in an identical manner (so as not to destroy the equality) until the desired term is isolated.

(a) $a = \dfrac{b}{c}$

To solve for a, divide b by c.
To solve for b, multiply both sides of the equation by c.

$$a \times c = \dfrac{b}{\cancel{c}} \times \cancel{c}$$

$$b = a \times c$$

To solve for c, multiply both sides of the equation by $\dfrac{c}{a}$.

$$\cancel{a} \times \dfrac{c}{\cancel{a}} = \dfrac{b}{\cancel{c}} \times \dfrac{\cancel{c}}{a}$$

$$c = \dfrac{b}{a}$$

(b) $\dfrac{a}{b} = \dfrac{c}{d}$

To solve for a, multiply both sides of the equation by b.

$$\dfrac{a}{\cancel{b}} \times \cancel{b} = \dfrac{c}{d} \times b$$

$$a = \dfrac{c \times b}{d}$$

To solve for b, multiply both sides of the equation by $\dfrac{b \times d}{c}$.

$$\dfrac{a}{\cancel{b}} \times \dfrac{\cancel{b} \times d}{c} = \dfrac{\cancel{c}}{\cancel{d}} \times \dfrac{b \times \cancel{d}}{\cancel{c}}$$

$$b = \dfrac{a \times d}{c}$$

(c) $a \times b = c \times d$

To solve for a, divide both sides of the equation by b.

$$\dfrac{a \times \cancel{b}}{\cancel{b}} = \dfrac{c \times d}{b}$$

$$a = \dfrac{c \times d}{b}$$

(d) $\dfrac{(b - c)}{a} = d$

To solve for b, first multiply both sides of the equation by a.

$$\dfrac{\cancel{a}(b - c)}{\cancel{a}} = d \times a$$
$$b - c = d \times a$$

Then add c to both sides of the equation.

$$b - \cancel{c} + \cancel{c} = d \times a + c$$
$$b = (d \times a) + c$$

When $a = 1.8$, $c = 32$, and $d = 35$,

$$b = (35 \times 1.8) + 32 = 63 + 32 = 95$$

---

9. Exponents, Powers of 10, Expression of Large and Small Numbers   In scientific measurements and calculations, we often encounter very large and very small numbers—for example, 0.00000384 and 602,000,000,000,000,000,000,000. These numbers are troublesome to write and awkward to work with, especially in calculations. A convenient method of expressing these large and small numbers in a simplified form is by means of exponents or powers of 10. This method of expressing numbers is known as **scientific** or **exponential notation**.

An *exponent* is a number written as a superscript following another number; it is also called a *power* of that number, and it indicates how many times the number is used as a factor. In the number 10^2, 2 is the exponent, and the number means 10 squared, or 10 to the second power, or $10 \times 10 = 100$. Three other examples are

$$3^2 = 3 \times 3 = 9$$
$$3^4 = 3 \times 3 \times 3 \times 3 = 81$$
$$10^3 = 10 \times 10 \times 10 = 1000$$

For ease of handling, large and small numbers are expressed in powers of 10. Powers of 10 are used because multiplying or dividing by 10 coincides with moving the decimal point in a number by one place. Thus, a number multiplied by 10^1 would move the decimal point one place to the right; 10^2, two places to the right; 10^{-2}, two places to the left. To express a number in powers of 10, we move the decimal point in the original number to a new position, placing it so that the number is a value between 1 and 10. This new decimal number is multiplied by 10 raised to the proper power. For example, to write the number 42,389 in exponential form (powers of 10), the decimal point is placed between the 4 and the 2 (4.2389), and the number is multiplied by 10^4; thus, the number is 4.2380×10^4.

$$\underset{4\ 3\ 2\ 1}{\underbrace{42{,}389}} = 4.2389 \times 10^4$$

The exponent (power) of 10 (4) tells us the number of places that the decimal point has been moved from its original position. If the decimal point is moved to the left, the exponent is a positive number; if it is moved to the right, the exponent is a negative number. To express the number 0.00248 in exponential notation (as a power of 10), the decimal point is moved three places to the right; the exponent of 10 is -3, and the number is 2.48×10^{-3}.

$$\underset{1\ 2\ 3}{\underbrace{0.00248}} = 2.48 \times 10^{-3}$$

Study the following examples.

$$1237 = 1.237 \times 10^3$$
$$988 = 9.88 \times 10^2$$
$$147.2 = 1.472 \times 10^2$$
$$2{,}200{,}000 = 2.2 \times 10^6$$
$$0.0123 = 1.23 \times 10^{-2}$$
$$0.00005 = 5 \times 10^{-5}$$
$$0.000368 = 3.68 \times 10^{-4}$$

Exponents in multiplication and division.   The use of powers of 10 in multiplication and division greatly simplifies locating the decimal point in the answer. In multiplication, first change all numbers to powers of 10, then multiply the numerical portion in the usual manner, and finally add the exponents of 10 algebraically, expressing them as a power of 10 in the product. In multiplication, the exponents (powers of 10) are added algebraically.

$$10^2 \times 10^3 = 10^{(2+3)} = 10^5$$
$$10^2 \times 10^2 \times 10^{-1} = 10^{(2+2-1)} = 10^3$$

Multiply: $40{,}000 \times 4200$

Change to powers of 10: $4 \times 10^4 \times 4.2 \times 10^3$

Rearrange: $4 \times 4.2 \times 10^4 \times 10^3$

$$16.8 \times 10^{(4+3)}$$
$$16.8 \times 10^7 \qquad \text{or} \qquad 1.68 \times 10^8 \quad \text{(Answer)}$$

Multiply: 380×0.00020

$$3.80 \times 10^2 \times 2.0 \times 10^{-4}$$
$$3.80 \times 2.0 \times 10^2 \times 10^{-4}$$
$$7.6 \times 10^{(2-4)}$$
$$7.6 \times 10^{-2} \qquad \text{or} \qquad 0.076 \quad \text{(Answer)}$$

Multiply: $125 \times 284 \times 0.150$

$1.25 \times 10^2 \times 2.84 \times 10^2 \times 1.50 \times 10^{-1}$

$1.25 \times 2.84 \times 1.50 \times 10^2 \times 10^2 \times 10^{-1}$

$5.325 \times 10^{(2+2-1)}$

5.32×10^3 (Answer)

In division, after changing the numbers to powers of 10, move the 10 and its exponent from the denominator to the numerator, changing the sign of the exponent. Carry out the division in the usual manner and evaluate the power of 10. The following is a proof of the equality of moving the power of 10 from the denominator to the numerator.

$$1 \times 10^{-2} = 0.01 = \frac{1}{100} = \frac{1}{10^2} = 1 \times 10^{-2}$$

In division, change the sign(s) of the exponent(s) of 10 in the denominator and move the 10 and its exponent(s) to the numerator. Then add all the exponents of 10 together. For example,

$$\frac{10^5}{10^3} = 10^5 \times 10^{-3} = 10^{(5-3)} = 10^2$$

$$\frac{10^3 \times 10^4}{10^{-2}} = 10^3 \times 10^4 \times 10^2 = 10^{(3+4+2)} = 10^9$$

Divide: $\dfrac{2871}{0.0165}$

Change to powers of 10: $\dfrac{2.871 \times 10^3}{1.65 \times 10^{-2}}$

Move 10^{-2} to the numerator, changing the sign of the exponent. This is mathematically equivalent to multiplying both numerator and denominator by 10^2.

$$\frac{2.871 \times 10^3 \times 10^2}{1.65}$$

$$\frac{2.871 \times 10^{(3+2)}}{1.65} = 1.74 \times 10^5 \quad \text{(Answer)}$$

Divide: $\dfrac{0.000585}{0.00300}$

$$\frac{5.85 \times 10^{-4}}{3.00 \times 10^{-3}}$$

$$\frac{5.85 \times 10^{-4} \times 10^3}{3.00} = \frac{5.85 \times 10^{(-4+3)}}{3.00}$$

$$= 1.95 \times 10^{-1} \quad \text{or} \quad 0.195 \quad \text{(Answer)}$$

Calculate: $\dfrac{760 \times 300 \times 40.0}{700 \times 273}$

$\dfrac{7.60 \times 10^2 \times 3.00 \times 10^2 \times 4.00 \times 10^1}{7.00 \times 10^2 \times 2.73 \times 10^2}$

$\dfrac{7.60 \times 3.00 \times 4.00 \times 10^2 \times 10^2 \times 10^1}{7.00 \times 2.73 \times 10^2 \times 10^2}$

$= 4.77 \times 10^1 \qquad \text{or} \qquad 47.7 \quad \text{(Answer)}$

10. Significant Figures in Calculations   The result of a calculation based on experimental measurements cannot be more precise than the measurement that has the greatest uncertainty. (See Section 2.5 for additional discussion.)

Addition and subtraction.   The result of an addition or subtraction should contain no more digits to the right of the decimal point than are contained in the quantity that has the least number of digits to the right of the decimal point.

Perform the operation indicated and then round off the number to the proper number of significant figures.

142.8 g	
18.843 g	93.45 mL
36.42 g	− 18.0 mL
198.063 g	75.45 mL
198.1 g (Answer)	75.5 mL (Answer)

Multiplication and division.   In calculations involving multiplication or division, the answer should contain the same number of significant figures as the measurement that has the least number of significant figures. In multiplication or division the position of the decimal point has nothing to do with the number of significant figures in the answer. Study the following examples:

	Round off to
$2.05 \times 2.05 = 4.2025$	4.20
$18.48 \times 5.2 = 96.096$	96
$0.0126 + 0.020 = 0.000252 \qquad$ or	
$\quad 1.26 \times 10^{-2} \times 2.0 \times 10^{-2} = 2.520 \times 10^{-4}$	$2.5 \ \times 10^{-4}$
$\dfrac{1369.4}{41} = 33.4$	33
$\dfrac{2268}{4.20} = 540$	540

11. Dimensional Analysis   Many problems of chemistry can be solved readily by dimensional analysis using the factor-label or conversion-factor method.

Dimensional analysis involves the use of proper units of dimensions for all factors that are multiplied, divided, added, or subtracted in setting up and solving a problem. Dimensions are physical quantities such as length, mass, and time, which are expressed in such units as centimeters, grams, and seconds, respectively. In solving a problem, we treat these units mathematically just as though they were numbers, which gives us an answer that contains the correct dimensional units.

A measurement or quantity given in one kind of unit can be converted to any other kind of unit having the same dimension. To convert from one kind of unit to another, the original quantity or measurement is multiplied or divided by a conversion factor. The key to success lies in choosing the correct conversion factor. This general method of calculation is illustrated in the following examples.

Suppose we want to change 24 ft to inches. We need to multiply 24 ft by a conversion factor containing feet and inches. Two such conversion factors can be written relating inches to feet.

$$\frac{12 \text{ in.}}{1 \text{ ft}} \quad \text{or} \quad \frac{1 \text{ ft}}{12 \text{ in.}}$$

We choose the factor that will mathematically cancel feet and leave the answer in inches. Note that the units are treated in the same way we treat numbers, multiplying or dividing as required. Two possibilities then arise to change 24 ft to inches:

$$24 \cancel{\text{ ft}} \times \frac{12 \text{ in.}}{1 \cancel{\text{ ft}}} \quad \text{or} \quad 24 \text{ ft} \times \frac{1 \text{ ft}}{12 \text{ in.}}$$

In the first case (the correct method), feet in the numerator and the denominator cancel, giving us an answer of 288 in. In the second case, the units of the answer are $\text{ft}^2/\text{in.}$, the answer being $2.0 \text{ ft}^2/\text{in.}$ In the first case, the answer is reasonable since it is expressed in units having the proper dimensions. That is, the dimension of length expressed in feet has been converted to length in inches according to the mathematical expression

$$\cancel{\text{ft}} \times \frac{\text{in.}}{\cancel{\text{ft}}} = \text{in.}$$

In the second case, the answer is not reasonable since the units ($\text{ft}^2/\text{in.}$) do not correspond to units of length. The answer is therefore incorrect. The units are the guiding factor for the proper conversion.

The reason we can multiply 24 ft times 12 in./ft and not change the value of the measurement is that the conversion factor is derived from two equivalent quantities. Therefore, the conversion factor 12 in./ft is equal to unity. And when you multiply any factor by 1, it does not change the value.

$$12 \text{ in.} = 1 \text{ ft} \quad \text{and} \quad \frac{12 \text{ in.}}{1 \text{ ft}} = 1$$

Convert 16 kg to milligrams. In this problem it is best to proceed in this fashion:

$$kg \longrightarrow g \longrightarrow mg$$

The possible conversion factors are

$$\frac{1000 \text{ g}}{1 \text{ kg}} \quad \text{or} \quad \frac{1 \text{ kg}}{1000 \text{ g}} \qquad \frac{1000 \text{ mg}}{1 \text{ g}} \quad \text{or} \quad \frac{1 \text{ g}}{1000 \text{ mg}}$$

We use the conversion factor that leaves the proper unit at each step for the next conversion. The calculation is

$$16 \text{ kg} \times \frac{1000 \text{ g}}{1 \text{ kg}} \times \frac{1000 \text{ mg}}{1 \text{ g}} = 1.6 \times 10^7 \text{ mg}$$

Many problems can be solved by a sequence of steps involving unit conversion factors. This sound, basic approach to problem solving, together with neat and orderly setting up of data, will lead to correct answers having the right units, fewer errors, and considerable saving of time.

12. Graphical Representation of Data   A graph is often the most convenient way to present or display a set of data. Various kinds of graphs have been devised, but the most common type uses a set of horizontal and vertical coordinates to show the relationship of two variables. It is called an $x-y$ graph because the data of one variable are represented on the horizontal or x axis (abscissa) and the data of the other variable are represented on the vertical or y axis (ordinate). See Figure I.1.

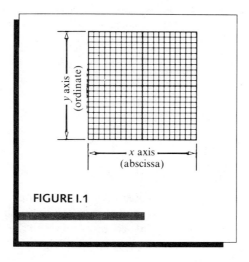

FIGURE I.1

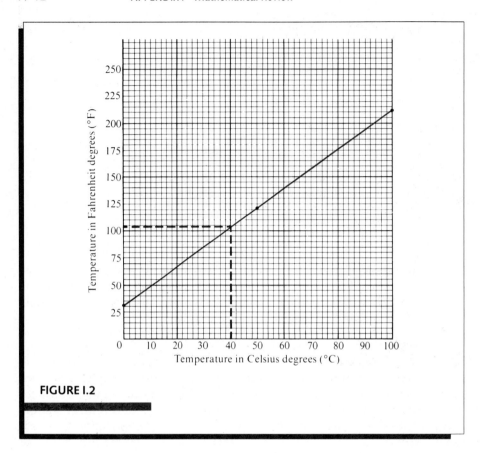

FIGURE I.2

As a specific example of a simple graph, let us graph the relationship between Celsius and Fahrenheit temperature scales. Assume that initially we have only the information in the following table.

°C	°F
0	32
50	122
100	212

On a set of horizontal and vertical coordinates (graph paper), scale off at least 100 Celsius degrees on the *x* axis and at least 212 Fahrenheit degrees on the *y* axis. Locate and mark the three points corresponding to the three temperatures given and draw a line connecting these points (see Figure I.2).

Here is how a point is located on the graph: Using the 50°C–122°F data, trace a vertical line up from 50°C on the x axis and a horizontal line across from 122°F on the y axis and mark the point where the two lines intersect. This process is called *plotting*. The other two points are plotted on the graph in the same way. [*Note*: The number of degrees per scale division was chosen to give a graph of convenient size. In this case there are 5 Fahrenheit degrees per scale division and 2 Celsius degrees per scale division.]

The graph in Figure I.2 shows that the relationship between Celsius and Fahrenheit temperature is that of a straight line. The Fahrenheit temperature corresponding to any given Celsius temperature between 0 and 100° can be determined from the graph. For example, to find the Fahrenheit temperature corresponding to 40°C, trace a perpendicular line from 40°C on the x axis to the line plotted on the graph. Now trace a horizontal line from this point on the plotted line to the y axis and read the corresponding Fahrenheit temperature (104°F). See the dashed lines in Figure I.2. In turn, the Celsius temperature corresponding to any Fahrenheit temperature between 32 and 212° can be determined from the graph by tracing a horizontal line from the Fahrenheit temperature to the plotted line and reading the corresponding temperature on the Celsius scale directly below the point of intersection.

The mathematical relationship of Fahrenheit and Celsius temperatures is expressed by the equation $°F = 1.8 \times °C + 32$. Figure I.2 is a graph of this equation. Because the graph is a straight line, it can be extended indefinitely at either end. Any desired Celsius temperature can be plotted against the corresponding Fahrenheit temperature by extending the scales along both axes as necessary. Negative, as well as positive, values can be plotted on the graph (see Figure I.3).

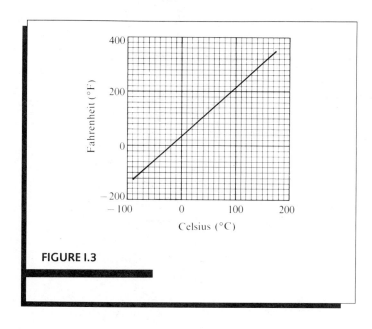

FIGURE I.3

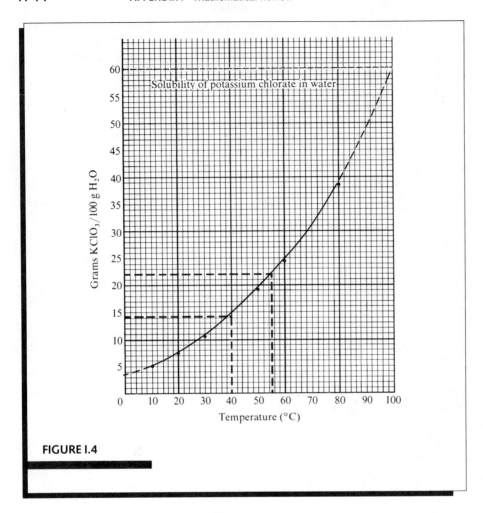

FIGURE I.4

Figure I.4 is a graph showing the solubility of potassium chlorate in water at various temperatures. The solubility curve on this graph was plotted from the data in the following table.

Temperature (°C)	Solubility (g $KClO_3$/100 g water)
10	5.0
20	7.4
30	10.5
50	19.3
60	24.5
80	38.5

In contrast to the Celsius–Fahrenheit temperature relationship, there is no known mathematical equation that describes the exact relationship between temperature and the solubility of potassium chlorate. The graph in Figure I.4 was constructed from experimentally determined solubilities at the six temperatures shown. These experimentally determined solubilities are all located on the smooth curve traced by the unbroken line portion of the graph. We are therefore confident that the unbroken line represents a very good approximation of the solubility data for potassium chlorate over the temperature range from 10 to 80°C. All points on the plotted curve represent the composition of saturated solutions. Any point below the curve represents an unsaturated solution.

The dashed-line portions of the curve are *extrapolations*; that is, they extend the curve above and below the temperature range actually covered by the plotted solubility data. Curves such as this one are often extrapolated a short distance beyond the range of the known data, although the extrapolated portions may not be highly accurate. Extrapolation is justified only in the absence of more reliable information.

The graph in Figure I.4 can be used with confidence to obtain the solubility of $KClO_3$ at any temperature between 10 and 80°C, but the solubilities between 0 and 10°C and between 80 and 100°C are less reliable. For example, what is the solubility of $KClO_3$ at 55°C, at 40°C, and at 100°C?

First draw a perpendicular line from each temperature to the plotted solubility curve. Now trace a horizontal line to the solubility axis from each point on the curve and read the corresponding solubilities. The values that we read from the graph are

55°	22.0 g $KClO_3$/100 g water
40°	14.2 g $KClO_3$/100 g water
100°	59 g $KClO_3$/100 g water

Of these solubilities, the one at 55°C is probably the most reliable because experimental points are plotted at 50°C and at 60°C. The 40°C solubility value is a bit less reliable because the nearest plotted points are at 30°C and 50°C. The 100°C solubility is the least reliable of the three values because it was taken from the extrapolated part of the curve, and the nearest plotted point is 80°C. Actual handbook solubility values are 14.0 and 57.0 g of $KClO_3$/100 g of water at 40°C and 100°C, respectively.

The graph in Figure I.4 can also be used to determine whether a solution is saturated or unsaturated. For example, a solution contains 15 g of $KClO_3$/100 g of water and is at a temperature of 55°C. Is the solution saturated or unsaturated? *Answer*: The solution is unsaturated because the point corresponding to 15 g and 55°C on the graph is below the solubility curve; all points below the curve represent unsaturated solutions.

APPENDIX II

Vapor Pressure of Water at Various Temperatures

Temperature (°C)	Vapor pressure (torr)	Temperature (°C)	Vapor pressure (torr)
0	4.6	26	25.2
5	6.5	27	26.7
10	9.2	28	28.3
15	12.8	29	30.0
16	13.6	30	31.8
17	14.5	40	55.3
18	15.5	50	92.5
19	16.5	60	149.4
20	17.5	70	233.7
21	18.6	80	355.1
22	19.8	90	525.8
23	21.2	100	760.0
24	22.4	110	1074.6
25	23.8	—	—

APPENDIX III

Units of Measurement

Physical Constants

Constant	Symbol	Value
Atomic mass unit	amu	1.6606×10^{-27} kg
Avogadro's number	N	6.022×10^{23}/mol
Gas constant	R	0.0821 L atm/K mol
Mass of an electron	m_e	9.11×10^{-31} kg
		5.5×10^{-4} amu
Mass of a neutron	m_n	1.675×10^{-27} kg
		1.00866 amu
Mass of a proton	m_p	1.673×10^{-27} kg
		1.00728 amu
Speed of light	c	2.997925×10^8 m/s

SI Units and Conversion Factors

Length

SI unit: meter (m)

1 meter	= 1.0936 yards
1 centimeter	= 0.3937 inch
1 inch	= 2.54 centimeters (exactly)
1 kilometer	= 0.62137 mile
1 mile	= 5280 feet
	= 1.609 kilometers
1 angstrom	= 10^{-10} meter

Mass

SI unit: kilogram (kg)

1 kilogram	= 1000 grams
	= 2.20 pounds
1 pound	= 453.59 grams
	= 0.45359 kilogram
	= 16 ounces
1 ton	= 2000 pounds
	= 907.185 kilograms
1 ounce	= 28.3 g
1 atomic mass unit	= 1.6606×10^{-27} kilograms

Volume

SI unit: cubic meter (m^3)

1 liter	= 10^{-3} m^3
	= 1 dm^3
	= 1.0567 quarts
1 gallon	= 4 quarts
	= 8 pints
	= 3.785 liters
1 quart	= 32 fluid ounces
	= 0.946 liter
1 fluid ounce	= 29.6 mL

Temperature

SI unit: kelvin (K)

$0 \text{ K} = -273.15°C$
$\quad\quad = -459.67°F$
$K = °C + 273.15$

$$°C = \frac{(°F - 32)}{1.8}$$

$°F = 1.8(°C) + 32$
$°F = 1.8(°C + 40) - 40$

Energy

SI unit: joule (J)

1 joule	= 1 kg m^2/s^2
	= 0.23901 calorie
1 calorie	= 4.184 joules

Pressure

SI unit: pascal (Pa)

1 pascal	= 1 kg/m^1 s^2
1 atmosphere	= 101.325 kilopascals
	= 760 torr (mmHg)
	= 14.70 pounds per square inch (psi)

APPENDIX IV

Solubility Table

	F^-	Cl^-	Br^-	I^-	O^{2-}	S^{2-}	OH^-	NO_3^-	CO_3^{2-}	SO_4^{2-}	$C_2H_3O_2^-$
H^+	S	S	S	S	S	s	S	S	s	S	S
Na^+	S	S	S	S	S	S	S	S	S	S	S
K^+	S	S	S	S	S	S	S	S	S	S	S
NH_4^+	S	S	S	S	—	S	(S)	S	S	S	S
Ag^+	S	I	I	I	I	I	—	S	I	I	I
Mg^{2+}	I	S	S	S	I	d	I	S	I	S	S
Ca^{2+}	I	S	S	S	I	d	I	S	I	I	S
Ba^{2+}	I	S	S	S	s	d	s	S	I	I	S
Fe^{2+}	s	S	S	S	I	I	I	S	s	S	S
Fe^{3+}	I	S	S	—	I	I	I	S	I	S	I
Co^{2+}	S	S	S	S	I	I	I	S	I	S	S
Ni^{2+}	s	S	S	S	I	I	I	S	I	S	S
Cu^{2+}	s	S	S	—	I	I	I	S	I	S	S
Zn^{2+}	s	S	S	S	I	I	I	S	I	S	S
Hg^{2+}	d	S	I	I	I	I	I	S	I	d	S
Cd^{2+}	s	S	S	S	I	I	I	S	I	S	S
Sn^{2+}	S	S	S	s	I	I	I	S	I	S	S
Pb^{2+}	I	I	I	I	I	I	I	S	I	I	S
Mn^{2+}	s	S	S	S	I	I	I	S	I	S	S
Al^{3+}	I	S	S	S	I	d	I	S	—	S	S

Key: S = soluble in water
    s = slightly soluble in water
    I = insoluble in water (less than 1 g/100 g H_2O)
    d = decomposes in water

APPENDIX V

Glossary

A-20

absolute zero   *See* Kelvin scale.

acid   (1) A substance that produces H^+ (H_3O^+) when dissolved in water. (2) A proton donor. (3) An electron-pair acceptor. A substance that bonds to an electron pair.

acid anhydride   A nonmetal oxide that reacts with water to form an acid.

acid ionization constant (K_a)   The equilibrium constant for the ionization of a weak acid in water.

activated complex   The intermediate high-energy species formed when reactants collide. The complex can decompose to form either reactants or products.

activation energy   The amount of energy needed to form the activated complex.

activity series of metals   A listing of metallic elements in descending order or reactivity.

addition reaction   In organic chemistry a reaction in which two substances join together to produce one substance.

alchemists   Practitioners of chemistry during the Middle Ages whose aims were to change the baser metals into gold and to discover the "philosopher's stone," which was thought to bring eternal youth.

alcohol   An organic compound consisting of an —OH group bonded to a carbon atom in a nonaromatic hydrocarbon group.

aldehyde   An organic compound that contains the —CHO group. The general formula is RCHO.

alkali metal   An element (except H) from Group IA of the periodic table.

alkaline earth metal   An element from Group IIA of the periodic table.

alkane or paraffin hydrocarbon   A straight- or branched-chain compound composed of carbon and hydrogen and having only single bonds between the atoms.

alkene   An unsaturated hydrocarbon whose molecules have at least one carbon–carbon double bond.

alkyl group   An organic group derived from an alkane by removal of one H atom. The general formula is C_nH_{2n+1} (for example, CH_3, methyl). Alkyl groups are generally indicated by the letter R.

alkyl halide   A halogen atom bonded to an alkyl group. The general formula is RX (for example, CH_3Cl).

alkyne   An unsaturated hydrocarbon whose molecules have at least one carbon–carbon triple bond.

allotropy   A phenomenon in which a substance exists in two or more molecular or crystalline forms. Graphite and diamond are two allotropic forms of carbon.

alpha particle   A particle emitted from a nucleus during radioactive decay; consists of a He-4 nucleus with a mass of 4 and a charge of $+2$.

amide   An organic compound containing the

$$\overset{\displaystyle O}{\underset{\displaystyle |}{\overset{\displaystyle \|}{-C}}}-N\diagdown \ \text{group.}$$

amino acid   An organic compound containing two functional groups—an amino group (NH_2) and a carboxyl group (COOH). Amino acids are the building blocks for proteins.

amorphous   A solid without definite crystalline form.

amphoteric substance   A substance that can react as either an acid or a base.

anion   A negatively charged ion.

anode   The electrode where oxidation occurs in an electrochemical reaction.

aqueous solution   A water solution.

aromatic compound   An organic compound whose molecules contain a benzene ring or that has properties resembling benzene.

artificial or induced radioactivity   Radioactivity produced during some types of transmutations. Artificial radioactive nuclides behave like natural radioactive elements in that they disintegrate in a definite fashion with a specific half-life.

atmospheric pressure   The pressure experienced by objects on the earth as a result of the layer of air surrounding our planet. A pressure of 1 atmosphere (1 atm) is the pressure that will support a column of mercury 760 mm high at $0°C$.

atom   The smallest particle of an element that can enter into a chemical reaction.

atomic mass   The average relative mass of the isotopes of an element referred to the atomic mass of C-12 as exactly 12 amu.

atomic mass unit (amu)   A unit of mass equal to one-twelfth the mass of a carbon-12 atom.

atomic number   The number of protons in the nucleus of an atom.

atomic theory   The theory that substances are composed of atoms, and that chemical reactions are explained by the properties and the interactions of these atoms.

Avogadro's law   Equal volumes of different gases at the same temperature and pressure contain equal numbers of molecules.

Avogadro's number 6.022×10^{23}; the number of formula units in 1 mole of whatever is indicated by the formula

balanced equation   A chemical equation having the same number and kind of atoms and the same electrical charge on each side of the equation.

barometer   A device used to measure pressure.

base   (1) A substance that produces OH^- when dissolved in water. (2) A proton acceptor. (3) An electron-pair donor.

basic anhydride   A metal oxide that reacts with water to form a base.

beta particle   A particle identical in charge and mass to an electron.

binary compound   A compound composed of two different elements.

biochemistry   The branch of chemistry concerned with the chemical reactions occurring in living organisms.

boiling point   The temperature at which the vapor pressure of a liquid is equal to the pressure above the liquid.

bond dissociation energy   The energy required to break a covalent bond.

bond length   The distance between two nuclei that are joined by a chemical bond.

Boyle's law   At constant temperature, the volume of a given mass of gas is inversely proportional to the pressure (PV = constant).

buffer solution   A solution that resists changes in pH when diluted or when small amounts of a strong acid or strong base are added.

---

calorie (cal)   One calorie is a quantity of heat energy that will raise the temperature of 1 g of water 1°C (from 14.5 to 15.5°C).

carbohydrate   A polyhydroxy aldehyde or polyhydroxy ketone, or a compound that upon hydrolysis yields a polyhydroxy aldehyde or ketone. Sugars, starch, and cellulose are examples.

carbonyl group   The structure $\diagdown$C$=$O

carboxyl group   The functional group of carboxylic acids:

$$-C\diagup^{\displaystyle O}_{\diagdown\, OH}$$

carboxylic acid   An organic compound having a carboxyl group.

catalyst   A substance that influences the rate of a reaction and can be recovered in its original form at the end of the reaction.

cathode   The electrode where reduction occurs in an electrochemical reaction.

cation   A positive charged ion.

Celsius scale (°C)   The temperature scale on which water freezes at 0°C and boils at 100°C at 1 atm pressure.

chain reaction   A self-sustaining reaction in which the products cause the reaction to continue or to increase in magnitude.

Charles' law   At constant pressure, the volume of a gas is directly proportional to the absolute (K) temperature (V/T = constant).

chemical bond   The attractive force that holds atoms together in a compound.

chemical change   A change producing products that differ in composition from the original substances.

chemical equation   An expression showing the reactants and the products of a chemical change (for example, 2 H$_2$ + O$_2$ → 2 H$_2$O).

chemical equilibrium   The state in which the rate of the forward reaction equals the rate of the reverse reaction for a chemical change.

chemical family   *See* groups or families of elements.

chemical formula   A shorthand method for showing the composition of a compound using symbols of the elements.

chemical kinetics   The study of reaction rates or the speed of a particular reaction.

chemical properties   Properties of a substance related to its chemical changes.

chemistry   The science dealing with the composition of matter and the changes in composition that matter undergoes.

colligative properties   Properties of a solution that depend on the number of solute particles in solution and not on the nature of the solute (for example, vapor pressure lowering, freezing point lowering, boiling point elevation).

combination reaction   A direct union or combination of two substances to produce one new substance.

combustion   In general, the process of burning or uniting a substance with oxygen, which is accompanied by the evolution of light and heat.

common ion effect   The shift of an equilibrium caused by the addition of an ion common to the ions in the equilibrium.

compound   A substance composed of two or more elements combined in a definite proportion by mass.

concentrated solution   A solution containing a relatively large amount of solute.

concentration of a solution   A quantitative expression of the amount of dissolved solute in a certain quantity of solvent or solution.

condensation   The process by which molecules in the gaseous state return to the liquid state.

conjugate acid-base   Two molecules or ions whose formulas differ by one H$^+$. (The acid is the species with the H$^+$, and the base is the species without the H$^+$.)

coordinate-covalent bond   A covalent bond in which the shared pair of electrons is furnished by only one of the bonded atoms.

copolymer   A polymer containing two different kinds of monomer units.

covalent bond   A chemical bond formed between two atoms by sharing a pair of electrons.

critical mass   The mass of a fissionable material required to sustain a nuclear chain reaction.

curie (Ci)   A unit of radioactivity indicating the rate of decay of a radioactive substance: $1 \text{ Ci} = 3.7 \times 10^{10}$ disintegrations per second.

Dalton's atomic theory   The first modern atomic theory to state that elements are composed of tiny, individual particles called atoms.

Dalton's law of partial pressures   The total pressure in a mixture of gases is equal to the sum of the partial pressures of each gas in the mixture.

decomposition reaction   A breaking down or decomposition of one substance into two or more different substances.

deliquescence   The absorption of water by a compound beyond the hydrate stage to form a solution.

density   The mass of an object divided by its volume.

diffusion   The process by which gases and liquids mix spontaneously because of the random motion of their particles.

dilute solution   A solution containing a relatively small amount of solute.

dipeptide   Two α-amino acids joined by a peptide linkage.

dipole   A molecule with separation of charge causing it to be positive at one end and negative at the other end.

disaccharide   A carbohydrate that yields two monosaccharide units when hydrolyzed.

disintegration series   The spontaneous decay of a certain radioactive nuclide by emission of alpha and beta particles from the nucleus, finally stopping at a stable isotope of lead or bismuth.

dissociation   The process by which a salt separates into individual ions when dissolved in water.

DNA   Deoxyribonucleic acid; a high molar-mass polymer of nucleotides, present in all living matter, that contains the genetic code and transmits hereditary characteristics.

double bond   A covalent bond in which two pairs of electrons are shared.

double-displacement reaction   A reaction of two compounds to produce two different compounds by exchanging the components of the reacting compounds.

efflorescence   The spontaneous loss of water of hydration by a compound when exposed to air.

effusion   The passage of gas through a tiny orifice from a region of high pressure to a region of lower pressure.

Einstein's mass-energy equation $E = mc^2$; the relationship between mass and energy.

electrolysis   The process whereby electrical energy is used to bring about a chemical change.

electrolyte   A substance whose aqueous solution conducts electricity.

electrolytic cell   An electrolysis apparatus in which electrical energy from an outside source is used to produce a chemical change.

electron   A subatomic particle that exists outside the nucleus and has an assigned electrical charge of -1.

electron affinity   The energy released or absorbed when an electron is added to an atom or an ion.

electron-dot structure   See Lewis structure

electronegativity   The relative attraction that an atom has for the electrons in a covalent bond.

electron shell   *See* energy levels of electrons.

element   A basic building block of matter that cannot be broken down into simpler substances by ordinary chemical changes.

empirical formula   A chemical formula that gives the smallest whole-number ratio of atoms in a compound.

endothermic reaction   A chemical reaction that absorbs energy from the surroundings as it proceeds from reactants to products.

energy   The capacity or ability of matter to do work.

energy levels of electrons   Areas in which electrons are located at various distances from the nucleus.

energy sublevels   The s, p, d, and f orbitals within a principal energy level occupied by electrons in an atom.

enzyme   A protein that catalyzes a biochemical reaction.

equilibrium   A dynamic state in which two or more opposing processes are taking place at the same time and at the same rate.

equilibrium constant (K_{eq})   A value representing the equilibrium state of a chemical reaction involving the concentrations of the reactants and the products.

equivalent mass   That mass of a substance that will react with, combine with, contain, replace, or in any other way be equivalent to 1 mole of hydrogen atoms or hydrogen ions.

essential amino acid   An amino acid that is not synthesized by the body and therefore must be supplied in the diet.

ester   An organic compound derived from a carboxylic acid and an alcohol. The general formula is

$$R-C{\overset{\displaystyle O}{\underset{\displaystyle OR'}{<}}}$$

ether   An organic compound having two hydrocarbon groups attached to an oxygen atom. The general formula is $R-O-R'$.

evaporation   The escape of molecules from the liquid state to the gas or vapor state.

exothermic reaction   A chemical reaction in which heat is released as a product.

Fahrenheit scale (°F)   The temperature scale on which water freezes at 32°F and boils at 212°F at 1 atm pressure.

fats and oils   Esters of fatty acids and glycerol.

fatty acids   Long-chain carboxylic acids present in lipids (fats and oils).

formula mass   The sum of the atomic masses of all the atoms in a chemical formula.

free radical   A chemical species having one or more unpaired electrons.

freezing or melting point   The temperature at which the solid and liquid states of a substance are in equilibrium.

functional group   An atom or group of atoms that characterizes a class of organic compounds. For example, —COOH is the functional group of carboxylic acids.

galvanic cell   *See* voltaic cell.

gamma ray   High-energy photons emitted by radioactive nuclei.

gas   The state of matter that is the least compact of the three physical states; a gas has no shape or definite volume and completely fills its container.

Gay-Lussac's law   At constant volume, the pressure of a fixed mass of gas is directly proportional to the absolute (K) temperature (P/T = constant).

Gay-Lussac's Law of Combining Volumes of Gases   At constant temperature and pressure, the ratios of the volumes of reacting gases are small whole numbers.

genes   Basic units of heredity that consist primarily of DNA and proteins and occur in the chromosomes.

Graham's Law of Effusion   The rates of effusion of different gases are inversely proportional to the square root of their molecular masses or densities.

groups or families of elements   Vertical groups of elements in the periodic table (IA, IIA, and so on). Families of elements have similar outer-orbital electron structures.

half-life ($t_{1/2}$)   The time required for one-half of a specific amount of a radioactive nuclide to disintegrate. For example, the half-life for $^{32}_{15}P$ is 14 days; the half-life for $^{14}_{6}C$ is over 5000 years.

halogen family   Group VIIA of the periodic table; consists of the elements fluorine, chlorine, bromine, iodine, and astatine.

halogenation   The substitution of a halogen atom for a hydrogen atom in an organic compound.

heat   A form of energy associated with the motion of small particles of matter.

heat of fusion   The amount of heat required to change 1 gram of a solid into a liquid at its melting point.

heat of reaction   The quantity of heat produced by a chemical reaction.

heat of vaporization The amount of heat required to change 1 gram of a liquid to a vapor at its normal boiling point.

heterogeneous Matter without uniform composition; two or more phases present.

homogeneous Matter having uniform properties throughout.

homologous series A series of compounds in which the members differ from one another by a regular increment. For example, each member of the alkane series of hydrocarbons differs by a CH_2 group.

hydrate A substance that contains water molecules as a part of its crystalline structure; $CuSO_4 \cdot 5H_2O$ is an example.

hydrocarbon A compound composed entirely of carbon and hydrogen.

hydrogen bond A chemical bond between polar molecules that contain hydrogen covalently bonded to the highly electronegative atoms F, O, or N:

hydrolysis A chemical reaction with water in which the water molecule is split into H^+ and OH^-.

hydrometer An instrument used to measure the specific gravity of a liquid. It consists of a weighted bulb at the end of a sealed calibrated tube.

hydronium ion H_3O^+; formed when an H^+ ion reacts with a water molecule.

hygroscopic substance A substance that readily absorbs and retains water vapor.

hypothesis A tentative explanation of the results of data to provide a basis for further experimentation.

---

ideal gas A gas that obeys the gas laws and the Kinetic-Molecular Theory exactly.

ideal gas equation $PV = nRT$; a single equation relating the four variables—P, V, T, and n—used in the gas laws. R is a proportionality constant known as the ideal or universal gas constant.

immiscible Incapable of mixing. Immiscible liquids do not form solutions with one another.

induced radioactivity *See* artificial radioactivity.

inorganic chemistry The chemistry of the elements and their compounds other than the carbon compounds.

ion An electrically charged atom or group of atoms. A positively charged ($+$) ion is called a *cation*, and a negatively charged ($-$) ion is called an *anion*.

ionic bond A chemical bond between a positively charged ion and a negatively charged ion.

ionization The formation of ions.

ionization constant (K_i) The equilibrium constant for the ionization of a weak electrolyte in water.

ionization energy The energy required to remove an electron from an atom, an ion, or a molecule.

ion product constant for water (K_w)
$K_w = [H^+][OH^-] = 1 \times 10^{-14}$ at 25°C

isomerism The phenomenon of two or more compounds having the same molecular formula but different molecular structures.

isomers Compounds having identical molecular formulas but different structural formulas.

isotopes Atoms of an element having the same atomic number but different atomic masses. Since the atomic numbers are identical, isotopes vary only in the number of neutrons in the nucleus.

IUPAC International Union of Pure and Applied Chemistry.

---

joule (J) The SI unit of energy; 4.184 J = 1 cal.

Kelvin (absolute) scale (K) Absolute temperature scale starting at absolute zero, the lowest temperature possible. Freezing and boiling points of water on this scale are 273 K and 373 K, respectively, at 1 atm pressure.

ketone An organic compound that contains a carbonyl group between two other carbon atoms. The general formula is $R_2C{=}O$.

kilocalorie (kcal) 1000 cal; the kilocalorie is also known as the nutritional or large Calorie, used for measuring the energy produced by food.

kilogram (kg) The standard unit of mass in the metric system.

kinetic energy (KE)   Energy of motion; $KE = \frac{1}{2}mv^2$.

Kinetic-Molecular Theory (KMT)   A group of assumptions used to explain the behavior and properties of ideal gas molecules.

---

law   A statement of the occurrence of natural phenomena that occur with unvarying uniformity under the same conditions.

Law of Conservation of Energy   Energy cannot be created or destroyed, but it may be transformed from one form to another.

Law of Conservation of Mass   There is no detectable change in the total mass of the substances in a chemical reaction; the mass of the products equals the mass of the reactants.

Law of Definite Composition   A compound always contains the same elements in a definite proportion by mass.

Law of Multiple Proportions   Atoms of two or more elements may combine in different ratios to produce more than one compound.

Le Chatelier's principle   If the conditions of an equilibrium system are altered, the system will shift to establish a new equilibrium system under the new set of conditions.

Lewis structure   A method of indicating the covalent bonds between atoms in a molecule or an ion such that a pair of electrons (:) represents the valence electrons forming the covalent bond.

limiting reactant   A reactant in a chemical reaction that limits the amount of product formed. The limitation is imposed because an insufficient quantity of the reactant, compared to amounts of the other reactants, was used in the reaction.

lipids   Organic compounds found in living organisms that are water insoluble, but soluble in such fat solvents as diethyl ether, benzene, and carbon tetrachloride. Examples are fats, oils, and steroids.

liquid   One of the three physical states of matter. The particles in a liquid move about freely while the liquid still retains a definite volume. Thus, liquids flow and take the shape of their containers.

liter (L)   A unit of volume commonly used in chemistry; 1 L = 1000 mL; the volume of a kilogram of water at 4°C.

logarithm (log)   The power to which 10 must be raised to give a certain number. The log of 100 is 2.

---

macromolecule   *See* polymer.

mass   The quantity or amount of matter that an object possesses.

mass defect   The difference between the actual mass of an atom of an isotope and the calculated mass of the protons and neutrons in the nucleus of that isotope.

mass number (A)   The sum of the number of protons and neutrons in the nucleus of a given isotope of an atom.

mass percent solution   The grams of solute in 100 g of a solution.

matter   Anything that has mass and occupies space.

mechanism of a reaction   The route or steps by which a reaction takes place. The mechanism describes the manner in which atoms or molecules are transformed from reactants into products.

meiosis   The process of cell division to form a sperm cell and an egg cell in which each cell formed contains half of the chromosomes found in the normal single cell.

metal   An element that is lustrous, ductile, malleable, and a good conductor of heat and electricity. Metals tend to lose their valence electrons and become positive ions.

metalloid   An element having properties that are intermediate between those of metals and nonmetals.

meter (m)   The standard unit of length in the SI and metric systems.

metric system   A decimal system of measurements.

miscible   Capable of mixing and forming a solution.

mitosis   Ordinary cell division in which a DNA molecule is duplicated by uncoiling to single strands and then reassembling with complementary nucleotides. Each new cell contains the normal number of chromosomes.

mixture   Matter containing two or more substances that can be present in variable amounts.

molality (m)   The number of moles of solute dissolved in 1000 grams of solvent.

molarity (M)   The number of moles of solute per liter of solution.

molar mass   The mass of Avogadro's number of atoms or molecules.

molar solution   A solution containing 1 mole of solute per liter of solution.

molar volume of a gas   The volume of 1 mol of a gas at STP, 22.4 L/mol.

mole   The amount of a substance containing the same number of formula units (6.022×10^{23}) as there are in exactly 12 g of C-12. One mole is equal to the molar mass in grams of any substance.

molecular formula   The true formula representing the total number of atoms of each element present in one molecule of a compound.

molar mass (weight)   The sum of the atomic masses of all the atoms in a molecule.

molecule   A small, uncharged individual unit of a compound formed by the union of two or more atoms.

mole ratio   A ratio of the number of moles of any two species in a balanced chemical equation. The mole ratio can be used as a conversion factor in stoichiometric calculations.

monomer   The small unit or units that undergo polymerization to form a polymer.

monosaccharide   A carbohydrate that cannot be hydrolyzed to simpler carbohydrate units; for example, simple sugars like glucose or fructose.

---

net ionic equation   A chemical equation that includes only those molecules and ions that have changed in the chemical reaction.

neutralization   The reaction of an acid and a base to form water plus a salt.

neutron   A subatomic particle that is electrically neutral and has an assigned mass of 1 amu.

noble gases   A family of elements in the periodic table—helium, neon, argon, krypton, and xenon—that contain a particularly stable electron structure.

nonelectrolyte   A substance whose aqueous solutions do not conduct electricity.

nonmetal   Any of a number of elements that do not have the characteristics of metals. They are located mainly in the upper right-hand corner of the periodic table.

nonpolar covalent bond   A covalent bond between two atoms with the same electronegativity value. Thus, the electrons are shared equally between the two atoms.

normal boiling point   The temperature at which the

vapor pressure of a liquid equals 1 atm or 760 torr pressure.

normality   The number of equivalents of solute per liter of solution.

nuclear binding energy   Energy equivalent to the mass difference between the theoretical sum of the masses of the protons and neutrons in a nucleus and the actual mass of the nucleus.

nuclear fission   The splitting of an atom with a heavy nucleus into two or more large fragments when the nucleus is struck by a neutron. As fission occurs, energy and two or three more neutrons are released.

nuclear fusion   The uniting of two light nuclei to form one heavier nucleus, accompanied by the release of energy.

nuclear power   The energy derived from a nuclear fission or fusion reaction.

nucleic acids   Complex organic acids essential to life and found in the nucleus of living cells. They consist of thousands of units called nucleotides. Includes DNA and RNA.

nucleons   A general term for the neutrons and protons in the nucleus of an atom.

nucleotide   The building-block unit for nucleic acids. A phosphate group, a sugar residue, and a nitrogenous organic base are bonded together to form a nucleotide.

nucleus   The central part of an atom where all the protons and neutrons of the atom are located. The nucleus is very dense and has a positive electrical charge.

nuclide   A general term for any isotope of any atom.

---

octet rule   An atom tends to lose or gain electrons until it has eight electrons in its outer shell.

oil   *See* fats and oils.

orbital   A cloudlike region around the nucleus where electrons are located. Orbitals are considered to be energy sublevels within the principal levels and are labeled *s*, *p*, *d*, and *f*.

organic chemistry   The branch of chemistry that deals with carbon compounds.

oxidation   An increase in the oxidation number of an atom as a result of losing electrons.

oxidation number (oxidation state)   A small number representing the state of oxidation of an atom. For an ion, it is the positive or negative charge on the ion; for covalently bonded atoms, it is a positive or negative number assigned to the more electronegative atom; in free elements, it is zero.

oxidation-reduction   A chemical reaction wherein electrons are transferred from one element to another.

oxidizing agent   A substance that causes an increase in the oxidation state of another substance. The oxidizing agent is reduced during the course of the reaction.

---

partial pressure   The pressure exerted independently by each gas in a mixture.

peptide linkage   The amide bond in a protein molecule; bonds one amino acid to another.

percent composition of a compound   The mass percent of each element in a compound.

percent yield $\dfrac{\text{Actual yield}}{\text{Theoretical yield}} \times 100$

periodic law   The properties of the chemical elements are a periodic function of their atomic numbers.

periodic table   An arrangement of the elements according to their atomic numbers, illustrating the periodic law. The table consists of horizontal rows or periods and vertical columns or families of elements. Each period ends with a noble gas.

periods of elements   The horizontal groupings of elements in the periodic table.

pH   A method of expressing the H^+ concentration (acidity) of a solution. $pH = -\log[H^+]$; $pH = 7$ is a neutral solution, $pH < 7$ is acidic, and $pH > 7$ is basic.

phase   A homogeneous part of a system separated from other parts by a physical boundary.

photosynthesis   The process by which green plants utilize light energy to synthesize carbohydrates.

physical change   A change in form (such as size, shape, physical state) without a change in composition.

physical properties   Characteristics associated with the existence of a particular substance. Inherent characteristics such as color, taste, density, and melting point are physical properties of various substances.

physical states of matter   Solids, liquids, and gases.

pOH   A method of expressing the basicity of a solution. $pOH = -\log[OH^-]$. $pOH = 7$ is a neutral solution, $pOH < 7$ is basic, and $pOH > 7$ is acidic.

polar covalent bond   A covalent bond between two atoms with differing electronegativity values resulting in unequal sharing of bonding electrons.

polyatomic ion   An ion composed of more than one atom.

polycyclic (fused) aromatic ring system   Aromatic systems that consist of multiple rings in which two carbon atoms are common to two rings.

polyhydroxy alcohol   An alcohol that has more than one —OH group.

polymer (macromolecule)   A natural or synthetic giant molecule formed from smaller molecules (monomers).

polymerization   The process of forming large, high-molar-mass molecules from smaller units.

polypeptide   A peptide chain containing up to 50 amino acid units.

polysaccharide   A carbohydrate that can be hydrolyzed to many monosaccharide units; cellulose, starch, and glycogen are examples.

positron   A particle with a $+1$ charge having the mass of an electron (a positive electron).

potential energy   Stored energy or the energy an object has because of its relative position.

pressure   Force per unit area; expressed in many units, such as mm Hg, atm, $lb/in.^2$, torr, pascal.

primary alcohol   An alcohol in which the carbon atom bonded to the —OH group is bonded to only one other carbon atom.

product   A chemical substance produced from reactants by a chemical change.

properties   The characteristics, or traits, of substances. Properties are classified as physical or chemical.

protein   A polymer consisting mainly of α-amino acids linked together; occurs in all animal and vegetable matter.

proton   A subatomic particle found in the nucleus of

all atoms; has a charge of $+1$ and a mass of about 1 amu. An H^+ ion is a proton.

----

quantum mechanics or wave mechanics The modern theory of atomic structure based on the wave properties of matter.

----

rad A unit of absorbed radiation indicating the energy absorbed from any ionizing radiation. 1 rad = 0.01 joule of energy absorbed per kilogram of matter.

radioactive decay The process by which an unstable nucleus emits particles or rays and is transformed into an atom of another element.

radioactivity The spontaneous emission of radiation from the nucleus of an atom.

rate of reaction The rate at which the reactants of a chemical reaction disappear and the products form.

reactant A chemical substance entering into a reaction.

redox An abbreviation for *oxidation–reduction*.

reducing agent A substance that causes a decrease in the oxidation state of another substance. The reducing agent is oxidized during the course of a reaction.

reduction A decrease in the oxidation number of an element as a result of gaining electrons.

rem A unit of radiation-dose equivalent taking into account that the energy absorbed from different sources does not produce the same degree of biological effect.

representative element An element in one of the A groups in the periodic table.

reversible reaction A chemical reaction in which products can react to form the original reactants. A double arrow ($\rightleftarrows$) is used to indicate that a reaction is reversible.

RNA Ribonucleic acid; a high-molar-mass polymer of nucleotides present in all living matter. Its main function is to direct the synthesis of proteins.

roentgen (R) A unit of exposure of gamma radiation based on the quantity of ionization produced in air.

rounding off numbers The process by which the value of the last digit retained is determined after dropping nonsignificant digits.

----

salts Ionic compounds of cations and anions.

saturated solution A solution containing dissolved solute in equilibrium with undissolved solute.

scientific laws Simple statements of natural phenomena to which no exceptions are known under given conditions.

scientific method A method of solving problems by observation; recording and evaluating data of an experiment; formulating hypotheses and theories to explain the behavior of nature; and devising additional experiments to test the hypotheses and theories to see if they are correct.

scientific notation A number between 1 and 10 (the decimal point after the first nonzero digit) multiplied by 10 raised to a power; for example, 6.022×10^{23}.

secondary alcohol An alcohol in which the carbon atom bonded to the —OH group is bonded to two other carbon atoms.

significant figures The number of digits that are known plus one that is uncertain are considered significant in a measured quantity.

simplest formula *See* empirical formula.

single bond A covalent bond in which one pair of electrons is shared between two atoms.

single-displacement reaction A reaction of an element and a compound to produce a different element and a different compound.

soap A salt of a long-carbon-chain fatty acid.

solid One of the three physical states of matter; matter in the solid state has a definite shape and a definite volume.

solubility An amount of solute that will dissolve in a specific amount of solvent.

solubility product constant (K_{sp}) The equilibrium constant for the solubility of a slightly soluble salt.

solute The substance that is dissolved in a solvent to form a solution.

solution A homogeneous mixture of two or more substances.

solvent The substance present to the largest extent in a solution. The solvent dissolves the solute.

specific gravity The ratio of the density of one substance to the density of another substance taken as a standard. Water is usually the standard for liquids and solids; air, for gases.

specific heat The quantity of heat required to change the temperature of 1 gram of any substance by 1°C.

spectator ion An ion in solution that does not undergo chemical change during a chemical reaction.

standard boiling point *See* normal boiling point.

standard conditions *See* STP.

stoichiometry The area of chemistry that deals with the quantitative relationships among reactants and products in a chemical reaction.

STP (standard temperature and pressure) 0°C (273 K) and 1 atm (760 torr).

strong electrolyte An electrolyte that is essentially 100% ionized in aqueous solution.

subatomic particles Mainly protons, neutrons, and electrons.

sublimation The process of going directly from the solid state to the vapor state without becoming a liquid.

substance Matter that is homogeneous and has a definite, fixed composition. Substances occur in two forms—as elements and as compounds.

substrate In biochemical reactions the substrate is the unit acted upon by an enzyme.

supersaturated solution A solution containing more solute than a saturated solution at a particular temperature. Supersaturated solutions tend to be unstable; jarring the container or dropping in a "seed" crystal will cause crystallization of the excess solute.

symbol In chemistry, an abbreviation for the name of an element.

---

temperature A measure of the intensity of heat or how hot or cold a system is.

tertiary alcohol An alcohol in which the carbon atom bonded to the —OH group is bonded to three other carbon atoms.

theoretical yield The maximum amount of product that can be produced according to a balanced equation.

theory An explanation of the general principles of certain phenomena with considerable evidence of facts to support it.

Thomson model of the atom The model that proposes that atoms are composed of smaller parts and therefore are not indivisible. The model states that atoms contain both positively and negatively charged particles embedded in the atomic sphere.

titration The process of measuring the volume of one reagent required to react with a measured mass or volume of another reagent.

torr A unit of pressure (1 torr = 1 mm Hg).

total ionic equation An equation that shows compounds in the form in which they actually exist. Strong electrolytes are written as ions in solution, whereas nonelectrolytes, weak electrolytes, precipitates, and gases are written in the un-ionized form.

transcription The process of forming RNA from DNA.

transition elements The metallic elements characterized by increasing numbers of d and f electrons in an inner shell. These elements are located in Groups IB through VIIB and in Group VIII of the periodic table.

transmutation The conversion of one element into another element.

transuranium element An element having an atomic number higher than that of uranium (>92).

triacylglycerol (triglyceride) An ester of glycerol and three molecules of fatty acids.

triple bond A covalent bond in which three pairs of electrons are shared between two atoms.

---

un-ionized equation (molecular equation) A chemical equation in which all the reactants and products are written in their molecular or normal formula expression.

unsaturated hydrocarbon A hydrocarbon whose molecules contain one or more double or triple covalent bonds.

unsaturated solution A solution containing less solute per unit volume than its corresponding saturated solution.

---

valence electrons Electrons in the outer energy

level of an atom. These electrons are primarily involved in chemical reactions.

vaporization   *See* evaporation.

vapor pressure   The pressure exerted by a vapor in equilibrium with its liquid.

vapor pressure curve   A graph generated by plotting the temperature of a liquid on the x axis and its vapor pressure on the y axis. Any point on the curve represents an equilibrium between the vapor and liquid.

vital force theory   A theory that held that organic substances could originate only from some form of living material. The theory was overthrown early in the 19th century.

volatile substance   A substance that evaporates readily; a liquid with a high vapor pressure and a low boiling point.

voltaic cell   An electrochemical cell that produces an electric current from a spontaneous chemical reaction.

volume   The amount of space occupied by matter.

volume percent solution   The volume of solute in 100 mL of solution.

———————

water of crystallization or hydration   Water molecules that are part of a crystalline structure, as in a hydrate.

weak electrolyte   A substance that is ionized to a small extent in aqueous solution.

weight   An extraneous property that an object possesses. The weight of an object depends on the gravitational attraction of the earth for that object. Therefore, an object's weight depends on its location in relation to the earth.

word equation   A statement in words, in equation form, of the substances involved in a chemical reaction.

———————

yield   The amount of product obtained from a chemical reaction.

APPENDIX VI

Answers to Selected Problems

Chapter 1

5. The following statements are correct:
   a, b, d, f, g, h
10. 16, 12, 4

Chapter 2

2. 135°F, 27°F, 212°F
3. 7.62 in.
15. The following statements are correct:
    a, c, d, e, g, h, i, j, l, n, p, q
17. (a) 93.2 (b) 8.87 (c) 8.0285 (d) 21.3
    (e) 4.64 (f) 130 (g) 34.3 (h) 2.00×10^6
18. (a) 2.9×10^6 (b) 4.56×10^{-2} (c) 5.8×10^{-1}
    (d) 4.0822×10^3 (e) 8.40×10^{-3}
    (f) 4.030×10^1 (g) 1.2×10^7 (h) 5.5×10^{-6}
19. (a) 14.4 (b) 58.5 (c) 1.08×10^8
    (d) 4.0×10^1 (e) 2.0×10^2 (f) 2009
    (g) 0.504 (h) 1.79×10^3 (i) 7.18×10^{-3}

(j) 2.49×10^{-4}
20. (a) 0.833 (b) 0.0429 (c) 0.750 (d) 0.500
21. (a) 1.9 (b) 86.7 (c) 2.1 (d) 51.3 (e) 8.93
    (f) 0.030
22. (a) 100°C (b) 72°F (c) 298 K (d) 4.6 mL
23. (a) 0.280 m (b) 1.000 km (c) 92.8 mm
    (d) 1.5×10^{-4} km (e) 6.06×10^{-6} km
    (f) 4.5×10^8 Å (g) 6.5×10^3 Å
    (h) 1.21×10^3 cm (i) 8.0×10^3 m
    (j) 31.5 cm (k) 2.5×10^7 mm
    (l) 1.2×10^{-6} cm (m) 5.20×10^4 cm
    (n) 38.84 nm (o) 107 cm (p) 4.1×10^4 in.
    (q) 811 km (r) 12.9 cm^2 (s) 117 ft
    (t) 10.2 mi (u) 7.4×10^4 mm
    (v) 1.25×10^{19} mm^3
24. (a) 1.068×10^4 mg (b) 6.8×10^{-2} kg
    (c) 0.00854 kg (d) 94.2 lb (e) 0.164 g
    (f) 6.5×10^5 mg (g) 5.5×10^3 g
    (h) 4.3×10^4 g

25. (a) 2.5×10^{-2} L   (b) 2.24×10^4 mL
    (c) 3.3×10^3 mL   (d) 1.3×10^3 m^3
    (e) 468 mL   (f) 9.41 gal
    (g) 9.0×10^{-3} mL   (h) 75.7 L
26. (a) 89 km/hr   (b) 81 ft/s
27. (a) 37 ft/s   (b) 25 mi/hr
28. 297 g NaCl remaining
29. (a) 7.5 mi/s   (b) 12.1 km/s
30. 77.2 kg
31. 0.32 g
32. 5.0×10^2 s
33. 3×10^4 mg
34. \$0.39/kg
35. \$2520
36. 3.0×10^3 times heavier
37. \$20.93
38. 55.6 L
39. 230 L
40. 76,000 drops
41. 160 L
42. 5.0×10^{-3} mL
43. 2.83×10^4 mL
44. 4×10^5 m^2
47. (a) 72°C   (b) -17.8°C   (c) 255.2 K
    (d) 0°F   (e) 90.°F   (f) -22.6°C   (g) 546 K
    (h) -61°C
45. 37.0°C
46. -100°C < -138°F
48. (a) -40°F   (b) -11.4°C
49. 1.565 g/mL
50. 3.12 g/mL
51. 7.1 g/mL
52. 595 g
53. 1.28 g/mL
54. 3.40×10^2 g
55. 27.6 g ethyl alcohol,
    76.0 g = (mass of cylinder + alcohol)
56. 1.74 g/mL Mg; 2.70 g/mL Al; 10.5 g/mL Ag
57. 3.57×10^3 g
58. 0.965 g/mL
59. 49.8 mL
60. H_2O = 50. mL; alcohol = 63 mL
62. \$1,033,718
63. 0.842 g/mL

Chapter 3

40. The following statements are correct:
    b, c, e, f, g, j, m, o, q, t, u, w, x, y, aa, dd
41. 2.67 g/mL
42. 54.5 mL

43. 18 carats
44. 75% C
45. 7.58×10^3 g Au
46. 1570 kg of alloy

Chapter 4

19. The following statements are correct: a, d, f, h, i
20. (a) 391 K   (b) 244.4°F
21. 30.3% Fe
22. 2.25 g copper(II)oxide
23. 21.9 g Cu
24. 60.2 g Hg
25. 8.5% fat
26. (a) 6.9 g oxygen   (b) 60.3% magnesium
27. 21.7 g chlorine
28. 1.7×10^3 J
29. 1.9×10^2 J
30. 0.301 J/g °C
31. 0.480 J/g °C
32. 40°C
33. 41.6°C
34. 3.0×10^3 cal
35. 29.1°C
36. 5.03×10^{-2} J/g °C
38. 29.2°C
39. 6.4 g coal
40. (a) 1.2×10^{14} cal   (b) 3.9×10^8 gal

Chapter 5

9. The following statements are correct: a, b, c, f, g
14. b, c, d, f, g, h, k
29. (a) 5.1×10^5 (b) 2.74×10^{-4} (c) 1×10^{-6}
    (d) 5.12×10^2
31. 131 amu
32. Silver = 107.87 amu
33. Magnesium = 24.31 amu
34. 6.03×10^{24} atoms
35. 69.7 amu = gallium

Chapter 6

29. The following statements are correct:
    a, d, f, g, i, j, l, m, o, p, r, u, w, x, y, aa

Chapter 7

9. The following statements are correct:
   a, b, c, d, h
12. a, d, f, g, j, l, m
13. (a) 119.0   (b) 142.0   (c) 331.2   (d) 46.0

(e) 60.0   (f) 231.4   (g) 342.0   (h) 342.3
(i) 132.0

14. (a) 40.0   (b) 275.8   (c) 152.0   (d) 96.0
    (e) 146.3   (f) 122.0   (g) 180.0   (h) 368.2
    (i) 244.3

15. (a) 0.344 mol Zn   (b) 2.83×10^{-2} mol Mg
    (c) 7.5×10^{-2} mol Cu   (d) 6.49 mol Co
    (e) 4.6×10^{-4} mol Sn   (f) 28 mol N atoms

16. (a) 0.625 mol NaOH   (b) 0.275 mol Br_2
    (c) 7.18×10^{-3} mol $MgCl_2$
    (d) 0.463 mol CH_3OH
    (e) 2.03×10^{-2} mol Na_2SO_4
    (f) 5.97 mol ZnI_2

17. (a) 108 g Au   (b) 284 g H_2O (c) 686 g Cl_2
    (d) 252 g NH_4NO_3 (e) 0.0417 g H_2SO_4
    (f) 11.5 g CCl_4

18. (a) 7.59×10^{23} molecules O_2
    (b) 3.4×10^{23} molecules C_6H_6
    (c) 6.02×10^{23} molecules CH_4
    (d) 1.65×10^{25} molecules HCl

19. (a) 3.441×10^{-22} g Pb
    (b) 1.791×10^{-22} g Ag
    (c) 2.99×10^{-23} g H_2O
    (d) 3.77×10^{-22} g $C_3H_5(NO_3)_3$

20. (a) 550 g Cu   (b) 24.6 kg Au
    (c) 1.7×10^{-23} mol C
    (d) 8.3×10^{-21} mol CO_2
    (e) 0.885 mol S   (f) 42.7 mol NaCl
    (g) 1.05×10^{24} atoms Mg   (h) 9.47 mol Br_2

21. (a) 6.022×10^{23} molecules CS_2
    (b) 6.022×10^{23} C atoms
    (c) 1.204×10^{24} S atoms
    (d) 1.806×10^{24} atoms

22. 8.43×10^{23} atoms P

23. 5.88 g Na

24. 10.8 g/atomic mass

25. 5.54×10^{19} m

26. 1.2×10^{14} dollars/person

27. (a) 8.3×10^{16} drops   (b) 7.3×10^6 ml^3

28. (a) 10.3 cm^3 (b) 2.18 cm

29. (a) 6.02×10^{23} atoms 0
    (b) 3.75×10^{23} atoms 0
    (c) 3.60×10^{23} atoms 0
    (d) 6.0×10^{24} atoms 0
    (e) 5.36×10^{24} atoms 0
    (f) 5.0×10^{18} atoms 0

30. (a) 14.4 g Ag   (b) 1.28 g Cl   (c) 266 g N
    (d) 6.74 g O   (e) 6.00 g H

31. 10.3 mol H_2SO_4

32. 1.62 mol HNO_3

33. (a) H_2O will contain the least molecules
    (b) CH_3OH will contain the least molecules

34. 6.022×10^{23} formula units

35. (a) 22.4% Na; 77.6% Br
    (b) 39.1% K; 1.0% H; 12.0% C; 48.0% O
    (c) 34.4% Fe; 65.6% Cl
    (d) 16.5% Si; 83.5% Cl
    (e) 15.8% Al; 28.1% S; 56.1% O
    (f) 63.5% Ag; 8.2% N; 28.3% O

36. (a) 47.9% Zn; 52.1% Cl
    (b) 18.2% N; 9.1% H; 31.2% C; 41.6% O
    (c) 12.3% Mg; 31.3% P; 56.5% O
    (d) 21.2% N; 6.1% H; 24.3% S; 48.4% O
    (e) 23.1% Fe; 17.4% N; 59.6% O
    (f) 54.4% I; 45.6% Cl

37. (a) 77.7%   (b) 69.9%   (c) 72.3%   (d) 15.2%

38. (a) 47.6%   (b) 34.1%   (c) 83.5%
    (d) 83.7% Highest %Cl is LiCl;
        lowest %Cl is $BaCl_2$

39. 43.7% P; 56.3% O

40. 24.2% C; 4.0% H; 71.72% Cl

41. 8.60 g Li

42. (a) 76.98% Hg   (b) 46.4% O   (c) 17.3% N
    (d) 2.72% Mg

43. (a) H_2O (b) N_2O_3 (c) equal   (d) $KClO_3$
    (e) $KHSO_4$ (f) Na_2CrO_4

44. (a) N_2O (b) NO   (c) N_2O_5 (d) Na_2CO_3
    (e) $NaClO_4$ (f) Mn_3O_4

45. (a) CuCl   (b) $CuCl_2$ (c) Cr_2S_3 (d) K_3PO_4
    (e) $BaCr_2O_7$ (f) PBr_8Cl_3

46. SnO_2

47. V_2O_5

48. There is insufficient Zn

49. $C_6H_6O_2$

50. $C_6H_{12}O_6$

51. $C_9H_8O_4$

52. 4.77 g O

53. GaAs

54. (a) CCl_4 (b) C_2Cl_6 (c) C_6Cl_6 (d) C_3Cl_8

Chapter 8

21. The following statements are correct:
    a, d, e, f, h, i, j, l, n, o, p, q

Chapter 9

1. (a) 0.247 mol KNO_3
   (b) 0.0658 mol $Ca(NO_3)_2$
   (c) 4.4 mol $(NH_4)_2C_2O_4$
   (d) 25.0 mol $NaHCO_3$

(e) 3.85×10^{-3} mol $ZnCl_2$

   (f) 0.056 mol NaOH   (g) 16 mol CO_2

   (h) 4.3 mol C_2H_5OH (i) 0.237 mol H_2SO_4

2. (a) 272 g $Fe(OH)_3$ (b) 1.31 g $NiSO_4$

   (c) 1.25×10^5 g $CaCO_3$ (d) 3.60 g $HC_2H_3O_2$

   (e) 179 g NH_3 (f) 373 g Bi_2S_3 (g) 2.6 g HCl

   (h) 1.35 g $C_6H_{12}O_6$ (i) 2×10^3 g Br_2

   (j) 18 g K_2CrO_4

3. (b) 2.84×10^{23} molecules $C_6H_{12}O_6$

6. 2.80 mol Cl_2

7. 500. g NaOH

8. 0.800 mol $Al(OH)_3$

9. 19.7 g $Zn_3(PO_4)_2$

10. (a) 0.500 mol Fe_2O_3 (b) 12.4 mol O_2

   (c) 6.20 mol SO_2 (d) 65.6 g SO_2

   (e) 0.871 mol O_2 (f) 332 g FeS_2

11. 4.20 mol HCl

12. (a) 2.22 mol H_2 (b) 162 g HCl

13. 87.4 kg Fe

14. 117 g H_2O, 271 g Fe

15. (a) 52.5 mol O_2 (b) 13.0 g CO_2

   (c) 2.20×10^2 g CO_2

16. (a) KNO_3 is limiting; KOH in excess

   (b) H_2O is limiting, NaOH in excess

   (c) Bi_2S_3 is limiting; $Bi(NO_3)_3$ in excess

   (d) H_2O is limiting; Fe in excess

17. (a) 2.0 mol Cu; 2.0 mol $FeSO_4$;

     1.0 mol $CuSO_4$ (unreacted)

   (b) 15.9 g Cu; 38.1 g $FeSO_4$;

     6.03 g (unreacted)

18. (a) 3.0 mol CO_2 (b) 9.0 mol CO_2

   (c) 1.8 mol CO_2

   (d) 6.0 mol CO_2; 8.0 mol H_2O; 4.0 mol O_2

   (e) 16.5 g CO_2 (f) 60.0 g CO_2

   (g) 60.0 g C_3H_8

19. 57.8%

20. 45.7 g CH_3OH; 4.29 g H_2 unreacted

21. 95.0% yield of Cu

22. (a) 3.2×10^2 g C_2H_5OH

   (b) 1.10×10^3 g $C_6H_{12}O_6$

23. 5.2×10^2 g C

24. 77.8% $CaCl_2$

25. $MgCl_2$ produces more AgCl than $CaCl_2$

26. 3.7×10^2 kg Li_2O

27. 14 kg concentrated H_2SO_4

28. 13 tablets

29. 67.2% $KClO_3$

30. The following statements are correct:
   a, c, d, f

31. The following statemets are correct: a, c, e

Chapter 10

34. The following statements are correct: a, c, d, e,
   f, h, i, j, k, l, n, p, q, r, u

Chapter 11

47. The following statements are correct: a, d, e, g,
   h, i, k, n, o, p, t, v

Chapter 12

40. The following statements are correct: a, b, e, h,
   i, j, l, o, q, r, s, t, u, v, w, x, y, z, bb, cc,
   gg, hh, jj

Chapter 13

25. The following statements are correct: b, d, f1,
   f4, g, h, o, p, q, t

26. (a) 0.941 atm   (b) 28.1 in.   (c) 13.8 lb/in^2.
   (d) 715 torr   (e) 953 mbar   (f) 95.3 kPa

27. (a) 0.037 atm   (b) 78.95 atm   (c) 1.05 atm
   (d) 0.0493 atm

28. (a) 263 mL   (b) 8.0×10^2 mL   (c) 132 mL

29. (a) 374 mm Hg   (b) 7.1×10^2 mm Hg

30. 1.0×10^2 atm

31. (a) 6.60 L   (b) 6.17 L   (c) 2.42 L   (d) 8.35 L

32. 600 K (327°C)

33. 1.48×10^3 torr

34. -62°C

35. 65 atm

36. 320°F

37. (a) 3.6×10^2 mL   (b) 7.80×10^2 mL

38. 6.1 L

39. 3.55 atm or 2.7×10^3 torr

40. 2.4×10^5 L

41. 33.4 L

42. 7.39×10^{21} molecules; 2.22×10^{22} atoms;
   275 mL at STP

43. 703 torr

44. 1.10×10^3 torr

45. 2.29 L

46. 3.70×10^2 torr

47. 56 L

48. (a) 34 mol   (b) 1.2×10^2 g H_2

49. 4.9 g CO_2

50. 43 g/mol

51. (a) 22.4 L   (b) 8.30 L   (c) 44.6 L

52. 2.69×10^{22} molecules

53. 44.6 mol Cl_2

54. 39.9 g/mol

55. (a) 3.74 g/L   (b) 0.18 g/L   (c) 3.58 g/L
   (d) 2.50 g/L

56. 1.54 g/L
57. 195 K ($-78°C$)
58. (a) 8.4 L H_2 (b) 8.20 g CH_4 (c) 8.47 g/L
     (d) 63.5 g/mol
59. 1.64×10^2 g/mol
60. 57 L Ne
61. 123 L
62. 2.81 K ($276°C$)
63. 0.13 mol N_2
64. 279 L H_2 STP
65. (a) 5.5 mol NH_3 (b) 5.6 mol NH_3
     (c) 9.6 L NO   (d) 640 mL   (e) 2.4 L NO
     (f) 1.1×10^2 g O_2 (g) 9.7 g NH_3
66. (a) 268 L O_2 (b) 153 L SO_2
67. 9.0 atm
68. He effuses 2.6 times as fast as N_2
69. (a) He/CH_4 = 2:1   (b) 67 cm from He end
70. C_4H_8
71. (a) 10. mol CO_2; 3.0 mol O_2; no CO
     (b) 29 atm
72. 76% $KClO_3$
74. (a) 250 torr   (b) 1060 torr
75. (a) 1.1×10^2 torr CO_2; 1.2×10^1 torr H_2
     (b) 120 torr
76. 0.72 g/L
77. Air enters the room
78. 60.4 atm

Chapter 14

59. The following statements are correct: a, b, c,
     f, h, l, m, o, p, s, t, u, w, y
60. (a) 0.420 mol $CoCl_2 \cdot 6H_2O$
     (b) 0.262 mol $FeI_2 \cdot 4H_2O$
61. (a) 2.52 mol H_2O (b) 1.05 mol H_2O
62. 51.1% H_2O
63. 48.6% H_2O
64. $Pb(C_2H_3O_2)_2 \cdot 3 H_2O$
65. $FePO_4 \cdot 4 H_2O$
66. 3.11×10^5 J
67. 5.5×10^4 J
68. 1.62×10^5 cal
69. 40.7 kJ/mol
70. Yes, sufficient ice
71. 74°C
72. 3.1×10^3 cal
73. Mixture of ice and water
74. 82 g ice remains
76. 2.29×10^6 J
77. 4.39×10^1 kJ

78. 40.2 g H_2O
79. 18.0 mL with volume 22.4 L
80. (a) 18.0 g H_2O (b) 36.0 g H_2O
     (c) 0.783 g H_2O (d) 18.0 g H_2O
     (e) 0.447 g H_2O (f) 0.167 g H_2O
81. 6.97×10^{18} molecules/sec
82. 54 g H_2SO_4
83. (a) yes, O_2 remains   (b) 20.0 mL O_2

Chapter 15

30. The following statements are correct: a, b, f,
     h, j, k, l, m, n, p, r, s, t, v
42. (a) 20.0% NaBr   (b) 10.7% K_2SO_4
     (c) 7.41% $Mg(NO_3)_2$
43. (a) 240 g solution   (b) 544 g solution
44. (a) 23.1% NaCl   (b) 22% $HC_2H_3O_2$
     (c) 15.3% $C_6H_{12}O_6$
45. (a) 3.3 g KCl   (b) 37.5 g K_2CrO_4
     (c) 6.4 g $NaHCO_3$
46. 455 g solution
47. (a) 4.5 g NaCl   (b) 449 g H_2O
48. 160 g H_2O needs to evaporate
49. (a) 22.0% CH_3OH (b) 33.6% NaCl
50. (a) 22% CH_3OH (b) 22% CCl_4
51. 214 mL solution
52. (a) 424 g HNO_3 (b) 1.18 L
53. (a) 0.40 M (b) 3.8 M NaCl   (c) 2.5 M HCl
     (d) 0.59 M $BaCl_2 \cdot 2H_2O$
54. (a) 0.327 M Na_2CrO_4 (b) 1.8 M $C_6H_{12}O_6$
     (c) 2.19×10^{-3} M $Al_2(SO_4)_3$
     (d) 0.172 M $Ca(NO_3)_2$
55. (a) 40. mol LiCl   (b) 0.0750 mol H_2SO_4
     (c) 3.5×10^{-4} mol NaOH   (d) 16 mol $CoCl_2$
56. (a) 8.8×10^3 g NaCl   (b) 13 g HCl
     (c) 4.6×10^2 g H_2SO_4 (d) 8.58 g $Na_2C_2O_4$
57. (a) 1.68×10^3 mL   (b) 3.91×10^4 mL
     (c) 1.05×10^3 mL   (d) 7.81×10^3 mL
58. 6.72 M HNO_3
59. (a) 6.0 M HCl   (b) 0.064 M $ZnSO_4$
     (c) 1.60 M HCl
60. (a) 2.0×10^2 mL 12 M HCl
     (b) 20. mL 15 M NH_3 (c) 16 mL 16 M HNO_3
     (d) 69 mL 18 M H_2SO_4
61. 8.2 L
62. (a) 0.48 M H_2SO_4 (b) 0.74 M H_2SO_4
     (c) 1.83 M H_2SO_4
63. 540. mL H_2O to be added
64. 0.32 M HNO_3
65. Take 31 mL of 5.00 M KOH and dilute with
     water to a volume of 250 mL

66. (a) 7.60 g $BaCrO_4$ (b) 15 mL of 1.0 M $BaCl_2$
67. (a) 33.3 mL 0.250 M Na_3PO_4
    (b) 1.10 g $Mg_3(PO_4)_2$
68. (a) 0.300 mol H_2 (b) 7.00 L H_2
69. 2.08 M HCl
70. (a) 0.67 mol KCl   (b) 0.33 mol $CrCl_3$
    (c) 0.30 mol $FeCl_2$
    (d) 69 mL of 0.060 M $K_2Cr_2O_7$
    (e) 35 mL of 6.0 M HCl
71. (a) 1.3 mol Cl_2 (b) 16 mol HCl
    (c) 1.3×10^2 mL 6.0 M HCl   (d) 3.2 L Cl_2
72. (a) HCl: 36.5 g/eq; NaOH: 40.0 g/eq
    (b) HCl: 36.5 g/eq; $Ba(OH)_2$: 85.7 g/eq
    (c) H_2SO_4: 49.1 g/eq; $Ca(OH)_2$: 37.1 g/eq
    (d) H_2SO_4: 49.1 g/eq; KOH 56.1 g/eq
    (e) H_3PO_4: 49.0 g/eq; LiOH 23.9 g/eq
73. (a) 4.0 N HCl   (b) 0.243 N HNO_3
    (c) 6.0 N H_2SO_4 (d) 5.55 N H_3PO_4
    (e) 0.250 N $HC_2H_3O_2$
74. 1.86 N H_2SO_4
75. $Mg(OH)_2$ tablet
76. (a) 9.943 mL NaOH   (b) 7.261 mL NaOH
    (c) 20.19 mL NaOH
77. (a) 4.38 m CH_3OH (b) 10 m C_6H_6
    (c) 5.6 m $C_6H_{12}O_6$
78. (a) CH_3OH (b) Both the same
79. 6.2 m H_2SO_4, 5.0 M H_2SO_4
80. Solution will freeze in 20°F temperature
81. 101.5°C
82. (a) 10.8 m (b) 105.53°C   (c) −20.1°C
83. (a) 0.545 m (b) 2.8°C   (c) 81.5°C
84. 153 g/mol
85. 163 g/mol
86. 500. g H_2O
87. (a) 8.04×10^3 g $C_2H_6O_2$
    (b) 7.24×10^3 mL $C_2H_6O_2$ (c) −4.0°F
88. 60. g 10% NaOH
89. 96.5 g NaOH
90. (a) 1.6×10^2 g sugar   (b) 0.47 M
    (c) 0.516 m
91. Molecular formula is $C_8H_4N_2$
92. 16.2 m HCl
93. 0.141 M

Chapter 16

28. The following statements are correct:
    a, c, f, i, j, k, l, n, p, r, t
29. (a) 0.015 M Na^+; 0.015 M Cl^-
    (b) 4.25 M Na^+; 4.25 M K^+; 4.25 M SO_4^{2-}

(c) 0.75 M Zn^{2+}; 1.5 M Br^-
(d) 4.95 M SO_4^{2-}; 3.30 M Al^{3+}
(e) 0.20 M Ca^{2+}; 0.40 M Cl^-
(f) 0.3 M K^+; 0.3 M I^-
(g) 0.6 M NH_4^+; 0.3 M SO_4^{2-}
(h) 0.0627 M Mg^{2+}; 0.125 M ClO_3^-
30. (a) 0.035 g Na^+; 0.053 g Cl^-
    (b) 9.78 g Na^+; 16.6 g K^+; 40.8 g SO_4^{2-}
    (c) 4.9 g Zn^{2+}; 12 g Br^-
    (d) 8.91 g Al^{3+}; 47.6 g SO_4^{2-}
    (e) 0.80 g Ca^{2+}; 1.4 g Cl^-
    (f) 1.04 g K^+; 3.36 g I^-
    (g) 1.23 g NH_4^+; 3.28 g SO_4^{2-}
    (h) 0.152 g Mg^{2+}; 1.04 g ClO_3^-
31. 0.260 M Ca^{2+}
32. (a) 1.0 M Na^+; 1.0 M Cl^-
    (b) 0.50 M Na^+; 0.50 M Cl^-
    (c) 1.0 M K^+; 0.5 M Ca^{2+}; 2.0 M Cl^-
    (d) 0.20 M H^+; 0.20 M K^+; 0.40 M Cl^-
    (e) No ions present in solution
    (f) 0.67 M Na^+; 0.67 M NO_3^-
33. 3.0×10^3 mL
34. (a) 0.684 M HCl   (b) 0.542 M HCl
    (c) 0.367 M HCl   (d) 0.147 M NaOH
    (e) 0.964 M NaOH   (f) 0.4750 M NaOH
35. 0.309 M $Ba(OH)_2$
44. (a) 40.8 mL of 0.245 M HCl
    (b) 1.57×10^3 mL of 0.245 M HCl
45. 0.201 M HCl
46. 0.1234 M HCl
47. 0.673 g KOH
48. 38.0% NaOH
49. 7.0% NaOH in sample
50. (a) 0.468 L H_2 (b) 0.932 L H_2
51. (a) 2   (b) 0   (c) 8.19   (d) 7   (e) 0.30
    (f) 4
52. (a) 3.4   (b) 2.6   (c) 4.3   (d) 10.5
53. 4
54. 13.9 mL of 18.0 M H_2SO_4
55. 0.025 mol NaOH remains
56. 0.3586 N NaOH
57. 22.7 mL of 0.325 M NaOH
58. 22.7 mL of 0.325 M H_2SO_4
59. 0.4536 N H_3PO_4; 0.1512 M H_3PO_4
60. 0.16 N acid
61. 1.2×10^2 mL
62. (a) 0.32 N H_2SO_4 (b) 0.392 g H_2SO_4
63. 4.8 N HCl; 9.6 N NaOH
64. 1.2×10^2 g/equiv
65. 57 g/equiv

Chapter 17

35. The following statements are correct:
c, d, g, i, j, k, l, n, p, q, s, t, u, v, x, z
36. 4.20 mol HI
37. (a) 3.16 mol HI
(b) 3.4 mol HI; 0.30 mol H_2; 0.57 mol I_2
(c) 3.16 mol HI; 0.42 mol H_2; 0.42 mol I_2
38. 2.75 mol H_2; 0.530 mol I_2; 0.500 mol HI
39. $K_{eq} = 29$
40. 128 times as fast
41. HOCl, 3.5×10^{-8}; $HC_3H_5O_2$, 1.3×10^{-5}; HCN, 4.0×10^{-10}
42. (a) $[H^+] = 2.1 \times 10^{-3}$ M   (b) pH = 2.7
(c) 0.84% ionized
43. $K_a = 2.7 \times 10^{-5}$
44. $K_a = 7 \times 10^{-10}$
45. (a) 0.42% ionized; pH = 2.4
(b) 1.3% ionized; pH = 2.9
(c) 4.2% ionized; pH = 3.4
46. $K_a = 1 \times 10^{-7}$
47. $K_a = 7 \times 10^{-6}$
48. $[OH^-] = 1.7 \times 10^{-15}$
49. (a) pH = 4.0; pOH = 10.0
(b) pOH = 12.0; pH = 2.0
(c) pOH = 2.6; pH = 11.4
(d) pH = 4.2; pOH = 9.8
(e) pOH = 4.9; pH = 9.1
50. (a) $[OH^-] = 1.0 \times 10^{-10}$
(b) $[OH^-] = 3.6 \times 10^{-9}$
(c) $[OH^-] = 2.5 \times 10^{-6}$
51. (a) $[H^+] = 1.7 \times 10^{-8}$ (b) $[H^+] = 1 \times 10^{-6}$
(c) $[H^+] = 2.2 \times 10^{-9}$
52. (a) 1.5×10^{-9} (b) 1.9×10^{-12}
(c) 1.2×10^{-23} (d) 2.6×10^{-13}
(e) 3.1×10^{-70} (f) 1.7×10^{-10}
(g) 2.4×10^{-5} (h) 5.13×10^{-17}
(i) 1.81×10^{-18}
53. (a) 4.5×10^{-5} mol/L   (b) 7.6×10^{-10} mol/L
(c) 1.6×10^{-2} mol/L   (d) 1.2×10^{-4} mol/L
54. (a) 8.9×10^{-4} g $BaCO_3$
(b) 9.3×10^{-9} g $AIPO_4$
(c) 0.50 g Ag_2SO_4 (d) 7.0×10^{-4} g $Mg(OH)_2$
55. (a) 2.1×10^{-4} mol Ca^{2+}/L;
4.2×10^{-4} mol F^-/L
(b) 8.2×10^{-3} g CaF_2
56. (a) Precipitate formed   (b) Precipitate formed
(c) No precipitate formed
57. (a) 3.0×10^{-8} M SO_4^{2-}
(b) 7.0×10^{-7} g $BaSO_4$

58. $BaSO_4$ will precipitate first
59. (a) 5.0×10^{-12} mol AgBr
(b) 2.5×10^{-12} mol AgBr
60. No precipitate of $PbCl_2$ will form
61. $K_{eq} = 3.0 \times 10^2$
62. $K_{sp} = 2.3 \times 10^{-5}$
63. $[NH_3] = 8.0$ M
64. $K_{sp} = 4.7 \times 10^{-7}$
65. $K_{sp} = 1.07 \times 10^{-27}$
66. (a) $[H^+] = 3.6 \times 10^{-5}$; pH = 4.4
(b) $[H^+] = 1.8 \times 10^{-5}$; pH = 4.7
67. (a) Change in pH = 5.3 units
(b) Change in pH = 0.02 units

Chapter 18

26. The following statements are correct:
a, c, e, g, j, k, m, p, q, r, s
27. 0.0772 mol NO
28. 20.2 L Cl_2
29. 17 g $KMnO_4$
30. 66.2 mL $K_2Cr_2O_7$
31. 10.0 mL $K_2Cr_2O_7$
32. 91.3% KI
33. 149 g Cu
34. 4.22 L NO
35. 5.56 mol H_2

Chapter 19

39. 0.0625 mg Sr-90 remains
40. $380
41. 2064 AD; one-eighth of the starting amount remains
42. 1.1×10^4 years old
43. 18 minutes/half-life
44. (a) 0.0424 g/mol   (b) 3.8×10^{12} J/mol
45. (a) 2.9×10^{-11} J/atom U-235
(b) 1.8×10^{13} J/mol   (c) 0.08185% mass loss
46. (a) 5.4×10^4 J/mol   (b) 0.195% mass loss
47. 7 alpha and 4 beta particles lost
49. 0.16 ng remains
51. The following statements are correct:
c, e, f, g, j, k, n, o, q, r, t

Chapter 20

61. The following statements are correct:
b, f, h, j, k, l, n, o, p, q, r, t, v, x, z

Chapter 21

59. The following statements are correct:
a, b, c, d, e, f, j, k, m, o, p, r, s, u, v

INDEX

Names and Formulas of Common Ions

Positive Ions		Negative Ions	
Ammonium	NH_4^+	Acetate	$C_2H_3O_2^-$
Copper(I)	Cu^+	Bromate	BrO_3^-
(Cuprous)		Bromide	Br^-
Hydrogen	H^+	Chlorate	ClO_3^-
Potassium	K^+	Chloride	Cl^-
Silver	Ag^+	Chlorite	ClO_2^-
Sodium	Na^+	Cyanide	CN^-
Barium	Ba^{2+}	Fluoride	F^-
Cadmium	Cd^{2+}	Hydride	H^-
Calcium	Ca^{2+}	Bicarbonate	HCO_3^-
Cobalt(II)	Co^{2+}	(Hydrogen carbonate)	
Copper(II)	Cu^{2+}	Bisulfate	HSO_4^-
(Cupric)		(Hydrogen sulfate)	
Iron(II)	Fe^{2+}	Bisulfite	HSO_3^-
(Ferrous)		(Hydrogen sulfite)	
Lead(II)	Pb^{2+}	Hydroxide	OH^-
Magnesium	Mg^{2+}	Hypochlorite	ClO^-
Manganese(II)	Mn^{2+}	Iodate	IO_3^-
Mercury(II)	Hg^{2+}	Iodide	I^-
(Mercuric)		Nitrate	NO_3^-
Nickel(II)	Ni^{2+}	Nitrite	NO_2^-
Tin(II)	Sn^{2+}	Perchlorate	ClO_4^-
(Stannous)		Permanganate	MnO_4^-
Zinc	Zn^{2+}	Thiocyanate	SCN^-
Aluminum	Al^{3+}	Carbonate	CO_3^{2-}
Antimony(III)	Sb^{3+}	Chromate	CrO_4^{2-}
Arsenic(III)	As^{3+}	Dichromate	$Cr_2O_7^{2-}$
Bismuth(III)	Bi^{3+}	Oxalate	$C_2O_4^{2-}$
Chromium(III)	Cr^{3+}	Oxide	O^{2-}
Iron(III)	Fe^{3+}	Peroxide	O_2^{2-}
(Ferric)		Silicate	SiO_3^{2-}
Titanium(III)	Ti^{3+}	Sulfate	SO_4^{2-}
(Titanous)		Sulfide	S^{2-}
Manganese(IV)	Mn^{4+}	Sulfite	SO_3^{2-}
Tin(IV)	Sn^{4+}	Arsenate	AsO_4^{3-}
(Stannic)		Borate	BO_3^{3-}
Titanium(IV)	Ti^{4+}	Phosphate	PO_4^{3-}
(Titanic)		Phosphite	PO_3^{3-}
Antimony(V)	Sb^{5+}		
Arsenic(V)	As^{5+}		

2.00

$H_2C_2O_4$

Atomic masses are based on carbon-12. Atomic masses in parentheses indicate the most stable or best-known isotope. Slight disagreement exists as to the exact electronic configuration of several of the high-atomic-number elements.

Noble gases(18)

			IIIA(13)	IVA(14)	VA(15)	VIA(16)	VIIA(17)	2 Helium **He** 4.00260
			5 Boron **B** 10.81	6 Carbon **C** 12.011	7 Nitrogen **N** 14.0067	8 Oxygen **O** 15.9994	9 Fluorine **F** 18.99840	10 Neon **Ne** 20.179

(10)	IB(11)	IIB(12)	13 Aluminum **Al** 26.98154	14 Silicon **Si** 28.086	15 Phosphorus **P** 30.97376	16 Sulfur **S** 32.06	17 Chlorine **Cl** 35.453	18 Argon **Ar** 39.948
28 Nickel **Ni** 58.71	29 Copper **Cu** 63.546	30 Zinc **Zn** 65.38	31 Gallium **Ga** 69.72	32 Germanium **Ge** 72.59	33 Arsenic **As** 74.9216	34 Selenium **Se** 78.96	35 Bromine **Br** 79.904	36 Krypton **Kr** 83.80
46 Palladium **Pd** 106.4	47 Silver **Ag** 107.868	48 Cadmium **Cd** 112.40	49 Indium **In** 114.82	50 Tin **Sn** 118.69	51 Antimony **Sb** 121.75	52 Tellurium **Te** 127.60	53 Iodine **I** 126.9045	54 Xenon **Xe** 131.30
78 Platinum **Pt** 195.09	79 Gold **Au** 196.9665	80 Mercury **Hg** 200.59	81 Thallium **Tl** 204.37	82 Lead **Pb** 207.2	83 Bismuth **Bi** 208.9804	84 Polonium **Po** (210)[a]	85 Astatine **At** (210)[a]	86 Radon **Rn** (222)[a]

Inner transition elements

63 Europium **Eu** 151.96	64 Gadolinium **Gd** 157.25	65 Terbium **Tb** 158.9254	66 Dysprosium **Dy** 162.50	67 Holmium **Ho** 164.9304	68 Erbium **Er** 167.26	69 Thulium **Tm** 168.9342	70 Ytterbium **Yb** 173.04	71 Lutetium **Lu** 174.97
95 Americium **Am** (243)[a]	96 Curium **Cm** (247)[a]	97 Berkelium **Bk** (249)[a]	98 Californium **Cf** (251)[a]	99 Einsteinium **Es** (254)[a]	100 Fermium **Fm** (253)[a]	101 Mendelevium **Md** (256)[a]	102 Nobelium **No** (254)[a]	103 Lawrencium **Lr** (257)[a]